Swine Production and Nutrition

ANIMAL SCIENCE TEXTBOOK SERIES

Swine Production and Nutrition

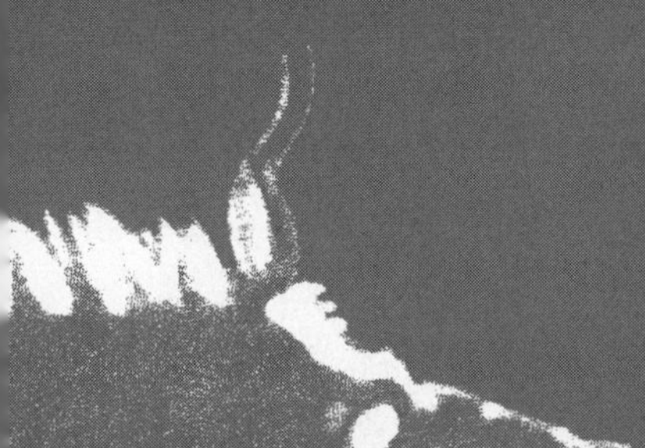

W. G. POND
USDA
Clay Center, Nebraska

J. H. MANER
IADS
Arlington, Virginia

avi
AVI Publishing Company, Inc. Westport, Connecticut

The frontispiece illustration is a reproduction of a painting of a galloping wild boar from the Altamira Caves in northern Spain. It is reproduced through the courtesy of The American Museum of Natural History.

Library of Congress Cataloging in Publication Data

Pond, Wilson G., 1930–
Swine production and nutrition.

Bibliography: p.
Includes index.
1. Swine. 2. Swine—Feeding and feeds. I. Maner, J. H., 1935– . II. Title.
SF395.P568 1984 636.4 83–25699
ISBN 0–87055–450–6

ABCDE 3210987654

Printed in the United States of America

Preface

The purpose of this text as stated in the preface to our book "Swine Production in Temperate and Tropical Environments," 1974, is to provide a technical basis for successful production of pork in both temperate and tropical environments. The competitive position of pork as a food source both in the developed and developing countries has continued to be strengthened during the past 10 years. In this book, we update the current knowledge and technology upon which pork production is based—covering extensively the new knowledge of feeds and their nutrient values.

The current growth of the world swine population is faster than that of the human population, reflecting the high demand for pork among consumers in all parts of the world. Advances in genetics, physiology, nutrition, and biotechnology have permitted continued improvements in efficiency of pork production, and prospects are bright for continued advances. The principles of modern production technology addressed in this book provide the basis for application by the swine industry in a broad range of environmental and economic settings. We have attempted to present a balanced coverage of the biological, economic, and husbandry aspects of swine production for use by college and university students, personnel in the feed and food industries, livestock extension specialists, and commercial swine producers. We hope that this book will contribute to the continued role of swine production in providing a means by which human nutrition can be enhanced by maximizing the use of surplus and noncompetitive feed resources in complex and varied plant–animal ecosystems. Innovative and balanced approaches to meeting the food needs of an expanding human population in the decades ahead will be required, and we believe pork production will play an important role in fulfilling this challenge.

We express our sincere appreciation to the U.S. Department of Agriculture for making available its office and library facilities and to the Rockefeller Foundation for providing the opportunity for one of us (JHM) to prepare this book. We are indebted to Martha Daza, Peggy Green, and Sherry Hansen for the stenographic work.

W. G. Pond
J. H. Maner

Contents

II. Biology of the Pig

4 Breeding and Selection

5 Prenatal Development

6 Postnatal Development

7 Growth

III. Feeds and Feeding

11 Nutrients and Feed Formulation

12 Grain and Grain By-product Energy Sources

13 Roots, Tubers, Fats, and Other Energy Sources

14 Animal Protein Sources

15 Plant Protein Sources

Part I

The Pork Industry

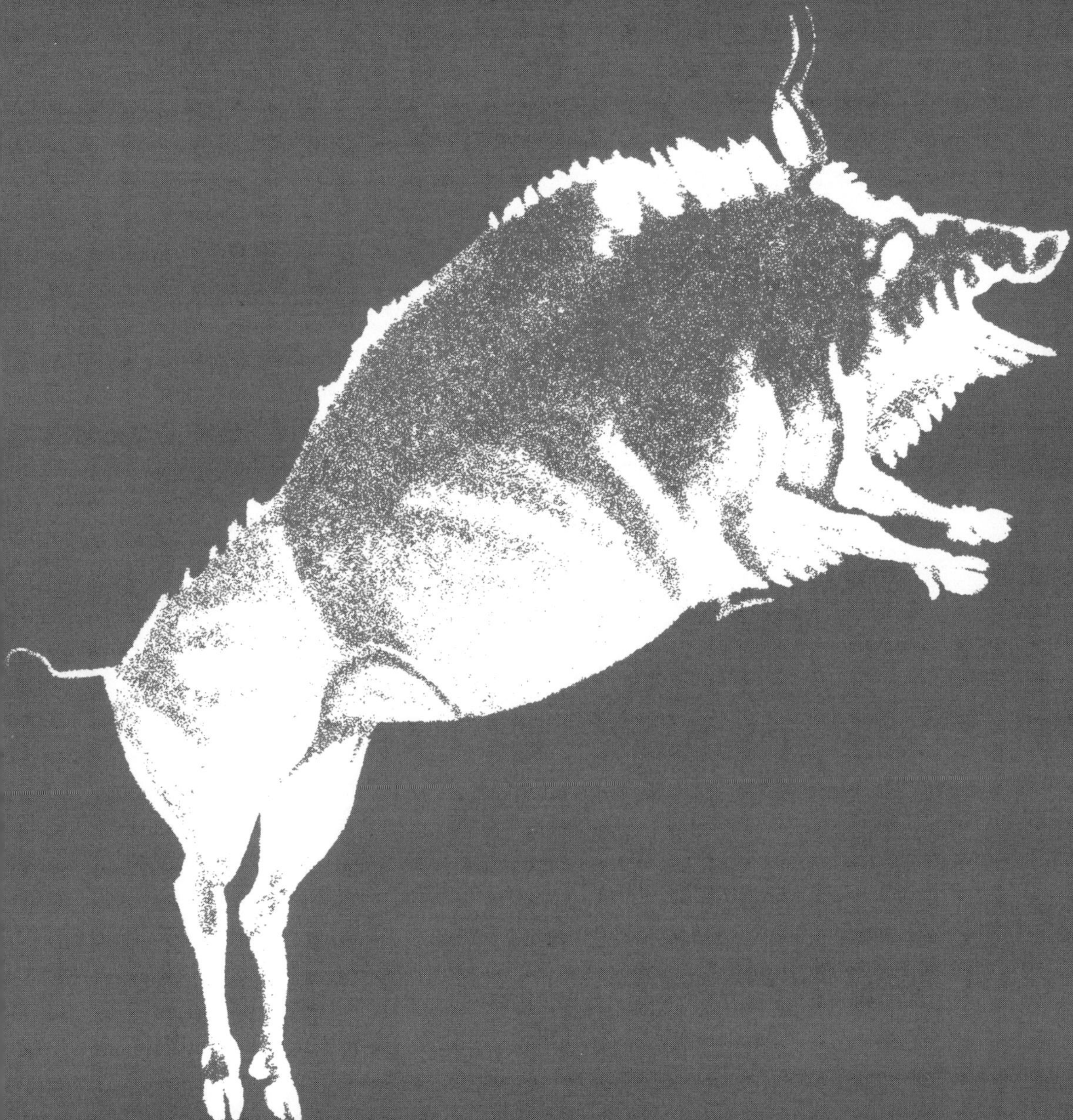

1

The Past, Present, and Future of Swine Production

INTRODUCTION

The technical knowledge of any subject cannot be applied most effectively without an appreciation of the magnitude of its importance. In this introductory chapter we paint a broad picture of the swine industry in the context of the past, present, and future. The objectives are to illustrate the scope of significance of the pig to human welfare and to provide the background information necessary for building a technical knowledge of swine production for application in all parts of the world.

EVOLUTION AND DOMESTICATION

The evolution and domestication of the pig has been described in several books (Towne and Wentworth 1950; Mellen 1952; Briggs 1969; Mason 1969), and the student interested in a detailed account of this fascinating story is referred to them. Recorded domestication appears in Biblical accounts as early as 2000 B.C. Despite some ancient cultural and religious taboos forbidding the consumption of pork, the domestication of the pig as a source of human food has persisted, and the continued increases in numbers of swine throughout the world up to the present time provide evidence of the contribution of the pig to human nutrition through the ages.

The pig has been traditionally a scavenger, and in early domestication it was raised as a means of utilizing human food wastes. In many parts of the world the pig still performs this function as a "backyard" inhabitant (see Figs. 1.1 and 1.2). Even in the developed countries some swine produced for commercial pork are fed kitchen wastes. However, such enterprises are closely controlled by local

FIG. 1.1. (a) Pigs on the north coast of Colombia. (b) Pigs tethered in the mountains of Ecuador.

and state government health regulations (such as those requiring that raw garbage be cooked) to prevent the spread of diseases to humans and other animals.

The proportion of commercially produced swine fed on garbage (kitchen wastes) in the developed countries is less than 1% of total production. Most commercial swine production in the developed countries is located in areas where cereal grains, maize (corn), and other high-energy feeds are grown in large quantities and are available at low cost. A classic example of the relatively recent development of commercial swine production on a large scale based on the availability of a source of high-energy feedstuffs is that of the "corn belt" of the United States. The corn belt, located in the north-central section of the United States, has soil and climate conditions that are ideal for the production of large amounts of high-yielding corn (*Zea mays*) hereafter referred to as maize (Fig. 1.3). The cultivation of hybrid maize and the use of fertilizer have resulted in yields of feed energy per acre far above anything realized before their introduction. The abundance of favorably priced feedstuffs provides an opportunity to convert this energy to pork for sale at a greater profit than could be obtained by selling the maize as a cash crop. Because of the profitability of swine production in the United States, the pig long ago became known among farmers as "the mortgage lifter."

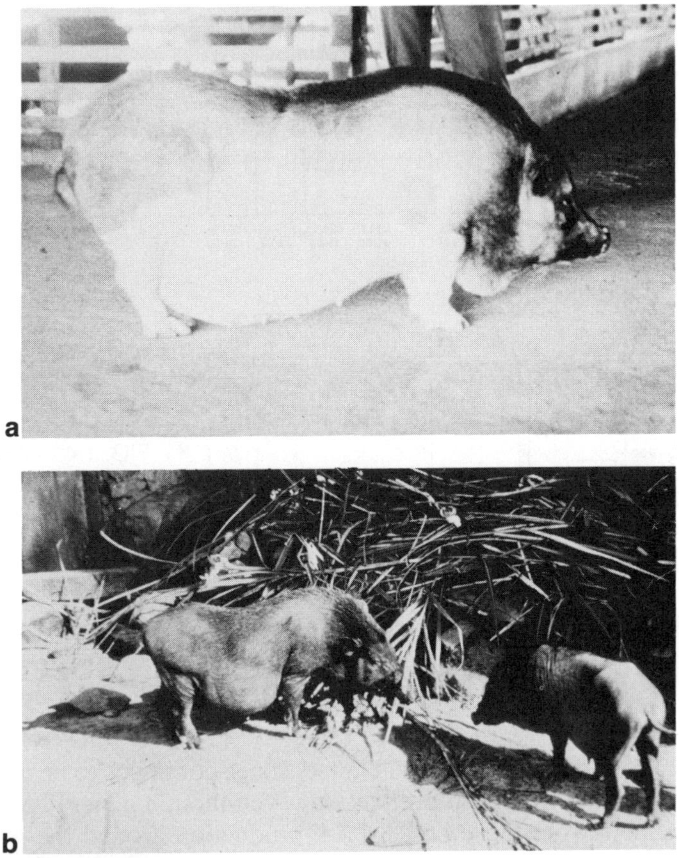

FIG. 1.2. (a) Pigs of Thailand; note the swayback typical of pigs in southeast Asia. (b) Pigs in North Sumatra.

FIG. 1.3. Well-filled ears of maize ready for harvest.
Photograph by J. C. Allen and Son.

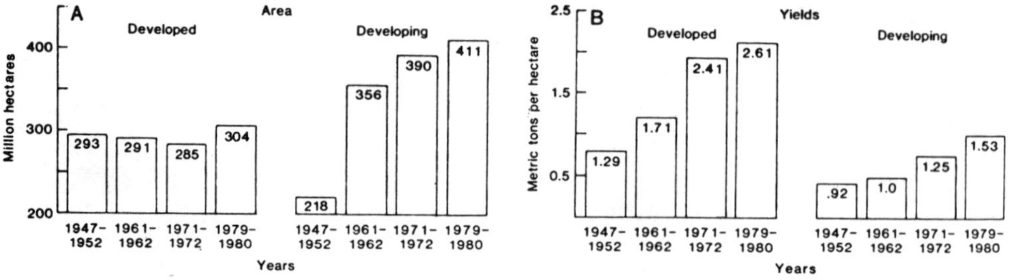

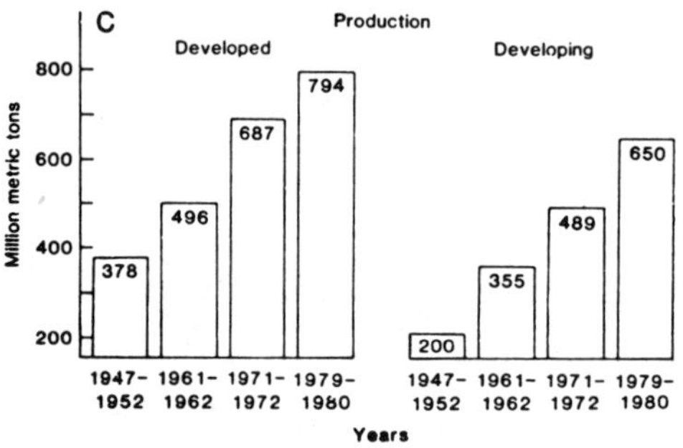

FIG. 1.4. World grain in developed and developing countries: (A) area under cultivation, (B) yields, and (C) production.
From Barr, 1981. Reprinted with permission of the author and the American Association for the Advancement of Science.

THE IMPORTANCE OF THE PIG IN SUPPLYING FOOD

Pork production in the United States continues to be concentrated in the areas of maize and sorghum production, even though generally the population centers are remote from these areas. Pork production around the world tends to be concentrated in grain producing areas and in areas where sufficient grain can be imported. The total land area devoted to grain production and the yield per hectare and the total tonnage of grain production in the developed and developing countries of the world in various periods from 1947 to 1980 are shown in Figs. 1.4A, B, and C (Barr 1981). World production of small grains by kind and by region is shown in Fig. 1.5. Table 1.1 (Global 2000 Report 1980) lists the amount of grain production, consumption, and trade in regions and countries of the world in 1969–1971 and 1973–1975 and projections for the year 2000, along with the percentage

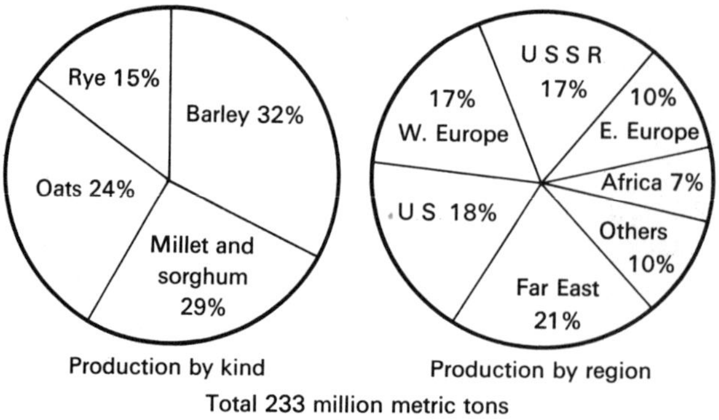

FIG. 1.5. Relative world production of some small grains.
From USDA 1964.

TABLE 1.1 Grain Production, Consumption, and Trade, Actual and Projected, and Percentage Increase in Total Food Production and Consumption

Locations	*Grain[a] (million metric tons)*			*Food (percent increase over 1970–2000 period)*
	1969–1971	*1973–1975*	*2000*	
Industrialized countries				
Production	401.7	434.7	679.1	43.7
Consumption	374.3	374.6	610.8	47.4
Trade	+32.1	+61.6	+68.3	
United States				
Production	208.7	228.7	402.0	78.5
Consumption	169.0	158.5	272.4	51.3
Trade	+39.9	+72.9	+129.6	
Other developed exporters				
Production	58.6	61.2	106.1	55.6
Consumption	33.2	34.3	65.2	66.8
Trade	+28.4	+27.7	+40.9	
Western Europe				
Production	121.7	132.9	153.0	14.6
Consumption	144.2	151.7	213.1	31.6
Trade	−21.8	−19.7	−60.1	
Japan				
Production	12.7	11.9	18.0	31.5
Consumption	27.9	30.1	60.1	92.8
Trade	−14.4	−19.3	−42.1	
Centrally planned countries				
Production	401.0	439.4	722.0	74.0
Consumption	406.6	472.4	758.5	79.9
Trade	−5.2	−24.0	−36.5	
Eastern Europe				
Production	72.1	89.4	140.0	83.2
Consumption	78.7	97.7	151.5	81.7
Trade	−6.1	−7.8	−11.5	
U S S R				
Production	165.0	179.3	290.0	72.7
Consumption	161.0	200.7	305.0	85.9
Trade	+3.9	−10.6	−15.0	
People's Republic of China				
Production	163.9	176.9	292.0	69.0
Consumption	166.9	180.8	302.0	71.4
Trade	−3.0	−3.9	−10.0	
Less-developed countries (LDCs)				
Production	306.5	328.7	740.6	147.7
Consumption	326.6	355.0	772.4	142.8
Trade	−18.5	−29.5	−31.8	
Exporters[b]				
Production	30.1	34.5	84.0	125.0
Consumption	18.4	21.5	36.0	58.0

(continued)

TABLE 1.1. (*Continued*)

Locations	Grain[a] (million metric tons)			Food (percent increase over 1970–2000 period)
	1969–1971	1973–1975	2000	
Trade	+11.3	+13.1	+48.0	
Importers[c]				
Production	276.4	294.2	656.6	149.3
Consumption	308.2	333.5	736.4	148.9
Trade	−29.8	−42.6	−79.8	
Latin America				
Production	63.8	72.0	185.9	184.4
Consumption	61.2	71.2	166.0	165.3
Trade	+3.2	+0.2	+19.9	
North Africa/Middle East				
Production	38.9	42.4	89.0	157.8
Consumption	49.5	54.1	123.7	167.3
Trade	−9.1	−13.8	−29.7	
Other African LDCs				
Production	32.0	31.3	63.7	104.9
Consumption	33.0	33.8	63.0	96.4
Trade	−1.0	−2.4	+0.7	
South Asia				
Production	119.1	127.7	259.0	116.8
Consumption	125.3	135.1	275.7	119.4
Trade	−6.2	−9.3	−16.7	
Southeast Asia				
Production	22.8	21.4	65.0	210.0
Consumption	19.3	17.9	47.0	163.6
Trade	+3.4	+3.7	+18.0	
East Asia				
Production	29.9	34.0	73.0	155.3
Consumption	38.3	42.9	97.0	164.9
Trade	−8.8	−9.7	−24.0	
World				
Production/Consumption	1108.0	1202.0	2141.7	91.0

Source: Global 2000 Report to the President, Table 6-5, p. 18, 1980.
[a]*Note:* In trade figures, plus sign indicates export, minus sign indicates import.
[b]Argentina and Thailand.
[c]All others, including several countries that export in some scenarios (e.g., Brazil, Indonesia, and Colombia).

increases in food production and consumption for the period 1970–2000. Among the industrialized countries, the United States is currently producing about 53% of the grain and will be producing about 59% by 2000. Countries with central planning (see Table 1.1) and less-developed countries, with the exception of Argentina and Thailand, are currently consuming more grain than they produce, and the deficit will be more severe by 2000. Most of the population growth (92%) will be in the less-developed countries (Global 2000 Report 1980). Of the projected 6.35 billion humans alive in 2000, 5 billion will live in these less-developed countries. As the human population of the world continues to increase [by a projected 86 million people per year during the 1980s (Table 1.2)], the proportion of plant products that can be fed to pigs for conversion to meat can be expected to decline, since the pig competes directly with humans for these feeds as a source of carbohydrate (energy) and protein. However, protein is relatively more limited than carbohydrates as a nutrient for humans and, in general, plant products, except some legume seeds, do not contain an adequate quantity or quality of protein to meet human requirements. The pig serves as an efficient synthesizer and storage depot for fat, proteins, vitamins, and minerals which, taken separately, would be very expensive. Furthermore, no one has yet produced a synthetic ham or pork chop equal in acceptability to the real thing at any price. Strong dietary preferences for traditional foods exist within individual regions of the world so that there is reluctance to accept esthetically inferior replacements. Since animal products when available at reasonable prices are readily consumed by people throughout the world, this potential for providing adequate nutrients cannot be overlooked. Thus, although the pig is considered by some to be a source of nutrients that make it too luxurious to survive in a world overpopulated with humans, there are both nutritional and esthetic grounds for arguing for its survival and in fact for its increased propagation at least until technology has been developed to the point at which other acceptable protein sources are available more cheaply. Nonetheless, there has been a dramatic increase in per capita meat consumption in all parts of the world since 1960, especially in those countries whose economic growth has been most rapid. Figure 1.6 (Barr 1981) shows the per capita meat consumption (in kilograms) in 34 countries plotted against percentage change for each country from 1961 to 1980. World meat consumption grew at an annual rate of nearly 3%, and poultry consumption grew more than 5% during that period (Barr 1981). Countries reporting the most rapid increase in pork consumption during the past decade were Greece, Bulgaria, and Japan, each of which doubled their intake, and Spain, Italy, Taiwan, and Venezuela, each of which recorded gains of 50% or more. This trend toward indirect consumption of grains (marketing grains through livestock) has resulted in increased demand for grains and oilseed meals and other by-product feedstuffs. The degree to which animal protein production can continue to increase in response to human demand will depend on economic as well as esthetic factors.

The amounts of pork needed to supply human demands and the number of animals required to produce it in 1965, 1985, and 2000 (as projected) are shown in Table 1.3. The rapid increase in human population ensures a substantial increase in demand for pork even with static per capita consumption. If per capita consumption of 30 kg, as currently recorded in the United States, were extended to the entire world, the number of pigs required to meet the demand would increase dramatically to about 3 billion (about 4.5 times the present number). Such an increase seems highly unlikely, but even if per capita consumption of pork remains at the current level of 9 kg, the total world demand will double by the year 2000 (Fig. 1.7).

The factors presented above do not include many unrealized potential applications. For example, in swine production large quantities of waste bananas, cassava, and other root crops could be used as sources of energy, and legume seeds could be used as sources of protein. This would provide large quantities of protein and other nutrients for human consumption at relatively low cost. In addition to these examples of tropical applications, there are many potential uses of temperate plant materials. The pig bred for production in temperate environments

TABLE 1.2. World Population, Actual and Projected

Region or country	*Compound annual growth rates (%)*			*Absolute population (millions)*			*Annual increase in people (millions)*		
	1960–1970	*1970–1980*	*1980–1990*	*1970*	*1980*	*1990*	*1960–1970*	*1970–1980*	*1980–1990*
Developed	1.06	0.80	0.68	728	788	843	7.3	6.0	5.5
United States	1.25	0.80	0.95	205	222	244	2.4	1.7	2.2
Canada	1.55	1.34	1.18	21	24	27	0.3	0.3	0.3
Western Europe	0.82	0.47	0.29	357	374	385	2.8	1.7	1.1
South Africa	2.61	2.80	2.74	22	29	38	0.5	0.7	0.9
Japan	1.02	1.10	0.59	104	116	123	1.0	1.2	0.7
Oceania	1.73	1.93	1.23	19	23	26	0.3	0.4	0.3
Centrally planned	1.68	1.29	1.07	1173	1334	1484	18.0	16.1	15.0
Eastern Europe	0.60	0.66	0.53	103	110	116	0.6	0.7	0.6
Soviet Union	1.32	0.90	0.90	244	267	292	3.0	2.3	2.5
China	1.93	1.48	1.18	826	957	1076	14.4	13.1	11.9
Developing	2.55	2.59	2.55	1776	2293	2949	39.6	51.7	65.6
Central and South America	2.79	2.66	2.65	283	368	478	6.8	8.5	11.0
East Asia (excluding Japan and China)	2.46	2.14	1.89	51	63	76	1.1	1.2	1.3
South Asia	2.50	2.51	2.40	1110	1422	1803	24.3	31.2	38.1
Africa (excluding South Africa)	2.55	2.86	3.01	332	440	592	7.4	10.8	15.2
World total	1.96	1.85	1.80	3677	4415	5276	65.0	73.8	86.1

Source: Barr 1981.

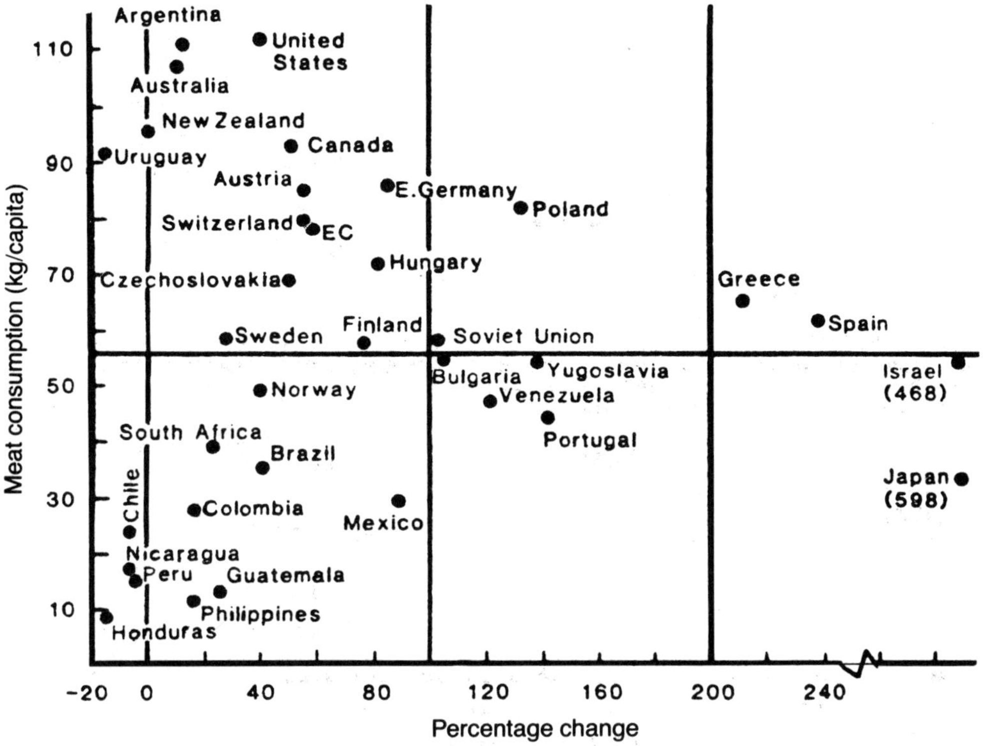

FIG. 1.6. Per capita meat consumption and percentage change, 1961–1980 (EC = European Economic Community).
From Barr 1981. Reprinted with permission of the author and the American Association for the Advancement of Science.

TABLE 1.3. Amounts of Pork Needed to Supply Human Demands in 1965, 1985, and 2000

	1965	1985[a]		2000[a]	
Human population, in billions	3281	4746		6351	
World per capita consumption of pork, kg	9	9	30	9	30
Amount of pork needed, in billions of kg	29.5[c]	42.8	143.3	57.2	190.5
Amount of pork protein needed, in billions of kg[b]	4.7	6.8	22.8	9.1	30.5
Number of animals needed to supply it, in millions[c]	468	679	2259	908	3024

Sources: FAO Production Yearbook 1967, reported in The World Food Problem 1967, Table 4-4, p. 256 (based on 1962–1964 data).

[a]Two amounts each are predicted for the years 1985 and 2000: the first is based on the assumption that the world per capita consumption stays at 9 kg/year; the second is that the total world consumption increases to the present U.S. per capita consumption of 30 kg/year.

[b]Based on the assumption that edible carcass contains 16% protein.

[c]Based on the assumption that the extraction rate is 100 and animals are slaughtered at 90 kg liveweight and have a dressing percentage of 70. (Actually there were closer to 600 million pigs in the world in 1965.)

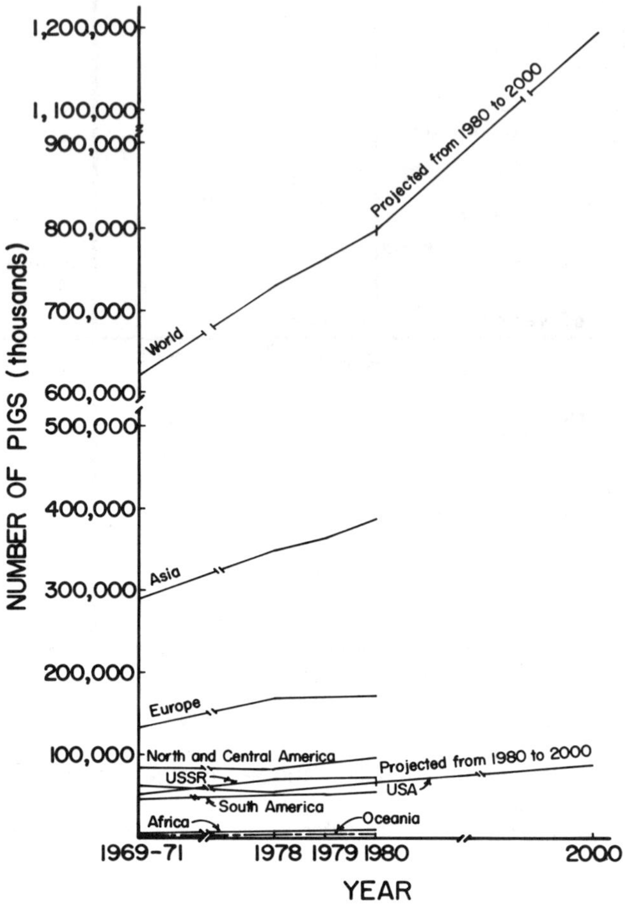

FIG. 1.7. World and regional pig population in 1969–1971, 1978, 1980, and projected to 2000, based on assumed constant annual per capita pork consumption of 9 kg worldwide. According to Food and Agriculture Organization (FAO) statistics, there were 783,485,000 pigs in the world in 1982. Asia had 47.4%, Europe (excluding USSR) 22.7%, USSR 9.4%, North and Central America 11.9%, South America 6.8%, Africa 1.3%, and Oceania 0.6% of the total. Total pork production (FAO statistics) in 1982 was 55,263,000 tons, of which Asia produced 37.7%, Europe 33.5%, USSR 9.2%, North and Central America 14.3%, South America 3.2%, Africa 0.8%, and Oceania 0.5%. *From FAO Production Yearbook 1982.*

adapts well to the tropics so that it is not necessary to incorporate genes of indigenous animal populations for maximum productivity, as it is with other food animals.

Tables 1.4 and 1.5, adapted from the FAO (Food and Agriculture Organization) Production Yearbook (1968, 1981) contain data on crop production by regions of the world during the 5-year period 1948–1952 and during 1967. The crops listed are some of those which are presently or could potentially be used for swine production. Many of these crops and their by-products are described as to nutritive value and special characteristics as feeds for swine in Chapters 11–17.

TABLE 1.4. World Production of Cereal Grains, Root Crops, Potatoes, Pulses, Sugar, Bananas, and Oilseeds, 1948–1981 (in millions of metric tons)

Crop	*1948–1952*	*1967*	*1977*	*1981*
Wheat	170.9	298.0	387.6	458.2
Rice (paddy)	167.3	275.9	371.7	413.8
Maize	139.4	264.5	370.0	451.7
Barley	59.2	118.3	160.3	158.5
Total cereals[a]	691.7	1137.5	1470.4	1663.8
Root crops[b]	119.2	418.6	568.2	561.6
Potatoes	247.4	307.9	268.2	257.0
Pulses[c]	29.3	39.6	42.4	42.4
Sugar	37.8	77.4	89.8	92.0
Bananas	12.3	26.0	36.6	39.9
Oilseeds (oil equivalent)	15.4[d]	29.4[d]	45.4	53.0
Total meat			129.1	142.3
Total pork[e]			50.1	55.0

Sources: FAO Production Yearbook 1968, 1981.

[a]The discrepancies between the sums of the cereal crops and the figures given for Total cereals are due to the fact that the total figure includes all cereal grains produced, including minor crops not listed in the table.

[b]Root crops include sweet potatoes, yams, cassava, and additional minor root crops.

[c]Pulses include dry beans, dry peas, dry broad beans, chick peas, lentils, pigeon peas, cow peas, lupins, and miscellaneous pulses.

[d]Calculated from oilseed production based on assumption that oilseeds contain 30% oil.

[e]From FAO Production Yearbook 1980, Table 87, pp. 220–222.

ALTERNATIVE PROTEIN SOURCES FOR HUMANS

The race to feed an exploding human population during the coming decades will be won or lost on the basis of the ability of science and technology to improve conventional methods and to develop new methods of food production. The chief conventional method of food production until now has been based on agriculture. Except in the developed countries, much of agriculture has been and still is at subsistence level with inefficient methods and primitive technology. The sophisticated agricultural technology of the more-developed countries is being applied with ever increasing pressure in the less-developed countries, although it is increasingly clear that such technology transfer is slow and under constraints of local cultures and resources.

Nonetheless, unconventional potential applications of agricultural technology to meeting human food needs include many exciting possibilities. Noncompetitive sources of feed and ways in which their usefulness can be extended have been reviewed (Altschul and Wilcke 1981). Such approaches include amino acid fortification of feeds and utilization of animal wastes and of microorganisms in animal feeds. Other examples are the domestication of selected species of wild life or the control of their ecology in their natural environment; the utilization of exotic plant materials; the employment of different methods to process existing plant materials; and the genetic manipulation of existing plant species by conven-

TABLE 1.5. Production of Cereal Grains, Root Crops, Potatoes, Pulses, Sugar, Bananas, and Oilseeds by Regions, 1977 and 1981 (in millions of metric tons)

Crop	North American developed			Western Europe			Oceania			Africa		
	1977	*1981*	*% of world,[a] 1981*	*1977*	*1981*	*% of world, 1981*	*1977*	*1981*	*% of world, 1981*	*1977*	*1981*	*% of world, 1981*
Wheat	75.5	100.5	21.9	53.6	64.8	14.1	9.7	16.8	3.7	3.8	4.3	8.9
Rice (paddy)	4.5	8.4	2.0	1.3	1.6	0.4	0.5	0.8	0.2	5.5	6.1	1.5
Maize	169.4	214.5	47.5	29.6	32.3	7.3	0.4	3.6	0.8	14.1	15.5	3.4
Barley	21.1	23.8	15.0	51.2	50.5	31.9	2.7	3.6	2.3	1.4	1.4	0.9
Total cereals	308.3	384.0	23.1	153.3	165.3	9.9	15.3	24.1	1.4	43.2	47.3	2.8
Root crops	19.2	18.3	3.3	55.0	48.2	8.6	1.0	1.2	0.2	75.7	82.6	14.7
Potatoes	18.6	17.7	6.9	54.9	48.0	18.7	1.0	1.2	0.5	2.5	3.2	1.2
Pulses	.9	1.8	4.3	1.7	1.6	3.8	0.1	0.2	0.5	4.4	4.6	10.8
Sugar	5.4	5.9	6.4	15.4	18.9	20.5	3.3	3.5	3.8	3.1	3.9	4.2
Bananas	0.003	0.003	<0.01	0.42	0.48	1.2	0.10	0.12	0.3	3.9	4.1	10.3
Oilseeds (oil equivalent)	11.8	13.4	25.3	2.6	2.8	5.3	0.09	0.13	0.2	3.7	4.1	7.7
Total meat	26.02	27.45	19.3	25.76	28.97	20.4	4.08	3.81	2.7	4.17	4.73	3.3
Total pork[b]	7.35	9.13	16.6	12.18	13.09	23.8	0.27	0.29	0.05	0.34	0.36	0.07

Crop	Latin America			Asia and Far East			Asia centrally planned			Eastern Europe and USSR centrally planned		
	1977	*1981*	*% of world, 1981*	*1977*	*1981*	*% of world, 1981*	*1977*	*1981*	*% of world, 1981*	*1977*	*1981*	*% of world, 1981*
Wheat	11.5	15.0	3.2	38.9	49.5	10.8	41.7	57.6	12.6	121.3	115.2	25.1
Rice (paddy)	15.1	15.6	3.8	171.4	196.3	47.4	149.3	164.7	39.8	2.4	2.6	0.6
Maize	43.7	55.8	12.4	15.4	19.9	4.4	51.8	64.4	14.3	30.9	29.3	5.7
Barley				3.3	3.2	2.0	3.4	3.8	2.4	67.0	58.8	37.1
Total cereals	86.1	105.2	6.3	251.9	292.1	17.6	264.8	309.5	18.6	266.0	247.4	14.9
Root crops	45.9	46.8	8.3	51.9	61.0	10.9	160.2	154.0	27.4	145.2	135.2	24.1
Potatoes	10.1	11.8	4.6	9.4	12.3	4.8	16.3	17.4	6.8	145.2	135.2	52.6
Pulses	4.6	5.6	13.2	13.8	13.3	31.4	6.4	7.3	17.2	8.2	5.9	13.9
Sugar	27.3	26.5	28.8	12.5	12.3	13.4	3.2	4.2	4.6	13.9	11.0	12.0
Bananas	18.5	19.1	47.9	11.3	13.4	33.3	1.0	1.2	3.0	—	—	—
Oilseeds (oil equivalent)	5.4	6.3	15.8	11.0	13.2	24.9	3.9	6.3	11.9	4.7	4.5	8.5
Total meat	13.17	14.82	10.4	4.35	5.34	3.8	20.78	24.18	17.0	23.90	25.07	17.6
Total pork[b]	2.20	2.53	4.6	1.30	1.46	2.7	15.10	17.04	31.0	11.00	10.76	19.6

Source: FAO Production Yearbook 1981.

[a]Percentage for each commodity does not always total 100.0 because of rounding errors and because some areas of the world (some developing countries in the Near East and other areas) are not included.

[b]From FAO Production Yearbook 1980, Table 87, pp. 220–222.

tional methods of breeding as well as by exciting new recombinant DNA (gene-splicing) techniques.

Man must depend mostly on agricultural technology as a major source of food (either by direct consumption of by feeding to animals) until nonagricultural techniques of food production become advanced to the point of making a major contribution. Innovative new food sources include single-cell protein (yeasts, bacteria, and algae), cultivated fish (in fresh and salt water), and petroleum by-products (which in combination with inorganic nitrogen can be used to produce single-cell protein). Petroleum by-products and other fossil energy sources obviously have restricted usefulness because they are available in limited quantities. Food sources (such as blue–green algae) generated by photosynthesis (solar energy), just as those produced by conventional agriculture except for the improved efficiency, offer greater potential application. Other energy sources (e.g., nuclear) undoubtedly also will make significant contributions to human food production in the future. Most nonagricultural food sources, while technologically sound, will be too expensive to compete for consumer purchasing power for many years, especially in the less-developed countries where foodstuffs of high nutritional value are in short supply. Crude preparations of many of these nonagricultural food sources first may find application as feedstuffs for livestock.

Thus, one can conclude that the war against hunger still has as a major participant conventional agricultural technology, including pork production.

FUTURE PROSPECTS FOR PORK PRODUCTION

Recent estimates (Global 2000 Report 1980) put the world human population by 2000 A.D. at approximately 6.35 billion (approximately 4.5 billion in 1983).

TABLE 1.6. Amounts of Energy and Protein Needed to Feed the Human Population of the World Adequately in 1970, 1985, and 2000

Requirements	*Year*		
	1970	*1985*	*2000*[a]
Protein needs per person[b] per day, grams			
FAO Reference Protein	35.4	35.1	35.1
Protein with NPU of 62[c]	57.1	56.6	56.6
Total world protein needs per day, metric tons			
FAO Reference Protein	127,922	168,139	211,325
Protein with NPU of 62	205,600	272,000	341,863
Energy needs per day, kilocalories			
Per person	2,354	2,393	2,393
Total world needs	7,787,000,000	11,118,000,000	13,974,000,000[d]

Source: Values for 1970 and 1985 taken from The World Food Problem 1967.

[a]Values for 2000 are computed on the basis of projections for human population of The World Food Problem 1967. The low estimate of 5.965 billion was chosen for the computations.

[b]Assumptions: Adult male weighs 65 kg; adult female weighs 55 kg; children weigh according to FAO standards.

[c]NPU is net protein utilization based on egg protein = 100. The different values listed for 1970 and 1985 represent changes in median age of the population.

[d]The appropriate figure would be 15,195,550,000 if the population were 6.35 billion as projected by Global 2000 Report (1980).

TABLE 1.7. Efficiency of Energy and Protein Utilization and the Energy Cost of Protein for Beef, Lamb, Poultry, Eggs, and Milk

Product	*Edible protein (g/100 g consumed)*	*Edible energy (kcal/100 kcal ME consumed)*[a]	*Edible protein (g/100 kcal ME consumed)*[a]
Beef	6.0	7.0	2.6
Lamb	3.0	3.0	1.3
Pork	12.0	23.0	6.0
Poultry	20.0	13.0	11.0
Eggs	16.0	15.0	11.0
Milk	23.0	21.0	10.0

Source: Robinson and Vohra (1976), adapted from Holmes (1970).
[a]ME = metabolizable energy.

The amounts of calories and protein needed to feed people properly in the year 2000 based on published nutritive requirements appear in Table 1.6, adapted from a report of the President's Science Advisory Committee (The World Food Problem 1967). Figure 1.7 shows the changes in pig numbers by continents and in the world with time since 1969–1971 (FAO Production Yearbook 1980) with projections to 2000 for the World and the United States. The amount of pork needed to meet global human demand and the amount of energy (calories) and protein required to meet these demands, based on current per capita pork consumption, where presented earlier (see Table 1.2). The required pork production could be achieved by increased land area devoted to these crops or by increased yield per unit land area or by both. Undoubtedly, both factors will be involved in increased future production, along with changes in proportions of what is produced and the introduction of other sources of feed.

The efficiency of utilization of feed for pork production depends on the provision of a balanced diet and on husbandry conditions and genetic selection conducive to high efficiency. The pig is by far the most efficient among the farm animals in conversion of feed energy to body energy and ranks only behind poultry and dairy cattle in efficiency of conversion of feed energy to protein. These comparisons are shown in Table 1.7 (Holmes 1970).

A useful measure of efficiency of swine production for a country or region is the *extraction rate* (the number of pigs slaughtered per year/the number of pigs on hand at a particular time). The developed countries of northern Europe and North America, for example, have high extraction rates (Belgium, 182; Denmark, 139; United Kingdom, 174; United States, 157) whereas some countries have extraction rates under 50 (i.e., Mexico, Brazil, Venezuela, Viet Nam). The low values reported in many developing countries reflect a need to improve efficiency of production, for as efficiency improves unit cost of production declines. The low efficiency is probably due to a variety of factors, but it seems clear that significant improvements can be made in most cases by appropriate adjustments in feeding, breeding, and management.

Although the values given previously (Table 1.2) for total protein needed to feed the human population indicate the magnitude of the problem, it is important to recognize that animal protein, while generally providing a better balance of amino acids (higher biological value) than plant proteins, also provides other needed nutrients not found in adequate amounts in plants. In the past, pork has

TABLE 1.8. Food Energy and Protein Values for Humans from Ruminants, Pigs, and Poultry in 1970 and Projections for the Year 2000

Product	*Livestock units[a] (millions)*			*Food energy[b] (billion Mcal)*		*Food protein[c] (million metric tons)*	
	1970	*2000*	*Increase (%)*	*1970*	*2000*	*1970*	*2000*
Ruminants	1195	1540	29	373	638	16	29
Pigs	154	190	23	144	215	2	3
Poultry	55	68	23	57	100	4	7
Total	1404	1798	28	574	953	22	39

Source: Fitzhugh et al. 1978, Table 1.1.

[a]Livestock unit for ruminants: cattle = 0.8; buffalo = 1.0; sheep and goats = 0.1. Livestock unit for nonruminants: pigs = 0.5; poultry = 0.01.

[b]Mcal/kg carcass weight: Meat—cattle and buffalo meat, 2.31; sheep and goat meat, 2.0; pig meat, 4.2; poultry meat, 1.4. Milk—cattle, 0.62; buffalo, 1.0; sheep, 1.12; goats, 0.75. Eggs—1.5.

[c]Net protein value (NPV), expressed as g/kg: Meat—cattle and buffalo meat, 105; sheep and goat meat, 89; pig meat, 60; poultry meat, 126 g/kg carcass weight. Milk—cattle, 28; buffalo, 32; sheep, 48; goat, 28 g/kg. Eggs—115 g/kg.

provided a major source of animal protein and can be expected to continue to do so during the rest of the present century. Data from the FAO Production Yearbook (1967) show that pork represents nearly one-half of the meat production from indigenous animals in the world and about 40% from those in the United States. It has been predicted (The World Food Problem 1967) that after 1990, animal proteins will decline in importance in contributing to total protein needs, owing to an increase in the direct competition of animals with humans for available feed supplies as well as to the appearance of protein sources from unconventional agricultural and nonagricultural production. However, this decline will not be dramatic as illustrated in Tables 1.8 and 1.9.

The top 20 countries in the world in 1983 for pig numbers, pig slaughter, and

TABLE 1.9. Present and Anticipated Animal Protein Content of Food Supplies per Person by Regions (g/person/day)

Region	*1970*	*1980*	*1990*	*2000*
World[a]	16.8	19.8	20.2	19.8
Europe	46.4	50.0	53.2	57.2
North America	67.1	71.3	80.1	82.4
Latin America	23.7	22.6	23.3	23.0
Near East	17.8	18.0	18.2	18.1
Far East	8.6	8.1	7.1	6.1
Africa	11.3	10.9	11.3	10.4
Oceania	71.3	81.9	84.5	88.7
United States	65.9	69.2	76.9	79.1

Source: The World Food Problem 1967.

[a]These means do not reflect the great variability among regions and countries of the world. For example, in 1965 supplies of animal protein per capita were 20 g/day, but poor distribution resulted in a value of 44 g for North America and Oceania versus only 9 g in the developing countries (FAO short-term goal is 15 g/person/day; long-term goal is 21 g/person/day).

TABLE 1.10. Top 20 Countries in Pig Numbers, Pig Slaughterings, and Pork Production, 1983

Pig numbers (thousands)		*Pig slaughterings (thousands)*		*Pork production (thousand tons)*	
USSR	73,900	USA	77,950	USA	6,078
USA	53,230	USSR	70,080	USSR	5,200
Brazil	33,500	West Germany	37,600	West Germany	2,680
West Germany	21,965	France	20,830	France	1,660
Poland	16,500	Japan	19,700	Japan	1,470
Mexico	16,000	Mexico	15,500	East Germany	1,251
East Germany	12,900	Spain	15,400	Poland	1,250
Romania	12,000	United Kingdom	15,240	Netherlands	1,180
France	11,513	Denmark	14,800	Mexico	1,132
India	10,600	Netherlands	14,300	Spain	1,125
Spain	10,500	Yugoslavia	14,000	Italy	1,050
Japan	10,050	Poland	13,817	Hungary	1,000
Netherlands	10,000	Brazil	13,800	United Kingdom	1,000
Denmark	9,875	Canada	13,450	Denmark	990
Italy	9,084	East Germany	12,600	Brazil	970
Canada	9,000	Philippines	11,726	Romania	954
Hungary	8,700	Romania	11,480	Czechoslovakia	850
Yugoslavia	8,281	Italy	10,650	Canada	825
United Kingdom	8,075	Belgium/Luxemburg	8,850	Yugoslavia	795
Philippines	7,950	Czechoslovakia	8,650	Belgium/Luxemburg	725

Source: Marbery 1983.

pork production are listed in Table 1.10. Discrepancies in ranking for pig numbers (census figures) and pigs slaughtered reflect differences in efficiency of production (number of pigs slaughtered per breeding female and age at which market weight is reached). The United States ranked first in total pork production and pigs slaughtered in 1983, but was second, behind the USSR, in pig numbers. The developed countries dominate the world in pork production, but as pig numbers increase in the developing countries changes can be expected in rankings for pork production. Total pork production of 36.813 million tons from 50 countries in 1983 (Marbery 1983) may be increased considerably in the coming decades, primarily by increased pork production in countries with rapidly expanding swine industries.

The pig will continue to contribute a significant proportion of energy and protein in the diets of developed and developing countries well into the twenty-first century (Pond 1983; Fredeen and Harmon 1983).

REFERENCES

ALTSCHUL, A. M., and WILCKE, H. L. 1981. New Protein Foods, Vol. 4: Animal Protein Supplies, Part B. Academic Press, New York.

BARR, T. N. 1981. Science *214*, 1087.

BRIGGS, H. M. 1969. Modern Breeds of Livestock, 3rd Edition. Macmillan Co., New York; Collier-Macmillan, London.

FAO PRODUCTION YEARBOOK. 1961. Food and Agric. Organization, Rome, Italy.
FAO PRODUCTION YEARBOOK. 1968. Food and Agric. Organization, Rome, Italy.
FAO PRODUCTION YEARBOOK. 1980. Food and Agric. Organization, Rome Italy.
FAO PRODUCTION YEARBOOK. 1981. Food and Agric. Organization, Rome, Italy.
FITZHUGH, H. A., HODGSON, H. J., SCOVILLE, O. J., NGUYEN, T. D., and BYERLY, T. C. 1978. The Role of Ruminants in Support of Man. Winrock Int. Livestock Research and Training Center, Morrilton, Arkansas.
FREDEEN, H., and HARMON, B. G. 1983. J. Anim. Sci. *57* (Suppl. 2), 100.
GLOBAL 2000 REPORT TO THE PRESIDENT. 1980. Entering the Twenty-First Century, Vol. I. Government Printing Office, Washington, D.C.
HOLMES, W. 1970. Proc. Nutr. Soc. *29,* 237.
MARBERY, S. 1983. Pig American, August, p. 34.
MASON, I. L. 1969. A World Dictionary of Livestock Breeds, Types and Varieties. Tech. Communication No. 8, Commonwealth Bureau of Animal Breeding and Genetics, Edinburgh. Commonwealth Agric. Bureau, Farnham Royal, Bucks., England.
MELLEN, I. M. 1952. The Natural History of the Pig. Exposition Press, New York, New York.
POND, W. G. 1983. Sci. Am. *248,* 96.
THE WORLD FOOD PROBLEM. 1967. A Report of the President's Science Advisory Committee, Vol. II, Report of the Panel on the World Food Supply. Superintendent of Documents, U.S. Government Printing Office Washington, D.C.
ROBINSON, D. W. and VOHRA P. 1976. Feedstuffs, February 23, p. 20.
TOWNE, C. W., and WENTWORTH, E. N. 1950. Pigs from Cave to Cornbelt. Unversity of Oklahoma Press, Norman, Oklahoma.
UNITED NATIONS. Department of Social Affairs, Population Division. 1966. Population Studies No. *41*. Washington, D.C.
USDA. 1964. Misc. Publ. *705*.

2

General Considerations about the Pig

INTRODUCTION

A sound knowledge of swine production requires, as a basis, a general appreciation for the types of pig that currently exist, a knowledge of some of the characteristics of the pig that make this species unique as an important provider of human food, and an appreciation for its behavior and other traits that must be considered in maximizing its usefulness to humans. These aspects of the pig are considered in this chapter.

BREEDS

The domestic pig[1] has evolved by natural selection and more recently by the application of the principles of population genetics by animal breeders. The product pork provides a valuable source of nutrients, by conversion from other less nutritious plant and animal products.

There are currently some 87 recognized "breeds" (Mason 1969) of domestic pigs in the world, most of them in Europe and North America. In addition, there are another 225 more "varieties" of pigs not recognized as breeds, but each having unique characteristics, appearance, or geographical location. The numbers of swine registered in each of the major breeds of the United States in recent years are shown in Table 2.1. Photographs of representatives of these and other breeds

[1]Phylum, Chordata; class, Mammalia; order, Artiodactyla; family, Suidae; genus, *Suis;* species, *domesticus* (*suis*).

TABLE 2.1. Purebred Swine Registrations in the United States in Recent Years

Breeds	*No. of animals registered*			
	1961	*1968*	*1970*	*1979*[a]
Berkshire	14,138	8,502	8,012	15,012
Chester White	17,112	17,115	19,934	62,679
Duroc	60,542	64,676	76,394	288,245
Landrace	27,307	8,123	8,810	79,331
Poland China	26,409	19,512	16,102	18,905
Spotted	13,020	12,367	13,974	130,023
Yorkshire	26,817	40,033	56,506	246,356
Inbred Registry[b]	—	708	805	—

[a]From Anon. 1980.
[b]This included the Minnesota Nos. 1–4, Montana No. 1, Maryland No. 1, Beltsville No. 1, Palouse, San Pierre, Conner Prairie Farms Nos. 1 and 2.

and varieties are shown in Fig. 2.1. Several commercial concerns produce and sell breeding stock and mating programs with various combinations of genetic stock representing the major breeds of U.S. swine as well as selected European breeds, including Large White and Landrace.

CHROMOSOMES

The domestic pig has 38 somatic chromosomes (19 pairs) (Jimenez *et al.* 1962). As in other mammals, the male determines the sex of the offspring by producing both X and Y chromosomes. The karyotype of the domestic pig is shown in Fig. 2.2.

A high proportion (up to 10%) of blastocysts from normal pigs have grossly detectable chromosome defects. The proportion of these defects associated with the high early embryonic mortality in pigs is not known. However, since they have not been reported in living adult pigs, many are probably lethal.

BODY SIZE

Much genetic variation exists among and within breeds in mature size and characteristics. The average mature weight of conventional pigs raised for meat is approximately 200–300 kg, whereas the breeds of miniature pigs developed for use in biomedical research have a mature weight of less than 100 kg. This wide variation makes it possible to breed selectively for small or large body size and is the basis for the development of miniature pigs. Average mature weights of some common breeds are given in Table 2.2.

TABLE 2.2. Average Mature Weights of Representative Breeds[a]

Breed	*Boars (wt in kg)*	*Sows (wt in kg)*
Poland China	205	180
Spot	200	160
Duroc	200	160
Chester White	195	150
Yorkshire	150	125
Minipig	75	75
Labco	70	70

Source: Briggs 1958.
[a]Figures in table (except for miniature pigs) represent weight in show condition. For breeding condition weights should be about 50 kg less than indicated for both boars and sows. Mature weight is reached by 3 years of age. Recent evidence suggests that in some breeds growth may persist beyond 3 years.

BEHAVIOR TRAITS

The behavior of swine has been described in detail (Pond and Houpt 1978; Hafez *et al.* 1962). The general behavior patterns described here are pertinent to commercial swine production. Pigs are notoriously curious. This trait is probably reflected in the tendency of the suckling pig to sniff and nibble at clean earth or feed materials to which it has access. Older pigs given access to pasture will spend hours rooting through the soil and will inevitably locate an open gate or a flaw in the fence. Pigs readily learn to open the lid of a self-feeder or push a valve for water. Even newborn pigs, whose suckling instinct is highly developed, can be taught readily to drink from a bowl when weaned from their dam. Their urge to suckle makes it highly desirable to cage pigs individually when weaned at a few days of age, since they constantly spend effort sucking the ears and navels of other pigs if kept together. Individual pigs within litters suckling the sow establish teat position within the first week of age so that possession is jealously guarded and attempted takeover of a teat by a littermate is met with resistance and a struggle for maintenance of position.

The pig is a "nibbler." That is, if feed is offered *ad libitum* from a self-feeder the total daily intake will be spread over many small meals throughout a 24-hr period. If given only one or two meals per day to satisfy appetite, daily food intake is reduced in young pigs, but in pigs approaching market weight it is maintained at a level equal to that achieved with continuous *ad libitum* feeding.

Pigs tend to be good "housekeepers" when given the chance. If the size of the nesting area is restricted they usually use a corner of the pen apart from the nesting area to defecate and urinate. Since the pig cannot efficiently dissipate body heat by sweating, high ambient temperatures encourage rooting in cool, moist earth or wallowing in mud, if available, in an effort to keep cool.

During gestation sows tend to become more docile, especially as they approach term. About one day before farrowing, the sow prepares a nest by rooting and

a

Poland China sow

b

Chester White gilt

c

Spotted gilt

d

Miniature sow

e
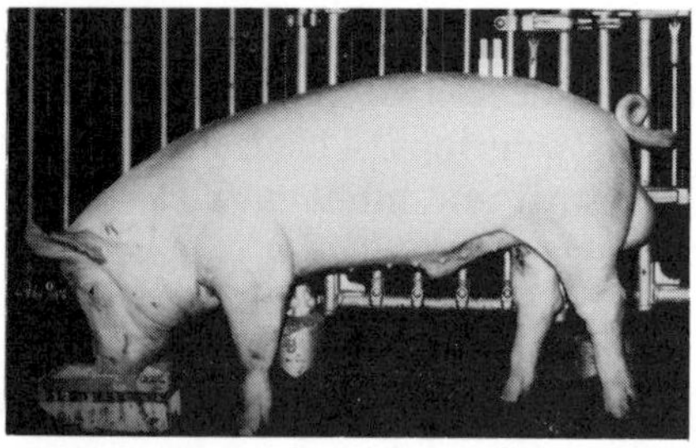
Yorkshire boar

f
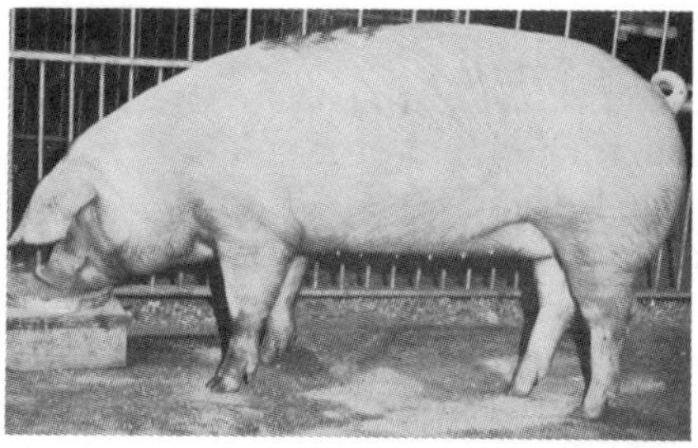
Danish Landrace gilt

g
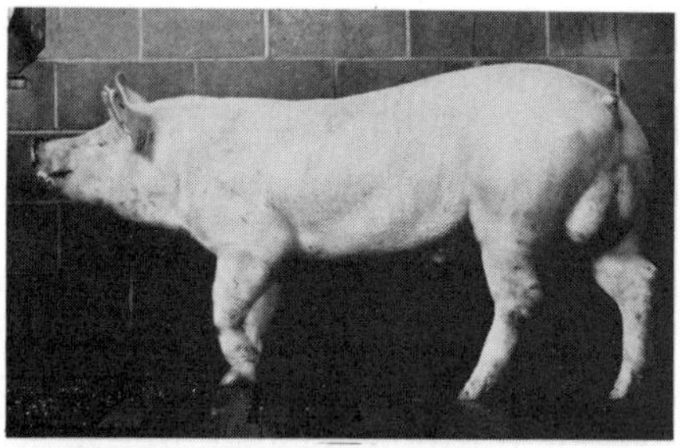
Babcock C-Line GP hybrid boar

h

Babcock $\frac{2}{3}$ F-Line-$\frac{1}{3}$ C-Line hybrid gilt

i

Hampshire gilt

j
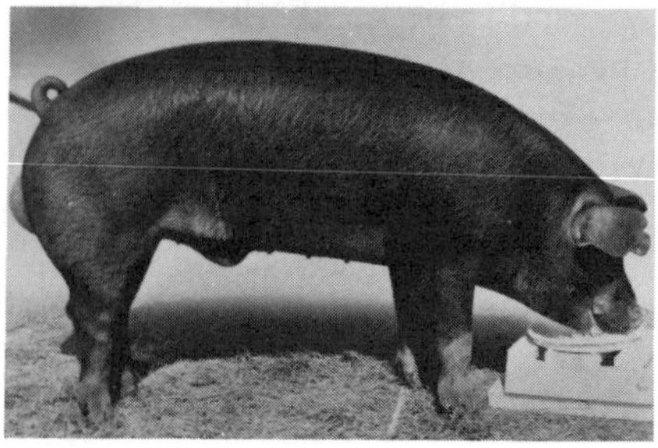
Duroc boar

k
Berkshire barrow

l
Synthetic-Line 24 boar

m
Belgian Pietrain pig

n
Camborough Blue gilt

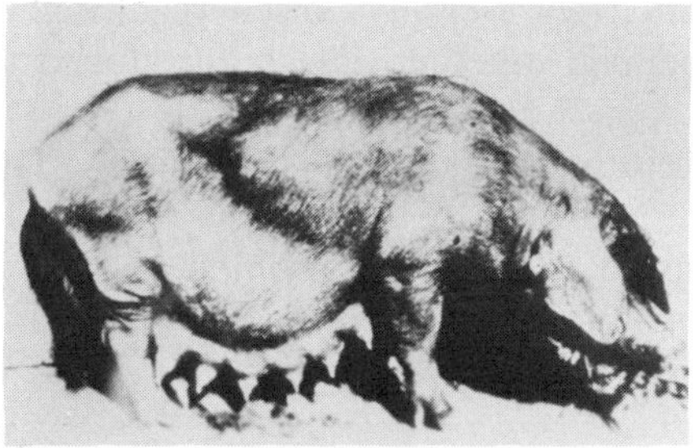

o
Ming sow, Northeast China

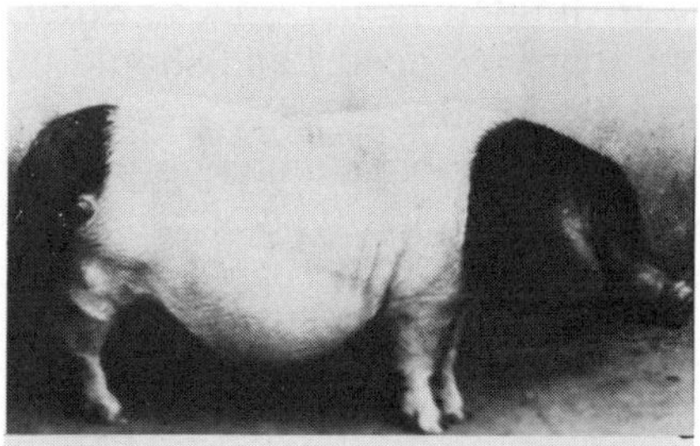

p
Jinhua gilt, Zhejiang Province, China

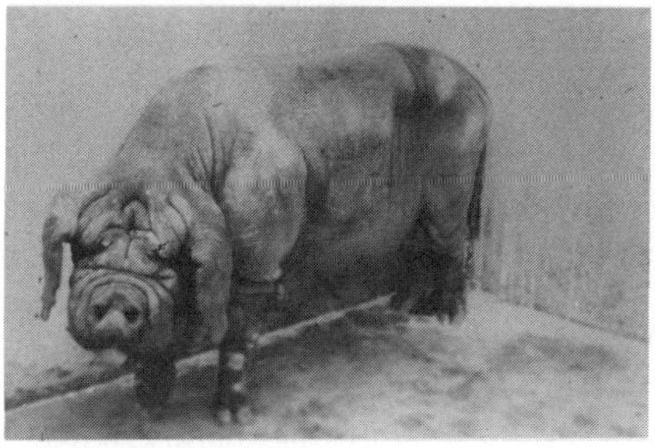

q
Taihu-type boar,
Eastern Coast Rivers area, China

r
Taihu-type sow,
Eastern Coast Rivers area, China

FIG. 2.1. See page 26 for legend.

pawing straw or other bedding into a small area of the pen. Colostrum usually can be milked from the teats at the time of nest-building. Nest-building is associated with distress and irritability and is followed, shortly before farrowing, by a period of up to several hours during which the sow lies quietly before labor begins. Labor includes restlessness, repeated rising and lying down, and straining. When farrowing begins, the sow usually remains lying down until the last pig is born, sometimes a period of several hours. Milk letdown is continuous during farrowing and the first-born pigs find their way to a teat before their littermates are born. The sow is usually indifferent to their presence. Occasionally sows are hostile to their pigs during farrowing and may bite them or throw them violently across the pen, causing serious injury or death. When this happens, pigs should be protected from the sow until farrowing is completed and then allowed free access to the sow only with caution.

A type of social dominance order, such as in the establishment of teat position, is evident very early in the life of the pig. This same social dominance order prevails when strange pigs are penned together after weaning. There is invariably fighting, manifested mainly by biting and head confrontation, which ultimately ends in a social hierarchy. The same pattern exists among sows kept together and fed with limited feeder space. Strange boars fight fiercely, often resulting in serious injury or death, and should not be penned together unless introduced as immature animals.

A common problem among growing pigs kept in confinement is tail biting. The cause of this peculiar type of cannibalism is not known, but hanging a rubber tire or similar object in the pen sometimes seems to alleviate the problem partially. Usually one or two pigs in a group are responsible for starting the tail biting and

---

FIG. 2.1. Swine breeds and types common in the United States and in other parts of the world. Pictured (from left to right on p. 24) are the following: (a) Poland China sow. (b) Chester White gilt. (c) Spotted gilt. (d) Miniature sow: Originally developed at Hormel Institute, Austin, Minnesota, for use in biomedical research. Shown with litter sired by Yorkshire boar. (e) Yorkshire boar. (f) Danish Landrace gilt. (g) Babcock C-Line GP hybrid boar: *Photo is courtesy of Babcock Swine, Inc., Rochester, Minnesota.* (h) Babcock $\frac{2}{3}$ F-Line-$\frac{1}{3}$ C-Line hybrid gilt: *Photo is Courtesy of Babcock Swine, Inc., Rochester, Minnesota.* (i) Hampshire gilt. (j) Duroc boar. Pictured (on p. 25) are the following: (k) Berkshire barrow. (l) Synthetic-Line 24 boar (Pig Improvement Co., Inc.): Original breed makeup was Duroc 50, Pietrain 25, and Large White 25%. (m) Belgian Pietrain pig. (n) Camborough Blue gilt: Original breed makeup was British Saddleback and British Landrace. (o) Ming sow from Northeast China: *Photo is courtesy of Professor Peilieu Cheng, Chinese Acad. Agric. Sciences, Beijing.* (p) Jinhua gilt from Zhejiang Province, China: *Photo is courtesy Professor Peilieu Cheng, Chinese Acad. Agric. Sciences, Beijing.* (q) Taihu-type boar from Eastern Coast Rivers area of China: *Photo is courtesy of L. D. Young, USDA/ARS.* (r) Taihu-type sow from Eastern Coast Rivers area of China: *Photo is courtesy of L. D. Young, USDA/ARS.*

Current information available (Legault, C. 1983. 34th Annu. Meet. of European Assoc. Anim. Prod., Madrid, Spain, October 3–6) supports the claim that some Chinese breeds of swine may reach puberty at less than 4 months of age, produce 17–20 live pigs per litter, yield more milk, and have several more functional nipples than U.S. and European breeds.

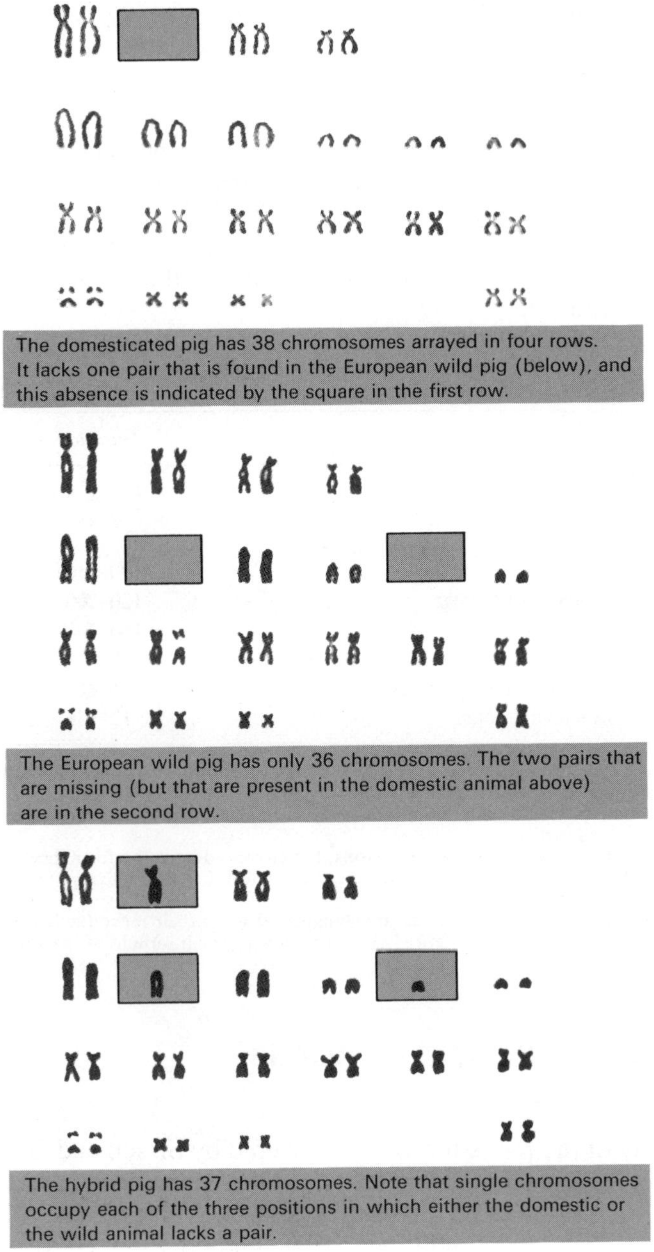

FIG. 2.2. Chromosomes of the domestic pig, European wild pig, and hybrid pig.
Courtesy of AEC Agricultural Research Laboratory, University of Tennessee, Knoxville.

the habit seems to spread to others. The infections and occasional deaths that result have led most commercial producers to the practice of removing the tail from each pig at birth to eliminate the problem.

LIFE CYCLE

The life cycle can be divided conveniently into the prenatal period, the suckling period, the growing-finishing period, and the mature or reproductive period. The time sequence of the life cycle is outlined in Table 2.3. Each has its own peculiar husbandry and management requirements. Suitable pen and housing arrangements that take into account these characteristics of the pig are described in detail in Chapters 18 and 19.

TABLE 2.3. The Life Cycle of the Pig

Period	*Duration*
Prenatal period, days	114 ± 4
Suckling period, wks	3–8[a]
Growing-finishing period (to 90-kg body wt), days	90–150[b]
Age at 90 kg, days	120–200[b]
Age at puberty	150–200
Optimum age at first service, months	7–8[c]
Reproductive longevity, years	4–8
Longevity, years	12–15[d]

[a]Pigs can be weaned to cow milk or synthetic diets at birth but it is more common in commercial production to wean at three weeks or older.
[b]Boars and male castrates grow slightly faster than females, so can be expected to reach 90 kg body wt several days earlier.
[c]Since puberty is reached before growth stops, better reproductive performance is obtained (large litters in females and more sperm produced in males) if reproductive service is not initiated before 7 months of age.
[d]Most female and male breeding stock are slaughtered when their reproductive efficiency begins to decline, so the longevity figure is based on a very small sample of the total swine population.

GROSS ANATOMY AND CARCASS MEASUREMENTS

The anatomy of the pig is thoroughly covered by Sisson and Grossman (1953), so the description of anatomical details need not be included here. Gross anatomy of the pig fetus is described in several dissection guides for students (Walker 1964; Holstad 1956; Stuart and Pinder 1948; Gilbert 1966; Harrison 1971). Some details of structure and functions of some of the major internal organs of the pig are presented in subsequent sections under the appropriate headings (see Chapters 4–9).

There is considerable variation among pigs in the number of ribs and thoracic and lumbar vertebrae. Results of extensive X-ray studies of live animals and examination of carcasses by several workers (Shaw 1930; Berge 1948; Freeman

TABLE 2.4. Carcass Traits of Winning Barrows at the National Barrow Show, Austin, Minnesota, 1981

Breed	*No. of animals*	*Length (cm)*	*10th rib back fat (cm)*	*Loin eye area (cm)*	*Age at 100 kg (days)*
Berkshire	31	79.5	2.90	29.2	175
Chester White	43	78.0	3.00	29.8	179
Duroc	113	79.3	2.53	29.6	177
Hampshire	63	79.5	2.00	32.5	182
Landrace	41	82.0	3.05	26.3	175
Poland China	47	78.3	2.38	32.0	177
Spotted	60	80.5	2.90	28.7	173
Yorkshire	73	81.5	2.73	26.9	176
Crossbred	107	80.5	2.85	32.4	170
Average, 1981	578	80.0	2.68	29.9	176

Source: Anon. 1981.

1939) show that lumbar and thoracic vertebrae are variable in number. With an increase in the number of ribs the number of lumbar vertebrae decreases. The heritability of pre-sacral vertebrae number is high and has been estimated at 0.74 (Shaw 1930). (For a definition and discussion of heritability, see Chapter 4.) There are breed but not sex differences in average vertebrae number, additional ribs being generally associated with breeds of larger size. Selection for carcass length (distance from first cervical vertebrae to aitch bone or from first rib to aitch bone) has probably resulted in an increase in the average number of vertebrae in selected populations.

A problem of practical importance in some strains of pigs is the tendency toward unequal development in the size of the members of digit pairs (toes) on the front and/or hind hoofs. Usually the inside toe is shorter than the outside toe so that the two do not make equal contact with the ground, and body weight is unequally distributed, sometimes resulting in lameness and soreness. When the pasterns (second and third phalanges) are long the defect is more noticeable than when the pasterns are short.

In commercial swine production, carcass merit is a major consideration. Although no breed has a monopoly on superior animals in any one trait, Jensen *et al.* (1967) have made comparisons in some economically important carcass traits among some important U.S. breeds of swine. The carcass measurements shown in Table 2.4 are those of winning carcass barrows from various breeds at the National Barrow Show, Austin, Minnesota, during the year 1981 (Anon. 1981). The animals in competition are generally from the superior herds of each breed. Therefore, values for back fat, ham weight, and loin eye area would be better than the average values for all pigs being slaughtered in the United States during the same period. The points of reference on the carcass where these measurements are taken are shown in Fig. 2.3. Variations from this standard procedure often are used, and exact measurements taken will be influenced by the relative economic value of parts of the carcass in a particular country or geographic area. Pictures of pigs representing each of the official USDA slaughter grades of swine are shown in Fig. 2.4 (USDA 1970).

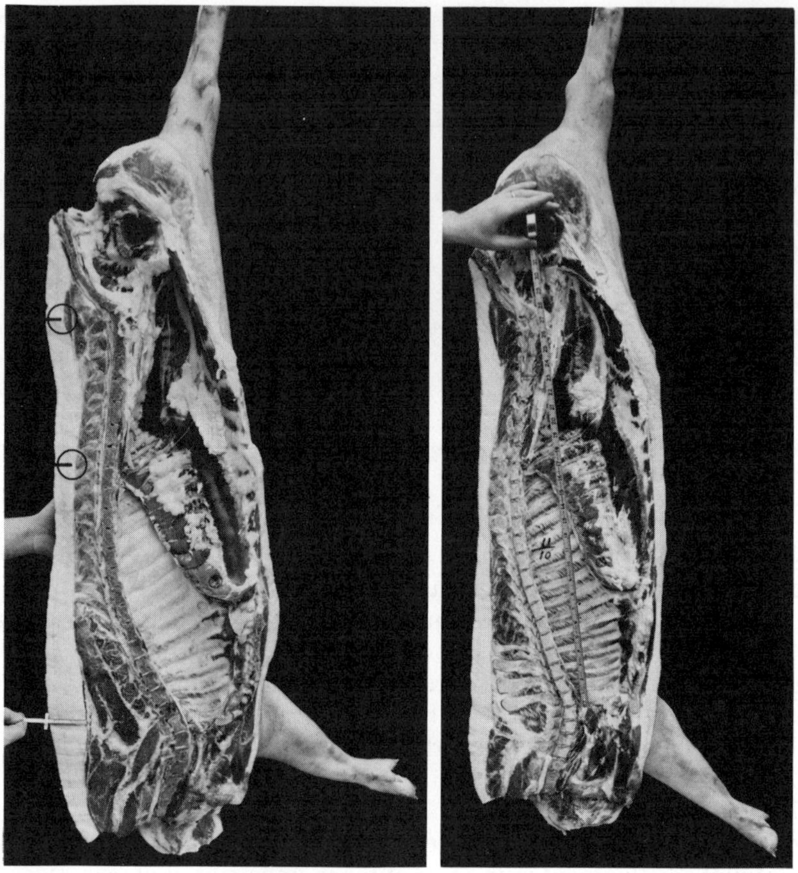

FIG. 2.3. Carcasses of fat and lean pigs showing points of reference for carcass measurements.

PLASTICITY OF THE PIG

Swine breeders have long recognized the fact that to meet changing consumer demands, changes in the general shape and type of pig can be made by selection based on visual appraisal. This is exemplified by the cyclic changes that have occurred in the type of pig produced in the United States during the past half-century. When lard was high priced, selection pressure was exerted for short, chuffy pigs yielding a high percentage of fat. From this extreme the pendulum swung to the long, narrow, slowly maturing hog. In World War II, the demand for lard was again increased and swine breeders responded by selecting fat pigs to meet this demand. With the present demand for lard greatly reduced in the United States by competition from vegetable oils and synthetic detergents, the goal is for pork with a minimum amount of fat. In some countries lard is highly valued and pigs are selected accordingly. Rapid changes can be made in the kind of pig produced as a result of the high degree of heritability of body type (Hetzer *et al.* 1944), the relatively short generation interval of pigs, and their high prolificacy.

For example, the selection of swine for small size can be accomplished in a

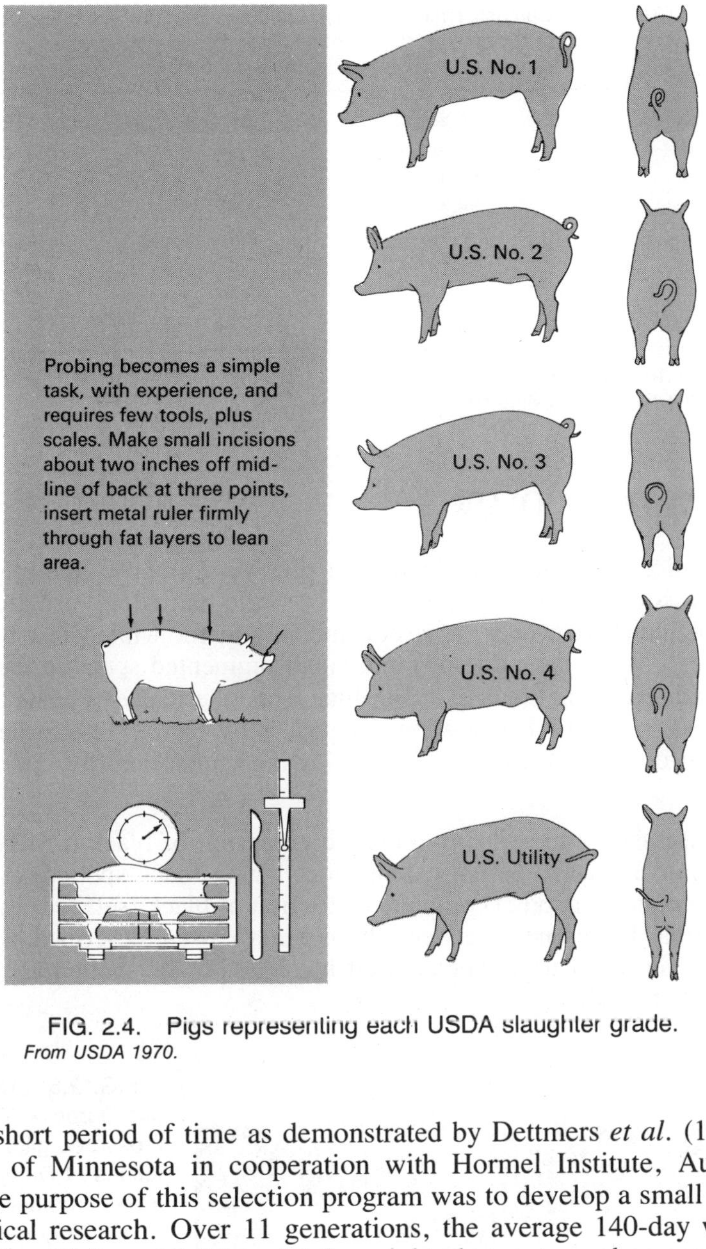

FIG. 2.4. Pigs representing each USDA slaughter grade.
From USDA 1970.

relatively short period of time as demonstrated by Dettmers *et al.* (1965) at the University of Minnesota in cooperation with Hormel Institute, Austin, Minnesota. The purpose of this selection program was to develop a small pig for use in biomedical research. Over 11 generations, the average 140-day weight was reduced from 39.2 to 17.5 kg. Birth weight, but not number per litter, was reduced as well, although the only criterion of selection was body weight at 140 days of age. Carcass traits can also be altered by selection as clearly shown by Clausen (1970) based on the results of the Danish progeny-testing program summarized in Table 2.5. Similar responses to selection for back fat have been reported elsewhere by Hetzer and Harvey (1967) and Berruecos *et al.* (1970). Extreme differences in back-fat thickness between individuals from populations selected for low or high fatness are evident not only from carcass measurements but from general appearance (Fig. 2.5).

TABLE 2.5. Changes in Carcass Traits of Danish Landrace Pigs, in Response to Selection

Year	*Body length* (*cm*)	*Back fat* (*cm*)	*Area of loin eye* (*sq cm*)
1926–27	88.9	4.05	—[a]
1936–37	92.8	3.49	—[a]
1946–47	93.4	3.36	—[a]
1957–58	94.8	3.05	29.3[b]
1965–66	96.3	2.46	30.1
1966–67	96.3	2.41	31.1
1967–68	96.0	2.36	32.5
1968–69	96.4	2.34	31.8

Source: Clausen et al. 1970.
[a]A dash indicates that the data were not available.
[b]1959/60.

HAIR AND SKIN COLOR

The hair color and skin pigmentation of pigs vary greatly. Skin pigmentation generally is associated with hair color; that is, white-haired pigs usually have very little skin pigment. However, it is not uncommon for white pigs, even within recognized breeds, to have one or more small pigmented spots on the skin even though the hair covering these spots is white. The inheritance of coat color in pigs was described in the first of a series of papers by Hetzer (1945) dealing with results of various breed crosses. The following categories of coat color were given:

1. Tortoise-shell patterns: (a) complete extension of black (self black) as in unbelted Hampshires, (b) partial extension of black (black spotted) as in Poland China and Berkshire, and (c) complete restriction of black (self red) as in Durocs and Tamworths. Differences between these patterns were described as being due primarily to a series of triple alleles. Self black and black spotted are apparently both dominant to red.

FIG. 2.5. Representative Duroc × Yorkshire crossbred gilts from populations of Durocs and Yorkshires selected for about 20 generations for low- or high-back-fat thickness. Note the lean, long appearance of the gilt on the left (low back fat) compared with the fat, short appearance of the gilt on the right (high back fat).
Photo: Courtesy of U.S. Department of Agriculture.

2. White patterns: (a) self white of the U.S., British, German, and Scandinavian breeds; (b) slaty or dirty white on dark skin found in the Mangalitza breed; (c) white belt of the Hampshire, Wessex Saddleback, Essex, and Hanovarian; and (d) white responsible for the white feet, nose, and tail of Poland China or Berkshire breeds.

Self white is dominant over most black and red breeds, although blue-gray or blue-roan individuals have been produced in some crosses. Also, the slaty white of the Mangalitza has been reported dominant over the self white of Norwegian Landrace. The width of the white shoulder belt of Wessex Saddleback (Donald 1951) and presumably of the Hampshire is apparently associated with three genotypes: homozygous belted pigs have a medium-to-wide belt; heterozygous belted pigs have a narrow-to-medium belt; and homozygous black pigs have no belt. Persistence of occurrence of solid black pigs apparently is due to the breeder's preference for a narrow or medium belt in selection.

The wide range of color patterns (white, black, red, gray, brown, and all combinations) among crossbred pigs on commercial farms in the United States is ample evidence for the complex nature of coat color inheritance in pigs.

The wild European hog has characteristic longitudinal stripes which are more conspicuous at birth than later. Occasionally the texture of the hair of newborn pigs of some domestic pigs also suggests the presence of this characteristic hair pattern.

A peculiar type of curly (woolly) hair occurs in Conastras (Brazilian) pigs, not associated with a particular color or sex, but apparently under the influence of a single completely dominant gene. The hair shaft of "woolly" hair is oval or flat in cross-section as compared to the round shaft of normal, straight hair.

Pigs in which hair very sparsely covers the body are represented by the Extremadura breed of Spain and Tatu breed of Brazil. The trait is apparently controlled by an incompletely dominant gene for hair, so that the heterozygous individual is midway between the two homozygous extremes in the amount of hair covering. Mating two hypotrichotic pigs results in 100% hypotrichotic offspring.

REFERENCES

ANON. 1980. National Hog Farmer, April 15, 1980, p. 24.

ANON. 1981. Hormel Farmer Magazine *36*(2),20. Geo. A. Hormel & Co., Austin, Minnesota.

BERGE, S. 1948. J. Anim. Sci. *7*,233.

BERRUECOS, J. M., DILLARD, E. U., and ROBISON, O.W. 1970. J. Anim. Sci. *30*,844.

BRIGGS, H. M. 1958. Modern Breeds of Livestock, 3rd Edition. Macmillan Co., New York.

CLAUSEN, H. 1970. Sammenlignende forsøg med. svin. *379*,74–75. Nat. Agric. Research Laboratory, Copenhagen, Denmark.

DETTMERS, A. E., REMPLE, W. E., and COMSTOCK, R. E. 1965. J. Anim. Sci. *24*,216.

DONALD, H. P. 1951. J. Agric. Sci. *41*,214.

FREEMAN, V. A. 1939. J. Hered. *30*,61.

GILBERT, S. G. 1966. Pictorial Anatomy of the Fetal Pig, 2nd Edition. Univ. of Washington Press, Seattle.

HAFEZ, E. S. E., SUMPTRON, L. J., and JAKWAY, J. S. 1962. *In* The Behavior of Domestic Animals, p. 334. E. S. E. Hafez (Editor). Williams and Wilkins, Baltimore.

HARRISON, B. M. 1971. Embryology of the Chick and Pig. William C. Brown, Dubuque, Iowa.

HETZER, H. O. 1945. J. Hered. *36*,121; *26*,187; *36*,255; *36*,308; *38*,120; *39*,122.

HETZER, H. O., DICKERSON, G. E., and ZELLER, J. H. 1944. J. Anim. Sci. *3*,390.

HETZER, H. O., and HARVEY, W. R. 1967. J. Anim. Sci. *26*,1244.

HOLSTAD, G. E. 1956. The Fetal Pig. An Introduction to the Anatomy of the Fetal Pig. Burgess, Minneapolis.

JENSEN, P., CRAIG, H. B., and ROBISON, O. W. 1967. J. Anim. Sci. *26*,1252.

JIMENEZ, M., LOPEZ-SAEZ, J. F., and MONGE, E. G. 1962. J. Hered. *53*,281.

MASON, I. L. 1969. A World Dictionary of Livestock Breeds, Types and Varieties. Technical Communication No. 8, Commonwealth Bureau of Animal Breeding and Genetics, Edinburgh. Commonwealth Agric. Bureau, Farnham House, Farnham Royal, Slough, Bucks., England.

NORDBY, J. E. 1939. J. Hered. *30*,307.

POND, W. G., and HOUPT, K. A. 1978. The Biology of Pig, pp. 1–371. Cornell University Press, Ithaca, New York.

SHAW, A. M. 1930. Sci. Agr. *10*,690.

SISSON, S., and GROSSMAN, J. D. 1953. The Anatomy of the Domestic Animals, 4th Edition. W. B. Saunders Co., Philadelphia.

STUART, R. R., and PINDER, D. E. 1948. The Anatomy of the Fetal Pig. Denoyer-Geppert, Chicago.

USDA. 1970. Grades for Slaughter Swine and Feeder Pigs. Marketing Bull. No. *51*, U.S. Govt. Printing Office, Washington, D.C.

WALKER, W. F., JR. 1964. Dissection of the Fetal Pig, Laboratory Studies in Biology. W. H. Freeman and Co., San Francisco.

3

Pork as Human Food

INTRODUCTION

Because pigs are produced mainly as a source of human food, it is important to consider the characteristics of pork in terms of its acceptance by humans and its nutritive value for humans. Those associated with the swine industry must not lose sight of the fact that the consumer ultimately determines the amount and kind of pork produced. The purpose of this chapter is to characterize pork as to nutrient composition and other properties that affect its acceptability to humans.

NUTRIENT COMPOSITION OF PORK

Pork is an excellent source of high-quality protein, B vitamins, and trace elements. Its chemical composition has been described by Leverton and Odell (1958), Watt and Merrill (1963), Kiernat *et al.* (1964), by Moss *et al.* (1983) and by USDA (K. Ono 1982, unpublished data) from analyses of carcasses from contemporary swine of considerably leaner composition than those used in earlier studies. The dramatic changes that have occurred in the fat content of pork since the 1960s are illustrated in Fig. 3.1. Rice (1971) drew on much of the earlier work in reviewing the nutritional value of pork and other meat products. Table 3.1 contains average percentages of water, protein, and fat in raw fresh retail cuts of pork, and the average calorie content of each cut from lean (U.S. Grade 1) moderate (U.S. Grade 2) and fat (U.S. Grade 3) pigs. It is clear that the overall gross chemical composition of these cuts is similar. Cooking reduces the water content so that percentages of protein, fat, and ash are increased appreciably and calorie concentration is increased. Roasted lean pork may contain as little as 82% of the water and up to 150% of the protein of raw pork (Kiernat *et al.* 1964).

The type of cooking will, of course, have an influence on the nature of the weight loss. Any cooking method, such as frying or charcoal broiling, that results in dripping losses of fat would result in a reduction in percentage of fat as well as of water in the final product. Table 3.2 contains data on percentages of connective

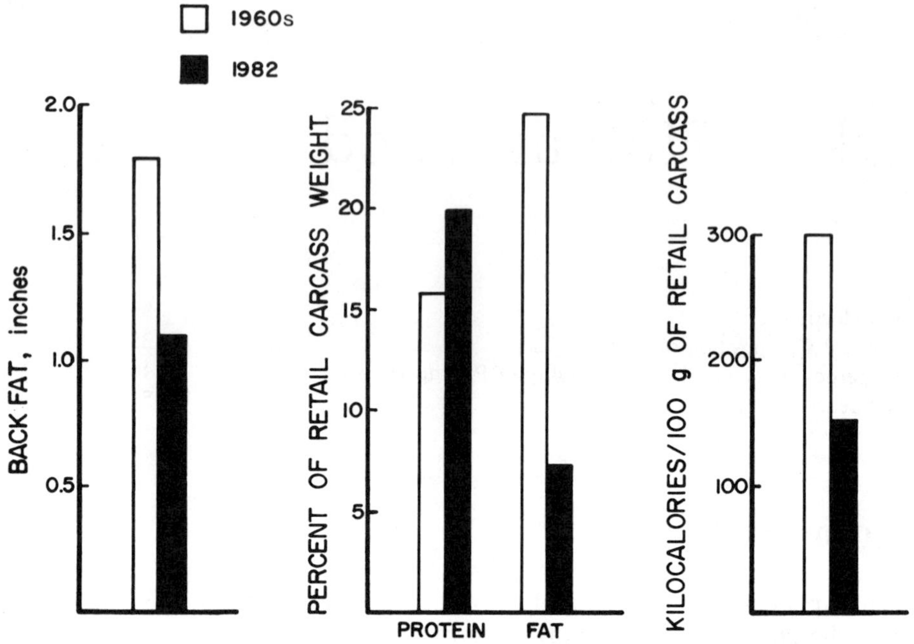

FIG. 3.1. Changes in fat content of pork since the 1960s as illustrated by back fat thickness, percentages of protein and fat in the retail carcass, and kilocalories of energy per gram of retail carcass.

TABLE 3.1. Water, Protein, Fat and Calorie Content of Raw and Cooked Fresh Retail Cuts of Pork

Retail cut	*USDA grade*	*Back fat (cm)*	*Protein (%)*	*Fat (%)*	*Water (%)*	*kcal/100g*
Picnic	1	2.78	19.9	7.2	73.6	150
	2	4.01	19.4	7.4	73.3	150
	3	4.51	19.2 (15.8)[a]	6.8 (24.7)	74.0 (58.9)	144 (290)
Ham	1	2.78	20.0	5.5	74.0	140
	2	4.01	19.4	5.6	73.9	135
	3	4.51	20.5 (15.9)	6.2 (26.6)	73.6 (56.5)	144 (308)
Loin	1	3.00	20.7	6.9	71.5	151
	2	4.09	20.6	8.1	70.7	161
	3	4.74	20.2 (17.1)	9.8 (24.9)	69.2 (57.2)	175 (298)
Spare rib	1	2.49	17.8	26.2	57.7	312
	3	4.23	17.0 (14.5)	25.2 (33.2)	58.1 (51.8)	300 (361)

Source: USDA (unpublished data supplied by K. Ono, 1982, Meat Science Research Laboratory, Agricultural Research Service, Beltsville, Maryland).

[a]Values in parentheses are from Kiernat et al. (1964) based on data from pork produced in the 1950s and early 1960s. Note the dramatic decrease in fat content of picnic, ham, and loin and the increase in protein content during the 20-year period. These dramatic changes have resulted from improved feeding and genetic selection for leanness.

TABLE 3.2. Percentages of Lean, Bone, Connective Tissue, and Intramuscular and Subcutaneous Fat in Retail Cuts of U.S. Grade 1 Pork Carcasses

Retail cut	*Lean*	*Connective tissue*	*Bone*	*Fat*		
				Intramuscular	*Subcutaneous*	*Total*
Picnic	55.6	8.9	16.8	9.1	10.4	19.5
Blade loin roast	53.6	4.9	20.7	10.6	10.2	20.8
Ham rump	69.5	4.9	12.8	4.6	8.3	12.9
Ham shank	62.5	7.7	11.6	5.0	13.1	18.1
Rib chop	57.9	3.7	18.5	6.8	14.8	21.6
Shoulder blade steak	68.0	5.5	9.2	10.8	9.0	19.8
Sirloin roast	57.8	5.9	16.6	4.0	15.7	19.7
Spare rib	58.4	0.7	40.4	7.6	0.03	7.6

Source: USDA (unpublished data supplied by K. Ono, 1982, Meat Science Research Laboratory, Agricultural Research Service, Beltsville, Maryland).

tissue, lean, bone and intramuscular and subcutaneous fat in retail cuts of U.S. Grade 1 pork carcasses (K. Ono 1982, unpublished data). The protein content of 20–30% on a fresh basis is higher than nearly all plant materials and similar to most foods of animal origin. The National Research Council Food and Nutrition Board (1980) recommends a daily protein intake of 56 g for young adult men (70 kg, 177-cm tall) and 44 g for young adult women (55 kg, 163-cm tall). Therefore, the entire daily protein requirement is met by a 150–280-g serving of lean pork. Organs such as brain, heart, and liver are also good sources of protein and energy (Table 3.3). The amino acid composition of the protein is of importance, as is total amount of protein. Most animal products, including pork, provide a better balance of essential amino acids than most plant products provide. This is illustrated in Table 3.4 which lists the essential amino acid composition of pork and casein, which is the major protein in milk (used to represent animal proteins), and that of maize and peanut meal (used to represent plant proteins). Both pork

TABLE 3.3. Water, Protein, Fat, Ash, Carbohydrate, and Calorie Content of Fresh Organs of Pigs

Organ	*Water* (%)	*Protein* (%)	*Fat* (%)	*Ash* (%)	*Carbohydrate* (%)	*Calories* (*no. per 100 g*)
Brain	77.2	11.5	8.5	1.3	1.5[a]	129
Heart	77.4	16.8	4.4	1.0	0.4	113
Kidney	81.0	15.1	2.7	1.1	0.1[a]	—
Liver	71.6	20.6	3.7	1.5	2.6	131
Lung	81.1	15.2	1.9	0.9	0.9[a]	82
Pancreas	69.2	20.7	8.6	1.4	0[a]	160
Spleen	76.8	19.0	2.8	1.4	0[a]	101
Tongue	66.1	16.8	15.6	1.0	0.5	215

Source: Kiernat et al. 1964.
[a]Determined by difference where actual values were not available.

TABLE 3.4. Amino Acid Composition of Fresh Pork, Casein (a Good-Quality Protein), Peanut Meal (a Poor-Quality Protein), and Maize (a Poor-Quality Protein)

Amino acid	*Percent of protein*			
	Pork	*Casein*	*Peanut meal*	*Maize*
Arginine	6.4	3.9	12.4	4.6
Histidine	3.2	3.2	2.4	2.9
Isoleucine	4.9	5.1	3.8	5.0
Leucine	7.5	7.6	6.8	16.1
Lysine	7.8	8.2	3.7	2.8
Methionine–cystine	3.8	3.1	2.8	4.9
Phenylalanine	4.1	4.1	6.3	5.4
Threonine	5.1	4.1	3.1	3.5
Tryptophan	1.4	1.0	1.0	0.6
Valine	5.0	6.0	4.2	7.2

Source: Schweigert and Payne 1956.

protein and casein are adequate in amino acid balance to support good growth, although casein is slightly deficient in sulfur-containing amino acids (methionine and cystine). Most plant proteins are deficient in lysine, tryptophan or methionine–cystine, or a combination of two or more of them. A comparison of the amounts of lysine and methionine–cystine in peanut meal and of lysine and tryptophan in corn with the amounts of these amino acids in pork protein and casein (Table 3.4) illustrates this characteristic of plant proteins. Thus, a diet consisting entirely of plant foodstuffs is almost certain to be deficient in at least one of the essential amino acids for man. Pork in the diet, even if only a small proportion, can measurably improve the protein quality of the diet.

Pork is also an excellent source of some of the mineral elements and a poor source of others. For example, it is high in phosphorus, but almost devoid of calcium; it is high in potassium, but rather low in sodium; it is a good source of iron, zinc, manganese, and magnesium. Thus, if pork were the only source of food for animals and man, a deficiency of calcium would most certainly occur, but the supply of many of the elements would be sufficient.

Fortunately, the biological availability of minerals in pork and in the skeleton (bone) is very high and, unlike the case with some plant sources of minerals, there is no problem with naturally occurring chelating agents which interfere with the utilization of the minerals present. Some organs are especially good sources of particular mineral elements. For example, pork liver is an excellent source of iron. The amounts of trace elements such as iron, copper, zinc, and manganese deposited in the liver of the pig are related to the diet content of each mineral. That is, pigs fed diets high in copper tend to accumulate copper in the liver. Gipp (1971) found levels of copper as high as 1400 ppm (0.14%) in the liver of pigs fed a diet containing 250 ppm copper. Unfortunately, toxic elements such as cadmium, lead, and mercury also tend to accumulate in the liver and kidneys of pigs ingesting them, so that a human health hazard can be created by consumption of organs from such animals. However, a steady diet of liver or kidneys or other organ meats is not likely, so that a single meal of organs from a pig exposed to

TABLE 3.5. Total Ash Content (%) and Selected Mineral Content (mg/100 g) of Pork and Organs of Pigs

Mineral	*Pork*	*Brain*	*Heart*	*Kidney*	*Liver*	*Pancreas*	*Spleen*	*Tongue*
Ash	1.2	1.4	1.0	1.2	1.5	1.1	1.4	1.0
Calcium	9	10	3	11	10	11	—	29
Phosphorus	175	312	131	218	356	282	298	186
Iron	2.3	2.4	3.3	6.7	19.2	1.0	29.4	1.4
Sodium	70	125	54	115	73	44	—	—
Potassium	285	219	106	178	261	217	—	—
Magnesium	18	—	—	—	16	—	—	—

Sources: Watts and Merrill 1963; Rice 1971.

high levels of a toxic element is of less potential danger. Average concentrations of several required elements present in various organs of pigs are shown for comparison in Table 3.5.

The water-soluble vitamin content of pork is high compared to most food products. Also, the organs commonly used for food, especially the liver, are unusually good sources of the vitamins. Average concentration of vitamins in raw and cooked fresh ham and loin and in cured ham are given in Table 3.6, and concentrations in various organs are in Table 3.7. Pork is especially high in thiamin and, as shown in Table 3.6, a 100-g serving of pork loin provides 70% of the daily requirement for adult man. The same 100-g serving supplies substantial proportions of the daily requirements of other vitamins as well. Liver is an excellent source of vitamin A. Vitamin B_{12} is a component of animal and microbial products only, so unless a meat, dairy, or poultry product is eaten, a vitamin B_{12} deficiency may result. For comparison, the concentrations of three important water-soluble vitamins, thiamin, riboflavin, and niacin, in a variety of plant and

TABLE 3.6. Water-Soluble Vitamin Content (mg/100g) of Raw and Cooked Pork[a,b] and Percentage of Daily Adult Human Requirement

Vitamin	*Raw pork*				*Cooked pork*			*% of daily adult requirement*
	Fresh ham	*Cured ham*	*Fresh loin*	*Fresh composite*[b]	*Fresh ham*	*Cured ham*	*Fresh loin*	*from 100 g of loin*[c]
Thiamin	0.77	0.72	0.98	—	0.51	0.47	1.13	70
Riboflavin	0.19	0.19	0.24	0.32	0.23	0.18	0.33	14
Niacin	4.1	4.1	5.2	—	4.6	3.6	6.8	29
Pantothenic acid	0.72	0.64	2.0	—	—	—	—	—
Vitamin B_6	0.42	0.39	0.50	0.52	—	—	—	25
Vitamin B_{12}	0.0009	0.001	—	0.0009	—	—	—	20
Folic acid	0.009	0.011	0.007	—	—	—	—	2
Biotin	0.0053	0.005	0.0055	—	—	—	—	—

[a]Leverton and Odell 1958; Watt and Merrill 1963; Kiernat et al. 1964.
[b]Moss et al. 1983.
[c]National Research Council 1980.

TABLE 3.7. Content of Water-Soluble Vitamins and Vitamin A in Organs of Pigs (mg/100 g)

Vitamin	*Brain*	*Heart*	*Kidney*	*Liver*	*Lung*	*Pancreas*	*Spleen*	*Tongue*
Thiamin	0.23	0.43	0.58	0.30	0.09	0.11	0.13	0.17
Riboflavin	0.26	1.24	1.73	3.03	0.27	0.46	0.30	0.29
Niacin	4.4	6.6	9.8	16.4	3.4	3.5	4.3	5.0
Pantothenic acid	2.8	2.5	—[a]	6.6	0.9	4.6	1.1	—
Vitamin B_6	—	0.35	—	0.5	—	—	—	—
Vitamin B_{12}	0.0028	0.0024	—	0.023	—	0.0065	0.0041	—
Folic acid	—	—	—	0.074	—	—	—	—
Biotin	—	0.018	—	0.085	—	—	—	—
Choline	—	231	262	552	—	329	208	137
Vitamin C	18	3	12	23	—	—	—	—
Vitamin A[b]	—	30	130	10,900	—	—	—	—

Sources: Watt and Merrill 1963; Kiernat et al. 1964.
[a]A dash indicates that the value is not available.
[b]International Units per 100 g.

animal products is listed in Table 3.8 along with the amount of each vitamin supplied by each product as a percentage of that supplied by an equal amount of lean pork. Pork is by far the best source of thiamin among both plant and animal products with the exception of peanuts, and of yeast, which is notable as a B-vitamin source. Beef and lamb contain only 9 and 17% of the amount of thiamin provided by pork. Only some of the seed legumes, including red beans, cowpeas and soybeans, and the cereal grain, wheat, contain half or more of the amount of thiamin supplied by pork. Pork is comparable to milk, beef, and lamb as a source of riboflavin and niacin and is considerably higher than almost all plant products in these two vitamins. Furthermore, the niacin of cereal grains and maize, although approaching the level of that found in pork, is almost completely unavailable to the animal as it is chemically bound in a form not easily broken down in digestion. Thus, the value of pork as a source of niacin is even greater than tabular values of food composition would indicate.

As pork contains a relatively high proportion of fat compared to most plant products, it is of importance to be aware of the composition of the fat. The controversial question of the relationship between dietary fat and heart disease has not been answered, but it is often assumed that animal fats may enhance the development of atherosclerosis. A large long-term study completed among humans in Framingham, Massachusetts, summarized by Kannel and Gordon (1971) has failed to incriminate diet as a major factor.

Other more recent studies suggest a probable relationship between high levels of dietary fat and heart disease in some population groups. In response to such a possible relationship, U.S. pork producers have reduced the average lard produced from about 14% of the carcass in 1960 to less than 5% in 1983 (Fredeen and Harmon 1983). Pork carcass composition can be estimated accurately by measuring back fat thickness and cross-sectional area of the loin muscle in the live animal or carcass (Fahey *et al.* 1977; Edwards *et al.* 1981).

Regardless of the eventual outcome of the diet–heart disease controversy, it is

TABLE 3.8. Water-Soluble Vitamin Content of Pork Compared to That of Some Other Common Foods

Food	Thiamin		Riboflavin		Niacin	
	mg/100 g	% of that supplied by lean pork	mg/100 g	% of that supplied by lean pork	mg/100 g	% of that supplied by lean pork
Pork						
Total edible	0.50	57	0.12	57	2.7	59
Separable lean	0.87	—	0.21	—	4.6	—
Separable fat	0.17	20	0.04	19	0.9	20
Apples	0.03	3	0.02	10	0.1	2
Bananas	0.05	6	0.06	29	0.7	15
Beans, red	0.51	59	0.20	95	2.3	50
Beans, snap (yellow or snap)	0.08	9	0.11	52	0.5	11
Beef	0.08	9	0.17	81	4.7	102
Cabbage	0.05	6	0.05	24	0.3	7
Cheese, cheddar	0.03	3	0.46	219	0.1	2
Cheese, cottage	0.03	3	0.25	119	0.1	2
Chicken, light meat without skin	0.05	6	0.09	43	10.7	233
Chicken, dark meat without skin	0.08	9	0.20	95	5.2	113
Clams	0.10	11	0.18	86	1.3	28
Coconut	0.05	6	0.02	10	0.5	11
Corn, yellow	0.15	17	0.12	57	1.7	37
Cowpeas	0.43	49	0.13	62	1.6	35
Eggs, whole	0.11	13	0.30	143	0.1	2
Grapefruit	0.04	5	0.02	10	0.2	4
Halibut	0.07	8	0.07	33	8.3	180
Herring	0.02	2	0.15	71	3.6	78
Ice cream (12% fat)	0.04	5	0.19	90	0.1	2
Lamb	0.15	17	0.20	95	4.8	104
Liver, beef	0.25	29	3.26	1,552	13.6	296
Liver, chicken	0.19	22	2.49	1,186	10.8	235
Liver, lamb	0.40	46	3.28	1,562	16.9	367
Liver, pork	0.30	34	3.03	1,443	16.4	357
Milk, whole fluid	0.03	3	0.17	81	0.1	2
Milk, skim fluid	0.04	5	0.18	86	0.1	2
Oranges	0.10	11	0.04	19	0.4	9
Peanuts	0.99	114	0.13	62	15.8	343
Potatoes	0.10	11	0.04	19	1.5	33
Rice, brown	0.34	39	0.05	24	4.7	102
Rice, white	0.07	8	0.03	14	1.6	35
Soybeans	0.44	51	0.16	76	1.4	30
Sugar, beet or cane	0.01	1	0.03	14	0.2	4
Sweet potatoes	0.10	10	0.06	29	0.6	13
Wheat	0.55	63	0.12	57	4.3	93
Yams	0.10	11	0.04	19	0.5	11
Yeast, dried or bakers'	2.33	267	5.41	2,576	36.7	798

Source: Watt and Merrill 1963.

of value to characterize the fat composition of pork and to clear up some often-stated but unfounded claims regarding pork. Among those claims are these two:

1. Pork is high in saturated fatty acids and cholesterol and low in unsaturated (polyunsaturated) fatty acids. *Answer:* False. Pork is higher in polyunsaturated fatty acids than beef, lamb, or milk and is lower in cholesterol than butter, cheddar cheese, eggs, organ meats, and many seafoods (Table 3.9) (Watt and Merrill 1963).

2. Pork fat is difficult to digest. *Answer:* False. Digestibility experiments with animals and man have failed to show differences between pork fat (lard) and other animal and plant fats in apparent digestibility.

Average fatty acid composition of typical pork fat taken from various body sites is shown in Table 3.10 (Austin *et al.* 1964; Rice 1971). The pattern of fatty acid distribution is similar among sites (loin, belly, ham, and shoulder) and between subcutaneous, intermuscular, and intramuscular fat within each site. Linoleic acid makes up from 8 or 9% to more than 13% of total fatty acids in a typical pork fat

TABLE 3.9. Cholesterol Content of Pork Compared to That of Some Other Common Foods

Food	*Cholesterol content per 100 g of edible portion (mg)*
Pork	61[a]
Beef	70
Brain	>2000
Butter	250
Caviar	>300
Cheese, cheddar	100
Cheese, cottage or creamed	15
Chicken	60
Corn (maize)	0
Eggs, whole	550
Egg, white	0
Egg, yolk	1500
Fish	70
Heart	150
Lamb	70
Margarine, all-vegetable	0
Lard	95
Liver	300
Milk, whole	11
Milk, skim	3
Orange	0
Oyster	>200
Rice	0
Shrimp	125
Veal	90

Source: Watt and Merrill 1963 (animal products only).
[a]Moss et al. 1983.

TABLE 3.10. Fatty Acid Composition of Pork Fat

Tissue	Fatty acids[a] (% of total contained in tissue)								
	Myristic ($C_{14:0}$)	*Palmitic* ($C_{16:0}$)	*Palmitoleic* ($C_{16:1}$)	*Stearic* ($C_{18:0}$)	*Oleic* ($C_{18:1}$)	*Linoleic* ($C_{18:2}$)	*Linolenic* ($C_{18:3}$)	*Arachidonic* ($C_{20:4}$)	*Miscellaneous*
Loin									
Subcutaneous	1.3	22.5	2.8	14.1	42.5	13.5	0.9	0.3	2.1
Intermuscular	2.0	22.5	2.2	14.2	42.9	11.9	0.9	0.4	2.0
Intramuscular	1.8	21.3	4.2	13.4	32.8	13.4	0.3	3.2	4.6
Belly									
Subcutaneous	1.1	23.3	3.5	8.7	50.2	10.8	0.8	0.2	1.4
Intermuscular	1.6	25.7	1.7	11.5	40.5	10.3	0.7	0.4	7.6
Intramuscular	1.3	22.7	4.3	10.9	46.5	9.4	0.7	0.5	3.7
Ham									
Subcutaneous	2.2	22.5	3.5	9.4	47.8	11.6	0.8	0.3	1.9
Intermuscular	2.2	21.3	3.1	12.6	36.9	13.3	1.1	0.7	8.8
Intramuscular	2.1	21.2	4.3	11.5	38.8	13.7	0.6	3.8	4.0
Shoulder									
Subcutaneous	1.7	21.9	3.4	11.5	45.8	12.8	1.0	0.4	1.5
Intermuscular	1.6	24.3	2.5	14.9	43.5	11.5	0.8	0.3	0.6
Intramuscular	2.1	22.9	3.8	13.4	43.5	10.0	0.7	1.2	2.4

Sources: Austin et al. 1964; Rice 1971.

[a]Number before colon refers to the length of the carbon chain; number after colon refers to the number of double bonds; for example, $C_{14:0}$ is a fatty acid with 14 carbons and no double bonds.

sample. Linolenic and arachidonic acids, the other two predominant polyunsaturated fatty acids, contribute an additional 1–4 or 5% of total fatty acids in the fat from a typical pig. Thus, pork fat can be considered a good source of polyunsaturated fatty acids. Fatty acid composition of back fat is affected by sex (boars have a higher content of linoleic acid than gilts), sire, and diet (Koch *et al.* 1967). As shown in Table 3.11 (Dugan 1957), the fatty acid composition of pork fat is affected to a great extent by diet. The fatty acid pattern of the carcass fat tends to resemble that of the dietary fat. For example, it is possible to increase the percentage of linoleic acid in the depot fat substantially by feeding a diet high in linoleic acid. Feeding a diet containing 3% safflower oil, which contains more than 50% linoleic acid, results in a carcass containing 17% or more of linoleic acid in the depot fat. This flexibility in the fatty acid composition of pork allows the production of pork to meet consumer demands for polyunsaturated fat. Although there is still some stigma attached to "soft pork" by meat packers and others, taste panel studies of pork produced from pigs that had been fed diets of whole maize and cooked soybeans, which resulted in a linoleic acid content of 14.1% in the back fat, failed to show any objectionable flavor or any decline in acceptability of the pork (Wahlstrom *et al.* 1971). There are potential problems with storage and rancidity of "polyunsaturated pork," but proper refrigeration and handling should avoid such problems.

The proportions of ash, protein, water, and fat in the pig carcass are controlled by genetics as well as by environmental factors. Suffice it to say here, that the proportions of fat and water in the edible carcass vary inversely with each other,

TABLE 3.11. Fatty Acid Composition (%) of Pork Back Fat from Pigs Fed Various Diets

Fatty acid	Composition in typical pork fat[a]	Composition in fat from pigs fed various diets Diet description			
		3% coconut fat[b]	*3% safflower oil*[b]	*Maize–soybean meal*[b]	*Maize–whole soybeans*[c]
Capric ($C_{10:0}$)	—[d]	0.18	0.20	0.11	—
Lauric ($C_{12:0}$)	—	0.63	0.12	0.12	—
Myristic ($C_{14:0}$)	—	3.85	2.00	1.76	1.2
Palmitic ($C_{16:0}$)	28	20.00	15.60	19.31	28.7
Palmitoleic ($C_{16:1}$)	3	16.47	13.01	12.68	1.5
Stearic ($C_{18:0}$)	13	55.93	50.58	55.76	9.4
Oleic ($C_{18:1}$)	46				
Linoleic ($C_{18:2}$)	10	0.93	17.15	8.32	14.1
Linolenic ($C_{18:3}$)	0.7	0.09	0.09	0.09	—
Eicosamonoenoic ($C_{20:1}$)	—	0.76	0.56	1.08	—
Eicosatrienoic ($C_{20:3}$)	—	0.11	0.46	0.49	—
Arachidonic ($C_{20:4}$)	2.0	0.07	0.19	0.23	—
Total polyunsaturated	—	1.96	1.30	10.21	—

[a]Dugan 1957.
[b]Babatunde et al. 1968.
[c]Wahlstrom et al. 1971.
[d]Value not available.

TABLE 3.12. Effect of Degree of Carcass Fatness on Water, Protein, Fat, Ash, and Calorie Content of Retail Cuts of Pork

Retail cut	*Water* (%)	*Protein* (%)	*Fat* (%)	*Ash* (%)	*Calories* (*no. per 100 g*)
Fresh ham					
Lean	60	17.2	22	0.9	270
Medium	53	15.2	31	0.8	340
Fat	46	13.2	40	0.7	410
Smoked ham					
Lean	49	19.5	25	5.8	300
Medium	42	16.9	35	5.4	380
Fat	36	14.6	44	5.1	460
Loin					
Lean	63	17.9	18	1.0	230
Medium	58	16.4	25	0.9	290
Fat	52	14.8	32	0.8	350
Shoulder butt					
Lean	52	14.4	33	0.8	360
Medium	45	12.5	42	0.7	430
Fat	37	10.3	52	0.6	410

Source: Chatfield and Adams 1940.

so that as fattening of the animal progresses, water content declines. Percentages of protein and ash remain rather constant, but tend to decline slightly in fatter animals. These variations are illustrated in Table 3.12, which gives the percentages for ham, loin, and shoulder of pigs, in lean, medium, and fat types (Chatfield and Adams 1940). These data were obtained more than 40 years ago, and as a result of selection for leanness, many pigs today produce carcasses with less fat and more protein and water than suggested by the values listed for lean pigs in Table 3.12.

The proportion of the live animal that is available as edible lean meat is often referred to as "percent lean cuts." These lean cuts are the shoulder (Boston butt plus picnic), loin, and ham. Other important cuts from the carcass are the jowl, clear plate, spareribs, belly, back fat, feet, and organs such as the brain, liver, and kidney. The wholesale cuts showing the location of each component on the carcass are indicated in Figs. 3.2 and 3.3. The "lean cuts" constitute 39% of the weight of a U.S. No. 1 hog but 73% of the value under U.S. marketing conditions. The fat trim of a U.S. No. 1 hog is 17% of the weight and 5% of the value, but as fatness increases the fat trim makes up 21% of the weight and 7% of the value of a U.S. No. 3 hog. A description and the specifications for official U.S. market grades of hogs are given in Chapter 19. The relative weights and values of wholesale cuts of pork from a U.S. No. 1 pig are shown in Table 3.13 (Pearson 1966). In fatter pigs (U.S. Nos. 2 and 3) the weights of the trimmed ham and liver as a percentage of liveweight tend to decline while the fat trim and leaf fat increase. The lean cuts therefore represent a smaller proportion of the total weight and total value of the carcass as the carcass increases in fatness.

TABLE 3.13. Weight and Value of Wholesale Pork Cuts from U.S. No. 1 Slaughter Hogs

Cut	*Weight (% of liveweight value)*	*Value (% of total carcass value)*
Ham, skinned	14.7	30.3
Loin	11.8	24.9
Boston butt plus picnic[a]	12.8	18.0
Belly (bacon)	10.0	14.1
Lean trim	1.8	1.8
Fat trim plus leaf fat	16.6	5.0
Spareribs	1.9	3.5
Jowls	2.0	1.1
Neckbones	1.3	0.6
Feet	1.9	0.6
Lean cuts[b]	89.3	73.2

Source: Pearson 1966.
[a]New York shoulder is Boston butt and picnic left unseparated.
[b]Ham, loin, and Boston butt and picnic combined.

PROPERTIES AFFECTING THE ACCEPTABILITY OF PORK TO HUMANS

Pork Quality Factors

In addition to the high nutritive value of pork as discussed above, there are some properties of pork that make it inherently a sought-after food when the consumer can afford to purchase it. These properties are shared with other animal products (beef, lamb, poultry), and the amount of pork consumed in relation to alternative meat sources is dependent on esthetic, economic, religious, and cultural factors. The esthetic aspects will be considered here. Those working in meat technology research generally divide palatability factors into tenderness, juiciness, color, aroma, and flavor. Pork is less variable in these criteria than are beef and lamb so that differences in age, breed, and environment have a relatively small effect on the quality of pork. Desirable quality pork has a firm, dry surface with a pinkish red color (Briskey and Kauffman 1971). Color, firmness, and water-holding capacity (exudation) are included together as a general appraisal of pork in quality ranging from pale, soft, exudative (PSE) pork to dark, firm, dry pork (DFD). Between these extremes is the most desirable. The amount of fat within the muscle is rather consistent and no differences have been observed in protein, fat and water content between PSE and normal muscle (Briskey *et al.* 1959; Briskey and Kauffman 1971.) Although PSE muscle shows very little visible intramuscular fat, ether extraction of the muscle fails to reveal differences in fat content of PSE versus normal or DFD muscle, indicating that the fat in PSE muscle is somehow masked.

The Pale, Soft, Exudative (PSE) Pork Problem

The pale color and excessive exudation of PSE pork creates merchandising problems in the retail store. As discussed in Chapter 18, the development of PSE pork is related to a more rapid than normal depletion of muscle glycogen

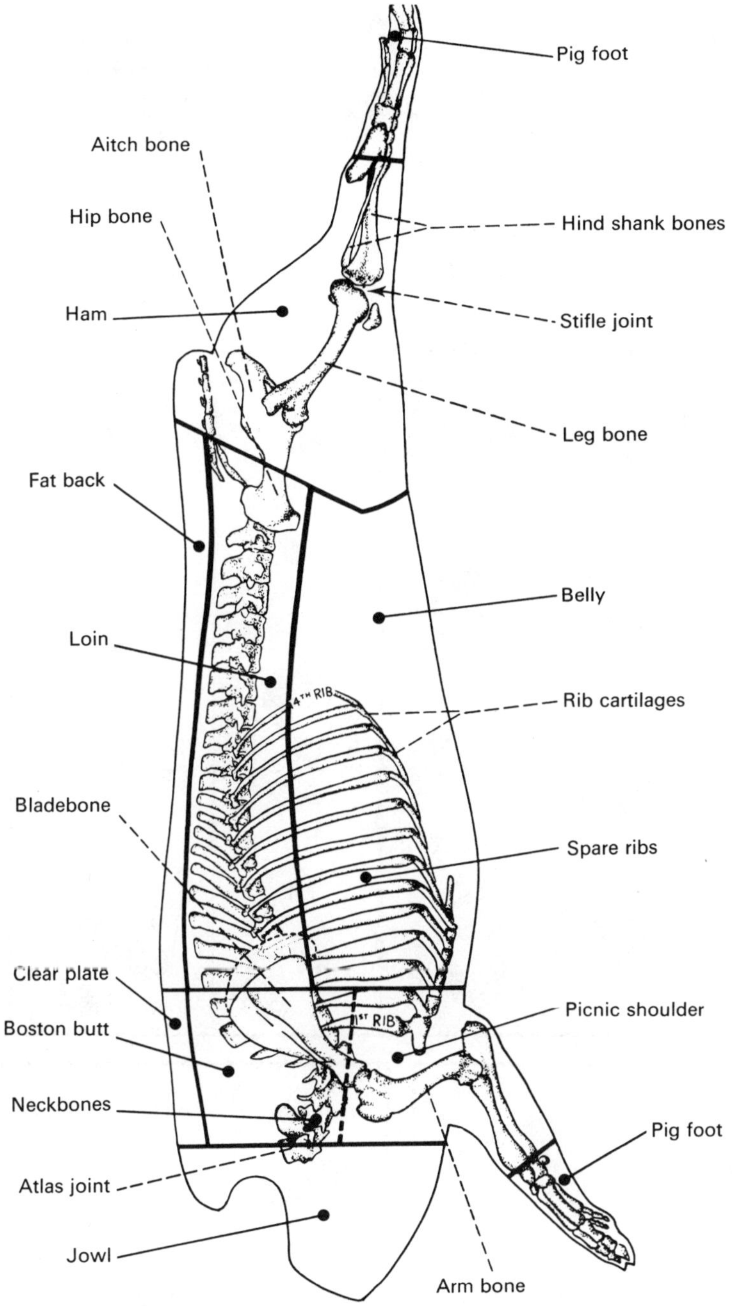

FIG. 3.2. Carcass diagram showing location of wholesale cuts.
Courtesy of J. R. Stouffer, Cornell University.

PORK CHART

RETAIL CUTS OF PORK — WHERE THEY COME FROM AND HOW TO COOK THEM

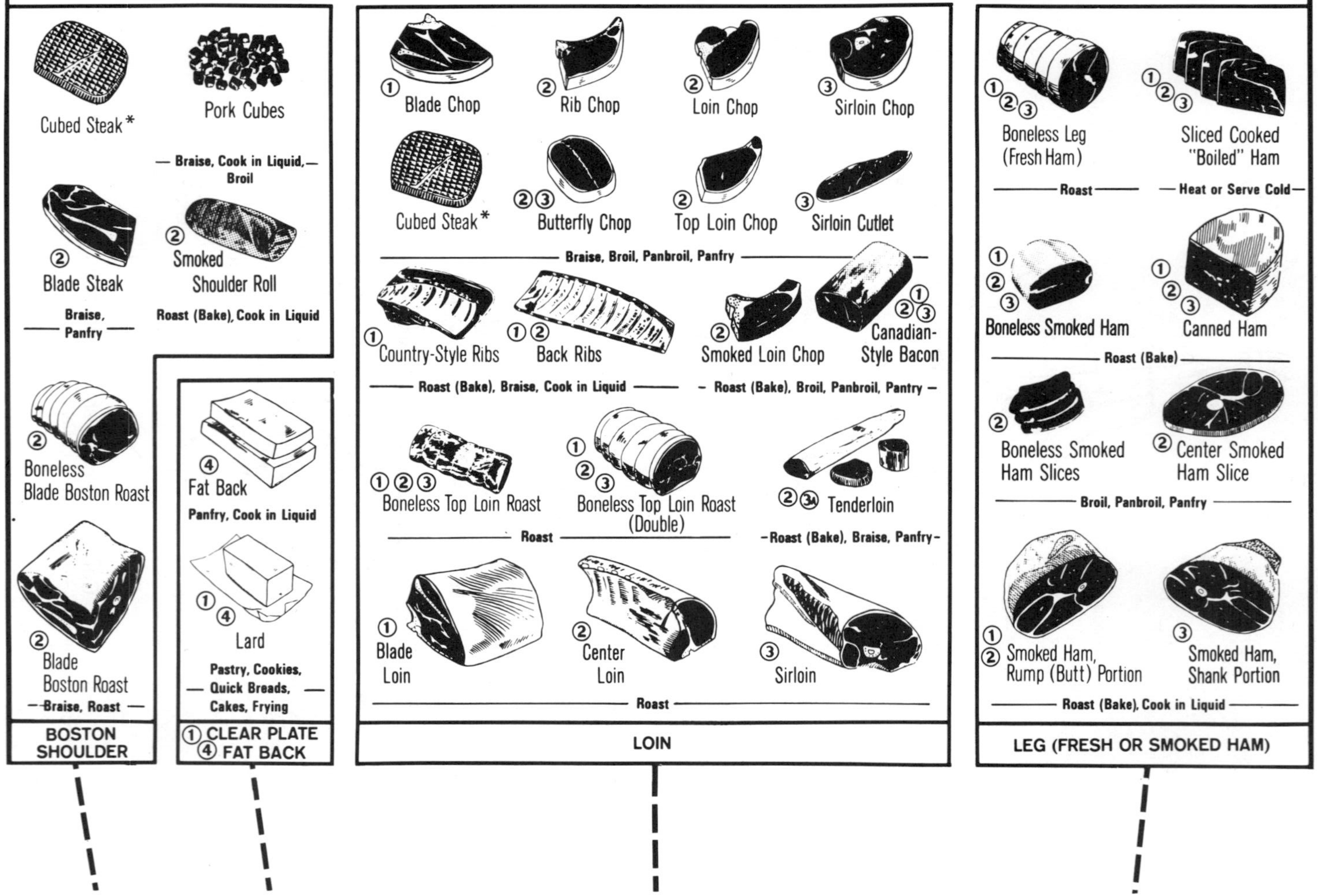

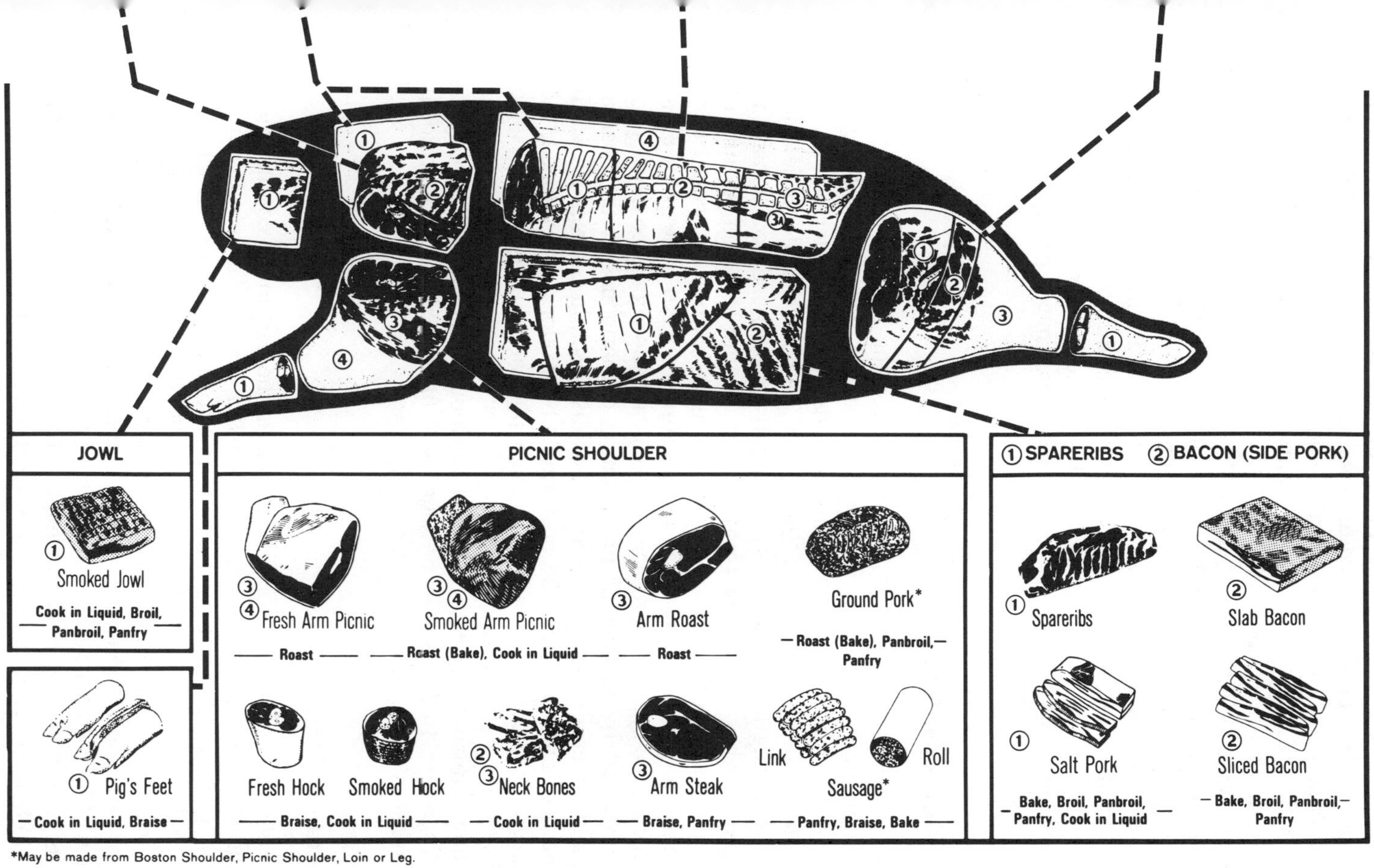

FIG. 3.3. Wholesale cuts of pork and their uses.
Courtesy of National Livestock and Meat Board, Chicago.

postmortem which results in a reduced pH of the muscle related to lactic acid production. Muscle pH is closely related to its water-holding capacity (Herring *et al.* 1971). This condition is apparently partly under genetic control and seems to be associated with stress resistance in live animals. Stress-susceptible pigs are prone to sudden death when exposed to excitement in a warm environment. The muscles of these animals show the typical PSE syndrome and the adrenal hormone secretions are reduced compared to those of normal animals. Thus, we see that physiological response to stress in pigs can have an influence on the acceptability of pork by humans. Much remains unknown concerning all of the factors involved in determining pork quality, the relative importance of each factor, and the degree to which pork acceptability can be controlled through husbandry practices, marketing and slaughter procedures, and meat processing and technology.

The Boar Odor Problem

Meat from mature boars has an objectionable odor and flavor when cooked. This odor is produced by at least one fat soluble substance, 5-α-androst-16ene-3-one, characterized by Craig *et al.* (1962) and later identified by Patterson (1968). In taste-panel studies designed to detect "boar odor" in pork, the odor is detected in most boars weighing more than 90 kg (6 months of age) and occasionally in gilts, sows, and barrows. Since boars grow faster, require less feed per unit of body weight gain, and produce leaner carcasses than barrows, it would be a distinct advantage to the swine producer and consumer alike if a means of eliminating or reducing the boar odor problem without castration could be developed. This is being studied by feeding or implantation of hormones in the boar at various ages and under a variety of conditions as discussed in Chapter 19. Until suitable methods of circumventing the problem of boar odor are developed, the current management practice of castrating male pigs at an early age must be followed. Meat from mature boars sold for slaughter after outliving their usefulness as breeding animals is perfectly safe and nutritious and is normally sold as a component of sausage. In this way its presence goes unnoticed and the nutritional value of the meat is salvaged.

The Trichinosis Problem

Perhaps no other factor has been as detrimental in the past to the consumption of pork as the threat of infection with the parasite, *Trichinella spiralis,* from eating infected pork which has been improperly cooked. Due to strict garbage cooking laws and better education of the public about trichinosis the incidence of this infection in humans has been drastically reduced. In the United States, the annual number of reported human cases has declined steadily in recent years so that fewer than 100 cases per year are now being reported. The life cycle of this parasite and its control are described in Chapter 10. Suffice it to say here, that the current threat to human health is not serious when one considers that only a small fraction of all swine is infected with trichinella and that heating pork to an internal temperature of 77° C or freezing for 20 days at −15° C destroys the parasite (Price and Schweigert 1971). Also, there is now a drug, thiabendazole, available for successful treatment of humans infected with trichinella. Thiabendazole can also be used to treat infected animals.

Thus, we see that the threat of trichinosis, while real, remains grossly exaggerated as a stigma attached to consumption of pork. The newly developed procedure for identifying infected pigs in the slaughter house will undoubtedly do much toward eliminating this parasite as a deterrent to pork consumption.

REFERENCES

AUSTIN, F. L., NG, C. S., OSTRANDER, J. A., and SLOVER, H. R. 1964. Proc. 16th Am. Meat Inst. Res. Conf., Am. Meat Inst. Found. Circ. No. *76*, Chicago.

BABATUNDE, G. M., POND, W. G., WALKER, E. F., JR., and CHAPMAN, P. 1968. J. Anim. Sci. *27*,1290.

BRISKEY, E. J., BRAY, R. W., HOEKSTRA, W. G., PHILLIPS, P. H. and GRUMMER, R. H. 1959. J. Anim. Sci. *18*,146.

BRISKEY, E. J., and KAUFFMAN, R. G. 1971. *In* The Science of Meat and Meat Products, 2nd Edition (J. F. Price and B. S. Schweigert, Editors). W. H. Freeman and Co., San Francisco.

CHATFIELD, C., and ADAMS, G. 1940. USDA Circ. No. *549*, Washington, D.C.

CRAIG, H. B., PEARSON, A. M., and WEBB, N. B. 1962. J. Food Sci. *27*,29.

DUGAN, L. R., JR. 1957. Am. Meat Inst. Found. Circ. No. *36*, Chicago.

FAHEY, T. J., SHAEFER, D. M., KAUFFMAN, R. G., EPLEY, R. J., GORIS, P. F., ROMANS, J. R., SMITH, G. C., and TOPEL, D. G. 1977. J. Anim. Sci. *44*,8.

EDWARDS, R. C., SMITH, G. C., CION, H. R., and CARPENTER, Z. L. 1981. J. Anim. Sci. *52*,703.

FREDEEN, H. T., and HARMON, B. G. 1983. J. Anim. Sci. *57*(Suppl. 2), 100.

GIPP, W. F. 1971. The effect of supplemental copper on growth, hematology and iron utilization in swine. Ph.D. Thesis, Cornell University, Ithaca, New York.

HERRING, H. K., HAGGARD, J. H., and HANSON, L. J. 1971. J. Anim. Sci. *33*,578.

KANNEL, W. B., and GORDON, T. 1971. Section 24: The Framingham diet study; diet and regulation of cholesterol. U.S. Dep. of Health, Education and Welfare, Public Health Service, Natl. Inst. of Health Publication. Bethesda, Maryland.

KIERNAT, B. H., JOHNSON, J. A., and SIEDLER, A. J. 1964. Am. Meat Inst. Found. Bull. No. *57*, Chicago.

KOCH, D. E., PARR, A. F., and MERKEL, R. A. 1967. J. Food Sci. *33*,1.

LEVERTON, R. M., and ODELL, G. V. 1958. Okla. Agric. Exp. Stn. Misc. Publ. *MP-49*, 90 pp.

MOSS, M., ONO, K., CROSS, R., SLOVER, H., BERRY, B., LANZA, E., THOMPSON, R., WOLF, W., VANDERSLICE, J., JOHNSON, H., HOLDEN, J. M. and STEWART, K. 1983. Nutrient composition of fresh retail pork. J. Food Sci. *48*, 1767.

NATIONAL RESEARCH COUNCIL. 1968. Recommended Dietary Allowances. 7th Edition. National Academy of Sciences. Publ. *1694*. Washington, D.C.

NATIONAL RESEARCH COUNCIL. 1980. Recommended Dietary Allowances. 9th Edition. Natl. Acad. Sci., Washington, D.C.

PATTERSON, R. L. S. 1968. J. Sci. Food Agric. *19*,31.

PEARSON, A. M. 1966. Meats. *In* Introduction to Livestock Production, 2nd Edition. H. H. Cole (Editor). W. H. Freeman and Co., San Francisco.

PRICE, J. F., and SCHWEIGERT, B. S., Editors. 1971. The Science of Meat and Meat Products. 2nd Edition. W. H. Freeman and Co., San Francisco.

RICE, E. E., 1971. *In* The Science of Meat and Meat Products, 2nd Edition. V. F. Price and B. S. Schweigert, Editors. W. H. Freeman and Co., San Francisco.

SCHWEIGERT, B. S., and PAYNE, B. J. 1956. Am. Meat Inst. Found. Bull. No. *30*, Chicago.

WAHLSTROM, R. C., LIBAL, G. W. and BERNS, R. J. 1971. J. Anim. Sci. *32*,891.

WATT, B. K., and MERRILL, A. L. 1963. Composition of Feeds—Raw, Processed, Prepared. Foods Agric. Handbook No. 8. Consumer and Food Economics Research Division, Agric. Res. Serv., USDA, Washington, D.C.

Part II

Biology of the Pig

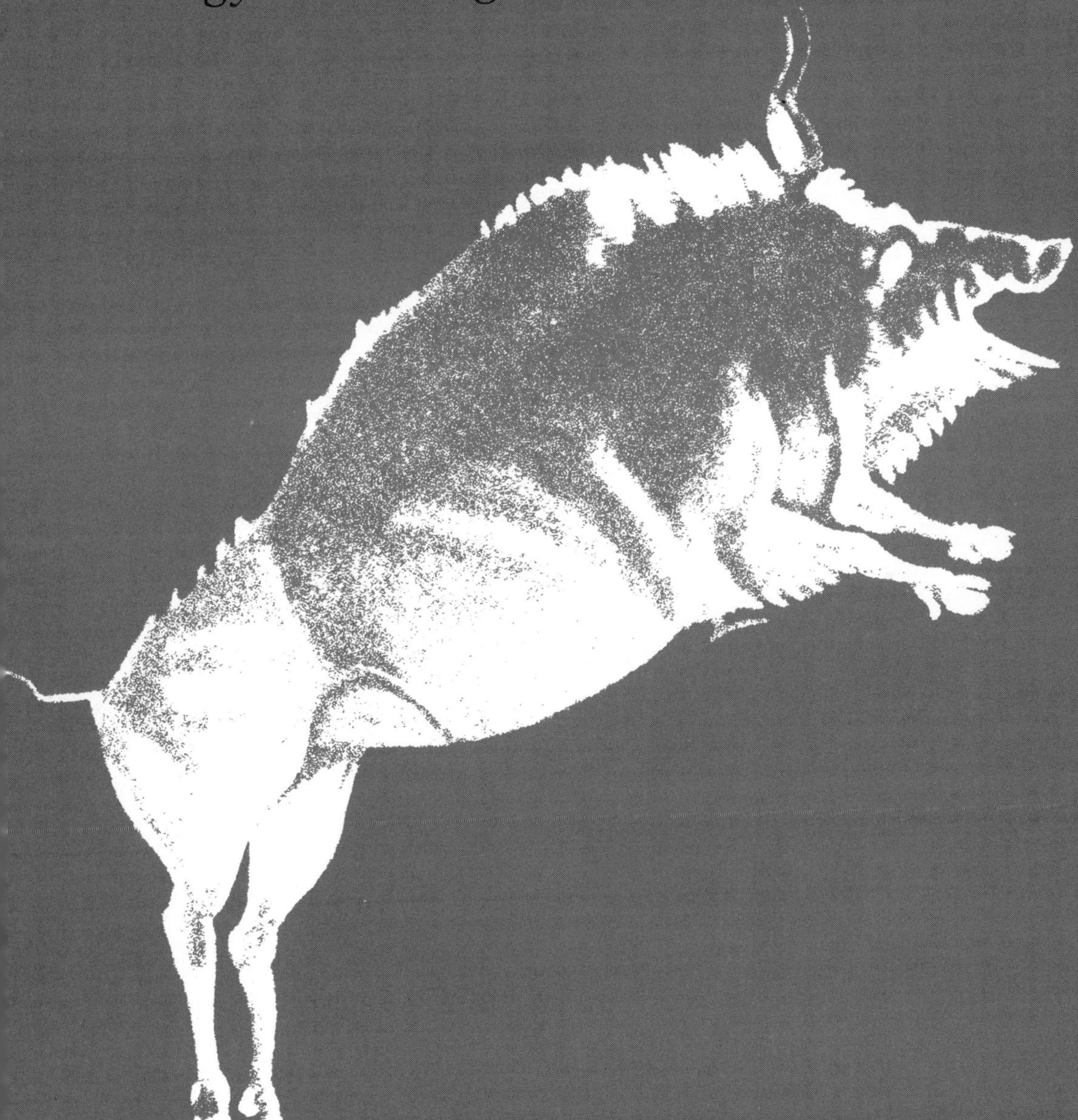

4

Breeding and Selection

INTRODUCTION

The basis for a successful swine production enterprise is a sound breeding and selection program. Changes in total efficiency can be made by altering feeding and management systems. Such changes are temporary and easily reversible; changes involving genetics are more permanent in nature. Decisions on culling and breeding made at any point in time may be felt and their effects sustained for several generations of swine. Thus, it is vital to understand thoroughly the principles of swine selection and the methods of applying these principles. Current knowledge of applied genetics and the theoretical bases on which modern tools of genetic engineering and other practices may be used have received recent review (Dickerson and Willham 1983; Rutledge and Seidel 1983). This chapter is devoted to providing the basis for intelligent swine breeding and selection.

GOALS

As indicated in Chapter 1, the changes that have taken place in swine since the beginning of domestication have been directed toward the production of an animal that would most effectively meet the demand existing at a particular time. At the present time, the type of pig in greatest demand differs between countries and even within countries. For example, in the United States and in most of Europe, the present market demands a lean carcass with a minimum of fat whereas in some parts of Latin America, Southern Europe, and in some other areas of the world a premium is paid for fat carcasses. Since the fat content of the carcass increases rapidly from birth to maturity, the weight of the pig at slaughter is an important factor in meeting the local demand. However, nutritional energetics, feed supplies, labor supply, and other factors related to economics must be considered as well.

A long-term goal of the swine breeder, regardless of current local demands, must be maximum efficiency of production. This must involve the propagation of pigs with the genetic capacity to respond favorably to a particular set of environmental conditions that are economically favorable in a particular setting. For example, if confinement feeding of growing swine on slotted floors is of importance as a means of saving labor in areas of the world where this represents a large proportion of production costs, it is important for the swine breeder to select swine that are able to thrive and produce pork efficiently under these conditions. Unfortunately, it is difficult to select swine on traits that do not lend themselves to objective measurement, as compared to traits such as daily weight gain, which is easily obtained. Nevertheless, if the industry is to thrive, the concept of *efficiency* of pork production must prevail in swine breeding and selection.

HERITABILITY

If one is to attain the goals specified in the foregoing discussion, an understanding of the basic role of genetics in swine selection is essential. Several textbooks (Lasley 1963; Lush 1945; Rice 1957) devoted specifically to animal breeding and genetics are available and the student interested in detailed theoretical and mathematical treatment can refer to one of them. Most of the economically important traits of swine are influenced to some degree by heredity. However, the degree to which genetics (heredity) affects performance depends on the particular trait concerned.

The classical effects of Mendelian genetics involving dominant and recessive genes that result in 3:1 or 1:2:1 phenotypes in the F_1 generation of the mating of two heterozygous individuals cannot be used as the basis of selection for most of the economically important traits. Most traits involve many genes (perhaps several hundred in some cases) so that the best approach to livestock improvement has been to study the gross effects of these multiple gene actions by various selection and mating systems.

Differences among individuals provide the basis on which the breeder makes changes in his herd in the particular trait of concern. The variation observed is due to variations in environment, to differences in heredity, and to joint effects of variation in environment and heredity that cannot be ascribed to either alone. The breeder can affect the environment by husbandry practices, but can affect heredity only by selection of parents to be used in forming the new generation. Most economically important traits (growth rate, feed efficiency, carcass measurements) involve more than one gene pair and are controlled by both heredity and environment. Thus, population genetics has grown out of and is based on Mendelian genetics, but is concerned with genetic changes in the total herd or in the average of the individuals in that herd.

The question of whether a trait is hereditary or environmental is a question of what proportion of the variation in a population is due to differences in heredity and what proportion is due to differences in environment. The answer will not be the same for all traits in a single population nor for a single trait in all populations. This is true because environment may be standardized more in some cases than in others. Lush (1945) has given the following formula to calculate the degree of heritability of a particular trait:

σ_0^2 = observed variance;

σ_H^2 = portion of variance caused by differences in heredity;

σ_E^2 = portion of variance caused by differences in environment.

Then

$$\sigma_H^2 + \sigma_E^2 = \sigma_0^2$$

and

$$\sigma_H^2/(\sigma_H^2 + \sigma_E^2) = \sigma_H^2/\sigma_0^2$$

= portion or observed variance for which differences in heredity are responsible.

When σ_H^2/σ_0^2 is large the trait is said to be highly heritable; when small it is said to be largely environmental. The symbol H is generally used to denote this fraction. Thus, $H = .80$ would indicate that 80% of the variance in that trait was due to heredity. This would be considered a highly heritable trait.

Population geneticists have developed mathematical procedures for estimating the relative importance of heredity and environment for various traits of economic importance. Briefly, the procedure consists of studying the performance in a particular trait of a large number of individuals in a population and comparing it with that of related animals. Heritability estimates have been made in this way by dam–daughter, sire–daughter, or paternal half sib comparisons among genetically related animals. Corrections must be made for effects due to environmental variations such as season, year, location, and any other identifiable sources of variation.

The heritability of several economically important traits obtained by animal geneticists working under a variety of experimental conditions was summarized by Craft (1958) and is listed in Table 4.1. It should be recognized that these are merely estimates, since the exact value obtained under a given set of conditions is subject to experimental error. However, for practical purposes, the traits can be classified according to degree of heritability (high, medium, and low). Fortunately, some useful heritable traits are positively correlated with each other, or at least are not antagonistic, so that selection for one trait leads to a gain in the others, or at least not a reduction. When there are several traits of economic importance, as with swine, in contrast to dairy cattle (in which milk production is the single most important trait), the individual breeder must decide on which trait to place the most emphasis in selection or select for more than one trait. Selection for more than one trait results in less progress in each trait than if selection is for only one trait. In either case, the progress that can be expected per generation will

TABLE 4.1. Heritability of Economically Important Traits in Swine

Trait	Heritability[a] (%) average	Range of observed values	Degree of heritability (arbitrary)
Performance Traits			
Litter size at birth (no. of pigs)	15	0–24	Low
Post-weaning rate of wt gain	29	14–58	Medium
Feed per unit of wt gain	31	8–72	Medium
Litter size at weaning (no. of pigs)	12	0–32	Low
Litter wt at weaning	17	3–37	Low
Anatomical traits			
Length of body	59	40–81	High
Length of legs	65	51–75	High
No. of vertebrae	75	—	High
No. of nipples[b]	32	10–42	Medium
Carcass traits			
Length	59	40–81	High
Loin eye area	48	16–79	High
Ham wt (% of carcass wt)	58	51–65	High
Shoulder wt (% of carcass wt)	47	38–56	High
Fat cuts (% of carcass wt)	63	52–69	High
Lean cuts (% of carcass wt)	31	14–76	High

Source: Craft 1958.

[a]Heritability $(H) = \frac{\text{genetic variance (additive + dominance + epistatic)}}{\text{genetic variance + environmental variance}} \times 100$. (Dominance variance and epistatic variance are not important and can be ignored in computing heritability.)

[b]Pumfrey et al. 1980.

depend on (1) the heritability of that trait and (2) selection differential (the amount by which those individuals selected to be parents exceeds the average of all pigs born in the herd in their generation).

Genetic superiority of selected parents depends on the fraction of potential breeding animals selected, the accuracy of evaluation of the trait, and the genetic variation (genetic standard deviation) for the trait.

An example of the *expected* progress in a single trait in one generation of selection is shown below:

Trait: Average daily gain, 35–170 days of age.

Heritability (*H*) of average daily gain: 29%.

Average daily gain of females in population from which a gilt was selected: 750 g.

Average daily gain of selected female: 900 g.

Expected average daily gain of female progeny of the gilt is computed as follows:

Step 1: *H* × (daily gain of the gilt minus daily gain of population from which selected).

$$29 \times (900 - 750) = 0.29 \times 150 \text{ (selection differential)}$$
$$= 43.50 \text{ g (expected genetic change).}$$

Step 2: Daily gain of population from which selected plus genetic change: 750 ± 43.5 = 793.5 g (assuming the gilt is mated to a boar of equal genetic potential).

If the gilt is mated to a boar that had an average daily gain of 1000 g selected from a population of boars with an average daily gain of 800 g, the expected average daily gain of the female progeny of this mating would be

$$750 + \tfrac{1}{2}[43.5 + (0.29 \times 1000 - 800)] = 750 + 50.75 = 801 \text{ g}.$$

Contrast this *expected* improvement with what might occur in random selection. In random selection, the average does not change, for the probability of a choice above the average is equal to that of a choice below the average. It is possible that a female whose average daily gain was less than that of her contemporaries might be selected by chance as a herd replacement, and the result would be a reduction in performance in the next generation. This is an even bigger danger in the selection of boars, because the influence of the boar may be spread across a large number of litters in a single breeding season. This represents a permanent setback in that trait in the herd involved.

SELECTION INDEX

In selection for more than one trait, one can devise a *selection index* that allows a rank order evaluation of potential breeding stock on the basis of the combined score on several traits. The weighting of the several traits selected can take into consideration the degree of heritability of each trait and its relative economic importance. An example of a selection index that was developed by animal geneticists at Iowa State University for use at the Swine Evaluation Centers in Iowa, and an illustration of its use, is given below:

$$\text{Index } (I) = 270 + (100 \times \text{av. daily gain in lb}) - (150 \times \text{back fat in in.}) - (35 \times \text{lb of feed/lb of gain})$$

	Av daily gain	*Back fat*	*Feed/gain*	*Index*
Performance of Boar A	1.8	1.5	3.2	113
Performance of Boar B	1.6	1.1	2.6	174

Decision: Even though Boar A grew faster than Boar B, Boar B had a higher index because of his superior back fat and feed efficiency. Select Boar B for use as a herd sire.

In the above example, Boar B had a higher index and was therefore saved for breeding. Had the index been weighted heavily toward selection for daily gain, boar A might have had the higher index. Thus, it is clear that the *selection index* serves to *rank individuals* according to a predetermined set of performance traits and that the ranking depends on the arithmetic emphasis placed on each trait.

In the 1970s, the National Swine Improvement Federation devised an index to rate boars tested in central stations. The index for test pens with more than one boar (full and half sibs) is

$$I = 100 - 60(G - \bar{G}) - 75(F - \bar{F}) - 70(B - \bar{B}),$$

where G = daily gain (lb), F = feed/gain, and B = back fat of the individual. $\bar{G}$, $\bar{F}$, and $\bar{B}$ are test group means for each trait.

CULLING RATES

The usefulness of the selection index in terms of genetic progress depends, of course, on the degree of heritability of the traits chosen for inclusion in the index, and also on the proportion of potential breeding stock that is saved for breeding. The more rigid the culling, the more rapid will be the expected progress. The decision as to what proportion of potential breeding stock to save for breeding must be based on the number necessary to maintain the size of the breeding herd. This number, in turn, will depend on the degree of culling due to infertility, death, and other similar factors contributing to productive longevity. If, in gilt selection, the goal is to select only the top 1% of potential breeding stock as herd replacements, it will be necessary to cull no more than 10% of the herd (assuming that each sow produces 10 gilts/year) if herd size is to be maintained. This could result in retaining in the herd some individuals which are actually inferior to some of the potential replacements that have been discarded on the basis of the previous decision to cull 99% of the potential herd replacements. Clearly, this decision would not be conducive to the most rapid herd improvement. A more realistic approach would be to save that proportion of herd replacements which shows promise, on the basis of performance information, of improving the production efficiency of the herd.

This may mean culling some sows from the herd that have not reached the end of their productive ability, if there is good evidence that their replacements will be superior. Thus, gilt herd replacements must include that number needed to maintain herd size through normal turnover plus that number showing clear evidence of production potential superior to that of the sows they are to replace.

In purebred boar selection, again, the decision to replace a herd sire should be based (aside from considerations of inbreeding) on the evidence that the replacement sire will be superior to the sire he is to replace. The evidence, in the case of young potential sires, must be on the basis of their own performance, as they have not had an opportunity to produce progeny. Data on progeny should be used in selection for traits of low heritability. For traits of high heritability, performance testing should be used, although outstanding individual performance does not ensure prepotency (a marked capacity to transmit a trait or traits to the progeny).

TESTING PROCEDURES

The usefulness of selection for an individual trait or for several traits depends on the use of accurate and consistent measurements. For some traits it is possible to use the *performance test*. In this test the performance of the individual is measured; growth rate and feed efficiency are examples of traits that lend themselves to this type of testing. For other traits, such as some carcass characteristics, the progeny test or sibling test must be used, as the individual being tested must be slaughtered before breeding. This would apply for such measurements as ham weight and percentage of lean cuts. Some carcass traits can be estimated with sufficient accuracy on the live animal, so that the performance test is valid. These include back fat thickness and, more recently, measurements of the size of individual muscles, obtained by ultrasonic devices. For maternal production traits, such as litter size and milk producing ability, performance testing can be employed in females, but obviously *progeny tests* or *sibling tests* are required to identify sires able to transmit superior maternal characteristics to their daughters.

MECHANICS OF PERFORMANCE AND PROGENY TESTING

Valid information on production traits depends on accurate and unbiased measurements. Two broad types of testing programs have evolved. These are *central test programs* and *on-farm test programs*. Each of these can use either performance or progeny testing or a combination of both. A classical example of the central test station is the Danish system of progeny testing, which involves an organized cooperative system of controlled testing of pigs from more than 200 recognized ''elite'' breeders. These breeders concentrate on improved production traits in their herds through a closely supervised program of progeny testing for several traits, including weight gain, efficiency of feed utilization, and several carcass measurements. The physical facilities for testing include several identically constructed buildings scattered in appropriate geographic locations in Denmark, each with pens for feeding pigs under standardized environmental conditions, and litters of pigs chosen by the breeder for testing on the basis of their pedigree and the performance of their ancestors.

The information obtained is used as the basis for selection of boars for propagation of these superior traits among the herds of Denmark. The progress made in the average performance of tested pigs during more than 60 years of use of the progeny testing system has been recorded by Clausen (1970) and by Thomson *et al.* (1972), and is illustrated in Fig. 4.1. It must be recognized that, concomitant with genetic changes, there have been improvements in feeding and management, so that it would be imprudent to assign all of the observed improvements to genetics.

In the United States many central swine testing stations have been developed since the late 1950s, starting with the Ohio Swine Evaluation Center. An example of the type of data obtained and the effect of sex on the measurements obtained is shown in Table 4.2 (Bruner *et al.* 1958). Most of these central testing stations are operated with state or regional support in combination with partial or complete financial support by the participating breeders. Some representative facilities of this kind are illustrated in Fig. 4.2.

The central test station has the following advantages; (1) Animals can be tested under standardized conditions of feeding, housing, and management. (2) Close daily unbiased supervision is available.

The following disadvantages are inherent in the central test station: (1) Limited

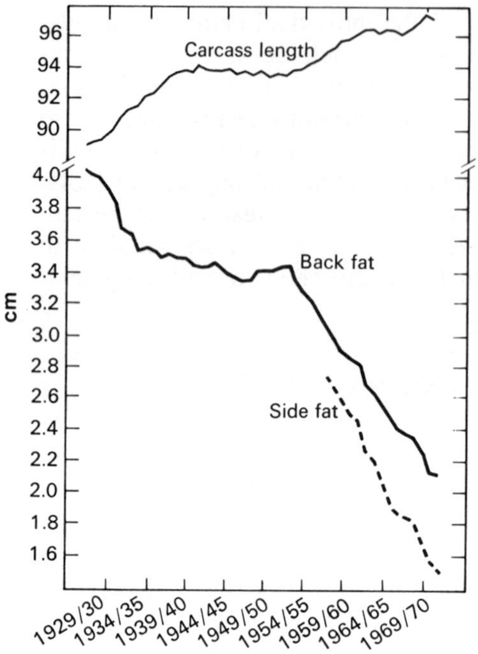

FIG. 4.1. Progress in performance of pigs in Danish progeny testing; note the reduction in back fat and side fat with little change in carcass length from 1929 to 1970.
Courtesy of H. Clausen and H. Staun, National Agricultural Research Laboratory Copenhagen, Denmark.

physical facilities permit the sampling and testing of only an extremely small percentage of the total swine population. (2) The congregation of animals from many farms presents the danger of spread of infectious diseases and parasites. (3) The test environment may be quite different from the normal environment under which animals are produced. This would not be a serious disadvantage if the *ranking* of animals did not change under the two environmental conditions. However, there is a possibility of a genetic *x* environment interaction such that animals that perform well under test conditions might not perform well under farm conditions or vice versa. (4) The sample of pigs from each farm is too small

TABLE 4.2. Performance Traits of Females and Castrated Males (Barrows) in the Ohio Swine Evaluation Station

Trait	*Females*[a]	*Barrows*[a]
No. of animals	385	385
Body wt, 63 days old, kg	18.4	18.3
Daily gain, g	744	804
Back fat, cm	3.90	4.15
Carcass length, cm	75.4	74.4
Loin eye area, cm^2	24.8	21.6
Trimmed loin weight, kg	10.6	10.1
Trimmed ham wt, kg	13.2	12.5
Lean cuts, % of carcass wt	51.6	49.3

Source: Bruner et al. 1958.
[a]Data obtained from full rib pairs tested in five seasons (fall 1954 through fall 1956).

FIG. 4.2. A central test station in the United States. The station in North Carolina is used by breeders to obtain performance data on their pigs under standardized conditions. There are more than 40 similar test stations scattered throughout the United States.
Courtesy of J. R. Jones, North Carolina State University at Raleigh.

to reflect differences between farms accurately. (5) The pretest environment is different for pigs from different farms, and carryover effects may affect performance during the test.

In avoiding some of these problems, the on-farm test program has achieved great importance in the United States. These programs are usually administered and partially financed by state and local extension services, but the physical facilities are provided on the farm premises of the participating breeder. Some of the larger purebred breeders in the United States now have developed their own private testing programs within their own herds. These testing programs (both central and on-farm), while usually obtaining data on performance, also provide a progeny test on each parent.

Another form of on-farm testing in use in the United States is illustrated in programs presently or formerly used by the individual purebred breed associations. These programs differ slightly among individual breed organizations but basically involve the following performance criteria:

Litter size: minimum of eight pigs alive at 3 (or 5) weeks of age.

Litter weaning weight: minimum of 48-kg total weight at 3 weeks of age (or higher minimum weight at older ages.).

Age at 90 kg: maximum of 170 days (correction factors are used for animals slaughtered at earlier or later ages).

Carcass data on pigs selected for slaughter at a standard weight of 90 kg: minimum of 72.5 cm long (first rib to aitch bone); maximum of 4.0 cm average back fat (average of three measurements taken opposite the first rib, last rib, and last lumbar vertebra); minimum of 25.0 cm^2 cross-sectional area of the longissimus (loin eye) muscle between the tenth and eleventh rib.

A litter that meets the litter size and litter weight requirements at 3 (or 5) weeks of age is designated a Production Registry (PR) litter or some similar designation, depending on the breed association. Litters that meet PR requirements are eligible to compete in the post-weaning weight gain and carcass phase of the testing program. Environmental factors play a major role in affecting the traits concerned (litter size and litter weight at birth and at weaning); but the heritabilities of these traits are sufficiently high to make such programs worthwhile. If pigs selected for slaughter meet all requirements, the litter is assigned special designation and is recognized as having fulfilled superior standards of performance. As continued genetic improvement results in a higher proportion of litters meeting these standards, individual breed associations will undoubtedly tighten their standards. Only by so doing can the most rapid genetic progress be made.

These on-farm testing programs make it possible to test a much higher proportion of the population than with only central testing. However, each system has advantages and both have a place in selection.

THE SHOW-RING

For many years local, regional, national, and international livestock shows have been an important phase of the swine business. Show-ring appraisal of breeding stock and slaughter hogs has served a useful purpose in standardizing type and appearance of purebred swine. While the correlation between productivity and show-ring appearance is not high, experienced judges can usually agree on how much individual animals deviate from an "ideal model." Show-ring appraisal has been the means by which changes in swine type (chuffy vs. rangy or lard type vs. meat type) have been accomplished to meet changing market demands. As more objective measures of animal merit have been developed, less attention is given to show-ring records by purebred and commercial swine breeders. However, the traditional livestock show provides a unique means of advertisement to other breeders as well as to the nonagriculturally oriented consumer of pork and for this reason this phase of the swine breeding and selection program will undoubtedly survive into the foreseeable future.

POTENTIAL SELECTION CRITERIA

Newer technology permits exploration of selection criteria more elaborate than conventional criteria used in the past, especially with respect to estimates of carcass traits based on measurements obtained with live animals.

1. Blood constituents. Selection of breeding animals at an early age on the basis of some blood constituent predictive of future performance has been an attractive idea to swine breeders for some time. This approach has not been highly successful to date. Blood volume (Austad 1969) as well as several blood constituents have been examined (Austad 1969; Heidenreich *et al.* 1964; Bowland and Hironaka 1957; Saffle *et al.* 1958; Tremere *et al.* 1966; Jensen *et al.* 1968; Smith *et al.* 1968) including serum lipids, cholesterol, creatinine, and protein. Although blood group systems might be expected to have predictive value because of their specific nature, they have not been found useful in this respect.

2. Indirect measures of body composition. Of great potential value is the estimation of lean tissue content of the live animal, by use of ^{40}K measurement of the body using a whole body radioactive counter (Stant *et al.* 1969; Lohman *et al.* 1968). This is based on the known close relationship between body K and body protein content. Since the radioactive isotope of potassium ^{40}K is distributed in an animal's body at a constant ratio of total K, it should be possible to estimate accurately the total protein content of the body by measuring ^{40}K. To date, problems of ^{40}K contamination of the body surface from the environment and of the gastrointestinal content from the feed and water and the failure of K to maintain a constant relationship with other chemical constituents in different tissue types have produced discouraging results. Also under study is the use of deuterium oxide (H_2O with the H replaced by heavy hydrogen ^2H) to estimate total body water. This method, as well as that of ^{40}K measurement, offers the advantage of no harm or discomfort to the animal and is preferable to methods involving the injection of dyes or radioisotopes that make the meat of the animal inedible. One method involving radioisotopes is the use of radioactive chromium (^{51}Cr) to determine red blood cell volume which in turn is closely related to protein content of the body (Doornenbal *et al.* 1962). This method is technically valid, but offers less appeal than a method that does not represent an environmental hazard. Recently developed electronic meat measuring equipment (EMME) has been shown (Mersmann *et al.* 1984) to be a poor predictor of carcass composition in living animals.

Specific gravity of the carcass is very highly correlated with body lean tissue content (Doornenbal *et al.* 1962; Whiteman and Hillier 1953). Attempts to measure specific gravity of live animals have not been successful, owing to the problem of underwater breathing and errors introduced by variable amounts of air in the lungs.

$$\text{Specific gravity} = \frac{\text{Weight in air}}{\text{Weight in air} - \text{weight in water}}.$$

Muscle fiber number and diameter as an index of total muscle mass of the body has been studied by Staun (1963). The relationship between number and diameter of individual muscle fibers in the longissimus muscle of two breeds of swine and in crossbred animals of these breeds is shown in Table 4.3.

The number of muscle and adipose tissue cells is believed to be determined at some point in late prenatal or early postnatal life in the pig. The diameter of individual muscle fibers is affected by age and body weight. It is inversely related to the number of fibers in a particular muscle. However, as the muscle cross-sectional area increases, the total number of cells may increase even though the average individual fiber diameter decreases. Heritability estimates by Staun (1963) have indicated a high heritability for muscle cell number within a single muscle (longissimus). If it is assumed that the same relationship holds for other muscles, it should be possible to select for lean carcasses successfully in this way. If future research shows that the number of adipose tissue cells is determined in prenatal or early postnatal life, it should be possible to select breeding animals at an early age on the basis of their future propensity to deposit body fat. Such selection would be based on a high correlation between adipose tissue cell number and degree of body fatness in the adult, a relationship that appears to hold in

TABLE 4.3. Number and Diameter of Muscle Fibers in Longissimus Muscle in Pigs

Number and diameter	Landrace		Pietrain		Landrace × Pietrain	
	Female (20 animals)	*Male castrates (20 animals)*	*Female (20 animals)*	*Male castrates (20 animals)*	*Female (20 animals)*	*Male castrates (20 animals)*
Fiber diameter (μ)	65.0	63.9	73.3	68.9	70.3	69.0
No. of fibers/ cm^2	36,734	37,116	31,409	32,896	33,343	35,091
Total no. of fibers (×1000)	1,124	988	1,078	984	1,005	934
Area of cross section of longissimus muscle (cm^2)	30.5	26.7	34.4	30.0	30.2	26.8

Source: Staun 1963.

humans. Muscle biopsy techniques, perhaps along with other indirect estimates of total muscle mass, would allow selection at an early age without slaughter.

None of the potential selection criteria just discussed are free from serious problems as they now exist, but future refinement and modifications may provide the needed improvements.

3. Live back fat probe and ultrasonic estimation of back fat and lean tissue areas. An effective tool for estimating body leanness of pigs before slaughter is the live back fat probe. The first such approach was a mechanical measurement of back fat thickness (Hazel and Kline 1952) using a scalpel to cut the skin at selected points on the back and measuring the thickness of back fat at each point with a metal ruler (Fig. 4.3a). More recently, the use of ultrasonic techniques (Fig. 4.3b) for estimating back thickness as well as lean tissue cross-sectional area is common in research and on farms. The mechanical back fat probe and the "lean meter," a device that relies on the principle that electrical energy passes through fat at a different efficiency than through lean tissue, were used experimentally to reduce back fat thickness within a herd (Berruecos *et al.* 1970) and to obtain "lean" lines and "fat" lines of Duroc and Yorkshire swine (Hetzer and Harvey 1967). The average back fat thickness of the Duroc line selected for high fat was increased after 10 generations by 68% of the original mean; the corresponding difference for Yorkshires after 8 generations was 44%. This dramatic response for selection was accomplished without a reduction in daily gain or reproduction efficiency (Hetzer and Miller 1970). This inexpensive and simple selection tool is perhaps the most useful investment of the presently available methods for selection of carcass traits based on live animal measure-

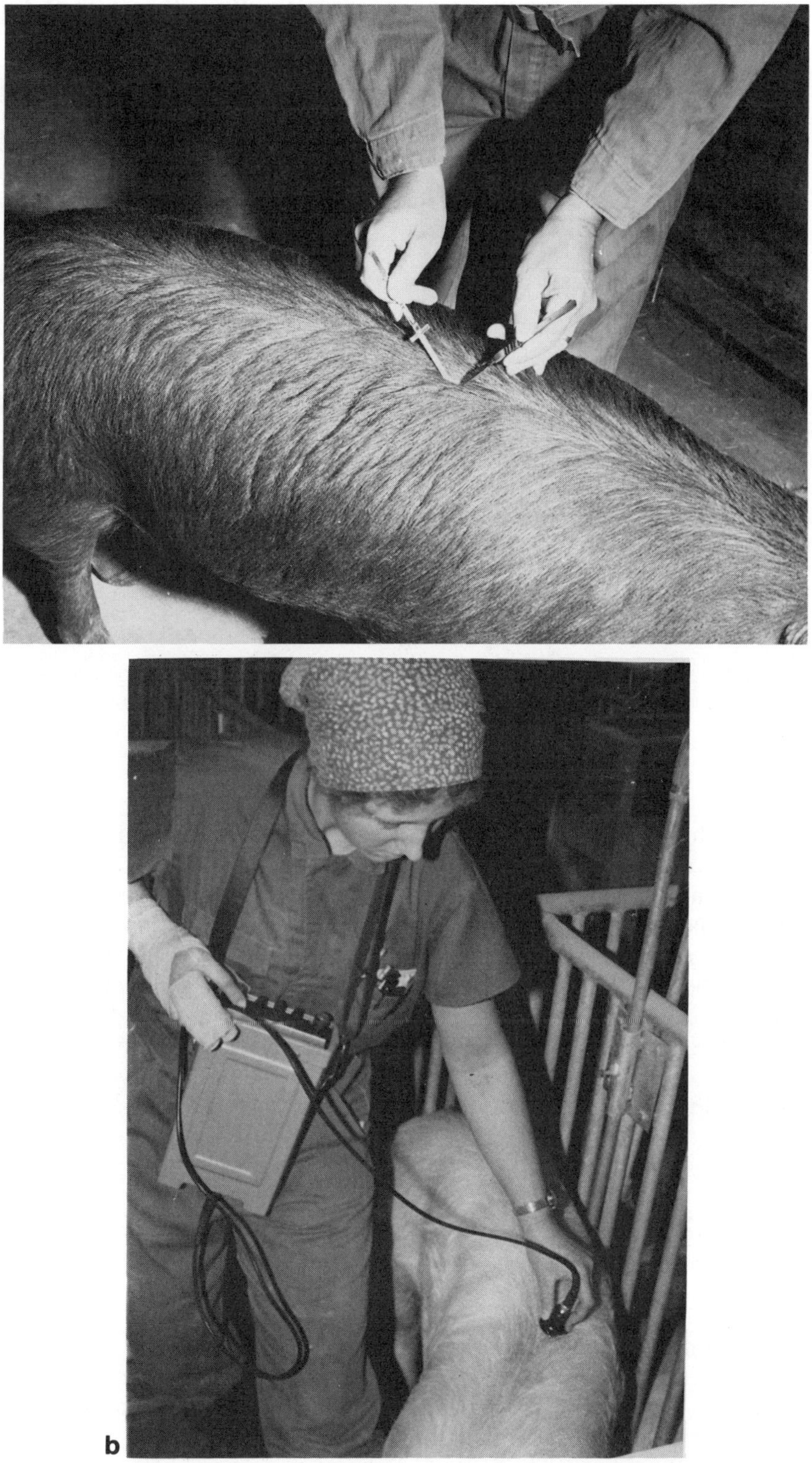

a

b

FIG. 4.3. Methods of estimating back fat thickness of live pigs: (a) use of metal ruler, and (b) use of ultrasonics.
Courtesy USDA.

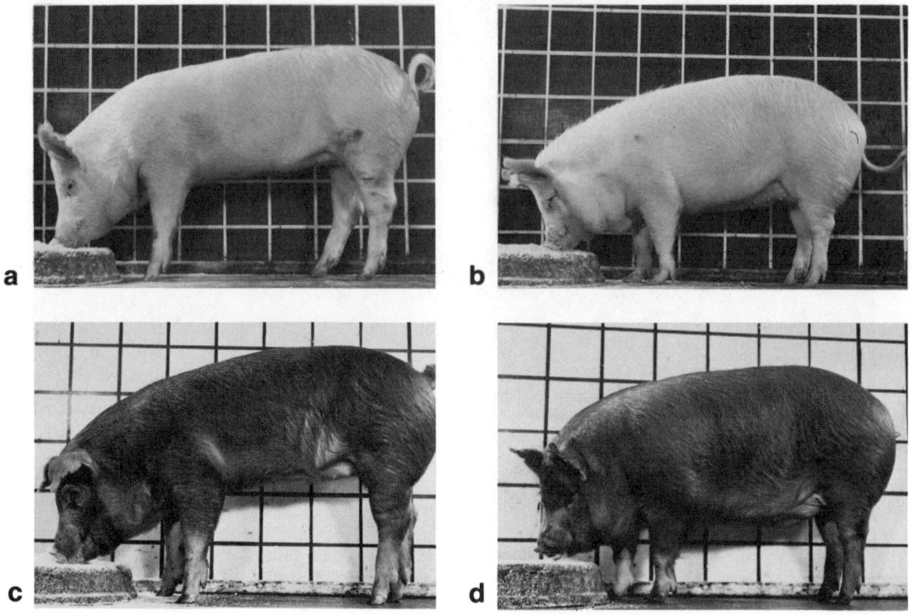

FIG. 4.4. Examples of lean and fat lines: (a) an eleventh generation low fat Yorkshire pig, weight 84 kg; (b) an eleventh generation high fat Yorkshire pig, weight 85 kg; (c) a thirteenth generation low fat Duroc pig, weight 85 kg; (d) a thirteenth generation high fat Duroc pig, weight 85 kg.

ments. Yorkshire and Duroc pigs representing the lean and fat lines selected in this manner are shown in Fig. 4.4.

The ultrasonic system of estimating back fat thickness and muscle cross-sectional area depends on the principle of differences among tissues (lean vs. fat) in the speed of travel of sound through them. These differences are reflected in "blips" on the oscilloscope which can then be expressed in quantitative terms after subjective interpretation. There has been considerable developmental work devoted to the application of this tool in swine selection. Techniques are now perfected and costs of instruments are reduced to the point at which ultrasonics play an important part in selection of swine for desired carcass traits (see Fig. 4.3b).

PUREBREEDING, CROSSBREEDING, INBREEDING, AND OUTBREEDING

These are all terms that describe systems of mating that refer to the degree of genetic relationship between animals. As discussed in Chapter 2, there are some 87 recognized "breeds" of swine in existence today. These breeds are populations that have some particular series of traits in common and matings within the breed produce progeny that maintain these same traits. Such traits are often referred to as "breed characteristics." Examples are hair color and ear carriage (erect or drooping). For purposes of this discussion, a purebred individual is one that is the progeny of purebred parents of the same breed. A registered purebred is one that has been assigned a permanent registration number or other designation in a purebred breed association. Thus, an animal can be a purebred without being a registered purebred.

A given breed has a gene pool different from that of other breeds. Therefore, mating of two individuals from different breeds introduces in the progeny a gene combination that is different from that existing in either parent or, in fact, in the breed of either parent. This mating of individuals of two breeds is called crossbreeding. In commercial swine production in the United States most swine slaughtered are crossbred. This can be accomplished by a two-way cross in which purebred gilts of one breed are mated to a boar of a second breed, or it can be a more complicated system involving three or more breeds. These systems of crossbreeding can be illustrated as follows:

Two-Way Rotational Crossbreeding

Generation	*Boar*	*Gilt*	*Result (%A, %B)*
1st	1, breed A	Breed B	50, 50
2nd	2, breed A	Crossbred AB	75, 25
3rd	1, breed B	Crossbred AAB	37.5, 62.5
4th	3, breed A	Crossbred BAAB	68.75, 31.25

With succeeding generations, the exact proportion of genes from each breed will vary depending on the sequence of boars.

Three-Way Rotational Crossbreeding

Generation	*Boar*	*Gilt*	*Result (%A, %B, %C)*
1st	1, breed A	Breed B	50, 50
2nd	1, breed C	Crossbred AB	25, 25, 50
3rd	2, breed A	Crossbred ABC	62, 5, 12.5, 25
4th	1, breed B	Crossbred ABC	31.25, 56.25, 12.5
5th	2, breed C	Crossbred ABC	15.625, 28.125, 56.25
6th	3, breed A	Crossbred ABC	57.8125, 14.0625, 28.125

As with the two-way cross, the exact proportion of genes originating from each breed will vary from one generation to the next, depending on the sequence of sires used.

A major purpose of crossbreeding is to take advantage of the observed improvement in performance of the progeny above the average of the parents. Another advantage of crossbreeding is to combine desirable traits of two or more breeds to produce a pig with more optimum characteristics. This phenomenon, termed "hybrid vigor" or "heterosis," is not well understood, but seems to be consistent enough to have encouraged its persistent use by commercial swine producers in the United States. An improvement of 5 to 15% in growth rate, efficiency of feed conversion, and reproductive traits is common in crossbred animals. Heterosis provides the basis for the fact that the vast majority of slaughter pigs produced in the United States are crossbred. Recent evidence (Johnson 1981) suggests that for sow productivity traits there is a large variation in heterosis in specific crosses. This point needs verification because of its important implications for animal breeders.

Crossbred boars exceed purebreds in conception rate, embryo number, and libido (Wilson *et al.* 1977) and may have larger testes and sperm production (Neely *et al.* 1980). Such advantages of crossbred boars have caught the imagination of commercial swine producers and suggest the possibility of increased future use of crossbred sires for pork production.

At this point in the discussion, it should be clear that the major objective of purebred swine production is to identify and propagate superior sires for use in commercial production primarily in crossbreeding programs as well as to propagate and identify superior females for maintaining valuable genetic material.

Improvements in commercial swine production will be only as rapid as improvements in the parent pure breeds. Furthermore, the practice of crossbreeding is not worthwhile unless superior purebred sires are used, for the improvement in performance of progeny of two average individuals from two different breeds may not be large enough to produce performance better than that obtained by a purebred mating.

Outbreeding and inbreeding both involve mating of animals within a breed. *Outbreeding* is the mating of individuals less closely related than the average of the breed. This is the most common kind of mating in normal purebred production, in which boars are often selected on the basis of performance or progeny testing and there is no common ancestor present for 4 or 5 generations back in the pedigree of the boar and the females with which he is mated.

Inbreeding is the mating of individuals more closely related than the average of the breed. The degree of inbreeding can range from the closest possible mating (full sibs or sire–daughter) to the mating of animals that have no common ancestors for several generations back. For practical purposes, the influence of a common ancestor more than 3 generations removed in the pedigree is probably negligible. However, if one computes the inbreeding coefficient of such an individual, the value will indicate that the animal is not the result of outbreeding. Thus, it becomes clear that the terms inbreeding and outbreeding have little meaning in practical terms except to indicate the probability of differences in relative *homozygosity*. That is, the mating of closely related animals (such as sire–daughter) has the effect of concentrating common genes in the progeny. This increase in frequency of homozygous gene pairs in inbred animals applies both to undesirable and desirable traits. Many of the undesirable traits, including sublethal anatomical defects (hernia, cryptorchidism) involve recessive genes so that in individuals that carry genes for these traits, mating with a relative will increase the odds of the trait being detected. Thus, if a potential sire is suspected of carrying genes for cryptorchidism, for example, it is advisable to mate him to full sibs or his own dam. If he carries the gene, the odds favor the occurrence of cryptorchidism in some of the progeny. Known carriers of undesirable genes such as this should be destroyed to avoid perpetuating the trait in the herd. (More detailed descriptions of undesirable inherited defects are given under Inherited Defects in this chapter.)

Formulas for *coefficient of inbreeding* and *coefficient of relationship* between two individuals and examples of their computation are as follows:

Coefficient of inbreeding (a measure of the percentage of homozygosis):

$$Fx = [(\tfrac{1}{2})^{n+n'+1}(1 + Fa)],$$

where Fx is the inbreeding coefficient for the individual; Fa, inbreeding coefficient for the common ancestor; n, number of generations between the common ancestor and the sire of the individual; and n', number of generations between the common ancestor and the dam of the individual.

Coefficient of relationship (a measure of the relationship of 2 individuals):

$$Rxy = \frac{\Sigma[(\tfrac{1}{2})^{n+n'} \times (1 + Fa)]}{\sqrt{(1 + Fx)(1 + Fy)}},$$

where Fa is the inbreeding coefficient of common ancestor; Fx and Fy, inbreeding coefficient of individuals whose relationship we want to measure; n and n', number of generations by which x and y, respectively, are removed from the common ancestor.

Closed populations of purebreds, such as the Danish Landrace, in which introduction of new genes is prevented by prohibiting importation of live pigs from other countries, show a gradual increase in the inbreeding coefficient for the breed with time. At the present time the value is reported to be about 9%. This is probably almost meaningless for populations as large as this because other factors such as selection and even mutations may completely offset the homozygosity that occurs in large populations. The rationale for the "closed herd" in Denmark has been the preservation and protection of the traits that are considered essential for best meeting market demand for pork (largely export).

As indicated earlier, selection for superiority in a specific trait (or traits) is most effective when there is a large variation among individuals. As degree of inbreeding increases within a population, the individual variation may decline. A good example is the uniformity of commercially available laboratory rats which have been inbred for many generations specifically to satisfy a market demand for animals that respond similarly to experimental treatments.

This concept of uniformity within lines of inbred swine led to the hypothesis that development of such inbred lines and subsequent crossing of these lines (in a fashion patterned after "hybrid corn" production) might offer a means of obtaining superior performance in swine in the same way but to a greater extent than was observed with crossbreeding. Considerable time and money were devoted to testing this hypothesis by the USDA in cooperation with several Land-Grant State Experiment Stations during the years between 1930 and 1970. These efforts resulted in several new "breeds" based on propagation of selected lines of inbred animals resulting from foundation stock representing two or more breeds in the initial generations. Some of the breeds developed in this way were the Minnesota Nos. 1, 2, 3: Montana No. 1 (Hamprace); Beltsville Nos. 1, 2. The Inbred Livestock Registry Association was organized to record purebred registrations in these and other similarly developed breeds. The performance of these breeds and that of progeny resulting from selected matings of animals from different breeds or from different inbred lines within breeds has not, on the average, been superior to that of crossbred animals from outbred populations or of purebred animals of the conventional breeds.

The question as to the best approach to selection to provide the greatest long-term gain to the industry, therefore, remains unanswered, but it appears that intensive selection within a purebred population to provide outstanding sires for use in commercial crossbreeeding programs may be as acceptable, or more so, than more elaborate selection schemes. Recent evidence indicates that the use of crossbred sires, produced under carefully managed systems, may provide a practical approach to applying the effects of heterosis in commercial pork production. The utilization of inbred lines for crossbreeding on a commercial basis would require vast government or other similar support to maintain the purity of the inbred lines under production systems of the size and scope presently existing even in the most advanced areas of swine production. Although some commercial swine farms (or perhaps more properly called pork production enterprises) now produce several thousand slaughter hogs annually, the number of breeding animals needed to maintain inbred lines for crossbreeding would preclude the application of such a program even if favorable from a genetic point of view. There are now several swine breeding programs of this kind in the United States financed by private industry. Whether or not commercial swine breeders in the areas of concentrated swine production in the United States and elsewhere will someday depend on such organizations to supply replacement breeding stock remains to be seen.

ARTIFICIAL INSEMINATION AND CONTROLLED ESTRUS

The technical aspects of artificial insemination (AI) and controlled estrus are covered in Chapter 8. The justification of artificial insemination in commercial swine production is mainly on the basis of the opportunity to make maximum use of superior sires. In natural service, a mature sire can be expected to mate with no more than 250 to 300 sows/year, whereas with artificial insemination the number can be increased to 2500 or more services with the present state of knowledge, and this may be increased further as the technology of semen preservation and dilution improves. Thus, the potential for genetic improvement in swine could be enhanced by artificial insemination, because it would help to identify sires out-

TABLE 4.4. Reproductive Performance of Gilts Fed Allyl Trenbolone, a Synthetic Progesterone at 15 mg/Animal for 18 Days

	Control			*Synchronized*[a]		
	Natural service	*Frozen semen*	*Combined*	*Natural service*	*Frozen semen*	*Combined*
No. of gilts bred	38	30	68	29	29	58
No. of gilts farrowed	33	17	50	27	14	41
Farrowing rate, %	86.8	56.7	73.5	93.1	48.3	70.7
Total pigs/litter	10.3	6.8	9.1	11.0	9.6	10.5
Live pigs/litter	10.1	6.4	8.8	10.3	9.2	9.9

Source: Pursel et al. 1981.
[a]All but two treated gilts returned to estrus 4–7 days after withdrawal of allyl trenbolone.

standing enough for continued heavy use. The relatively high cost per insemination, if only a few animals per day are inseminated, and the difficulty in accurately detecting estrus discourage the widespread use of AI in most large commercial herds. Experimental results with oral progesterone derivatives as a means of controlling the estrous cycle enhance the potential utilization of AI. Some studies have shown good synchronization of estrus by feeding progesterone derivatives to gilts. Estrus can be blocked regardless of the stage of the estrous cycle at which the hormone feeding is started. An example of the kinds of results obtained in this way in combination with frozen semen is shown in Table 4.4 (Pursel *et al.* 1981). Further efforts should result in fewer problems with "silent heat" (ovulations without complete estrus) and cystic ovarian follicles. When these problems are solved, AI and controlled estrus can be used concurrently to control the production schedule carefully throughout the year.

INHERITED DEFECTS

There are many anatomical defects in pigs, most of them associated with recessive genes.

Lethal defects of genetic origin in pigs include hydrocephalus, hind leg paralysis, atresia ani, cleft palate, thickened forelimbs, muscle contracture (rigid forelimbs), split ears, and amputated (legless) condition. The following description of congenital abnormalities includes these lethal defects as well as sublethal defects considered to be of genetic origin. Examples of such defects are shown in Fig. 4.5. It is important to keep in mind the possibility of nutritional and/or other environmental factors contributing to the occurrence of these defects. Blindness and microphthalmia in newborn pigs, originally thought to have a genetic basis, have been experimentally produced in sows fed vitamin A deficient diets by Hale (1933) and more recently in Denmark by Palludan (1966). Viral and other infections (Young *et al.* 1955) as well as ingestion of drugs such as thalidomide (Palludan 1966) during pregnancy can also seriously alter the normal development of the pig fetus.

The inherited defects are listed here in alphabetical order.

Atresia ani. This condition involves a closed rectum so that death results within the first few days after birth, owing to inability to void feces. In some cases corrective surgery is possible. In females, the anus is closed, but usually an opening exists on the ventral side of the rectum with communication to the vagina so that defecation can occur. Such females can survive to maturity and are usually fertile. Atresia ani has a genetic basis (Berge 1941) but the exact mode of inheritance has not been determined. No differences have been found between normal females and those with atresia ani in chromosome counts or karyotypes (Vogt 1967).

Cleft palate. This condition has been reported several times in pigs and is often associated with other defects. For example, a peculiar ear defect involving a bilobed outer ear has been found (Annett 1938) in conjunction with cleft palates.

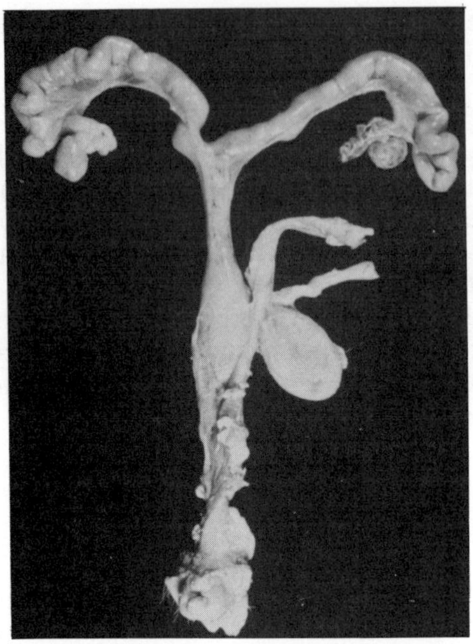

a

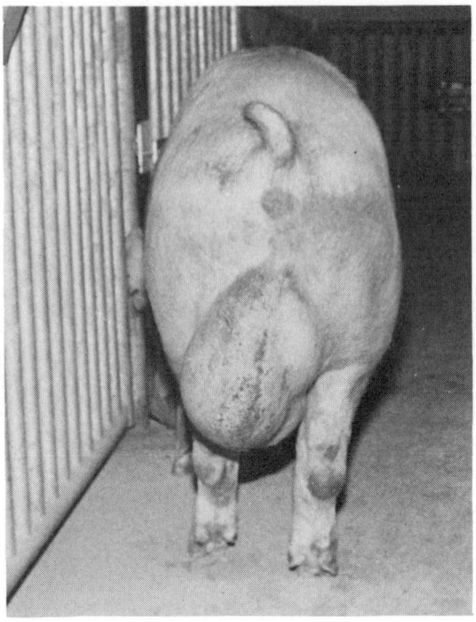

b

c

FIG. 4.5. Inherited defects: (a) the reproductive tract of a true hermaphrodite: note the general appearance of the female tract, but the testes and epididymis at the tip of the uterine horn; (b) boar with scrotal hernia; (c) barrow with umbilical hernia.

The incidence of cleft palate is not high enough to establish the mode of inheritance. Its expression may, in fact, be of nutritional origin, in some cases.

Cryptorchidism. This defect involves the failure of one or both testes to descend from the body cavity through the inguinal ring into the scrotum during late embryonic life. They are then retained in the body cavity throughout life. This inhibits spermatogenesis, owing to the higher temperature of the body cavity. Testosterone production may be appreciably reduced (Hanes and Hooker 1937), although the development of secondary sex characters is not greatly affected. When only one testis fails to descend, fertility is usually not destroyed, but when neither descends infertility almost always results. Bilateral cryptorchids are occasionally reported to be temporarily fertile, but this is probably due to temporary descent of the testes. Cryptorchidism is thought to be under the control of a sex-limited recessive gene (McPhee and Buckley 1934), although there is some evidence that several genes are involved. Boars and sows producing cryptorchids should be destroyed to avoid perpetuating the trait.

Epithelial defect. There is a comparatively rare condition (Bentinck-Smith 1951) involving an absence of the epithelium and hair follicles over a part of the back. In such cases there is a bare, raw patch in the affected area. Affected animals sometimes survive to slaughter weight, but are susceptible to infections by the entrance of disease organisms through the unprotected area.

Heart and circulatory system abnormalities. Several abnormalities occur in the anatomy of major blood vessels and of the heart itself (Kitchell *et al.* 1957). Often affected pigs are stillborn or die within a few days after birth. Abnormalities include interventricular septal defects and arterial abnormalities. The incidence of defects of the heart and arteries in normal populations of swine is not known, nor is the heritability of such defects.

A rare hemophilia-like condition in an inbred line of pigs has been reported in Missouri (Bogart and Muhrer 1942), which is apparently due to a recessive gene and one or more modifying genes. The condition expresses itself in both sexes at about 2 months of age and increases in severity with age. Surgery, including castration, and minor skin abrasions result in prolonged bleeding.

Hermaphrodites and pseudohermaphrodites. The intersex condition is a common occurrence in pigs and has been suggested (Johnston *et al.* 1958) to involve one sex-linked recessive gene as well as one or more additive genes. The distinction between true and pseudohermaphrodites is that in the true hermaphrodite, both ovarian and testicular tissue are present, whereas in the pseudohermaphrodite, testicular but not ovarian tissue is found in the presence of female secondary sex organs, including vulva, vagina, and uterus. Most intersexed pigs that have been studied cytologically have been genetic females (Pond *et al.* 1961; Gerneke 1964; Hard and Eisen 1965). The degree to which these anomalies deviate from normal structure ranges from very minor abnormalities in some pseudohermaphrodites to the presence of almost all of the secondary sex organs of both sexes in true hermaphrodites. A reproductive tract from a true hermaphrodite (Pond *et al.* 1961) is shown in Fig. 4.5a. In one pig with only one testis descended, that one testis contained spermatozoa, whereas the one that had not descended contained no spermatozoa and resembled a testis from a typical cryptorchid.

Hernias of the intestine. Warwick (1926) categorized hernias of the intestinal tract into scrotal, umbilical, inguinal, and ventral. Ventral hernias occur between abnormally arranged or weak muscles of the abdomen. The other three types are described in more detail in this section. (See Figs. 4.5b and c.)

Hydrocephalus. There are two types of hydrocephalus (Blunn and Hughes 1938). One is internal, resulting in fluid accumulation in the ventricles, distension of the cerebral hemispheres, and an increase in the size of the head; the other type is external involving fluid accumulation outside the brain in the subarachnoid spaces. Fluid pressure causes a partial collapse of the brain so that it occupies the anterior and ventral part of the cranium. The second, external type is most common in pigs. This is due to a simple recessive gene (Gilman 1956) exhibiting the typical ratio of three normal to one defective in matings of two heterozygous individuals. The defect is lethal and has been noted in several breeds. Externally there is a protuberance in the center of the forehead about midway between the top of the skull and a line drawn horizontally between the eyes. The size of this protuberance varies and the amount covered with hair varies. There is a thick membrane covering the affected area instead of a true outer epidermis. There is lack of contact along the mid-line of the skull, always associated with the parietal bone. In markedly affected animals the skull is dome shaped, the eye sockets are flattened, and the eyes are smaller than normal.

Affected pigs have been known to live for as long as a few days. It has been suggested that some cases of the first type may result from penetrance of a pair of recessive genes that causes defective development of the forebrain under the influence of a maternal dietary deficiency, but the evidence is only fragmentary.

The occurrence of the external type seems to be always associated with a short tail, and in some strains, with a lighter coat color. In one instance (Blunn and Hughes 1938) even in animals whose head shape appeared normal, if the coat color were light and the tail short, fluid was found in the cranium on autopsy.

Inguinal hernia. This is a condition in females parallel to scrotal hernia in boars, involving the escape of the intestine through the inguinal ring. Its occurrence is believed to be an inherited weakness, but evidence is meager. Its presence has been associated with animals found to be true or pseudohermaphrodites (Gregory 1959), but whether the occurrence of inguinal hernia depends on the presence of the intersex condition is not known.

Kinky tail. A condition involving the unilateral fusion of adjacent caudal vertebrae results in a crooked or "kinky tail." Postmortem examination (Donald 1949) of "kinky tailed" pigs revealed a high incidence of abnormalities of kidneys, ureters, and genitalia. Inbred "kinky-tailed" pigs were difficult to rear and survivors were usually sterile.

Leglessness. The birth of pigs without legs apparently involves a simple recessive gene (Johnson 1940). The condition is rare and affected pigs are apparently otherwise normal.

Malformed limbs. Malformations of feet and legs have been reported, but some of these may be partly or wholly a result of faulty maternal nutrition or of ingestion during pregnancy of substances having effects similar to those of thalidomide (Palludan 1966). These have involved club foot, thick limbs (involving excessive fibrous connective tissue and hypoplasia of muscle groups) (Kaye

1962), missing fibula, and abnormal projections resembling claws on one or both hind legs.

Mammary defects and variations. The number of functional mammary glands in pigs varies greatly, averaging approximately 12 to 14 (6 or 7 pairs), but more are not uncommon. Glands usually but not always occur in pairs, so that an odd number is not uncommon. The desired number, of course, is that which will adequately support the nutritional needs of a large litter of suckling pigs, normally one gland per pig. The number of glands is not highly heritable (Allen *et al.* 1959, Nordby 1934, Enfield and Rempel 1961). There is an inherited condition involving "inverted" nipples in which the nipple fails to project out from the surface of the gland, making it impossible to suckle. Usually the glandular tissue is normal, but lactation soon ceases when no milk is removed. This defect can involve one or all of the nipples on an individual sow and can usually be detected before sexual maturity. The center glands are more frequently affected than the pectoral or inguinal glands.

Monsters. Several reports (Duckworth 1908; Baumgartner 1927; Carey 1917; Williams and Rauch 1917, Gilbert and Thurley 1967; King 1969; Ludvigsen *et al.* 1963) have described cases of pigs afflicted with a variety of teratologic deviations from normal. Siamese twin pigs have been also reported (McKay 1957). Gross defects such as this may be, in some cases, of genetic origin; the defect may be induced by teratogenic agents. Much remains unknown concerning the causes of these bizarre births.

Myoclonia congenita ("shakers"). Pigs born with nervous incoordination commonly described as "shakers" or "tremblers" have been reported in several breeds. This condition varies greatly in severity, ranging from slight shivering with no apparent effect on performance to shaking so severe that starvation results because the affected pig cannot grasp the nipple to nurse. Severity of shaking generally decreases with age and in some cases disappears altogether. It is generally considered that the condition is an inherited defect: but present evidence suggests that it may be a manifestation of hog cholera virus infection in the pregnant sow. Some believe that a nutritional deficiency is involved, but many cases are found in otherwise normal litters (Hughes and Hinman 1936), and affected pigs are often as large and vigorous at birth as their normal littermates. Histological studies show retarded development of the myelin sheath (Christensen and Christensen 1936) and degeneration of Purkinje fibers in some areas of the cerebellar cortex.

Polycystic kidneys. The condition is characterized by single or multiple fluid-filled cysts on one or both kidneys; cysts are often accompanied by fissures in the surface of the normal smooth, bean-shaped kidney. The pathogenesis of the condition in humans has been studied intensively and it is clearly under genetic control. In adult-onset polycystic kidney disease, the abnormality usually appears in young adulthood and seems to be produced by a single recessive gene; therefore, affected individuals can be homozygous or heterozygous for the trait. The mode of inheritance of polycystic kidneys in swine has not been established clearly, even though the condition has been reported frequently. Littermates sharing the abnormality have been identified at the time of slaughter at 10 weeks or at 20 weeks of age and show no evidence of adverse effects on performance unless the reduction in proportion of normal kidney tissue is extensive.

Polydactyly. A rare condition, more of a curiosity than of practical importance, has been reported in a herd of pigs in which animals of both sexes have extra toes on one or both front feet (Hughes 1935).

Reproductive organ abnormalities. Since only a small percentage of females are normally kept for breeding purposes the exact incidence of anatomical defects of the reproductive organs serious enough to interfere with normal reproduction is not known with certainty. However, a wide variety of abnormalities have been found (Nalbandov 1952; Perry and Pomeroy 1956). These include hydrosalpinx, pyosalpinx, unilateral blind uterine horn, unilateral missing segment in one uterine horn, blind uterine body, and infantilism. These account for more than one-half of all reproductive failures involving anatomical abnormalities. Cystic ovaries, abnormal uterine histology, and other endocrine aberrations account for most of the remainder. In a survey by Wiggins *et al.* (1950) of abnormalities of the reproductive tract in more than 5000 sows and gilts slaughtered in a meat-packing plant, 1.4% were found to have tubular abnormalities, 1.7% had cystic follicles with or without corpora lutea and 8.9% had rudimentary male ducts.

Scrotal hernia. This involves a weakening of the musculature of the inguinal ring and other structures, allowing the intestine to pass into the scrotum of the male pig. Successful castration of affected pigs requires surgery to retain the intestine in the body cavity. This weakness is inherited ($H = 0.15$) (Magee 1951), though the exact mode of inheritance is not firmly established. Boars or sows producing litters containing affected pigs should be destroyed to avoid perpetuating the trait in a given strain. In one study (Berge 1941) the incidence of scrotal hernia in males was 2% and in another it was a 5.1% (Magee 1951). The incidence doubtlessly varies among strains and depends on the degree of inbreeding practiced.

Skeletal anomalies. Congenital anomalies of swine involving the skeleton and skull have been summarized by Eaton (1937) and Lerner (1944). The incidence of this type of anomalies in swine was found to be 0.28% in a study of Canadian swine (Freeden and Jarmoluk 1963). These defects included fused vertebrae, incomplete vertebrae, headlessness, and missing ribs.

Splayleg. Affected newborn pigs are unable to stand or walk because the legs (especially the hind legs) tend to splay out sideways or forward. This is a muscular weakness associated with too few myofibrils per muscle fiber. The condition usually disappears within 2 weeks of birth. Analysis of herd records (Thurley and Gilbert 1967, Dobson 1968) suggests a hereditary disposition. Kohler *et al.* (1969) found that newborn pigs kept on a slippery surface became splaylegged, while littermates kept on a straw-covered surface remained normal. Postmortem examination of affected pigs showed edema on the medial surface of both thighs and more space than normal between muscle bundles and fibers of the abductor muscle. Provision of a surface not conducive to slipping is recommended for pigs prone to the splayleg condition.

Umbilical hernia. This involves a weakened supportive musculature at the umbilicus, allowing the intestines to protrude through the orifice. The size varies from that of a golf ball or smaller to that of a volley ball, or larger, depending on the amount of intestine displaced. This condition does not always manifest itself at birth and it may not appear until several weeks later. It apparently has little, if any, adverse effect on the general health of the animal, since seriously affected individuals have been marketed for slaughter along with their normal littermates. The weakness is apparently inherited but the mode of inheritance is not well established. The incidence in one study (Berge 1941) was 0.6%, but as with scrotal hernia, this doubtlessly varies. Boars and sows producing litters containing affected pigs should be destroyed.

Wattles. These are harmless projections hanging in pairs from the neck area; the wattle is covered by normal skin. The trait is due to a simple dominant gene; 75% of the offspring from the mating of two heterozygous individuals would be expected to possess wattles.

REFERENCES

ALLEN, A. D., TRIBBLE, L. F., and LASLEY, J. F. 1959. Mo. Agric. Exp. Stn. Res. Bull. *684*.
ANNETT, H. E. 1938. J. Hered. *29*,469.
AUSTAD, D. 1969. Acta Scandinavica *19*,181; *20*,65.
BAUMGARTNER, W. J. 1927. Anat. Rec. *37*,303.
BENTINCK-SMITH, J. 1951. Cornell Vet. *41*,47.
BERGE, S. 1941. J. Hered. *32*,271.
BERRUECOS, J. M., DILLARD, E. U., and ROBISON, O. W. 1970. J Anim. Sci. *30*,844.
BLUNN, C. T., and HUGHES, E. H. 1938. J. Hered. *29*,203.
BOGART, R., and MUHRER, M. E. 1942. J. Hered. *33*,59.
BOWLAND, J. P., and HIRONAKA, R. 1957. J. Anim. Sci. *16*,62.
BRUNER, W. H., CAHILL, V. R., ROBISON, W. L., and WILSON, R. F. 1958. J. Anim. Sci. *17*,875.
CAREY, E. 1917. Anat. Rec. *12*,177.
CHRISTESTER, F. E. 1914. Anat. Rec. *8*,355.
CHRISTENSEN, E., and CHRISTENSEN, N. O. 1936. Nord. Veterinaermed. *8*,921.
CLAUSEN, H. 1970. Annu. Rep. Dep. of Anim. Sci., Natl. Agric. Res. Lab. Copenhagen, Denmark.
CRAFT, W. A. 1958. J. Anim. Sci. *17*,960.
DICKERSON, G. E., and WILLHAM, R. L. 1983. J. Anim. Sci. *57*(Suppl. 2), 248.
DOBSON, K. J. 1968. Aust. Vet. J. *44*,26.
DONALD, P. P. 1949. J. Agric. Sci. *39*,164.
DOORNENBAL, H., WELLINGTON, G. H., and STOUFFER, J. R. 1962. J. Anim. Sci. *21*,464.
DUCKWORTH, W. L. H. 1908. Proc. Comb. Phil. Soc. *14*,447.
EATON, O. N. 1937. J. Hered. *28*,320.
ENFIELD, F. D., and REMPEL, W. E. 1961. J. Anim. Sci. *20*,876.
FREDEEN, H. T., and JARMOLUK, L. 1963. Can. J. Anim. Sci. *43*,143.
GERNEKE, W. H. 1964. S. Afr. J. Sci. *69*,347.
GILBERT, F. R., and THURLEY, D. C. 1967. Vet. Rec. *80*,594.
GILMAN, J. P. 1956. Cornell Vet. *46*,487.
GREGORY, D. W. 1959. J. Am. Vet. Med. Assoc. *135*,624.

HALE, F. 1933. J. Hered. *24*,105.
HANES, F. M., and HOOKER, C. W. 1937. Proc. Soc. Exp. Biol. Med. *35*,349.
HARD, W. L., and EISEN, J. D. 1965. J. Hered. *56*,255.
HAZEL, L. N., and KLINE, E. A. 1952. J. Anim. Sci. *11*,313.
HEIDENREICH, C. J., GARWOOD, V. A., and HARRINGTON, R. B. 1964. J. Anim. Sci. *23*,496.
HETZER, H. O., and HARVEY, W. R. 1967. J. Anim. Sci. *26*,1244.
HETZER, H. O., and MILLER, R. H. 1970. J. Anim. Sci. *30*,481.
HUGHES, E. H. 1935. J. Hered. *26*,415.
HUGHES, E. H., and HINMAN, R. 1936. J. Am. Vet. Med. Assoc. *89*,96.
JENSEN, E. L., SMITH, C., BAKER, L. N., and COX, D. F. 1968. J. Anim. Sci. *27*,856.
JOHNSON, L. E. 1940. J. Hered. *31*,239.
JOHNSON, R. K. 1981. J. Anim. Sci. *52*,906.
JOHNSTON, E. F., ZELLER, J. M., and CANTWELL, G. 1958. J. Hered. *49*,254.
KAYE, M. M. 1962. Can. J. Comp. Med. Vet. Sci. *26*,218.
KING, G. J. 1969. J. Reprod. Fert. *20*,551.
KITCHELL, R. L., STEVENS, C. E., and TURBES, C. C. 1957. J. Am. Vet. Med. Assoc. *130*,453.
KOHLER, E. M., CROSS, R. F., and FERGUSON, L. C. 1969. J. Am. Vet. Med. Assoc. *155*,139.
LASLEY, J. F. 1963. Genetics of Livestock Improvement. Prentice-Hall, Englewood, New Jersey.
LERNER, I. M. 1944. J. Hered. *35*,219.
LOHMAN, T. G., COFFMAN, W. J., TWARDOCK, A. R., BREIDENSTEIN, B. C., and NORTON, H. W. 1968. Body Composition in Animals and Man, p. 291. National Academy of Science, Washington, D.C.
LUDVIGSEN, J., BASSE, A., CLAUSEN, H., and JONSSON, P. 1963. Annu. Rep. Natl. Agric. Res. Lab., Copenhagen, Denmark.
LUSH, J. L. 1945. Anim. Breeding Plans. Iowa State College Press. Ames.
MAGEE, W. T. 1951. J. Anim. Sci. *10*,516.
McKAY, W. M. 1957. Vet. Rec. *69*,1123.
McPHEE, H. C., and BUCKLEY, S. S. 1934. J. Hered. *25*,295.
MERSMANN, H. R., BROWN, L. J., CHAI, E. Y., and FOGG, T. J. 1984. J. Anim. Sci. *58*, 85.
NALBANDOV, A. V. 1952. Fert. and Ster. *3*,100.
NEELY, J. D., JOHNSON, B. H., and ROBISON, O. W. 1980. J. Anim. Sci. *51*,1070.
NORDBY, J. E. 1934. J. Hered. *25*,499.
PALLUDAN, B. 1966. *In* Swine in Biomedical Research. (L. K. Bustad, R. O. McClellan, and M. P. Burns, Editors). Frayn Printing Co., Seattle.
PERRY, J. S., and POMEROY, R. W. 1956. J. Agric. Sci. *47*,238.
POND, W. G., ROBERTS, S. J., and SIMMONS, R. 1961. Cornell Vet. *51*,394.
PUMFREY, R. A., JOHNSON, R. K., CUNNINGHAM, P. J., and ZIMMERMAN, D. R. 1980. J. Anim. Sci. *50*,1057.
PURSEL, V. G., ELLIOTT, D. L., NEWMAN, C. W., and STAIGMILLER, R. B. 1981. J. Anim. Sci. *52*,130.
RICE, V. A. 1957. Breeding and Improvement of Farm Animals. McGraw-Hill, New York.
RUTLEDGE, J. J., and SEIDEL, G. E., JR. 1983. J. Anim. Sci. *57*(Suppl. 2),265.
SAFFLE, R. L., ORME, L. E., SUTTON, D. D., ULLREY, D. E., and PEARSON, A. M. 1958. J. Anim. Sci. *17*,480.
SMITH, C., JENSEN, E. L., BAKER, L. N., and COX, D. F. 1968. J. Anim. Sci. *27*,848.
STANT, E. G., JR., MARTIN, T. G., and KESSLER, W. V. 1969. J. Anim. Sci. *29*,547.
STAUN, H. 1963. Acta Agric. Scand. *13*,293.
THOMSON, R. N., PEDERSEN, O. K., and CHRISTENSEN, A. 1972. Annu. Report, p. 93. Agric. Res. Lab., Copenhagen, Denmark.
THURLEY, D. C., and GILBERT, F. R. 1967. Vet. Rec. *80*,302.
TREMERE, A. W., OWENS, B. D., and BELL, J. M. 1966. J. Anim. Sci. *25*,138.
VOGT, D. W., 1967. J. Anim. Sci. *26*,1002.
WARWICK, B. L. 1926. Wis. Agric. Exp. Stn. Res. Bull. *69*.
WHITEMAN, J. V., and HILLIER, J. C. 1953. J. Anim. Sci. *12*,859.
WIGGINS, E. L., CASIDA, E., and GRUMMER, R. H. 1950. J. Anim. Sci. *9*,269.
WILSON, E. R., JOHNSON, R. K., and WETTEMAN, R. P. 1977. J. Anim. Sci. *44*,939.
YOUNG, G. A., KITCHELL, R. L., LUEDKE, A. J. and SAUTTER, J. H. 1955. J. Am. Vet. Med. Assoc. *126*,165.

5

Prenatal Development

INTRODUCTION

Prenatal development is a critical part of the life cycle of the pig, because the events of this period have an influence on subsequent growth and development and on the productive life of the pig. Complicated and fascinating, yet still incompletely understood, cell multiplication and differentiation occur early in prenatal life to form tissues and organs of specialized structure and function.

This discussion includes only the events of embryonic and fetal development considered to be of special developmental significance in swine production. For detailed embryology of the pig, the student should consult Patten (1948) or Marrable (1971). Some of these events are assembled chronologically in Table 5.1 (Patten 1948; Altman and Dittmer 1968). Many details of embryonic and fetal development beyond the scope of our discussion were reviewed in a symposium (Book and Bustad 1974; Brinster 1974; Ulberg and Rampacek 1974) and in two books (Dawes 1968; Hafez 1975). Walker (1964) described the dissection of the fetal pig.

FERTILIZATION

Fertilization (the union of the male and female pronuclei) includes the penetration of the surface of the ovum (zona pellucida) by a sperm. This results in a decrease in permeability and an increase in thickness of the ovum surface (Green and Winters 1946; Goldstein 1926; Heuser 1927), so that only one sperm enters the ovum. Cell division begins immediately following fertilization. The nutrient reserve of the fertilized ovum is extremely small, so almost from the start development depends on drawing nutrients from the surrounding fluid. Fertilization normally occurs in the oviduct a few hours after mating. The fertilized ova pass

TABLE 5.1. Chronological Events in Prenatal Development

Event	*Days*
Passes into uterus (4 cells)	2
Morula (16 cells)	3–4
Blastocyst	4.5
Differentiation of germ layers	7–8
Completed primitive streak, attachment to endometrium	11–12
Primitive gut	11–12
First somite differentiation	14–20 (anterior first)
Auditory pit	15–17
Allantoic arteries and veins; umbilical veins	15–17
Arteriosus, truncus arteriosus, dorsal aorta	15–17
Fusion of chorioamniotic fluids	16
Heart beat recognizable	16
Pancreas	16.5–17.5
Lung buds	16.5–17.5
Esophagus, stomach, intestine	16.5–18
Allantois prominent	16–17
Thyroid primordia	17
Amnion complete	17–18
Limb buds recognizable	17–19
Implantation	18
Optic cup and lens	19
Adrenal cortex	20
Olfactory pit	20
Umbilical veins routed through liver	20
Bronchi branching	20–21
Spleen	20–21
Bulbous arteriosus divides into aortic and pulmonary trunks	20–21
Thymus	21–22
Nostrils and eyes differentiated	21–28
Cecum	22
Cross striations in skeletal muscle	22
Parathyroids primordia	22
Allantois replaces all of exocoelom	25–28
Dental lamina	28
Digits differentiated	28
Epiphysis	28
Glomeruli and convoluted tubules in metanephros	28
Hair follicles recognizable	28
Mammary gland primordia	28
Testes differentiated	28
Ovaries differentiated	28
Activity of blood cell forming centers	30
Heart myoblasts become spindle shaped	30
Elongation of smooth muscle nuclei in longitudinal muscle of esophagus and stomach	32.5
Mandible ossifies	32.5
Adrenal medulla	34.5

(continued)

TABLE 5.1. (*Continued*)

Event	*Days*
Coils of large intestine	34.5
Humerus, femur and ribs ossify	34.5
Nasal chamber, nasal septum complete	34.5
Cerebellum	34–35
Skeletal muscle bundles	35
Vertebrae ossify	36
Liver glycogen storage	40
Secretion of epinephrine	40
Scrotum in male, clitoris in female	44
Serum protein synthesized in appreciable amount	50
Secretion of pituitary hormones	50
Secretion of thyroxin	50
Seminal vesicles, prostate, Cowper's glands in male; oviducts, uteri and vagina in female	51
Spinoid processes of vertebrae ossify	63
Fibrinogen synthesis	90
Testes enter inguinal canal in male	95
Birth	114

Source: Patten 1948; Altman and Dittmer 1968.

down the oviduct and into the uterus by the third day after mating. Cell specialization and rearrangement begin by the sixth day.

IMPLANTATION

Eleven-day embryos begin to show signs of attachment to the endometrium. The uterine lining shows a slight indentation at the site of attachment. However, true implantation of the blastocoele to form the placenta does not occur until about day 18 (embryo length of 5–10 mm). The three germ layers (the ectoderm, mesoderm, and entoderm) are by now clearly formed and cell specialization continues. From the ectoderm arise the outer epithelium of the body, the mammary, sebaceous, and sweat glands, the hair and hoofs, intestinal epithelium, teeth enamel, and all of the nervous system. From the entoderm arise the digestive tract (except epithelial covering), pancreas, liver, thyroid, pharynx, trachea, lungs, middle ear, and parathyroids. From the dorsal mesoderm arise the skeleton, skeletal muscle, and connective tissue; from the lateral mesoderm the blood vessels, heart, smooth muscle, connective tissue of blood vessels, blood cells, cortex of the adrenal gland, visceral pleura and peritoneum, and the pericardium; and from the intermediate mesoderm the primary reproductive organs and accessory glands and the kidney tubules and ureters.

By the 20th day the embryo has reached a crown–rump length of about 10 mm and can be successfully used for gross laboratory study. Fetuses at this and subsequent stages of development are diagrammed and photographed in Figs. 5.1 and 5.2.

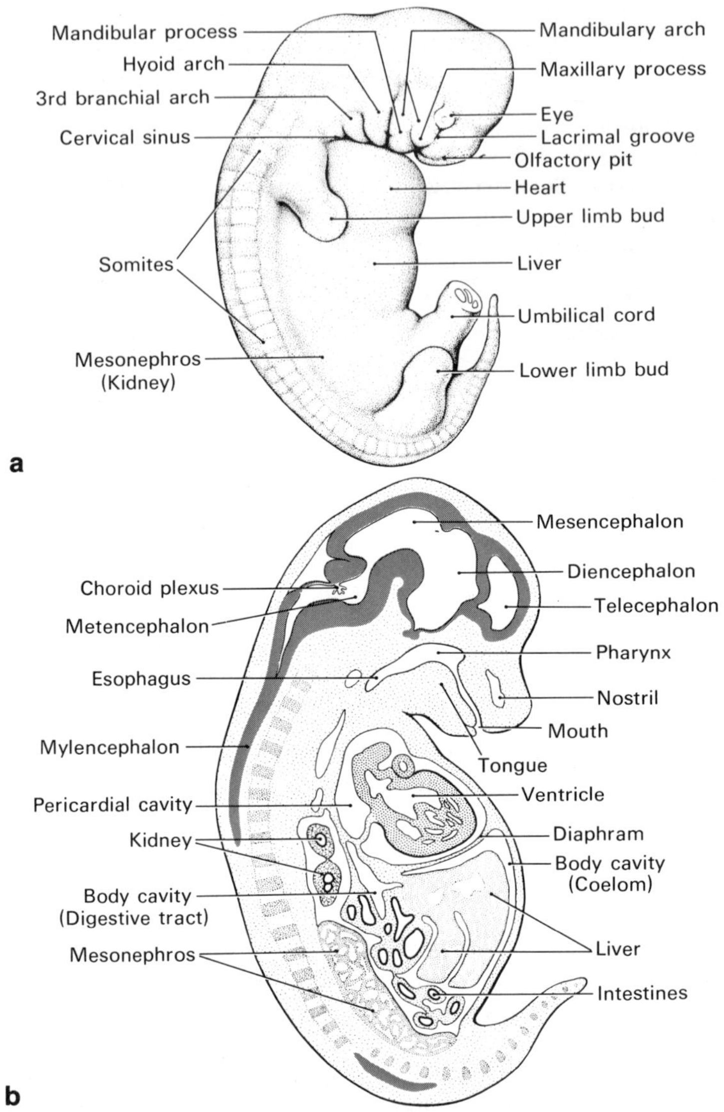

FIG. 5.1. Stages of development of pig embryo: (a) 10–12-mm embryo (12–15 days); (b) 35-mm embryo (20–25 days).
From Harrison 1971.

DIFFERENTIATION OF ORGANS

Many of the major organs can be seen by day 20. These include the heart, pharynx, trachea, esophagus, stomach, lung, liver, eye socket, nasal pit, and anterior and posterior appendage buds. The elongation of the intestines has just begun at this stage. The digestive tube protrudes into the belly stalk and communicates with the yolk stalk which is now greatly enlarged and largely replaces the

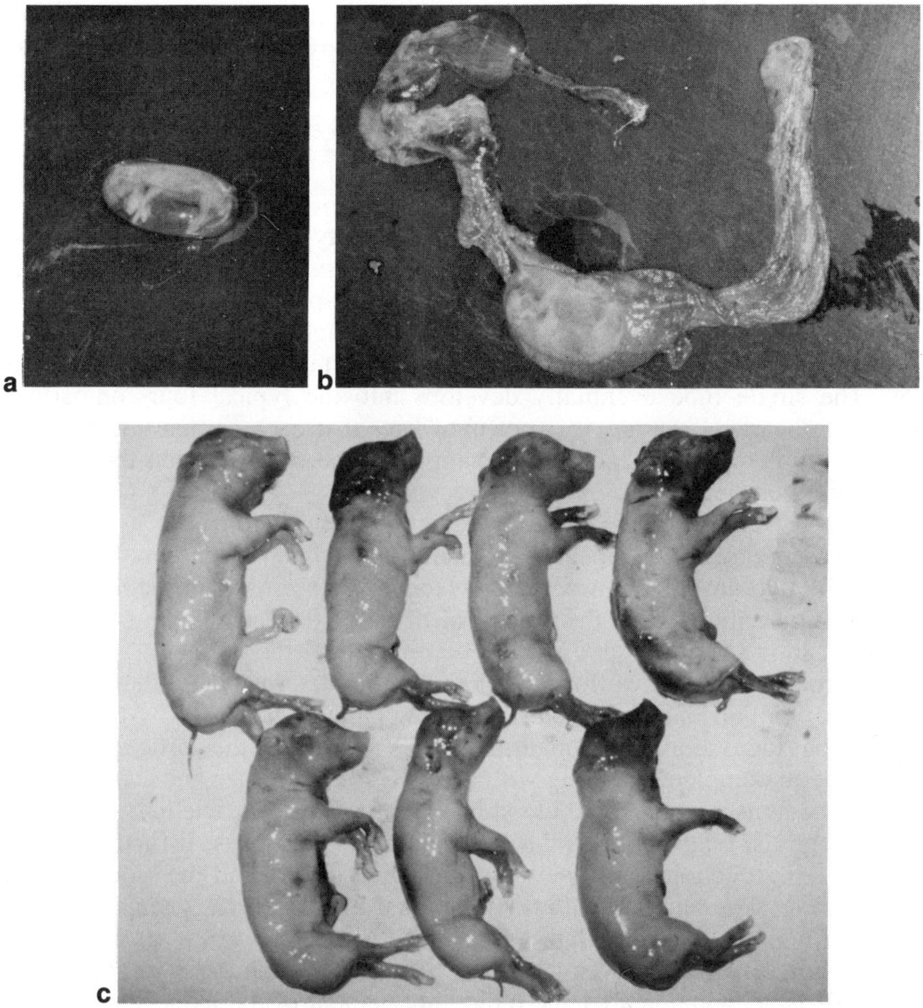

FIG. 5.2. Development of the fetal pig: (a) 30-day fetus in amnion; (b) 35-day fetus in placenta; (c) 80-day fetuses, where the general appearance is similar to newborn pigs.

yolk sac. The caudal end of the digestive tube is the cloaca which ruptures at about this stage, providing the opening to the outside.

The body cavity has not yet divided into the pericardial, pleural, and peritoneal cavities, but there is by now an incomplete division between the liver and heart, the septum transversum. Later this septum fuses with the pleuroperitoneal folds to complete the diaphragm.

The urinary system at this stage is composed of a large network of tubules, the mesonephros, and represents a great mass of tissue in relation to total embryo size, making it a conspicuous organ.

The brain has already reached the five-divisions stage to include the myelencephalon, metencephalon, mesencephalon, diencephalon, and telencephalon as distinguishable entities and the 12 cranial nerves have already developed.

DEVELOPMENT OF THE CIRCULATORY SYSTEM AND BLOOD CELLS

The embryonic development of the pig heart is typical of that of most mammals, but certain features are of special interest in swine production. At day 20, the heart is a tubular organ pumping blood from its posterior and out through its anterior end into a pair of aortas which extend through the length of the developing embryo. The blood is eventually returned to the heart via the common cardinal veins. The single tube eventually develops into the typical four-compartment mammalian heart. However, it is not until later that each compartment is separated by complete septa. Openings remain for some time between the two atria and the two ventricles.

The one between the two atria is the foramen primarum. As it approaches closure, a second opening appears, the interatrial foramen secundum. The septum between the two atria is soon complete except for the foramen secundum. This single opening remains until the time of birth as the foramen ovale, covered by a flap-like tissue to form a valve system between the atria.

Blood from the vena cava flows through the foramen ovale into the left atrium. Meanwhile, the left and right ventricles are separated gradually by the eventual fusing of the ridges along the truncus arteriosus to separate the aortic and pulmonary circulation completely. Along with the increase in the size and capacity of the lungs, as the fetus develops, the blood leaving the right ventricle passes more and more through the lungs and less through the ductus arteriosus. It finally becomes completely nonfunctional and the lungs are able to handle all the blood from the right ventricle. The lungs remain nonfunctional until birth. Oxygenated blood is introduced into the fetal circulation across the placenta.

The umbilical (allantoic) veins become rerouted through the liver at about day 20 and this is accompanied by the rapid enlargement of the liver. The umbilical veins become fused together just before entering the body so that the left umbilical vein takes over the transport of placental blood to the liver. As the vessels through the liver develop, almost all the placental blood soon passes through the liver. Thus, the liver has become the common organ through which all of the extra embryonic circulation and most of the intraembryonic circulation must pass.

The first embryonic blood cells are produced in the yolk sac. However, by the 30th day there is considerable blood-cell formation, including erythrocytes (red blood cells, RBC), within the developing fetus. Later, the kidney and liver take over followed by the lymphoid organs and finally the bone marrow. The changes occurring in blood cell population during prenatal life are summarized in Table 5.2 (Waddill *et al.* 1962; Wintrobe and Shumacker 1935, 1936).

By day 51, the number of RBC has increased fourfold to about two million per cubic millimeter with a corresponding hemoglobin of about 8 g/100 ml and a hematocrit of about 24. The number of erythrocytes and the hematocrit and hemoglobin concentration continue to increase throughout the remainder of the fetal period (Dittmer and Grebe 1959).

TABLE 5.2. Changes in Blood Cell Population and Serum Protein Concentrations during Fetal Life

Blood constituent	*Day after conception*				
	30	*51*	*72*	*93*	*Newborn*
Red blood cells, million/cu mm	0.6	2.1	3.4	4.0	5.6
Hemoglobin, g/100 ml	6.4	8.7	8.6	—	11.7
Hematocrit	20	24	30	32	38
Mean corpuscular volume, cu micra	—	117	91	80	67
Mean corpuscular hemoglobin, picograms	—	31	26	22	21
Leukocytes per cu mm	—	1,430	2,230	2,770	6,270
Total serum protein, g/100 ml	—	2.8	2.3	2.5	2.9
Albumin, % of total protein	—	23	22	19	21
Alpha globulin, % of total protein	—	34	40	50	48
Beta globulin, % of total protein	—	35	27	23	20
Gamma globulin + fibrinogen, % of total protein	—	9	10	9	10

Source: Waddill et al. 1962; Wintrobe and Shumacker 1935, 1936.

At 30 days most erythrocytes are nucleated, but later the proportion rapidly diminishes and at term almost all are mature cells. Male and female fetuses are similar in RBC morphology and in blood hemoglobin concentration.

The total white blood cell (leukocyte) concentration increases throughout the embryonic period. At term, males appear to have higher counts than females, and purebreds higher than crossbreds.

At 50 days the total serum protein level is just under 3 g/100 ml consisting of about one-third each of α- and γ-globulins, 9–10% β-globulin, and 20–25% albumin. The concentration of total serum protein then declines somewhat, followed by a gradual rise to term. There are apparently no fetal sex or breed differences either in total serum protein concentration or in proportions of the fractions. A comparison of serum protein values at term with values for older pigs demonstrates the relative physiological immaturity in this respect in newborn pigs. The α- and γ-globulin fractions are bolstered largely by ingestion of colostral proteins shortly after birth. The synthesis of serum proteins by the young pig is not maximum until several weeks after birth.

PLACENTA AND FETAL MEMBRANES (Fig. 5.3)

The placenta of the pig is termed *epitheliochorial.* The extra-embryonic membranes can be separated from the uterus with no damage to or destruction of the uterine epithelium and, unlike in other species, the blood must pass through the chorion epithelium to reach the fetus. The following six tissues separate the maternal and fetal blood in the pig: maternal tissues composed of (1) uterine endothelium, (2) uterine connective tissue, and (3) uterine epithelium, and the fetal chorion composed of the (1) trophoblast, (2) connective tissue layer, and (3) endothelium.

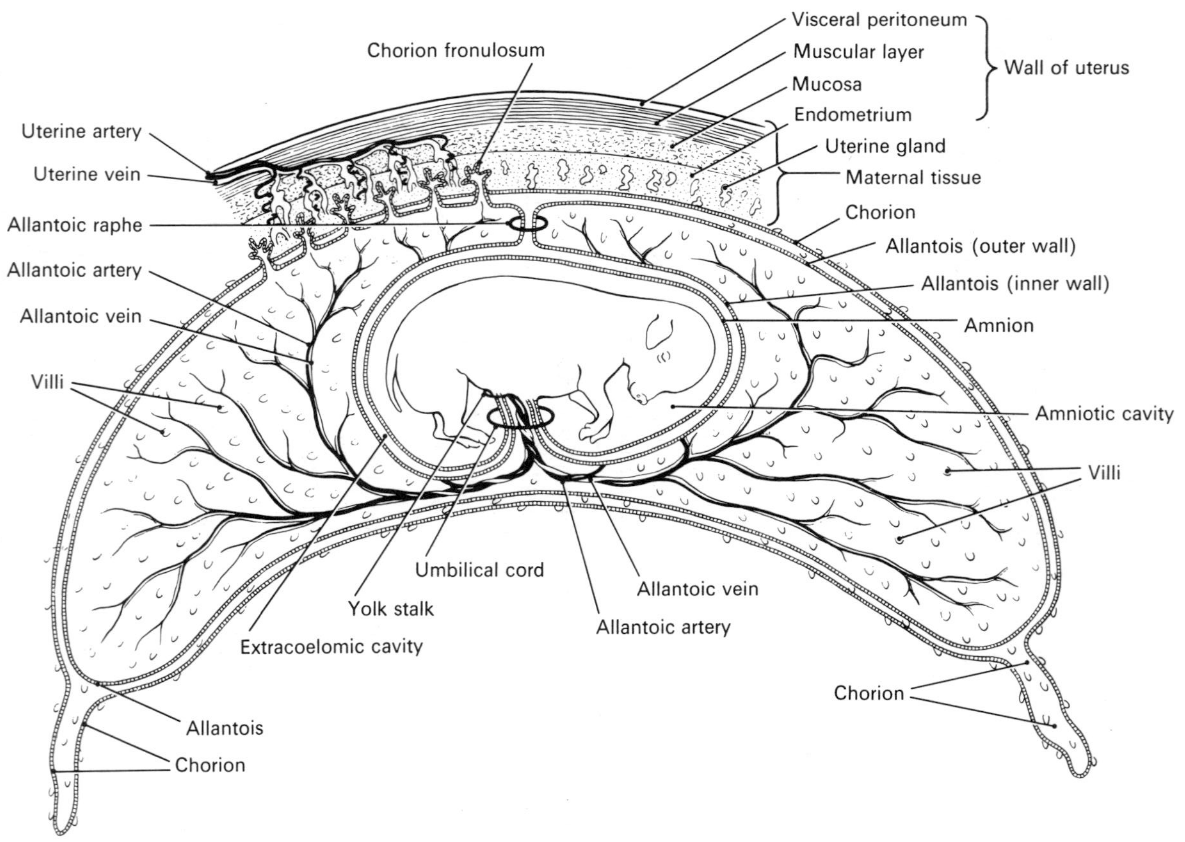

FIG. 5.3. View of fetal membranes and placenta.
From Harrison 1971.

The transfer of large organic molecules across the placenta is limited in the pig so that prenatal passive immunity, provided by maternal blood supply, is almost completely unavailable to the fetus (Hemmings and Brambell 1961). The blood of the newborn pig is therefore almost devoid of immune globulins (gamma globulins).

Each embryo has separate covering membranes, except for identical twins. The incidence of identical twins in swine is not known, but is probably quite low. Thus, after implantation each fetus lives within its own protective membranes and depends on the maternal blood circulation flowing across the six-layered epitheliochorial placenta for its growth and development.

Several membranes are associated with the developing pig fetus. These extra-embryonic membranes are (1) the yolk sac, (2) the amnion, (3) the serosa, and (4) the allantois.

1. **Yolk sac.** It serves mainly in very early embryonic life by carrying nutrients to the embryo. The surface of the yolk sac comes in contact with the uterine epithelium, making transfer of nutrients and oxygen from the uterus to the embryo very efficient. The thick network of blood vessels in the yolk sac which carry these nutrients to the developing embryo is termed the vitelline circulation. This function of the yolk sac disappears as the allantois develops and the yolk sac eventually becomes a shriveled vestige.

2. **Amnion.** This is a single-layered membrane that surrounds the embryo and when completed is filled with fluid which engulfs the fetus and protects it against mechanical injury and adhesions between the membrane and the body surface. The amnion first begins to form at about one week, and by the 17th or 18th day the amnion is complete and the fetus is completely surrounded with amniotic fluid. It is attached where the body opens ventrally in the region of the yolk stalk, and the opening eventually becomes the umbilical ring, through which nutrients pass via the umbilical artery and umbilical vein.

3. **Serosa (trophoderm or somatopleure).** This membrane has several names, depending on what stage of embryonic development is involved, but the term serosa is used here. The serosa encloses the other extra-embryonic cavities with their respective surrounding membranes and, together with the splanchnopleure of the allantois, represents the chorion, the fetal part of the placenta.

4. **Allantois.** This is a single-layered membrane that arises from the hind gut. It rapidly enlarges and dilates, except at the point where it leaves the body cavity (belly-stalk) where it becomes a constricted connection with the hind gut. The allantois has an abundant blood supply and its rapid growth and vascularity results in a rapid increase in its relative importance as a vehicle for supplying nutrients and oxygen to the fetus. It supplants the yolk sac in this function and eventually occupies an enormous space within the cavity of the serosa and becomes fused with it to form the chorion.

The chorion furnishes a double-layered contact directly with the uterine epithelium and provides the means of nutrient exchange from dam to fetus and excretory exchange from fetus to dam.

TABLE 5.3. Relative Weights of Extra-embryonic Membranes, Their Fluids, and the Uterus at Various Stages of Fetal Development

Days after conception	No. fetuses	Membranes		Fluids		Empty uterus	
		g	*% of 47 day value*	*g*	*% of 47 day value*	*g*	*% of nonpregnant uterus*
47	12	800		1,350		1,300	
63	11	2,100	263	5,050	374	2,450	189
66	10	1,570		4,220		2,450	
81	12	2,550	319	5,650	419	2,600	200
86	11	2,150		2,800		2,550	
96	10	2,500	313	2,250	207	3,411	265
102	10	2,500		1,250		3,200	
108	9	2,500	313	1,890	140	3,770	290

Source: Absolute weights of membranes, fluids and empty uterus taken from Moustgaard 1962.

The development of the fetus is accompanied, as an important counterpart, by the growth and development of the surrounding membranes and changes in the uterine epithelium. This increase in size and mass of the membranes involves storage of nutrients obtained from maternal circulation, so the net positive balance of nutrient intake and outgo of the pregnant sow represents that amount required for embryonic growth plus that required for the membranes and for the uterus itself. The relative increase in mass of the membranes and of the fetus during the first few weeks of development heavily favors the former. Most of the growth of fetal membranes is completed by day 70 of gestation (Warnick 1928; Pomeroy 1960) while the fetus has its greatest increase in weight after this time. These relative changes in mass of the fetal membranes, fluids, and empty uterus are depicted in Table 5.3 (Moustgaard 1962). Changes in length and weight of the fetus to term are given in Table 5.4 (Ullrey *et al.* 1965).

The maximum volume of the amniotic and allantoic fluids is highly variable but may reach several hundred milliliters for each fetus at about 65–80 days followed

TABLE 5.4. Length and Weight of the Pig Fetus at Various Ages

Days after conception	Length[a]		Weight	
	cm	*Multiple of 30-day length*	*g*	*Multiple of 30-day wt*
30	2.5 ± 0.3	—	1.5 ± 0.05	—
51	9.8 ± 1.0	3.9	49.8 ± 1.4	33.2
72	16.3 ± 2.0	6.5	220.5 ± 7.3	147.0
93	22.9 ± 2.0	9.2	616.9 ± 15.0	411.3
114	29.4 ± 8.6	11.8	1,040.9 ± 42.7[b]	693.9

Source: Ullrey et al. 1965.

[a]Thirty days measured from crown to rump; 51 days and older measured from most anterior portion of frontal crest to anus.

[b]This mean was obtained from first litter dams and represents a mean approximately 100 g less than expected for litters from mature dams.

TABLE 5.5. Composition of Amniotic and Allantoic Fluids at Various Fetal Ages (per Liter of Fluid)

Constituents	Age						
	20 days	41–46 days		55–65 days		90 days	
	Allantoic	*Amniotic*	*Allantoic*	*Amniotic*	*Allantoic*	*Amniotic*	*Allantoic*
No. of samples	32	13	16	5	5	—	—
Volume, ml	5	26	110	60	350	—	—
Total osmolar conc. mmoles	250	290	120	300	90	270	240
Urea, mmoles	3	4	8	5	10	—	—
Creatinine, mg	12	8	37	3	—	—	—
Inorganic phosphorus, mmoles	2	1	2	2	1	—	1
Chlorine, meq	70	104	30	90	20	8	75
Sodium, meq	115	125	13	120	14	110	4
Potassium, meq	14	19	8	9	6	8	8
Magnesium, meq	1.5	1.5	1.0	—	0.5	—	5.0
Calcium, meq	3	5	19	—	15	—	35

Sources: McCance and Dickerson 1957; McCance and Widdowson 1960; Economou-Mavrou and McCance 1958.

by a constant decline to term. The concentration of cations and anions decreases whereas that of urea and creatinine increases with time. The allantoic fluid is produced first by secretion of the allantoic membranes but later by metabolic products of the fetus as well (McCance and Dickerson 1957).

Compiled data (McCance and Dickerson 1957; Economou-Mavrou and McCance 1958; McCance and Widdowson 1960) on composition of allantoic and amniotic fluid are given in Table 5.5.

The uterus itself also increases in weight throughout pregnancy. Its empty weight increases from approximately 0.5–1.5 kg in the nonpregnant female to 3.0–4.0 kg at term, depending on the number and size of the fetuses. The ratio of placental weight to fetal weight at parturition is about 1:10.

GROWTH OF THE FETUS

The number of developing fetuses is inversely related to the ultimate size of individual fetuses at term. The size of individual fetuses is directly related to the size of the placental (chorionic) membranes associated with them. This is probably related to the degree of the vascular connection to the fetus and in turn to the amounts of nutrients available for deposition in the fetus. However, the greater weight of fetuses at the ends of the uterine horn than in middle and at the ovarian end than at the cervical end cannot be explained on the basis of vascular architecture of the uterus (Perry and Rowell 1969).

The size of the developing fetus increases tremendously during the final half of the prenatal period when the bulk of nutrients are stored. The *relative* rate of increase in size is greatest in the very young fetus and declines gradually in late

TABLE 5.6. Relative Weights of Various Organs at Various Stages of Development

Organ	Organ wt (% of body weight)							
	Prenatal Crown-rump length, mm						Postnatal	
	15[a]	18	25	58	84	97	Newborn	Adult
Brain	8	9	7	7	5	—	4	0.09
Spinal cord	—	1.9	1.3	0.8	0.4	—	0.33	0.04
Eyeballs	—	0.2	0.4	0.9	1.15	—	0.4	0.01
Heart	4.6	3.5	—	1.4	—	0.9	1.0	0.32
Lungs[b]	—	—	0.6	2.9	3.5	—	2.1	0.7
Liver	—	12	16	15	10	3.2	3.1	1.4
Kidneys	—	—	0.3	2.6	2.4	1.1	1.0	0.25
Spleen	—	—	—	0.04	0.05	0.17	0.17	0.13
Pancreas	—	—	—	0.11	0.10	—	0.16	0.14
Thymus	—	—	—	0.1	0.1	—	0.37	0.0001
Adrenals	—	—	—	0.13	0.07	—	0.02	0.005
Stomach + intestine	—	0.3	1.2	2.9	2.8	—	5.9	6.8
Thyroid	—	—	—	0.045	0.035	—	0.025	0.005

Sources: Pomeroy 1960; Lowrey 1911; Ullrey et al. 1965.
[a]Approximately 24 days old. [b]Right is larger than left by ratio of 1.4:1.

prenatal and in postnatal life, even though the greatest daily *absolute* increase in size occurs during late prenatal life and in postnatal life. The comparative rate of growth of various organs during prenatal life is summarized in Table 5.6 (Pomeroy 1960; Ullrey *et al.* 1965; Lowrey 1911–1912).

Viable and strong young have been known to weigh as little as 400 g and more than 2000 g at birth after a normal gestation period. The size of individual pigs varies inversely with the number of pigs in the litter, and directly with the size of their extra-embryonic membranes. The amounts of the major nutrients (protein, energy, and minerals) deposited in the fetus increase by just more than 3% per day, so that the absolute amounts of each are extremely large during the terminal stages.

The gross chemical composition (water, lipid, protein, ash) of the pig fetus at

TABLE 5.7. Gross Chemical Composition of the Pig Fetus at Various Ages

Fetal age (days)	Water	Percent of fresh wt		
		Lipid	Protein	Ash
30	94.7	0.5	3.6	0.9
60	89.5	0.9	6.2	1.9
100	85.3	1.3	9.1	3.1
107	83.6	1.4	9.7	3.2
Term	80.2(73.5–83.8)	0.6(0.2–1.0)	10.9(5.4–13.0)	4.0(3.2–5.3)

Sources: Salmon-Legagneur 1965 (values for fetal pigs); Curtis et al. 1967 (values for 72 newborn pigs from 12 litters, with ranges in parentheses).

TABLE 5.8. Mineral Composition of the Pig Fetus at Various Ages

Constituents	*Fetal age (days)*[a]				
	51	*74*	*97*	*108*	*113*
Body weight, g	55.6	234	823	937	1,127
Water, %	89	89	84	81	81
Total nitrogen, %	1.15	1.07	1.43	1.64	1.84
Calcium, mg/100 g	218	440	800	854	1,070
Magnesium, mg/100 g	14.5	13.5	22.7	27.6	27.6
Phosphorus, mg/100 g	209	297	490	540	628
Sodium, mg/100 g	202	206	202	176	214
Potassium, mg/100 g	185	158	191	214	191
Nitrogen–potassium ratio	6.2	6.8	7.5	7.7	9.6

Source: Pomeroy 1960.
[a]In fetuses ranging from largest to smallest in litter, very little difference was observed in percent composition of each component at any stage.

various stages of development is shown in Table 5.7 (Salmon-Legagneur 1965; Curtis *et al.* 1967). Changes in water, nitrogen, and major mineral concentration of the fetus from day 51 to term are shown in Table 5.8 (Pomeroy 1960). It is of interest to note the wide variation in chemical composition among newborn pigs. Yet, considering fetuses within litters, very little variation was encountered in percentage of mineral composition between the largest and smallest pig (Nielsen 1975). Gestation length apparently has no effect on birth weight, body composition of the newborn pig or thermostability at birth (Curtis *et al.* 1969).

ADAPTATIONS AT BIRTH

The sudden need for the lung to begin functioning at birth is met by the presence of the foramen ovale and the ductus arteriosus as they remain anatomically open at birth. The anatomical closure of the ductus arteriosus is not complete until 6–8 weeks after birth, although it becomes functionally closed at birth. The foramen ovale also ceases to function within a few days after birth, though it may not actually close until much later.

When the umbilical cord is severed at birth, the umbilical vein and artery involute as they have no further function. There is a great danger of hemorrhaging through the umbilical arteries at birth. It is important to make sure that blood flow through them has ceased. This can be done by pinching the cord with blunt scissors or by tying a knot in the cord close to the umbilicus.

Several hours often elapse between the birth of the first and the last pig. This long delay provides the possibility of suffocation should the umbilical cord be severed sometime before birth. The injection of oxytocin into the sow early in parturition in an effort to hasten parturition has given variable results. Most swine producers who have used oxytocin for this purpose can cite success in individual cases. Apparently, time and amount injected are important.

Breed differences exist in the incidence of stillbirths. The frequency of stillbirths may range from less than 3% to more than 8% among different breeds handled under similar conditions. Stillborn pigs are smaller, on the average, than live littermates at birth, so that probably other causative factors in addition to suffocation are involved. It seems possible that poor adaptation of the blood flow through the heart and lungs associated with closure of the foramen ovale and ductus arteriosus at birth are responsible for some stillborn pigs, but this has not been established. Certainly one of the opportunities for improved efficiency in pig production lies in reducing this high frequency of stillbirths among apparently normal pigs. Experience of veterinary practitioners with hysterectomy-derived pigs suggest that virtually all pigs except mummified fetuses are alive at 112 days.

REFERENCES

ALTMAN, P. L., and DITTMER, D. S. 1968. Metabolism. Fed. Am. Soc. of Exper. Biol., Bethesda, Maryland.

BOOK, S. A. and BUSTAD, L. K. 1974. The fetal and neonatal pig in biomedical research. J. Anim. Sci. *38*,997.

BRINSTER, R. L. 1974. Embryo development. J. Anim. Sci. *38*,1003.

CURTIS, S. E., HEIDENREICH, C. J., and MARTIN, T. G. 1967. J. Anim. Sci. *26*,749.

CURTIS, S. E., ROGLER, J. C., and MARTIN, T. G. 1969. J. Anim. Sci. *29*,335.

DAWES, G. S. 1968. Foetal and Neonatal Physiology. Yearbook Medical Publishers, Chicago.

DITTMER, D. S., and GREBE, R. M. 1959. Handbook of Circulation. W. B. Saunders, Philadelphia.

ECONOMOU-MAVROU, C., and McCANCE, R. A. 1958. Biochem. J. *68*,573.

GILBERT, S. G. 1966. Pictorial Anatomy of the Fetal Pig, 2nd Edition. Univ. of Washington Press, Seattle.

GOLDSTEIN, S. R. 1926. Anat. Rec. *34*,25.

GREEN, W. W., and WINTERS, L. M. 1946. J. Morphol. *78*,305.

HAFEZ, E. S. E. 1975. The Mammalian Fetus. Thomas Publishers, Springfield, Illinois.

HARRISON, B. M. 1971. Embryology of the Chick and Pig. Wm. C. Brown, Dubuque, Iowa.

HEMMINGS, W. A., and BRAMBELL, F. W. R. 1961. Br. Med. Bull. *17*,96.

HEUSER, C. H. 1927. Carnegie Inst. Contrib. to Embryol. *19*,229.

LOWREY, L. G. 1911–1912. Am. J. Anat. *12*,107.

MARRABLE, A. W. 1971. The Embryonic Pig: A Chronological Account. Pitman, New York.

McCANCE, R. A., and DICKERSON, J. W. T. 1957. J. Embryol. Exp. Morphol. *5*,43.

McCANCE, R. A., and WIDDOWSON, E. M. 1960. J. Physiol. *151*,484.

MOUSTGAARD, J. 1962. *In* Nutrition of Pigs and Poultry (J. T. Morgan and D. Lewis, Editors), p. 191. Butterworth's, London.

NIELSEN, H. E. 1975. *In* Festskrift til Hjalmar Clausen (H. Staun, Editor) p. 165. Danske Landhusholdingsseiskab, Copenhagen, Denmark.

PATTEN, B. M. 1948. Embryology of the Pig, 3rd Edition. McGraw-Hill, New York.

PERRY, J. S., and ROWELL, J. G. 1969. J. Reprod. Fert. *19*,527.

POMEROY, R. W. 1960 J. Agric. Sci. *54*,31.

SALMON-LEGAGNEUR, E. 1965. Chemical composition of the pig foetus. Ph.D. Thesis, Faculty of Science, Univ. of Paris.

ULBERG, L. C., and RAMPACEK, G. B. 1974. Embryonic and fetal development: uterine components and influence. J. Anim. Sci. *38*,1013.

ULLREY, D. E., SPRAGUE, J. I., BECKER, D. E., and MILLER, E. R. 1965. J. Anim. Sci. *24*,711.

WADDILL, D. G., ULLREY, D. E., MILLER, E. R., SPRAGUE, J. I., ALEXANDER, E. A., and HOEFER, J. A. 1962. J. Anim. Sci. *21*,583.

WALKER, W. F., JR. 1964. Dissection of the Fetal Pig. W. H. Freeman and Co., San Francisco.

WARWICK, B. L. 1928. J. Morphol. *46*,59.

WINTROBE, M. M., and SHUMACKER, H. B., JR. 1935. J. Clin. Invest. *11*,837.

WINTROBE, M. M., and SHUMACKER, H. B., JR. 1936. Amer. J. Anat. *58*,313.

6

Postnatal Development

INTRODUCTION

The newborn pig is rather physiologically immature, and many important changes take place during its early postnatal life to equip it for later life. Some of these changes, and factors affecting them are discussed here.

BIRTH WEIGHT AND LITTER SIZE

The average birth weight of all improved domestic pigs is probably very near 1.2 kg (smaller for ''miniature'' pigs and many indigenous local strains). A survey by Vestal (1936) of a sample of U.S. pigs showed this to be the mean of the population studied, with a close association between birth weight and survival. Of pigs that weighed 1.1–1.35 kg at birth, approximately 75% survived to weaning while of those that weighed 0.57 kg or less at birth, less than 2% survived to weaning. The frequency of pigs farrowed in each 114-g weight group and the relationship between body weight and stillbirths in this study is shown in Fig. 6.1. First-born piglets tend to have larger birth weights than last-born within the same litter.

An increase in litter size (number of live pigs) at birth reduces the chance of survival slightly owing to smaller individual pigs in large litters (an inverse relationship between total litter weight and weight of individual pigs) (Friend and Cunningham 1966). This is depicted in Danish data (Table 6.1).

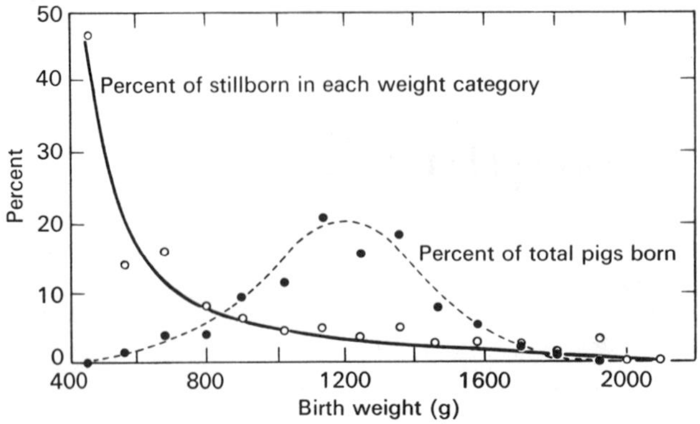

FIG. 6.1. The relationship between the body weight and the incidence of stillbirths. *From Vestal 1936.*

The physiological reasons for the lower survival rate of smaller pigs are not completely known. The phenomenon is probably related to the genetic ability of the individual to withstand extrauterine stress, to differences in the placental supply of nutrients, to other spacial or endocrine differences in the environment, and to the relative physiological immaturity of the newborn pig. The survival rate of small pigs can be improved greatly by raising them in individual cages, thus eliminating the stress of competition with larger littermates (Lecce 1971).

Holub (1968) reviewed the functional changes that occur in early postnatal life of the pig and emphasized the low body energy reserves and immature thermoregulatory and circulatory systems as important factors in neonatal mortality.

If the inherited capacity to attain a large birth weight is not different between individual pigs representing litters of different sizes, there should be a possibility of producing large litters with large individual pigs. This might be achieved by altering the growth pattern of the placenta, the transfer of nutrients across it, or some related aspect of the intrauterine environment. Whether this can be done experimentally or on a practical basis, remains to be seen. The maximum total litter birth weight one can expect to attain at the present time appears to be about 20 kg.

A high proportion of stillbirths occurs during the last third of farrowing (Friend and Cunningham 1966). In a survey of reproductive performance of three breeds

TABLE 6.1. Influence of Birth Weight on Mortality

Weight intervals, kg at birth	*No. of pigs*	*Mortality % of liveborn*
Up to 0.90 kg	123	56.1
0.91–1.20 kg	416	20.4
1.21–1.40 kg	550	11.3
1.41–1.60 kg	637	9.1
1.61–1.80 kg	431	6.5
1.81–2.00 kg	159	6.9
Above 2.00 kg	46	8.7

Source: From Nielsen 1975.

in the same herd (Pond *et al.* 1960), the breed with the smallest average litter size (8.4 live pigs) had the largest average live pig birth weight (1315 g), and the breed with the largest litter size (10.7) had the smallest live pig birth weight (1108 g). The third breed was intermediate in both criteria. Second and subsequent litters tend to have more and larger pigs than do first litters. This is at least partly due to the smaller size of first-litter gilts as compared to mature sows. The average litter size reaches a maximum at the third litter and then plateaus until the sixth or seventh litter followed by a decline.

By far the greatest rate of mortality in young pigs is during the first day after birth (13% including stillbirths in a study by Cox 1962) with smaller losses during the second and third days (3%), and less during the fourth day (2%) followed by a gradual decline to 0.02% during the period 28 to 154 days of age. This clearly shows the effect of age on mortality rate. In a more recent report from Denmark (Neilsen *et al.* 1974), stillbirths accounted for 6.1% of pigs born; while deaths due to starvation, trauma, and gastrointestinal diseases accounted for 3.5, 4.0, and 2.0%, respectively, of pigs born; and miscellaneous causes, including infectious diseases and malformations, were responsible for the remainder of the 22.3% total preweaning mortality. This loss is typical of that experienced in the U.S. swine industry and represents a challenge to the swine producer for improved management practices.

It is important to recognize that prenatal nutrition affects the size and viability of the newborn pig. This is discussed further in Chapter 11.

Several physiological parameters are of importance in the development of the young pig. Each of the following has economic implications in commercial swine production.

BLOOD GLUCOSE

The newborn pigs has a very limited supply of readily available energy stored as liver glycogen and, to a lesser extent, heart and skeletal muscle glycogen. The fat content of the body is less than 1% by weight at birth. Because of this small energy reserve, the pig must nurse shortly following birth, or the blood glucose level falls dramatically. The normal value of 100 mg of glucose/100 ml of blood at birth may drop to 10 mg/100 ml or less within the first two days (Hanawalt and Sampson 1947) in the fasted newborn pig.

When this low blood glucose is reached, lethargy, coma, and death result unless the condition is promptly corrected by nursing or by oral or intravenous glucose administration. Morrill (1952) completed a series of detailed studies designed to characterize baby pig mortality. The following knowledge was gained from his work.

Providing distilled water or physiological saline *ad libitum* to fasted pigs fails to delay the hypoglycemia but may delay dehydration and aid in renal excretion of nitrogenous wastes and in temperature regulation.

Fasting also results in reduced body temperature (less than 34.5° vs. 39°C (normal) during the first 2 or 3 days. The hypoglycemia and reduced body temperature are accompanied by a loss of 30% or more of body weight and finally death. The liver loses proportionately more weight than the remainder of the body during fasting. The environmental temperature during fasting affects the rate of decline in blood glucose concentration. Exposure of fasting newborn pigs to low temperatures (<21°C) results in a rapid drop in blood glucose level so that within 24 hr coma and death result. Pigs fasted in the same way at a temperature of above 27° C survived without coma for 2 to 3 days in a controlled experiment.

Hypoglycemic coma can occur even among pigs allowed to nurse. These cases are usually associated with a cold or drafty environment in combination with an inadequate milk supply. Injection of glucose reverses hypoglycemic coma and the related symptoms if the condition is of short duration, but coma of 1 hr or longer causes irreversible damage and death.

Newton and Sampson (1951) showed that carbohydrates other than glucose (such as sucrose, fructose, and galactose) are not utilized by the baby pig for energy when given intravenously. Resistance to fasting hypoglycemia greatly increases as the pig gets older. Blood glucose of 10-day-old pigs was maintained at a concentration of 60 mg/100 ml after 2 weeks of fasting in a study by Hanawalt and Sampson (1947), and some pigs survived for more than two weeks when allowed only water.

This resistance to hypoglycemia after the first week of postnatal life is probably related to the amount of reserve energy stored before fasting, as well as to metabolic changes associated with development.

Wide individual differences exist in the ability of pigs to metabolize an oral load of glucose (Bunding *et al.* 1956; Hanawalt *et al.* 1947; Meyer *et al.* 1962). A normal blood glucose concentration of 60–80 mg/100 ml of blood in weanling pigs shows a sharp rise to 200 mg/100 ml 1 to 2 hr after administration of such a dose in some animals but only a slight rise in others. This is mostly associated with individual differences, although young animals tend to have a higher glucose tolerance (lower peak blood glucose concentration and faster removal after dosage). Animals with high glucose tolerance tend to have higher muscle glycogen levels. Whether this is related to the problem of PSE (pale, soft, exudative) pork, a condition associated with postmortem changes in muscle acidity, is not known.

BODY TEMPERATURE CONTROL AND CLIMATIC PHYSIOLOGY

For a detailed treatment of the climatic physiology of the pig the student is referred to a book by Mount (1968). The present discussion covers only the salient features of the subject pertaining to development.

The newborn pig is not able to control body temperature efficiently (Newland *et al.* 1952; Pomeroy 1953). This works against the survival of newborn pigs exposed to a stressful environment or a limited milk supply. The poor response to low environmental temperatures would be expected because of the newborn pig's sparse hair coat and limited subcutaneous fat (Mount 1962A; Widdowson 1950). Small pigs are less resistant to cold than are larger pigs because of their greater relative surface area. Maintenance of body temperature in the presence of reduced environmental temperatures requires a high level of heat production (high metabolic rate). Below the critical temperature (that environmental temperature below which an increase in metabolic rate is necessary to maintain body temperature), the newborn pig, in fact, is able to increase heat production (as measured by oxygen consumption) to a surprisingly efficient extent while maintaining body temperature essentially constant (Mount 1959). The metabolic rate (oxygen consumption) increases considerably during the first day after birth.

There is an increase in body temperature from about 38°C at birth to 39° C later, so that the observed increase in oxygen consumption is not surprising (Mount 1962B).

The newborns of wild pigs, which are found roaming free in the southern United States, are more cold resistant than newborn domestic pigs during the first two days postpartum. This advantage in thermostability is partly due to a higher oxygen consumption during exposure to cold in wild than in domestic piglets, and partly to a more dense hair cover in the former (Foley *et al.* 1971).

Heat loss in a cold environment may be lessened by skin vasoconstriction and also by the fact that individuals in a litter huddle together, reducing their effective surface area, and therefore, their rate of heat loss (Mount 1964). Weight is of greater importance than age in affecting heat production after the first few days. Development of chemical thermoregulation has been described (Holub *et al.* 1957) as consisting of a period of poor thermoregulation (first 6 days), a period of gradual improvement (days 7 to 20), and a fully developed system (after 20 days). Table 6.2 shows the effect of environmental temperature and body weight on metabolic rate (Cairnie and Pullar 1957). As ambient temperature falls, metabolic rate increases to maintain body temperature within normal limits. The increase in energy expenditure in a cold environment becomes smaller as the animal becomes larger, owing to decreased relative surface area and increased insulation with subcutaneous fat.

The baby pig can rebound quickly to warming and feeding following exposure to near freezing temperatures for several hours, even when rectal temperature drops below 33° C and the pig has entered coma (Pomeroy 1953).

TABLE 6.2. Effect of Body Weight and Ambient Temperature on Metabolic Rate (kcal of Heat Produced per Day) in Young Pigs

	Ambient temperature (°C)			
Body weight (kg)	*15*	*20*	*25*	*30*
	Metabolic rate (kcal/pig/day)			
4.1	530	430	370	330
5.9	700	680	630	610
8.2	990	920	890	880
10.0	1,220	1,160	1,140	1,150
11.4	1,450	1,400	1,400	1,420

Source: Cairnie and Pullar 1957.

TABLE 6.3. Protein Profile of Newborn Colostrum-Deprived Pig Serum as Compared to That of Mature Pig Serum

Serum protein profile	*Newborn*	*Mature*
Total protein, g/100 ml	2.8	6.3
Percentage of total protein:		
Albumin	12	40
Alpha globulin	74	22
Beta globulin	14	20
Gamma globulin	NM[a]	18

Source: Lecce and Matrone 1960.
[a]NM = not measurable.

DIGESTION AND ABSORPTION OF FOOD

Colostrum and immune antibodies: The colostrum of the sow is very high in gamma globulin and its associated antibodies, so that the newborn pig is provided with an immediate passive immunity if allowed to nurse. The gamma globulin content of the colostrum gradually declines but considerable amounts persist for several days. During the first 24 hr the intestinal wall is permeable to almost any protein materials, including infectious agents. Following the ingestion of colostrum (or milk), the ability of the pig to absorb intact proteins is lost ("gut closure"). This is due partly to a developmental response of the intestinal tract, as evidenced by histological studies (Payne and Marsh 1962A, 1962B) and partly to the presence of a trypsin inhibitor present in sow colostrum (Laskowski *et al.* 1957). Colostrum-acquired antibodies persist in the serum for at least 6 weeks after birth (Brown *et al.* 1961). A typical serum protein profile of a newborn colostrum-deprived pig as compared to the mature profile is given in Table 6.3 (Lecce and Matrone 1960). Active antibody production in the young pig begins at about 3 weeks of age, but production is small until 4 or 5 weeks. It is generally considered that 3 weeks of age is the most critical immunological period for the pig, since colostrum-acquired antibodies (passive immunity) are low and active antibody production (active immunity) has not yet attained full development. Depending on the disease or agent involved, the pig's immunity to it is controlled to varying degrees by passively and actively acquired antibodies. New research (Whiting *et al.* 1983) indicates that feeding of immunoglobulins removed from blood serum to colostrum-deprived low birth weight pigs improves survival markedly when given for the first 3 weeks after birth compared with the improvement achieved when given for only the first day. These immunoglobulins are not absorbed after "gut closure," but do protect the pig from infection by pathogenic microbes in the gastrointestinal tract.

SPECIAL CONSIDERATIONS RELATED TO DIGESTIVE SYSTEM DEVELOPMENT

The newborn pig adapts well to a liquid diet for the first few days after birth, and can be trained to survive on dry feed beginning at a few days of age. Synthetic milk and cow's milk can be used to replace sow's milk, even for the newborn pig,

a

b

FIG. 6.2. Automated feeding device for baby pigs: (above) view of animals at feedings; (right) internal view of the device showing the arrangement of the feeding conveyors.
From Lecce 1971.

and can be supplied either from a bottle and nipple or from a bowl. Figure 6.2 (Lecce 1971) shows an automated feeding system for metered dispensing of liquid milk replacers on a preset time schedule to a large number of individually caged pigs.

Pigs suckling sows normally nurse about every hour around the clock. When early-weaned pigs are fed manually, labor requirements usually limit the feeding frequency to three or four times daily. Pigs fed as infrequently as twice daily convert feed to body weight and gain more efficiently than littermates fed 24 times daily at hourly intervals (Braude *et al.* 1970), but their daily feed intake is reduced and slower body weight gains are obtained. More frequent feeding allows more rapid growth, but there is a danger of digestive upsets and death if intake at a single feeding is excessive. Therefore, it is advisable to limit the food offered to the amount that is consumed in a few minutes.

It is advisable to keep weaned pigs separated to avoid injury and continued agitation caused by persistent sucking of ears and nuzzling of other pigs when kept together.

The utilization of nutrients depends on the ability of the animal to release them in a form available for absorption. For example, the amino acids must be released from intact protein, the fatty acids from triglycerides, the glucose from the disaccharide, lactose from milk, or from other carbohydrates in other feeds. Each major nutrient (carbohydrate, fat, protein) will be discussed briefly, with respect to its utilization by pigs of various ages.

Carbohydrates. The newborn pig readily absorbs glucose, but cannot digest sucrose (Becker *et al.* 1954) or polysaccharides (Braude *et al.* 1958; Dollar *et al.* 1957) efficiently. Lactose is efficiently utilized by the newborn pig, but becomes less well utilized in the older pig, because of the decline in secretion of lactase (Bailey *et al.* 1956), which splits the disaccharide lactose into its component monosaccharides, glucose and galactose. Galactose is converted to glucose as a source of energy.

The young pig is not able to utilize xylose (Wise *et al.* 1954) or fructose (Kidder *et al.* 1963). Unlike the ruminant, the young pig lacks the capacity to utilize large amounts of some of the complex carbohydrates found in forages (cellulose). The pig does not possess the necessary digestive enzymes to hydrolyze these materials. There is some microbial cellulase activity in the cecum of pigs, but this is not an important nutritional contribution to the host, as it has been shown (Lloyd *et al.* 1958) that cecectomy results in no apparent reduction in nutrient utilization by growing pigs. Feeding large amounts of cellulose to growing pigs increases the size of the stomach and cecum (Horszxzaruk 1962; Bohman *et al.* 1953), increases the rate of food passage through the tract, and has a laxative effect. Microbial fermentation of fibrous feeds in the colon of finishing pigs and pregnant sows may provide 10–20% or more of the energy requirements for maintenance (Kass *et al.* 1980).

Fat. The pig is able to hydrolzye and absorb large amounts of fat very efficiently from birth, as would be expected, since approximately one-third of the dry matter of sow's milk is fat. Most of the fat hydrolysis occurs in the small intestine, in the presence of substantial secretion of pancreatic lipases (Gjessing and Harthelt 1960; Hartman *et al.* 1961). The apparent absorbability of dietary fat is usually more than 80% in pigs of all ages. Increased chain length of fatty acids decreases absorbability in young pigs (Hartman *et al.* 1961; Lloyd and Crampton 1957) as does high unsaturation, but this effect tends to disappear with age. In dry milk replacers for early weaned pigs, lard and beef tallow are utilized with approximately equal efficiency, while absorbability of corn oil and other unsaturated fats is less. In addition to lipases, dietary lipids are prepared for absorption by the emulsification of fats and reduction in particle size through the action of bile secretions. Bile secretion is active even in the newborn pig.

Protein. The hydrolysis of protein to its constituent amino acids is inhibited during the first 36 to 48 hr by the presence of a trypsin inhibitor in sow colostrum (Laskowski *et al.* 1957). Subsequently "gut closure" occurs and proteins must be hydrolyzed before they can be used. Apparent absorbability of milk protein by the newborn pig is very high (92–95%). There is a distinctly higher efficiency of digestibility of casein than of soybean protein (95 vs. 89%) by the very young pig (Maner *et al.* 1961; Hays *et al.* 1959). The secretions of pepsin by the stomach and proteinases from the intestinal mucosa and pancreas are low at birth and gradually increase during the first 8 weeks. The inferior performance of pigs less than 3-weeks old that are fed soybean protein and other plant proteins, rather than casein, may be related to the low level of proteinase secretions by the very young pig, but this theory has not been substantiated. The difference in utilization of milk protein and plant proteins disappears by about 5 weeks of age.

By about 5 weeks of age the pig can utilize a wide variety of foods other than milk. The lack of response generally obtained by experimental digestive enzyme supplementation of nonmilk-containing diets for pigs weaned before 3 weeks of age indicates that enzyme secretions are not limiting factors in utilization of the feed. Perhaps unknown factors in development of digestive physiology are overriding the beneficial effects of the supplemental enzymes.

RATE OF PASSAGE OF FEED RESIDUES

In the 4-week-old pig fed synthetic milk, the rate of passage of food residues through the alimentary tract varies from 12 hr with soybean protein to 24 hr with casein as the protein source. By 10 weeks of age approximately 35 hr are required, and there is no effect of protein source (Maner *et al.* 1962). The ratio of dry feed to water plays little, if any, part in 22–90-kg pigs nor does the digestibility of the feed (Castle and Castle 1956). Dry matter content of the feces is remarkably constant over a wide range of feed dry matter. Pelleted feed passes through faster than nonpelleted feed; mean retention time in full-fed growing pigs in one study (Seerley *et al.* 1962) was 47.8 hr for meal feed and 25.3 hr for pelleted feed. Diurnal variation exists in the composition of feces; nitrogen and ash are in higher concentrations in the A.M. and nonnitrogenous organic matter higher concentrations in the P.M., probably due to differences in rate of passage of various feed fractions through the gastointestinal tract. This phenomenon probably has little or no practical importance in commercial swine production.

INTESTINAL MICROFLORA

Even though the pig depends mostly on its own digestive secretions to prepare its food for absorption, the normal gastrointestinal tract contains a wide variety of microorganisms within 24 hr after birth (Fuller 1962; Wilbur *et al.* 1960; Willingale and Briggs 1955). A typical microbial population in the feed, feces, and various segments of the gastrointestinal tract of growing pigs is shown in Table 6.4 (Horvath *et al.* 1958).

With the demonstration that dietary antibiotics stimulate growth of pigs, much attention has been given to alterations in the intestinal microflora that might account for this response. One such study by Larson and Hill (1960) showed that the metabolic activities of the microflora of the ileum are markedly reduced by chlortetracycline supplementation in young pigs. This reduction was associated with less amine formation from catabolism of amino acids by *Eschericia coli.*

TABLE 6.4. Microflora of Feed, Gastrointestinal Contents, and Feces in 90-kg Pigs[a]

	Coliforms	*Enterococci*	*Lactobacilli*	*Yeasts*	*Molds*
Feed	5.4	4.9	3.0	2.8	3.3
Stomach	7.7	6.8	7.9	4.6	3.5
Small intestine	9.3	6.7	7.8	4.6	3.5
Large intestine	9.2	8.0	8.2	4.4	4.0
Feces	9.9	7.6	8.5	4.1	4.0

[a]Average log of the number of viable cells per gram of dry matter; a mean of 8 animals (except for enterococci with mean of 7).

Amine formation has been suggested as being related to inhibition of growth of young pigs (Larson and Hill 1960; Michel and Francois 1956).

The inhibition of urease-producing microflora by antibiotics has also been suggested by Visek (1968, 1978) as responsible for the growth stimulation obtained by feeding antibiotics. The growth stimulation by antibiotics has also been suggested to be related to the inhibition of intestinal microflora that reduce nutrient utilization. It has been observed that the thickness of the intestinal wall is reduced by antibiotics. This may favor nutrient absorption.

The significance of the microorganisms in the alimentary tract of the pig and the mechanism of antibiotics in promoting growth await further study. Controlling factors must be at work in maintaining a suitable balance of microbes to prevent the overwhelming of the host by the several species that include pathogenic strains but are ever present as normal inhabitants.

BLOOD AND CIRCULATION

Blood volume of the pig expressed as a percentage of body weight, decreases steadily with age (Hansard *et al.* 1951; Bush *et al.* 1955). The ingestion and absorption of milk during the first few hours after birth tend to increase the total volume and probably contribute to the decline in hematocrit and blood hemoglobin concentration that occur in piglets during the first few days after birth. The mean blood volume of a newborn pig is approximately 80–90 ml/kg body weight and increases quickly after nursing to 95–100 ml/kg, where it remains for a few days before gradually declining, so that at two or three weeks of age it is 70–75 ml/kg body weight. Mature obese swine weighing more than 275 kg have a blood volume of only 30–35 ml/kg of weight. There is a gradual decline between these two extremes as the pig becomes larger and fatter.

Systolic blood pressure increases linearly with body weight from birth to maturity (50–60 mm Hg at birth, 140–160 m Hg or more at maturity) (Engelhardt 1966). Diastolic pressure also rises with increased body weight but is 30–45 mm Hg lower than systolic pressure.

Heart rate has been reported in various studies reviewed by Stone and Sawyer (1966) to be as low as 60 and as high as 200 beats per minute. Excitement increases heart rate 40–50% which could account for some of the higher values reported. Heart rate declines as body weight and age increase. At birth, a heart rate of 200 beats per minute is normal, with a gradual decline to 80 or below at maturity.

Cardiac output is approximately 150 ml/kg body weight per minute in pigs from 7 to 50 kg body weight, and with a steady decline to approximately 30–40 ml/kg at maturity (Hoernicke 1966). Iron deficiency anemia increases cardiac output several-fold. The inability of the pig to withstand severe exertion is probably related not only to its limited ability to dissipate heat efficiently, due to its sparse sweat glands and large subcutaneous fat insulation, but also to its relatively small heart, small blood volume, and unstable circulatory physiology. Recent evidence (Wegner 1970) indicates wide variation in heart size among individuals. This may be related to differences in susceptibility to stress. Domestication and selection have probably been factors in the characteristic cardiovascular physiology of present breeds of pigs, as wild pigs have a higher blood volume, higher oxygen-carrying capacity of the blood, and better ability to increase blood volume and hemoglobin in response to exercise than do domestic swine.

HEMATOLOGY

Cells

Red blood cells (RBC) (erythrocytes). Hematopoiesis begins during fetal life. The pig RBC is a biconcave disc of approximately 5–6 μ in diameter. The amounts of blood platelets and leukocytes and the changes in hematocrit (packed cell volume as percentage of total blood volume), in the young pig during the first 2 weeks after birth, are shown in Table 6.5 (Lie 1968).

The bone marrow is the major site of erythrocyte and leukocyte formation. The RBC during maturation becomes smaller and extrudes its nucleus; its cytoplasm changes from basophilic to eosinophilic and finally takes on the characteristic red color of the mature erythrocyte.

During the suckling period, the pig must depend on a source of supplemental iron for hemoglobin synthesis since poor placental transfer limits body reserves at birth, and milk iron is deficient. Iron deficiency in the young pig results in microcytic hypochromic anemia. Hemoglobin and hematocrit (packed cell volume) are both good indices of iron nutrition. The life of the RBC in pigs is approximately 70 days (Bush *et al.* 1955; Talbot and Swenson 1963; Jensen *et al.* 1956; Withrow and Bell 1969). The RBC lifespan may be shortened by external factors but age-dependent destruction is not subject to environment. Formation may be impaired in various nutritional deficiencies. These are described in Chapter 11.

TABLE 6.5. Platelets, Leukocytes, and Packed RBC Volume of Pigs from Birth to 2 Weeks of Age

Day after birth	*Platelets (thousand/cu mm)*[a]	*Leukocytes (thousand/cu mm)*[b]	*Packed RBC volume (%)*[c]
0	337	16.9	40.9
1	276	15.3	36.1
2	241	9.4	32.9
3	279	10.2	34.0
4	309	11.8	32.2
5	378	11.6	32.4
6	410	10.2	33.0
7	465	10.8	31.5
8	535	9.1	34.2
9	525	8.0	33.9
10	578	9.4	34.4
11	495	8.3	35.0
12	553	8.6	36.8
13	532	8.0	38.5
14	527	9.4	36.1

Source: Lie 1968.
[a]51 to 141 observations at each age.
[b]26 to 106 observations at each age.
[c]21 to 95 observations at each age.

White Blood Cells (Leukocytes)

The white blood cells are fewer in number than the RBC, but with a wide normal range and a lower average concentration in the newborn animal. Infections and stress increase the leukocyte count. Lymphocytes make up the largest fraction of the leukocyte population (about 50%), while neutrophils account for about 38%; esosinophils, 5%; and basophils, 2% of the total. Leukocyte counts are important in clinical diagnosis of disease (Dunne 1970).

Serum Proteins

The newborn pig is poorly supplied with blood serum proteins until the first colostrum is ingested. Before nursing, the total plasma proteins are only about 2.4 g/100 ml of blood. Since fibrinogen, which is converted to fibrin when blood clots, is present only as a component in the blood clotting mechanism and represents only 3 or 4% of the total serum proteins at birth, it is more meaningful and convenient to discuss the blood proteins in terms of serum proteins (minus fibrin).

Serum proteins, while initially provided in large amounts by colostrum, are soon produced in ample quantities by the liver and to a lesser extent by the spleen and bone marrow. A moderate deficiency of a single amino acid or of total protein in the diet results in a decline in serum protein. Even in extreme and prolonged protein deficiency, the pig is able to maintain serum gamma globulin at about the same absolute level, while albumin is reduced to almost zero. The ratio of albumin to globulin is, therefore, a good index of protein nutrition.

By 8 or 10 weeks the total serum protein level plateaus at approximately 7 g/100 ml where it remains until and during mature life.

Many physiological studies with swine have resulted in the recording of normal parameters for various blood constituents. Tabular data of this kind have been reported (Dittmer 1961; Altman and Dittmer 1962, 1968; Frandson 1965; Bustad *et al.* 1966). Hematological data collected from those sources are summarized in Table 6.6. The values represent a variety of ages, breeds, and types of pigs.

The pig has at least 14 blood group systems (Andresen 1962; Moustgaard and Hesselholt 1966) similar to blood groups in man. The blood type of pigs is detected by means of blood typing reagents. It is potentially possible to characterize individual pigs on the basis of a blood type, as this is a heritable and permanent character of that individual. Though the ability to identify traits of this kind is often predicted to be of value, whether or not these traits themselves may eventually have a place in selection of animals in commercial swine breeding remains to be seen.

DENTITION

The dental pattern of the pig is illustrated in Fig. 6.3 (Jump and Weaver 1966). At birth the pig has eight teeth, including four incisors and four canines. The full complement of temporary teeth is 28, and consists of 12 incisors (three pairs on upper and lower jaws), four canine (one pair on upper and lower jaws), and 12 molars (three pairs on upper and lower jaws). The permanent teeth erupt at

TABLE 6.6. Normal Hematological Values for Pigs of Various Ages

Blood constituent	*Age or weight*	*Concentration*	
		Whole blood	*Serum (or plasma)*
Amino acids	50 kg, 15 wk		
Essential			
Arginine	50 kg, 15 wk	—	2 (1–4) mg/100 ml
Histidine	50 kg, 15 wk	—	2 (1.5–2.5) mg/100 ml
Isoleucine	50 kg, 15 wk	—	2.5 (2–4) mg/100 ml
Leucine	50 kg, 15 wk	—	3 (2.5–5) mg/100 ml
Lysine	50 kg, 15 wk	—	5 (4–6) mg/100 ml
Methionine	50 kg, 15 wk	—	0.5 (0.1–1.0) mg/100 ml
Cystine	50 kg, 15 wk	—	2 (0.5–3) mg/100 ml
Phenylalanine	50 kg, 15 wk	—	2 (1.5–3.0) mg/100 ml
Threonine	50 kg, 15 wk	—	5 (3–6) mg/100 ml
Tryptophan	50 kg, 15 wk	—	0.5 (0.1–0) mg/100 ml
Valine	50 kg, 15 wk	—	4 (3–5) mg/100 ml
Total		—	25 (20–30) mg/100 ml
Nonessential			
Alanine	50 kg, 15 wk	—	4 (3–6) mg/100 ml
Aspartic acid	50 kg, 15 wk	—	0.5 (0.3–1.2) mg/100 ml
Glutamic acid	50 kg, 15 wk	—	5 (4–8) mg/100 ml
Glycine	50 kg, 15 wk	—	4 (2–7) mg/100 ml
Hydroxyproline	50 kg, 15 wk	—	0.9 (0.8–1.0) mg/100 ml
Norleucine	50 kg, 15 wk	—	5 (4–6) mg/100 ml
Proline	50 kg, 15 wk	—	4 (3–6) mg/100 ml
Serine	50 kg, 15 wk	—	3 (2–4) mg/100 ml
Tyrosine	50 kg, 15 wk	—	2 (1–3) mg/100 ml
Total	50 kg, 15 wk	—	23 (18–35) mg/100 ml
Ammonia	3.5–4.5 kg	50 (10–100) μg/100 ml	—
	50–90 kg	—	500 (425–750) μg/100 ml
Ceruloplasmin	90 kg	—	35 mg/100 ml
Cholesterol, total	Newborn, 10 wk	—	68 (50–86) mg/100 ml
	90 kg		150 (100–200) mg/100 ml
Choline-esterase	12–32 kg	0.3 (0.1–0.7) delta pH units/hr	146 (121–171) μg/100 ml
Corticosterine, 17-α-hydroxy	Newborn	—	44 (33–53) μg/100 ml
	4–6 wk	—	12 (6–18) μg/100 ml
	1 yr	—	5 (4.5–55) μg/100 ml
Creatinine	1 yr	—	1.0 (0.7–1.3) mg/100 ml
Fatty acids, free	27 kg	—	520 (280–750) μeq/l
Glucose	Newborn	100 (80–120) mg/100 ml	—
	48–52 kg	110 (90–140) mg/100 ml	—
Glutamic oxalacetic transaminase	Newborn	—	19 (16–22) units/ml
	5–10 wk	—	23 (21–24) units/ml
	36 wk	—	18 (17–19) units/ml
Growth hormone	50 kg	—	6 (5–7) mμg/ml
Inorganic elements			
Calcium	145–185 days	—	10 (8–12) mg/100 ml
Chlorine	145–185 days	—	355 (335–375) mg/100 ml

(*continued*)

TABLE 6.6. (*Continued*)

Blood constituent	*Age or weight*	*Concentration*	
		Whole blood	*Serum (or plasma)*
Copper	145–185 days	1.25 (0.9–1.6) mg/liter	—
Iodine, protein bound	—	—	5.0 (3–7) μg/100 ml
Iron, total	5–14 wk	—	140 (100–160) μg/100 ml
Total iron-binding capacity	5–14 wk	—	350 (300–400) μg/100 ml
Magnesium	145–185 days	—	2.5 (2–5) mg/100 ml
Manganese	90 kg	—	2.5 (2–3) mg/100 ml
Phosphorus, inorganic	145–185 days	—	6 (3–9) mg/100 ml
Potassium	145–185 days	—	6 (5–7) mg/l
Sodium	145–185 days	—	325 (295–355) mg/100 ml
Zinc	145–185 days	0.55 (2–0.8) mg/100 ml	—
Bicarbonate	145–185 days	—	20 (15–27) meq/l
Insulin	38–42	—	5.4 (3–7) μU/ml
Ketones, total	4 wk	0.5 (0.2–0.7) mg/100 ml	—
Lactate	4 wk	10 (5–20) mg/100 ml	—
Lipids, total	10 wk	—	300 (250–400) mg/100 ml
Pentoses	4 wk	20 (15–30) mg/100 ml	—
Phosphatase, alkaline	Newborn	—	40 (30–60) sigma units/ml
	10 wk	—	5 (3–6) sigma units/ml
Protein, total	Newborn	—	2.2 (2–3) g/100 ml
	24 hr	—	5.2 (5.0–5.4) g/100 ml
	1 wk	—	4.8 (4.6–5.0) g/100 ml
	6 wk	—	5.3 (5.1–5.5) g/100 ml
	3 mo	—	5.7 (5.6–5.8) g/100 ml
	6 mo	—	6.8 (6.7–6.9) g/100 ml
	1 yr	—	7.5 (7.4–7.6) g/100 ml
Albumin	Newborn	—	16.7 (15.4–18.0) g/100 g ser. prot.[a]
	24 hr	—	8.9 (8.4–9.4) g/100 g ser. prot.
	1 wk	—	31.6 (30.7–32.5) g/100 g ser. prot.
	6 wk	—	47.0 (46–48) g/100 g ser. prot.
	3 mo	—	45.3 (44.5–46.1) g/100 g ser. prot.
	6 mo	—	48.6 (47.9–49.3) g/100 g ser. prot.
	1 yr	—	52.7 (52–53) g/100 g ser. prot.
Alpha globulin	Newborn	—	60.8 (58.9–62.7) g/100 g ser. prot.
	24 hr	—	23.4 (21.8–25.0) g/100 g ser. prot.
	1 wk	—	25.6 (24.6–26.6) g/100 g ser. prot.
	6 wk	—	25.3 (24.7–25.9) g/100 g ser. prot.
	3 mo	—	28.7 (28.0–29.4) g/100 g ser. prot.
	6 mo	—	19.4 (19.1–19.7) g/100 g ser. prot.
	1 yr	—	18.2 (17.5–18.9) g/100 g ser. prot.
Beta globulin	Newborn	—	15.9 (15.0–16.8) g/100 g ser. prot.
	24 hr	—	21.3 (20.4–22.2) g/100 g ser. prot.

(*continued*)

TABLE 6.6. (*Continued*)

Blood constituent	*Age or weight*	*Concentration*	
		Whole blood	*Serum (or plasma)*
	1 wk	—	17.5 (17.0–18.0) g/100 g ser. prot.
	6 wk	—	18.6 (18.2–19.0) g/100 g ser. prot.
	3 mo	—	16.0 (15.5–16.5) g/100 g ser. prot.
	6 mo	—	12.6 (12.4–12.8) g/100 g ser. prot.
	1 yr	—	12.6 (11.9–13.3) g/100 g ser. prot.
Gamma globulin	Newborn	—	6.5 (6.0–7.0) g/100 g ser. prot.
	24 hr	—	46.3 (44.9–47.7) g/100 g ser. prot.
	1 wk	—	25.2 (24.1–26.3) g/100 g ser. prot.
	6 wk	—	9.1 (8.7–9.5) g/100 g ser. prot.
	3 mo	—	20.0 (19.2–20.8) g/100 g ser. prot.
	6 mo	—	19.3 (18.7–19.9) g/100 g ser. prot.
	1 yr	—	16.4 (15.5–17.3) g/100 g ser. prot.
Albumin:Globulin	Newborn	—	0.2
	24 hr	—	0.11
	1 wk	—	0.44
	6 wk	—	0.90
	3 mo	—	0.82
	6 mo	—	0.96
	1 yr	—	1.1
Red blood cells (RBC)	16 wk	8 (3–10) million/cu mm	—
RBC size	8–28 wk	5 μ	—
Hemoglobin	Newborn	9 (8–11) g/100 ml	—
	20 wk	11 (9–14) g/100 ml	—
Mean corpuscular volume	5–14 wk	50–60 cu μ	—
Mean corpuscular Hb concentration	5–14 wk	30%	—
Mean corpuscular Hb	5–14 wk	14–18 picograms	—
Vol. of packed red cells (hematocrit)	New born	30 (25–35) ml/100 ml	—
	5–14 wk	42 (35–45) ml/100 ml	—
White blood cells (leukocytes)	16 wk	19 (6–23) thousand/cu mm	—
Basophils	16 wk	1%	—
Eosinophils	16 wk	5%	—
Lymphocytes	16 wk	50%	—
Monocytes	16 wk	4%	—
Neutrophils	16 wk	40%	—
Platelets	2 yr	350 (50–650) thousand/cu mm	—
Vitamins			
A	145–185 days	—	120 (65–175) IU/100 ml
Carotene	145–185 days	—	30 (13–50) μg/100 ml
E (tocopherol)	90 kg	160 mg/100 ml	—

Sources: From Dittmer 1961; Altman and Dittmer 1962, 1968; Frandson, 1965; Bustad et al. 1966.
[a]Serum protein = ser. prot.

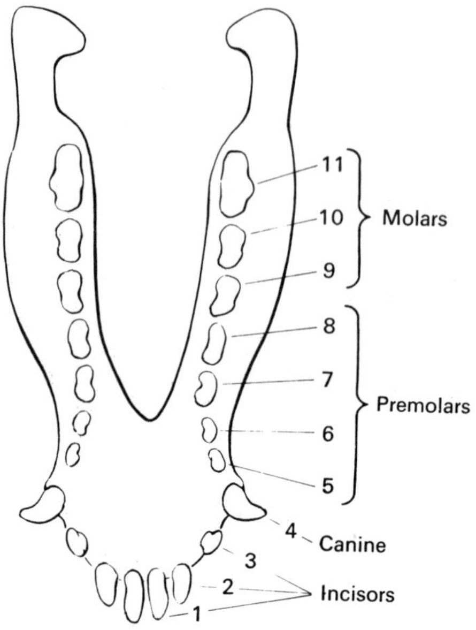

FIG. 6.3. Dental pattern of the pig.

varying intervals from 3 or 4 months (first molar) to 18 or 20 months (third molar). During the preceding time, the dentition shows a mixture of deciduous and permanent teeth. The mature dental pattern is three pairs of incisors, one pair of canines, four pairs of premolars, and three pairs of molars, upper and lower, for a total of 44 teeth.

The eight teeth present at birth are very sharp and have been descriptively termed needle teeth. It is common commercial practice to clip these temporary teeth at the gum line shortly after birth to avoid injury to other pigs and to the nipples of the dam during suckling. The canines of the adult grow very long and protrude out of the mouth as tusks. In the male, it is common practice to clip them to avoid injury to humans, or to other pigs.

TABLE 6.7. Organ Weight Changes in Late Prenatal and Early Postnatal Life

Organ	*Days from birth*						
	−16	*−12*	*−8*	*−4*	*+2*	*+4*	*+6*
Whole body, kg	0.89	1.11	1.18	1.35	1.45	1.71	1.96
Carcass (minus internal organs), kg	0.40	0.52	0.54	0.64	0.72	0.81	0.93
Liver, g	22.8	27.0	28.5	37.5	37.8	51.8	64.3
Stomach (empty), g	6.0	6.8	7.0	6.8	8.5	10.5	12.5
Heart, g	7.3	9.0	10.0	9.8	10.5	12.8	14.0
Kidneys, g	10.0	10.0	9.5	10.3	12.5	12.8	15.0
Spleen, g	2,8	2.0	2.8	1.1	2.0	3.0	4.0
Brain, g	22.8	24.5	25.5	29.3	32.8	35.3	36.8
Lungs, g	28.9	31.5	32.5	22.3	20.3	25.8	29.0

Source: Brooks and Davis 1969.

TABLE 6.8. Body Composition of Pigs During Early Postnatal Life

Constituent	Postnatal period (age, days)				
	0	*2*	*7*	*14*	*28*
Livewt, g	1,520	1,815	3,221	5,563	9,928
Empty livewt, g	1,450	1,741	3,044	5,284	9,651
Fat-free tissue, g	1,432	1,701	2,738	4,488	7,888
Fat, g	18	40	306	796	1,763
Fat, %	1.1	—	—	—	—
Water, g	1,198	1,398	2,207	3,557	6,138
Water, %	84.1	—	—	—	—
Protein, g	174	237	437	770	1,427
Protein, %	11.3	—	—	—	—
Ash, g	60.4	66.8	94.1	160.8	323.9
Calcium, g	16.7	17.9	24.1	41.7	87.6
Phosphorus, g	9.3	10.8	16.3	28.1	56.3
Potassium, g	2.8	3.8	6.8	11.5	19.9
Sodium, g	2.8	3.1	4.3	6.9	12.3
Magnesium, g	0.42	0.48	0.83	1.53	3.01
Iron, mg	54	49	77	118	153
Zinc, mg	26	33	60	114	236
Copper, mg	3.9	5.4	10.9	18.6	23.2
Manganese, mg	0.69	0.75	0.70	0.98	1.94

Source: Manners and McCrea 1963

CHANGES IN BODY COMPOSITION WITH TIME

Rapid changes occur in relative and absolute weights of organs from a few days before birth, to a few days after birth (Brooks and Davis 1969). The stomach, liver, and kidneys increase in weight proportionately faster than the whole body immediately after birth. Nitrogen-free extract (presumably largely glycogen) concentration of liver and carcass increases sharply during the last 10 days before birth and then declines rapidly during the first few hours after birth. This probably reflects the importance of provision being made during fetal life for a source of available energy during the early postnatal period, until the pig suckles. The high efficiency of fat synthesis by the newborn pig is indicated by the twofold increase in carcass fat during the first 48 hr after birth. The sharp rise in concentration of linoleic acid in liver and carcass during this same period suggests that there is more efficient transfer of linoleic acid to the pig via milk than via the placenta. The changes in weight of several organs from 98 days after conception (approximately 2 weeks before birth) to 6 days after birth are illustrated in Table 6.7.

The newborn pig is much lower in fat content and higher in water than most other newborn mammals. The changes that occur in gross composition and in mineral element content during the first 4 weeks after birth, are summarized in Table 6.8 (Manners and McCrea 1963). It is clear that rate of fat deposition exceeds rate of deposition of other organic and inorganic components.

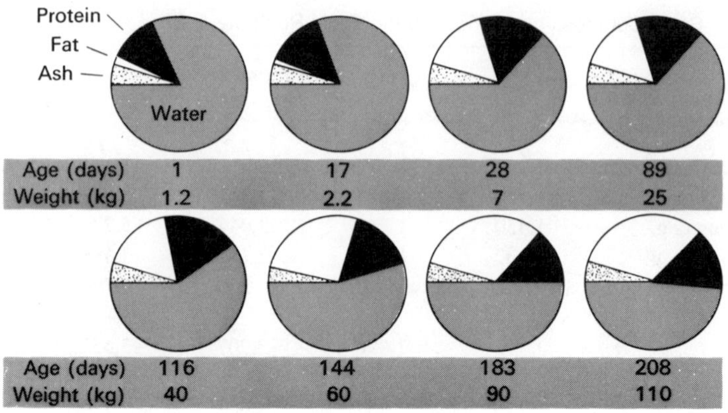

FIG. 6.4. Changes in body composition with age and weight.
Courtesy of Oslage 1963.

These changes in proportions of ash, fat, protein, and water with increased age and weight are depicted in Fig. 6.4 from the data of Oslage (1963). They depict edible carcass composition, but probably closely represent changes in total body composition, since the inedible and edible components change similarly in composition as weight increases (Babatunde *et al.* 1967). Boars are leaner than females at a given weight during growth, and both sexes are leaner than castrated males at the same weight.

As fat deposition increases with increased body weight, the percentage of saturated acids tends to increase while unsaturated acids tend to decrease (Table 6.9) (Sink *et al.* 1964). Fatty acid composition of the depot fat is highly responsive to diet. Unsaturated fat, such as safflower oil, results in a soft depot fat whereas a saturated fat such as hydrogenated coconut oil results in a firm depot fat when fed to growing pigs (Table 6.10) (Babatunde *et al.* 1968).

TABLE 6.9. Fatty Acid Composition of Depot Fat of Growing Swine as Affected by Body Weight

Body wt (kg)	Fatty acid composition (%)						
	Saturated fatty acids						
	$C_{12:0}$	$C_{14:0}$	$C_{15:0}$	$C_{16:0}$	$C_{17:0}$	$C_{18:0}$	*Total*
34	0.34	1.75	0.09	22.25	0.52	9.79	34.89
59	0.25	1.72	0.05	22.69	0.46	10.67	35.91
86	0.21	1.98	0.03	23.81	0.23	10.93	37.26
	Unsaturated fatty acids						
	$C_{16:1}$	$C_{17:1}$	$C_{18:1}$	$C_{18:2}$	$C_{18:3}$	$C_{20:4}$	*Total*
34	4.87	0.75	44.05	14.47	0.96	0.16	65.11
59	4.00	0.54	45.79	12.71	1.06	0.07	64.09
86	3.68	0.26	45.20	12.80	0.76	0.08	62.74

Source: Sink et al. 1964.

TABLE 6.10. Effect of Diet on Fatty Acid Composition of Depot Fat of Growing Pigs

	Diet[a]			
Fatty acid	*Fat-free*	*3% safflower oil*	*3% hydrogenated coconut oil*	*Maize–soybean meal*
	Percent of total fatty acids			
$C_{10:0}$	0.13	0.20	0.18	0.11
$C_{12:0}$	0.10	0.16	0.63	0.12
$C_{14:0}$	1.79	2.00	3.85	—
$C_{16:0}$	18.71	15.60	20.99	—
$C_{16:1}$	16.85	13.01	16.47	12.68
$C_{18:0+18:1}$	60.16	50.58	55.93	55.76
$C_{18:2}$	1.28	17.15	0.93	8.32
$C_{18:3}$	0.18	0.09	0.09	0.09
$C_{20:1}$	0.98	0.56	0.76	1.08
$C_{20:3}$	0.17	0.46	0.11	0.49
$C_{20:4}$	0.07	0.19	0.07	0.23

Source: Babatunde et al. 1968.
[a]Amount of linoleic acid ($C_{18:2}$) in diet was 0, 2.4, 0.06 and 0.8% for fat-free, 3% safflower, 3% hydrogenated coconut oil and maize–soybean meal diets, respectively.

REFERENCES

ALTMAN, P. L., and DITTMER, D. S. 1962. Growth. Fed. Am. Soc. of Experimental Biol., Bethesda, Maryland.

ALTMAN, P. L., and DITTMER, D. S. 1968. Metabolism. Fed. Am. Soc. of Exp. Biol., Bethesda, Maryland.

ANDRESEN, E. 1962. Annu. N.Y. Acad. Sci. *97*,203.

BABATUNDE, G. M., POND, W. G., VANVLECK, L. D., KROENING, G. H., and REID, J. T. 1967. J. Anim. Sci. *26*,718.

BABATUNDE, G. M., POND, W. G., WALKER, E. F., JR., and CHAPMAN, P. 1968. J. Anim. Sci. *27*,1290.

BAILEY, C. B., KITTS, W. D. and WOOD, A. J. 1956. Can. J. Agric. Sci. *36*,51.

BECKER, D. E., ULLREY, D. E., TERRILL, S. W., and NORTZOLD, R. A. 1954. Science *120*,345.

BOHMAN, V. R., KIDWELL, J. F., and McCORMICK, J. N. 1953. J. Anim. Sci. *12*,876.

BRAUDE, R., DOLLAR, A. M. MITCHELL, K. G., PORTER, J. W. G., and WALKER, D. M. 1958. Proc. Nutr. Soc. *17*,XLIX.

BRAUDE, R., MITCHELL, K. G., NEWPORT, M. J., and PORTER, J. W. G. 1970 Brit. J. Nutr. *24*,501.

BRAUDE, R., MITCHELL, K. G., and SUFFOLK, S. F. 1969. J. Inst. Anim. Tech. *20*,43.

BROOKS, C. C., and DAVIS, J. W. 1969. J. Anim. Sci. *29*,325.

BROWN, H., SPEER, V. C., QUINN, L. Y., HAYS, V. W., and CATRON, D. V. 1961. J. Anim. Sci. *20*,323.

BUNDING, I. M., DAVENPORT, M. E., JR., and SCHOOLEY, M. A., 1956. J. Anim. Sci. *15*,234.

BUSH, J. A., BERLINE, N. I., JENSEN, W. N., BRILL, A. B., CARTWRIGHT, G. E., and WINTROBE, M. M. 1955. J. Exp. Med. *101*,451.

BUSH, J. A., JENSEN, A. N., CARTWRIGHT, G. E., and WINTROBE, M. M. 1955. Am. J. Physiol. *181*,9.

BUSTAD, L. K., McCLELLAN, R. O., and BURNS, M. P. 1966. Swine in Biomedical Research, Frayn Printing Co., Seattle.

CAIRNIE, A. B., and PULLAR, J. D. 1957. J. Physiol. *139*,15.

CASTLE, E. J., and CASTLE, M. E. 1956. J. Agric. Sci. *47*,196.

COX, D. F. 1962. Iowa State Univ. Res. Bull. *500*,339.

DITTMER, D. S. 1961. Blood and Other Body Fluids. Fed. Am. Soc. of Exp. Biol., Washington, D.C.
DOLLAR, A. M., MITCHELL, K. G., and PORTER, J. W. G. 1957. Proc. Nutr. Soc. *16*,XII.
DUNNE, H. W. 1970. Diseases of Swine, 2nd Edition. Iowa State Univ. Press, Ames.
ENGELHARDT, W. 1966. *In* Swine in Biomedical Research (L. K. Bustad, R. O. McClellan, and M. P. Burns, Editors). Frayn Printing Co., Seattle.
FOLEY, C. W., SEERLEY, R. W., HANSEN, W. J., and CURTISS, S. E. 1971. J. Anim. Sci. *32*,926.
FRANDSON, R. D. 1965. Anatomy and Physiology of Farm Animals. Lea and Febiger, Philadelphia.
FRIEND, D. W., and CUNNINGHAM, H. M. 1966. Can. J. Comp. Med. Vet. Sci. *30*,179.
FULLER, R. 1962. Studies on the gut flora of the pig and chick. Ph.D. Thesis. Univ. of Reading, England.
GJESSING, E., and HARTHELT, J. C. 1960. Fed. Proc. *19*,49.
HANAWALT, V. M., and SAMPSON, J. 1947. Am. J. Vet. Res. *8*,235.
HANAWALT, V. M., LINK, R. P., and SAMPSON, J. 1947. Proc. Soc. Exp. Biol. Med. *65*,41.
HANSARD, S. L., SAUBELICH, H. E., and COMAR, C. L. 1951. Proc. Exp. Biol. Med. *78*,544.
HARTMAN, P. A., HAYS, V. W., BAKER, R. D., NEAGLE, L. H., and CATRON, D. V. 1961. J. Anim. Sci. *20*,114.
HAYS, V. W., SPEER, V. C., HARTMAN, P. A. and CATRON, D. V. 1959. J. Nutr. *69*,179.
HOERNICKE, H. 1966. *In* Swine in Biomedical Research (L. K. Bustad, E. O. McClellan, and M. P. Burns, Editors), p. 419. Frayn Printing Co., Seattle.
HOLUB, A. 1968. Functional Periodization of the Early Postnatal Period in Piglets. Acta Universitatis Agriculturae, Fakulta Veterinarni. Brno, Cechoslovenia (English summary), 125 pp.
HOLUB, A., FORMAN, Z., and JIZKOVA, D. 1957. Nature *180*,858.
HORSZXZARUK, F. 1962. Rocz. Nauk. Roln. Ser. B *80*,115.
HORVATH, D. J., SEELEY, H. W., WARNER, R. G., and LOOSLI, J. K. 1958. J. Anim. Sci. *17*,714.
JENSEN, W. N., BUSH, J. A., ASHENBRUCKER, H., CARTWRIGHT, G. E., and WINTROBE, M. M. 1956. J. Exp. Biol. Med. *103*,145.
JUMP, E. B., and WEAVER, M. E. 1966 *In* Swine in Biomedical Research (L. K. Bustad, R. O. McClellan, and M. P. Burns, Editors), p. 543. Frayn Printing Co., Seattle. 1963. Res. Vet. Sci. *4*,145.
KASS, M. L., VANSOEST, P.J., and POND, W. G. 1980. J. Anim. Sci. *50*,192.
KIDDER, D. E., MANNERS, M. J., McCREA, M. R., and WEAVER, B. W. 1963. Res. Vet. Sci. *4*,145.
LARSON, N. L., and HILL, E. G. 1960. J. Bacteriol. *80*,188.
LASKOWSKI, M., KASSELL, B., and HAGERTY, G. 1957. Biochem Biophys. Acta *24*,300.
LECCE, J. G. 1971. J. Anim. Sci. *33*,47.
LECCE, J. G., and MATRONE, G. 1960. J. Nutr. *70*,13.
LIE, H. 1968. Acta Vet. Scand. *9*,105.
LLOYD, L. E., and CRAMPTON, E. W. 1957. J. Anim. Sci. *16*,377.
LLOYD, L. E., DALE, D. G., and CRAMPTON, E. W. 1958. J. Anim. Sci. *17*,684.
MANER, J. H., POND, W. G., and LOOSLI, J. K. 1961. J. Anim. Sci. *20*,614.
MANER, J. H., POND, W. G., LOOSLI, J. K., and LOWREY, R. S. 1962. J. Anim. Sci. *21*,49.
MANNERS, M. J., and McCREA, M. R. 1963. Br. J. Nutr. *17*,495.
MEYER, J. A., BRISKEY, E. J., HOEKSTRA, W. G., and BRAY, R. W. 1962. J. Anim. Sci. *21*,543.
MICHEL, M., and FRANCOIS, A. C. 1956. Compt. Rend. *242*,1770.
MORRILL, C. C. 1952. Am. J. Vet. Res. *13*,164,322,325,327.
MOUNT, L. E. 1959. J. Physiol. *147*,333.
MOUNT, L. E. 1962A. *In* Nutrition of Pigs and Poultry (J. T. Morgan and D. Lewis, Editors), Chap. 6. Butterworths, London.
MOUNT, L. E. 1962B. J. Physiol. *164*,274.
MOUNT, L. E. 1964. J. Physiol. *170*,286.
MOUNT, L. E. 1968. Climatic Physiology of the Pig. Williams and Wilkins, Baltimore.
MOUSTGAARD, J., and HESSELHOLT, M. 1966. *In* Swine in Biomedical Research (L. K. Bustad, R. O. McClellan, and M. P. Burns, Editors), p. 25. Frayn Printing Co., Seattle.
NIELSEN, H. E. 1975. *In* Festskrift til Hjalmer Clausen (H. Staun, Editor), pp. 165–181. Danske Landhusholdingsselskab, Copenhagen, Denmark.
NEILSON, N. C., CHRISTENSEN, K., BILLE, N., and LARSEN, J. L. 1974. Nord. Veterinaermed. *26*,137.
NEWLAND, H. W., McMILLAN, W. N., and REINEKE, E. P. 1952. J. Anim. Sci. *11*, 118.
NEWTON, W. C., and SAMPSON, J. 1951. Cornell Vet. *41*, 377.
OSLAGE, H. J. 1963. Z. Tierphysiol. Tierernaehr. Futtermittelkd. *17*,357.

PAYNE, L. C., and MARSH, C. L. 1962A. J. Nutr. *76*,151.
PAYNE, L. C., and MARSH, C. L. 1962B. Fed. Proc. *21*,909.
POMEROY, R. W. 1953. J. Agric. Sci. *43*,182.
POND, W. G., ROBERTS, S. J. DUNN, J. A., and WILLMAN, J. P. 1960. J. Anim. Sci. *19*,881.
SEERLEY, R. W., MILLER, E. G., and HOEFER, J. A. 1962. J. Anim. Sci. *21*,834.
SINK, J. D., WATKINS, J. L., ZIEGLER, J. H., and MILLER, R. C. 1964. J. Anim. Sci. *23*,121.
STONE, H. L., and SAWYER, D. C. 1966. *In* Swine in Biomedical Research. (L. K. Bustad, R. P. McClellan, and M. P. Burns, Editors), p. 411. Frayn Printing Co., Seattle.
TALBOT, R. B., and SWENSON, M. J. 1963. Proc. Soc. Exp. Biol. Med. *112*,573.
VESTAL, C. M. 1936 Ind. Agric. Exp. Stn. Bull. *413*.
VISEK, W. J. 1968. J. Dairy Sci. *51*,1.
VISEK, W. J. 1978. J. Anim. Sci. *46*,1447.
WEGNER, W. 1970. Fleischwirtschaft *50*,72.
WHITING, R., OWEN, B. D., ELLIOT, J. I., and BEAMES, R. M. 1983. Can. J. Anim. Sci. *63*,993.
WIDDOWSON, E. M. 1950. Nature *166*,626.
WILBUR, R. D., CATRON, D. V., QUINN, L. Y., SPEER, V. C., and HAYS, V. W. 1960. J. Nutr. *71*,168.
WILLINGALE, J. M., and BRIGGS, C. A. E. 1955. J. Appl. Bacteriol. *18*,284.
WISE, M. B., BARRICK, E. R., WISE, G. H., and OSBORNE, J. C. 1954. J. Anim. Sci. *13*,365.
WITHROW, G., and BELL, M. C. 1969. J. Anim. Sci. *28*,240.

7

Growth

INTRODUCTION

Some aspects of growth are covered in Chapters 2, 4–6, 10, 11, 18, and 19, but since it is such an important topic, it is discussed in general more directly, if briefly, here. The characteristic S-shaped growth curve of all biological systems, whether it be an increase in weight gain of a plant or animal or increase in number of yeast cells in a population (Brody 1945) is shown in Fig. 7.1. The term growth, as applied to animal production, is often considered to be synonymous with increase in body weight of the animal from conception to maturity. Brody (1945) has defined growth as "the constructive or assimilatory synthesis of one substance at the expense of another (nutrient) which undergoes dissimilation." The energetic efficiency of growth as just defined is, therefore, the energy stored in the animal per unit of energy consumed. The pig is especially efficient in growth, in comparison with other farm animals. Detailed discussion of the energetic efficiency of growth in meat producing animals has been provided by Webster (1980). The complex metabolic and endocrine aspects of tissue and animal growth are exciting areas of animal biology and have been reviewed by Trenkle and Marple (1983).

METHODS OF EXPRESSING GROWTH

Growth, as indicated by increase in body weight, can be expressed as absolute gain in a given period of time or as relative gain (or percentage gain). The absolute growth rate

$$\frac{\text{larger weight} - \text{smaller weight}}{\text{final time} - \text{initial time}}$$

can be misleading from the standpoint that the average value may not represent the actual value at any one point in time. For example, a pig whose average daily

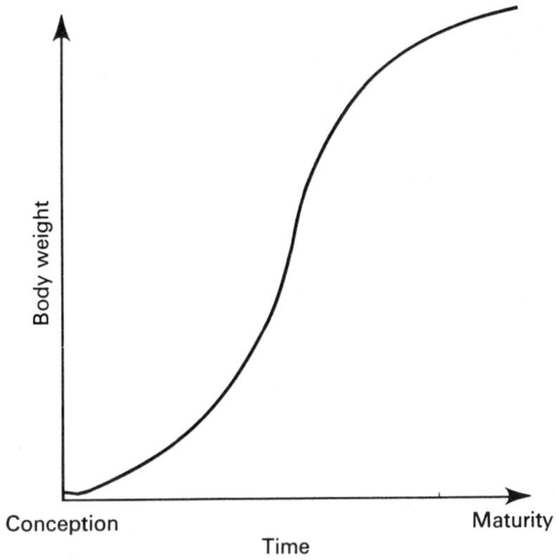

FIG. 7.1. S-shaped growth curve typical of all biological growth systems.
From Brody 1945.

weight gain from birth to slaughter was 500 g,

$$\frac{91\ \text{kg} - 1\ \text{kg}}{180\ \text{days}},$$

might never have gained exactly 500 g on any one day (see S-shaped growth curve in Fig. 7.1). The absolute gain the day after conception might have been a few micrograms, followed by a gain of 50 g on the first day after birth, and 1000 g on the last day before slaughter. Nevertheless, from a practical standpoint, the absolute average daily gain to slaughter is an important measure of growth as used in the broad sense.

Relative growth rate

$$\frac{\text{larger weight} - \text{smaller weight}}{\text{smaller weight}}$$

declines steadily as the animal approaches maturity.

The growth rate of individual parts of the pig is not identical. McMeekan (1940) showed, by dissection and carcass studies, that the head and shoulders reach mature size before the posterior parts of the body. That is, early in life, the head and shoulders represent a higher proportion of the total body weight than they do later. Similarly, individual organs and tissues grow at different rates, and indeed, individual cell types within a single organ grow at different rates. The changes in weight of various tissues of the pig from birth to 100-kg body weight are illustrated in Table 7.1 (McMeekan 1940).

HYPERPLASIA AND HYPERTROPHY

Growth of tissues and organs and of the whole organism (the pig, in this discussion) occurs in two phases: increase in *number* of cells (*hyperplasia*) and increase in *size* of cells (*hypertrophy*). Immediately after conception, most growth is by hyperplasia. During late prenatal life and early postnatal life, the two processes occur concurrently. Finally, at some point in postnatal life, cell division ceases

TABLE 7.1. Organ Weights of Pigs from Birth to 100-kg Body Weight

Body weight[a] (kg)	*1.34*	*5.9*	*13.2*	*36.1*	*52.1*	*71.4*	*100.0*
Organ			*Organ weight (% of body weight)*				
Liver	3.5	3.2	3.2	2.8	2.7	2.9	1.7
Heart	0.76	0.82	0.47	0.46	0.38	0.36	0.27
Thyroid	—[b]	—	—	—	0.082	—	0.0061
Gall bladder	0.03	0.03	0.09	0.04	0.05	0.05	0.04
Kidneys	0.81	0.69	0.08	0.47	0.41	0.31	0.23
Adrenals	—	—	—	—	0.0041	—	0.0032
Spleen	0.10	0.15	0.16	0.10	0.12	0.11	0.10
Lungs and trachea	1.9	1.5	1.5	1.2	1.1	0.9	0.8
Diaphragm	0.50	0.41	0.40	0.46	0.45	0.48	0.41
Mesenteric and kidney fat	0.10	0.35	0.38	0.66	1.00	1.41	2.30
Pancreas	0.12	0.24	0.23	0.26	0.25	0.15	0.14
Esophagus	0.12	0.08	0.11	0.08	0.08	0.08	0.06
Stomach	0.44	0.66	1.04	1.02	0.86	0.84	0.57
Small intestine	1.6	3.7	2.4	2.8	2.5	2.2	1.40
Cecum	0.07	0.09	0.11	0.21	0.22	0.17	0.13
Large intestine and rectum	0.56	0.68	1.43	1.71	1.69	1.26	1.02
Mesentery	0.54	0.64	0.81	1.02	1.14	1.15	1.71
Urinary bladder	0.19	0.05	0.08	0.04	0.05	0.06	0.05
Hoofs	0.22	0.12	0.13	0.11	0.10	0.10	0.10

Source: McMeekan 1940.
[a]Body weights correspond to pigs at birth and at 4, 8, 16, 20, 24, and 28 weeks of age.
[b]Dashes indicate that no data were available.

(except in some tissues) and growth occurs only by hypertrophy. This is illustrated diagrammatically in Fig. 7.2. Nervous tissue and muscle tissue perhaps fit this concept most closely. Other tissues, such as the epithelium of skin and digestive tract, and the blood have continuous cell turnover throughout life. For example, the life of the RBC of the pig is about 70 days; that of the intestinal mucosal cell is only a few days. For those tissues that reach mature size by proceeding through the three phases (hyperplasia, hyperplasia–hypertrophy, and hypertrophy) the ultimate size is presumably determined genetically; that is, the upper limit imposed by genetics cannot be exceeded, although environmental factors can have an effect in reducing ultimate size. There is considerable research effort underway in several laboratories at the present time to determine possible effects of nutritional regimes imposed in early life on subsequent growth of body parts of animals. Work of this kind with pigs suggests permanent effects of prenatal malnutrition on postnatal growth and development (Pond *et al.* 1969; Atinmo *et al.* 1974; Buitrago *et al.* 1974). The exact nature of this "stunting" is not known, but probably is associated with such factors as reduced cell number,[1]

[1]Estimates of cell number and cell size can be obtained by measuring nucleic acid concentrations of tissues. The nucleic acid DNA (deoxyribonucleic acid) is found only in cell nuclei and remains constant throughout the life of the cell. The nucleic acid RNA (ribonucleic acid) varies with the size of the cell. Therefore DNA, RNA, and the DNA:RNA ratio can be used to estimate cell number and size.

endocrine changes, and protein synthetic activity in the tissues affected. Pigs from dams severely restricted in gestation protein intake appear to be permanently retarded in growth rate while those from energy-restricted dams appear to recover even though birth weight of progeny of protein-restricted and of energy-restricted dams may be only about 80% of that of pigs from adequately fed dams (Atinmo *et al.* 1974). Starvation of the dam (access to water and salt only) for 40 days during mid- or late gestation reduces pig birth weight and early postnatal growth (Hard and Anderson 1979) while starvation for 7–21 days prepartum does not effect fetal survival or piglet birth weight (Ezekwe 1981). Additional research is needed to identify the metabolic bases for these contrasting responses. Small birth weight pigs (<1000 g) grow at a slower rate postnatally than their larger littermates and have a higher proportion of fat and less muscle at comparable slaughter weights (Powell and Aberle 1980).

The time sequence of cell proliferation and cell growth in individual tissues and organs of pigs is not well mapped. It seems probable that most cell division (hyperplasia) in muscle is complete at birth or shortly thereafter and that most postnatal body growth is associated with an increase in cell size (hypertrophy). Studies of number and cross-sectional diameter of individual muscle cells taken from the loin eye (longissimus) muscle of pigs (Staun 1963; Chrystall *et al.* 1969) suggest differences among individuals in *cell number* in this tissue, but they indicate that most of the increase in cross-sectional area of the loin eye as the pig grows is related primarily to an increase in the diameter of individual muscle

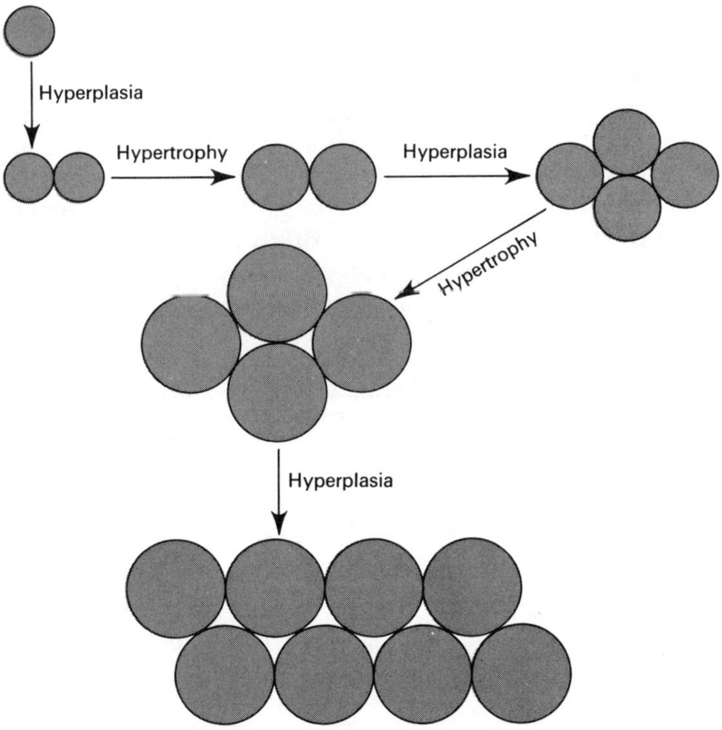

FIG. 7.2. Diagram of growth by hyperplasia and hypertrophy.

fibers rather than to an increase in number of fibers. From birth to 25 days of age fiber diameter increases 100% but from 100 to 125 days of age the increase is only 10%; about 95% of maximum diameter is achieved by 150 days of age (Chrystall *et al.* 1969). Subsequent increases in muscle cross-sectional area, therefore, mainly reflect increases in intramuscular fat and connective tissue. There is a high correlation between cross-sectional area of the loin eye of a pig at slaughter and individual muscle fiber diameter. During the late stages of growth (82–110-kg liveweight) the loin eye increases linearly at a rate of about 0.14 cm^2/kg liveweight (Miller and Wheat 1970).

The ontogeny of fat cell hyperplasia is also not clear. It is considered by some that the fat cell number is established prenatally in most animals. If hyperplasia does stop at or before birth in the pig, it should be possible, by biopsy of fat depots in the newborn pig, to predict whether or not a particular individual is likely to become very fat if allowed free access to food in later life. Several groups of scientists (Mersmann *et al.* 1975, 1981; Steffen *et al.* 1981; Hausman *et al.* 1981, 1982; Etherton 1982) are working to unlock the secrets of the fattening process in swine. Factors affecting rates of lipogenesis (fat formation) and lipolysis (fat breakdown) are under investigation; more complete knowledge of these fundamental processes will allow appropriate adjustments in production practices to improve efficiency of pork production and to meet consumer demands. The significance of these relationships from the standpoint of the animal breeder is discussed briefly in Chapter 4. Exciting research now underway in several laboratories (Richer 1981; Goll *et al.* 1978) on the regulation of muscle protein synthesis and degradation rates promises to provide new knowledge applicable to lean pork production.

It is well known that hormones (thyroid hormone, growth hormone, etc.) affect growth of young animals. Newer knowledge of endocrinology and biochemistry shows that these effects of the endocrine system on growth are mediated through effects on enzyme systems that control anabolic and catabolic reactions of nutrients and their metabolites. Thus, we see that growth of the pig, in keeping with the broad definition of growth cited earlier (Brody 1945), proceeds from a fertilized zygote to a mature adult through a process of cell hyperplasia and hypertrophy (accompanied by differentiation of cells, tissues, and organs) under the

TABLE 7.2. Absolute and Relative Weight of the Pig at Various Stages of Prenatal and Postnatal Growth

Stage of development	*Weight (g)*	*No. of times ovum wt*
Ovum	0.000003	
21 days	0.25	83,000
112 days (term)	1,000	333,000,000
135 days (21 days postnatal)	5,000	1,667,000,000
274 days (160 days postnatal)	90,000	30,000,000,000

Source: Lowrey 1911.

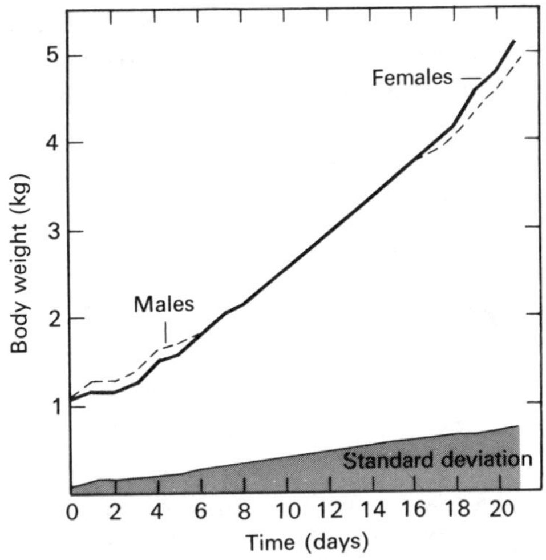

FIG. 7.3. Growth curve of Yorkshire pigs from birth to 3 weeks.
From Gipp et al. 1969.

influence of genetically and environmentally controlled hormone and enzyme systems within the body.

BODY WEIGHT GAIN AS A MEASURE OF GROWTH

In practical swine production, the most meaningful and useful measure of growth is increase in body weight gain. Table 7.2 (Lowrey 1911) illustrates the rapid increase in absolute and relative body weight from conception to slaughter at 90 kg (160 days postnatal age). Growth curves representing birth to 3 weeks of age (Gipp *et al.* 1969) and 8–26 weeks of age (Headley *et al.* 1961) are shown in Figs. 7.3 and 7.4 respectively. It should be recognized that these curves are derived from limited numbers of animals of specific genetic background and raised under specific environmental conditions. Genetics as well as environment will affect the shape of

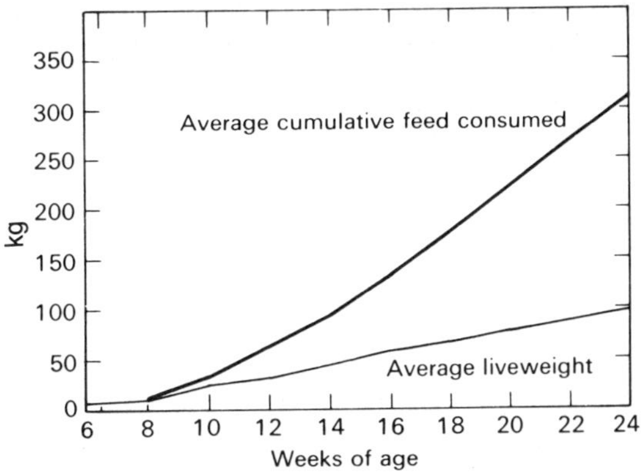

FIG. 7.4. Growth curve of pigs 8–26 weeks of age.
From Headley et al. 1961.

the curve, but it can be used as a guideline. The concept of individual variation and its application in swine breeding is discussed in Chapter 4.

Topel (1968) has prepared a two-volume collection of abstracts related to growth and body composition of pigs and chemical and physical properties of pork. The reader interested in breaking into the literature on this broad subject is referred to this work.

REFERENCES

ATINMO, T., POND, W. G., and BARNES, R. H. 1974. J. Anim. Sci. *39*,703; J. Nutr. *104*,1033.

BRODY, S. 1945. Bioenergetics and Growth. Hafner, New York.

BUITRAGO, J. A., MANER, J. H., GALLO, J. T., and POND, W. G. 1974. J. Anim. Sci. 39,47.

CHRYSTALL, B. B., ZOBRISKY, S. E., and BAILEY, M. E. 1969. Growth *33*,351.

ETHERTON, T. 1982. J. Anim. Sci. *54*,58.

ETHERTON, T. 1982. J. Anim. Sci. *54*,58.

EZEKWE, M. O. 1981. J. Anim. Sci. *53*,1504.

GIPP, W. F., POND, W. G., VANVLECK, L. D., and MILLER, P. D. 1969. J. Anim. Sci. *29*,330.

GOLL, D. 1978. *In* Protein Turnover and Lysosome Function. (H. Segal and D. J. Doyle, Editors). pp. 587–588. Academic Press, New York.

HARD, D. L. and ANDERSON, L. L. 1979. Am. J. Physiol. *237*,E273.

HAUSMAN, G. J. and MARTIN, R. J. 1982. J. Anim. Sci. *54*,1286.

HAUSMAN, G. J., CAMPION, D. R., McNAMARA, J. P., RICHARDSON, R. L., and MARTIN, R. J. 1981. J. Anim. Sci. *53*,1634.

HEADLEY, V. E., MILLER, E. R., ULLREY, D. E., and HOEFER, J. A. 1961. J. Anim. Sci. *20*,312.

LOWREY, L. G. 1911. Am. J. Anat. *12*,107.

MANER, J. H., and GALLO, J. T. 1971. Unpublished data.

McMEEKAN, C. P. 1940. J. Agric. Sci. *30*,301,511.

MERSMANN, H. J., ALLEN, C. D., CHAI, E. Y., BROWN, L. J., and FOGG, T. J. 1981. J. Anim. Sci. *52*,1298.

MERSMANN, H. J., GOODMAN, J. R. and BROWN, L. J. 1975. J. Lipid Res. *16*,269.

MILLER, L. R., and WHEAT, J. D. 1970. J. Anim. Sci. *30*,732.

POND, W. G., STRACHAN, D. N., SINHA, Y. N., WALKER, E. F., JR., DUNN, J. A., and BARNES, R. H. 1969. J. Nutr. *99*,61.

POWELL, S. E. and ABERLE. E. D. 1980. J. Anim. Sci. *50*,860.

RICHER, L. L., EWERT, D. H., and GRAF, G. 1981. Fed Proc. *40*,1625 (Abstr. 495).

STAUN, J. 1963. Acta Agric. Scand. *13*,292.

STEFFEN, D. G., ARAKELIAN, M. C., PHILLEY, G., BROWN, L. J. and MERSMANN, H. J. 1981. J. Anim. Sci. *52*,1306.

TOPEL, D. G. 1968. Bib liography of Porcine Research. Vol. 1: Growth and body composition. Vol. 2: body composition. Iowa State Dept. of Agric. and Agric. and Home Econ. Exp. Stn., Iowa State Univ., Ames.

TRENKLE, A. H. and MARPLE, D. N. 1983. J. Anim. Sci. *57*(Suppl. 2),273.

WEBSTER, A. J. F. 1980. The Energetic Efficiency of Growth. Livestock Prod. Sci. *7*,243.

8

Reproduction

INTRODUCTION

A high reproductive rate is essential in a successful purebred or commercial swine enterprise. A knowledge of the normal anatomy and physiology of the female and male reproductive tract is fundamental to the achievement of a high reproductive rate of swine. In this chapter the following topics are discussed: anatomy of the female and male reproductive tracts, puberty, the estrous cycle, factors affecting ovulation rate and time of estrus, movement and survival of sperm in the female reproductive tract, conception rate, duration of gestation, parturition, composition of semen, sperm survival and storage, artificial insemination, and potentially important techniques, though presently in limited use, such as ova transfer, artificial control of the estrous cycle, superovulation, and storage of sperm by freezing. Each of the above subjects is of significance in the total picture of reproduction. Recent overviews of current knowledge of male and female reproduction in swine are recommended (Dziuk 1977; Dziuk and Bellows 1983; Bazer and First 1983; Amann and Schanbacher 1983).

THE FEMALE

Reproductive physiology of the sow has been reviewed thoroughly by Boda (1959). Only the important aspects related to application to commercial swine production are considered here.

ANATOMY OF THE REPRODUCTIVE TRACT

The reproductive tract of the sow includes the ovaries (right and left), the oviducts (fallopian or uterine tubes), uterus, consisting of the body and two uterine horns, cervix, the vagina, and the vulva. A normal reproductive tract is illustrated in Fig. 8.1. The uterine horns are long and tortuous to accommodate numerous developing fetuses. In the mature sow they may be 1 m or more long when extended, while the body of the uterus is about 5-cm long, and the cervix and vagina each about 10-cm long. The muscular nature of the cervix is also shown in Fig. 8.1. In natural mating, the penis penetrates the first few cervical folds or beyond. The ovaries are lobular, owing to follicles in varying stages of development, and vary in weight from 3 to 7 g and in diameter from 2 to 4 cm. There may be 10–25 individual mature follicles each 8–12 mm in diameter. The size and weight of the reproductive organs are shown in Table 8.1 (Reddy *et al.* 1958). A normal ovary during diestrus is shown in Fig. 8.2. The fallopian tubes are 15–30-cm long with a large ovarian end.

PUBERTY

Puberty (onset of estrous cycle) occurs at 6 or 7 months of age. The number of ova released per estrus increases gradually over the first several estrous cycles. There is a large range even among normal animals kept under adequate husbandry conditions. Breed is an important factor in the onset of first estrus. Christenson and Ford (1979) reported that the percentage of Landrace gilts that showed regular estrous cycles at 6 months of age was higher than that of Duroc, Hampshire, Large White, and Yorkshire gilts. Puberty is also influenced by season of year (Christenson 1981), social environment, and confinement (Christenson 1981; Christenson and Ford 1979). The percentage of gilts cycling at 9 months of age is higher during the winter than during the summer season. Gilts raised in confinement often show delayed puberty and a higher incidence of behavioral anestrus (silent estrus). Behavioral anestrus is influenced by breed and season of year (higher incidence in summer) as well as by confinement. Inbreeding delays onset of first estrus by several weeks (Warnick 1951), and it is retarded by excessive fatness (Self *et al.* 1955). Isolation of gilts from boars tends to delay puberty

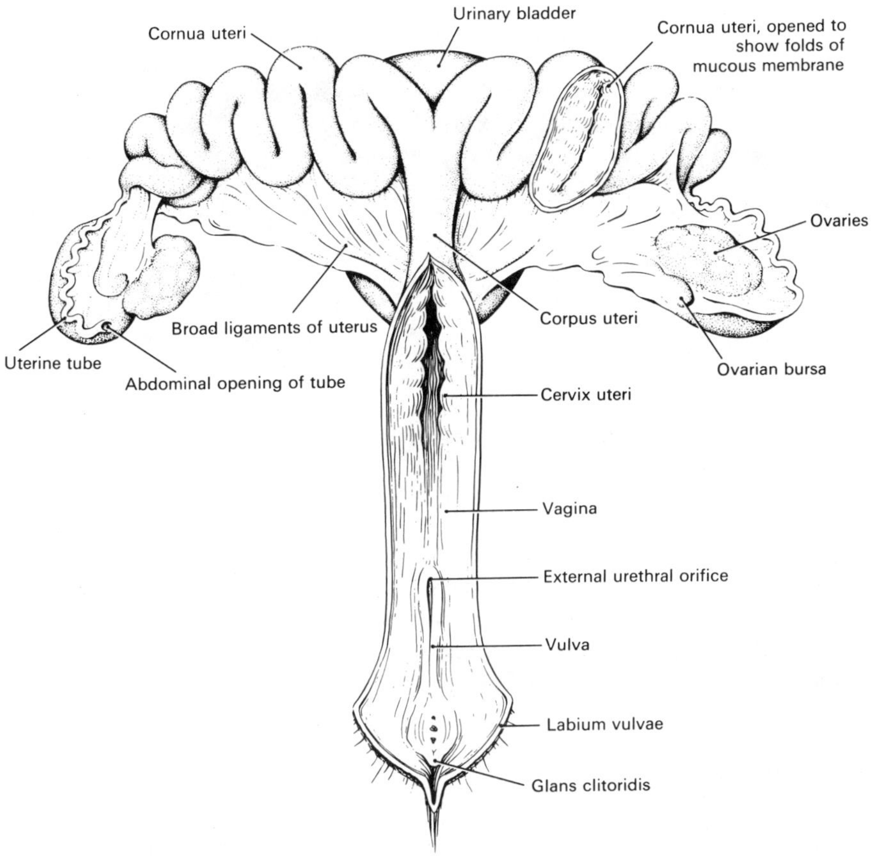

FIG. 8.1. Normal reproductive tract of the female pig showing muscular folds of cervix.
From Sisson and Grossman 1938.

TABLE 8.1. Size and Weight of Reproductive Organs of Sows

	Before puberty	*After puberty*	*Increase (%)*
Age, days	169	186	—
Length of vagina, mm	292	318	9
Average length of uterine horns, mm	383	605	58
Average length of fallopian tubes, mm	217	241	11
Average weight of uterus, g	153	263	72
Average weight of ovaries, g	214	283	32

Source: Reddy et al. 1958.

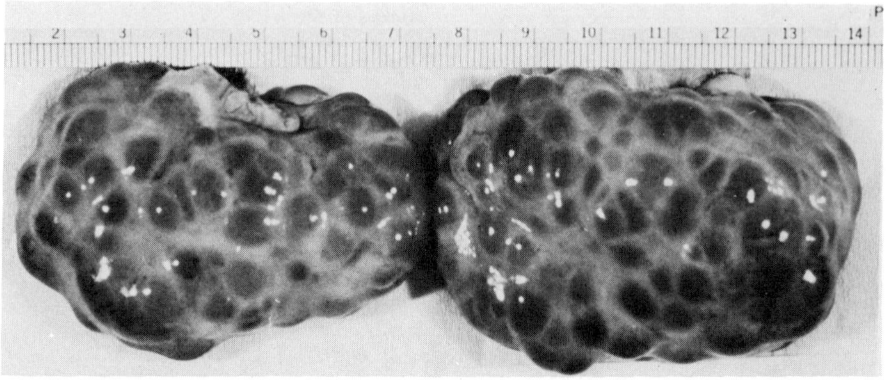

FIG. 8.2. Normal ovaries from sow at 35 days pregnancy. Note numerous follicles over entire ovary surface. Scale in centimeters.
Photo courtesy R. K. Christenson.

(Brooks and Cole 1970). Introduction of a boar to gilts 165–190 days of age that have been previously isolated from boars seems to induce estrus. Although age at puberty is several months, the number of ova available for release in the course of a lifetime, determined in prenatal life as oogenesis, is complete by day 100 after conception (Block and Erickson 1968).

ESTROUS CYCLE

The estrous cycle length is generally discussed in terms of the estrous periods since it is the important phase from the standpoint of planning of breeding dates. The length of the cycle in gilts and sows (onset of one estrus to the onset of the next) averages approximately 21 days (the range is about 18–24 days). The length of the cycle does not seem to be related to the duration of estrus. Ovulation without external manifestations of estrus (silent heat) is uncommon in swine. Gilts reared in total confinement have a high incidence of abnormalities in the estrous cycle, including delayed puberty and behavioral anestrus (''silent heat''). Christenson (1981) reported differences in breed, housing (confinement increases), and season (hot weather increases) in these phenomena. The underlying physiological bases for these observations are not clearly understood, but the results illustrate the ineffectiveness of pen mating behaviorally anestrous gilts. This phenomenon is of concern in attempts at synchronization of estrus by feeding progestational agents.

The cycle can be divided into proestrus, estrus, and diestrus. During proestrus (1–3 days), females are alert to the approach of the boar, will mount other females and accept mounting by diestrus females, but will not tolerate mounting by a boar. During estrus, the swollen vulva and vaginal discharge first observed during late proestrus are accompanied by restlessness (fence-walking, agitation), mounting of other animals, frequent mounting by other females, and frequent urination. When pressure is applied to the back or rump, the animal takes on a rigid stance and has a characteristic ear carriage. The external signs of estrus may last for a period of 3–4 days, or longer, but the female will mate only for a period of 2–3 days.

Sex pheromones synthesized in the testes of the boar and transferred to the salivary gland, from which they are secreted during premating behavior, are probably important in enhancing receptivity of the female to the boar. The steroid 5-androst-6-en-3-ol is one such hormone.

Attempts have been made to relate vaginal smears to stage of estrus to utilize more effectively artificial insemination. These attempts have been largely unsuccessful.

OVULATION

Ovulation occurs late in estrus, usually on the second day after the onset of estrus. There seem to be genetic differences in the length of the estrous period and in the time in which ovulation occurs. Ovulation is delayed with increases in the duration of estrus (Burger 1952). The time span for all ova to be released in a given estrus may be up to at least 7 hr (Burger 1952; Ito *et al.* 1959).

The pig ovum is easily visible under a low-power microscope (120–170-μ diameter). Figure 8.3 shows (a) an unfertilized ovum, (b) a fertilized ovum in the first division, and (c) the first two blastomeres with spindle remnants between them (Thibault 1959). It is possible to retrieve ova from the flushed oviduct by scanning the fluid collected in a petri dish under low magnification. The ova are not motile but are moved down the oviduct by its peristaltic action. When the number of ova released from the two ovaries is unequal at a given estrus, there is a tendency for fertilized ova to migrate to the uterine horn corresponding to the ovary that released the fewest ova (Lasley *et al.* 1963).

Although the mechanism of ovulation is not completely understood, it is closely governed by delicate interactions among hormones of the pituitary, ovary, and uterus (Anderson 1966). Plasma progesterone concentration rises during the first 14 to 15 days of the estrous cycle, then declines precipitously in the course of a 48-hour period and remains low for 4 or 5 days to the next ovulation (Stabenfeldt *et al.* 1969). Follicle-stimulating hormone (FSH) and luteinizing hormone (LH) contents of the pituitary are related to the stage of the estrous cycle. Follicle-stimulating hormone increases and LH decreases during pregnancy. Corpora lutea persist during pregnancy and ovulation is prevented. The signal for maintenance of the corpora lutea apparently is mediated by some local uterine effect or by the presence of the embryo (Niswender *et al.* 1970). This signal is initiated at about day 12 of pregnancy. The pituitary appears to be essential for extended life of the corpora lutea but not for their formation as ovulation and corpus luteum formation occur in hypophysectomized pigs (Anderson *et al.* 1967). The uterus and pituitary apparently regulate the life span of the corpus luteum systemically by both glands as well as locally by the uterus. During pregnancy, progesterone derivatives excreted in the urine increase during the first 54 days, and then decline until term (Edgerton and Erb 1971). This is a rough index of blood progesterone levels. Levels of plasma progesterone and LH are not closely related to number of embryos present during pregnancy (Tillson and Erb 1970).

Superovulation can be induced by injection of gonadotrophins (Tanabe *et al.*

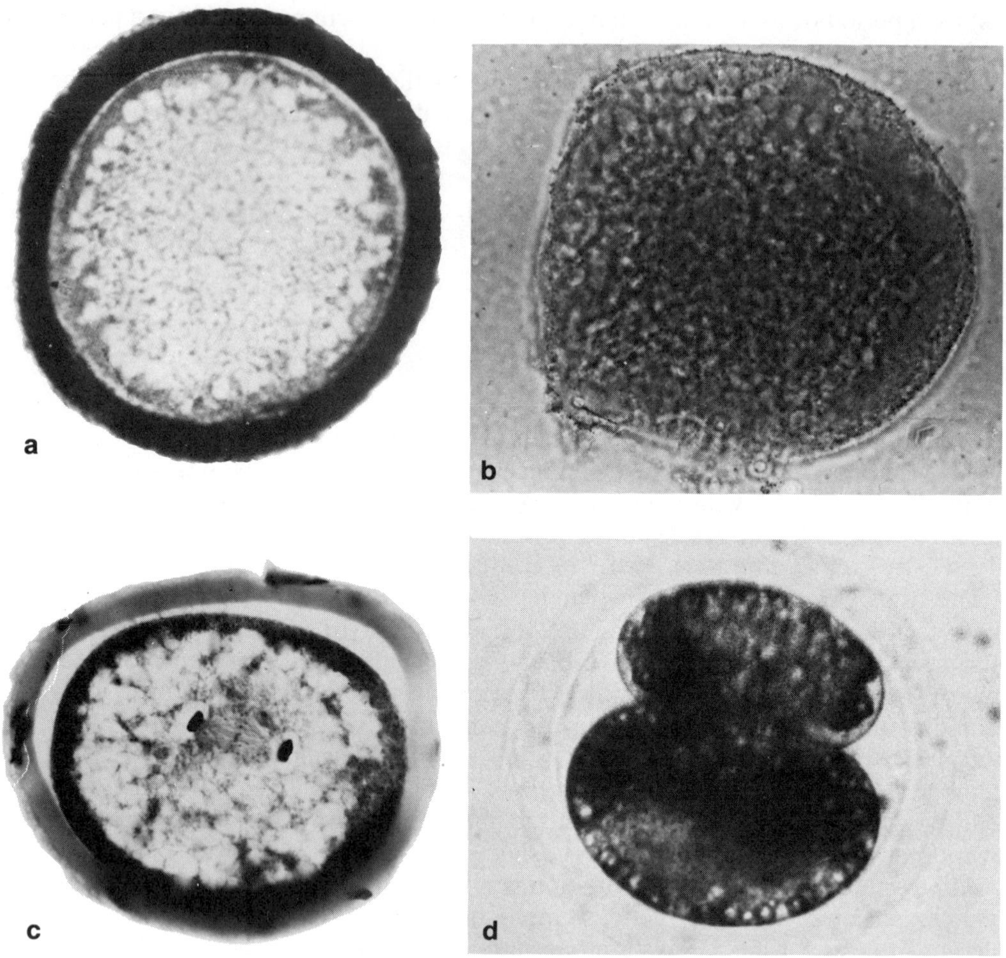

FIG. 8.3. Ova under low power magnification: (a) Unfertilized ovum. (b) Ovum with spermatozoa pentrating zona pellucida (lower left) and nucleus on periphery of ovum (bottom). (c) Fertilized ovum in first division. (d) Fertilized ovum with two blastomeres.
(a) and (c): From Thibault 1959.
(b) and (d): Photos courtesy of R. K. Christenson, Roman L. Hruska U.S. Meat Animal Research Center, Clay Center, Nebraska.

1949; Hunter 1964, 1966; Longenecker and Day 1968), but the practical value of this is in question at the present time because of the smaller birth weight of pigs from large litters and the accompanying reduction in vigor and survival rate. The number of available nipples for sow-nursed pigs is also a factor.

Fertilization occurs within 6 to 10 hr after natural mating and may occur within 2 hr after artificial insemination. It can be defined as the penetration by a sperm of the outer covering (zona pellucida) of the ovum. Normally, a single sperm enters the ovum but occasionally polyspermy does occur (Hancock 1959), especially if ovulation has taken place considerably earlier (Thibault 1959; Dziuk and Polge 1962; Hunter 1967).

Ova remain fertile as long as 24 hr after release so that artificial insemination with fresh semen the day following ovulation can result in conception. However, insemination should be done 6–10 hr before ovulation for highest conception rates. Age of sperm at the time of insemination affects embryo survival (First *et al.* 1963; Hancock 1961). Storage of boar semen before insemination results not only in reduced fertilization but in increased early embryonic mortality. The first division occurs very shortly after the time of fertilization.

Two to three days before the onset of estrus the follicle begins to enlarge (Corner 1921). The nucleus at this time moves toward the periphery where it undergoes mitosis and the polar body is expelled. At the time of rupture the granulosa remains intact (Spalding *et al.* 1955). The granulosa cells then hypertrophy and are changed to the luteal cells of the mature corpus luteum. By the seventh day after the onset of estrus, the corpus luteum is fully developed. About day 15 of the cycle, regression of the corpora lutea begins unless conception has taken place; if it has, the corpora lutea are maintained. The progesterone content of the corpora lutea increases to about day 14 when it declines if conception has not occurred or persists if it has occurred.

After ovulation, the epithelial cells of the oviduct and uterus rapidly hypertrophy and become high columnar-type cells by the end of the first week following ovulation (Steinbach and Smidt 1970). About 10 days after ovulation the columnar epithelium persists and the invasion of eosinophilic leukocytes is at its peak. If conception does not occur, the surface epithelium again reverts to low columnar after day 10 and the eosinophilic leukocytes disappear.

In approximately 21 days the cycle repeats itself. The vaginal epithelium increases in height at estrus (Steinbach and Smidt 1970) and then progressively decreases to a minimum at day 12 to 16 of the cycle. The superficial layers slough beginning at about the fourth day and continuing to day 16. Leukocytes invade the epithelium during late diestrus and reach a maximum just after estrus.

A phenomenon called arborization of the cervical mucus occurs when mucus is allowed to dry on an unstained slide. This consists of the appearance of a crystalline pattern of leafy projections resembling fern leaves. Although in humans this occurs at a specific phase of the estrous cycle, in sows it is not predictable enough to be of value in conjunction with artificial insemination. Change in pH of the vaginal smear has also been used as a means of attempting to identify estrus in sows, but, again, results have not been adequate for predictive use.

FACTORS AFFECTING OVULATION RATE AND TIME OF ESTRUS

The number of ova released at a single estrus gradually increases through the first four or five estrous cycles. Therefore, gilts bred at the first estrus have a smaller average litter size than those bred at a subsequent estrus. Because of increased ovulation rate with progressive maturity, first-litter gilts usually farrow fewer pigs than they do subsequently.

Generally, the litter size at birth represents only about 60–70% of the number of ova originally released. In mature sows the average total number of ova released at a single estrus (as measured by counting corpora lutea) is approximately 17 while that of first-litter gilts is about 13 (Ito *et al.* 1959). Breed differences exist in ovulation rate and in prolificacy. An increase in the energy (glucose) intake during the 2 weeks prior to breeding ("flushing") is associated with an increase of two to three in total number of ova released. This has also been observed in sows fed increased energy just prior to postlactational estrus (Hardy and Lodge 1969). Extreme restrictions in feed intake during the breeding season do not entirely inhibit estrus and, in fact, it is difficult to interfere with the estrous cycle by underfeeding except through its delaying effect on age at puberty as discussed earlier. Ovulation rate does not seem to be affected by continuous light, unlike the effect for the hen (Waddill *et al.* 1968). High environmental temperatures (33° C) reduce ovulation rate, but have no effect on estrous cycle length (d'Arce *et al.* 1970).

Sows do not normally ovulate during lactation, although they often show estrus during the first day or two after parturition. Estrus usually occurs within a few days after the pigs are weaned. When pigs are weaned at 5 weeks of age, estrus usually occurs in less than a week; when pigs are weaned earlier there is usually a longer interval. The initiation of estrus and ovulation after weaning appears to be associated with changes in production of the gonadotrophin-releasing factor from the hypothalamus of the sow (Cox and Britt 1982).

Estrus and ovulation can be induced during weeks 3–6 of lactation by injection of 1000–1500 IU of pregnant mare serum gonadotrophin (PMSG) followed by 500 or 1000 IU human chorionic gonadotrophin (HCG) 72–96 hr later (Hausler *et al.* 1980). Hodson *et al.* (1981) found that best results were obtained with multiparous sows treated with PMSG 25 days or more postpartum.

However, when the pigs are removed from the sow at birth or within 2 days after farrowing most sows will come into estrus within 2 weeks and the ova are usually fertile. There is apparently no relationship between number of pigs suckling the sow and the time of return of estrus following weaning (Burger 1952).

ARTIFICIAL CONTROL OF THE ESTROUS CYCLE

Since the estrous cycle is under endocrine control, it is possible to alter it by administering exogenous hormones. Several ways are available:

1. *Parenteral gonadotrophin administration after weaning*. This includes injection of a follicle-stimulating material (pregnant mare serum gonadotrophin, PMSG) at weaning (Longenecker and Day 1968) or PMSG followed 56–72 hr later with human chorionic gonadotrophin (HCG), an ovulating hormone (Christenson and Teague 1975). Combining PMSG and HCG is satisfactory for synchronizing estrus and ovulation in sows after weaning and permits artificial insemination of groups of synchronized sows at one time without relying on estrus detection.
2. *Parenteral gonadotrophin administration to the cycling female* (Day *et al.* 1959). This involves injection of PMSG followed by HCG 96 hr later. The dosage level is critical; improper levels result in a high proportion of gilts failing to show estrus. Dosages of 1200 IU of PMSG and 500–1000 IU of HCG have given good results.

3. *Oral administration of estrogens followed by progesterone derivatives* (Pond *et al.* 1965) *or of a progestogen, allyl trenbolone* (Davis *et al.* 1979), *to the cycling female.*

4. *Oral administration of a dithiocarbamoylhydrazine derivative that affects the hypothalamus so as to inhibit ovulation* (Polge 1965). This drug (sometimes called methallibure or ICI 33828) has received widespread attention as a means of controlling estrus without problems with cystic ovarian follicles. However, it produces teratogenic effects when ingested during pregnancy, so probably will not find commercial use.

5. *Injection of 10 mg of prostaglandin $F_2\alpha$ ($PGF_2\alpha$) on days 13 and 14 after inducing accessory corpora lutea with PMSG–HCG administration in cycling sows.* Employing an injection sequence of 1000 IU of PMSG at the time of the second $PGF_2\alpha$ injection followed by 500 IU of HCG 96 hr later is effective in synchronization of ovulation, but suffers from the high labor intensity required as compared to that associated with the use of orally active compounds. As more is learned about the physiological effects of prostaglandins, further refinements can be expected in ovulation control methods.

Orally active progestational compounds and nonsteroid inhibitors of anterior pituitary hormone activity hold promise for application in controlled estrous programs to be used in conjunction with artificial insemination. This also offers a practical means of implementing a closely planned farrowing schedule, whereby groups of sows farrow together. The application is foremost with first litter gilts, as the estrous cycle can be controlled effectively in older sows by a systematic weaning program. It has been postulated that the suppression of ovulation in sows during lactation is due to failure of FSH release and of LH synthesis during lactation (Crighton and Lamming 1969).

OVA TRANSFER

Transplantation of pig ova from the uterus of one female to that of another has been achieved (Kvasnicki 1951; Hancock and Hovell 1962; Vincent *et al.* 1964; Smidt *et al.* 1965; Dziuk *et al.* 1964). Blastocysts recovered on days 7 or 8 by flushing the uterus with 40 ml of pig serum have been transferred with pregnancy resulting in some recipients (Hunter *et al.* 1967). Thus, the groundwork is laid for refinement in conditions and techniques that will allow the routine use of ova transfer. This technique could have special use as a means of increasing the number of progeny obtained from genetically superior females, by using common or inferior recipient mothers to incubate the developing fetuses.

MOVEMENT AND SURVIVAL OF SPERM IN THE FEMALE REPRODUCTIVE TRACT

Studies of sperm transport in the female reproductive tract (Ito *et al.* 1944; Mann *et al.* 1956; Pitkjanen and Subin 1961; du Mesnil du Buisson and Dauzier 1955; Lovell 1958) have established that sperm reach the uterotubular junction within 10 min but that passage into the oviducts may not occur for more than $1\frac{1}{2}$ hr after mating. There is no difference in the number of spermatozoa recovered from the oviducts of sows inseminated with live or dead spermatozoa (First *et al.* 1968A) or in the number of sperm in the oviducts of sows inseminated during estrus or on day 7 postestrus (First *et al.* 1968B). Spermatozoa can maintain viability in the oviduct for 25–30 hr or longer. More than half of the spermatozoa have disappeared from the uterus within 15 min after insemination and almost all are gone after 24 hr. Motility of spermatozoa in the uterus is drastically reduced after 2 hr and phagocytosis has been observed at 8 hr (First *et al.* 1968A).

Sperm age has an effect on fertilization rate. Semen stored for 54 hr after collection produced a lower rate of fertilization in artificially inseminated gilts than semen stored for 6 hr. The number of sperm in the zona pellucida of 3-day cleaved eggs is also significantly lowered with older sperm (First *et al.* 1963).

CONCEPTION RATE

Conception rate (percentage of matings, either artificial or natural, that result in development of viable fetuses) may refer to first service or to first and second services or to total services. An acceptable first service conception rate for gilts and sows is 70%. Failures to attain 100% conception are due to many factors. Some of these factors are abnormalities in the anatomy of the female reproductive tract, cystic follicles, the improper timing of mating, and infertility of the boar. Environmental temperature exerts an effect on conception rate in swine. Cold stress on the day of mating (−25° C) does not interfere with conception (Swierstra 1970), but heat stress as indicated by summer versus winter conception rates clearly reduces reproductive efficiency (Teague 1970; Omtvedt *et al.* 1971).

In swine, it is generally considered that conception is an all or nothing phenomenon since several billion sperm are competing to fertilize fewer than 25 ova. However, there is some evidence that smaller litters may be associated with marginal fertility in the boar. Double mating of the sow (two services a day apart during estrus) is associated with a higher conception rate and with one or two extra pigs per litter, probably because the duration of ovulation may be several hours.

DURATION OF GESTATION

The gestation period of swine averages about 114 days (Kenneth 1953; Joubert and Bonsma 1957). There are small genetic differences (Joubert and Bonsma 1957). There is no important effect of litter size or of first pregnancy versus subsequent pregnancy on gestation length (Rathnasabapathy *et al.* 1956).

PRENATAL DEATHS

It is common to have a 30–40% loss of developing young between the first and 114th day of gestation (Pond *et al.* 1960). More than half of these losses occur during the first 25 days (before or shortly after implantation). Mummified fetuses

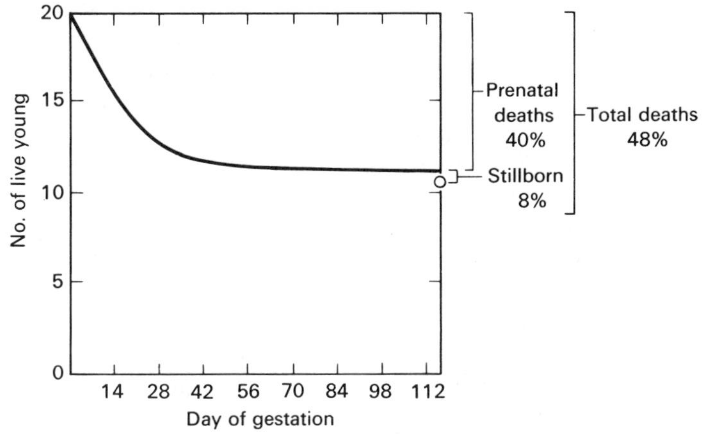

FIG. 8.4. Sequence and proportion of prenatal deaths.

also account for a surprisingly large total. These represent deaths that occur during the last part of gestation and are differentiated from stillborn pigs by partial resorption. Typical sequence and proportion of losses are illustrated diagrammatically in Fig. 8.4.

Litter size depends on the number of ova produced, on the percentage of fertilization, and on the number of prenatal deaths and stillbirths.

It has been suggested that early embryonic mortality may be due to an excess secretion of progesterone by the corpora lutea, but no clear data are available to substantiate this (Baker *et al.* 1958; Perry 1954; Lerner *et al.* 1957; Reddy *et al.* 1958; Spies *et al.* 1959).

The high embryo mortality rate in swine has also been proposed as being related to uterine crowding. However, negative results were obtained by ligating portions of the uterine horns to provide uterine spaces of varying capacity (Dziuk 1968). Uterine crowding may affect fetal survival when the number of fetuses exceeds about 14 but it is probably not an important factor until after day 30 of gestation (Dhindsa and Dziuk 1968B; Fenton *et al.* 1970). From embryo transfer studies, in which color inheritance was used as a code for identifying fetuses in late gestation, evidence has been presented to suggest that the barrier imposed by uterine capacity must be removed somehow before a marked increase in reproductive rate of sows can be realized (Bazer *et al.* 1969).

A large environmental effect on embryonic mortality is related to energy intake. A high plane of nutrition immediately following breeding results in an increase in embryonic mortality (Waldorf *et al.* 1957; Gossett and Sorenson 1959A,B). The metabolic basis for this is not known. Reciprocal embryo transfer studies using gilts fed high or low energy diets have shown that embryos from gilts of each group are equally viable (Bazer *et al.* 1968). Therefore, it appears that the controlling factor is some uterine factor rather than any effect of the embryo per se. Although a marked decline in body weight occurs in gilts subjected to severe energy restriction or inanition in mid-gestation, fetal survival rate is unaffected. Hard and Anderson (1982) have found that uterine blood flow is maintained at a high level during late pregnancy of energy-restricted gilts despite changes in plasma volume. They have suggested that this special adaptation allows normal fetal development during prolonged maternal starvation.

The incidence of stillbirths is greater with large litters than with small ones (Carmichael and Rice 1920; Asdell and Willman 1941). Virtually all pigs, except mummified fetuses, are alive at 112 days in the experience of veterinarians involved with hysterectomy-derived pigs. The simultaneous occurrence of stillbirths and mummified fetuses does not have an influence on frequency of stillbirths. In one study (Pond *et al.* 1960), only 29 out of 70 litters that contained stillbirths contained mummified fetuses. Single mummified fetuses occurred more frequently than two or more in a litter. Mummified fetuses were more prevalent in large than in small litters.

Normal 25-day embryos are approximately 20 mm long and by 40 days they have reached approximately 40 mm (Gossett and Sorenson 1959B). Most mummified fetuses are at least 50-mm long and often nearly full term, indicating that total resorption probably occurs when death precedes the 40th day.

PARTURITION

The duration of parturition may vary from 1 to 2 to as many as 24 hr in a normal pregnancy. Attempts to hasten parturition by injection of oxytocin are not always successful, probably because dosage level and timing of the injection are critical. Progesterone in maternal blood plasma declines before the onset of natural parturition. Since corpora lutea are the primary source of progesterone for maintenance of pregnancy, cessation of their function is necessary for initiation of parturition. Prostaglandin $F_2\alpha$ ($PGF_2\alpha$) is a natural luteolytic agent in swine at term, producing a decline in progesterone and a release of relaxin from the corpora lutea. Exogenous $PGF_2\alpha$ administration results in loss of corpora luteal function and a drop in plasma progesterone and has been found to be effective in inducing rapid parturition in sows (Nara and First 1981).

When an appropriate amount of $PGF_2\alpha$ is administered intramuscularly on day 112 of gestation, parturition can be expected to begin 20 to 24 hr later. This phenomenon provides a practical tool for adjusting parturition time to the convenience of the producer, rather than allowing farrowing to be spread randomly over a 24-hr period.

Parturition can be delayed for several days beyond the normal gestation by subcutaneous administration of progesterone (Minar and Schilling 1976; Wilson *et al.* 1981) or a progesterone derivative, 6-methyl-17-acetoxy-progesterone (MAP) (Nellor *et al.* 1975). However, the delayed parturition may be associated with abnormal patterns of plasma progesterone, estrogen, and relaxin, which in turn may contribute to the high incidence of stillbirths in delayed parturition. Further research is needed to eliminate these increased stillbirths and thereby to create the conditions under which artificially delayed parturition can be used as a management tool for controlling the time of farrowing.

Just before parturition the sow shows general uneasiness and attempts to make a nest with the bedding if available. The presence of colostrum in the mammary glands is a signal of imminent parturition. The birth of the first pig is usually accompanied by the placental membranes but occasionally one or more pigs are born before any placenta is expelled. The umbilical cord is usually still attached when the pig is expelled but breaks soon afterward as the pig struggles to stand and explore his new environment. Presentation may be with the front feet first or the hind feet first and there appears to be no chronological pattern within a given litter. Occasionally, lack of uterine motility or the inability of a large pig to pass the pelvic region necessitates assistance, but usually this is unnecessary. Occasionally cesarean section becomes necessary, or the failure of some of the placental membranes to be expelled results in uterine infection. This can be avoided or alleviated by oxytocin injection and antibiotic administration. Within the first few hours after the completion of parturition the sow will be mobile and will accept water and feed. Pigs born earliest soon find their way to a nipple and may suckle from several glands before the last pig is born.

THE MALE

ANATOMY OF THE REPRODUCTIVE TRACT

The anatomy of the reproductive organs of the boar was thoroughly described by Mackenzie *et al.* (1938). A normal reproductive tract is shown in Fig. 8.5. The size and weight of the primary and accessory reproductive organs are shown in Table 8.2 (Blom 1962).

All the typical accessory organs of the mammal are present. The testes are carried in the external body cavity in the scrotum, where they become located after migration out of the body cavity by approximately the 100th day of prenatal life. The testes, covered by a tough capsule (tunica albuginea), are carried in a near vertical position with the head of the epididymis on the ventral end. The tail of the epididymis forms a cap over the dorsal end of the testis. In normal castration procedure the epididymis and the attached tunica vaginalis are removed with the testes. Stored sperm leave the epididymis via the vas deferens. The seminal vesicle, Cowper's (bulbourethral) gland, and prostate gland all contribute to semen volume. The semen leaves the reproductive tract through the urethra and the penis at ejaculation.

A microscopic section through the testis (Fig. 8.6) shows the relative proportions of seminiferous tubules and interstitial cells. The interstitial tissue in boars is more plentiful than in most other animals. The epididymis consists of a long, highly convoluted tube which increases in diameter as it reaches the point of transition into the vas deferens. On each side of the pelvic urethra lie the seminal vesicles. They cover the neck of the bladder, the Cowper's, the prostate, and the vas deferens. The openings from the vas deferens and the seminal vesicles into the urethra represent four separate openings located in close proximity to each other. The Cowper's glands lie parallel to the urethra and are attached to it on the ventral side. The prostate gland is a multilobular gland composed of the body (bulb) attached on the dorsal surface of the urethra and of the pars disseminata attached to the body of the gland itself. The pars disseminata of prostate is conspicuous at its anterior end because of its yellow color.

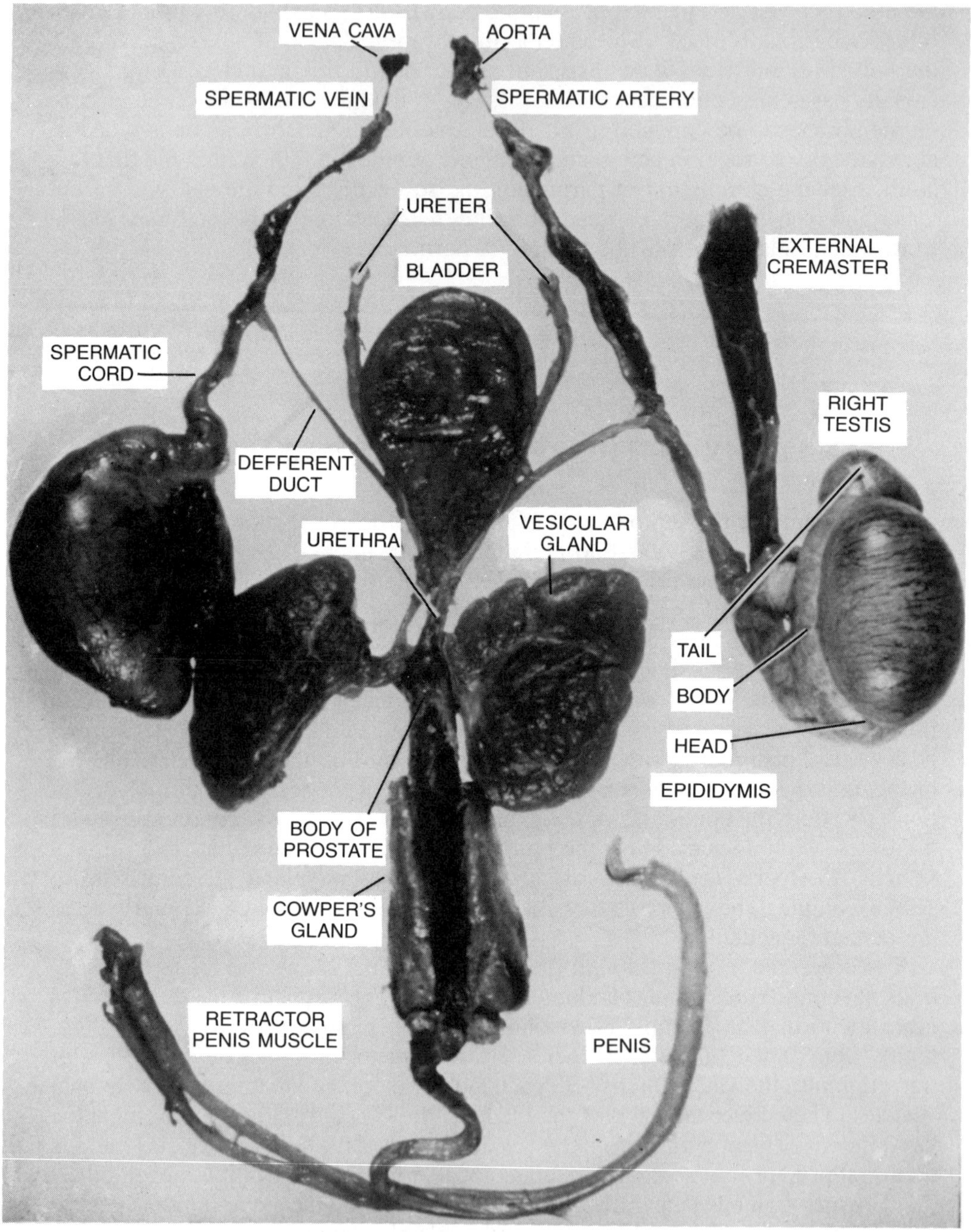

FIG. 8.5. Normal reproductive tract of the boar.
Photo courtesy of R. K. Christenson, Roman L. Hruska U.S. Meat Animal Research Center, Clay Center, Nebraska.

TABLE 8.2. Size of Boar Reproductive Organs

Organ	*Size*
Testis plus epididymis	13 × 7 × 7 cm 250–300 g (single)
Epididymal duct length	55–65 m
Vas deferens	25–30 cm long × 2 mm outside diameter
Seminal vesicle	13 × 7 × 5 cm
Prostate	3 × 3 × 1 cm (15–25 g)
Cowper's gland	17 × 3 × 3 cm
Penis	50–75 cm long 2 cm diameter (erect)

Source: Blom 1962.

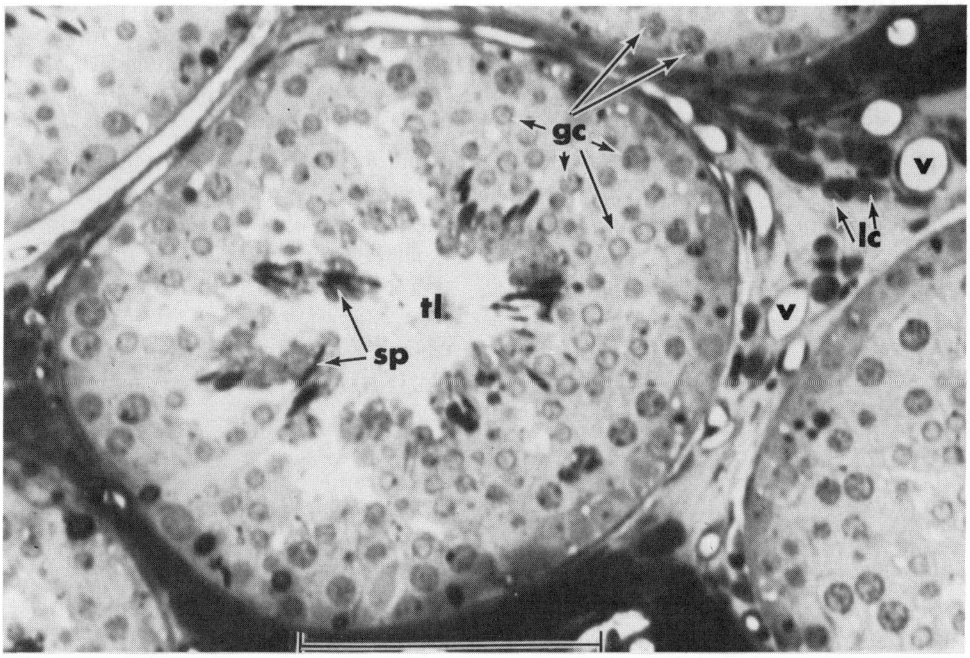

FIG. 8.6. Microscopic appearance of boar testes showing interstitial tissue with Leydig cells, blood and lymph vessels, and seminiferous tubules with spermatozoa in various stages of development. Key: sp = elongated spermatids; tl = tubules lumen; gc = germ cells in different stages of development; lc = Leydig cells; v = blood and lymph vessels; and bar = 100 microns.
Photo courtesy of D. D. Lunstra, Roman L. Hruska U.S. Meat Animal Research Center, Clay Center, Nebraska.

PUBERTY AND SEXUAL BEHAVIOR

Sexual maturity in the boar is a gradual process in which sperm production and sexual desire begin concurrently in increasing intensity at about 4 months of age. However, it is generally considered that boars should not be used in service before 6–8 months of age. Semen has been collected as early as 5 months of age, but volume and percentage of normal sperm are less than in semen from mature boars. Extending daylight by artificial light may hasten onset of sexual interest and lipids in young boars but does not appear to affect semen quality (Berger *et al.* 1980). Sexual behavior of males has been described by Hafez *et al.* (1963), and by Houpt and Wolski (1982). It includes phases of courtship, mounting, and intromission and ejaculation. Mature boars should not be allowed together if not accustomed to being penned together, as the fighting instinct will result in injury or death to one or both boars and possible injury to attendants. When boars are raised together there is no problem in maintaining them in the same pen at maturity. Shy breeders or those introduced to a strange environment will often serve better when the female to be bred is brought to them rather than vice versa. The duration of copulation is 3–20 min or longer. Competition among boars for a female in estrus usually results in an establishment of order of dominance among two or more boars that influences the frequency of siring of litters by a particular boar. There are apparently genetic differences in pattern and intensity of sexual behavior.

Mature size is not reached until more than 1 yr of age so that increases in total semen and sperm production accompany both the development of sexual maturity and the increase in body size until about 1 yr of age. The size of the testis tends to increase proportionately with total body size and is directly related to total sperm output. Severe restriction of feed intake delays puberty. It results in a reduction in the diameter of the seminiferous tubules of the testis as well as a reduction in its total growth (Phillips and Andrews 1936). The effect is reversible by returning the animal to a normal level of intake.

ARTIFICIAL INSEMINATION AND SEMEN PRODUCTION

The practice of artificial insemination on a commercial basis exists in some countries of the world. For economic reasons the practice, even though technically feasible, has not achieved widespread use in North and South America. Two problems that still exist are difficulty in detecting estrus in the gilt and the poor survival of boar sperm stored beyond about 3 days. The first problem can be alleviated by keeping a vasectomized boar in the presence of the gilts to be bred. The volume of total fluid (semen) inseminated and the total number of motile sperm are both important factors in successful insemination. The total number of motile sperm required to result in a high conception rate is believed to be in the order of three billion and the mimimum volume about 50 ml (Self 1959). Assuming 60% motility, this would require a total of five billion sperm in 50 ml of inseminating fluid. Since boar semen is more concentrated than this, it is possible to use a diluent. None of the diluents (extenders) in use to date will promote viability of sperm beyond a few days of storage at 5° or 15°C. Freezing kills boar sperm. However, research has resulted in viable litters of pigs from semen frozen under special conditions (Graham *et al.* 1971; Johnson *et al.* 1981). This is a technical breakthrough that will allow eventual commercial application. When conditions are favorable, the rate of conception with artificial insemination is equal to that obtained by natural service (Self 1959; Hancock 1960). The development of extenders and of thawing procedures for frozen semen (Pursel 1983) allows storage of semen over longer periods of time and will encourage the use of outstanding sires on a commercial basis in artificial insemination. The optimum time for insemination of fresh, whole semen is 12–16 hr before ovulation (Dziuk 1970); in contrast, maximum fertility with frozen–thawed spermatozoa is obtained when pigs are inseminated at about the time of ovulation (Pursel 1980).

The total volume of semen obtained in a single ejaculate is affected by size of boar and frequency of semen collection. In a study involving Yorkshire and Chester White boars (Turkheimer *et al.* 1958), from which semen was collected weekly, the total volume per ejaculate for boars 18–33 weeks of age was 125 ml, whereas that of boars 47–52 weeks of age was 184 ml. Therefore, one can expect a sizable increase in total semen volume between initial semen collection at puberty and at 1 yr of age. In boars housed together, there is a tendency toward pederasty which makes it advisable to house boars individually for utilization of semen production for either artificial insemination or natural service.

The total volume per ejaculate is decreased, but the total sperm collected is increased if collections are daily rather than less frequent. Motility is apparently not affected appreciably. In commercial production it is common to use a mature boar in natural service to breed three sows in a single day. The total semen production of a mature boar is, therefore, probably considerably greater than the amount measured during semen collection studies. Total semen production (as measured by number of fertile matings) remains at a high level for several years after the boar reaches sexual maturity. Boars are often kept in heavy service to an age of 5 or 6 yr, or more, and are usually disposed of for other reasons, such as an ill temper (which represents a danger to the safety of those handling the boar) or physical disability, rather than a failure in semen production. Low ambient temperature (−15°−−20° C) does not interfere with testicular development, sperm production, or semen quality in healthy boars. (Swierstra 1970).

On the other hand, acute heat stress (exposure to environmental temperatures of 33.3° C for 72 hr) reduces total number of sperm per ejaculate, reduces sperm motility, increases abnormal sperm per ejaculate, and reduces pregnancy rates of gilts inseminated with semen from boar postexposure (Christenson *et al.* 1972). Detrimental effects on semen quality and fertility were first evident at 2–3 weeks postexposure, continued to decline until 4.5 weeks, and then returned gradually to pre-exposure conditions by 8–9 weeks. Likewise, chronic heat stress (exposure to an environmental temperature of 34.5° C for 8 hr and 31° C for 16 hr during each 24-hr period for 11 weeks) reduces daily sperm production by nearly one-half and reduces the conception rate significantly (Wettemann *et al.* 1979). Furthermore, the reduced sperm motility associated with heat stress persists for at least 5 weeks after the heat stress is removed. Results of such studies under controlled laborato-

TABLE 8.3. Characteristics of a Typical Single Ejaculate of Boar Semen[a]

Criterion	*Amount*
Total volume, ml	300
Strained volume, ml	210
Gelatinous volume, ml	90
Spermatozoa per ml of strained semen	200×10^6
Total spermatozoa per ejaculate	42×10^9
Motile sperm, %	70
Total motile sperm per ejaculate	29.4×10^9
Number of services possible	10

Source: Foote et al. 1959; Kennelly and Foote 1964.

[a]Minimum volume of fluid and number of motile sperm needed for 1 service: 3×10^9 motile sperm in 50 ml of fluid.

Maximum number of services from above ejaculate:

$$\frac{29.4 \times 10^9}{3 \times 10^9} = 9.8 = \text{approximately } 10$$

500 ml of fluid needed for 10 services

210 ml of semen + 290 ml of extender = 500 ml of diluted semen for 10 services = 50 ml per service.

ry conditions simulating temperature fluctuations that normally occur in many environments in which swine are produced suggests the importance of providing a means of artificial cooling for boars during summer months. Overhead misting or sprinkling systems or temperature-modified housing may prove to be a worthwhile investment for boars in many environments.

In large mature boars it is often possible to collect up to a liter or more of semen at a single ejaculate. This represents a tremendous fertility potential for outstanding boars that remain in service for any length of time. However, more common sperm production figures and the number of services that might be expected from them are shown in Table 8.3 (Foote *et al.* 1959; Kennelly and Foote 1964). Only about two-thirds of the total sperm produced are actually obtained when semen is collected by an artificial vagina because of masturbation, resorption of sperm before ejaculation, losses in the urine, and adherence to surfaces of collection equipment (Kennelly and Foote 1964).

COMPOSITION OF SEMEN

The composition of boar semen is summarized in Table 8.4 (Westmoreland *et al.* 1967; White, 1958) and that of seminal vesicle secretion in Table 8.5 (Mann 1954, 1959). The total composition is a reflection of the secretions from the secondary sex glands as well as from the testes. The composition of a single ejaculate is conveniently divided into three components: (1) pre-sperm fraction, often high in bacterial contamination and containing mainly secretions of the accessory sex glands; (2) sperm-rich fraction containing the largest concentration of sperm and contributing the largest volume; (3) the post-sperm fraction consisting mainly of gelatinous material. The three fractions can be collected in separate

TABLE 8.4. Chemical Composition of Boar Semen (per 100 ml of fluid)

Constituent	*Mean*[a] (*per 100 ml of fluid*)
Water, g	95(94–98)
Sodium, mg	650(290–850)
Potassium, mg	240(80–390)
Calcium, mg	5(2–6)
Magnesium, mg	11(5–14)
Chloride, mg	330(260–430)
Zinc, μg/ml	31.8(7.8–78)[b]
Fructose, mg	13(3–50)
Sorbitol, mg	12(6–18)
Citric acid, mg	130(30–330)
Inositol, mg	530(380–630)
Glycerylphorphoryl choline (GPC), mg	(110–240)
Ergothioneine, mg	(6–23)
Protein, g	(3–7)

Source: White 1958.
[a]Mean, with range in parentheses.
[b]Westmoreland et al. 1967.

containers. In practice, it is common to collect the entire ejaculate in one container and strain the semen through a sterile cheesecloth to separate the gelatinous mass. This is to avoid problems with clogging of the inseminating tube and to obtain a sample of uniform sperm concentration. Generally, the higher the sperm concentration, the lighter the color of the semen (Polge 1956B).

TABLE 8.5. Composition of Seminal Vesicle Secretions in Boars (per Liter of Fluid)

Constituent	*Seminal vesicle secretion*[a]
Water (g)	814
Total N (g)	15.5
Sodium (meq)	30
Potassium (meq)	55
Chloride (meq)	3
Magnesium (meq)	—
Calcium (meq)	6
Citric acid (mmole)	30
Total phosphorus (mmole)	10
Inorg. phosphorus (mmole)	1
Acid sol. phosphorus (mmole)	9

Source: Mann 1954, 1959.
[a]The concentration of nitrogen, citric acid and calcium is higher than in whole semen; that of chloride, sodium and total phosphorus is much lower; that of other listed constituents is similar.

SPERM SURVIVAL AND STORAGE

Removal of any one accessory gland apparently does not alter libido nor does it change the activity of the remaining accessory glands or the process of ejaculation, but it does cause a proportional reduction in semen volume. Fertility apparently is not reduced when the seminal vesicles and Cowper's glands are removed and, in fact, the number of abnormal sperm may be reduced and the duration of sperm motility increased. Regeneration of tissue from completely removed glands does not take place. Motility is possible both in aerobic and anaerobic conditions; in the presence of oxygen the spermatozoa oxidize lactic acid and in its absence they utilize glycolizable sugar, mainly fructose.

The viability of boar sperm after ejaculation is only a few days. Unlike bull semen, the semen from boars cannot be readily frozen without loss of motility and fertility. Much effort has been exerted to determine the conditions required to maintain motility of boar semen by various storage procedures, but the longest storage of diluted boar semen thus far achieved until very recently without appreciable loss of motility is 4 or 5 days. Conception has been accomplished in gilts inseminated with frozen boar semen at the USDA (Pursel and Johnson 1971; Johnson *et al.* 1981), in Scandinavia (Crabo and Einarsson 1971), and at the University of Minnesota (Graham *et al.* 1971). A variety of extenders have been tried, including solutions of sodium sulfate and potassium sulfate, calcium chloride, peptone (Milovanov, 1932), egg yolk–phosphate mixtures (Lasley and Bogart 1944), cow's milk (du Mesnil du Buisson 1956), glucose tartrate (Lasley and Bogart 1944) and antibacterial agents such as penicillin and streptomycin (Polge 1956A; Aamdal and Hoyset 1957; Foley *et al.* 1967). In view of the high concentration of inositol, citrate, glucose, and glycine in semen, each of these has also been used. Citrate and phosphate extenders, which are capable of prolonging the life of bull sperm, are toxic to boar sperm under the same conditions (Noll 1949; Roy 1955; Polge, 1956B). Extenders containing 0.33% citrate maintain the highest motility and as the concentration of citrate increases, livability of sperm decreases (Young 1958). An extender which maintains the motility for 4–5 days consists of a combination of 20% egg yolk, 0.33% citrate, 0.33% glycine, and 1% glucose. Glycerol, inositol, potassium chloride, and sodium bicarbonate seem to be detrimental to sperm motility whereas carbon dioxide has no effect. Osmotic pressure appears to be important in spermatozoa survival and fertility (Stevermer *et al.* 1964). Glass-sealed tubes produce higher sperm survival, as estimated by motility, than corked tubes. Storage at 15° C is superior to storage at either 5° or 25° C (Young 1958). However, motility of stored spermatozoa is not necessarily a good indication of fertility (Stevermer *et al.* 1964).

Boar spermatozoa are subject to "cold shock" when cooled suddenly to 6° C or less for storage. Attempts to prevent the loss of potassium and other ions from sperm exposed to quick cooling by addition of substances such as glycerol and dimethysulfoxide (DMSO) to the diluent thus far have been unsuccessful (Hood *et al.* 1970).

Boar semen exhibits anabiosis—that is, quick loss of motility and settling out on short-term storage (Ito *et al.* 1948)—but reactivation is possible by warming to 37° and shaking in the presence of oxygen. Therefore, it is necessary to take this into consideration in studies of boar sperm motility. As scientists continue to search for improved methods of freezing boar spermatozoa for long periods of storage, they are obtaining greater knowledge of sperm physiology, which will be of use in application of artificial insemination in commercial swine production in the future.

REFERENCES

AMANN, R. P., and SCHANBACHER, B. D. 1983. J. Anim. Sci. *57*(Suppl. 2), 380.
ANDERSON, L. L. 1966. J. Reprod. Fert., Suppl. *1*,21.
ANDERSON, L. L., DYCK, G. W., MORI, H., HENRICKS, D. M., and MELAMPY, R. M. 1967. Am. J. Physiol. *212*,1188.
ASDELL, S. A., and WILLMAN, J. P. 1941. J. Agric. Res. *63*,345.
BAKER, L. N., CHAPMAN, A. B., GRUMMER, R. H., and CASIDA, L. E. 1958. J. Anim. Sci. *17*,612.
BAZER, F. W., and FIRST, N. L. 1983. J. Anim. Sci. *57*(Suppl: 2), 425.
BAZER, F. W., CLAWSON, A. J., ROBISON, O. W., and ULBERG, L. C. 1969. J. Reprod. Fert. *18*,121.
BAZER, F. W., CLAWSON, A. J., ROBISON, O. K., VINCENT, C. K., and ULBERG, L. C. 1968. J. Anim. Sci. *27*,1021.
BERGER, T., MAHONE, J. P., SVOBODA, G. S., METZ, K. W., and CLEGG, E. D. 1980. J. Anim. Sci. *51*,672.
BLOCK, J. L., and ERICKSON, B. J. 1968. Anat. Rec. *161*,45.
BLOM, E. 1962. *In* Reproduction in Farm Animals (E. S. E. Hafez, Editor). Lea and Febiger, Philadelphia.
BODA, J. M. 1959. *In* Reproduction in Domestic Animals. (H. H. Cole and P. T. Cupps, Editors). Academic Press, New York.
BRINKLEY, H. J., WICKERSHAM, E. W., FIRST, N. L., and CASIDA, L. E. 1964. Endocrinology *74*,462.
BROOKS, P. H., and COLE, D. J. A. 1970. J. Reprod. Fert. *23*,435.
BURGER, J. F. 1952. Onderstepoort J. Vet. Res. Suppl. *2*,218.
CARMICHAEL, W. J., and RICE, J. B. 1920. Ill. Agric. Exp. Stn. Bull. *226*.
CHRISTENSON, R. K. 1981. J. Anim. Sci. *52*,821.
CHRISTENSON, R. K., and FORD, J. J. 1979. J. Anim. Sci. *49*,743.
CHRISTENSON, R. K., and TEAGUE, H. S. 1975. J. Anim. Sci. *41*,560.
CHRISTENSON, R. K., TEAGUE, H. S. GRIFO, A. P., JR., and ROLLER, W. L. 1972. The effect of high environmental temperature on the boar. Ohio Research Summary *61-4*, p. 19. Dep. Anim. Sci., Ohio State Univ., Columbus.
CORNER, G. W., 1921. Am. J. Anat. *26*,117.
COX, N. M., and BRITT, J. H. 1982. Biol. Reproduction *27*,70.
CRABO, B., and EINARSSON, S. 1971. Acta Vet. Scand. *12*,1.
CRIGHTON, D. B., and LAMMING, G. E., 1969. J. Endocrinology *43*,507.
D'ARCE, R. D., TEAGUE, H. S., ROLLER, W. L., GRIFO, A. P., JR., and PALMER, W. M. 1970. J. Anim. Sci. *30*,374.
DAVIS, D. L., KNIGHT, J. W. KILLIAN, D. B., and DAY, B. N. 1979. J. Anim. Sci. *49*,1506.
DAY, B. M., ANDERSON, L. L., HAZEL, L. N., and MELAMPY, R. M. 1959. J. Anim. Sci. *18*,909.
DHINDSA, D. S., and DZUIK, P. J. 1968A. J. Anim. Sci. *27*,122.
DHINDSA, D. S., and DZUIK, P. J. 1968B. J. Anim. Sci. *27*,668.
DZIUK, P. J. 1968. J. Anim. Sci. *27*,673.
DZUIK, P. J. 1970. J. Reprod. Fertil. *22*,277.
DZIUK, P. J. 1977. Reproduction in pigs. *In* Reproduction in Domestic Animals (H. H. Cole and P. T. Capps, Editors), pp. 455–474. Academic Press, New York.
DZIUK, P. J., and BELLOWS, R. A. 1983. J. Anim. Sci. *57* (Suppl. 2), 355.
DZIUK, P. J. and POLGE, C. 1962. J. Reprod. Fert. *4*,207.
DZIUK, P. J., POLGE, C., and ROWSON, L. E. 1964. J. Anim. Sci. *23*,37.
EDGERTON, L. A., and ERB, R. E. 1971. J. Anim. Sci. *32*,515.

FENTON, F. R., BAZER, F. W., ROBISON, W., and ULBERG, L. C. 1970. J. Anim. Sci. *31*,104.
FIRST, N. L., SHORT, R. E., PETERS, J. B., and STRATMAN, F. W. 1968A. J. Anim. Sci. *27*,1032.
FIRST, N. L., SHORT, R. E., PETERS, J. B., and STRATMAN, F. W. 1968B. J. Anim. Sci. *27*,1037.
FIRST, N. L., STRATMAN, F. W., and CASIDA, L. E. 1963. J. Anim. Sci. *22*,135.
FOLEY, C. W., MARCH, H. M., HEIDENREICH, C. J., GARWOOD, V. A., and ERB, R. E. 1967. J. Anim. Sci. *26*,1072.
FOOTE, R. H., YOUNG, D. C., TURKHEIMER, A. R., and BRATTON, R. W. 1959. *In* Colloquium on Reproduction and A. I. of the Pig, p. 27. Institut National de la Recherche Agronomique, Paris.
GARBERS, D. L., and FIRST, N. L. 1969. J. Reprod. Fert. *20*,465.
GOSSETT, J. W., and SORENSON, A. M. 1959A. J. Anim. Sci. *18*,48.
GOSSETT, J. W., and SORENSON, A. M. 1959B. J. Anim. Sci. *18*,40.
GRAHAM, E. F., RAJAMANNAN, A. H. J., SCHMEHL, M. K. L., MAKI-LAURILA, M., and BOWER, R. E. 1971. A. I. Digest XIX, No. 6, June.
GUTHRIE, H. O. 1979. J. Anim. Sci. *49*,158.
HAFEZ, E. S. E., SUMPTION, L. J., and JAKWAY, J. S. 1963. *In* Behavior of Domestic Animals (E. S. Hafez, Editor), p. 334. Williams and Wilkins, Baltimore.
HANCOCK, J. L., 1959. Anim. Prod. *1*,103.
HANCOCK, J. L. 1960. Stud. Fertil. *9*,146.
HANCOCK, J. L. 1961. J. Reprod. Fertil. *2*,307.
HANCOCK, J. L., and HOVELL, G. J. R. 1962. J. Reprod. Fertil. *4*,195.
HARD, D. L., and ANDERSON, L. L. 1982. Biol. Reprod. *27*,79.
HARDY, B., and LODGE, G. A. 1969. J. Reprod. Fertil. *19*,555.
HAUSLER, C. L., HODSON, H. H., JR., KAO, D. C., KINNEY, T. J., RAUWOLF, V. A., and STRACK, L. E. 1980. J. Anim. Sci. *50*,773.
HODSON, H. H., JR., HAUSLER, L. L., SNYDER, D. H., WILKENS, M. A., and ARTHUR, R. D. 1981. J. Anim. Sci. *52*,688.
HOOD, R. D., FOLEY, C. W., and MARTIN, T. G. 1970. J. Anim. Sci. *30*,91.
HOUPT, K. A., and WOLSKI, I. R. 1982. Domestic Animal Behavior for Veterinarians and Animal Scientists, pp. 1–356. Iowa State Univ. Press, Ames.
HUNTER, R. H. F. 1964. Anim. Prod. *6*,189.
HUNTER, R. H. F. 1966. Anim. Prod. *8*,457.
HUNTER, R. H. F. 1967. J. Reprod. Fertil. *13*,133.
HUNTER, R. H. F., POLGE, C., and ROWSON, L. E. 1967. J. Reprod. Fertil. *14*,501.
ITO, S., KUDO, A., and NIWA, T. 1944. Zootech. Exp. Stn. Res. Bull. *49*.
ITO, S., KUDO, A., and NIWA, T. 1959. *In* Colloquium on Reproduction and A. I. of the Pig, p. 104. Institut National de la Recherche Agronomique, Paris, France.
ITO, S., NIWA, T., KUDO, A., and MIZUHO, A. 1948. Zootech. Exp. Stn. Res. Bull. *55*.
JENSEN, A. H., YEN, J. T., GEHRING, M. M., BAKER, D. H., BECKER, D. E., and HARMON, B. G. 1970. J. Anim. Sci. *31*,745.
JOHNSON, L. A., AALBERS, J. G., WILLEMS, J. T. and SYBESMA, W. 1981. J. Anim. Sci. *52*, 1130.
JOUBERT, D. M., and BONSMA, J. C. 1957. S. Afr. J. Anim. Sci. *53*,340.
KENNELLY, J. J., and FOOTE, R. H. 1964. J. Anim. Sci. *23*,160.
KENNETH, J. H. 1953 Gestation Periods. A table and bibliography. Tech. Communications, The Commonwealth Bureau of Animal Breeding, Genetics No. 5, Farnham House, Farnham Royal, Slough, Bucks., England.
KIRKPATRICK, R. L., LAUDERDALE, J. W., FIRST, N. J., HAUSER, E. R., and CASIDA, L. E. 1965. J. Anim. Sci. *24*,1104.
KVASNICKI, A. V. 1951. Sov. Zootech. *1*,36. Anim. Breed. Abstr. *19*,224.
LASLEY, J. F., and BOGART, R. 1944. Am. J. Physiol. *141*,619.
LASLEY, J. F., DAY, B. N., and MAYER, D. T. 1963. J. Anim. Sci. *22*,422.
LERNER, E. H., MAYER, D. T., and LASLEY, J. F. 1957. Mo. Agric. Exp. Stn. Res. Bull. *629*.
LONGENECKER, D. E., and DAY, B. N. 1968. J. Anim. Sci. *27*,709.
LOVELL, J. E., 1958. Cytomorphosis of porcine spermatozoa as related to artificial insemination. Ph.D. Thesis, Iowa State University. Dissert. Abstr. *19*,8.
MacKENZIE, F. F., MILLER, J. C. and BAUGUESS, L. C. 1938. Mo. Agric. Exp. Stn. Res. Bull. *279*.
MANN, T. 1954. The Biochemistry of Semen. Methuen, London.
MANN, T. 1959. *In* Reproduction in Domestic Animals (H. H. Cole and P. T. Cupps, Editors). Academic Press, New York.

MANN, T., POLGE, C., and ROWSON, L. E. 1956. J. Endocrinology *13*,133.
du MESNIL du BUISSON, F. 1956. Anim. Breed. Abstr. *25*,292.
du MESNIL du BUISSON, F., and DAUZIER, L. 1955. Anim. Breed. Abstr. *24*,302.
MILOVANOV, V. K. 1932. Anim. Breed. Abstr. *1*,112.
MINAR, V. M., and SCHILLING, E. 1976. Dtsch. Tieraerztl. Wochenschr. *77*,428.
NARA, B. S., and FIRST, N. L. 1981. J. Anim. Sci. *52*,1360.
NELLOR, J. W., DANIELS, R. W., HOEFER, J. A., WILDT, D. E., and DUKELOW, W. 1975. Theriogenology *4*,23.
NISWENDER, G. D., DZIUK, P. J., KALTENBACH, C. C., and NORTON, H. W. 1970. J. Anim. Sci. *30*,225.
NOLL, D. C. 1949. Philipp. J. Anim. Ind. *10*,247.
OMTVEDT, I. T., NELSON, R. E., EDWARDS, R. L., STEPHENS, D. F., and TURMAN, E. J. 1971. J. Anim. Sci. *31*,312.
PERRY, J. S. 1954. J. Embryol. Exp. Morphol. *2*,308.
PHILLIPS, R. W., and ANDREWS, F. N. 1936. Mass. Agric. Exp. Stn. Bull. *331*.
PITKJANEN, I. G., and SUBIN, A. A. 1961. Anim. Breed. Abstr. *30*,84.
POLGE, C. 1956A. Proc. 2nd Int. Congr. Anim. Reprod. 3:59.
POLGE, C. 1956B. Vet. Rec. *68*,62.
POLGE, C. 1965. Vet. Rec. *77*, 232.
POLGE, C. 1968. Proc. 6th Int. Congr. Anim. Reprod. (Abstr. 294).
POND, W. G., ROBERTS, S. J., DUNN, J. A., and WILLMAN, J. P. 1960. J. Anim. Sci. *19*, 881.
POND, W. G., HANSEL, W., DUNN, J. A., BRATTON, R. W., and FOOTE, R. J. 1965. J. Anim. Sci., *24*, 536.
PURSEL, V. G. 1983. J. Anim. Sci. *56*,1161.
PURSEL, V. G. 1980. J. Anim. Sci. *51*(Suppl. 1),316.
PURSEL, V. G., and JOHNSON, L. A. 1971. Agric. Res. Service, USDA ARS *44*,227.
RATHNASABAPATHY, V., LASLEY, J. E., and MAYER, D. T. 1956. Mo. Agric. Exp. Stn. Res. Bull. *615*.
REDDY, V. B., LASLEY, J. F., and MAYER, D. T. 1958. Mo. Agric. Stn. Res. Bull. *666*.
REDDY, V. B., MAYER, D. T., and LASLEY, J. F. 1958. Mo. Agric. Stn. Res. Bull. *667*.
ROY, A. 1955. Vet. Rec. *67*,330.
SELF, H. L. 1959. *In* Coloquium on Reproduction and A. I. of the Pig. Institut National de la Recherche Agronomique, p. 121. Paris, France.
SELF, H. L., GRUMMER, R. H., and CASIDA L. E. 1955. J. Anim. Sci. *14*,573.
SISSON, S., and GROSSMAN, V. D. 1938. The Anatomy of Domestic Animals, W. B. Saunders, Philadelphia.
SMIDT, D., STEINBACH, J., and SCHEVEN, B. 1965. J. Reprod. Fert. *10*,153.
SPALDING, J. F., BERRY, R. O., and MOFFIT, J. G. 1955. J. Anim. Sci. *14*,609.
SPIES, J. G., ZIMMERMAN, D. R., SELF, H. L., and CASIDA, L. E. 1959. J. Anim. Sci. *18*,163. 1969. J. Reprod. Fert. *20*,443.
STABENFELDT, G. H., AKINS, E. L., EWING, L. L., and MORRISETTE, M. C. 1969. J. Reprod. Fert. *20*, 443.
STEINBACH, J., and SMIDT, D. 1970. J. Anim. Sci. *30*,573.
STEVERMER, E. J., FIRST, N. L., and HOEKSTRA, W. G. 1964. J. Anim. Sci. *23*,67.
SWIERSTRA, E. E. 1970. *In* Effect of Disease and Stress on Reproductive Efficiency in Swine, p. 8. Proc. Symposium, Extension Service, Univ. of Nebraska, College of Agric., Lincoln, and USDA, Washington, D.C.
TANABE, T. V., WARNICK, A. C., CASIDA, L. E., and GRUMMER, R. H. 1949. J. Anim. Sci. *8*,550.
TEAGUE, H. S. 1970. *In* Effect of Disease and Stress on Reproductive Efficiency in Swine, p. 21. Proc. Symposium, Extension Service, Univ. of Nebraska, College of Agric., Lincoln, and USDA, Washington, D.C.
TILLSON, S. A., and ERB, R. E. 1970. J. Anim. Sci. *31*,1179.
TURKHEIMER, A. R., YOUNG, D. G., and FOOTE, R. H. 1958. Cornell Vet. *58*,291.
VINCENT, C. K., ROBISON, O. W., and ULBERG, L. C. 1964. J. Anim. Sci. *23*,1084.
WALDORF, D. P., FOOTE, W. C., SELF, H. L., CHAPMAN, A. B., and CASIDA, L. E. 1957. J. Anim. Sci. *16*,965.
WARNICK, A. C., WIGGINS, E. L., CASIDA, L. E., GRUMMER, R. H., and CHAPMAN, A. B. 1951. J. Anim. Sci. *10*,479.

WESTMORELAND, N., FIRST, N. L., and HOEKSTRA, W. G. 1967. J. Reprod. Fert. *13*,223.
WETTEMANN, R. P., WELLS, M. E., and JOHNSON, R. K. 1979. J. Anim. Sci. *49*,1501.
WHITE, I. G. 1958. Anim. Breed. Abstr. *26*,109.
WILSON, M. E., EDGERTON, L. A., CROMWELL, G. L., and STAHLY, T. S. 1981. J. Anim. Sci. *52*,323.
YOUNG, D. C. 1958. Sperm production, sperm preservation, heat detection and artificial insemination in swine. M.S. Thesis. Cornell Univ. Ithaca.

9

Lactation

INTRODUCTION

Milk production in the sow is important for the survival of the young under present production practices. It is possible to wean pigs at birth, but sow's milk must still be considered the ideal food for the very young pig. A knowledge of the yield and composition of sow's milk is needed as the basis for making decisions on sow and pig feeding and management.

YIELD

Milk yield is more difficult to measure in the sow than in the cow, since oxytocin must be injected to induce milk letdown when the milk is obtained by hand or by machine (Hartman and Pond 1960; Hughes and Hart 1935). It can be estimated by weighing the pigs immediately before and after nursing, or by machine-milking (see Fig. 9.1), or by hand-milking of the sow at predetermined intervals after oxytocin injection. The peak daily yield is reached during the fourth or fifth week (Hartman and Pond 1960; Hughes and Hart, 1935; Allen and Lasley 1960). By the ninth or tenth week, daily production is negligible. The average daily yield over an 8-week lactation, may range from 5 to 8 kg or more (Smith 1952, 1959; Salmon-Legagneur, 1958; Hartman *et al.* 1962), depending on breed, body size, nutrition, and other management factors. Milk production efficiency in swine (milk energy per unit feed energy) is very high (45%) (Bowland 1966). Environmental temperature, number of young suckled, body fatness of the sow at the beginning of lactation, and level of energy intake during lactation all affect the apparent gross efficiency of milk production.

Lactation curves representative of data reported in the literature are shown in Fig. 9.2. The daily milk yield tends to be directly related to the number of pigs sucking, but inversely related to the yield per pig (Perrin 1955). That is, as litter size increases, total milk yield increases, but not sufficiently to accommodate maximal milk intake by each pig. By about week 4 of lactation, daily pig nutrient

FIG. 9.1. Milking machine for sows; it is useful for studying milk composition. Machinery designed by authors.

requirements begin to exceed the amount supplied by milk so supplemental feed must be made available. The relationship between milk supply and pig nutrient demands during the first 6 weeks of postnatal life is illustrated in Fig. 9.3.

Individual pigs identify themselves with a particular teat during the first few days of life and jealously protect their access to it (Perrin 1955).

In small litters it is common for a pig to claim more than one nipple. On the other hand, in large litters, the weakest pigs may starve because they cannot compete for a nipple. The dam will normally not accept pigs other than her own, unless transfer of orphan pigs is made during the first few days of lactation. However, when two or more sows and litters are penned together it is not uncommon for some pigs to nurse more than one sow. It is advisable to disguise odors and other cues by allowing the adopted pig to mingle with the others before introduction to the foster dam. This procedure allows a means of equalizing the size of litters if litters of disparate size are born during a short time interval.

The normal nursing behavior pattern is for a uniform time interval between each feeding over a 24-hr period. The average nursing interval is less than 1 hr day and night, so that the suckling pig normally receives more than 24 feedings daily. Each feeding consists of only a few minutes.

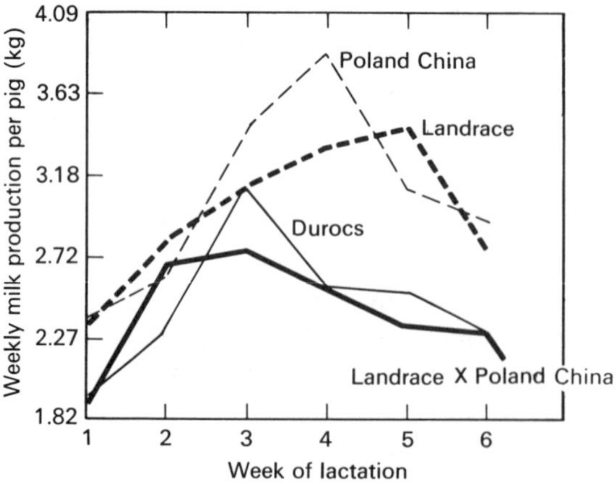

FIG. 9.2. Lactation curves for sows.
From Allen and Lasley 1960.

FIG. 9.3. Milk supply in relation to pig nutrient needs in 6 weeks.

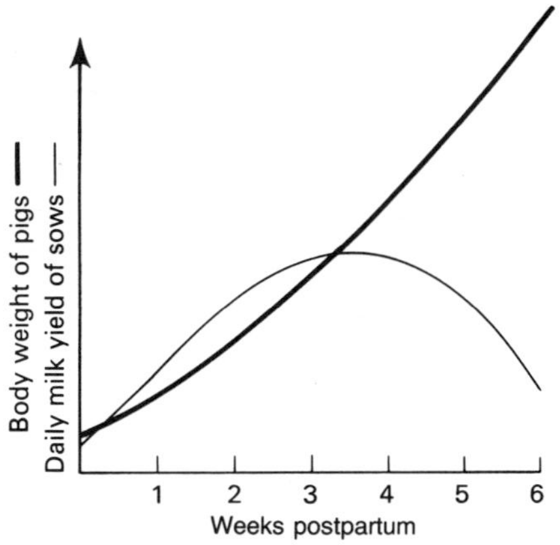

Nursing frequency tends to decrease with the advance of lactation as indicated in Table 9.1, which summarizes the results of a study of the nursing interval using time-lapse cinematography (Hartman *et al.* 1962).

COMPOSITION

The gross composition of sow milk is highly variable. These data are summarized in Table 9.2. The large variation is probably due partly to genetics, stage of lactation, prepartum diet, and other environmental factors.

The composition of colostrum during the first day is considerably different from that of later milk, as shown in Table 9.3 (Perrin 1955). The gradual change to milk occurs in the course of a 2–3-day transition period. Colostrum is higher in percentage of total solids and in protein than milk, but lower in ash, fat, and lactose. Recent data (Klaver *et al.* 1981) obtained with Dutch Landrace sows agrees well with earlier published data on sow milk yield and composition. Milk production at day 5 of lactation averaged 4.55 kg daily; energy concentration was 1.202 kcal/kg and dry matter, ash, protein, fat, and lactose were; 19.2, 0.74, 6.41, 6.88, and 5.16%, respectively. Such values are similar to those reported 15 yr ago for different breeds of swine (Bowland 1966). Sow condition affects milk production in early lactation; a high level of intake by thin sows may reduce fat

TABLE 9.1. Nursing Interval of Litters through 6 Weeks of Lactation

Time	*Week*						*Average*
	1	*2*	*3*	*4*	*5*	*6*	
Day	48[a]	54	51	58	54	56	53
Night	39	46	53	52	56	60	51
Average	44	50	52	55	55	58	52

Source: Hartman et al. 1962.
[a]Minutes from initial phase of one nursing to initial phase of next.

TABLE 9.2. Fat, Protein, Lactose, Vitamin, Calcium, Phosphorus, and Ash Content of Swine Colostrum and Milk

Constituents	Milk		Colostrum
	Unweighted mean	*Range of means*	*Range of means*
Total solids, %	19.4	17.1–25.8	22.0–33.1
Fat, %	7.2	3.5–10.5	2.7–7.7
Protein, %	6.1	4.4–9.7	9.9–22.6
Lactose, %	4.8	2.0–6.0	2.0–7.5
Ash, %	0.96	0.78–1.30	0.59–0.99
Calcium, %	0.21	0.12–0.35	0.05–0.08
Phosphorus, %	0.14	0.10–0.19	0.08–0.11
Vitamin A, μg/100 ml	—	15–255	44–144
Vitamin D, IU/100 ml	10	—	—
Vitamin E, mg/100 ml	0.14	—	—
Vitamin C, mg/100 ml	14.6	11.0–24.6	—
Thiamin, μg/100 ml	65	60–71	56–97
Riboflavin, μg/100 ml	355	137–820	45–650
Pantothenic acid, μg/100 ml	405	190–568	130–680
Niacinamide, μg/100 ml	741	430–903	165–167
Vitamin B_6, μg/100 ml	20.0	—	2.5
Biotin, μg/100 ml	1.4	—	5.3
Vitamin B_{12}, μg/100 ml	0.14	—	0.15
Folic acid, μg/100 ml	390	—	—

Source: Bowland 1966.

content without increasing total milk production (Klaver *et al.* 1981). In normal sows, high feed intake in early lactation increases milk production and tends to prevent excessive lactation weight loss. Considerable evidence (Boyd *et al.* 1978A; Seerley *et al.* 1978A,B) shows increased pig survival by feeding the sow a high fat diet during late gestation and early lactation. This effect appears to be the result of increased milk fat which provides a greater piglet calorie intake during the critical perinatal period (Pettigrew 1981). However, the practicality of using high fat diets for sows is limited by difficult mixing and pelleting and a tendency for the mixed feed to bridge in bulk bins and feeders.

The higher protein content of colostrum is due to its high immune globulin content. The ingestion of colostrum is especially critical to the survival of the newborn pig as this represents the chief source of immunity to disease during early life. The gamma globulin fraction is nearly one-half of total colostral protein at parturition and declines gradually during the first 24 hr and then plateaus (Table 9.4; Bourne *et al.* 1969). Gamma globulins are absorbed intact by the newborn pig but are not absorbed in appreciable proportions after about 24 hr.

Other differences (Salmon-Legagneur 1965) that exist between colostrum and milk are these: (a) Phospholipids, unsaturated fatty acids, vitamins A, C, and E, and thiamin are all higher in colostrum. (b) Nonprotein nitrogen, pantothetic acid, and niacin are lower in colostrum. (c) The amino acid distribution of

TABLE 9.3. Changes in Gross Composition of Swine Colostrum and Milk with Time following Parturition

Constituents	*Time after parturition*								
	Hours								*Weeks*
	0	*3*	*6*	*9*	*12*	*15–24*	*27–48*	*72–120*	*2–8*
Total solids, %	30.2	28.7	26.6	23.6	20.8	19.6	21.2	21.8	21.2
Fat, %	7.2	7.3	7.8	7.8	7.2	7.7	9.5	10.4	9.3
Protein, %	18.9	17.5	15.2	11.7	10.2	7.2	6.9	6.8	6.2
Lactose, %	2.5	2.7	2.9	3.0	3.4	3.7	4.0	4.6	4.8
Ash, %	0.63	0.62	0.62	0.63	0.63	0.66	0.72	0.77	0.95
Calcium, %	0.05	0.04	0.05	0.05	0.06	0.07	0.11	0.16	0.25
Phosphorus, %	0.11	0.11	0.11	0.11	0.11	0.12	0.13	0.14	0.15

Source: Perrin 1955.

colostrum is different from that of milk, although the difference probably is not of practical importance.

The diet fed during gestation and lactation also has an effect on the levels of other nutrients in the milk (Thomas *et al.* 1947; Whiting and Loosli 1948; Rook and Witter 1968; Bowland *et al.* 1949). Diets fortified with large quantities of some of these nutrients increase milk levels by increased transfer to the mammary

TABLE 9.4. Concentrations (g/100 ml) of Certain Proteins in Sow Colostrum and Milk

Hour postpartum	*Total protein*	*Beta globulin*	*Gamma globulin*
0	19.6	3.7	8.6
1	18.8	3.4	8.2
2	15.7	3.0	6.8
3	14.3	2.7	6.0
4	12.7	2.6	5.2
6	10.5	2.1	3.9
8	8.6	1.9	2.9
11	6.1	1.5	2.0
13	5.1	1.2	1.6
16	4.3	1.0	1.2
24	4.1	1.0	0.9
Day postpartum			
2	3.5	0.9	0.7
7	3.3	0.9	0.5
14	3.3	0.9	0.5
21	3.3	0.9	0.5
28	3.3	0.9	0.5

Source: Bourne 1969.

TABLE 9.5. Mineral Content of Swine Milk

Element	*Content*
Calcium	2.13 g/kg
Phosphorus	1.54 g/kg
Potassium	1.00 g/kg
Sodium	0.34 g/kg
Magnesium	0.20 g/kg
Chloride	1.0 g/kg
Iron	1.33 mg/liter
Zinc	4.94 mg/kg
Manganese	0.38 mg/kg (dry basis)

Sources: Perrin et al. 1955; Guegen, Salmon-Legagneur 1959; Pond et al. 1965; Plumlee et al. 1956.

gland. Thus, milk constituents are derived partly from synthesis within the mammary gland (lactose, fatty acids, some of the milk protein fractions) and partly by filtration or active transport from the blood.

There appears to be little variation among glands on the same sow in the composition of the milk produced (Pond *et al.* 1962), with the exception of the percentage of fat, which seems to be more variable. The level of vitamins in milk is related both to dietary intake during lactation, and to the amounts stored in liver and other tissues at the onset of lactation. For example, a large liver storage of vitamin A could temporarily mask the effect of a dietary insufficiency of vitamin A during lactation.

The same applies for some of the inorganic elements, such as, calcium and phosphorus, but milk volume probably suffers before striking changes occur in concentration in the milk.

The content of several mineral elements in swine milk is summarized in Table 9.5 (Guegen and Salmon-Legagneur 1959; Pond *et al.* 1965; Plumlee *et al.* 1956). Most mineral levels of milk appear to be resistant to dietary levels, while zinc (Earle and Stevenson 1965; Pond and Jones 1964) and manganese (Plumlee *et al.* 1956) can be increased in the milk by increasing dietary intake. Iron and copper, unlike zinc and manganese, are not appreciably increased in the milk by feeding (Veum *et al.* 1965) or even by injection (Pond *et al.* 1961). As a result, milk is characteristically deficient in iron and copper regardless of gestation or lactation diet.

The fatty acid composition of the milk is sensitive to changes in the composition of the dietary fat. This relationship is shown in Table 9.6 (de Man and Bowland 1963; Tollerz and Lindberg 1965). The unsaturated fatty acids, linoleic and linolenic, are more closely related to the dietary levels than are the other fatty acids.

Total fat concentration in sow milk is quite resistant to changes in level of dietary fat, although increasing the dietary fat level does increase milk fat level slightly (de Man and Bowland 1963, Salmon-Legagneur 1963). There is no

TABLE 9.6. Fatty Acid Composition of Swine Colostrum and Milk

Fatty acid	*Colostrum*	*Standard milk*	*6.3% cottonseed oil[a] milk*	*6.3% linseed oil[a] milk*
Capric ($C_{10:0}$)	—	0.2	—	—
Lauric ($C_{12:0}$)	—	0.3	—	—
Myristic ($C_{14:0}$)	1.4	3.3	3.2	2.9
Palmitic ($C_{16:0}$)	22.5	30.3	28.3	27.8
Palmitoleic ($C_{16:1}$)	5.0	9.9	6.6	8.1
Stearic ($C_{18:0}$)	5.7	4.0	5.4	3.2
Oleic ($C_{18:1}$)	41.7	35.3	34.3	27.1
Linoleic ($C_{18:2}$)	20.9	13.0	20.5	11.8
Linolenic ($C_{18:3}$)	2.4	2.5	0.9	18.3
Unidentified	0.4	1.2	—	—

Sources: de Man and Bowland 1963; Tollerz and Lindberg 1965.
[a]Cottonseed oil contains 51.9% $C_{18:2}$, 0.8% $C_{18:3}$; linseed oil contains 14.1% $C_{18:2}$, 58.2% $C_{18:3}$.

evidence to indicate the mechanism of milk fat secretion in swine is different from that of other mammals (Keenan *et al.* 1970).

MAMMARY GLANDS

The number of mammary glands varies from six to 20 (3–10 pairs), arranged in two parallel rows, one on each side of the ventral median line and extending from the pectoral to the inguinal region.

There may be an even or an odd number on each side and in total; frequency distribution of even or odd is about 60:40 (Parker and Bullard 1913). Each gland is separate from adjacent glands in secretory tissue.

Most breeds have a mean of approximately 12–14 nipples in females and a slightly higher number in males (Turner 1939). A relatively low heritability coefficient of 0.10–0.20 (Enfield and Rempel 1961, Allen *et al.* 1959) suggests that it is impractical to select for high nipple number. Most breed associations require that individual animals have a minimum of 12 functional nipples for purebred registration.

A normal nipple has two external duct openings. Supernumerary (nonfunctional) nipples are often present between normal ones, usually between the third and fourth pairs. Most of these are underdeveloped, even though they may be in association with a functional gland. Rudimentary nipples also sometimes occur on the posterior end of one or both rows of normal nipples. "Inverted" nipples occur occasionally in which the nipple fails to project out from the surface of the gland, making it impossible to suckle.

The blood supply to the mammary glands comes from two branches of the arterial system; the common carotid artery supplies the anterior glands and a branch of the abdominal aorta supplies the posterior glands. However, there is considerable anastomosis and some of the blood supplying the pectoral (anterior) glands may pass posteriorly to the posterior glands.

The lymphatic system of the mammary glands of swine is more complex than that of other farm animals. The details of structure have been described by Turner (1939).

In animals with red or black skin there is pigmentation within the nipple. Presence of this pigment in the nipples of slaughter animals, has led to the term "seedy belly" in the bacon trade. A red pigment in the nipples of some females of all breeds is found at estrus and is referred to as "red seedy cut" in carcasses from slaughter pigs. The mammary system appears early in prenatal life and, at birth, the nipples are easily visible. By 5 months of age a well-developed duct system is formed, and by puberty (6 months) the extent of development is still more advanced. During early pregnancy there is a further increase in the duct system, and by mid-pregnancy the lobules and alveoli are fully developed. Secretion into the lumen of the ducts and alveoli begins by the 75th day of pregnancy and by 90 days the enlargement of the gland is easily visible. By about 114 days (a few hours before parturition) the glands are full of colostrum and lactation begins when the first pig is born.

REFERENCES

ALLEN, A. D., and LASLEY, J. F. 1960. J. Anim. Sci. *19*,150.

ALLEN, A. D., TRIBBLE, L. F., and LASLEY, J. F. 1959. Mo. Agric. Exp. Stn. Res. Bull. *694*.

BOURNE, F. J. 1969. Anim. Prod. *11*,337.

BOWLAND, J. P. 1966. *In* Swine in Biomedical Research (L. K. Bustad, R. G. McClellan, and M. P. Burns, Editors). Frayn Printing Co., Seattle.

BOWLAND, J. P., GRUMMER, R. H., PHILLIPS, P. H., and BOHSTEDT, G. 1949. J. Anim. Sci. *8*,98.

BOYD, R. D., MOSER, B. D., PEO, E. R., JR., and CUNNINGHAM, P. J. 1978A. J. Anim. Sci. *47*,874.

BOYD, R. D., MOSER, B. D., PEO, E. R., JR., and CUNNINGHAM, P. J. 1978B. J. Anim. Sci. *47*,833.

EARLE, I. P., and STEVENSON, J. W. 1965. J. Anim. Sci. *24*,325.

ENFIELD, F. D. and REMPEL, W. E. 1961. J. Anim. Sci. *20*,876.

GUEGEN, L., and SALMON-LEGAGNEUR, E. 1959. C. R. Seances Acad. Sci. *249*,784.

HARTMAN, D. A., and POND, W. G. 1960. J. Anim. Sci. *19*,780.

HARTMAN, D. A., LUDWICK, T. M., and WILSON, R. F. 1962. J. Anim. Sci. *21*,883.

HUGHES, E. H., and HART, H. G. 1935. J. Nutr. *9*,311.

KEENAN, T. W., KING, J. L. and COLENBRANDER, V. F. 1970. J. Anim. Sci. *30*,806.

KLAVER, J., van KEMPEN, G. J. M., de LANGE, P. G. B., VERSTEGEN, M. W. A., and BOER, H. 1981. J. Anim. Sci. *52*,1091.

de MAN, J. M., and BOWLAND, J. P. 1963. J. Dairy Res. *30*,339.

PARKER, H. J., and BULLARD, C. 1913. Proc. Am. Acad. Arts Sci. *49*,399.

PERRIN, D. R. 1955. J. Dairy Res. *22*,103.

PETTIGREW, J. E., JR. 1981. J. Anim. Sci. *53*,107.

PLUMLEE, M. P., THRASHER, D. M., BEESON, W. M., ANDREWS, F. N., and PARKER, H. E. 1956. J. Anim. Sci. *15*,352.

POND, W. G., and JONES, J. R. 1964. J. Anim. Sci. *23*,1057.

POND, W. G., LOWREY, R. S., MANER, J. H., and LOOSLI, J. K. 1961. J. Anim. Sci. *20*,747.

POND, W. G., VANVLECK, L. D., and HARTMAN, D.A. 1962. J. Anim. Sci. *21*,293.

POND, W. G., VEUM, T. L., and LAZAR, V. A. 1965. J. Anim. Sci. *24*,668.

ROOK, J. A. F., and WITTER, R. C. 1968. Proc. Nutr. Soc. *27*,71.

ROOK, J. A. F., and WITTER, R. C. 1963. C.R. Acad. Sci. (Paris) *256*,1372.

SALMON-LEGAGNEUR, E. 1958. Ann. Zootech. *2*,143.

SALMON-LEGAGNEUR, E. 1963. C.R. Acad. Sci. (Paris) *256*,1372.

SALMON-LEGAGNEUR, E. 1965. Ann. Zootech. *14*,1.

SEERLEY, R. W., GRIFFIN, F. M. and McCAMPBELL, H. C. 1978A. J. Anim. Sci. *46*,1009.

SEERLEY, R. W., MAXWELL, J. S., and McCAMPBELL, H. C. 1978B. J. Anim. Sci. *47*,1114.

SMITH, D. M. 1952. N. Z. J. Sci. Technol. *A34*,65.

SMITH, D. M. 1959. N. Z. J. Agric. Res. *2*,1057.

THOMAS, J. W., LOOSLI, J. K. and WILLMAN, J. P. 1947. J. Anim. Sci. *6*,141.

TOLLERZ, G., and LINDBERG, P. 1965. Acta Vet. Scand. *6*,118.

TURNER, C. W., 1939. The Comparative Anatomy of the Mammary Glands. University Coop. Store, Univ. of Missouri, Columbia.

VEUM, T. L., GALLO, J. T., POND, W. G., VANVLECK, L. D., and LOOSLI, J. K. 1965. J. Anim. Sci. *24*,1169.

WHITING, F., and LOOSLI, J. K. 1948. J. Nutr. *36*,721.

10

Infectious and Noninfectious Diseases

INTRODUCTION

Many infectious, metabolic, and nutritional diseases affect swine. Some of the infectious diseases are of special significance because the organisms are pathogenic to humans. Examples are brucellosis, leptospirosis, erysipelas, and tuberculosis. Several metabolic diseases of swine have parallel etiology or pathogenesis in humans and provide the basis for the use of the pig as a model for study of the corresponding disorder in man. Examples are atherosclerosis, heart disease, and gastric ulcers.

The subject of swine diseases is thoroughly covered in Leman *et al.* (1981) and in the proceedings of a colloquium on infectious diseases of swine (*J. Am. Vet. Med. Assoc.*, Vol. 160, No. 4, Part 2, 1971). (These reports are cited throughout this chapter.) Two symposium proceedings dealing with the effects of disease and stress on reproductive efficiency (Lucas and Wagner 1970) and common diseases of baby pigs (Leman 1970) also contain useful information. Nutritional diseases of swine are described in detail in other sections of this book and by Whitehair and Miller (1981), Pond and Houpt (1978), and the National Research Council (1979). No attempt is made here to describe in detail the etiology, pathogenesis, treatment, or prevention of diseases of swine. The objective is simply to provide an introduction to the nature and scope of the infectious and noninfectious disease problem with swine, upon which a more sophisticated approach to knowledge in the field can be based. Only the most common diseases affecting the commercial production of swine are included at length. Others are merely listed. Major infectious diseases of swine and the causative agents are listed in Table 10.1.

TABLE 10.1. Major Infectious Diseases of Swine

Disease	*Agent*
Bacterial	
Brucellosis	*Brucella suis, B. abortus, B. melitensis*
Bordetellosis (Atrophic rhinitis)	*Bordetella bronchiseptica*
Colibacillosis	*Escherichia coli*
Edema disease (ED) (a form of enterotoxemic colibacillosis)	*Escherichia coli* (hemolytic)
Erysipelas	*Erysipelothrix rhusiopathiae*
Haemophilus infections	
Polyserositis and arthritis (Glasser's disease)	*Haemophilus parasuis*
Pleuropneumonia	*Haemophilus pleuropneumonia, H. parahaemalyticus*
Leptospirosis	*Leptospira pomona* (and other serovars in some countries)
Pasteurellosis (hemorrhagic septicemia)	*Pasteurella multocida*
Salmonellosis	*Salmonella choleraesuis* and *S. typhimurium*
Streptococcal	
Cervical lymphadenitis (jowl abscess)	*Streptococcus beta hemolytic* (Group E *Streptococcus*)
Septicemia, arthritis	Group C (*S. equisimilis*) and Group L (*Streptococcus*)
Arthritis, septicemia, meningitis	*Streptococcus suis* (Group D)
Swine dysentery	*Treponema hyodysenteriae*
Viral	
African swine fever	*Pestes africana suum*
Congenital tremors	Infectious congenital tremors virus
Foot and mouth disease	Foot and mouth disease virus
Hog cholera (swine fever)	*Togaviridae Pestivirus suis* or hog cholera virus
Porcine parvovirus infection	*Parvovirus* species
Porcine rotavirus infection	*Rotavirus* species
Pseudorabies (Aujeszky's disease)	Pseudorabies virus, a herpes virus
Swine influenza	Type A *influenza suis*
Swinepox	Swinepox virus
Transmissible gastroenteritis (TGE)	TGE virus
Mycoplasmal	
Arthritis	*Mycoplasma hyosynoviae*
Polyserositis and arthritis	*Mycoplasma hyorhinis* (see also *Haemophilus parasuis*)

(*continued*)

TABLE 10.1. (*Continued*)

Disease	*Agent*
Mycoplasma pneumonia	*Mycoplasma hyopneumoniae*
Mycotic	
Aspergillosis	*Aspergillus flavus*
Moniliasis	*Candida albicans, Oidium albicans, Monilia albicans*
Ergot	*Claviceps purpurea*
Mold toxicity	*Fusarium (Gibberella) roseum, F. tricinctum, Gibberella zeae*
Ochratoxicosis	*Aspergillus ochraceas* and *Penicillium* species
Miscellaneous	
Atrophic rhinitis	*Bordetella bronchiseptica, Mycoplasma hyorhinis,* noninfectious irritants, nutritional factors
Enterotoxemia (edema disease, gut edema, intestinal edema, gastric edema)	Toxin from *E. coli*
MMA-complex (metritis, mastitis, agalactia)	*Streptococcus* species, *Staphylococcus* species, *Actinomyces bovis, Actinobacillus lignieresi, Corynebacterium pyogenes, Mycobacterium tuberculosis, Spherophorus necrophorus,* possible endocrine factors

Internal parasites	*Parasite*	*Residence of adult*
Ascariasis	*Ascaris suum* (ascarid)	Small intestine
Trichinosis	*Trichinella spiralis*	Small intestine
Whipworm infection	*Trichuris suis*	Cecum and large intestine
Nodular worm infection	*Oesophagostomum* species (*Strongylid nematodes*)	Cecum and large intestine

External parasites	
Sarcoptic mange	*Sarcoptes scabiei* var. *suis*
Lice	*Haematopinus suis*

INFECTIOUS DISEASES

BACTERIAL DISEASES

Brucellosis (agent: *Brucella suis*). There are nine known biotypes of *B. suis*. Swine in the United States are susceptible to biotypes 1 and 3; those in Europe to biotype 2. *Brucella abortus, melitensis, neotomae, ovis,* and *canis* cause brucellosis in other animal species, but only *B. suis* causes systemic infection leading to reproductive failure in swine. The etiology, clinical signs, diagnosis, treatment, and control of brucellosis have been described by Devoe (1981). The main symptom in swine is abortion in pregnant females and intermittent fever in affected animals. However, there are frequently no outward signs of the disease in affected swine. In swine, there is sometimes, but not always, fever associated with the disease, and the presence of the organism in the blood (bacteremia) is intermittent. There is no prevention except by sanitation and avoidance of contact between affected and susceptible animals. Copulation is a common mode of transmission, but entrance of the organism through the alimentary tract is also common. Hemagglutination tests detect affected animals and carriers with fair accuracy, but blood testing should be done on a herd basis rather than on an individual basis owing to the uncertain diagnostic value of a single blood sample. Newer diagnostic methods such as complement-fixation and enzyme-linked immunosorbent tests show promise of greater future application. Safe and reliable vaccines against brucellosis have not been developed. Currently, there is a joint state–federal livestock brucellosis eradication program active in the United States. When herds are found to be infected, (1) the entire herd may be depopulated; (2) breeding stock may be sold for slaughter and weanling pigs blood tested later; (3) only serologic reactors may be removed and the remainder of the herd retested. Entire herd depopulation is the most effective approach. Human exposure to infected animals should be avoided whenever possible. The public health aspects of brucellosis may be important as well as direct losses to swine producers with infected swine.

Bordetellosis (agent: *Bordetella bronchiseptica*). Earlier, a bronchopneumonia caused by *B. bronchiseptica* was recognized as widespread in young pigs in the United States. During the past ten years, this organism has been shown to be the major cause of atrophic rhinitis (AR) in swine. Switzer (1971) has summarized the etiology, pathogenesis, clinical signs, lesions, treatment, and prevention of AR caused by *B. bronchiseptica*. Other microorganisms, including *Pasteurella multocida, Haemophilus* species, and *Pseudomonas aeruginosa,* are often isolated with *B. bronchiseptica* from nasal turbinates of AR-affected swine. Switzer (1971) estimated that 25–40% of U.S. swine have turbinate atrophy and that more than one-third of swine farms have AR present, with 14% of the pigs showing clinical signs.

When virulent *B. bronchiseptica* infects the piglet nasal cavity, organisms attach to the epithelial surface of the nasal cavity, proliferate, and destroy the cells in the region, and metabolic products diffuse into the nasal turbinate tissue and inhibit bone formation. The nasal turbinate bones become distorted and shrunken due to degeneration of osteoblasts (precursors of bone cells, osteocytes). The first clinical signs are sneezing and sniffing followed in about 3 weeks by turbinate atrophy. In severe cases, turbinates may be destroyed completely and nasal bleeding may occur sporadically. There is no cure for the nasal turbinate destruction and facial distortion associated with severe AR. Cats, rodents, and other animals are carriers of *B. bronchiseptica*. Noninfectious nasal irritants such as dilute acetic acid infused into the nasal cavity can produce turbinate atrophy in young swine free of *B. bronchiseptica* (Logomarsino *et al.* 1974). High levels of atmospheric ammonia in poorly ventilated swine buildings may aggravate turbinate atrophy in pigs with *B. bronchiseptica* AR (Curtis 1981).

Diagnosis of AR is made by postmortem examination of a cross section of the nasal cavity. *Bordetella bronchiseptica* infection is identified in the live animal by nasal culture. Antibodies induced by *B. bronchiseptica* have been detected by agglutination testing, but there is no routine serologic diagnostic test available.

Treatment is by sulfamethazine medication in the feed or sodium sulfathiazole medication in the water. Resistant strains develop against sulfa treatment, reducing the long-term effectiveness of this therapy.

An inactivated *B. bronchiseptica* bacterin has been developed for use in control of turbinate atrophy. Two doses of bacterin to the sow in late gestation yield high circulating antibody titers and high colostral antibody titers in most sows. This passive immunity provided the newborn piglet by colostrum ingestion protects for the first few weeks of life. A dose of bacterin to the piglet at 7 and again at 28 days of age affords active protection. Administration of bacterin does not protect exposed swine from *B. bronchiseptica* infection but may reduce the number of organisms that colonize the epithelium.

Vaccination with bacterin as a means of controlling *B. bronchiseptica* AR is a useful tool, but cannot be expected to eradicate the disease from a herd.

An often-used method for breaking the chain of infection is that used in specific pathogen-free (SPF) pig production in which pregnant sows are hysterectomized at term and the piglets are raised in isolation from infected swine and precautions are taken to minimize the risk of exposure to *B. bronchiseptica*.

Enteric colibacillosis (agent: *Escherichia coli*). This disease involves severe diarrhea and dehydration and is responsible for about one-half of cases of neonatal diarrhea in pigs. Other causes (Wilson 1981) include coccidiosis, rotavirus, transmissible gastroenteritis (TGE), enteroviruses, and salmonellosis. Clinical signs are white-gray feces, diarrhea, and dehydration (Wilson 1981); it often affects pigs during the first few days of life. When it affects lactating sows and their suckling pigs, it quickly spreads to several pigs in a litter. Its severity and rapidity of onset sometimes make it appear similar to TGE, but it can be differentiated by the fact that the disease responds favorably to administration of antibiotics. Oral treatment of lactating sows with antibiotics is generally impractical because of the difficulty of repeated administration, but younger pigs can be treated orally or by intramuscular injection. Some antibiotics are not effective in controlling colibacillosis, but dihydrostreptomycin as well as furazolidone and other nitrofurans are effective. Since spread of the disease is often by ingestion of contaminated feces, good sanitation is needed to control widespread outbreaks.

Milk from sows vaccinated with a live, formalin-treated vaccine affords protection against experimentally induced *E. coli* enteritis. These immunoglobulins are more effective against *E. coli* if ingested by the baby pig than if they function merely as circulating antibodies in the animal (Wilson 1971).

Kohler's (1971) description of pathogenesis is summarized as follows: Pathogenic strains of *E. coli* proliferate in the upper intestinal lumen; enterotoxins are produced and released; and the enterotoxins affect the cells of the intestinal mucosa causing a net loss of fluids and electrolytes into the lumen of the intestine, diarrhea, dehydration, acidosis, inanition, and failure of normal defense mechanisms to control microbial invasion.

Erysipelas (agent: *Erysipelothrix rhusiopathiae*). It has been estimated (Wood 1981) that 30–50% of healthy swine harbor this organism in their lymphoid tissues. The disease can occur in the acute form in which the clinical signs include anorexia, apathy, and high fever (40°–41°C, rectal temperature). These signs may persist for several days, but can be controlled effectively by treatment with antibiotics. After several days in the acute stage, large diamond-shaped red blotches may appear on some animals, accounting for the name "diamond skin disease," sometimes given to erysipelas. The acute form may be followed by chronic erysipelas involving inflammation and swelling of the joints and stiffness and soreness on moving. This disease, like brucellosis and leptospirosis, can be spread to humans. The disease can be treated and cured with antibiotics. Immunization against erysipelas is possible, and if adult animals are immunized twice a year, the degree of protection is high (Schuman and Wood 1970, Ose 1971).

Leptospirosis (agent: *Leptospira pomona* and other serovars or species). Leptospirosis has an incidence of as much as 20% or more in some areas of the United States (Hanson and Tripathy 1981) and is of worldwide significance. It is similar to brucellosis in clinical signs produced, and it can infect humans. Several serovars in addition to *L. pomona* have been isolated from swine, all of which produce a similar disease pattern (Hanson and Tripathy 1981). The most common manifestation in breeding animals is abortion, which usually occurs later in gestation than with brucellosis (Ferguson 1970). Often incidence of stillbirths and weak pigs is higher than normal (Huhn 1971). Only a small percentage of infected swine show clinical signs. Transmission is by coitus, by ingestion of contaminated feed and water supplied with excreta, by eating of infected rodents or their tissues, or by entrance of the organism through skin abrasions or conjunctiva. Many species of wildlife including deer, raccoons, opossums, skunks, woodchucks, foxes, and voles may be infected. Serum agglutination tests are used for diagnosis. Commercial bacterins containing the known serovars of *Leptospira* are used commonly for control. Since immunity lasts less than 1 yr, breeding stock should be vaccinated at each breeding period.

Pasteurellosis (hemorrhagic septicemia) (agent: *Pasteurella multocida*). This disease is rarely primary but usually occurs as a secondary infection following debilitation from mycoplasma pneumonia or from stress such as that occurring during shipment of animals from one place to another. The clinical signs are respiratory inflammation, dypsnea, anorexia, fever, apathy, and often a purulent nasal discharge (Carter 1970; Farrington 1981). In advanced cases, the septicemia is associated with extreme debilitation, incoordination, and death. There is no satisfactory treatment, but oral administration of sulfa drugs reduces death losses. Antibiotics found by *in vitro* sensitivity discs to be effective in a specific situation may alleviate clinical signs when given parenterally. Administration of antiserum may give temporary passive immunity. However, this procedure is not often used in normal management.

Salmonellosis (agents: *Salmonella choleraesuis* and *S. typhimurium*). Current knowledge of salmonellosis has been summarized by Wilcock (1981). The disease affects pigs of all ages, especially young growing swine, and occurs in humans as a result of food-borne contamination with the organism. It involves enteritis accompanied by anorexia and high fever and may also afflict the lungs. The organism has been isolated from numerous birds and animals, including swine and humans. The specific lesion in the intestinal tract is greatest in the duodenum and ileum and is sometimes referred to as necrotic enteritis. The lesion resembles that produced by dietary niacin deficiency. However, niacin supplementation of the diet does not alleviate the enteritis caused by *S. choleraesuis*. Oral administration of broad spectrum antibiotics and of furazolidone is effective in treatment and control, but as with all infectious diseases, prevention can often be attained by good sanitation. Animal feeds may be contaminated with *Salmonella*, especially those containing packing house by-products. As the awareness of such sources of infection grows, control of Salmonellosis in swine and other animals should be more effective.

Clostridial enteritis (hemorrhagic enteritis, infectious necrotic enteritis) (agent: *Clostridium perfringens* Type C). The disease is a hemorrhagic disease which is highly fatal in pigs less than 1-week old. Current knowledge of clostridial infections including clostridial enteritis has been summarized by Bergeland (1981). It is seen in young pigs in many parts of the world. It is characterized by bloody diarrhea and necrosis of the mucosa of the jejunum and ileum. The disease can be induced in colostrum-deprived pigs by oral administration of whole broth cultures or bacteria-free toxin. The pathogenesis is poorly understood. Such factors as stomach pH, trypsin inhibitors in colostrum, and injury or irritation of the intestinal mucosa may affect susceptibility. Colostrum-deprived pigs are more susceptible than colostrum-fed pigs and a high stomach pH predisposes a subject to the disease.

Immunization of young pigs with a vaccine or of the sow before farrowing may reduce the incidence. Antibiotics, arsenicals, and nitrofurans may reduce the severity, but the most effective control to date is sanitation and good management.

Streptococcal infections (agent: *Streptococcus beta hemolytic, S. equisimilis, S. zooepidemicus* and other *Streptococcus* species) The subject of diseases caused by *Streptococcus* species is covered in detail by Ross (1981B). The following are some of the important streptococcal diseases affecting swine.

Cervical abscess (jowl abscess, cervical lymphadenitis). The agent is a Group E streptococcus. Cervical abscesses occur mostly in the lymph nodes of the head and neck but other lymph nodes may be affected as well (Armstrong 1971). The abscesses are often found in the ventral areas of the body, especially in the lower ham, belly, and lower jowl. They can be single or multiple in an individual animal and may vary in size from a few millimeters to several centimeters in diameter. The incidence seems to increase gradually within a herd, starting with one or two affected sows or growing pigs and spreading in the course of a period of several months until one-half or more of the herd show visible superficial abscesses. The location can also be deep within the skeletal muscle so that it is not visible on the live animal. Presence of these deep-seated abscesses in the ham and other valuable parts of the carcass results in large losses in salable meat due to condemnation of the affected part at slaughter. In 1978, for example, more than 1.5 million carcasses in the United States were either condemned or passed for food after removal of affected parts (Ross 1981); this represents more than 2% of hogs slaughtered under federal inspection. Whereas the death losses are small, the losses in efficiency of feed conversion and growth rate in severely affected animals, although not known, are probably substantial. The incidence in a herd can be reduced and eliminated by continuous dietary supplementation with a large quantity of broad spectrum antibiotics in the feed. The invasion of *Streptococcus* species through superficial skin wounds can be minimized by eliminating mechanical hazards such as sharp protrusions from feeders and pen partitions.

A modified live-culture vaccine has been shown to be effective in preventing cervical abscesses.

Streptococcal arthritis. Invasion of *Streptococcus equisimilis* and other Group C streptococci into the joints of young or adult pigs is a frequent cause of lameness. Antibiotics such as penicillin and tetracyclines are effective in treatment. Vaccination of sows or of young pigs with *S. equisimilis* bacterin appears to offer some protection (Ross 1981).

Streptococcal invasion of central nervous system. Occasionally, the cerebrum and meninges are overwhelmed by *Streptococcus suis* (Group D). This results in encephalitis, fever, and anorexia, followed by convulsions and death within 2–3 days.

Streptococcal septicemia. Invasion of *Streptococcus suis* (Group D) to produce septicemia results in high fever, numerous cutaneous petechial hemorrhages, and often endocarditis followed by sudden death due to thrombus formation. Death losses may constitute one-half or more of an affected herd. Antibiotics or sulfa drugs are effective if used continuously for several days to avoid recurrence.

Swine dysentery (vibrionic dysentery, bloody scours, bloody dysentery; hemorrhagic dysentery, black scours, mucohemorrhagic diarrhea) (agent: *Treponema hyodysenteriae*). Current knowledge of the disease has been summarized by Harris and Glock (1981). Swine dysentery usually affects weanling pigs and occurs in most swine-producing areas of the world. The clinical signs are severe diarrhea, dehydration, and often, but not always, anorexia. The feces are sometimes black, owing to the presence of blood. The lesion is present mainly in the large intestine and the organism is found only in the cecum, colon, and rectum. The incubation period is about 2 weeks and death losses may run as high as 25–30% in growing pigs. The first indication of the presence of the disease in a herd may be the sudden death of one or more animals with no previous symptoms. Growth rate of survivors may be reduced considerably. There is often no fever associated with the disease, although intermittent fever may occur. Oral administration of organic arsenicals such as arsanilic acid or of streptomycin and bacitracin has been partially effective in curing the disease. There is apparently very little immunity acquired by infected animals, as the signs often recur within a few weeks or months. There is no reliable method available to identify carriers among previously affected swine.

Polyserositis and arthritis (Glasser's disease) (agent: *Haemophilus parasuis*). This syndrome usually affects young pigs postweaning; it is found worldwide and appears to be more prevalent in herds with specific-pathogen-free (SPF) swine. The etiology, clinical signs, and control of polyserositis and arthritis have been reviewed by Nicolet and Scholl (1981). Onset is sudden; a high fever (105°–108°F, 40.5°–42°C) is accompanied by apathy, anorexia, and generalized signs of discomfort. Involvement of the central nervous system, with muscle tremors and incoordination, is usually present. Mortality may reach 50% if therapeutic doses of antibiotics are not given early. A large array of antibiotics is effective against *H. parasuis*. The syndrome can be prevented by subcutaneous vaccination with killed *H. parasuis* vaccine.

Pleuropneumonia (agent: *Haemophilus pleuropneumonia; H. parahaemalyticus*). This respiratory disease is widely distributed throughout the world. It is spread primarily by direct contact of infected and susceptible animals. Transmission by small rodents or birds probably does not occur and *H. pleuropneumonia* organisms have never been isolated from the nose and throat of a human working with infected swine. Transmission of the disease via clothes or boots contaminated with mucus from affected animals is possible (Nicolet and Scholl 1981). Clinical signs include a sudden fever, apathy, anorexia, diarrhea, respiratory distress, and circulatory failure. Mortality is variable and has been reported as high as 100%, although is often less than 1% in affected animals. The acute phase is followed by a chronic cough with no fever and by a reduced appetite and growth rate. Treatment with antibiotics is effective if started early, as in the case of polyserositis and arthritis. Vaccination with a killed vaccine is effective in protecting susceptible animals and animals exposed to the disease. Nicolet and Scholl (1981) have suggested that eradication of pleuropneumonia from swine herds might be feasible by destroying breeding stock with high antibody titers and repopulating with seronegative pigs.

VIRAL DISEASES

African swine fever (ASF) (agent: *Pestes africana suum*). This disease is almost identical to hog cholera in clinical signs. Death losses are almost 100% in affected animals, and it is extremely contagious. The history, geographical distribution, etiology, diagnosis, treatment, and control of ASF are described in detail by McDaniel (1981). African swine fever virus is immunologically distinct from hog cholera virus so that animals immune to hog cholera are susceptible to ASF. There is, as yet, no effective vaccine against ASF. The disease is widely distributed in Africa and southern Europe and was reported for the first time in the Western Hemisphere in 1971 in Cuba and in 1978 in as widely scattered areas as Brazil, Dominican Republic, Malta, and Sardinia. It is believed that garbage containing uncooked pork was the means of transport (McDaniel 1981). The danger of spread of ASF to other parts of the world remains a serious threat, owing to its virulent nature and to the lack of a satisfactory vaccine against it. Countries free of ASF have strict importation restrictions on animals from countries in which ASF is present, just as is the case for hog cholera. With increasing human international travel it seems a wonder that the disease has been kept from the North American continent, Asia, and Australia. McDaniel (1981) has cited data illustrating the devastating financial effect its entry into the United States would have on the swine industry and on consumers.

Foot and mouth disease (FMD; aftosa) (agent: FMD virus; aftosa virus). This is a serious disease widespread in geographical distribution. Its etiology, transmission, diagnosis, and control have been described by Callis and McKercher (1981). Losses from FMD (often called aftosa in many countries) are sufficiently great that countries free of the disease enforce strict quarantine and importation restriction on animals coming from areas in which the disease is known to exist. For example, the United States has sophisticated control programs on importations of animals and fresh meat specifically to prevent the entrance of FMD into the country. The virus can occur in milk, in semen, and in meat and meat by-products such as hides and glands; it is inactivated at 148°C for 2.5 sec. Briefly, the symptoms are sores and erosions in and around the mouth and on the hooves and mammary glands accompanied by fever, depression, and lameness. The lesions are almost identical to those of vesicular exanthema and vesicular stomatitis. Since the disease also affects cattle and is contagious between cattle and swine, the problem of control through isolation of susceptible swine is increased.

There is no good treatment of infected animals. To control the spread of FMD where outbreaks occur it is customary to destroy all affected animals and burn or bury them. The disease can affect animals of all ages.

Vaccines against FMD have been developed (Bachrach and McKercher 1971) which are only moderately effective. A new vaccine against FMD has just been developed by Dr. H. Bachrach and other U.S. Department of Agriculture scientists who have used gene-splicing techniques that allow mass production of a safe and effective supply of the vaccine. This scientific breakthrough is a major contribution toward infectious disease control in animals and humans.

Since the FMD virus can be transported on contaminated boots of attendants or on wheels of vehicles, it is important to maintain high standards of sanitation and to restrict both human and vehicular traffic in areas where swine are kept.

Hog cholera (swine fever, Schweinepest) (agent: *Togaviridae Pestivirus suis* or hog cholera virus). Hog cholera was for many years one of the most serious diseases in North America in terms of annual economic losses to the industry. This is primarily because the disease is nearly 100% fatal in growing pigs, and its virulence, resistance to antibiotics and other drugs, and highly contagious nature make it difficult to control. An organized, federally supported eradication program was successful in eradicating hog cholera from the United States in 1978. Other countries free of this disease include Great Britain, Germany, Canada, the Scandinavian countries, Australia, and New Zealand (Stewart 1981).

The incubation period is approximately 1 week, and the onset of clinical signs is sudden and violent. The first signs are high fever, apathy, and anorexia. Blotchy reddening of the skin, especially on the ears and neck, often occurs, as do severe diarrhea, ataxia and anxiety, and finally convulsions and death in a few days (Stewart 1981).

The only control is by strict isolation of susceptible animals from infected animals or by immunization of susceptible animals. Because the virus, like that of TGE and FMD, can be transported mechanically on boots or wheels of vehicles, strict sanitation and control of human traffic as well as of movement of pigs are essential to prevention. For many years, measures used in the United States included simultaneous immunization of growing pigs with serum containing antibodies against hog cholera collected from affected animals (passive immunity) and with modified live virus or killed virus (active immunity). The serum gave immediate but temporary protection (3 or 4 weeks), while the animal began to produce its own antibodies in response to the challenge from the administered virus. The production of antibodies in this manner is theoretically adequate, after 3 or 4 weeks, to provide protection for many months. However, the use of hog cholera immunization (except for serum) is prohibited in the United States.

Young pigs are provided with passive immunity to hog cholera by ingestion of colostrum. The duration of colostral immunity is directly related to the antibody titer of the plasma of the dam which is, in turn, proportional to that of the colostrum. Therefore, suckling pigs from dams vaccinated against hog cholera during pregnancy have a much higher antibody titer than do pigs from unvaccinated sows, and the duration of their passive immunity is proportionally greater. Although passive immunity is high, active antibody protection in response to vaccination, in countries where vaccination is permitted, must be postponed until 10 or 12 weeks of age to induce active immunity in pigs that have a high antibody titer resulting from passive immunity transferred from their dams.

If the control program is based on the use of hog cholera vaccination, there is always a chance of some animals responding to the challenge by becoming affected with a full-blown case of hog cholera, thus acting as a reservoir for further spread of the disease. These ''breaks'' were of serious enough consequence in the long-term federal program of hog cholera eradication in the United States that the use of either modified (attenuated) live virus or killed virus was prohibited. The program was based on the assumption that if cases of hog cholera were identified and affected animals destroyed, the spread of the virus could be eliminated. The experience gained in this successful eradication program should be useful in application to control of other diseases.

Several major pork-producing countries of the world have been free of hog cholera for many years. This has been possible by maintaining strict quarantine practices, and in some countries, by completely eliminating import of animals from countries where the disease is present.

Swine influenza (agent: *Hemophilis influenza suis*). This disease resembles the human common cold and was thought, for a time, to have originated in man and spread to the pig. There is no evidence that the virus as it now exists is highly contagious from humans to pigs or vice versa, although interest in this possibility was reviewed in 1976 when swine influenza virus was isolated from humans in the United States, and it was shown that humans in contact with infected swine could develop acute respiratory disease as a result. However, there appears to be no serious threat to humans exposed to swine infected with swine influenza. Its etiology, diagnosis, and prevention have been described by Easterday (1981). The symptoms include fever, anorexia, prostration, and a typical abdominal breathing; recovery is rapid. The disease is associated with low mortality; it affects baby pigs most severely but is most frequent in older pigs, of from

46 to 60 kg (Easterday 1971); the main effect is to reduce growth and feed consumption of the affected animal during the 2- or 3-week course of the disease. Maternal antibodies provide some passive immunity, so that the course of the disease may be modified but prevention is not complete (Easterday 1971). There is no known treatment for the disease; antibiotics, sulfa drugs, and other drugs have no effect except in controlling secondary bacterial infections. Prevention and treatment lie mainly in good sanitation and freedom from stress. Affected animals develop temporary immunity to subsequent influenza.

Swinepox (agent: swinepox virus and vaccinia virus). This disease causes skin eruptions that persist on the entire body for 1 or 2 weeks and then disappear, usually without complications (Kasza 1981). The etiology of swinepox caused by swinepox virus or by vaccinia virus (VV) is similar. Vaccinia virus is a laboratory strain and in humans is used for immunization against smallpox (Kasza 1981). Animals with swinepox may show reduced appetite and growth reduction for a few days, but death is rare unless septicemia or other complications develop. As with chicken pox in humans, a high level of permanent immunity usually follows the disease.

Transmissible gastroenteritis (TGE) (agent: coronavirus species; TGE virus). Transmissible gastroenteritis is an extremely serious disease, causing severe diarrhea and dehydration, fever, anorexia, and in a large proportion of cases death (Doyle and Hutchings 1946; Bohl 1981; Haelterman 1971). In suckling pigs, the mortality is often 100% and in older pigs and mature sows, although the mortality rate is less, the disease causes diarrhea, dehydration, and reduced performance. In young pigs, death occurs within a few hours or days after the onset of the disease, and in many survivors, subsequent growth rate and productivity are retarded. The incubation period is 24–48 hr. Haelterman's (1971) description of the pathogenesis is summarized as follows: Ingested virus infects the epithelial cell of the small intestine; infected cells are sloughed into the lumen; and after a few hours most of the susceptible cells in the small intestine are destroyed leading to "villous atrophy" in which villi of each epithelial cell are small and distorted. The destruction of these cells leads to reduced absorption of nutrients and loss of fluids and electrolytes from the body.

Antibiotics and other drugs are ineffective in treatment and prevention. The only treatment available is to alleviate dehydration, acidosis, and starvation (Bohl 1981). Therefore, the best approach is prevention through good sanitation and management. Because the organism is carried mechanically on boots and vehicle wheels contaminated with excreta of affected animals, strict isolation of susceptible animals from carrier sources is essential. A good management practice is to place footbaths containing a cresol or other strong disinfectant at the entrance to buildings that house susceptible animals.

Survivors develop long-term immunity. There appears to be active antibody production, as well as passive maternal antibody protection (Bohl 1981). This has been the basis for development of vaccines for use against the disease. Considerable work is currently underway in veterinary research to develop safer and more effective immunization procedures against TGE. It is of interest in retrospect that perhaps one of the first oral vaccines in practical use was used empirically by swine producers who found that a degree of immunity against TGE in newborn pigs could be attained by feeding to pregnant sows the ground carcasses of pigs that had died of TGE. Antibodies against TGE are produced by the sow and passed *in utero* to the fetus and into the milk to the newborn pig so that the young pig is passively immune when challenged by the live virus in postnatal life. Sows that recover from TGE may remain immune for several months and pass the immunity on to their suckling pigs. Active and passive immunity to TGE and current immunization procedures against TGE have been reviewed by Bohl (1981). Although much is known about control of TGE, many unanswered questions remain about the immunological control of this dreaded swine disease.

Pseudorabies (Aujeszky's disease) (agent: pseudorabies virus, a herpes virus). Pseudorabies virus (PrV) is widespread in the United States and throughout most of the world. Gustafson (1981) reviewed the etiology, epidemiology, clinical signs, pathogenesis, diagnosis, and control of pseudorabies. The disease may have been present in the United States as early as 1813, but only during the past 10–15 years has its importance been generally recognized in the United States and in Western and Eastern Europe. Pseudorabies is caused by a member of the herpes virus groups and, unlike most herpes viruses whose host range is limited to a few species, PrV affects most mammals and many birds. However, swine are the major reservoir for the virus, as most other animals, including cattle, sheep, dogs, cats, and mice, are killed by the infection. Mortality in swine due to PrV infection is highest in baby pigs, but usually is nil in mature swine. The incubation period for pseudorabies is about 30 hr; the first signs in susceptible growing pigs are sneezing, coughing, and slight fever, followed by excessive salivation, spontaneous muscle spasms, convulsions and paddling, and finally death after 4–9 days. Pigs less severely affected may recover after a few days of fever and lassitude. Pseudorabies virus crosses the placenta; if infection occurs prior to 30 days of gestation, embryos are resorbed, infection after 30 days may result in death of some or all fetuses or birth of stillborn or weak infected pigs.

A large array of diagnostic procedures is available. Such procedures include demonstration of the virus in tissues (fresh or frozen) and several serum tests: virus neutralization, immunofluorescence, enzyme-linked immunosorbent assay, agar gel immunodiffusion, complement fixation, and indirect hemagglutination. In addition, a skin test and biological tests with other animals (rabbits, mice, chick embryos) injected with pig tissue suspensions are sometimes used.

Treatment of swine showing clinical signs of pseudorabies is unavailable, but intraperitoneal injection of anti-PrV-serum within 12 hr after exposure may offer protection in weanling pigs.

Prevention of pseudorabies is enhanced by careful control of the human and vehicular traffic and of feral animal traffic on swine premises. A live attenuated vaccine is available commercially. However, vaccinated animals may shed virulent virus if exposed to infection and may harbor a latent infection which may be shed later as a virulent virus. Thus, vaccination programs do not ensure herd eradication of pseudorabies and, in fact, as in the case of hog cholera in the United States before its eradication, vaccination with attenuated, live virus may result in persistence of occasional outbreaks of the disease.

The high level of concern in the United States about the escalation of the incidence of pseudorabies in swine herds has led the USDA to consider plans to attempt an eradication program. The future of such a program is uncertain due to the high cost and controversial probability of success.

Porcine parvovirus infection (agent: *Parvovirus* species). This virus causes reproductive failure in swine exposed during the first one-half of gestation. Fetuses are infected transplacentally, and embryonic and fetal death occurs without any clinical signs in the dam. The virus is ubiquitous in swine. Apparently, the only clinical sign of parvovirus infection is reproductive failure in females; the virus is found in many tissues and organs of apparently healthy growing swine. There is no treatment of infected pregnant females; a vaccine is available for use in gilts before conception to protect against the virus. Mengeling (1981) has reviewed the etiology, clinical signs, pathogenesis, and control of parvovirus infection.

Porcine rotavirus infection (agent: *Rotavirus* species). The significance of rotaviruses in swine is not fully clarified. Woods and Bohl (1981) reviewed current knowledge of the disease. Diarrhea is the chief clinical sign; the signs resemble those of transmissible gastroenteritis. Mortality is highest in colostrum-deprived pigs and may reach 100% when infected at 0–5 days of age; it is usually less than 10% in weaned pigs. Most, if not all adult swine, have serum antibody titers to rotavirus and pass immunity to newborn in colostrum. Rotavirus diarrhea appears to be more severe in the presence of other pathogens, especially *E. coli,* and with sudden adverse changes in environment, including ambient temperature and crowding and change to solid feed. There is no effective vaccine for control of porcine rotavirus infection. The best methods for prevention and control are enhancement of the immune status of the dam by exposure to natural infections and, as with any infectious disease, avoidance of contact of susceptible piglets to high doses of infection.

Congenital tremors (agent: infectious congenital tremors virus). The condition, characterized by shaking and tremors, occurs on newborn pigs and may persist for weeks, months, or years. Several factors have been suggested as causative agents for congenital tremors, including heredity, viral infection, nutritional deficiency, hyperthyroidism, muscle fiber abnormality, chilling, and nonrelated disease stress (Gustafson 1981).

The basis for an infectious origin of congenital tremors is the identification by electron microscopy of a virus in affected animals. Gustafson (1981) has summarized the knowledge of infectious congenital tremors. The importance of other infectious and noninfectious causative factors in contributing to this abnormality in baby pigs is unknown.

MYCOPLASMAL DISEASES

The organisms of this group are smaller than bacteria and almost devoid of a cell wall. Mycoplasmas, like viruses, are rather resistant to antibiotics and other chemotherapeutic agents. They survive only a short time in adverse environments and most members of the species show poor antigenicity.

The mycoplasmas that are pathogenic to swine are described in detail by Ross (1981A). Some of the mycoplasmal diseases are briefly described here.

Arthritis (agent: *Mycoplasma hyosynoviae*). The nasal passage harbors this organism but clinical signs of disease in the nasal passage caused by this organism have not been noted. The arthritis produced occurs in most pigs during the post-weaning stage of growth and is not necessarily associated with mild fever. It is responsive to antibiotics given parenterally and to cortisone. There is no immunization available, but protective immunity results from infection.

Polyserositis and arthritis (agent: *Mycoplasma hyorhinis*). The organism is commonly present in the nasal passage of swine. Most infected pigs show no evidence of clinical disease. Occasionally in colostrum-deprived pigs and in swine under some form of stress, such as early weaning or unusual environmental exposure, a small percentage develops polyserositis. This disease is characterized by sudden high fever, anorexia, and death within 24 hr after onset in a high proportion of affected pigs. Autopsy reveals the presence of large amounts of fibrin strands within the pleural and peritoneal cavities. In survivors, the acute phase abates in 10–14 days, and lameness and swollen joints, which may persist for several weeks or months, develop. There is evidence that other factors such as pneumonia or stress may act synergistically with *M. hyorhinis* in precipitating the disease syndrome. The clinical signs appear so suddenly and the course of the disease is so rapid that alert husbandry and keen observation are important to minimize losses. Parenteral administration of antibiotics usually is beneficial. The incidence of the disease seems to be higher in primary specific-pathogen-free (SPF) pigs, presumably owing to the absence of immune serum gamma globulins as a result of colostrum deprivation. Good sanitation and avoidance of environmental stress are recommended to minimize the incidence.

Mycoplasma pneumonia or enzootic pneumonia (agent:*Mycoplasma hyopneumoniae* or *M. suipneumoniae*). This disease was formerly believed to be caused by an unidentified airborne virus and thus was given the name virus pneumonia of pigs (VPP). This is a misnomer since the causative agent is a mycoplasma. It has been isolated and characterized (Friis 1969). It is worldwide in distribution and may be the most important single disease of pigs in terms of economic losses (Ross 1981). The pig is the only animal affected. Etiology, epidemiology, clinical signs, pathogenesis, treatment, and prevention of mycoplasma pneumonia have been reviewed in detail by Ross (1981). The most typical clinical sign is a chronic, deep-seated cough that may persist throughout the growing-finishing period. This disease is believed to be spread to suckling pigs from their dam. Mature animals usually do not show clinical signs but remain carriers. Certain diagnosis is verified by postmortem examination of the lungs. The lesion is a distinctive purple, plum-colored discoloration occurring in one or more lobes, usually the cardiac lobe first. The affected area is hard and nonpliable and its pulmonary function is impaired or destroyed. Usually no fever accompanies the cough, except at the onset. Mortality rate is low in uncomplicated cases. The main adverse effect is a reduction in appetite, weight gain, and efficiency of feed utilization.

Since the organism is airborne and apparently viable only on immediate transfer from affected to susceptible animals, control of the disease is based on the isolation of susceptible animals from those affected. Efforts are underway in veterinary research to develop serological tests capable of identifying carrier animals (Ross 1981).

A system of performing the hysterectomy of affected sows and raising their pigs in isolation under sterile conditions was employed by Young and co-workers (Young and Underdahl 1953; Young *et al.* 1955) to produce mycoplasma pneumonia-free pigs experimentally. (Atrophic rhinitis, AR, control was also advocated by this procedure; see the discussion of AR, under Miscellaneous Diseases.) This approach gained rapid acceptance, and state and regional programs of control came into existence in the United States. Pigs so derived were designated specific-pathogen-free (SPF) pigs. Details of procedures and progress to date in SPF-pig production have been reviewed by Twiehaus and Underdahl (1970).

Hysterectomy-derived pigs are designated primary SPF; second-generation pigs produced by normal parturition are designated secondary SPF; in third and subsequent generations the prefix is generally omitted and the pigs are referred to simply as SPF pigs. The success of an SPF-pig production program depends on strict adherence to isolation of SPF pigs from others. This precludes the introduction of new breeding stock into a herd unless they are free of mycoplasma pneumonia. If this precaution is followed, mycoplasma pneumonia-free status often can be maintained, although breakdowns in some herds have occurred.

It is also possible to eradicate mycoplasma pneumonia from an individual swine herd by removing pigs from affected dams at birth and raising them in isolation (Pond *et al.* 1961; 1967). Maintenance of strict sanitation and a schedule of frequent feeding (four or more times per day) of a limited supply of milk or liquid milk-replacer are essential for high survival rate in these colostrum-deprived pigs. Parenteral administration of porcine gamma globulin also helps survival rate.

In the early years of SPF-pig production it was difficult to obtain acceptable breeding stock because of the high proportion of available breeding stock within a herd that was retained solely to propagate as rapidly as possible. Now, however, there are many herds producing purebred SPF pigs of high quality.

There is no effective treatment for mycoplasma pneumonia. Antibiotics and other drugs are not effective, except in controlling secondary bacterial infections. Limited evidence suggests the possibility of clearing mycoplasma pneumonia from swine by oral administration of specific antibiotics, but this work is not definitive. Attempts at immunological control of mycoplasma pneumonia have not been generally successful to date.

MYCOTIC DISEASES

The importance of mycotoxins in swine production was reviewed recently by Carson and Lloyd (1981). Mycotoxins produced by fungi and molds are numerous (Shull and Cheeke 1978); those of major concern are discussed briefly here.

Aspergillosis (agent: *Aspergillus flavus*). This fungus is common in grain, peanuts, and maize harvested under wet conditions and stored at high mositure content. It may also be present in fish meal when the moisture content exceeds 18% (Love 1968). Clinical signs of *A. flavus* toxicity are anorexia, diarrhea, slow weight gain, pulmonary distress, and death in severe cases (Hintz 1967). Symptoms are due to absorption from the gastrointestinal tract of mycotoxins produced by *A. flavus* (aflatoxins). Aflatoxins have been associated with liver cancer in humans.

Aspergillosis can be prevented by avoiding the use of feed contaminated or suspected of being contaminated with *A. flavus*.

Moniliasis (agent: *Candida albicans, Oidium albicans, Monilia albicans*). These agents are widespread in the environment and are pathogenic mainly in malnourished animals. The symptoms are inflammation and hyperkeratosis of the epithelium of the oral cavity, esophagus, and stomach. In severe cases, the mouth becomes so sore the animal refuses to eat. The lesion has been observed in young pigs fed diets deficient in vitamin A (Hajsig *et al.* 1962) or zinc (Norrdin 1969).

Orally administered copper sulfate (Qain 1952) or narrow spectrum antibiotics have been reported to be effective in prevention and cure of affected animals. The best prevention is adequate nutrition and a clean environment.

Mold toxicity (agent: *Fusarium (Gibberella) roseum, Fusarium tricinctum, Gibberella zeae*). The feeding of moldy maize and other feeds to swine may cause poor growth and reproductive problems. Molds of the genus *Fusarium* produce zearalenone (F-2 toxin) which causes an estrogenic syndrome involving the genital tracts of both sexes of swine (Nelson *et al.* 1969). Metabolites produced by *Gibberella zeae* have been implicated in uterine hypertrophy, vomiting (therefore, the name vomitoxin), and poor consumption of contaminated feeds (Curtin and Tuite 1966). Maize that has been experimentally innoculated with cultures of *Fusarium (Gibberella) roseum* when fed to swine (Sharda *et al.* 1971; Mirocha 1977; Young *et al.* 1981) reduces weight gain, severely depresses feed intake and efficiency of feed utilization, and impairs reproduction (Young *et al.* 1982). When fed to laboratory animals, the maize produces hemorrhages in the liver and lungs. *Fusarium roseum* and *F. tricinctum,* present in significant amounts in the feed of swine, can cause major economic losses.

Ochratoxicosis (agent: *Aspergillus ochraceas* and *Penicillium* species). Concentrations of as little as 200 parts per billion (ppb) of ochratoxin A in feeds have been shown to produce kidney lesions after several weeks of feeding. Ochratoxicosis is prevalent in Denmark among pigs fed barley containing ochratoxin A isolated from fungal streams belonging to species of *Aspergillus* and *Penicillium* (Madsen *et al.* 1982). Others signs of toxicity include diarrhea, dehydration, and liver necrosis; and with levels of ochratoxin above 1400 ppb, there is reduction in weight gain and feed utilization. Residues of ochratoxin are found in the kidney, muscle, and liver of pigs fed diets containing as little as 200 ppb for several weeks before slaughter.

Ergot (agent: *Claviceps purpurea*). This fungus invades cereal grains, especially rye, and its toxic alkaloids produce reproductive problems in sows and gangrene of extremities (feet, ears, nose, and tail). The effects of ergot disappear quickly on withdrawal of infected grain from the diet.

MISCELLANEOUS DISEASES

Atrophic rhinitis (AR) (agents: *Bordetella bronchiseptica, Mycoplasma hyorhinis,* nutritional factors, noninfectious irritants). This was first described in Germany about a century ago but was not recognized in the United States until about 1940. Switzer (1981) reported *Bordetella bronchiseptica* to be the causative infectious agent of AR (see Bordetellosis under Bacterial Diseases, page 159, in this chapter). A confusing element is the fact that the organism is not always found in affected animals and is sometimes found in animals without clinical signs of AR.

The clinical signs include frequent sneezing, occasional nose bleeding, and, in severe cases, distortion of the snout to the left or right in accordion-like fashion. The commonly accepted positive diagnosis, when the animal is examined by cross section of the snout at slaughter, is absence or shrinkage of the nasal turbinates. Subjective evaluation of turbinates in this way is of limited value. A more objective, though still limited, approach used in some SPF-pig production programs is to measure the space between the turbinates and the nasal passage (in millimeters). A space in excess of some prescribed acceptable maximum is considered to be a manifestation of shrinkage of the turbinate indicative of AR. The method takes no account of histological changes that must be associated with bone to allow distortion of the face and snout in advanced cases.

Since the disease obviously involves changes in bone metabolism, the possibility of a nutritional basis for AR has been examined. Clinical signs of AR were produced in growing pigs by dietary calcium levels below the requirement or by an excess of dietary phosphorus in relation to dietary calcium (Brown *et al.* 1966). These signs included atrophied nasal turbinates, distorted facial bones, and lameness in some animals fed imbalanced diets. Histological examination of turbinates and other bones from affected pigs revealed lesions identical to those observed in field cases of AR. The bone lesions are characterized by excessive bone resorption and replacement of osseous tissue by fibrous tissue (fibrous osteodystrophy). Dietary calcium insufficiency or calcium–phosphorus imbalance produces hyperparathyroidism as a result of the inability of the parathyroid to maintain serum calcium homeostasis in dietary calcium insufficiency. Levels of dietary calcium sufficient for normal weight gains are not necessarily sufficient for normal bone development.

There is no cure for the turbinate destruction and facial distortion associated with severe atrophic rhinitis. The addition of high levels of antibiotics or of sulfa drugs or a combination of them to the diet eliminates *Bordetella bronchiseptica* from the nasal passage. A detailed account of the control and treatment of AR caused by *B. bronchiseptica* is given earlier in this chapter under Bacterial Diseases, page 159. The fact that cats, rodents, and other animals are carriers of *B. bronchiseptica* creates a practical problem in control of AR by culling of swine whose nasal passage harbors *B. bronchiseptica.* Logomarsino *et al.* (1974) showed that atrophy and distortion of the nasal turbinate could be induced in *B.*

bronchiseptica-free young pigs by a single intranasal infusion of a dilute solution of acetic acid, illustrating that noninfectious irritants can cause changes in the nasal turbinates. Good sanitation, avoidance of gastrointestinal disturbance and nasal irritants, and an adequate diet, as well as adherence to the recommendations for controlling *B. bronchiseptica,* are needed for control of AR.

Enterotoxemia (edema disease, gut edema, intestinal edema, gastric edema). The disease is produced by a toxin released from toxic strains of *E. coli;* nonpathogenic strains reside as a normal constituent of the microflora of the gastrointestinal tract. The toxin is produced by the same organism that is responsible for enteric colibacillosis, discussed under Bacterial Diseases, page 160, in this chapter. Nielsen (1981) reviewed the etiology, pathogenesis, clinical signs, treatment, and prevention of edema disease. It is a serious disease in the sense that it produces sudden symptoms of distress, anorexia, and often death in growing-finishing pigs, sometimes with no rise in body temperature or other warning signs. It is believed by some that the development of enterotoxemia is a manifestation of anaphylactic shock as a result of absorption of toxins of *E. coli* from the gastrointestinal tract.

The disease usually occurs in growing-finishing pigs and often, but not always, follows a sudden change in diet or in management. Severely affected animals may exhibit circulatory distress manifested by dypsnea (difficult breathing) and cyanosis (poor oxygen supply to tissues).

There is no satisfactory treatment. Antibiotics and other drugs are without effect except to control secondary bacterial infections. There is some evidence that fasting or reduction in feed intake aids in recovery. Good sanitation and avoidance of stresses are recommended for prevention.

MMA complex (metritis, mastitis, agalactia) (agents: endocrine factors, *Streptococcus* species, *Escherichia coli, Enterobacter, Klebsiella;* several other organisms including *Staphylococcus, Actinomyces, Actinobacillus, Corynebacterium, Aerobacter, Citrobacter, Clostridia, Proteus, Mycoplasma, Chlamydia*). These three maladies occur in association with each other so often that they are usually treated together (Martin and Elmore 1981) and are so discussed here. Metritis (uterine infection) is characterized by a purulent discharge from the vulva late in gestation or in early lactation accompanied by fever and anorexia. This syndrome is usually, though not always, accompanied by mastitis (mammary gland infection) early in lactation. The udder becomes hot and hard and the milk ejected is often abnormal. The mastitis may affect one or several glands, and the effect on the suckling pigs will depend on the severity of infection and on the number of glands affected. Pigs nursing infected glands may develop diarrhea. Usually the mastitis is followed or accompanied by a decrease in milk production of the affected glands, so that the growth of pigs nursing these glands is depressed by the hypolactia. Agalactia or hypolactia may occur as a general phenomenon associated with metritis, so that the poor performance of the suckling pigs may be related to agalactia caused by both metritis and mastitis in varying combinations. The lack of milk often results in one or more pigs in a litter dying of starvation within 1 or 2 days following birth if agalactia occurs early.

The complicated and variable nature of the MMA complex makes prevention and treatment difficult. High levels of antibiotics and arsenicals in the gestation diet for 2 or 3 weeks prepartum and in the lactation diet seem to reduce the incidence and decrease the severity of the syndrome in affected animals. Some of these drugs are secreted in the milk in sufficient quantities to offer some protection to the suckling pigs. It has been suggested that endocrine factors are involved in the MMA complex (Martin and Threlfall 1970A, 1970B). This could be through endogenous hormone imbalances or excessively high estrogen content of feedstuffs such as alfalfa meal or pasture. Administration of corticosteroids to the lactating sow may alleviate lactation failure. Synthetic estrogen injections also may be beneficial.

Prevention is aided by good sanitation in the farrowing facilities. It is preferable to allow a period of several days between groups of litters, when the facilities are unoccupied, for thorough cleaning (and fumigation) when the syndrome persists. Also, management techniques aimed at reducing stress at the time of farrowing are important, including control of temperature, humidity, and noise and provision of a laxative diet.

PARASITES

There are many parasites that affect swine but most of these do not survive in man. Several of the internal parasites of swine require a secondary host to complete their life cycle. Thorough coverage of internal and external parasites of swine appears elsewhere (Corwin *et al.* 1981; Dobson 1981), and the reader interested in details of each parasite, its life cycle, and methods of control is referred to one of these sources. The present discussion includes only the salient features of identification and control of several of the most common parasites affecting swine and a description of the clinical signs and the effects on the host.

External Parasites

Sarcoptic mange mite (*Sarcoptes scabiei*). The sarcoptic mange mite burrows under the outer layers of epidermis and causes irritation and thickening of the skin. (The subspecies of *S. scabiei* that affects swine apparently does not affect man and vice versa.) The advanced lesion resembles the parakeratosis caused by dietary zinc deficiency, but differential diagnosis is made easily by microscopic examination of epidermis for the presence of the mite. Heavy infestation is accompanied by extreme itching and discomfort manifested by the animal rubbing itself against pen partitions and feeders in an attempt to alleviate the discomfort. In severe cases the entire body surface becomes affected with thick scurfy skin and alopecia is common. Appetite is reduced and body weight gains are severely depressed.

Infestation can be controlled by spraying the animal with an aqueous suspension of malathion, toxaphene, lindane, polysulfides, trichlorfon, diazinon or of lime sulfur. The suspension should be applied under high pressure (30 kg/cm^2) to allow penetration under the outer layers of epidermis where contact is made with the parasite. The entire body must be exposed for complete elimination of the mite.

A new antibiotic, ivermectin, given in a single subcutaneous injection to infected sows 8–37 days prepartum, completely eliminated mange mites from the sows (Courtney *et al.* 1983); this is a major breakthrough in external parasite control.

Demodectic mange mite (*Demodex phylloides* or *D. folliculorum*). This species is less common in swine than *Sarcoptes scabiei* but occurs frequently in man, dog, sheep, and other animals. They are easier to control because they do not penetrate so deeply under the outer layers of epidermis. Control is the same as for that of *S. scabiei*.

Louse. One species of lice *Haematopinus suis* occurs on swine. The continual skin punctures induced to suck blood result in irritation and in rubbing and scratching by the pig against any available object for relief. Control is easy compared to that of other external parasites. Dry powder containing such substances as 5.0% ronnel can be applied on the entire body surface or, preferably,

TABLE 10.2. Treatment of External Parasites of Swine[a]

Chemical	*Concentration and method of application*	*Parasites affected*	*Preslaughter interval*
Ciodrin	0.25%; spray with up to 4 gal/pig	Lice	None
Coumaphos (Co-Ral)	0.06% solution; spray	Lice, horn flies	None
	0.12% solution; spray	Tick, mites	None
	0.24% solution; spray	Screwworms, blowflies	None
	1%; dust	Lice	None
	5%; dust	Ear ticks	None
Diazinon	0.05% emulsion; spray or dip	Lice, mites	None
Dioxathion (Delnev)	0.15% solution; spray or dip	Lice, ticks	None
Lindane	0.06% emulsion; dip or spray	Lice, sarcoptic mites	30 days after spray; 60 days after dip
	1%; dust	Fleas	30 days
	3%; paste or aerosol	Screwworms, blowflies	30 days
Malathion	0.5% emulsion; spray	Lice, ticks, mites	None
	6%; dust	Fleas, lice	None
	2.5% emulsion; spray	Houseflies, stable flies, fleas	None
Polysulfides	2%; spray	Sarcoptic mange mites	None
Rabon	2%; spray	Houseflies, stable flies	None
Ronnel (Korlan)	0.25% emulsion; spray	Lice	None
Rotenone	1%; dust	Fleas	None
Toxaphene	0.5% emulsion; spray	Lice, ticks, mites	28 days
Trichlorfon (Neguvon)	0.125% spray; spray buildings	Houseflies, stable flies, mites	None

[a]Adapted from Dobson 1981, based on compilation from Bennett 1975.

the animal can be immersed in or sprayed with a suspension of such aqueous substances as 0.25% malathion, 0.25% ronnel, 0.6% Co-Ral, 0.2% Ciodrin–0.05% Vapona combination, 0.5% carbaryl or 0.15% dioxathion (Matthysse 1971). Legal requirements for application and withdrawal before slaughter should be followed carefully.

Effective control is possible by the same means used for control of mange.

Other external parasites. A variety of other external parasites and pests of swine, including fleas, flies, and ticks, is described and discussed by Griffiths (1970). Some of these are of importance because they act as vectors for transport of pathogens to man and other animals. Their control is by the same means used for control of lice.

Dobson (1981) tabulated from the compilation of Bennett (1975) information on insecticides for control of external parasites of swine. This information is summarized in Table 10.2.

Internal Parasites

Ascarid (*Ascaris suis or A. lumbricoides*). There are several parasites that inhabit the gastrointestinal tract of swine and enter one or more body tissues during their life cycle. By far the most important of these is the intestinal ascarid (Fig. 10.1). The cycle has been reviewed (Corwin *et al.* 1981) and is diagrammatically shown in Fig. 10.2. Briefly, the important features are as follows. The adult lives in the upper intestinal tract where it mates. The fertilized eggs are excreted in the feces and go through a period of maturation on the ground, the time depending on environmental temperature and humidity. The larvae, if ingested by a pig several weeks later, are swallowed and after dissolution of the protective covering by the digestive juices of the host, are absorbed into the blood from the gastrointestinal tract. The developing larvae lodge in the liver where scar tissue is formed at the local site of infestation. Heavy infestation results in severe liver damage seen at slaughter as white fibrous tissue scars on the surface of the liver and within the liver (Fig. 10.3). Severe scarring results in veterinary condemnation of the liver for human consumption. The adverse effects of ascarid infestation are manifested both by the competition between the adult ascarid and the host for nutrients entering the gastrointestinal tract of the host and by the liver

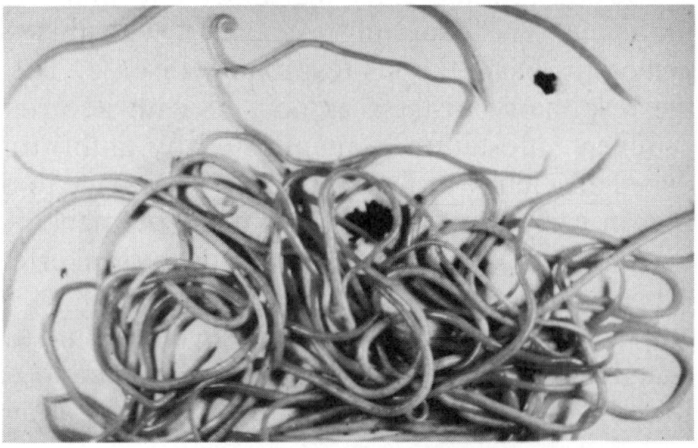

FIG. 10.1. Adult ascarids in the small intestine.

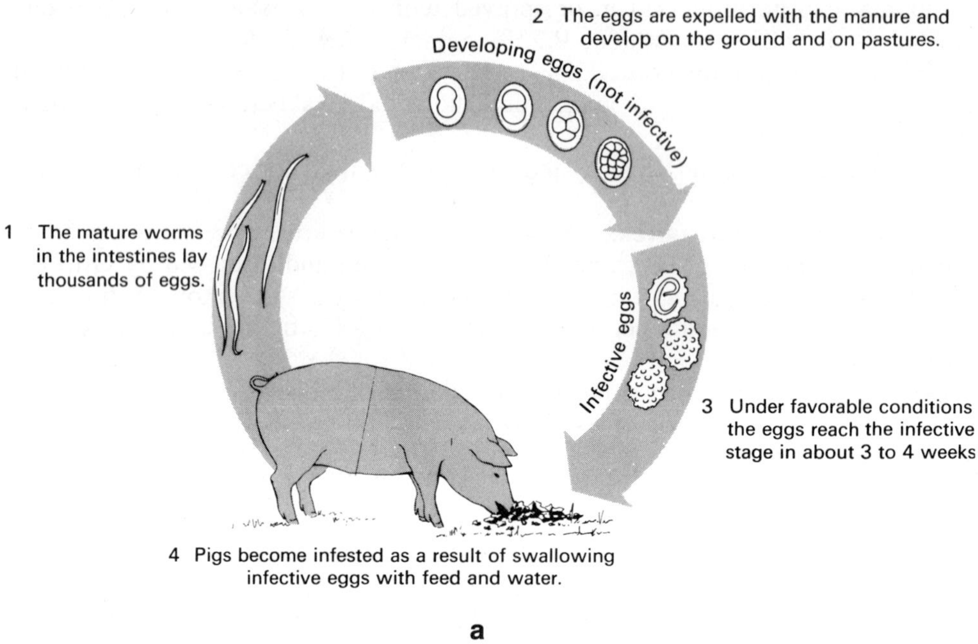

FIG. 10.2. (a) Life cycle of the ascarid; (b) journey of the ascarid through the pig.
From USDA 1963.

damage caused by migration of the larvae. Recent evidence (Stephenson *et al.* 1980) suggested reduced nitrogen absorption and retention in growing pigs infected with ascarids when dietary protein intake is marginal; small intestinal weight was increased proportionally with ascarid burden, the increased weight being due mainly to hypertrophy of the muscle layers surrounding the mucosal lining. The physiological importance of the intestinal response to the presence of adult ascarids in the lumen is unclear, but the data do emphasize the complex interactions between host and parasite and suggest that the adverse effects of adult ascarid load on pig performance may be more important than generally recognized. From the liver, the maturing larvae are pumped by the heart to the lungs where they are expelled into the bronchioli and coughed up into the mouth. They are swallowed and enter the upper intestinal tract where they mature and repeat the cycle. The coughing associated with this migration is often confused with symptoms of infectious respiratory disease. The migration through the lungs and the respiratory distress associated with it, usually persist for only a few days.

Severe infestation is manifested by unthriftiness, poor weight gains, and reduced efficiency of feed utilization. Damage to the liver caused by severe infestation in early life probably produces permanent damage to performance, so that even in animals freed of ascarids by an effective anthelmintic, performance subsequent to treatment may be inferior.

It is not uncommon to find several adult ascarids in the upper intestine of slaughter pigs even when good sanitation and parasite control programs are followed. A population of only a few adults at any point in time probably has no

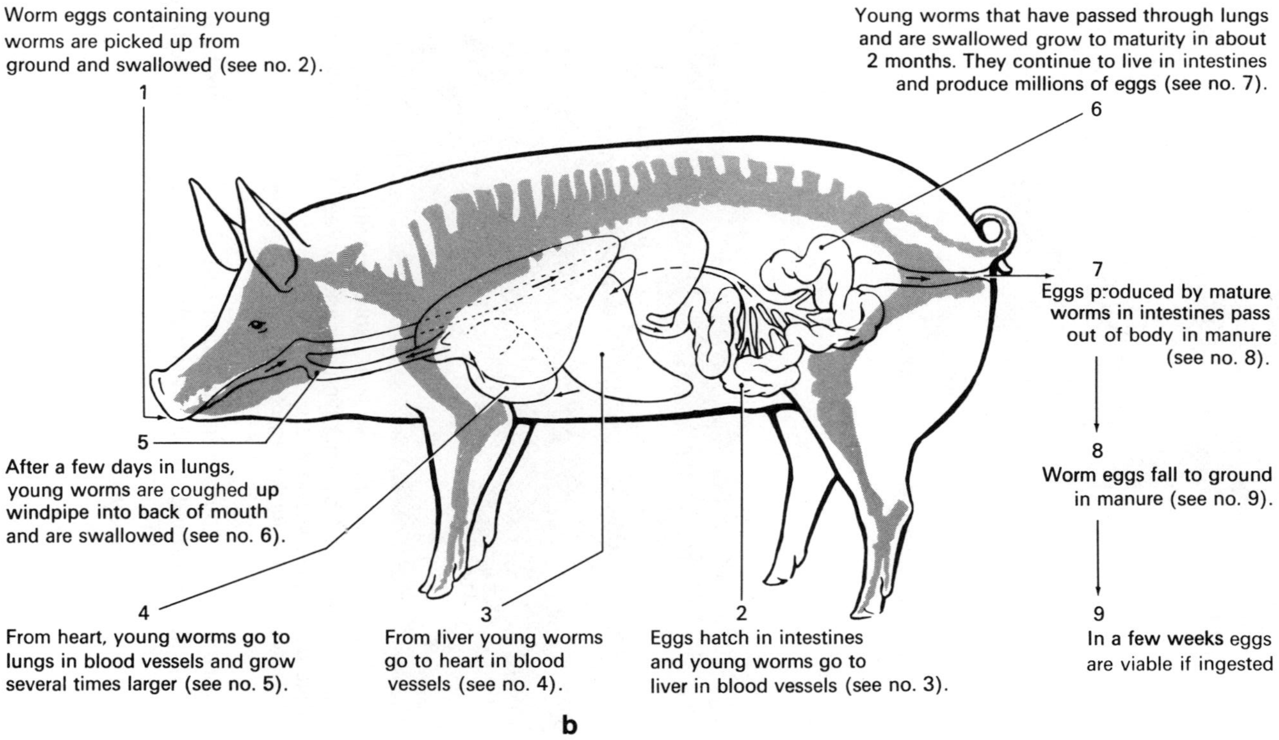
Worm eggs containing young worms are picked up from ground and swallowed (see no. 2).
1
Young worms that have passed through lungs and are swallowed grow to maturity in about 2 months. They continue to live in intestines and produce millions of eggs (see no. 7).
6
7
Eggs produced by mature worms in intestines pass out of body in manure (see no. 8).
8
Worm eggs fall to ground in manure (see no. 9).
9
In a few weeks eggs are viable if ingested
5
After a few days in lungs, young worms are coughed up windpipe into back of mouth and are swallowed (see no. 6).
4
From heart, young worms go to lungs in blood vessels and grow several times larger (see no. 5).
3
From liver young worms go to heart in blood vessels (see no. 4).
2
Eggs hatch in intestines and young worms go to liver in blood vessels (see no. 3).
b

FIG. 10.2b.

a 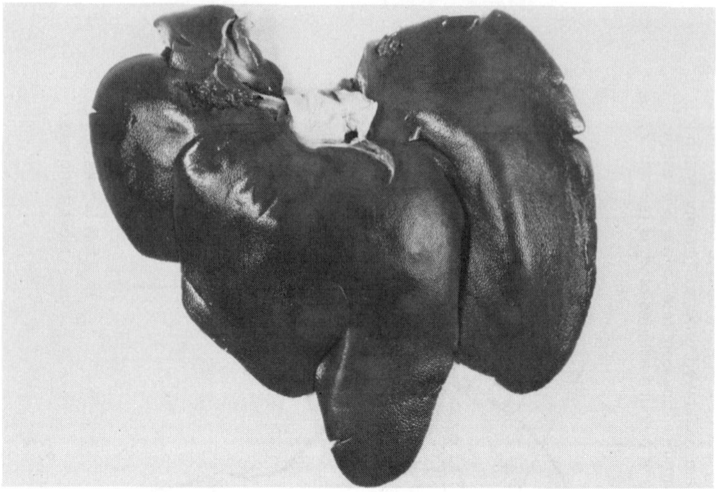

b 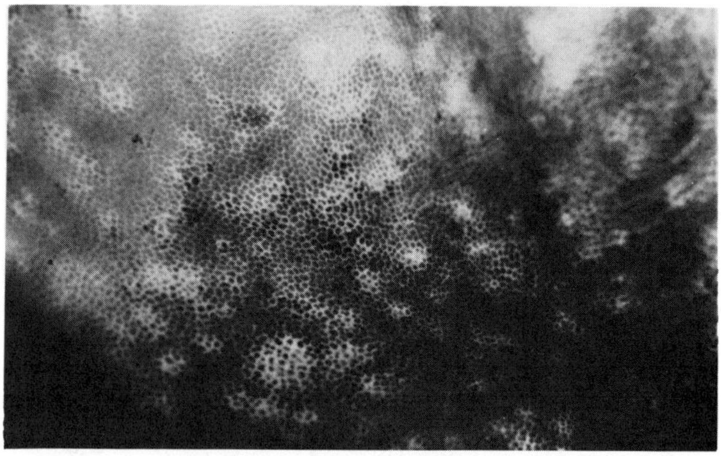

FIG. 10.3. (a) Normal pig liver. Note smooth, dark surface. (b) Pig liver damaged by migrating ascarid larvae. Note numerous white connective tissue scars.
(a) Courtesy of USDA. (b) Courtesy of D. L. Ferguson, Univ. Nebraska.

effect on performance in young pigs but controlled observations have indicated that a large adult population at slaughter weight is associated with reduced weight gain to slaughter and with profound changes in the morphology of the small intestine described previously (Stephenson *et al.* 1980).

Complete eradication of ascarids from individual swine herds is impractical, as the ascarid is transferred to the fetus *in utero*. Therefore, unless the dam is completely free of ascarids, the progeny are already infected at birth.

Control of ascarid infestation to a level compatible with optimum performance is achieved by good sanitation and by a systematic program of administration of anthelmintics. There are several effective treatments available for control of ascarids: piperazine, hygromycin B, dichlorvos, levamisole HC1, pyrantel tartrate, fenbendazole, and oxfendazole. Most of these anthelmintics are effective against other internal parasites as well. Corwin *et al.* (1981) tabulated the broad spectrum anthelmintic compounds currently available to control important internal parasites of swine. This information is listed in Table 10.3 for parasites residing as adults in the alimentary tract. Piperazines can be given to pregnant sows before being brought into the farrowing facilities and to growing pigs immediately after weaning and 2 weeks later either in the feed or water. The second treatment is needed as piperazines are only effective against adults. Larvae not killed by the first treatment are vulnerable as adults. Hygromycin can be fed continuously to growing pigs of up to 40-kg body weight. Dichlorvos is effective against both adults and larvae, so that a single dose in the feed over a 24-hr period is sufficient. A new injectable antibiotic, ivermectin, shows promise in controlling ascarids and other internal parasites (van Veen and Gibson 1983).

***Trichinella spiralis* (trichinosis).** This parasite is highly infectious to man and for this reason has probably been an important deterrent to consumption of pork in both the developing and developed countries. The life cycle has been described in detail elsewhere (Brown 1969; Gould 1970) but is summarized here briefly as follows: Adults live and mate in the intestinal tract and fertilized ova are deposited in the mucosa. The larvae pass into the lymphatics and blood and become encysted in the diaphragm and skeletal muscle of the host. The time required from original ingestion of infected meat to encysting of the larvae in skeletal muscle of the host is about 1–4 weeks. The encysted larvae can remain viable in the skeletal muscle of the host for several years. The life cycle of *Trichinella spiralis* is diagrammed in Fig. 10.4. The ingestion of infected pork by man results in severe nausea, anorexia, and digestive upset, followed by muscle pain as the larvae migrate to the skeletal muscles. Severe infestation may cause death.

It has been estimated from U.S. slaughter house surveys (consisting of microscopic examination of skeletal muscle from representative animals) that most of the infected pork sold in the United States is from garbage-fed pigs. The incidence of trichina infestation among garbage-fed pigs is estimated at 0.5% (Jeffries 1966). The percentage of pigs fed garbage in the United States is less than 0.1%. About 73% of human cases of trichinosis diagnosed in the United States in 1975 were attributed to pork products, mostly sausage (Schantz *et al.* 1977). The reduction that has occurred since a similar survey 25 yr ago is significant. Therefore, the odds of trichina infestation of human beings in the United States are extremely small. The parasite is destroyed by heat so that cooking pork at an internal temperature of 77°C for 30 min is sufficient to avoid trichinosis. Freezing pork for 20 days at −15°C or below also destroys trichinae. Rapid cooking methods, including microwaving, are not dependable to destroy live trichinae.

A rapid laboratory test for the presence of trichina larvae in pork (USDA 1969) is now available, so that identification of sources of infestation is possible. This should lead to progress in eradication of the parasite from swine.

Other internal parasites. Compared to ascarid and trichina infestation other internal parasites are of less economic importance. These other parasites include the whipworm (*Trichuris suis*) and nodular worm (*Strongyloides ransomi* and *S. luis*), both intestinal parasites with no entry into the body of the host during the life cycle; the lung worm (*Metastrongylus* species); the liver fluke (*Fasciola hepatica*), which is the most common in lowland coastal planes of the United

TABLE 10.3. Anthelmintic Compounds for Control of Internal Parasites of Swine[a]

Part of alimentary tract inhibited by adult	Parasite	Compound						
		Piperazine	*Hygromycin B*	*Dichlorvos*	*Levamisole HCl*	*Pyrantel tartrate*	*Thiabendazole*	*Fenbendazole, oxfendazole*
Stomach	*Hyostrongylus,*			×		×	×	
	Ascarops			×				
Small intestine	*Ascaris,*	×	×	×	×	×		×
	Strongyloides (nodular worm)				×		×	
Large intestine	*Trichuris* (whipworm),		×	×	×			×
	Oesophagostomum	×	×	×	×	×	×	×

[a]Adapted from Corwin et al. 1981.

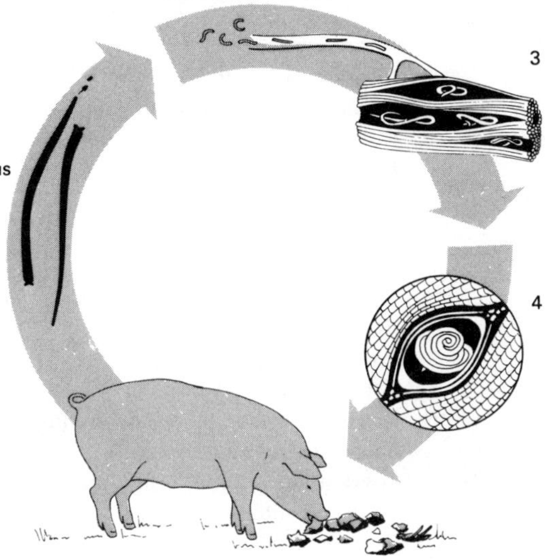

FIG. 10.4. Life cycle of the trichina.
From USDA 1963.

States and in tropical countries; kidney worm (*Stephanuses dentatus*) and *Taenia solium,* the larvae of which invade the cardiac muscle, diaphragm, tongue, and other muscles. Whipworm and nodular worms can be controlled by continuous feeding of an antibiotic, hygromycin, or by several other compounds (Table 10.3). The lung worm can be controlled by levamisole added to feed or water (Poeschel and Emro 1972) or by fenbendazole. The platyhelminths, including *F. hepatica* and *T. solium,* are best controlled by strict sanitation. Kidney worms can be controlled by levamisole HC1, and by fenbendazole and oxfendazole.

OTHER INFECTIOUS DISEASES

Table 10.4 contains a list of infectious diseases that affect swine, but because they are relatively rare, they are not discussed here. Detailed information is provided on each one in Leman *et al.* (1981).

NONINFECTIOUS DISEASES

POISONOUS PLANTS

A large number of poisonous plants are of potential danger to swine. Kingsbury (1964) has treated this subject in detail in a textbook, which contains the necessary botanical information as well as a comprehensive description of symptoms of poisoning in swine. The naturally occurring toxic principles of poisonous plants

TABLE 10.4. Other Infectious Diseases of Swine

Disease	*Agent*	*Type*
Anthrax	*Bacillus anthracis*	Bacterial
Botulism	*Clostridium botulinum* (*Bacillus botulinus*)	Bacterial
Corynebacterial infections	*Corynebacterium* species	Bacterial
Listerosis	*Listeria monocytogenes*	Bacterial
Malignant edema	*Clostridium septicum*	Bacterial
Necrotic rhinitis	*Spherophorus necrophorus*	Bacterial
Tetanus	*Clostridium tetani*	Bacterial
Tuberculosis	*Mycobacterium tuberculosis avium, bovis, hominus*	Bacterial
Actinomycosis	*Actinomyces bovis*	Mycotic
Encephalomyocarditis	*Cardiovirus* species	Viral
Eperythrozoonosis	*Eperythrozoon suis*	Rickettsial
Hemagglutinating encephalomyelitis	Hemagglutinating encephalomyelitis virus	Viral
Porcine adenoviruses	*Adenovirus* species	Viral
Porcine enteroviruses	Teschen virus, Virus T80, Virus F7, Virus Californianos	Viral
Porcine cytomegalovirus infection (inclusion body rhinitis)	*Cytomegalovirus* species	Viral
Rabies	*Rhabovirus* species	Viral
Reovirus infection	Reovirus	Viral
Vesicular exanthema	Vesicular exanthema virus	Viral
Vesicular stomatitis	Vesicular stomatitis virus	Viral
Coccidiosis	*Eimeria* and *Isopora* species	Protozoan
Toxoplasmosis	*Toxoplasma gondii*	Protozoan

as outlined by Kingsbury (1964) and as described by Carson and Lloyd (1981) and Kasali *et al.* (1977) are listed as follows: alkaloids; glycosides and glucosides (cyanogenic glycosides, goitrogenic substances, irritant oils, coumarin glycosides, steroid and triterpenoid glycosides, and cardiac glycosides); nitrates and nitrites; oxalates; photosensitivity-causing compounds (primary photosensitivity and hepatogenic photosensitization compounds); polypeptides; resins and resinoids; saponins; toxalbumens (phytotoxins); and vitamin D metabolites.

In addition to plants that contain toxic compounds listed above, many plants, especially the seed legumes such as soybeans, contain heat-labile inhibitors, which limit their usefulness in human and animal feeding unless heated. The development of cooking of plant products is regarded as a major evolutionary advance for humans (Leopold and Ardrey 1972). Processing of specific feeds by this and other methods as a means of improving usefulness for swine is discussed in depth in Chapter 17. The following plants are of special significance as potential hazards for swine because they contain one or more of the above toxic substances.

***Xanthium strumarium* (cocklebur).** This plant is widely distributed throughout the world. It contains a toxic glucoside which, when ingested, produces anorexia, vomiting, ataxia, rapid respiration and pulse rate, depressed body temperature, and in severe poisoning, death within a few hours of ingestion.

Administration of mineral oil immediately after ingestion retards absorption of the glucoside. Anticholinesterases can also be injected to counteract the effects.

***Solanum nigrum* (black nightshade).** This species of *Solanum* contains a variety of toxic alkaloids. Ingestion causes anorexia, vomiting, muscular tremors, incoordination, convulsions, coma, and, in severe cases, death. The pupils become dilated, and breathing and pulse are accelerated.

Treatment consists of a parasympathetic stimulant such as pilocarpine given parenterally.

***Crotalaria* species.** Several species are common in various parts of the world. The plant is an annual legume that produces increased respiration and pulse rates, diarrhea, anemia, and loss of coat pigments. In severe poisoning, death occurs as a result of petechial hemorrhages, congestion, and necrosis of stomach, liver, heart, and other tissues. There is no satisfactory treatment.

Cottonseed (gossypol poisoning). Cottonseed, and some samples of processed cottonseed meal are high in a toxic complex hydrocarbon, gossypol. Modern methods of processing cottonseed meal remove most of the gossypol so that the danger of gossypol poisoning is less than it formerly was. Clinical signs of gossypol toxicity include dyspnea, generalized edema, muscular weakness, and in severe cases, death. It has been shown that high levels of dietary iron or lysine tie up much of the gossypol and prevent its absorption (Clawson *et al.* 1961; Kornegay *et al.* 1961). Performance is superior with soybean or peanut meal to that with cottonseed meal in the presence of high levels of gossypol. Prevention of gossypol poisoning is attained by using degossypolized cottonseed meal and/or feeding a large quantity of either iron or lysine in the diet.

***Leucaena leucocephala* (Lam.) de Wit poisoning.** This plant is a shrub legume common in the tropical and subtropical regions of the world. When consumed by pregnant gilts at levels greater than 15% of the dry matter in the diet, it produces fetal death and malformed limbs of some of the surviving fetuses. It has been recommended that this legume be removed from the diet of sows at least 2 weeks before breeding (Wayman *et al.* 1970).

***Amaranthus retroflexus* (pigweed).** Swine given sudden access to pastures or fence rows containing redroot pigweed develop a syndrome called perirenal edema in 5–10 days. Signs include trembling, incoordination, and finally nearly complete paralysis of the rear legs. Coma and death occur in up to 75–80% of affected pigs within 48 hr of onset of the clinical signs. The syndrome is associated with severe renal disease and with changes in the electrocardiogram which probably is the cause of death (Carson and Lloyd 1981).

***Cestrum diurnum* (day-blooming jessamine, wild jasmin, king-of-the-day).** The ingestion of leaves of this plant, which is a shrub found in Florida and perhaps elsewhere in the southern United States, has been implicated in a chronic lameness, hypercalcemia, osteonecrosis, and calcification of soft tissues in horses and cattle. In swine experimentally fed a corn–soybean-meal-type diet containing 3% *Cestrum diurnum,* plasma Ca increased to twice the normal level within 4 weeks and soft tissue calcification, parathyroid atrophy, and retarded cell differentiation of growth cartilages occurred even earlier. The biochemical and anatomical changes are similar to those in vitamin D toxicity. The basis for the lesion is the presence in *Cestrum diurnum* of the vitamin D metabolite, 1,25-dihydroxycholecalciferol, or a similar compound, which is many times more active than vitamin D itself.

Plants known to accumulate nitrates. Carson and Lloyd (1981) have listed more than 20 common crop plants and weeds that accumulate nitrate in their leaves to levels that are toxic to animals. When animals consume plants high in nitrate and concurrently drink nitrate-contaminated water, there is potential for accumulation of methemoglobin in the blood. Nitrate itself is not harmful, but it is reduced to nitrite which oxidizes ferrous iron in hemoglobin to ferric iron, forming methemoglobin which cannot transport oxygen, resulting in death if methemoglobin levels reach 80%. Pigs given single doses of greater than 20 mg nitrate N/kg body weight show clinical signs, including increased respiratory rate, salivation, and anorexia; blood becomes a chocolate brown color due to the methemoglobin. Pigs have been fed up to 18 mg nitrate N/kg body weight for many weeks with no serious effects. It is highly unlikely for plants to accumulate levels of nitrate that would yield such levels of nitrite. Therefore, even though such common plants as oats, soybeans, sudangrass, wheat, and corn do accumulate nitrate in the leaves, toxic levels are rare.

TOXICANTS THAT CAUSE DISEASE

Modern agriculture makes use of a wide variety of substances—rodenticides, insecticides, herbicides, and fungicides—developed to control pests. Many of these are not only toxic to the pest they were developed to control, but also to the animal for which protection is desired. Some of the most common of these toxicants are discussed briefly here.

Rodenticides

Sodium fluoroacetate. This compound is fatal to swine in extremely small quantities (0.3 mg/kg body weight) (McGirr and Papworth 1955), and in even smaller amounts, within 30 min of ingestion, it produces severe symptoms of vomiting, muscular spasms, alteration of cardiac function, and cyanosis due to stimulation of the central nervous system.

Treatment consists of intravenous administration of pentobarbital or glycerol monoacetate to control ventricular fibrillation.

Warfarin (3-phenyl-acetyl-4-hydroxycumarin). This is an antimetabolite of vitamin K and produces hemorrhages as a result of interference with the function of vitamin K in blood coagulation. It was developed as a rat poison. A single oral dosage of 3 mg/kg body weight or repeated oral doses of 0.05 mg/kg daily for 7 days produces toxic effects (Carson and Lloyd 1981).

Signs of poisoning are the appearance of subcutaneous hemorrhages, especially on the face and ventral parts of the body, dypsnea, and death within 2 or 3 days of ingestion. Treatment consists of blood transfusions or intravenous injection of vitamin K.

Insecticides

Organophosphates. This group of compounds produces immediate toxicity upon ingestion. The symptoms include salivation, dyspnea, and holding of the mouth open, followed quickly by convulsions and death (Link 1970). Atropine sulfate is the treatment in doses of 0.44 mg/kg body weight given intravenously.

Carbamates. The mechanism of action of this group of compounds is similar to that of organophosphates. Under normal conditions, acetylcholine released at the synapses of nerves is quickly hydrolyzed; organophosphates and carbamates inhibit this enzyme hydrolysis and result in the convulsions and death seen in affected pigs and other animals. Intramuscular injection of atropine sulfate at 0.5 mg/kg body weight should be given immediately; oximes, often used in human therapy, are effective but very expensive (Carson and Lloyd 1981).

Chlorinated hydrocarbons (methoxychlor, dieldrin, and others). The symptoms of naphthenate and pentachlorophenone toxicity include hyperirritability, vomiting, salivation, cervical muscle spasms, convulsions, and death (Walters 1952). Treatment is by intramuscular injection of chlorohydrate, atropine sulfate, or pentobarbital. Dieldrin, when given experimentally to pregnant sows (Uzoukwu and Sleight 1971), failed to cause abortion, and tissue levels of dieldrin in fetuses were lower than in dams, indicating inefficient placental transfer. Much more information is needed about the toxicity and tissue-residue problem with this class of compounds for swine and about the implications for public health. Chlorinated hydrocarbons are concentrated in the fat of animals that have ingested them, so tissue residues represent a potential human health hazard; concerted efforts are underway by product manufacturers, marketers, and regulatory agencies to minimize this risk.

Herbicides

Dinitrocresols and dinitrophenols. These compounds are often used as wood preservatives. Symptoms of toxicity include elevation of body temperature, tachycardia, convulsions, coma, and death (Schipper 1961). Diagnosis of the toxic agent can often be made by the yellow appearance of the mouth and other tissues with which the material has come in contact. There is no good treatment, but oral administration of sodium bicarbonate solution by stomach tube may alleviate symptoms.

Dipyridal compounds. Paraquat is widely used as a herbicide in no-till cropping. Accidental and malicious dosing of swine causes death at 75 mg/kg body weight. Acute signs include necrosis of the oral and stomach lining, while more classic effects involve pulmonary congestion and occur 7–10 days after ingestion (Carson and Lloyd 1981).

Phenoxy compounds. Phenoxy compounds including 2,4-D; 2,4,5-T; and silvex are used widely in crop production. These compounds, though toxic for the dog, are apparently not toxic to the pig in the quantities normally encountered (Carson and Lloyd 1981). Experimental administration of large doses produces anorexia, muscular weakness, and incoordination.

Arsenicals. Various arsenic-containing compounds, some organic, others inorganic, are present in some herbicides. Organic forms are less toxic and in fact are often used as feed additives in the control of enteric diseases of bacterial origin (Gleason *et al.* 1957). However, an excess of arsenic ingestion allows absorption of quantities sufficient to cause irritability, muscular spasms and tremor, and excessive nasal secretion. The pulse becomes weak and tachycardia is common. Chronic changes include inflammation and necrosis in the stomach and intestinal linings followed by diarrhea and dehydration. Treatment is administration of raw egg white or other protein that tends to tie up arsenic and prevent its absorption.

Arsenic interferes with the absorption of selenium from the gastrointestinal tract and has been used experimentally to prevent or alleviate selenium toxicity in animals fed diets containing toxic levels of selenium (Wahlstrom and Olson 1959). Thus, as with several other mineral elements, minerals can be either toxic or beneficial depending on the amount ingested over a given period of time.

Fungicides

Captan. Field seed corn produced in the United States is generally treated with 1000 ppm of captan; there is little danger of severe poisoning because the lethal dose is several-fold higher (>250 mg/kg body weight).

Organomercurials. Mercuric compounds, including phenyl mercuric chloride, phenyl mercuric acetate, and ethyl mercuric chloride are seed treatments that can produce mercury poisoning if ingested in large quantities.

Pentachlorophenol. This compound has been used for many years as a wood preservative and fungicide. Acute poisoning is unlikely to result from contact with treated wood, but stillborn pigs have been produced in high incidence from sows housed in pens containing treated wood. Chronic exposure may lead to levels of 10–1000 ppb in whole blood of pigs.

METABOLIC DISEASES

Swine are affected by a variety of metabolic disorders of importance in commercial production. Some of these have their counterpart in humans or in other farm animals. The use of swine as the model animal for studying these disorders has been reviewed (Bustad *et al.* 1966; Pond and Houpt 1978). They include atherosclerosis, thrombosis, osteoporosis, and gastric ulcers. The economic losses to the swine industry as a result of degenerative diseases such as atherosclerosis are probably not great, since in today's husbandry practices the animals are usually slaughtered before they develop the disorder enough to show clinical signs. However, losses due to death or crippling of breeding animals from osteoporosis are significant in commercial herds.

Some metabolic diseases of major economic importance to the swine industry are discussed here.

Gastric ulcers. The occurrence of stomach ulcers in growing-finishing pigs has apparently increased substantially in recent years. The reasons for this are not clear but it appears that stress is an important factor. O'Brien (1981) summarized the current knowledge of gastric ulcers in swine and reported a widespread incidence around the world; 5% in United States, 29% in England, 22% in Sweden, and even higher levels in several other countries. The secretion of gastric juice with high acidity and high pepsin concentration increases the susceptibility to ulceration, but the mechanism of the effect of diet is not clear. Heated (gelatinized) maize is associated with a higher incidence of ulcers than unheated maize (Nuwer *et al.* 1965); in comparison, oats tend to protect against ulcers. An alcohol-soluble substance extracted from oats has been shown to be a potent inhibitor of gastric acid secretion and to protect against ulcers in the pig (Seether *et al.* 1971). The process of ulceration can occur within a 2- or 3-day period. Stress associated with marketing (fasting, long duration of shipment) and crowding (Pickett *et al.* 1969) increases the incidence. Finely ground feeds increase the incidence (Baustad and Nafstad 1969). Other dietary components and nutritional factors such as the inclusion of cheese whey, corn starch, and unsaturated fats have been implicated (O'Brien 1981). The extent of economic loss to the industry is difficult to evaluate but probably losses are restricted mainly to deaths among affected pigs, as there is no good evidence that growth rate or feed utilization are adversely affected; presumably deaths are due either to perforation of the stomach wall, resulting in peritonitis, or to such severe ulceration that digestive function is halted. Until more is known concerning etiology and pathogenesis of gastric ulcers, the preventions seem to be avoidance of stresses during growing and at marketing, and avoidance of feeding finely ground diets.

Osteoporosis. Although documentation is not extensive, there is sufficient indirect evidence showing that many of the cases of lameness and spontaneous bone fractures in extremities in older sows result from excessive bone resorption due to suboptimal dietary calcium intake in the course of extended periods of time. The calcium demands during lactation are great, and insufficient dietary calcium during this stage of production may result in excessive bone resorption with no noticeable change in composition or quantity of milk produced. In growing pigs kept in close confinement, or in sows kept continuously in stalls during gestation and in farrowing crates throughout lactation, a potential exists for development of osteoporosis severe enough to produce clinical signs and economic losses through crippling or bone fractures. Present knowledge of etiology and pathogenesis of osteoporosis does not permit solid recommendations, but adequate dietary calcium and at least moderate exercise are probably helpful.

Ketosis. An abnormally high level of ketone bodies in the blood during late gestation or early lactation is occasionally a problem in swine (Newton 1961). Metabolically, this disease is probably identical to or similar to ketosis in dairy cattle, in which it is a much more serious problem. The basic causes of the disease have not been determined although it is now recognized that it involves a defect in energy metabolism or in energy intake relative to current metabolic needs. It is seen most commonly in lactating sows in thin condition and manifests itself by anorexia, body weight loss, and the obvious presence of acetone and other ketones in the expired air. In severe ketosis the animal may become comatose and death may occur following a period of unconsciousness. Treatment is by injection of sodium propionate or of glucose. This is merely an expediency, as the basic metabolic defect is not corrected. Adequate nutrition, especially during late gestation and in lactation, is important for minimizing the incidence.

Parturient paresis (milk fever). This disease is related to calcium metabolism and is precipitated by hypocalcemia induced by dietary or metabolic means. Metabolically it is probably identical or similar to the disease as it occurs in dairy cattle, in which it is much more common. It almost always occurs during lactation when large amounts of calcium and other minerals are needed for milk production. The inability of the diet to supply adequate calcium or of the body to mobilize sufficient calcium from bone to maintain serum calcium homeostasis results in a decline in serum calcium concentration to a level low enough to cause the clinical signs of parturient paresis. The signs include anorexia, apathy, ataxia, tetany, unconsciousness, and death. Treatment is to inject a calcium salt such as calcium gluconate intravenously. The response is dramatic. Animals nearly comatose are able to rise and appear normal within a few minutes after calcium administration. Adequate dietary calcium, phosphorus, and vitamin D intake, especially during lactation, minimize the incidence. There are undoubtedly endocrine factors of importance in addition to the nutritional factors. Parathyroid hormone injection can be used as a means of mobilizing bone calcium, but can be considered only as an expediency. There is some evidence that excess calcitonin production from the thyroid gland may be a factor in inhibiting bone resorption to a sufficient degree to induce parturient paresis. There is also the possibility that exercise during gestation acts through the endocrine system to produce an effect on bone mineral mobilization. Research on parturient paresis has been centered on dairy cattle, although similar mechanisms probably control the syndrome in swine. Current knowledge of the disease has been reviewed by Littledike *et al.* (1981).

Photosensitization. This is a problem in white pigs or those with small amounts of skin pigments. It involves sensitization of the skin to ultraviolet light following the ingestion of plants containing photosensitizing agents (Bourne 1953). Since both ultraviolet light exposure and ingestion of certain plant materials are needed to cause photosensitization, the problem exists only in swine kept outdoors or otherwise exposed to ultraviolet light. Several legumes, including alfalfa forage and other common legumes as well as buckwheat, contain these photodynamic agents. The symptoms begin approximately one week after the animal begins to consume the diet and include severe reddening, swelling, and exudation of parts of the body exposed to the sun, especially the ears. This swelling and reddening is associated with hemolysis of red blood cells and is followed by tissue necrosis of affected areas. Behavioral changes may also be observed, including staggering and repeated falling.

Osteochondrosis. This is a disturbance in ossification of endochondral tissue and in epiphyses of long bones, especially ulna and femur. The syndrome involves lameness and discomfort and appears to have a higher incidence in pigs raised on concrete or on slatted floors. Lameness develops when a weakened area of the articular-epiphyseal cartilage separates from the underlying bone. Scandinavian work (Grondalen 1974; Reiland *et al.* 1978) has shown a relationship between visual scoring of locomotion in young pigs and the severity of osteochondrosis, but the causal relationship between the abnormal shape of the articular surfaces and long bones and the development of osteochondrosis is controversial.

The syndrome appears to be noninfectious but attempts to identify nutrient deficiencies or imbalances (vitamins A and D, protein, calcium, phosphorus) as causative factors have been unsuccessful (Nielsen *et al.* 1971; Kornegay *et al.* 1981). There is evidence (Reiland *et al.* 1978; Grondalen 1974) that severity of osteochondrosis has some degree of heritability, so that selection against the syndrome may have some merit.

The endocrine and metabolic bases for osteochondrosis have not been thoroughly studied and it appears that until the syndrome is more completely characterized, the major means of control will be by culling affected animals from breeding and providing comfortable housing and opportunity for exercise.

Rectal and vaginal prolapse. Cases of prolapse are sporadic on individual farms and often the cause is obscure. Culling rates for rectal and/or vaginal prolapse in breeding swine have been reported to be as high as 8.9% in a single herd (Smith and Penny 1981), but more commonly the incidence is less than 1%. Rectal prolapse is most common in pigs 3- to 5-months old, usually at an incidence of less than 1%. The physiological basis for vaginal and rectal prolapses is unclear, although such factors as nutrition (dietary constituents that lead to constipation) drugs (estrogenic substances), genetics (inbreeding may increase incidence), and lack of exercise have been implicated. Treatment must be cost effective; often a simple purse-string suture is sufficient to allow reversal of the prolapse.

Porcine stress syndrome (PSS). Stress susceptibility in swine is a commonly observed syndrome. Reports of stress susceptible swine have come from all of the major swine producing countries of the world. The incidence, inheritance, detection, and prevention of PSS have been summarized by Topel and Christian (1981). Affected animals are recognized by muscle and tail tremors, dypsnea, irregular breathing, blanched and reddened areas of the skin, increased body temperature, cyanosis, and death (Topel and Christian 1981). Blood and muscle tissue accumulation of lactic acid results in acidosis which, in turn, leads to the production of pale, soft, exudative (PSE) pork at slaughter in affected animals. Porcine stress syndrome-susceptible swine have an elevated turnover of adrenal cortical hormones and generally a hypofunction of the adrenal cortex.

Susceptible animals are identified by their adverse response to halothane anesthesia. This relationship between PSS and halothane has led to the use of halothane as a screening tool to identify stress-susceptible swine. Other tools for identifying stress susceptibility include measurement of blood serum levels of the enzyme, creatine phosphokinase (CPK), which is elevated in PSS swine; blood typing to identify PSS-susceptible and resistant swine; and visual appraisal. Visual indicators include extreme muscling, small stature, and tight skin in the jowl region; visual appraisal is too subjective to be highly useful in identifying stress-susceptible swine.

Prevention is approached most effectively by genetic selection against PSS. Topel and Christian (1981) have recommended elimination of all positive reactors (by halothane test) and of their parents, sibs, and half sibs. Animals known to be stress susceptible should be subjected to minimum stress such as avoidance of mixing with strange pigs and maintenance at a thermally comfortable environment. Marketing of PSS swine is discussed in Chapter 19.

NUTRITIONAL DISEASES

Marginal dietary levels of one or more nutrients are probably common in swine fed homegrown feeds not fortified with commercial sources of vitamins. Mineral and protein supplements are tailored to meet specific needs. These marginal (subclinical) deficiencies of a general nature are manifested as reduced body weight gain or reduced efficiency of feed utilization, often so slight that they are not even recognized.

However, over and above these marginal deficiences, usually involving more than one nutrient, there are several specific nutrient deficiencies that manifest themselves by typical clinical signs (National Research Council 1979; Whitehair and Miller 1981; Cunha 1977; Pond and Houpt 1978). As pointed out by Whitehair and Miller (1981), nutritional deficiencies result from not only specific dietary deficiencies, but also from substances that interfere with nutrient utilization, from infections that lower the intake or utilization of nutrients, and from diseases that increase nutritional requirements (for example, infections that cause fever and thereby increase the metabolic rate). Only simple deficiencies of some nutrients of special concern are described here.

Zinc deficiency (parakeratosis). The specific clinical sign is a typical skin lesion characterized by excess keratinization of the epidermis accompanied by a brown exudate and alopecia. The lesion resembles that produced by the mange mite (*Sarcoptes scabiei*), but is distinguished from mange by the rapid and dramatic disappearance of the lesion when the diet is supplemented with zinc. The skin lesion is accompanied by reduced growth rate and efficiency of feed utilization, and the serum alkaline phosphatase concentration is depressed. All of these effects are dramatically reversed by zinc supplementation.

Tucker and Salmon (1955) first published evidence that zinc would cure or prevent the lesion and these workers are generally given credit for the discovery, although R. F. Sewell and co-workers at the University of Georgia had previously obtained firm evidence of a relationship between parakeratosis and zinc.

The biological availability of zinc is affected by several dietary factors. A high level of calcium increases the dietary zinc requirement so that supplemental zinc is required to prevent parakeratosis when the diet is high in calcium, but may not be required with lower levels of calcium. Phytic acid (the hexaphosphoric acid ester of inositol), present in high concentrations in soybean meal and other plant proteins, ties up zinc in the gastrointestinal tract and thereby increases the dietary requirement. Parakeratosis can be prevented by the addition of 100 ppm of zinc as an inorganic salt to the diet of growing pigs. The toxic level is not reached until the diet contains more than 1000 ppm of zinc (Brink *et al.* 1959).

Iron deficiency (baby pig anemia). Iron deficiency anemia is of practical importance mainly in suckling pigs as a result of inefficient placental and mammary transfer of iron. The newborn pig has very limited body reserves of iron and must get iron in addition to that obtained from milk to avoid anemia. Pigs allowed access to soil or pasture are exposed to enough iron from environmental contamination to meet their needs, but those born and raised in concrete-floor pens develop anemia within a few days after birth if not given a supplemental source. Oral tablets and pastes or parenteral administration of commercial iron–dextran preparations will cure or prevent anemia (Maner *et al.* 1959). Despite the availability of iron sources, baby pig anemia remains as a major nutritional disease of swine, primarily because many management systems do not ensure that oral sources of iron offered to suckling pigs are actually consumed in sufficient amounts. Prevention is assured by parenteral administration of nontoxic iron–dextran preparations. Iron deficiency is not a serious problem after weaning, because normal feedstuffs usually contain more iron than required.

Selenium and vitamin E deficiency (hepatosis dietetica, mulberry heart disease). The exact metabolic interrelationships between selenium and vitamin E are not well understood, but it appears that one can substitute for the other as a dietary component to prevent the liver or heart lesions that result in development of deficiency signs. The deficiency signs are seen almost entirely on postmortem examination, following sudden death of apparently healthy animals. The liver shows massive necrosis and the heart often has grossly visible muscle dystrophy, verifiable by histology. Skeletal muscles may also show dystrophy, although this is not as commonly seen in pigs as in sheep that are deficient in vitamin E. Depot fats may have a brown discoloration.

Although vitamin E (tocopherol) is stored in the liver and other tissues, it is possible to produce deficiency signs by long periods of dietary vitamin E inadequacy. For many years it was considered that swine fed diets composed mainly of maize or cereal grains had no danger of developing vitamin E deficiency. However, in recent years many cases of vitamin E deficiency have been reported in swine fed diets composed of feedstuffs grown over a wide geographic area. D. E. Ullrey and associates at Michigan State University (Ku *et al.* 1972) have shown a close relationship between body tissue concentration of selenium in pigs and selenium content of the diet fed.

It is now well known from soil surveys (Kubota *et al.* 1967) that selenium-deficient soils exist in many parts of the world. Crops grown on these soils tend to be low in selenium and when they are fed in diets also deficient in vitamin E, the above syndrome is produced. The reason for the apparent increase in incidence of selenium and vitamin E deficiency is swine in recent years is not clear but may be related to methods of processing feeds during and following harvest and the increasing practice of confinement rearing with limited access to fresh forages.

Calcium, phosphorus, or vitamin D deficiency (rickets). Failure to provide adequate calcium, phosphorus, or vitamin D in the diet of the growing pig results in abnormalities in bone structure induced by inadequate mineralization. A deficiency of any one or all three of these nutrients produces rickets in growing animals. Vitamin D can be supplied as such in the diet or can be synthesized in skin by ultraviolet light activating 7-dehydrocholesterol to form the active vitamin or its precursor (current evidence indicates that the active form of vitamin D is 1,25-dihydroxycholecalciferol). Thus, pigs given exposure to as little as 30 min of direct sunlight daily have no need for dietary vitamin D (Johnson and Palmer 1941). If adequate vitamin D is supplied either in the diet or by sunlight, dietary calcium and phosphorus utilization are adequate. Therefore, if rickets develops in the presence of adequate vitamin D, dietary calcium and/or phosphorus deficiency is indicated. Symptoms are reduced mineralization of the protein matrix and the outward manifestations are bowing of the legs, lameness, and occasionally a spontaneous fracture of an extremity or of the ribs. Growth rate and efficiency of feed utilization may be reduced. Rickets can be prevented by feeding adequate calcium and phosphorus and by feeding vitamin D or its metabolites or allowing exposure to ultraviolet light.

Calcium–phosphorus imbalance (fibrous osteodystrophy). A dietary calcium–phosphorus ratio of less than 1 : 1 may result in excessive bone resorption in relation to deposition with replacement of bone by fibrous tissue (Brown *et al.* 1966). This is a manifestation of nutritional secondary hyperparathyroidism induced by continued low serum calcium concentration.

The lesion is generalized throughout the skeleton and, in severe cases, results in spontaneous fracture of long bones and of ribs as in rickets. However, the histological picture is different in fibrous osteodystrophy than in rickets. The high phosphorus content of many plant materials probably contributes to the frequent occurrence of this nutritional imbalance. Prevention is achieved by dietary calcium adequacy and by maintenance of a calcium–phosphorus ratio greater than 1 : 1.

Protein deficiency (hypoproteinemia, fatty liver, edema, osteoporosis). Dietary protein deficiency is most easily produced in young rapidly growing animals. Older animals have a lower protein requirement and are more resistant to deficiency signs. In young pigs fed corn or cereal grain as the only source of protein, body weight gain may be reduced to zero, liver fat is increased, and serum total protein and serum albumin concentrations are severely reduced (Gallo and Pond 1968). Failure of normal formation of protein matrix for ossification of skeletal tissue results in reduced activity of the growth plate of long bones and decreased density and thickness of bone cortices (osteoporosis) (Lowrey *et al.* 1962). These signs are aggravated when the energy concentration of the diet (and consequently the calorie–protein ratio) is increased by adding fat (Pond *et al.* 1965). This increased calorie–protein ratio produces edema, evident by puffiness around the umbilical area, neck, and jowl and in the scrotum of males. The syndrome produced by low-protein–low-fat diets (maize alone) resembles the condition of marasmus in human infants whereas that produced by low-protein–high–fat diets resembles kwashiorkor in human infants (Pond *et al.* 1966). Protein deficiency can be prevented by the addition of a protein supplement to low-protein,high-energy feedstuffs. Some of the new varieties of maize, which contain a superior balance of amino acids and, in some cases, a higher protein content than common varieties, can be used with less protein supplement.

Niacin deficiency (necrotic enteritis). A suboptimum level of niacin in the diet, in addition to reducing growth rate (as does any B-vitamin deficiency) results in ulceration and necrosis of the epithelial lining of the gastrointestinal tract (Luecke *et al.* 1949). These lesions are associated with severe diarrhea, dehydration, and weight loss. The lesions and the clinical signs resemble those produced by *Salmonella cholerasuis* (see the discussion on salmonellosis) but are not responsive to antibiotics or furazolidone, except in cases in which secondary infections have occurred.

The biological availability of niacin in maize and cereal grains is very low so supplementation is necessary even in diets for which analytical data indicate an adequate intake (Luce *et al.* 1967). An intake of niacin in excess of the requirement is merely excreted in the urine. Since there is no evidence for niacin toxicity, deficiency can be prevented by dietary supplementation.

NUTRIENT TOXICITY DISEASES

In the broad sense, any excess of a single nutrient has potential danger to health, including even fat, in which the threat to health is obesity. However, under most conditions only a few nutrients cause serious ill effects if fed continuously in excess. Some of the most common of these maladies are included here.

Hypervitaminosis A. Massive oral doses of vitamin A during the first half of pregnancy result in numerous teratological effects in the offspring of swine (Palludan 1966). These include histological abnormalities in the eyes, and a variety of abnormalities in the heart and circulatory system. There are no data available on hypervitaminosis A in growing pigs, but the observed effects on the pregnant sows in the Palludan experiment suggest that the greatest effect might be discomfort and lameness as a result of skeletal changes in the joints of extremities. Hypervitaminosis A is likely to occur only as a result of experimental or accidental administration of an excess either parenterally or in the feed.

Hypervitaminosis D. Massive doses of vitamin D given parenterally or orally produce excessive calcification (Cunha *et al.* 1968). Vitamin D synthesis in skin, resulting from exposure to sunlight, is inversely related to the amount of exposure, so that hypervitaminosis D is not likely as a result of this. As with hypervitaminosis A, the only likely means of producing hypervitaminosis D is by experimental or accidental excesses given parenterally or by errors in feed mixing. Plants containing compounds that mimic or that are active metabolites of vitamin D may also create signs of vitamin D toxicity (see Poisonous Plants, p. 185, in this chapter).

Selenium toxicity. In some areas of the United States and other parts of the world the selenium content of soil is sufficiently high to result in excessive concentration of selenium in the plant. Swine fed cereal grains or other feeds grown on these soils develop alopecia and crooked hoofs, and weight gains are severely depressed (Wahlstrom *et al.* 1955). Excesses of dietary selenium in pregnant sows result in reduced birth weights and increased mortality in progeny (Wahlstrom and Olson 1959). The addition of arsanilic acid to high-selenium diets counteracts some of the signs of selenium toxicity, presumably by interfering with absorption of selenium from the gastrointestinal tract (Wahlstrom and Olson 1959).

Molybdenum toxicity. Pigs are more resistant to molybdenum toxicity than are ruminants. Levels that would be toxic for ruminants have been fed to swine throughout the growing-finishing period with no toxic signs except a slight reduction in average daily gain (Gipp *et al.* 1967). The adverse effect of molybdenum is apparently the result of its ability to tie up copper in the gastrointestinal tract so that it induces a copper deficiency. When larger than adequate amounts of copper are fed, the adverse effects of excessive molybdenum are prevented.

Copper toxicity. The addition of copper salts to diets for growing swine as a means of stimulating weight gains has become a common practice in some parts of the world (Barber *et al.* 1955). Isolated cases of copper poisoning have been reported from Europe and more frequently from experimental studies in the United States. The main signs of acute toxicity are internal hemorrhages and generalized icterus throughout the body fluids and tissues. Chronic toxicity is manifested by depressed hemoglobin concentration and reduced body weight gain. Sudden death occurs in some affected animals, presumably as a result of hemorrhage and/or impaired liver function. Other dietary factors affect the biological availability of copper. High-protein diets depress absorption (Wallace 1968); casein induces more severe copper-toxicity symptoms than does soybean meal (Gipp 1971). The anemia produced by high-copper diets is due to induced iron deficiency probably caused by interference with iron absorption from the gastrointestinal tract (Gipp 1971). More work is needed to elucidate the factors affecting the response to high levels of copper.

Sodium (sodium chloride or salt) toxicity. Pigs deprived of water and ingesting diets containing 1–2% sodium chloride may show clinical signs of salt toxicity including severe thirst, convulsions, and death. Levels of sodium in blood serum above 160 meq/liter indicate excessive sodium retention. Free access to water allows sodium to be excreted by the kidney and protects against the effects of excess salt ingestion.

Toxicity of other minerals. An excess of any required nutrient or ingestion of harmful levels of minerals in the environment but not required by the animal for normal metabolism may result in adverse effects on normal body processes. Tolerance levels for required and toxic levels of minerals for farm animals, including swine, have been described (National Research Council 1980) and signs of toxicity of other nutrients required by swine have been reviewed (Carson and Lloyd 1981; Church and Pond 1982) but are not discussed here.

REFERENCES

ARMSTRONG, C. H. 1971. J. Am. Vet. Med. Assoc. *160*,655.

BACHRACH, H. L., and McKERCHER, P. D. 1971. J. Am. Vet. Med. Assoc. *160*,521.

BARBER, R. S., BRAUDE, R., and MITCHELL, K. G. 1955. J. Nutr. *9*,378.

BAUSTAD, B., and NAFSTAD, I. 1969. Pathol. Vet. *6*,546.

BENNETT, D. G. 1975. External Parasites. *In* Diseases of Swine, 4th Edition. Iowa State Univ. Press, Ames.

BERGELAND, M. E. 1981. Clostridial infections. *In* Diseases of Swine. (A. D. Leman, R. D. Glock, W. L. Mangeling, R. H. C. Penney, E. Scholl, and B. Straw, Editors), 5th Edition, Chap. 40, p. 418. Iowa State Univ. Press, Ames.

BOHL, E. H. 1981. Transmissible gastroenteritis. *In* Diseases of Swine (A. D. Leman *et al.*, Editors), 5th Edition, p. 195. Iowa State Univ. Press, Ames.

BOURNE, R. F. 1953. North Am. Vet. *34*,173.

BRINK, H. F., BECKER, D. E., TERRILL, S. W., and JENSEN, A. H. 1959. J. Anim. Sci. *18*,836.

BROWN, H. W. 1969. Basic Clinical Parasitology, 3rd Edition. Appleton-Century-Crofts, New York. (Life cycle reproduced in J. F. Price and B. S. Schweigert, 1971. The Science of Meat Products, p. 278. W. H. Freeman and Co., San Francisco.)

BROWN, W. R., KROOK, L., and POND, W. G. 1966. Cornell Vet. *56*(Suppl. 1),108.

BUSTAD, L. K., McCLELLAN, R. A., and BURNS, M. P., Editors. 1966. Swine in Biomedical Research. Frayn Printing Co., Seattle, Washington,

CALLIS, J. J., and McKERCHER, P. D., 1981. Foot-and-Mouth Disease. *In* Diseases of Swine. (A. D. Leman *et al.*, Editors), 5th Edition, p. 278. Iowa State Univ. Press, Ames.

CARSON, T. L, and LLOYD, W. E. 1981. Toxic chemicals, plants, metals and mycotoxins. *In* Diseases of Swine (A. D. Leman *et al.*, Editors), 5th Edition, Chap. 57, p. 603. Iowa State Univ. Press, Ames.

CARTER, G. R. 1970. Pasteurellosis. *In* Diseases of Swine (H. W. Dunne, Editor), 3rd Edition. Iowa State Univ. Press, Ames.

CHURCH, D. C. and POND, W.G. 1982 Basic Animal Nutrition and Feeding, 2nd Edition. John Wiley & Sons, New York.

CLAWSON, A. J., SMITH, F. H., OSBORNE, J. C., and BARRICK, E. R. 1961. J. Anim. Sci. *20*,547.

CORWIN, R. M., McDOWELL, A. E., and TALENT, N. K. 1981. Internal parasites. *In* Diseases of Swine (A. D. Leman, Editors), 5th Edition, Chap. 53., p. 560. Iowa State Univ. Press, Ames.

COURTNEY, C. H., INGALLS, W. L., and STITZLEIN, S. L. 1983. Am. J. Vet. Res. *44,* 1220.

CUNHA, T. J. 1977. Swine Feeding and Nutrition. Academic Press, New York.

CUNHA, T. J., BECKER, D. E., BOWLAND, J. P., CONRAD, J. H., HAYS, V. W., and POND, W. G. 1968. Nutrient Requirement of Swine Natl. Res. Council Publ. *1599*. Natl. Acad. of Sci., Washington, D.C.

CURTIN, T. M., and TUITE, J. 1966. Life Sci. *5*,1937.

CURTIS, S. E. 1981. Environmental Management in Animal Agriculture, p. 24–3. Animal Environment Services, Mahomet, Illinois.

DEVOE, B. L. 1981 Brucellosis. *In* Diseases of Swine, (A. D. Leman *et al.*, Editors), 5th Edition, Chap. 39, p. 410.

DOBSON, K. J. 1981. External parasites. *In* Diseases of Swine (A. D. Leman *et al.*, Editors), 5th Edition, p. 579. Iowa State Univ. Press, Ames.

DOYLE, L. P., and HUTCHINGS, L. M. 1946. J. Am. Vet. Med. Assoc. *108*,257.

EASTERDAY, B. C. 1981. Swine influenza. *In* Diseases of Swine (A. D. Leman *et al.*, Editors) 5th Edition, p. 184. Iowa State Univ. Press, Ames.

EASTERDAY, B. C. 1971. J. Am. Vet. Med. Assoc. *160*,645.

FARRINGTON, D. O. 1981. Pasteurellosis. *In* Diseases of Swine (A. D. Leman *et al.*, Editors), 5th Edition, p. 378. Iowa State Univ. Press, Ames.

FERGUSON, L. C. 1970. Leptospirosis. *In* Diseases of Swine (H. W. Dunne, Editor), 3rd Edition. Iowa State Univ. Press, Ames.

FRIIS, N. F. 1969. Acta Vet. Scand. *10*,295.

GALLO, J. T., and POND, W. G. 1968. J. Anim. Sci. *27*,73.

GIPP, W. F. 1971. The effect of supplemental copper on growth, hematology and iron utilization of swine. Ph.D. Thesis, Cornell University, Ithaca, New York.

GIPP, W. F., POND, W. G., and SMITH, S. E. 1967. J. Anim. Sci. *26*,727.

GLEASON, M. N., GOSSELIN, R. E., and HODGE, H. C. 1957. Clinical Toxicology of Commercial Products. Williams and Wilkins, Baltimore, Maryland.

GOULD, S. E. 1970. Trichinosis in Man and Animals. Thomas, Springfield, Illinois.

GRIFFITHS H. J. 1970. *In* Diseases of Swine (H. W. Dunne, Editor), 3rd Edition. Iowa State Univ. Press, Ames.

GRONDALEN, T. 1974. Acta Vet Scand. *15*,1,2,6,43,47,61,555,574.

GUSTAFSON, D. P. 1981. Pseudorabies. *In* Diseases of Swine (A. D. Leman *et al.*, Editors), 5th Edition, Chap. 14, p. 209. Iowa State Univ. Press, Ames.

GUSTAFSON, D. P. 1981. Congenital tremors. *In* Diseases of Swine (A. D. Leman *et al.*, Editors), 5th Edition, Chap. 29, p. 335. Iowa State Univ. Press, Ames.

HAELTERMAN, E. O. 1971. J. Am. Vet. Med. Assoc. *160*,534.

HAJSIG, M., RIZNAS, S., and MARZAN, B. 1962. Vet. Arch. *32*,276.

HANSON, L. E., and TRIPATHY, D. N. 1981. Leptospirosis. *In* Diseases of Swine (A. D. Leman, *et al.*, Editors), 5th Edition, Chap. 36, p. 236. Iowa State Univ. Press, Ames.

HARRIS, D. L., and GLOCK, R. D. 1971. J. Am. Vet. Med. Assoc. *160*,561.

HESS, G. W., MOULTHROP, J. I., and NORTON, H. R. 1970. J. Am. Vet. Med. Assoc. *157*,1975.

HINTZ, H. F. 1967. Proc. Soc. Exp. Biol. Med. *124*,266.

HUHN, R. G. 1971. J. Am. Vet. Med. Assoc. *160*,634.

JEFFRIES, J. C., BEAL, V., JR., MUTRISHAW, T. R., and ZIMMERMAN, W. J. 1966. *In* Proc. 70th Annu. Mtg. U.S. Livestock Sanitary Assoc. p. 349.

JOHNSON, D. W., and PALMER, L. S. 1941. J. Agric. Res. *63*,639.

KASALI, O. B., KROOK, L., POND, W. G., and WASSERMAN, R. H. 1977. Cornell Vet. *68*,190.

KASZA, L. 1981. Swine Pox. *In* Diseases of Swine (A. D. Leman *et al.*, Editors), 5th Edition, Chap. 18, p. 254. Iowa State Univ. Press, Ames.

KINGSBURY, J. M. 1964. Poisonous Plants of the United States and Canada. Prentice-Hall, Englewood Cliffs, New Jersey.

KOHLER, E. M. 1971. J. Am. Vet. Med. Assoc. *160*,567.

KORNEGAY, E. T., CLAWSON, A. J., SMITH, F. H., and BARRICK, E. R. 1961. J. Anim. Sci. *20*,597.

KORNEGAY, E. T., THOMAS, H. R., CARTER, J. H., ALLEN, L. B., BROOKS, C. C., and HINKLEMANN, K. H. 1981. J. Anim. Sci. *52*,1085.

KU, P. K., ELY, W. T., GROCE, A. W., and ULLREY, D. E. 1972. J. Anim. Sci. *34*,208.

KUBOTA, J. W., ALLAWAY, W. H., CARTER, D. L., CARY, E. E., and LAZAR, V. A., 1967. J. Agric. Food Chem. *15*,448.

LEMAN, A. D. 1970. Swine Health. Common Diseases Affecting Baby Pigs—A Research Review. College Vet. Med. Univ. Illinois, Urbana.

LEMAN, A. D., GLOCK, R. D., MENGELING, W. L., PENNY, R. H. C., SCHOLL, E., and STRAW, B. (Editors). 1981. Diseases of Swine, 5th Edition. Iowa State Univ. Press, Ames. 832 pp.

LEOPOLD, A. C., and ARDREY, R. 1972. Science *176*,512.

LINK, R. P. 1970. *In* Diseases of Swine (H. W. Dunne, Editor), 3rd Edition. Iowa State Univ. Press, Ames.

LITTLEDIKE, E. T., YOUNG J. W., and BEITZ, D. C. 1981. J. Dairy Sci. *64*,1465.

LOGOMARSINO, J. V., POND, W. G., KROOK, L., and KIRTLAND, D. 1974. J. Anim. Sci. *39*,544.

LOVE, T. D. 1968. Fish. Ind. Res. *4*,139.

LOWREY, R. S., POND, W. G., BARNES, R. H., KROOK, L., and LOOSLI, J. K. 1962. J. Nutr. *78*,245.

LUCAS, L. E. and WAGNER, W. C. 1970. Effect of Disease and Stress on Reproductive Efficiency in Swine. Proc. Symp. Ext. Ser. Univ. of Nebraska and USDA, Washington, D.C. 128 pp.

LUCE, W., G., PEO, E. R., JR., and HUDMAN, D. B. 1967. J. Anim. Sci. *26*,76.
LUECKE, R. W., THORP, F., JR., McMILLEN, W. N., DUNNE, H. W., and STAFSETH, H. J. 1949. Mich. Agric. Exp. Stn. Tech. Bull. *211*.
McDANIEL, H. A. 1981. African Swine Fever. *In* Diseases of Swine (A. D. Leman *et al.*, Editors), 5th Edition, Chap. 16, p. 237. Iowa State Univ. Press, Ames.
McGIRR, J. L., and PAPWORTH, D. S. 1955. Vet. Rec. *67*,124.
MADSEN, A., MORTENSEN, H. P., and HALD, B. 1982. Acta Agric. Scand. *32*,225.
MANER, J. H., POND, W. G., and LOWREY, R. S. 1959. J. Anim. Sci. *18*,1373.
MARTIN, C. E., and ELMORE, R. G. 1981. Mammary glands. *In* Diseases of Swine (A. D. Leman *et al.*, Editors), 5th Edition, Chap. 10, p. 155. Iowa State Univ. Press, Ames.
MARTIN, C. E., and THRELFALL, W. R. 1970A. Vet. Rec. *87*,768.
MARTIN C. E., and THRELFALL, W. R. 1970B. *In* Symposium on Effect of Disease and Stress on Reproductive Efficiency in Swine. Univ. Nebraska Ext. Ser. and USDA, Lincoln.
MATTHYSSE, J. G. 1971. Livestock external parasite control recommendations for New York State Dep. of Entomology Mimeo, Cornell Univ., Ithaca, New York.
MAUER, F. D., and GRIESEMER, R. A. 1970. *In* Diseases of Swine (H. W. Dunne, Editor), 3rd Edition. Iowa State Univ. Press, Ames.
MENGELING, W. L. 1981. Procine parvovirus infection. *In* Diseases of Swine (A. D. Leman *et al.*, Editors), 5th Edition, Chap. 33, p. 352. Iowa State Univ. Press, Ames.
MIROCHA, C. J. 1977. Cornell Nutr. Conf. Proc., p. 126, Ithaca, New York.
NATIONAL RESEARCH COUNCIL. 1980. Mineral Tolerance of Domestic Animals. Natl. Acad. Sci., Washington, D.C.
NATIONAL RESEARCH COUNCIL. 1979. Nutrient Requirements of Swine. Natl. Acad. Sci. Washington, D.C.
NELSON, G. H., CHRISTENSON, C. M., and MIROCHA, C. J. 1969. Proc. 20th Annu. Hybrid Corn Inc. Res. Conf., p. 86. Univ. Minnesota, St. Paul.
NEWTON, W. M. 1961. Ill. Vet. *3*,33.
NICOLET, J. and SCHOLL, E. 1981. Polyserositis and Arthritis. *In* Diseases of Swine (A. D. Leman *et al.*, Editors), 5th Edition, Iowa State Univ. Press, Ames.
NIELSEN, N. C., ANDERSON, S. A., MADSEN, A., and MORTENSEN, H. P. 1971. Acta Vet. Scand. *12*,202.
NIELSEN, N. O. 1981. Edema disease. *In* Diseases of Swine (A. D. Leman *et al.*, Editors), 5th Edition, Chap. 45, p. 478. Iowa State Univ. Press, Ames.
NORRDIN, R. W. 1969. Experimental zinc deficiency in weanling pigs. Ph.D. Thesis. Cornell Univ. Ithaca, New York.
NUWER, A. J., PERRY, T. W., PICKETT, R. A., CURTIN, T. M., FEATHERSTON, W. R., and BEESON, W. M. 1965. J. Anim. Sci. *24*,113.
O'BRIEN, J. J. 1981. Gastric ulcers. *In* Diseases of Swine (A. D. Leman *et al.*, Editors), 5th Edition, Chap. 59, p. 632. Iowa State Univ. Press, Ames.
OSE, E. E. 1971. J. Am. Vet. Med. Assoc. *160*,603.
PALLUDAN, B. 1966. *In* Swine in Biomedical Research, (L. K. Bustad, R. O. McClellan, and M. P. Burns, Editors.). Frayn Printing Co., Seattle, Washington.
PICKETT, R. A., FUGATE, W. H., HARRINGTON, R. B., PERRY, T. W., and CURTIN, T. M. 1969. J. Anim. Sci. *28*,837.
POESCHEL, G. P., and EMRO, J. E. 1972. J. Am. Vet. Med. Assoc. *160*,1637.
POND, W. G., and HOUPT, K. A. 1978. Biology of the Pig. Cornell Univ. Press. Ithaca, New York.
POND, W. G., ROBERTS, S. J., DUNN, J. A., KING, J. M., MANER, J. H., LOWREY, R. S., and OLAFSON, P. 1961. J. Anim. Sci. *20*,88.
POND, W. G., BARNES, R. H., BRADFIELD, R. B., KWONG, E., and KROOK, L. 1965. J. Nutr. *85*,57.
POND, W. G., BARNES, R. H., REID, J., KROOK, L., KWONG, E., and MOORE, A. U. 1966. *In* Swine in Biomedical Research (L. K. Bustad, R. O. McClellan, and M. P. Burns, Editors). Frayn Printing Co., Seattle, Washington.
POND, W. G., TODD, G. C., DUNN, J. A., KROOK, L., ROBERTS, S. J., and JONES, J. R. 1967. Vet. Med./Small Anim. Clin. *62*,67.
QAIN, A. H. 1952. Can. J. Comp. Med. *16*,265.
REILAND, S., ORDELL, N., LUNDEHEIM, N., and OLSSON, S. E. 1978. Acta Pathol (Suppl.) *358*,123.

RICHARDS. W. P. C., and FRASER, C. M. 1961. Cornell Vet. *51*,245.
ROSS, R. F. 1981A. Mycoplasmal diseases. *In* Diseases of Swine (A. D. Leman *et al.*, Editors), 5th Edition, Chap. 51, p. 535. Iowa State Univ. Press, Ames.
ROSS, R. F. 1981B. Streptococcal diseases. *In* Diseases of Swine (A. D. Leman *et al.*, Editors), 5th Edition, Chap. 52, p. 550. Iowa State Univ. Press, Ames.
SCHANTZ, P. M., JURANEK, D. D., and SCHULTZ, M. G. 1977. J. Infect. Dis. *136*,712.
SCHIPPER, I. A. 1961. Amer. J. Vet. Res. *22*,401.
SCHUMAN, R. D., and WOOD, R. L. 1970. *In* Diseases of Swine (H. W. Dunne, Editor). 3rd Edition, p. 508. Iowa State Univ. Press., Ames.
SEETHER, K. A., MJUA, T. S., PERRY, T. W., and BOEHM, P. N. 1971. J. Anim. Sci. *32*,1160.
SHARDA, D. P., WILSON, R. F., WILLIAMS, L. E., SWIGER, L. A., and CROSS, R. F. 1971. J. Anim. Sci. *32*,1169.
SHULL, L. R., and CHEEKE, P. R. 1983. J. Anim. Sci. *57* (Suppl. 2), 330.
SMITH, W. J., and PENNY, R. H. C. 1981. Rectal and vaginal prolapse. *In* Diseases of Swine (A. D. Leman, Editors), 5th Edition, Chap. 63, p. 681. Iowa State Univ. Press, Ames.
STEPHENSON, L. S., POND, W. G., NESHEIM, M. C., KROOK, L. P., and COMPTON, D. W. T. 1980. Exp. Parasitol. *49*,15.
STEWART, W. C. 1981. Hog cholera. *In* Diseases of Swine (A. D. Leman *et al.*, Editors), 5th Edition, Chap. 15, p. 224. Iowa State Univ. Press, Ames.
SWITZER, W. P. 1970. *In* Diseases of Swine (H. W. Dunne, Editor), 3rd Edition. Iowa State Univ. Press, Ames.
SWITZER, W. P. 1971. J. Am. Vet. Med. Assoc. *160*,651.
THARP, V. L. 1970. *In* Diseases of Swine (H. W. Dunne, Editor), 3rd Edition. Iowa State Univ. Press, Ames.
TOPEL, D. G., and CHRISTIAN, L. G. 1981. Porcine stress syndrome. *In* Diseases of Swine (A. D. Leman *et al.*, Editors, 5th Edition, Chap. 60, p. 647. Iowa State Univ. Press, Ames.
TUCKER, H. F., and SALMON, W. D. 1955. Proc. Soc. Exp. Biol. Med. *88*,613.
TWIEHAUS, M. J., and UNDERDAHL, N. R. 1970. *In* Diseases of Swine (H. W. Dunne, Editor), 3rd Edition. Iowa State Univ. Press, Ames.
ULLREY, D. E. 1970. Proc. Minn. Nutr. Conf. Univ. Minnesota, St. Paul, p. 15.
USDA 1983. USDA Farmer Bull. *1787*.
USDA 1969. Laboratory Diagnosis of Trichina in Pork. USDA Tech. Bull.
UZOUKWU, M., and SLEIGHT, S. D. 1971. J. Am. Vet. Med. Assoc. *160*,1641.
van VEEN, T. W. S. and GIBSON, C. D. 1983. Am. J. Vet. Res. *44*, 1732.
WAHLSTROM, R. C., and OLSON, O. E. 1959. J. Anim. Sci. *18*,141,578.
WAHLSTROM, R. C., KAMSTRA, L. D., and OLSON, O. E. 1955. J. Anim. Sci. *14*,105.
WALLACE, H. D. 1968. Feedstuffs *40*(27),22.
WALTERS, C. S. 1952. Proc. Amer. Wood Press. *48*,302.
WAYMAN, O. I., IWANGA, I., and HUGH, W. I. 1970. J. Anim. Sci. *30*,583.
WHITEHAIR, C. K. and MILLER, E. R. 1981. Nutritional Deficiencies. *In* Diseases of Swine (A. D. Leman *et al.*, Editors), 5th Edition, Chap. 61, p. 656, Iowa State Univ. Press, Ames.
WILCOCK, B. P. 1981. Salmonellosis. *In* Diseases of Swine (A. D. Leman *et al.*, Editors), 5th Edition, Chap. 42, p. 445. Iowa State Univ. Press, Ames.
WILSON, M. R. 1971. J. Amer. Vet. Med. Assoc. *160*,585.
WILSON, M. R. 1981. Enteric colibacillosis. *In* Diseases of Swine (A. D. Leman *et al.*, Editors), 5th Edition, Chap. 44, p. 471. Iowa State Univ. Press, Ames.
WOOD, R. L. 1981. Erysipelas. *In* Diseases of Swine (A. D. Leman *et al.*, Editors), 5th Edition, Chap. 43, p. 457. Iowa State Univ. Press, Ames.
WOODS, G. N., and BOHL, E. H. 1981. Porcine rotavirus infection. *In* Diseases of Swine (A. D. Leman *et al.*, Editors), 5th Edition, Chap. 26, p. 310. Iowa State Univ. Press, Ames.
YOUNG, G. A., and UNDERDAHL, N. R., 1953. Am. J. Vet. Res. *14*,571.
YOUNG, G. A., UNDERDAHL, N. R., and HINZ, R. W. 1955. Am. J. Vet. Res. *16*,123.
YOUNG, G. A., UNDERDAHL, N. R., WELCH, L. C., and CALDWELL, J. D. 1962. J. Am. Vet. Med. Assoc. *140*,1196.
YOUNG, L. G., VESONDER, R. F., FUNNEL, H. S., SIMMONS, J., and WILCOCK, B. 1981. J. Anim. Sci. *52*,1312.
YOUNG, L. G., KING, G. J., McGIRR, L., and SUTTON, J. C. 1982. J. Anim. Sci. *54*,976.

Part III

Feeds and Feeding

11

Nutrients and Feed Formulation

INTRODUCTION

Feed represents 55–85% of the total cost of commercial swine production, depending mainly on the relative costs of feed, labor, and housing in a particular situation. For this reason, it is highly important that economical as well as nutritionally balanced diets are provided during all phases of the life cycle. The economics of swine feeding are largely dependent on local conditions of feedstuff availability and competition for the same foodstuff for use by either humans or other animals. The pig has a simple digestive system with limited ability to utilize large quantities of forage so that, like the chicken, it is in direct competition with man for available food supplies. The degree of competition is related to cultural differences in food preferences. For example, wheat and potatoes are not usually fed to swine in the United States as the demand for human consumption holds the price too high, but in many other parts of the world these crops are commonly fed to swine. Similar relationships exist for other crops in other parts of the world.

PRACTICAL SWINE FEEDING

Practical swine feeding consists of meeting biological needs with appropriate combinations of feed ingredients and pure sources of limiting nutrients to yield an economical and nutritionally balanced diet. The nutrient requirements of swine are similar within each phase of the life cycle except for genetic differences in utilization or quantitative requirements for individual nutrients and for environmental effects such as ambient temperature, altitude, and season. Therefore, specific nutrients and signs of deficiency are discussed first, quantitative nutrient requirements for each phase of the life cycle are discussed next, and finally, practical aspects of diet formulation and feeding are considered. Since available feedstuffs are quite different in different parts of the world, practical feeding must take this into account. The nutrient composition and feeding value of the common feedstuffs available for formulation of swine diets in temperate and tropical environments are discussed in detail in Chapters 12–16. Average composition of some common feed ingredients for swine is given by the National Research Council (1982A) and in the National Research Council (1979) publication on nutrient requirements of swine.

NUTRIENTS REQUIRED BY SWINE

A nutrient, for the purpose of this discussion, is any chemical entity required by the animal to meet metabolic needs. It can be supplied by feeding, or in some cases by parenteral administration. Quantitative requirements and clinical signs of deficiency are given in publications by the National Research Council (1979) and by the Agricultural Research Council (1981). Practical feeding recommendations and current knowledge of digestion and absorption of nutrients by swine are reviewed by Low and Partridge (1980) and provided in several textbooks (Cunha 1977; Morrison 1956; Krider *et al.* 1981; Pond and Houpt 1978). In addition, the effects of environment on nutrient requirements have been reviewed (National Research Council 1982B). In this second edition, most of the specific references on which current knowledge of nutrient requirements of swine are based have been deleted to save space. Literature on early research in swine nutrition can be found in Swine Production in Temperate and Tropical Environments (Pond and Maner 1974).

The nutrients known to be required by swine follow:

Water. Water is of course necessary.

Fatty acids. Linoleic acid and perhaps arachidonic acid are required and some (0.06%) fat is needed for absorption of fat-soluble vitamins.

Vitamins. Vitamins include the fat soluble—A, D, E, and K; and the water soluble—biotin, choline (replaceable by methionine), folacin, niacin (nicotinic acid), pantothenic acid, riboflavin (vitamin B_2), thiamin (vitamin B_1), vitamin B_{12}, and vitamin B_6 (pyridoxine, pyridoxamine, or pyridoxal).

Protein. Proteins are probably not required as such, but they are necessary as a source of essential amino acids and N for nonessential amino acid synthesis.

Amino acids. Amino acids include arginine (for growth only), histidine, isoleucine, leucine, lysine, methionine (50% replaceable by cystine), phenylalanine (30% replaceable by tyrosine), threonine, tryptophan, and valine.

Inorganic elements. Inorganic elements include calcium, chlorine, cobalt (as component of vitamin B_{12}), copper, iodine, iron, magnesium, manganese, sodium, phosphorus, potassium, selenium, sulfur (as component of sulfate and of thiamine, methionine, cystine, and several other organic compounds), and zinc.

Carbohydrates. There is no proof that carbohydrates are required as a source of energy, but since all natural feedstuffs contain some carbohydrate, it becomes an academic question. Certainly, by far the greatest source of energy commonly available in swine feeding is carbohydrate and if a dietary requirement exists, it is as glucose since the ultimate utilization of energy by the cell is as glucose from dietary intake or from glucose formation from amino acids and fat.

Several of the vitamins are required only under specialized conditions. Biotin deficiency is normally encountered only when large quantities of raw egg white are fed. This contains a protein, avidin, which ties up biotin in the gastrointestinal tract. Recent field reports indicate that biotin deficiency may occur under special conditions of confinement feeding in commercial production. Vitamin K deficiency has been observed in pigs kept in wire- or slotted-floor pens which prevent coprophagy. Apparently sufficient vitamin K is synthesized by microflora in the lower intestinal tract to meet the metabolic requirement of the pig when given access to the feces. As confinement feeding of swine on slotted floors becomes more ccmmon (see Chapter 18), it behooves scientists and swine producers to be alert to deficiencies of vitamins and other nutrients induced by prevention of coprophagy.

CLINICAL SIGNS OF NUTRIENT DEFICIENCIES

Under practical feeding conditions, in which natural feedstuffs are used, only some of the required nutrients are likely to be deficient. These include the following: total protein; the amino acids, lysine and tryptophan and occasionally others; the vitamins A, D, E, riboflavin, pantothenic acid, niacin, choline, vitamin B_{12}; and the minerals calcium, phosphorus, iodine, sodium, chlorine, zinc, and selenium.

A chart adapted from a National Research Council publication (1979) summarizing clinical signs of deficiency of nutrients is shown in Table 11.1A. A similar chart of signs of dietary excesses is shown in Table 11.1B. Clinical signs of deficiency of some of these nutrients are recorded in several textbooks on nutrition and in the scientific literature. Brief descriptions of typical nutrient deficiency and toxicity signs are given in this book. Gross and microscopic pathology of nutrient deficiencies are illustrated in books by Follis (1958) and King *et al.* (1976).

The preparation of tables of nutrient requirements as done by the National Research Council (1979) and the Agricultural Research Council (1981) for use by feed manufacturers, swine producers, and scientists requires compilation and evaluation of results of research obtained under a wide variety of experimental conditions and environments. A diet that is nutritionally adequate for one population of pigs might not be entirely adequate for another population or for the same population under different environmental conditions. Therefore, a clear distinction must be made between minimum requirements and minimum recommended allowances. The minimum requirement can be considered to be that amount of a single nutrient needed to allow normal body function of all animals in a population under ideal conditions. The minimum recommended allowance can be considered the minimum requirement plus some arbitrary increment of that nutrient added as a "safety factor" to allow for increased metabolic needs induced by

TABLE 11.1A. Signs of Nutrient Deficiencies

Nutrient	Signs of nutrient deficiency	
	Clinical	*Subclinical*
Energy	Weakness, low body temperature, loss of weight, coma, and death	Hypoglycemia Loss of subcutaneous fat Elevated hematocrit and serum cholesterol Reduced blood glucose, calcium, and sodium
Protein: amino acid	Impaired growth Unthriftiness Reduced resistance to bacterial infection	Kwashiorkor-like signs in baby pigs, including reduced serum protein and serum albumin, anemia, gross edema, and increased lipid liver concentration
Fat: linoleic acid	Scaly dermatitis may appear	Small gall bladder Elevated triene/tetraene in tissue lipids
Vitamin A	Incoordination Lordosis Paralysis of rear limbs Night blindness Congenital defects	Retarded bone growth Increase in cerebrospinal fluid pressure Degeneration of sciatic and femoral nerves Minimal visual purple Atrophy of epithelial layers of genital tract
Vitamin D	Rickets Osteomalacia Low calcium tetany	Lack of bone calcification and proliferation of epiphyseal cartilage Rib and vertebra fracture Low plasma calcium, magnesium, and inorganic phosphorus levels Elevated serum alkaline phosphatase levels
Vitamin E—selenium	Edema Sudden death	Generalized edema Liver necrosis (hepatosis dietetica) Microangiopathy Cardiac muscle degeneration (mulberry heart) Pale, dystrophic muscle
Vitamin K	Pale newborn pigs with loss of blood from umbilical cord Sudden death following dicoumarin intake	Increased prothrombin time Increased blood-clotting time Internal hemorrhage Anemia due to blood loss
Thiamin	Poor appetite Poor growth Sudden death	Cardiac hypertrophy Bradycardia First and second degree auriculoventricular block Elevated plasma pyruvate
Riboflavin	Slow growth Seborrhea Impaired sow reproductivity	Lens cataracts Increase in neutrophilic leukocytes Birth of weak pigs with skeletal anomalies
Niacin	Poor appetite Poor growth	Necrotic lesions of intestine

TABLE 11.1A. (*Continued*)

Nutrient	*Signs of nutrient deficiency*	
	Clinical	*Subclinical*
	Severe diarrhea	
	Dermatitis	
Pantothenic acid	Poor appetite	Inflammation of colon
	Poor growth	Degeneration of sciatic and peripheral nerves
	Diarrhea	
	Unusual gait (goose-stepping)	Reduced blood pantothenic acid level
	Impaired sow reproductivity	Reduced free pantothenic acid level in milk
Vitamin B_6	Poor growth	Microcytic hypochromic anemia
	Epileptic seizures	Elevated serum iron
		Fatty infiltration of liver
		Elevated urinary xanthurenic acid
		Elevated gamma globulin-like blood protein fraction
Vitamin B_{12}	Depressed growth	Reduced serum and tissue B_{12} levels
	Hypersensitivity	
	Reduced sow reproductivity	
Choline	Slow growth	Fatty infiltration of liver
	Reduced litter size	Reduced conception rate
Biotin	Dermatosis	Reduced urinary biotin excretion
	Spasticity of hind legs	
Folacin	Poor growth	Normocytic anemia
	Weakness	
Calcium	Rickets	Lack of bone calcification
	Osteomalacia	Bones easily fractured
	Low calcium tetany	Low plasma calcium level
		Elevated serum inorganic phosphorus and alkaline phosphatase
Phosphorus	Poor growth	Lack of bone calcification
	Rickets	Bones easily fractured
	Osteomalacia	Low serum inorganic phosphorus level
		Elevated serum calcium and alkaline phosphatase
		Enlarged costochondral junction (beading)
Magnesium	Poor growth	Low serum magnesium and calcium
	Stepping syndrome	Reduced bone magnesium
	Weakened carpo-metacarpo-phalangeal and tarso-metatarso-phalangeal joints	
	Tetany	
Potassium	Anorexia	Reduced heart rate
	Rough hair coat	Increased *PR*, *QRS*, and *QT* intervals on electrocardiogram
	Emaciation	
	Ataxia	Reduced serum potassium

(*continued*)

TABLE 11.1A. (*Continued*)

Nutrient	Signs of nutrient deficiency	
	Clinical	*Subclinical*
Sodium	Poor appetite Low water consumption Unthriftiness	Negative sodium balance Elevated serum potassium Elevated plasma urea nitrogen Reduced chlorine retention
Chlorine	Poor growth	Reduced plasma chlorine Reduced sodium and potassium retention
Iron	Poor growth Rough hair coat Pallor Anoxia	Hypochromic microcytic anemia Enlarged heart and spleen Enlarged fatty liver Ascites Clumping of erythroblastic cells in bone marrow Reduced serum iron and percent transferrin saturation
Copper	Leg weakness Ataxia	Microcytic hypochromic anemia Reduced serum copper and ceruloplasmin Aortic rupture Cardiac hypertrophy
Zinc	Poor growth Poor appetite Parakeratosis	Reduced serum, tissue, and milk zinc Reduced serum albumin–globulin ratio Reduced serum alkaline phosphatase Reduced thymus weight Retarded testicular development Impaired reproductivity of sows
Iodine	Goiter Myxedema Sows farrow weak, hairless pigs	Enlarged hemorrhagic thyroid Hyperplasia of follicular epithelium of thyroid Reduced plasma protein-bound iodine
Manganese	Lameness in growing pigs Increased fat deposition in pregnant gilts with birth of weak pigs with poor sense of balance	Replacement of cancellous bone with fibrous tissue Early closure of distal epiphyseal plate Low serum manganese and alkaline phosphatase Negative manganese balance
Water	Poor appetite Dehydration Loss of body weight Possible salt poisoning Death	Elevated hematocrit Elevated plasma electrolytes Loss of temperature regulation Tissue dehydration

Source: National Resource Council 1979.

TABLE 11.1B. Signs of Dietary Excesses

Nutrient	*Toxic dietary level*[a]	*Age*	*Signs of dietary excess*
Calcium	1% (with limited zinc)	Immature	Depressed appetite, reduced rate of gain, parakeratosis
	1% (with adequate zinc and limited phosphorus)	Immature	Reduced rate of gain and reduced bone strength
Copper	300–500 mg/kg (in absence of higher levels of dietary iron and zinc)[b]	Immature	Reduced growth, lower hemoglobin, icterus, and death[c]
Iodine	800 mg/kg	Immature	Depressed feed intake and rate of gain, lowered hemoglobin,[d] and eye lesions
Iron	5000 mg/kg	Immature	Depressed feed intake and rate of gain, reduced serum inorganic phosphorus and femur ash, rickets[e]
Manganese	4000 mg/kg	Immature	Depressed feed intake, reduced growth rate, stiffness, and stilted gait
Selenium	5–8 mg/kg[f]	Immature	Anorexia, hair loss, separation of hoof and skin at coronary band, degenerative changes in liver and kidney
	10 mg/kg	Breeding (sows)	Reduced conception; pigs small, weak, or dead at birth
Sodium chloride and other sodium salts	1–8% (with severe water restriction)	All ages	Nervousness, weakness, staggering, epileptic seizures, paralysis and death
Zinc	2000 mg/kg	Immature	Growth depression, arthritis, hemorrhage in axillary spaces, gastritis, and enteritis
Arsenic	990 mg/kg	Immature	Poor growth, erythema, ataxia, posterior paralysis, quadraplegia, and blindness; myelin degeneration of optic and peripheral nerves
Cadmium	50 mg/kg	Immature	Reduced gain and hematocrit
	150 mg/kg	Immature	Severe depression of gain and hematocrit
	450 mg/kg	Immature	Severe depression of gain and hematocrit and appearance of dermatitis
Cobalt	400 mg/kg	Immature	Anorexia, growth depression, stiff-legged, humped back, incoordination and muscle tremors, anemia[g]
Fluorine:			
Soluble fluorides	100 mg/kg	Mature	Mottled enamel, enamel hypoplasia, softening of teeth, osteomalacia, excessive loss of weight by lactating sows
Rock phosphate F	200 mg/kg	Mature	

(*continued*)

TABLE 11.1B. (*Continued*)

Nutrient	*Toxic dietary level*[a]	*Age*	*Signs of dietary excess*
Gossypol[h]	200 mg/kg	Immature	Muscular weakness, dyspnea, generalized edema, death; myocarditis, hepatitis, and nephritis
Lead	660 mg/kg	Immature	Squealing as if in pain, diarrhea, salivation, grinding of teeth, depressed appetite, reduced growth rate, muscular tremors, ataxia, increased respiratory rate, decreased heart rate, enlarged carpal joints, impaired vision, clonic seizures, death
Mercury	Single oral dose of 5–15 mg methyl mercury dicyandiamide per kilogram of body weight	Immature	Anorexia, body weight loss, central nervous system depression, weakness, gagging, vomiting, diarrhea, ataxia, cyanosis, muscular tremors, postural and gait abnormalities, polyuria
Nitrate	1800 mg NO_3/kg	Immature	Growth depression, dyspnea and cyanosis, elevated methemoglobin, lymphocytosis, reduced serum vitamin A and E levels
Nitrite	400 mg NO_2/kg	Immature	
Urea	2.5%	Immature	Reduced feed intake and growth rate; increased plasma urea nitrogen level

Source: National Research Council 1979.
[a]The toxic dietary levels listed are those that have experimentally produced the signs indicated and are not necessarily minimum toxic or maximum tolerant levels.
[b]In a few instances, a dietary level of 250 mg/kg has resulted in signs of excess.
[c]In some instances, 500 mg/kg of copper has been fed without icterus or death occurring.
[d]Anemia of iodine toxicity alleviated with supplemental iron.
[e]Rickets from excessive dietary iron alleviated by increasing dietary phosphorus.
[f]Selenium toxicity partially alleviated with arsenic.
[g]Cobalt toxicity alleviated by supplemental methionine, iron, zinc, and manganese.
[h]Gossypol toxicity alleviated by increasing dietary iron to equal the weight of free gossypol.

environmental stresses. This concept is illustrated graphically in Fig. 11.1. The following discussion of individual nutrient deficiency signs pertains to results obtained with diets that failed to meet the minimum requirement of the specified nutrient. The same deficiency signs could be expected in animals in which the diet would have contained the minimum requirement if it had been fed under ideal conditions, but in which the metabolic requirement had been increased by some environmental stress. Examples of such stresses are cited under individual nutrients. From a practical standpoint, a reduction in rate of body weight gain, in efficiency of feed utilization, or in other productive function is the most important and usually the most common clinical sign of a nutrient deficiency. The following descriptions of individual deficiency signs, therefore, emphasize more specific changes. Practical aspects are emphasized in more detail in Chapters 14–16.

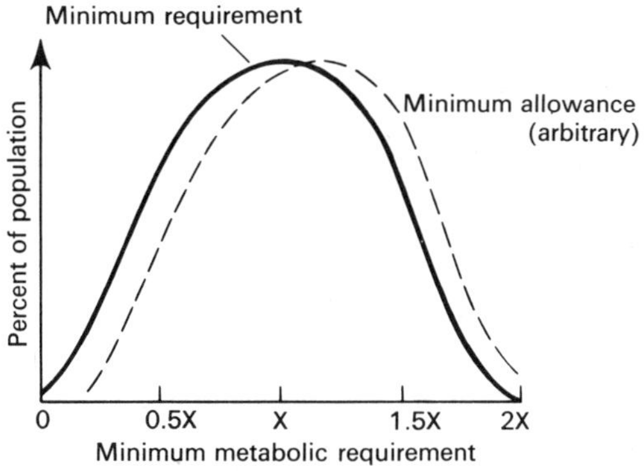

FIG. 11.1. Minimum requirements versus recommended allowance of nutrients.

Protein

Inadequate dietary protein is probably the most frequent nutrient deficiency, primarily because the common energy sources (maize, cereal grains, root and tuber crops) are low in protein, and protein supplements are expensive. The main signs of protein deficiency in growing pigs are slow growth and inefficient feed utilization accompanied in severe cases by unthriftiness. Blood serum protein is depressed, the liver becomes infiltrated with fat, and in severe deficiency there may be fluid accumulation (edema) in the jowl and scrotum.

The protein requirements for gestation and lactation have not been studied extensively but it appears that lactation is considerably more critical. Protein deficiency during lactation results in reduced milk yield and in reduced protein concentration of the colostrum but not of milk. The pregnant dam has the ability to protect the developing fetus partially against protein deficiency by using her own protein reserves to meet the needs of the fetus. For maximum nitrogen retention during late pregnancy in the gilt or sow, approximately 285 g of protein per day are required. This corresponds to a 15–18%-protein diet fed at accepted levels of feed intake during pregnancy. That this high level of protein is not required seems clear, because opaque-2 maize (containing 9.7% protein and approximately 0.3% lysine) as the only protein source allows normal reproduction although subsequent lactation requires a higher protein level. Amino acid balance is also apparently of importance since common maize is inferior to opaque-2 maize and to maize soybean meal diets. Complete deprivation of protein throughout pregnancy may cause the birth of smaller pigs but apparently has no effect on number born or livability. The long-term effects of protein deficiency on reproductive performance over more than two or three reproductive cycles is not known.

Lysine

This is the first or second limiting amino acid in maize and in most of the cereal grains so that a deficiency is common in pigs that are fed all-plant diets, unless the diet is maize and soybean meal, a combination which provides an excellent balance of amino acids. A deficiency of lysine results in anorexia and an unthrifty appearance in young pigs, but in older pigs a moderate deficiency may result in less protein deposition in the carcass and slightly less efficiency of feed utilization with no effect on weight gain. The requirement for gestation and lactation is apparently less than that for growth, but it is probably important to supplement low protein diets with protein sources high in lysine for optimum reproductive performance.

Tryptophan

Along with lysine, this amino acid is likely to be deficient in diets composed mainly of common maize. The signs of deficiency are slow growth, unthriftiness, and eye cataracts. Tryptophan can be converted by the tissues of the pig to the vitamin, niacin, so it is important to provide adequate niacin to minimize the tryptophan requirement. The hydroxy and keto analogues of tryptophan are utilized to some extent for growth, indicating that either the pig or the microflora of the gastrointestinal tract or both have the ability to add ammonia to form the amino acid. This may have future economic importance in amino acid fortification of swine diets.

Calcium

It is inappropriate to discuss mineral requirements independent of one another because of their interrelationships in metabolism. For example, phosphorus and vitamin D are both related to calcium metabolism, and levels of these nutrients in the diet affect the calcium requirement. An optimum ratio of calcium to phosphorus is between 1 : 1 and 2 : 1, depending on the biological availability of each in the feedstuffs used. When the ratio is less than 1 : 1, even a nearly adequate absolute amount of calcium in the diet may be associated with signs of calcium deficiency. Absorption of calcium from the gastrointestinal tract is increased by dietary vitamin D, decreased by high dietary fat, decreased by acid pH in the contents, decreased by phytin phosphorus, and probably affected by other poorly understood factors.

A deficiency in growing animals results in rickets (crooked legs, spontaneous bone fractures) due to failure of the protein matrix of bone to ossify. The corresponding condition in adults is osteomalacia. A dietary imbalance of calcium to phosphorus (ratio of less than 1 : 1) results in excessive bone resorption (fibrous osteodystrophy) due to secondary nutritional parathyroidism (Brown *et al.* 1966). The low level of calcium in the blood serum resulting from the calcium–phosphorus imbalance triggers the parathyroid to hypersecrete parathyroid hormone in an attempt to maintain blood serum calcium homeostasis. Calcium is lost from the bone in excessive amounts and replaced by fibrous connective tissue. This is a generalized phenomenon involving the entire skeleton. Clinical signs include distortion and loss of the nasal turbinates sometimes accompanied by a crooked snout and spontaneous fractures of ribs and long bones.

Calcium deficiency in gestation may result in spontaneous fractures of long bones, parturient paresis, or lameness of the dam. The fetus is partially protected by transfer of labile maternal sources of calcium across the placenta.

Phosphorus

A deficiency of phosphorus results in classical rickets. Adequate phosphorus supplementation to pig diets is especially important since there is a wide variation in the biological availability of the phosphorus from various sources. Plant sources of phosphorus (phytin phosphorus) are rather poorly utilized by the pig. Dicalcium phosphate is superior to soft phosphate with colloidal clay (colloidal phosphate) in promoting growth, feed utilization, femur ash, and inorganic serum phosphorus.

Phosphoric acid and di- and monocalcium phosphate are equal as sources of phosphorus while steamed bone meal and defluorinated rock phosphate are second in acceptability followed by Curaçao Island phosphate and colloidal phosphate. An average availability of phosphorus of 50% is believed to be typical for most diets fed to growing pigs (Agricultural Research Council 1981).

Zinc

The role of zinc in preventing skin lesions (parakeratosis) was first reported by Tucker and Salmon (1955). Since then, it has been established that a zinc deficiency can be precipitated by a high level of dietary calcium. Diets containing more than 1% calcium predispose to zinc-deficiency symptoms which can be prevented by the addition of zinc salts (carbonate, chloride, oxide, or sulfate) to the diet.

Since parakeratosis is not entirely absent even in diets containing less than 1.0% calcium, zinc should be added to most diets for swine. Hand-feeding a wet diet results in a lower incidence of parakeratosis than *ad libitum* feeding of a similar diet in the dry form.

The reduced growth rate and efficiency of feed utilization associated with parakeratosis are not entirely a reflection of food intake as shown by improved weight gains and feed conversion in the presence of added zinc at equalized energy intake. Serum alkaline phosphatase activity and tissue concentration of zinc are decreased in zinc deficiency. Phytic acid present in plant protein depresses zinc absorption and increases the incidence of parakeratosis. For this reason, the zinc requirement of pigs fed an all-plant protein diet is higher than that of pigs fed a milk-protein diet. Sex also affects dietary zinc requirement: females and intact males have a higher requirement than castrated males.

A wide margin of safety exists in zinc intake. Only levels as high as 1000 ppm of zinc (as zinc carbonate) in a maize–soybean meal diet produce toxicity symptoms. The addition of calcium to the diet has no effect on the level of zinc tolerance. Injection of zinc sulfate in parakeratotic animals causes local irritation, although it does relieve skin lesions.

Sodium and Chlorine

These two nutrients are considered together since the major dietary source is common salt (sodium chloride). While the required levels of sodium and chloride have reported to be approximately 0.10 and 0.13% of the diet, respectively, in practice, it is common to add 0.4% sodium chloride to the diet. Signs of deficiency include anorexia, muscular weakness, and a depraved appetite. An excess of salt can be induced by a diet containing as little as 2% of salt if water intake is restricted. Water restriction prevents normal urinary excretion of excess sodium and results in acute toxicity and death. There is very little danger of salt toxicity if water is not restricted.

Iron

Iron deficiency is common in the newborn pig because placental and mammary transfer are inadequate. Therefore, unless a supplemental source of iron is available to the baby pig, iron deficiency anemia (microcytic hypochromic) results within 2 to 3 weeks. Oral iron can be provided by tablet or drench (ferrous form is more efficiently absorbed than ferric form), but a more efficient way is by intramuscular injection of an iron–dextran complex. A level of approximately 150 mg of iron injected at 2–3 days of age is sufficient to meet the needs until about 3 weeks of age, when dry feed consumption provides adequate iron. Administration of iron to sows either orally or by injection during gestation or lactation does not prevent anemia in the suckling pigs. This reflects the poor placental and mammary transfer of iron. The slight increase in pig hemoglobin that has been observed by feeding sows high levels of iron during gestation is due to fecal contamination rather than metabolic transfer of iron through the milk.

Copper

The copper requirement of pigs is extremely small and symptoms of copper deficiency are obtained only by careful control of environmental contamination. In copper deficiency, there is microcytic hypochromic anemia, bowing of the legs, and spontaneous bone fractures.

Of far greater practical importance in swine production is the use of high levels of copper in the diet of growing pigs to stimulate growth. Levels of copper far above the metabolic requirement (200–250 ppm of diet) sometimes improve growth.

The mechanism of action is unknown. There is a danger of copper toxicity when dietary level of copper exceeds 200 ppm. The future of copper as a replacement for antibiotics as a growth stimulant for swine remains uncertain but the practice is widely accepted in Europe.

Iodine

Iodine deficiency in swine is manifested mainly by the birth of hairless pigs, often weak or stillborn. The thyroid of the progeny of sows fed iodine-deficient diets is increased in size and decreased in iodine content. The addition of 0.5% iodized salt containing 0.007% iodine (0.035 ppm in diet) to the diet of growing and adult animals prevents all signs of iodine deficiency and results in normal thyroid function.

Vitamin E and Selenium

A deficiency of vitamin E and selenium causes liver necrosis, a yellowish brown discoloration of body fat, and sudden death in young pigs. Whenever both the level of vitamin E and the level of selenium in the diet are marginal, the symptoms may occur. Apparently either one present in an adequate amount prevents the appearance of the lesions.

Common feedstuffs contain several tocopherols, the most biologically available of which is α-tocopherol. It has not always been considered important to be concerned with the level of vitamin E, when maize or cereal grains make up a significant proportion of the diet, but frequent field reports of vitamin E and/or selenium deficiency suggest that some methods of harvesting and storage (heat drying, for example) may destroy vitamin E or make it sufficiently biologically unavailable to precipitate deficiency signs when selenium is also limiting.

Some areas of the United States and other parts of the world are known to have soils of low selenium content (Kubota *et al.* 1967). Crops grown on these soils, in turn, are low in selenium.

Of practical concern in some locations is selenium toxicity, which occurs in areas where the soil levels and, in turn, plant levels, of selenium are excessive. Toxicity is manifested by poor growth, poor reproduction, and death. The reproductive failure associated with selenium toxicity can be prevented by feeding organic arsenic, which apparently reduces the amount of selenium absorbed from the gastrointestinal tract.

Cobalt

The only known metabolic requirement of cobalt is as a constituent of vitamin B_{12}. Some synthesis of vitamin B_{12} occurs in the lower intestinal tract, but absorption from this part of the tract is limited. Some reports have indicated a beneficial response to high cobalt feeding which would depend on vitamin B_{12} synthesis by the microflora of the gastrointestinal tract and absorption directly or after ingestion by coprophagy.

Sulfur

The requirement for sulfur is as inorganic sulfate and as a component of several organic nutrients, including the amino acids, methionine and cystine, and the vitamin thiamin. The normal diet provides adequate sulfur to meet these needs.

Manganese

The quantitative requirement of the pig is not clear, probably owing to differences among animals in the amounts stored in body tissues. Most feedstuffs contain manganese in amounts generally considered to exceed the dietary requirement. However, because of the widely diverse estimates of requirement, it is desirable to provide supplemental manganese in diet. The toxic level is several times the requirement.

Potassium and Magnesium

These are both required in large amounts, but are present in most common feedstuffs in amounts exceeding nutritional requirements. Therefore, they are not considered to be of practical importance in balancing diets.

Other Elements

Several elements normally present in animal tissues may have metabolic significance, but no requirement has been demonstrated for them in swine. These include aluminum, antimony, arsenic, barium, boron, bromine, cadmium, cesium, fluorine, germanium, lead, lithium, molybdenum, nickel, niobium, radium, rubinium, strontium, tellurium, tin, and zirconium. Their significance in practical swine nutrition remains unknown. Most of them are of more concern because of their toxic properties than as possible required nutrients. Others have been found to be required for specific animal species, but have not yet been shown to be required by swine. Such elements include vanadium, chromium, and silicon (National Research Council 1980; Church and Pond 1982).

Vitamin A

With the exception of yellow maize and forages, most common feedstuffs for swine are low in carotene and vitamin A. Liver storage on an adequate diet prevents immediate appearance of vitamin A deficiency symptoms when animals are fed a deficient diet; however, after tissue stores are depleted, growth is reduced and other deficiency signs appear. These include incoordination, emaciation and loss of weight, paralysis, xerophthalmia, and reproductive failure. In addition, several teratological effects have been noted; including malformations of the brain and spinal cord and of the palate, eyes, limbs, heart, lungs, diaphragm, liver, and kidneys.

Since vitamin A is stored in large quantities in the liver, hypervitaminosis A can be produced by excessive feeding. Effects include anomalies of the eyes, lungs, and major blood vessels in progeny of females fed an excess during pregnancy.

The pig can efficiently convert dietary carotene to vitamin A, so that blood levels of carotene are normally low while those of vitamin A are normally high. Thus, pork fat is low in carotene and has a characteristic white color.

Vitamin D

Vitamin D (vitamins D_2 and D_3 are equally utilized by the pig) is closely related to dietary calcium and phosphorus metabolism through its effect on their absorption and utilization for bone metabolism. Vitamin D deficiency results in reduced absorption of dietary calcium and phosphorus.

The source of protein, also, is related to the vitamin D requirement. The requirement is increased with soybean protein as compared with casein; this may be related to the presence of high levels of phytate in soybean protein. Dietary vitamin D is not required if the pig is allowed exposure to ultraviolet light from the sun which activates 7-dehydrocholesterol to vitamin D_3 which can then be utilized for metabolic purposes. Signs of deficiency include stiffness and lameness even in the presence of adequate calcium, phosphorus, and other minerals. Lameness is followed by appetite failure and growth depression; it can be corrected quickly by injection or feeding of vitamin D. As the trend toward complete confinement of swine throughout the life cycle continues, the importance of adequate dietary vitamin D fortification increases, especially in temperate climates where completely enclosed housing is used.

Riboflavin

Signs of riboflavin deficiency in baby pigs include dermatitis, eye lens cataracts, diarrhea, vomiting, and inflammation of the colon and anal mucosa. Baby pigs from sows fed riboflavin-deficient diets are incoordinated and weak, and mortality rate is high.

Niacin

The niacin present in most cereal grains and maize is not biologically available for the pig, so that it must be added to diets which, on an analytical basis, contain adequate quantities. Niacin deficiency results in unthriftiness, diarrhea, and necrotic enteritis. The lesions in the gastrointestinal tract resemble those produced by *Salmonella choleraesuis* (see Chapter 10, p. 161). Since the pig can convert dietary tryptophan to niacin, the level of tryptophan has an influence on the response to added niacin.

Pantothenic Acid

Pantothenic acid deficiency results in a characteristic gait involving elevation of the hind limbs (''goose-stepping''), alopecia, diarrhea, and abscesses and ulcers in the colon. In the adult female deficiency results in the birth of weak pigs or complete reproductive failure.

Vitamin B_{12}

Tissue storage is quite considerable and accounts for the earlier reference to vitamin B_{12} as animal protein factor (APF).

Vitamin B_{12} deficiency results in macrocytic anemia in growing animals and reproductive failure in adults. Because of the trend in many parts of the world toward all-plant diets for swine, the addition of vitamin B_{12} is of special importance in diet formulation, as tissue depletion occurs within a few weeks when a deficient diet is fed.

Vitamin B_1, Vitamin B_6, Folacin, Biotin, Choline, and Vitamin K

None of these is likely to be deficient in practical diets for swine. Vitamin B_1 (thiamin) deficiency, in addition to its adverse effect on normal growth and reproduction, causes bradycardia and damage to the heart. The small quantitative dietary requirement for thiamin and its relatively high tissue storage by pigs account for the infrequency of thiamin deficiency in pigs. The high tissue storage makes pork a good source of thiamin in human nutrition.

A deficiency of vitamin B_6 (pyridoxine, pyridoxal, and pyridoxamine are all active forms) results in microcytic hyperchromic anemia, epileptic-like convulsions, reduced growth, and impaired reproduction.

A folacin deficiency results in macrocytic anemia but a deficiency has never been reported in pigs fed diets composed of common feedstuffs.

Biotin is not usually considered of practical concern as it is normally synthesized in adequate amounts by the microflora of the intestinal tract. However, there have been field reports of deficiency signs under some conditions of confinement. These symptoms are dermatitis, alopecia, and cracked hoofs.

Choline is present in most feedstuffs in excess of the dietary requirement. However, as with other nutrients, the use of certain processed feedstuffs (such as cane sugar) as a major part of the diet may result in suboptimum choline intake. Therefore, a supplemental source should be provided.

Choline deficiency results in fatty infiltration of the liver, kidney lesions, birth of weak and uncoordinated pigs, and reduced growth. Dietary methionine at a level above that needed for tissue protein synthesis allows normal growth in choline-deprived pigs. Therefore, a high protein diet, unless severely deficient in methionine, would probably prevent choline deficiency, but this would not be an economical way to meet the choline requirement.

Vitamin K is normally synthesized in adequate amounts by the microflora of the intestinal tract. However, some field reports have suggested that a hemorrhagic syndrome in growing pigs, causing sudden deaths and subcutaneous hemorrhages, may be the result of vitamin K deficiency that has been induced by consumption of moldy feeds containing a vitamin K antagonist. Therefore, under certain conditions there may be a valid reason for adding vitamin K (menadione) to practical swine diets.

Fat and Fatty Acids

The pig has not been shown to require fat except as a source of possible essential fatty acids and as a vehicle for absorption of fat-soluble vitamins. The effect of dietary unsaturated fatty acids on growth remains controversial. Although early data indicated growth impairment with fat-free diets, more recent data have failed to demonstrate an effect on growth, although skin lesions develop. It is not clear whether linoleic acid deficiency or that of another fatty acid or an imbalance of fatty acids is responsible for the skin lesions.

A level of linoleic acid of 1% of the dietary calories has been suggested as a minimum requirement. Fat deficiency is not likely when diets contain appreciable amounts of maize or the cereal grains. However, when roots and tubers (such as potatoes and cassava) or processed carbohydrates (such as cane sugar) make up a major part of the energy source and when solvent-extracted protein supplements are used exclusively, there is potential danger of skin lesions related to fat (or fatty acid) deficiency.

QUANTITATIVE NUTRIENT REQUIREMENTS IN LIFE CYCLE FEEDING

Estimates of quantitative dietary requirements of each nutrient required by swine during each phase of the life cycle have been compiled by several groups of researchers (National Research Council 1979; Agricultural Research Council 1967, 1981). The data of the National Research Council are in Tables 11.2A,B

TABLE 11.2A. Nutrient Requirements of Growing-Finishing Swine Fed *Ad Libitum* (Percent or Amount per Kilogram of Diet)[a]

Liveweight (kg)		*1–5*[a]	*5–10*	*10–20*	*20–35*	*35–60*	*60–100*
Expected daily gain (g)		*200*	*300*	*500*	*600*	*700*	*800*
Expected efficiency (g gain/kg feed)		*800*	*600*	*500*	*400*	*350*	*270*
Expected efficiency (feed/gain)		*1.25*	*1.67*	*2.00*	*2.50*	*2.86*	*3.75*
Digestible energy[b]	kcal	3700	3500	3370	3380	3390	3395
Metabolizable energy[b]	kcal	3600	3400	3160	3175	3190	3195
Crude protein[c]	%	27	20	18	16	14	13
Indispensable amino acids							
Lysine	%	1.28	0.95	0.79	0.70	0.61	0.57
Arginine	%	0.33	0.25	0.23	0.20	0.18	0.16
Histidine	%	0.31	0.23	0.20	0.18	0.16	0.15
Isoleucine	%	0.85	0.63	0.56	0.50	0.44	0.41
Leucine	%	1.01	0.75	0.68	0.60	0.52	0.48
Methionine + cystine[d]	%	0.76	0.56	0.51	0.45	0.40	0.30
Phenylalanine + tyrosine[e]	%	1.18	0.88	0.79	0.70	0.61	0.57
Threonine	%	0.76	0.56	0.51	0.45	0.39	0.37
Tryptophan[f]	%	0.20	0.15	0.13	0.12	0.11	0.10
Valine	%	0.85	0.63	0.56	0.50	0.44	0.41
Mineral elements							
Calcium	%	0.90	0.80	0.65	0.60	0.55	0.50
Phosphorus[g]	%	0.70	0.60	0.55	0.50	0.45	0.40
Sodium	%	0.10	0.10	0.10	0.10	0.10	0.10
Chlorine	%	0.13	0.13	0.13	0.13	0.13	0.13
Potassium	%	0.30	0.26	0.26	0.23	0.20	0.17
Magnesium	%	0.04	0.04	0.04	0.04	0.04	0.04
Iron	mg	150	140	80	60	50	40
Zinc	mg	100	100	80	60	50	50
Manganese	mg	4.0	4.0	3.0	2.0	2.0	2.0
Copper	mg	6.0	6.0	5.0	4.0	3.0	3.0
Iodine	mg	0.14	0.14	0.14	0.14	0.14	0.14

(*continued*)

TABLE 11.2A. (*Continued*)

Liveweight (kg)		*1–5*[a]	*5–10*	*10–20*	*20–35*	*35–60*	*60–100*
Expected daily gain (g)		*200*	*300*	*500*	*600*	*700*	*800*
Expected efficiency (g gain/kg feed)		*800*	*600*	*500*	*400*	*350*	*270*
Expected efficiency (feed/gain)		*1.25*	*1.67*	*2.00*	*2.50*	*2.86*	*3.75*
Selenium	mg	0.15	0.15	0.15	0.15	0.15	0.10
Vitamins							
Vitamin A or β-carotene	IU	2200	2200	1750	1300	1300	1300
	mg	8.8	8.8	7.0	5.2	5.2	5.2
Vitamin D	IU	220	220	200	200	150	125
Vitamin E	IU	11	11	11	11	11	11
Vitamin K (menadione)	mg	2.0	2.0	2.0	2.0	2.0	2.0
Riboflavin	mg	3.0	3.0	3.0	2.6	2.2	2.2
Niacin[h]	mg	22	22	18	14	12	10
Pantothenic acid	mg	13	13	11	11	11	11
Vitamin B_{12}	μg	22	22	15	11	11	11
Choline[i]	mg	1100	1100	900	700	550	400
Thiamin	mg	1.3	1.3	1.1	1.1	1.1	1.1
Vitamin B_6	mg	1.5	1.5	1.5	1.1	1.1	1.1
Biotin[j]	mg	0.10	0.10	0.10	0.10	0.10	0.10
Folacin[j]	mg	0.60	0.60	0.60	0.60	0.60	0.60

Source: National Research Council 1979.

[a]Requirements reflect the estimated levels of each nutrient needed for optimal performance when a fortified grain–soybean meal diet is fed, except that a substantial level of milk products should be included in the diet of the 1–5 kg pig. Concentrations are based upon amounts per unit of air-dry diet (i.e., 90% dry matter).

[b]These are not absolute requirements but are suggested energy levels derived from diets containing corn and soybean meal (44% crude protein). When lower energy grains are fed, these energy levels will not be met; consequently, feed efficiency would be lowered.

[c]Approximate protein levels required to meet the need for indispensable amino acids when a fortified grain–soybean meal diet is fed to pigs weighing more than 5 kg.

[d]Methionine can fulfill the total requirement; cystine can meet at least 50% of the total requirement.

[e]Phenylalanine can fulfill the total requirement; tyrosine can meet at least 50% of the total requirement.

[f]It is assumed that usable tryptophan content of corn does not exceed 0.05%.

[g]At least 30% of the phosphorus requirement should be provided by inorganic and/or animal product sources.

[h]It is assumed that most of the niacin present in cereal grains and their by-products is in bound form and thus unavailable to swine. The niacin contributed by these sources is not included in the requirement listed. In excess of its requirement for protein synthesis, tryptophan can be converted to niacin (50 mg tryptophan yields 1 mg niacin).

[i]In excess of its requirement for protein synthesis, methionine can spare dietary choline (4.3 mg methionine is equal in methylating capacity to 1 mg choline).

[j]These levels are suggested. No requirements have been established.

TABLE 11.2B. Nutrient Requirements of Growing-Finishing Swine Fed *Ad Libitum*[a]

Liveweight (kg)		*1–5*[a]	*5–10*	*10–20*	*20–35*	*35–60*	*60–100*
Air-dry feed intake (g)		*250*	*500*	*1000*	*1500*	*2000*	*3000*
Digestible energy[b]	kcal	925	1750	3370	5055	6740	10110
Metabolizable energy[b]	kcal	900	1700	3160	4740	6320	9480
Crude protein[c]	g	67.5	100	180	240	280	390
Indispensable amino acids							
Lysine	g	3.2	4.8	7.9	10.5	12.2	17.1
Arginine	g	0.8	1.3	2.3	3.0	3.6	4.8
Histidine	g	0.8	1.2	2.0	2.7	3.2	4.5
Isoleucine	g	2.1	3.2	5.6	7.5	8.8	12.3

(*continued*)

TABLE 11.2B. (*Continued*)

Liveweight (kg)		*1–5*[a]	*5–10*	*10–20*	*20–35*	*35–60*	*60–100*
Air-dry feed intake (g)		*250*	*500*	*1000*	*1500*	*2000*	*3000*
Leucine	g	2.5	3.8	6.8	9.0	10.4	14.4
Methionine + cystine[d]	g	1.9	2.8	5.1	6.8	8.0	9.0
Phenylalanine + tyrosine[e]	g	3.0	4.4	7.9	10.5	12.2	17.1
Threonine	g	1.9	2.8	5.1	6.8	7.8	11.1
Tryptophan[f]	g	0.5	0.8	1.3	1.8	2.2	3.0
Valine	g	2.1	3.2	5.6	7.5	8.8	12.3
Mineral elements							
Calcium	g	2.3	4.0	6.5	9.0	11.0	15.0
Phosphorus[g]	g	1.8	3.0	5.5	7.5	9.0	12.0
Sodium	g	0.25	0.5	1.0	1.5	2.0	3.0
Chlorine	g	0.33	0.7	1.3	2.0	2.6	3.9
Potassium	g	0.75	1.3	2.6	3.5	4.0	5.1
Magnesium	g	0.10	0.2	0.4	0.6	0.8	1.2
Iron	mg	38	70	80	90	100	120
Zinc	mg	25	50	80	90	100	150
Manganese	mg	1.0	2	3	3	4	6
Copper	mg	1.5	3	5	6	6	9
Iodine	mg	0.04	0.07	0.14	0.21	0.28	0.42
Selenium	mg	0.04	0.08	0.15	0.22	0.30	0.30
Vitamins							
Vitamin A or β-carotene	IU	550	1100	1750	1950	2600	3900
	mg	2.2	4.4	7.0	7.8	10.4	15.6
Vitamin D	IU	55	110	200	300	300	375
Vitamin E	IU	2.8	5.5	11	17	22	33
Vitamin K (menadione)	mg	0.50	1.1	2.2	3.3	4.4	6
Riboflavin	mg	0.75	1.5	3.0	3.9	4.4	7
Niacin[h]	mg	5.5	11	18	21	24	30
Pantothenic acid	mg	3.3	6.5	11	17	22	33
Vitamin B_{12}	μg	5.5	11	15	17	22	33
Choline[i]	mg	275	550	900	1050	1100	1200
Thiamin	mg	0.33	0.65	1.1	1.7	2.2	3.3
Vitamin B_6	mg	0.38	0.75	1.5	1.7	2.2	3.3
Biotin[j]	mg	0.03	0.05	0.10	0.15	0.20	0.30
Folacin[j]	mg	0.15	0.30	0.60	0.90	1.2	1.8

Source: National Research Council 1979.

[a]Requirements reflect the estimated levels of each nutrient needed for optimal performance when a fortified grain–soybean meal diet is fed, except that a substantial level of milk products should be included in the diet of the 1–5 kg pig. Concentrations are based upon amounts per unit of air-dry diet (i.e., 90% dry matter).

[b]These are not absolute requirements, but are suggested energy levels derived from diets containing corn and soybean meal (44% crude protein). When lower energy grains are fed, these energy levels will not be met; consequently, feed efficiency would be lowered.

[c]Approximate protein levels required to meet the need for indispensable amino acids when a fortified grain–soybean meal diet is fed to pigs weighing more than 5 kg.

[d]Methionine can fulfill the total requirement; cystine can meet at least 50% of the total requirement.

[e]Phenylalanine can fulfill the total requirement; tyrosine can meet at least 50% of the total requirement.

[f]It is assumed that usable tryptophan content of corn does not exceed 0.05%.

[g]At least 30% of the phosphorus requirement should be provided by inorganic and/or animal product sources.

[h]It is assumed that most of the niacin present in cereal grains and their by-products is in bound form and thus unavailable to swine. The niacin contributed by these sources is not included in the requirement listed. In excess of its requirement for protein synthesis, tryptophan can be converted to niacin (50 mg tryptophan yields 1 mg niacin).

[i]In excess of its requirement for protein synthesis, methionine can spare dietary choline (4.3 mg methionine is equal in methylating capacity to 1 mg choline).

[j]These levels are suggested. No requirements have been established.

and 11.3A,B. It must be recognized that the values presented must be considered only as guidelines since values were obtained under experimental conditions using, in most studies, better than average physical facilities and management. Such factors as differences in genetic background and climatic or physical environment might be expected to affect the quantitative requirements (see Fig. 11.1).

TABLE 11.3A. Nutrient Requirements of Breeding Swine (Percent or Amount per Kilogram of Diet)[a]

		Bred gilts and sows; young and adult boars[b]	*Lactating gilts and sows*
Digestible energy	kcal	3400	3395
Metabolizable energy	kcal	3200	3195
Crude protein[c]	%	12	13
Indispensable amino acids			
Arginine	%	0	0.40
Histidine	%	0.15	0.25
Isoleucine	%	0.37	0.39
Leucine	%	0.42	0.70
Lysine	%	0.43	0.58
Methionine + cystine[d]	%	0.23	0.36
Phenylalanine + tyrosine[e]	%	0.52	0.85
Threonine	%	0.34	0.43
Tryptophan[f]	%	0.09	0.12
Valine	%	0.46	0.55
Mineral elements			
Calcium	%	0.75	0.75
Phosphorus[g]	%	0.60	0.50
Sodium	%	0.15	0.20
Chlorine	%	0.25	0.30
Potassium	%	0.20	0.20
Magnesium	%	0.04	0.04
Iron	mg	80	80
Zinc	mg	50	50
Manganese	mg	10	10
Copper	mg	5	5
Iodine	mg	0.14	0.14
Selenium	mg	0.15	0.15
Vitamins			
Vitamin A or β-carotene	IU	4000	2000
	mg	16	8
Vitamin D	IU	200	200
Vitamin E	IU	10	10
Vitamin K (menadione)	mg	2	2
Riboflavin	mg	3	3
Niacin[h]	mg	10	10
Pantothenic acid	mg	12	12
Vitamin B_{12}	μg	15	15
Choline	mg	1250	1250

(*continued*)

TABLE 11.3A. (*Continued*)

		Bred gilts and sows; young and adult boars[b]	*Lactating gilts and sows*
Thiamin	mg	1	1
Vitamin B_6	mg	1	1
Biotin[i]	mg	0.1	0.1
Folacin[i]	mg	0.6	0.6

Source: National Research Council 1979.

[a]Requirements reflect the estimated levels of each nutrient needed for optimal performance when a fortified grain–soybean meal diet is fed. Concentrations are based upon amounts per unit of air-dry diet (i.e., 90% dry matter).

[b]Requirements for boars of breeding age have not been established. It is suggested that the requirements will not differ significantly from that of bred gilts and sows.

[c]Approximate protein levels required to meet the need for indispensable amino acids when a fortified grain–soybean meal diet is fed. The true digestibilities of the amino acids were assumed to be 90%.

[d]Methionine can fulfill the total requirement; cystine can meet at least 50% of the total requirement.

[e]Phenylalanine can fulfill the total requirement; tyrosine can meet at least 50% of the total requirement.

[f]It is assumed that usable tryptophan content of corn does not exceed 0.05%.

[g]At least 30% of the phosphorus requirement should be provided by inorganic and/or animal product sources.

[h]It is assumed that most of the niacin present in cereal grains and their by-products is in bound form and thus unavailable to swine. The niacin contributed by these sources is not included in the requirement listed. In excess of its requirement for protein synthesis, tryptophan can be converted to niacin (50 mg tryptophan yields 1 mg niacin).

[i]These levels are suggested. No requirements have been established.

TABLE 11.3B. Daily Nutrient Requirements of Breeding Swine[a]

Air-dry feed intake (g):		*Bred gilts and sows; young and adult boars*	*Lactating gilts and sows*		
		1,800[b]	*4,000*	*4,750*	*5,500*
Digestible energy	kcal	6,120[c]	13,580	16,130	18,670
Metabolizable energy	kcal	5,760[c]	12,780	15,180	17,570
Crude protein	g	216	520	618	715
Indispensable amino acids					
Arginine	g	0	16.0	19.0	22.0
Histidine	g	2.7	10.0	11.9	13.8
Isoleucine	g	6.7	15.6	18.5	21.4
Leucine	g	7.6	28.0	33.2	38.5
Lysine	g	7.7	23.2	27.6	31.9
Methionine + cystine[d]	g	4.1	14.4	17.1	19.8
Phenylalanine + tyrosine[e]	g	9.4	34.0	40.4	46.8
Threonine	g	6.1	17.2	20.4	23.6
Tryptophan[f]	g	1.6	4.8	5.7	6.6
Valine	g	8.3	22.0	26.1	30.2
Mineral elements					
Calcium	g	13.5	30.0	35.6	41.2
Phosphorus[g]	g	10.8	20.0	23.8	27.5
Sodium	g	2.7	8.0	9.5	11.0
Chlorine	g	4.5	12.0	14.2	16.5
Potassium	g	3.6	8.0	9.5	11.0
Magnesium	g	0.7	1.6	1.9	2.2

(*continued*)

TABLE 11.3B. *(Continued)*

Air-dry feed intake (g):		*Bred gilts and sows; young and adult boars*	*Lactating gilts and sows*		
		1,800[b]	*4,000*	*4,750*	*5,500*
Iron	mg	144	320	380	440
Zinc	mg	90	200	238	275
Manganese	mg	18	40	48	55
Copper	mg	9	20	24	28
Iodine	mg	0.25	0.56	0.66	0.77
Selenium	mg	0.27	0.40	0.48	0.55
Vitamins					
Vitamin A or β-carotene	IU	7,200	8,000	9,500	11,000
	mg	28.8	32.0	38.0	44.0
Vitamin D	IU	360	800	950	1,100
Vitamin E	IU	18.0	40.0	47.5	55.0
Vitamin K	mg	3.6	8.0	9.5	11.0
Riboflavin	mg	5.4	12.0	14.2	16.5
Niacin[h]	mg	18.0	40.0	47.5	55.0
Pantothenic acid	mg	21.6	48.0	57.0	66.0
Vitamin B_{12}	μg	27.0	60.0	71.2	82.5
Choline	mg	2,250.0	5,000.0	5,940.0	6,875.0
Thiamin	mg	1.8	4.0	4.8	5.5
Vitamin B_6	mg	1.8	4.0	4.8	5.5
Biotin[i]	mg	0.18	0.4	0.48	0.55
Folacin[i]	mg	1.08	2.4	2.8	3.3

Source: National Research Council 1979.

[a]Requirements reflect the estimated levels of each nutrient needed for optimal performance when a fortified grain–soybean meal diet is fed. Concentrations are based upon amounts per unit of air-dry diet (i.e., 90% dry matter).

[b]An additional 25% should be fed to working boars.

[c]Individual feeding and moderate climatic conditions are assumed. An energy reduction of about 10% is possible when gilts and sows are tethered or individually penned in a stall in environmentally controlled housing. An energy increase of about 25% is suggested for cold climatic (winter) conditions.

[d]Methionine can fulfill the total requirement; cystine can meet at least 50% of the total requirement.

[e]Phenylalanine can fulfill the total requirement; tyrosine can meet at least 50% of the total requirement.

[f]It is assumed that usable tryptophan content of corn does not exceed 0.05%.

[g]At least 30% of the phosphorus requirements should be provided by inorganic and/or animal product sources.

[h]It is assumed that most of the niacin present in cereal grains and their by-products is in bound form and thus unavailable to swine. The niacin contributed by these sources is not included in the requirement listed. In excess of its requirement for protein synthesis, tryptophan can be converted to niacin (50 mg tryptophan yields 1 mg niacin).

[i]These levels are suggested. No requirements have been established.

Gestation and Prenatal Nutrition

The nutrient requirements for gestation are influenced by two separate productive functions; first, the needs of maintaining the pregnant animal per se and, second, the provision of an adequate environment and nutrient supply for the developing fetus. The first necessitates more than just meeting the maintenance requirements of an adult, since the endocrinological changes during pregnancy include increased basal metabolic rate and increased size and vascularity of the uterus as well as of the mammary glands in preparation for lactation. It is clear, then, that it is necessary to provide sufficient nutrition of the pregnant sow to allow normal reproduction and mothering ability (including subsequent lactation) and the marketing of normal progeny that have in no way been stunted in postnatal development due to prenatal malnutrition. It is possible, at least with some nutrients, to obtain normal reproduction through one reproductive cycle on a diet that is clearly not adequate for the dam. This possibility has been demonstrated experimentally for protein, and the results clearly show that the dam is able to draw on her own reserves to meet the needs of the fetus. However, such a diet cannot be considered adequate since it does not satisfy the long-term requirements of both dam and fetus.

The nutrition of the fetus depends on the transfer of nutrients across the placenta from the maternal blood. It is clear that adequate fetal nutrition, therefore, depends on adequate levels of circulating nutrients in the blood of the dam. With some nutrients the dam's blood serum concentration is maintained within a fairly normal range by homeostatic processes. For example, calcium can be resorbed from bone if there is insufficient dietary intake to maintain normal blood concentration. In contrast, a deficiency of the B vitamin, riboflavin, for example, causes a drop in blood riboflavin level of the dam because of limited tissue storage, and the amount available for transfer to the fetus is reduced accordingly. Therefore, in discussing fetal nutrition, it is important to specify the nutrient under consideration.

The nutrient requirements for fetal growth are hidden under the umbrella of pregnancy plus fetal nutrient requirements so that, until technical skills in gynecology allow the "test tube" rearing of pig fetuses, the question of the nutritive requirement of the fetus per se is academic. Therefore, these requirements are presented as requirements for gestation and are at best imprecise, as illustrated in the foregoing discussion.

Placental transfer of nutrients must occur or there would be no fetal growth. Some nutrients are transferred more efficiently than others. All of the fat-soluble vitamins and B vitamins are readily transferred to the fetus. Intact protein does not pass the placental barrier except in negligible amounts but individual amino acids do, and these form the main source of tissue protein synthesis by the fetus. Nonprotein nitrogen is used by the fetus to synthesize nonessential amino acids probably as early as 30 days of gestation. Individual fatty acids, but not triglycerides, apparently are transferred across the placenta, but the main source of energy for the fetal pig is glucose transported from maternal blood.

Transfer of water and the inorganic elements to the fetus increases several-fold during the latter part of gestation. The mineral composition of maternal plasma and the products of conception in swine at 110 days of gestation are summarized in Table 11.4. The mineral transfer to the fetus involves several factors which have been outlined by Hansard (1966) as follows: (a) availability of the mineral to the dam for transfer; (b) placental type and degree of permeability; (c) physiological age and size of the fetus; (d) size, weight, and charge of the particles being transferred. In general, the greatest placental transfer of minerals occurs during the last one-third of pregnancy. In comparison with the transfer of other mineral elements, the efficiency of iron transport to the fetus is very low and is unresponsive to maternal diet. This inefficient transport of iron across the placenta is not fully understood.

TABLE 11.4. Mineral Composition of Maternal Plasma and Placenta, Placental Fluids, and Fetus at 110 Days after Conception (per Kilogram Fresh Weight)

Mineral	*Maternal Plasma*	*Placental Fluids*	*Placenta*	*Fetus*
Calcium, g	0.118	0.620	0.589	9.9
Chloride, meq	106	87	88	50
Copper, mg	1.02	0.3	1.6	2.5
Iron, mg	1.15	0.8	30.5	34.8
Magnesium, g	0.026	0.022	0.085	0.3
Phosphorus, g	0.053	0.148	0.620	5.3
Potassium, meq	6	15	20	50
Sodium, meq	145	108	97	92
Sulfur (inorg.), g	0.012	0.037	0.032	1.0
Zinc, mg	0.92	0.4	5.8	6.8

Source: Hansard 1966.

Lactation

The nutrient output in milk during a 5-week lactation period is much greater than the nutrient deposition in fetuses and placental membranes during a 114-day gestation period. Therefore, the nutrient requirements of lactation are much more demanding than those of gestation. While the sow can draw on her own body reserves to some extent for milk production, the extent of this is limited, and complete or partial lactation failure results if nutrient restriction is severe or prolonged. The degree to which reserves of the dam can be drawn upon for lactation when dietary intake is insufficient is related to the specific nutrient concerned. For example, as discussd in the foregoing paragraphs, the transfer of iron into the milk is extremely inefficient so that even a very high dietary concentration does not appreciably increase the concentration in the milk. In general, a deficiency of a particular nutrient in the diet is manifested more by a reduction in total milk production than by a decreased concentration of that nutrient in the milk. The most striking effect of level of intake on total milk production is with energy. If the lactating sow is not allowed *ad libitum* or near *ad libitum* intake especially after the first week, milk production is decreased. Milk production increases as litter size increases, up to the point at which genetic capacity of the sow for milk production is reached. Mature sows raising large litters will often consume 8–10 kg or more of high concentrate diet (3600 kcal of metabolizable energy per kilogram of diet) per day at the peak of lactation. The amount of diet consumed and the changes in body weight during lactation depend on the obesity of the sow at the beginning of lactation. Obese sows tend to consume less feed during lactation and lose more weight than thin sows. The energetic efficiency of lactation is higher when milk is produced by current energy intake than by dependence on body fat depots. Therefore, in practical feeding, the greatest efficiency of energy utilization is achieved by controlled energy intake during gestation to minimize fattening and by full-feeding during lactation to minimize mobilization of depot fats for milk production.

The peak of lactation is reached at 4–5 weeks postpartum so that the quantitative daily nutrient requirement for lactation coincides with this peak. The daily milk production may be as high as 8 kg or more in individual sows nursing large litters at the peak of lactation. Since the caloric density of sow milk is considerably higher than that of cow milk, it represents a large supply of energy to individual pigs during a period of extremely rapid growth.

Quantitative nutrient requirements for lactation, assuming *ad libitum* energy intake, are given in Table 11.2A,B. It should be recognized that these values (expressed as units per kilogram of diet) will vary depending on the density of the diet and previous nutritional history of the sow.

Preweaning

The nutrient requirements (except iron) of the suckling pig are met by sow milk during the first 2 to 3 weeks. After this time, the rapid growth of the pig with the concomitant increase in daily nutrient requirement, combined with the decline in milk yield, necessitates the provision of supplemental feed. These relationships are illustrated diagrammatically in Fig. 9.3.

A supplemental source of feed (creep feed) must be supplied beginning at 3 weeks of age or earlier to allow maximum weight gain during the suckling period. The protein content, as a percentage of diet, does not have to be as high as that of sow milk, as the energy concentration is less, so that the ratio of calorie to protein is less. The carbohydrate source must be palatable and low in fiber to encourage consumption. For this reason, sucrose, dried molasses, glucose, or some other simple sugar is often added to make up 5 or 10%, or even more, of the diet. Nonnutritive sweeteners such as saccharine or monosodium glutamate are also used, but at a much lower concentration. While energy concentration of the creep diet is not critical, the addition of 5 or 10% fat to the diet has been shown to improve palatability (probably by reducing dustiness) and to encourage early consumption. Thus the quantitative nutrient requirements during the early preweaning period are met mostly by sow milk, but energy becomes limiting as the daily requirements increase in relation to the amount of milk available. The introduction of a dry, well-balanced diet early enough so that the pig consumes sufficient quantities for maximum weight gain is, therefore, a major consideration during the preweaning period, even though the total amount consumed per pig is negligible, compared with the total amount required to reach market weight.

Removal of the pig from the sow earlier than at 5 or 6 weeks of age can be considered early weaning for the purposes of this discussion. The younger the pig at weaning, the more critical are the dietary requirements. Primary specific pathogen-free (SPF) pigs, weaned at birth, are of the earliest possible weaning age. Because of deprivation of colostrum, not only must these pigs be fed a highly fortified diet, but also they must be kept in a sanitary environment and given parenteral antibody protection. The quantitative nutrient requirements of the newborn pig probably coincide closely with the composition of sow milk (for gross composition see Chapter 9). The digestive enzyme capacity of the newborn pig is limited when nonmilk proteins are fed. This is illustrated by the data shown in Table 11.5 (Maner *et al.* 1961), which demonstrates a difference in protein digestibility and in utilization of absorbed protein between pigs fed casein and

TABLE 11.5. Effect of Protein Source on Performance of Baby Pigs Fed Liquid Sow-Milk Replacers

Constituents	*Diet composition (g per l of finished milk)*	
	Pigs fed casein	*Pigs fed soybean protein*
Casein	44.3	—
Isolated soybean protein	—	44.3
Glucose	44.1	44.1
Lard	33.0	33.0
Soya lecithin	2.0	2.0
Minerals	10.6[a]	10.6[a]
Vitamins	+[b]	+[b]
Water	*to one liter*	*to one liter*
Criteria	*Performance*	
Av initial wt, kg	1.1	1.0
Av daily gain, g	113.0	100.0
Digestibility of dry matter, %	94.9	88.5
Digestibility of protein, %	96.6	89.8

Conclusion: The baby pig utilizes milk protein more efficiently than soybean protein. This is probably due to inability of the digestive system to hydrolyze soybean protein as efficiently for absorption. In older pigs this difference in utilization of the two protein sources disappears.

Source: Maner et al. 1961.
[a]Minerals added to meet recommendations of National Research Council.
[b]Vitamins added at levels at least twice those recommended by National Research Council.

those fed soybean protein. The liquid diets as used in this work must be fed at frequent intervals (a minimum of four times daily) to provide adequate intake for normal growth. When they are fed at this frequency, intake must be kept at less than full-feed to minimize problems with diarrhea. The digestive capacity of the young pig is not sufficient to handle large meals at infrequent intervals. Therefore, it is advisable to feed more frequently in smaller quantities per feeding to achieve best results. A commercially available automated feed dispenser that allows feeding more than 100 individually caged pigs 24 feedings per day at 1-hr intervals is shown in Fig. 6.2. Pigs can be weaned to dry diets at a few days of age. This reduces problems with sanitation.

Other experimentally developed automated feeders for baby pigs also have been developed (Lecce 1969; Braude *et al.* 1969). A less elaborate feeding system involves individually caged pigs fed four times daily from porcelain feeding bowls. It is important to house pigs individually when not allowed frequent feeding as the strong suckling instinct results in undue suckling and nuzzling of the ears and navels of other pigs when several pigs are kept together. It is important to increase intake gradually, as rapidly as freedom from digestive upsets will permit. Rate of weight gain is closely related to feed intake during this early period. The energy concentration of cow milk is not high enough to support maximum growth rate even when fed at nearly *ad libitum* intake. Therefore, most liquid sow milk replacers contain a higher dry matter and higher fat concentration than cow milk (Schneider and Sarett 1969). Young pigs can survive long periods on a daily energy intake so restricted that weight gain is almost nil. Although the pig has a tendency to compensate by extra rapid growth when it is transferred to a dry diet offered *ad libitum,* the effect of early deprivation on body size is probably never fully overcome (Barnes *et al.* 1967).

Growing Period (Early Post-Weaning)

This stage of the life cycle is arbitrarily set as the period from weaning (usually 5 or 6 weeks of age) to the point at which the pig has attained approximately 45-kg body weight. This arbitrarily assigned period represents a stage of nutrient requirements less critical than earlier stages, but more critical than that of the finishing period (from 45 kg to slaughter at 90 kg). The changes in nutrient requirements, as the pig matures, are related to changes in growth rate and body composition. Fat concentration of the body increases rapidly at the expense of water as the pig matures, and although the percentage of protein remains rather stable, the ratio of calories to protein in the body increases steadily to market weight. Feeding a protein deficient diet during the growing period results in an increase in this ratio (a fatter carcass); feeding protein in excess of the dietary requirement, on the contrary, does not yield a leaner carcass than that obtained when protein equal to the requirement is supplied. Clausen (1965) advocated high dietary protein for maximum carcass leanness but emphasized that the pig cannot be forced to increase lean meat deposition beyond its genetic capacity. Genetics influences carcass composition, but the effect of genetics on the response to diet remains uncertain. Experimental data with growing pigs suggest that the total protein required for normal performance is less than previously recognized. The addition of certain crystalline amino acids (lysine and methionine), in appropriate amounts and combinations to low protein diets, is now economically feasible, so that when the essential amino acid requirements are met in this way, the total amount of nitrogen needed for normal performance is reduced. Protein and amino acid requirements are usually expressed as a percentage of diet, but since energy concentration of the diet is important in affecting total daily feed intake, the expression of protein and of amino acid requirements as grams per unit of feed energy would be better. A clear-cut effect of energy concentration of the diet of growing pigs on performance can be observed by the addition of fat to the diet. Feed intake is thereby reduced, and if other nutrients are present in marginal or suboptimum amounts, the deficiency is aggravated (Pond *et al.* 1960) (see Table 11.6).

There are other examples of dietary interactions among nutrients. These include calcium and zinc (excess calcium increases the zinc requirement), methionine and choline (methionine can provide labile methyl groups used in metabolism that are normally supplied by choline, when choline is deficient in the diet), leucine and isoleucine (excess leucine inhibits the utilization of isoleucine), and tryptophan and niacin (tryptophan can be converted to niacin in the body). This last example

TABLE 11.6. Dietary Protein and Energy Interactions in Growing Pigs

Dietary protein, %:	*10*	*10*	*20*	*20*
Dietary fat, %:	*3*	*13*	*3*	*13*
Constituents	*Diet composition, %*			
Yellow maize (ground)	88.50	77.50	65.00	53.85
Soybean meal	5.00	5.00	28.90	28.90
Meat scraps	4.00	5.50	4.00	5.50
Dicalcium phosphate	1.50	1.00	1.10	0.75
Salt (trace mineralized)	0.50	0.50	0.50	0.50
Beef tallow (with antioxidant)	—	10.00	—	10.00
Antibiotic supplement	0.50	0.50	0.50	0.50
Vitamin supplement	+	+	+	+
Total, %	100.00	100.00	100.00	100.00
Criteria	*Performance*			
No. of pigs	6	6	6	6
Daily gain, g	554	500	667	790
Daily feed, g	2,025	1,671	2,193	2,147
Daily digestible energy, kcal	6,920	6,460	7,460	8,200

Conclusion: Dietary protein deficiency (10.0%) is aggravated by added fat (3% versus 13%); feed intake tends to be related to energy needs rather than protein needs when protein is deficient.

Source: Pond et al. 1960.

is worthy of further consideration because it has considerable practical significance in diets for growing swine, especially when maize or cereal grains make up a major part of the diet. This is for two reasons: tryptophan is a limiting amino acid in diets of this kind, and niacin, although present, is biologically unavailable. Therefore, the inadequate niacin absorption diverts the already short supply of tryptophan away from use for tissue protein synthesis, and growth depression results (Harmon *et al.* 1969). These interrelationships are shown in Table 11.7.

In addition to interrelationships among nutrients, antagonism exists between some nonnutritive constituents and certain nutrients. An example is the relationship between a protein source and quantitative zinc requirement whereby soybean protein increases the quantitative zinc requirement. Subsequent research has demonstrated that phytic acid and perhaps other organic constituents of plant materials are capable of altering the biological availability of zinc and other trace elements for growing pigs. As the trend in swine production moves more toward the use of all-plant diets, it seems likely that other such examples of antagonisms will be encountered. The swine producer, nutritionist, and feed manufacturer should be alert to these occurrences.

In addition to strictly nutritional factors affecting performance of growing swine, the physical form of the diet is important. For example, pelleting some types of diets has an effect on performance. The improved performance may be

TABLE 11.7. Dietary Niacin–Tryptophan Interrelationships in the Growing Pig

Level of niacin (ppm)[a]	*Daily gain for each level of maize (%)*[b]	
	0	*40*
0	327	74
4.4	490	527
8.8	663	531
13.2	658	704
17.6	690	672
22.0	645	690

Conclusion: The niacin in maize is biologically unavailable for growing pigs and the requirement appears to be increased in the presence of maize, possibly due to less tryptophan availability in maize than in casein or to other nutrient antagonisms.

Source: Harmon et al. 1969.
[a]2.5 and 10.8 ppm in 0 and 40% maize diets, respectively.
[b]The diet containing no maize was composed of corn starch, casein, dextrose, cellulose, corn oil and vitamin and mineral premixes and contained 0.10% tryptophan and 2.51 ppm niacin. The diet containing 40% maize was similar to the other diet except for the replacement of maize for corn starch and the reduction of casein from 14 to 12% of the diet. The 40% maize diet contained 0.12% tryptophan and 10.78 ppm of niacin.

due to improved utilization of nutrients or to less feed wastage or both. Other factors, including fineness of grind (particle size) and dustiness, contribute to variations in performance. Such factors are discussed in more detail in Chapter 17.

Finishing Period (45–90-kg Slaughter Weight)

The same principles and problems in meeting nutrient requirements during the growing period apply during the finishing period. However, the quantitative requirements for nutrients, other than energy, are less (in terms of percentage of diet) during the finishing period. The total daily energy requirement is considerably greater during finishing, not only because of the larger body size, but also because of the higher feed requirement per unit of body weight gain. This correspondence between decreased efficiency and increasing weight is related to the increased fat deposition and the high energy requirement associated with it.

Since energy is quantitatively the most important nutrient during the finishing period, it is important to consider the acceptability of available energy sources. Although the gross energy concentration of common cereal grains and root and tuber crops and plant by-products may be similar, the utilizable energy as well as the palatability varies widely. There are also differences in amino acid pattern that may have an effect when the protein supplement is eliminated or limited. An example of the effect of energy source, amino acid supplementation, and form (meal vs. pellets) on performance of finishing pigs is illustrated in Table 11.8 (Jensen *et al.* 1969). Table 11.9 shows that weight gain is not necessarily associ-

TABLE 11.8. Effect of Maize, Opaque-2 Maize, Wheat, Milo, and Lysine Supplementation on Performance of Finishing Pigs Fed Meal or Pelleted Diets

Constituents	*Composition of diets, %*				
	Maize	*Opaque-2*		*Milo*	*Wheat*
		No lysine	*Plus lysine*		
Maize	89.3	—	—	—	—
Opaque-2	—	93.0	97.0	—	—
Milo	—	—	—	96.6	—
Wheat	—	—	—	—	96.8
Lysine (50% L-form)	—	—	0.3	0.7	0.5
Soybean meal	8.0	4.3	—	—	—
Dicalcium phosphate	1.0	1.0	1.0	1.0	1.0
Ground limestone	1.0	1.0	1.0	1.0	1.0
Salt (trace mineralized)	0.5	0.5	0.5	0.5	0.5
Vitamin mix	0.1	0.1	0.1	0.1	0.1
Antibiotics	0.1	0.1	0.1	0.1	0.1
Total, %	100.0	100.0	100.0	100.0	100.0
Total protein, %	12.0	11.1	9.3	8.2	11.0
Lysine, %	0.44	0.45	0.49	0.48	0.50
Form of feed	*Performance*[a]				
	Av daily gain, g				
Meal	820	800	790	750	730
Pellet	870	820	850	850	850
	Gain/Feed				
Meal	0.255	0.282	0.260	0.249	0.266
Pellet	0.300	0.275	0.291	0.288	0.288

Conclusion: Finishing pigs can perform satisfactorily on less than 12% protein if amino acid balance is optimal and if all other nutrients are adequate; pelleting the diet improves feed utilization.

Source: Jensen et al. 1969.
[a]Nine pigs per group, av initial wt, 53 kg.

ated with maximum carcass leanness (Gallo and Pond 1968). Thus, it is important to consider nutrients in terms of their effects on efficiency of lean meat production, not only in terms of their effects on weight gain.

DIET FORMULATION

Adequate diet formulation depends on the use of valid measures of nutrient adequacy in the feedstuffs available for use. A chemical characterization of feeds as accomplished by *proximate analysis* is given as follows.

Nitrogenous compounds (analyzed as N in Kjeldahl procedure): These are protein N, nonprotein N, di-, tri-, polypeptides, amino acids, creatine and

TABLE 11.9. Effect of Protein Deficiency during the Finishing Period on Carcass Leanness

Constituents	*Composition of diets, %*	
	Maize–soybean meal	*Maize–glucose*
Maize (No. 2 yellow)	71.50	59.00
Glucose	—	34.32
Soybean meal	23.00	—
Decalcium phosphate	2.50	3.00
Ground limestone	1.50	1.25
Salt (trace mineralized)	0.50	0.50
Vitamin zinc supplement	0.50	0.50
Antibiotic supplement	0.50	0.50
Potassium phosphate	—	0.50
L-Lysine-HCl	—	0.37
L-Tryptophan	—	0.06
Total, %	100.00	100.00
Total protein, %	20.50	7.60
Criteria	*Performance*[a]	
Daily gain, g	670.00	680.00
Loin eye area, cm^2	29.40	19.90
Back fat, cm	3.30	3.90
Leaf fat, kg	1.36	1.91
Serum protein, g/100 ml	7.10	5.50

Conclusion: Diets for finishing pigs that provide adequate protein for normal weight gains may not be sufficient for normal carcass leanness.

Source: Gallo and Pond 1968.
[a]Five pigs per group; av initial wt, 47 kg; av final wt, 91 kg.

creatinine, purines, pyrimidines, urea, uric acid, amines, nucleotides, nucleic acids, ammonia, and some water-soluble vitamins.

Fat-soluble compounds (analyzed as ether extract): These are tri-, di-, and monoglycerides; free fatty acids; waxes; phospholipids; glycolipids; cholesterol, vitamin D and other sterols; vitamin A and carotenoids; vitamin E (tocopherols); and vitamin K.

Poorly digested carbohydrates (analyzed as crude fiber): These are cellulose and lignin.

Ash (determined by weight loss on ignition of fresh sample): This is all inorganic constituents including essential major and trace elements.

Well-digested carbohydrates (calculated by difference): These are hemicellulose, starch, dextran, glycogen, tri-, di-, and monosaccharides, and some water-soluble vitamins.

This is a widely used procedure that incorporates the analysis of protein (nitrogen × 6.25), ash, ether extract, moisture, crude fiber, and—by subtracting the sum of these from 100—the analysis of nitrogen-free extract (NFE). The specific classes of compounds contained in each fraction are indicated. The wide variety of compounds present in each fraction demonstrates the rather imprecise nature of the proximate analysis. It allows description of feeds only according to amounts of major nutrient groups and gives no information on amino acid content (protein quality), mineral composition, vitamin content, or such factors as palatability and toxicity. Also, the determination of crude fiber directly and of NFE (by subtracting the sum of all other components from 100) does not give an accurate prediction of the biological availability of the carbohydrates. It is now recognized that values obtained for crude fiber do not accurately describe the feed as an energy source for the pig, owing to the imprecise chemical characterization of the carbohydrates contained in it. Simpler techniques now available, utilizing acid and neutral detergent fiber analysis, are of greater value in characterizing the complex carbohydrates of plants (Van Soest 1963, 1967; Van Soest and Wine 1967). The cell wall of the plant contains the hemicellulose, cellulose, and lignin that are responsible for the low utilization of forages and seed hulls by swine. Hemicellulose is more digestible than cellulose by swine (Keys *et al.* 1970), and lignin is essentially nondigestible by swine as well as by ruminants. Thus, the overall digestibility of fibrous feeds by swine depends on the relative amounts of these cell-wall constituents present.

The crude fiber measurement obtained by proximate analysis is not completely without merit as a guide to the energy value of a feed, since levels of more than 5% in the diet generally reduce growth rates. Therefore, crude fiber content can be used as a general guide in selecting energy constituents for swine diets. The digestibility of crude fiber differs according to source; for example, the fiber of corn cobs has a lower energy value for pigs than fiber of barley.

The energy value of a feed for growth or other productive functions can be determined by using information obtained from the proximate analysis of the feed and of excreta of animals to which it is fed. The apparently digestible energy is computed by measuring the difference between the amount of energy consumed in a specified period of time and the amount of energy excreted in the feces in the same period of time. This is called the total digestible nutrient (TDN) value of the feed. The procedure for determining TDN of a feedstuff is as follows:

1. Determine the nitrogen, ether extract, crude fiber, and NFE content of the dried feed by proximate analysis.
2. Feed known quantities of the feed at constant daily intake for a specified time period (usually 5–10 days).
3. Collect the feces from the animal during the same time period.
4. Determine the nitrogen, ether extract, crude fiber, and NFE content of the dried feces by proximate analysis.
5. Calculate the amount of each nutrient absorbed (feed minus feces).
6. Calculate the percentage of each nutrient absorbed (amount absorbed divided by amount in feed). This is the *digestion coefficient*.
7. Multiply the value obtained for ether extract by 2.25 to account for the higher energy value of fat, which is 2.25 times that of carbohydrates.
8. Multiply the percentage of each nutrient in the diet by the percentage of each nutrient absorbed (digestion coefficient).
9. Add the percentage of each nutrient and divide the sum by 100. This gives TDN as a percentage of diet.

TABLE 11.10. Composition of Total Digestible Nutrients (TDN) from Data Obtained in a Digestion Trial[a]

Nutrient	*Consumed (g)*	*Excreted (g)*	*Amount apparently absorbed (g)*[b]	*Percent absorbed*	*Percent in diet*	*Percent absorbed (per 100 g diet)*
Protein	2,000	400	1600	80	20	16.0
Ether extract	500	50	450	90(×2.25) = 202.5	5	10.1
Crude fiber	500	450	50	10	5	0.5
NFE	5,000	500	4,500	90	50	45.0
Ash	800	—[c]	—	—	8	—
Water	1,200	—[c]	—	—	12	—
	10,000				100	71.6% TDN

[a]Fed to a 45-kg pig for 10 days at 1.0 kg of feed per day (10 kg total).

[b]The term, apparently absorbed, more accurately describes the process, since the nutrients excreted in the feces represent not only nutrients not absorbed, but also nutrients of endogenous origin (digestive secretions, sloughed cells from the intestinal lining, etc.).

[c]These values are ignored since they contribute no energy.

The mechanics are illustrated in Table 11.10.

A more direct approach to energy utilization, is to consider *calorie*[1] absorption and utilization. The calories contained in any material can be determined by measurement of the heat produced by complete ignition in a bomb calorimeter. This *gross* energy content can be expressed as calories per gram of material. Feedstuffs widely different in utilizable energy (such as corn vs. corn cobs) will yield similar values for gross energy. *Digestible* energy can be determined by completely combusting a sample of feed and of feces obtained in the same way as for the determination of TDN just described. In fact, TDN is really an indirect means of estimating digestible energy of a feed, because each nutrient (protein, fat, and carbohydrate) contains a specific approximate calorie concentration. For practical purposes, carbohydrate and protein are both considered to contain 4000 cal/g of pure nutrient,[2] while fat is considered to contain about 9000 cal/g (2.25 times as much). One gram of TDN contains approximately 4400 digestible calories (1 kg contains 4400 kcal).

Although digestible energy (or TDN) provides useful information about the value of feedstuffs, it does not take into account losses of energy through the urine, expired air, and so on. More refined techniques take these losses into account. The energy remaining for use by the body after these losses are accounted for is *metabolizable* energy. After a meal has been eaten, there is extra heat produced associated with assimilation of the meal. This is called the *heat increment*. The energy remaining for use after the heat increment has been deducted is the *net energy* of the feed and represents the most refined expression of

[1]A calorie is that amount of heat required to raise the temperature of 1 g of water 1°C (from 15.5° to 16.5°C).

[2]Protein contains about 5000 cal/g, but about 20% is unavailable to the body for use as energy because nitrogen from its metabolism is lost in the urine as urea.

the energy value of a feed. A flow sheet showing these relationships is shown as follows:

Energy value of feeds

Gross energy − fecal losses

↓

Digestible energy − urinary, expired air, and other losses

↓

Metabolizable energy − heat increment

↓

Net energy

Protein Value of Feeds

The protein value is basically the amino acid balance. The importance of proper amounts and proportions of amino acids has been amply demonstrated in the discussion of nutrient requirements of growing and finishing pigs. Methods of assessing amino acid adequacy, aside from analytical determination of the composition of a feedstuff, include the growth trial (measurement of growth of animals full-fed the test diet adequate in all other nutrients), protein digestibility (as described under TDN), nitrogen balance [feed nitrogen minus (fecal + urinary nitrogen)], and body composition studies. The terms PER (protein efficiency ratio) and NPU (net protein utilization) are sometimes used to describe experimentally determined values for protein adequacy of a feedstuff. Detailed descriptions of their derivation and application have been given by Loosli *et al.* (1977) and Church and Pond (1982). For practical purposes the most indicative index of amino acid adequacy of a feedstuff is the growth trial, as the appetite of the animal and its body weight gain are sensitive to a poorly balanced amino acid mixture.

Modern Feed Formulation

For the formulation of diets of swine that meet the requirements for growth, maintenance, reproduction, and lactation, a wide variety of sources of nutrients, from both plants and animals, is available. Since the nutrient requirements of the pig are similar to those of man, theoretically the feedstuffs used for swine are available only at the expense of food supplies for man. This is potentially true, but many of the feedstuffs used for swine are unsuitable for man or unacceptable to him on esthetic grounds. By-products of the milling industry, the meat-packing industry, the fisheries industry, and the plant oil industry all provide important feeds for swine that are not well suited for direct human use.

The efficient production of pork depends on equating the nutrient requirements of the pig to the available feed supply by using feedstuffs that complement one another in providing all nutrients in proper amounts and proportions at lowest economic cost.

The explosive growth of computer technology has brought linear programming and least-cost diet formulation to the farm. Now, not only commercial feed companies but individual pork producers use computers to capture optimum combinations of available feedstuffs for economical diet formulation. The nutrient composition and other attributes of a vast array of feed resources are used to make intelligent choices in feed formulation, using computer technology to process appropriate inputs of information.

CLASSIFICATION OF FEEDSTUFFS

Feedstuffs can be classified on the basis of origin (plant or animal) and major nutrient contributions. An exhaustive catalog of potential feedstuffs for swine is not given here. Rather, the most commonly used feedstuffs and those that seem to show promise of greater future use are described in terms of general characteristics, nutrient contents, and factors limiting their use. The swine industry has tended to be concentrated in areas in which production of these crops is high. The composition of many feedstuffs for swine has been listed by the National Research Council (1982). The feedstuffs listed here include some that are contained therein, but also several that are not. Other tables of nutrient composition of feedstuffs have been prepared by Morrison (1956) and Schneider (1947). Amino acid composition of a wide variety of feeds has been reported by Eggum (1968).

The feedstuffs in common use or of potential value in swine feeding are discussed in detail in the succeeding five chapters (energy sources in Chapters 12 and 13, protein sources in Chapters 14 and 15, and mineral and vitamin sources in Chapter 16. Most of the feeds considered as protein or energy sources also supply significant amounts of minor constituents (vitamins and minerals). Therefore, when diets for swine are being formulated to meet all nutrient requirements, the amounts of supplemental vitamins and minerals needed to fulfill the deficits existing in natural sources may vary from zero to considerable, depending on the nature of the feedstuffs used in the formulation. One must, therefore, consider the contributions of feedstuffs not only in terms of energy and protein but also in terms of their vitamin and mineral content. These considerations are included in the descriptions of individual feedstuffs that follow in Chapters 12–16.

Feedstuffs that are normally considered as energy sources also contribute protein to the diet. The contribution of cereal grains to protein nutrition for man and animals is emphasized in a reference list prepared by Coons (1968) and published by USDA. That list was prepared "to facilitate awareness and evaluation of available research data on cereal proteins. . ." and, though not recent, is still pertinent for those deeply involved with feed formulation.

The digestible energy value of some of the most common energy and protein sources in use in North America has been summarized in Table 11.11. It should be noted, first, that a wide variation exists in digestible energy content of feeds which are normally considered to be good energy sources for swine (for example, the difference in energy content between maize and oats) and, second, that many of the protein supplements are good sources of digestible energy. Of course, the prices of most protein supplements preclude their use as energy sources.

TABLE 11.11. Digestible and Metabolizable Energy Values of Some Common Feedstuffs for Swine (kcal/kg, as fed)

Feedstuff	*Energy value: DE*	*ME*
Alfalfa meal, dehydrated, 17% protein	2580	2270
Alfalfa meal, dehydrated, 17% protein[a]	2870	2610
Barley	3086	2870
Corn, dent yellow, grain	3525	3325
Corn and cob meal	3086	2500
Cottonseed meal, solvent	2689	2555
Fish meal, herring	3086	2500
Fish meal, Menhaden	2734	2230
Hominy feed[a]	4020	3840
Meat meal, 55% protein	2998	2540
Molasses, cane	2469	2343
Oats	2866	2668
Oat groats (dehulled oats)	3690	3400
Rice[b]	3700	3580
Rice bran[b]	3880	3710
Rice, broken	2513	2360
Rice, polishings	3792	3000
Skim milk, dried	3792	3360
Sorghum, grain, milo	3439	3229
Soybeans, full-fat	4056	3540
Soybean meal, dehulled, solvent	3860	3485
Soybean meal, solvent	3350	3090
Wheat, soft, red winter	3659	3416
Wheat, hard, red winter	3483	3220
Wheat bran	2513	2320
Whey, dried	3439	3190
Yeast, brewers' dried	3135	2707

Source: National Research Council 1979; [a]Stanley and Ewan 1982; [b]Robles and Ewan 1982.

Modern feed formulation laboratories use highly specialized equipment to determine vitamin, mineral, amino acid, and fatty acid analyses of feedstuffs. These analyses are expensive, but with modern automated equipment it is possible to obtain a complete nutrient analysis on a wide variety of feedstuffs, so that diets can be balanced within a few hours or days after the sample is obtained. As a result, computer formulation of diets, whereby prescribed limitations can be imposed on acceptable levels of nutrients (upper and lower limits) as well as of feedstuffs, has become increasingly convenient. It allows "least-cost" diet formulation which provides for frequent substitutions and alterations in proportion of feedstuffs, depending on current prices. Feed manufacturers and many individual swine producers now balance diets with the use of computers for least-cost formulation. The relatively low cost and easy programming of home computers now available encourages their use on the farm for diet formulation as well as the total business management program. The mechanics of least-cost formulation require that the following information be supplied to the machine: (1) tolerance (upper and lower) of specific ingredients, (2) accurate feedstuff composition data, and (3) current ingredient costs. If these three pieces of information are used, diets that meet all mathematical requirements can be formulated, but the final measure of acceptability is the performance of the animal to which the diet is fed. Least-cost formulation does not account for factors, such as palatability, that may have an important effect on animal response. In the final analysis, the most sophisticated computer cannot take the place of the animal itself for which the diet was formulated in determining the acceptability of the diet.

REFERENCES

AGRICULTURAL RESEARCH COUNCIL. 1967. The Nutrient Requirements of Farm Livestock, No. 3—Pigs. Agric. Res. Council, London, England.

AGRICULTURAL RESEARCH COUNCIL. 1981. The Nutrient Requirements of Pigs. Commonwealth Agricultural Bureaux, Farnham Royal, Slough. United Kingdom.

BARNES, R. H., MOORE, A. U., REID, I. M., and POND, W. G. 1967. Am. Diet. Assoc. *51*,34

BRAUDE, R., MITCHELL, K. G., and SUFFOLK, S. F. 1969. J. Inst. Anim. Technicians *20*,43.

BROWN, W. R., KROOK, L., and POND, W. G. 1966. Cornell Vet., Suppl. 1 *56*,108.

CHURCH, D. C., and POND, W. G. 1982. Basic Animal Nutrition and Feeding, 2nd Edition. John Wiley & Son, New York.

CLAUSEN, H. 1965. World Rev. Anim. Prod. *1*,28.

COONS, C. M. 1968. Selected References on Cereal Grains in Protein Nutrition. Human and Experimental Animal Studies of Major and Minor Cereals, 1910–1966. ARS 61-5, USDA, Washington, D.C.

CUNHA, T. J. 1977. Swine Feeding and Nutrition. Academic Press, New York.

EGGUM, B. O. 1968. Aminosyrekoncontration og proteinkvalitet. Stougaards Forlag, Copenhagen, Denmark.

FOLLIS, R. H., JR. 1958. Deficiency Disease, p. 43. Charles C. Thomas, Springfield, Illinois.

GALLO, J. T., and POND, W. G. 1968. J. Anim. Sci. *27*,73.

HANSARD, S. L. 1966. *In* Swine in Biomedical Research (K. L. Bustad, R. O. McClellan, and M. P. Burns, Editors), p. 79. Frayn Publishing Co., Seattle, Washington.

HARMON, B. G., BECKER, D. E., JENSEN, A. H., and BAKER, D. H. 1969. J. Anim. Sci. *28*,848.

JENSEN, A. H., BAKER, D. H., and HARMON, B. G. 1969. J. Anim. Sci. *29*,16.

KEYS, J. E., JR., VAN SOEST, P. J., and YOUNG, E. P. 1970 J. Anim. Sci. *31*,1172.

KING, J. M., HSU, F. S., HONG, C. B., and LEE, R. C. T. 1976. An Atlas of General Pathology. Pig Research Institute of Taiwan and Joint Commission on Rural Reconstruction, Taipei, Taiwan. 300 pp.

KRIDER, J. L., CONRAD, J. H., and CARROLL, W. E. 1981. Swine Production, 5th Edition. McGraw-Hill, New York.

KUBOTA, J., ALLAWAY, W. H., CARTER, D. L., COREY, E. E., and LAZAR, V. A. 1967. J. Agric. Food Chem. *15*,448.

LECCE, J. G. 1969. J. Anim. Sci. *28*,27.

LOW, A. G. and PARTRIDGE, I. G. 1980. Current concepts of digestion and absorption in pigs. Tech. Bull. *3*. Natl. Inst. Res. Dairying. Reading, England.

LOOSLI, J. K., WARNER, R. G., HINTZ, H. F., and MAYNARD, L. A. 1977. Animal Nutrition, 7th Edition. McGraw-Hill, New York.

MANER, J. H., POND, W. G., and LOOSLI, J. K. 1961. J. Anim. Sci. *20*,614.

MORRISON, F. B. 1956. Feeds and Feeding, 22nd Edition, Morrison, Ithaca, New York.

NATIONAL RESEARCH COUNCIL. 1979. Nutrient Requirements of Swine, *No. 2*. National Acad. Sci. Washington, D.C.

NATIONAL RESEARCH COUNCIL. 1980. Mineral Tolerance of Domestic Animals. Natl. Acad. Sci., Washington, D.C. 577 pp.

NATIONAL RESEARCH COUNCIL. 1982A. Joint U.S.–Canadian Tables of Feed Composition Natl. Acad. Sci., Washington, D.C.

NATIONAL RESEARCH COUNCIL. 1982B. Effect of Environment on Nutrient Requirements of Domestic Animals, NRS. Natl. Acad. Sci., Washington, D.C.
POND, W. G., and HOUPT, K. A. 1978. Biology of the Pig. Cornell Univ. Press, Ithaca, New York. 371 pp.
POND, W. G., and MANER, J. H. 1974. Swine Production in Temperate and Tropical Environments. W. H. Freeman and Co., San Francisco, California. 646 pp.
POND, W. G., KWONG, E., and LOOSLI, J. K. 1960. J. Anim. Sci. *19*,1115.
ROBLES, A., and EWAN, R. C. 1982. J. Anim. Sci. *54*,1175.
SCHNEIDER, B. H. 1947. Feeds of the World Agric. Exp. Stn., West Virginia Univ., Morgantown.
SCHNEIDER, D. L., and SARETT, H. P. 1969. J. Nutr. *98*,279.
STANLEY, D. L., and EWAN, R. C. 1982. J. Anim. Sci. *54*,1175.
TUCKER, H. F., and SALMON, W. D. 1955. Proc. Soc. Exp. Biol. Med. *88*,613.
VAN SOEST, P. J. 1963. J. Assoc. Off. Agric. Chem. *46*,829.
VAN SOEST, P. 1967. J. Anim. Sci. *26*,119.
VAN SOEST, P. J., and WINE, R. H. 1967. J. Assoc. Off. Agric. Chem. *50*,50.

12

Grain and Grain By-product Energy Sources

INTRODUCTION

Carbohydrates are the most abundant form of energy in plant materials and as such are the most widely available source of energy for swine feeding. Grains and their by-products are by far the most important sources of carbohydrates. Although the grains and their by-products discussed in this chapter also supply important amounts of proteins, vitamins, and minerals, they are included here because of their usefulness in supplying available carbohydrates to swine of all ages. In addition to the array of energy feedstuffs discussed here, many by-products of agriculture and industry are produced but have been underutilized as feed resources despite their potential value. Such products, including industrial food processing wastes, nonfood industrial wastes, forest residues, animal wastes, crop residues, and aquatic plants, are featured in a recent review to encourage their greater exploitation for animal feed use (National Research Council 1983). The future should bring increased use of these valuable resources in pork production.

TABLE 12.1. Feed Grains Produced in the United States and the World in 1980 (Millions of Metric Tons)

Grains	*United States*	*World*
Maize	164.1	379.2
Wheat[a]	64.3	431.1
Rice[a]	6.6	391.7
Barley	7.6	167.6
Grain Sorghum	14.9	58.4
Oats	6.5	45.4
Rye	0.4	24.8

Sources: USDA 1981; FAO Production Yearbook 1980.

[a]Most of wheat and rice produced is sold for human consumption; only a small fraction is fed to swine.

CORN (MAIZE)

Corn (*Zea mays*) is the most common grain used for feeding swine. Although in the United States it is referred to as corn, it is called maize in most parts of the world. It is hereafter called maize in this chapter. It is the basis for swine diets throughout the Americas, and in the United States alone about one-half of all maize produced is marketed for feeding to swine. The amounts of the most abundant feed grain sources produced in the United States and the world in 1980 are shown in Table 12.1 (USDA 1981). In the tropical areas of the Americas, although other energy sources are available, maize is the traditional feed for swine. In the more temperate environments of Canada and Europe, maize is commonly replaced by barley and wheat.

Although common maize predominates in yield and area planted, plant geneticists, in their search for maize of improved nutritional value, have identified mutants with interesting nutritional qualities. Even though commercially these mutants are of little importance, the nutritional characteristics of several are discussed to indicate their unique potential.

COMMON MAIZE

Common maize is an important grain in the food supply of swine. Although maize is commonly used for swine feeding and many times as the only source of high energy feed, it should never be fed as the sole source of protein in swine diets because of its relatively low protein content (Table 12.2) and poor balance of essential amino acids (Table 12.3). It was known as early as 1914 (Osborne and Mendel 1914) that the quality of maize protein was poor because of a deficiency of several of the essential amino acids, especially lysine and tryptophan. Without adequate supplementation with protein and/or essential amino acids, maize is an inadequate protein source for swine of all ages.

The chemical composition of the maize grain has been studied in detail (Mertz

TABLE 12.2. Composition of Maize (International Feed Number 4-02-935)

Component	*Amount*
Dry matter	89.0%[a]
Digestible energy	3399 kcal/kg
Metabolizable energy	3300 kcal/kg
Crude protein	9.6%
Total digestible nutrients	80.0%
Plant cell wall constituents	
Cell walls	8.0%
Cellulose	2.0%
Hemicellulose	5.0%
Lignin	1.0%
Acid detergent fiber	3.0%
Crude fiber	2.6%
Ether extract	3.8%
Ash	1.3%
Calcium	0.03%
Phosphorus	0.26%
Sodium	0.03%
Chlorine	0.04%
Potassium	0.33%
Magnesium	0.12%
Sulfur	0.11%
Iron	27 mg/kg
Copper	4 mg/kg
Manganese	5 mg/kg
Selenium	0.07 mg/kg
Zinc	13 mg/kg
Carotene (provitamin A)	2 mg/kg
Vitamin E	22 mg/kg
Choline	502 mg/kg
Niacin[b]	25 mg/kg
Pantothenic acid	5.9 mg/kg
Riboflavin	1.2 mg/kg

Source: National Research Council 1982.

[a]The dry matter may vary between 75 and 90%; for storage without spoilage in the absence of a preservative such as propionic acid a minimum of 85% is needed.

[b]Almost completely unavailable.

1963; Mertz and Bressani 1957; Mertz *et al.* 1958; Lloyd and Mertz 1958; Bressani and Mertz 1958). Extraction of the protein fractions with a copper reagent at different pHs has divided the protein of the endosperm and germ into four fractions; the zein, glutelin, acid-soluble, and residual fractions (Table 12.4). The importance of the proportions of these fractions in the grain is obvious from a study of the amino acid composition of each fraction (Table 12.5). The reason behind the poor protein quality of common maize is that zein, the protein that is soluble in alcohol, contributes as much as 50% of the total protein of the kernel in

TABLE 12.3. Essential Amino Acid Composition of Maize

Amino acids	*Amount (%)*
Arginine	0.43
Cystine	0.22
Histidine	0.26
Isoleucine	0.35
Leucine	1.21
Lysine	0.25
Methionine	0.17
Phenylalanine	0.48
Threonine	0.35
Tryptophan	0.08
Tyrosine	0.38
Valine	0.44

Source: National Research Council 1982.

most varieties. Zein has been studied extensively and has been shown to contain very low levels of both lysine and tryptophan (Moss 1966; Geiger *et al.* 1952; Rathman 1954). In fact, when compared to glutelin, zein contains lower levels of all of the essential amino acids except methionine, isoleucine, leucine, and phenylalanine.

Numerous studies (Bressani and Mertz 1958; Frey *et al.* 1949; Hamilton *et al.* 1951; Sauberlich *et al.* 1953; Tello *et al.* 1965; Bressani *et al.* 1962) on the factors affecting the protein quality of maize indicate that both environment and variety have a significant effect on lysine content. It has also been shown that nitrogen fertilization increases protein content and decreases protein quality. This is due to a total increase in the zein fraction which causes protein quality to be reduced (Sauberlich *et al.* 1953; Hansen *et al.* 1946; Schneider *et al.* 1952;

TABLE 12.4. Nature of Protein Fractions of the Endosperm and Germ of a Normal Dent Maize

Fraction	*Portion of kernel*	*Protein class*	*Percent of total nitrogen*
Zein	Endosperm	Prolamine	41–60
Zein	Germ	Unknown	5–10
Glutelin	Endosperm	Glutelin	17–31
Glutelin	Germ	Globulins, nucleoprotein	49–54
Acid-soluble	Endosperm	Albumin, globulins	16–26
Acid-soluble	Germ	Albumin, globulins	30–40
Residual	Endosperm	—	2–11
Residual	Germ	—	0–11

TABLE 12.5. Amino Acid Composition of the Protein Fractions of Normal Maize

Amino acid	Endosperm: Acid-soluble fraction	Endosperm: Alcohol-soluble fraction (zein)	Endosperm: Alkali-soluble fraction (glutelins)	Germ
Lysine	1.8	0.3	3.6	6.1
Histidine	1.1	1.6	4.2	2.9
Arginine	2.8	2.5	5.3	9.1
Threonine	2.4	3.4	3.4	3.9
Tryptophan	—	0.1	0.9	—
Valine	2.4	4.5	5.6	5.3
Methionine	1.4	2.5	1.1	1.7
Cystine	—	1.0	3.1	1.0
Isoleucine	3.0	4.2	3.4	3.1
Leucine	10.1	20.3	8.6	6.5
Phenylalanine	3.2	7.1	3.8	4.1

Source: Adapted from Mertz 1963.

MacGregor *et al.* 1961). This increase in the zein fraction accompanying the increase in the protein content in maize was also observed by Mitchell *et al.* (1952) who found that high-protein maize had a lower biological value. Pigs pair-fed 15% protein diets based on 12.8% protein maize produced significantly less gains and required more feed per unit gain than pigs fed similar diets based on 9.1% protein maize (Dobbins *et al.* 1950). No differences were found when the diets were based on 8.2 and 11.8% protein maize, and each variety supplied the same amount of protein in a 15% protein diet.

The energy value of maize is generally used as a standard with which other energy sources are compared. Thus, if the relative energy value of maize is taken as 100, the energy value of other available plant energy sources is generally less. This is mainly due to the low fiber content of the maize kernel and the high digestibility of its starch.

White and yellow maize are of similar composition except that yellow maize is high in carotene, a vitamin A precursor. Both are fair sources of vitamin E, but low in vitamin D and in the B vitamins. They are very deficient in calcium, and although they contain a reasonably high level of phosphorus, much of it is present in the biologically poor form of phytic acid phosphorus. Trace elements are present only in marginal quantities in maize.

Numerous studies with growing pigs have dealt with crystalline amino acid supplementation of corn protein. Both lysine and tryptophan are severely limiting and neither elicits a favorable response without supplementation with the other; tryptophan is slightly more limiting than lysine (Hays *et al.* 1957; Cromwell *et al.* 1967; Clawson and Matrone 1963; Picket and Beeson 1969; Halck *et al.* 1964; Gallo *et al.* 1968; Baker *et al.* 1970; Ilori and Conrad 1977). Data on finishing pigs (60 kg) also indicate that a response from the addition of lysine to a corn diet is not obtained unless the diet contains supplemental tryptophan (Gallo and Pond 1968). The third and fourth limiting amino acids in a fortified corn diet containing

0.221% lysine and 0.029% tryptophan were determined in gilts (30 kg) in a metabolism trial by Grosbach *et al.* (1980). The changes in urinary urea excretion obtained with the dietary addition of 0.054% L-threonine and/or 0.07% L-isoleucine indicate that these amino acids are, respectively, the third and fourth limiting amino acids in maize.

Pregnant swine exhibit a remarkable capacity to provide nutriment to the developing fetuses during periods of severe protein deficiency (Clawson *et al.* 1963; Rippel *et al.* 1965; Pond *et al.* 1968A,B, 1969A,B). However, Baker *et al.* (1970) demonstrated that unsupplemented maize diets were inadequate as the only protein source for gestating gilts. Even when fed 16% protein–maize–soybean meal diets during lactation, gilts fed the maize diet during the entire gestation weaned fewer pigs with lesser weaning weights than those fed either 12, 16, or 20% protein diets based on a mixture of maize and soybean meal. Although performance through farrowing was not affected by the all-maize diet, the severe amino acid deficiency during gestation causes maternal tissues to be sacrificed to the extent that subsequent lactation performance is impaired. For the gravid gilt, maize protein is first limiting in lysine and second limiting in tryptophan (Rippel *et al.* 1965; Allee and Baker 1970). In gravid gilts, maize supplemented with both lysine and tryptophan resulted in a nitrogen retention of 11.8 g/day (Allee and Baker 1970), a value not greatly different from the 12.30 g/day retention reported by Miller *et al.* (1969) when a standard 12% protein maize–soybean meal diet was fed. Maize diets containing supplemental lysine and tryptophan have been shown to be adequate for swine during gestation.

In balancing a practical diet for swine, based on maize as the energy source, supplemental protein sources rich in lysine and tryptophan are combined with maize at a level and proportion to satisfy the needs of the class of swine being fed. Some commonly used protein sources and their supplemental value are discussed in Chapters 14 and 15.

When soybean meal is used to supplement maize-based diets, it is generally acknowledged that lysine is first limiting. Crystalline lysine is available at economically competitive cost, and can be used to substitute for a portion of the supplemental protein and thereby reduce crude protein concentration of the diet. The requirement for lysine, expressed as a percentage of the diet, is directly related to the dietary protein concentration. The lysine requirement, expressed as a percentage of the diet, decreases by 0.02 percentage units with each 1% reduction in dietary protein level (Easter and Baker 1980; Baker *et al.* 1975).

OPAQUE-2 MAIZE

The existence of opaque-2 genes in maize (*Zea mays*) has been known since 1935 (Emerson *et al.* 1935). However, it was not until 1963 (Mertz 1963, 1968) that Drs. E. T. Mertz and O. E. Nelson, along with graduate student L. S. Bates determined that the opaque-2 mutant contained much more lysine than normal maize kernels. Common and opaque-2 kernels are shown in Fig. 12.1. Subsequent studies by these workers demonstrated that the endosperm of opaque-2 kernels contains twice as much lysine and tryptophan, 50% more arginine, aspartic acid, and glycine and 30% less alanine and leucine than other isogenic lines

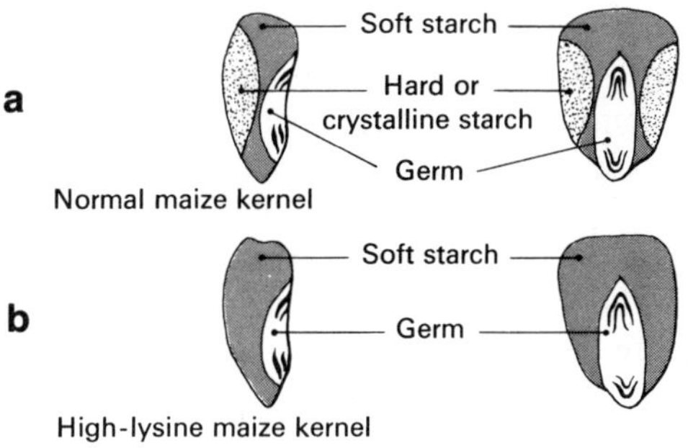

FIG. 12.1. (a) Common maize and (b) opaque = 2 maize; note the difference in seed structure.

that do not contain the opaque-2 mutant gene (Lloyd and Mertz 1958). A portion of the change in amino acid composition is attributed to a reduction in the ratio of zein to glutelin (Mertz *et al.* 1964). Whereas normal lines of maize contain 41–52% zein and 17–28% glutelin (Bressani and Mertz 1958), the maizes containing the opaque-2 gene contain approximately 16% zein and 42% glutelin. If the discussion on protein quality of normal maize is recalled (see the preceding section on Common Maize), the significance of this change can be well appreciated. Other reasons for the improvement in protein quality are related to the increase in lysine in both the acid-soluble protein fraction and in the zein present (Mertz *et al.* 1964). The changes resulting from the incorporation of the mutant opaque-2 gene provide a maize with a superior balance of amino acids. (Table 12.6). Later, Murphy and Dalby (1971) showed that the opaque-2 gene suppresses the synthesis of zein, especially during early development, resulting in

TABLE 12.6. Amino Acid Composition of Normal and Opaque-2 Maize

Amino acid	*Percent of protein*	
	Normal	*Opaque-2*
Argininc	5.1	6.8
Histidine	3.0	3.5
Isoleucine	3.4	3.2
Leucine	12.1	8.4
Lysine	2.6	4.2
Methionine	1.6	1.4
Cystine	1.5	1.7
Phenylalanine	4.9	4.4
Threonine	3.5	3.3
Tryptophan	1.0	1.3
Valine	4.6	4.9
Aspartic acid	6.7	10.0
Glutamic acid	20.8	18.7
Serine	3.5	4.3
Proline	9.7	8.6
Glycine	4.0	4.8

Source: Cromwell et al. 1967.

lower zein and higher lysine and tryptophan content of opaque-2 maize at maturity. The opaque-2 gene has little if any effect as a zein suppressant when present in one or two doses in the triploid endosperm. The lysine-raising effect is observed, only with three doses of the gene. When they have been properly used, the opaque-2 genes have consistently produced favorable changes in protein quality of maize lines into which they have been incorporated.

The original studies to demonstrate the nutritional value of opaque-2 maize were reported by Mertz and co-workers in 1965 (Mertz *et al.* 1965A, 1965B). Weanling rats fed opaque-2 maize were compared to rats fed a standard hybrid maize. The rats fed the opaque-2 maize gained an average of 97 g in 28 days as compared to only 27 g by rats fed the normal maize. Since that time extensive research has been conducted with rats (Von Zucher *et al.* 1968; Ocio and Treviño 1968; Bressani *et al.* 1969; Maner *et al.*, 1971; Rivera *et al.* 1978), chicks (Cromwell *et al.* 1968; Santos and Maner 1971; Von Zucher *et al.* 1968; Ocio and Treviño 1968), humans (Bressani 1966; Clark 1966) and pigs to demonstrate the superior values of maizes containing the opaque-2 gene.

The value of opaque-2 maize for pigs depends upon its amino acid adequacy in supplying the requirements of animals of different ages and performing different functions. It has been well established that young growing pigs require more of a high-quality protein than older pigs that are being finished for market. Therefore, in practice a high-quality 18–20% protein diet is fed to young pigs (5–18 kg), a 16% protein diet is fed to growing pigs (18–50 kg), and a 12 or 13% protein ration is used for finishing pigs (50–100 kg). Although gestating sows are less demanding in their protein requirement, lactating sows producing large quantities of milk also require a good-quality 15 or 16% protein ration to meet their daily nutrient requirements. These requirements are adequately furnished by normal maize supplemented with soybean meal, fish meal, or combinations of these with other high protein supplements. Therefore, the value of opaque-2 maize is associated with the ability of this grain to substitute for a portion of the supplemental protein or, in other words, to require less supplemental protein to produce results equal to those obtained with normal maize adequately supplemented. A limitation of opaque-2 maize is its susceptibility to insect damage.

Growth

The initial opaque-2 maize work with pigs was reported by Beeson and coworkers (Beeson *et al.* 1966) and Pickett (1966) at Purdue University. It was shown that growing pigs between 13.8 and 25.7 kg grew 3.6 times faster on opaque-2 maize than on normal maize. Those fed opaque-2 grew at a rate equal to that of pigs fed normal maize and soybean meal balanced to supply the same quantity of crude protein (Table 12.7). Finishing pigs with an average initial body weight of 59 kg gained weight 50% faster on opaque-2 maize than on normal maize and at about the same rate as pigs fed diets of normal maize and soybean meal containing 13.0% protein. Since the time of these initial pig trials many studies have been reported from many countries to demonstrate the value of opaque-2 maize in swine diets.

In subsequent studies at Purdue (Cromwell *et al.* 1967) 10–13-kg pigs were fed opaque-2 maize containing 104% more lysine and 67% more tryptophan (see

TABLE 12.7. Nutritional Value of Opaque-2 Maize for Growing and Finishing Swine

Criteria	*Treatment* 1 *Normal maize*	2 *Opaque-2 maize*	3 *Normal maize–soybean meal*	4 *Opaque-2 maize–soybean meal*
Weanling pigs				
Protein level, %	8.6	11.6	11.6	—
Av initial wt, kg	14.0	13.8	13.7	—
Av final wt, kg	17.3	25.8	25.6	—
Av daily gain, kg	0.12	0.43	0.42	—
Av daily feed, kg	.82	1.41	1.37	—
Feed/gain ratio	6.88	3.32	3.23	—
Finishing pigs				
Protein level, %	8.9	11.2	11.2	13.0
Av initial wt, kg	59.5	59.4	59.0	60.4
Av final wt, kg	73.8	81.0	64.3	83.2
Av daily gain, kg	0.68	1.03	.88	1.09
Av daily feed, kg	3.21	3.70	3.29	3.97
Feed/gain ratio	4.71	3.59	3.74	3.65

Source: Adapted from Beeson et al. 1966.

Table 12.6) than a normal maize. When these two maizes were compared, pigs consuming opaque-2 maizes as the sole source of protein and energy gained 4.3 times faster (0.26 vs. 0.06 kg/day) than pigs receiving normal maize. Since the opaque-2 maize contained more total protein (11.6 vs. 9.1%) than the normal maize, nonessential nitrogen in the form of diammonium citrate, glutamic acid, and glycine was added to the normal maize diet to make it isonitrogenous with the opaque-2 maize diet. The nonessential nitrogen was without effect in improving the performance of pigs fed normal maize. Neither lysine nor tryptophan supplementation significantly affected performance of pigs fed normal maize. However, either a combination of the two amino acids added to the normal maize diets at levels calculated to provide total lysine and tryptophan equal to that present in the opaque-2 maize diets, or the addition of soybean meal at a level to provide a quantity of total protein equal to that present in the opaque-2 maize diet, improved the performance of the pigs to a level similar to that obtained with opaque-2 maize. These studies demonstrated that higher levels of both lysine and tryptophan present in opaque-2 maize were responsible for the improved protein quality.

The baby pig, because of its requirement for higher levels and quality of protein, provides the most sensitive and critical test for assessing the value of the opaque-2 maize protein. Thirty-five-day-old pigs with an average body weight of 8.9 kg were fed diets in which all the protein was supplied by either common maize or opaque-2 maize (Maner *et al.* 1969). Groups fed only maize were compared with groups that had all been fed standard control diets containing a constant proportion of maize and soybean meal but at differing levels of protein.

This variation in protein level was accomplished by diluting the protein with sucrose.

The difference in pig response to each diet was striking (Table 12.8). Each decrease in level of protein from 16 to 7% in the control diet was accompanied by a corresponding decrease in daily feed consumption and in average daily gain and by an increase in feed required to produce a kilogram of gain.

The baby pigs fed common maize as the sole source of protein performed very poorly, gained an average of only 21 g/day during the entire 130-day experimental period and required 35.19 kg of feed to produce a kilogram of gain. The very poor balance of essential amino acids in the common maize was associated with a low level of feed consumed (0.74/day). As the experiment progressed, the pigs fed only common maize demonstrated the classical symptoms of protein deficiency and some died a few days after the experiment was terminated at 130 days. All animals were necropsied and their tissues examined histologically by both light and electron microscopy. Changes of the stomach, duodenum, and jejunum were characterized by atrophy of the epithelium. Study of the ribs and long bones indicated cessation of skeletal system development. The livers of all the pigs of this group were fatty and friable, and microscopically, lobular fatty accumulations were observed.

In contrast, the growth rate of pigs fed the opaque-2 hybrid diet was increased to 254 g/day which was a 12-fold increase when compared to those fed common maize. These pigs consumed 53% more feed but required only 4.43 kg/kg of body weight gain. The performance of these pigs fed on opaque-2 was, however, inferior to that of pigs fed an equal level of protein supplied by the mixture of maize and soybean meal. This inferior performance indicates that, although the level and balance of essential amino acids of the opaque-2 maize are much improved when compared to those of common maize, the proportion and level of these essential amino acids are not adequate to support a level of growth produced by the maize–soybean meal control diet containing an equal level of protein.

Although neither the growth rate nor feed conversion ratio resulting from the opaque-2 diet was equal to that produced by the control diet with equal protein, the pigs fed the opaque-2 maize showed no gross or microscopic protein-deficiency symptoms except for some fatty infiltration of the liver.

TABLE 12.8. Performance of Pigs Fed Opaque-2 Maize, Common Maize, and Common Maize–Soybean Meal Diets at Different Protein Levels from 35 to 165 Days of Age

Treatment	*Av daily gain (kg)*	*Feed efficiency*	*Daily feed cons. (kg)*
Common maize–soybean meal, 16%	.722	3.27	2.36
Common maize–soybean meal, 10%	.506	3.64	1.84
Common maize–soybean meal, 7%	.158	5.85	0.92
Common maize, 10%	.021	35.19	0.74
Opaque-2 maize, 10%	.254	4.43	1.13
Opaque-2 maize, 7%	.076	8.80	0.67
Common maize + L + T, 10%[a]	.143	5.29	0.76

Source: Maner et al. 1969.
[a]L = lysine, T = tryptophan

TABLE 12.9. Performance of Pigs Fed Opaque-2 and Common Maize During the Finishing Period

Treatment	*Protein* (%)	*Av daily gain* (*kg*)	*Feed/gain*
Common soya	16	.79[y]	3.35[y]
Common maize	10	.62[z]	4.58[z]
Opaque-2 maize	10	.81[y]	3.67[y]
Common maize + lysine + L-tryptophan	10	.81[y]	3.56[y]

Source: Gallo et al. 1968.
[y,z]Means in the same column with a different superscript are significantly different from each other.

The finishing pig (50–90 kg) requires less protein and has a lower requirement for amino acids than the growing pig. Opaque-2 maize fed as the only source of protein in the diet is adequate to support optimal growth (Gallo *et al.* 1968) (Table 12.9). The pigs fed this diet grew as rapidly and as efficiently as those receiving the control 16% maize–soya diet. However, those fed common maize as the protein source had slower and less efficient gains. These gains were improved to a level equal to those of the control or opaque-2 maize diet when 0.28% lysine and 0.04% tryptophan were used to supplement the common maize diet. Therefore, it appears that opaque-2 maize alone can serve as the only source of protein and energy for the finishing pig.

Pond *et al.* (1976) compared common maize and opaque-2 maize grown on the same farm and harvested at two stages of maturity. Although stage of maturity (90 and 102 days) of either common maize or opaque-2 maize had no significant effect on growing-finishing pig performance, pigs fed opaque-2 maize grew significantly faster on less feed per kilogram of gain than those fed the common maize. When both were included in a 12% protein diet containing soybean meal, pig performance and carcass quality were significantly better for pigs receiving opaque-2 maize. Using young pigs (4.5 and 20 kg), Rivera *et al.* (1978) demonstrated that neither stage of maturity nor drying temperature had any significant effect on amino acid availability or true nitrogen digestibility of either common or opaque-2 maize. There was a definite pattern, however, toward higher availability of all amino acids as well as true nitrogen digestibility of the immature maizes dried at the lowest temperature (40°C). All amino acid availabilities were slightly higher in opaque-2 maize compared to normal maize, although the differences were significant only in the case of threonine. Lysine and isoleucine were the least available and tryptophan the most available amino acid in opaque-2 maize. Amino acid availability of opaque-2 maize compares favorably with values for triticale, wheat, and barley reported by Sauer *et al.* (1974).

In studies with growing pigs (12.7 kg initial weight), based on growth rate and feed conversion efficiency, Wahlstrom *et al.* (1977) presented data to suggest that lysine is the most limiting amino acid in opaque-2 maize and that methionine and tryptophan did not significantly improve pig performance. Oestemer *et al.* (1970) were unable to demonstrate improved gains or feed to gain ratio in growing pigs fed opaque-2 maize supplemented with methionine. Gipp and Cline (1972) found that supplying diets of opaque-2 maize with lysine, methionine, and tryptophan did not significantly alter gain or feed-to-gain ratio of rats from those obtained when lysine alone was added. In the studies by Wahlstrom *et al.* (1977) a small improvement in gain and feed utilization was obtained when threonine was added to the opaque-2 maize diet supplemented with lysine, methionine, and tryptophan. The addition of isoleucine to the diet did not improve feed efficiency over that of the diets containing lysine, methionine, tryptophan, and threonine.

Using a complete protein evaluation procedure, Schelling *et al.* (1979) reported that the complete order of limiting amino acids in opaque-2 maize was lysine, threonine, tryptophan, and isoleucine equally and first-limiting followed in order by methionine, valine, phenylalanine, histidine, leucine, and arginine.

Gestation

It has been shown that the litter size, pig birth weight, and weaning weight at different ages are not responsive to different levels of protein fed during the complete gestation (Lenkeit 1947; Livingstone *et al.* 1966; Holden *et al.* 1968; Hesby *et al.* 1970A,B). Pregnant swine exhibit a remarkable capacity to provide nutriment to the developing fetuses during periods of severe protein restriction. No deleterious effect on offspring, except a 20% reduction in birth weight, was observed after feeding protein-free diets from mating to parturition (Pond *et al.* 1968, 1969).

Opaque-2 maize has been fed as the only source of protein and energy to gestating sows without altering normal reproduction (Hesby *et al.* 1970A,B; Baker *et al.* 1970). When these sows were compared to control sows that had been fed only common maize or maize–soybean meal diets containing from 11.2 to 20% crude protein, it was shown that opaque-2 maize fed during gestation had no effect on number or weight of pigs farrowed. However, number of pigs weaned per liter, litter gain, and total litter weaning weight were superior from gilts fed the opaque-2 maize diets as compared to those fed the normal maize diet (Baker *et al.* 1970).

Lactation

Mahan and co-workers (1971A) at the University of Illinois fed opaque-2 maize supplemented only with vitamins and minerals to first and second litter sows and compared this performance to that of other sows fed maize–soybean meal diets containing dietary protein levels from 10 to 18%. Gains of litters from sows fed opaque-2 maize were similar to those of litters from sows fed the 10% protein diet (Table 12.10). When the litter size of the first and second litter sows was maintained constant at 8 and 10 pigs per litter, respectively, and feeding level was controlled to ensure approximately equal intake of all sows, litter gains of sows fed opaque-2 corn were inferior to the litter gains obtained when 16 and 18% dietary protein levels were fed. The sows fed opaque-2 maize were in negative nitrogen balance during lactation as were sows fed only 10% dietary protein from maize and soybean meal (Mahan *et al.* 1971B). A marked decrease in milk production resulted when sows were fed either the opaque-2 maize diet or a maize–soybean meal diet containing 10% protein (Mahan *et al.* 1971C).

Maner and Colombian co-workers (1971) compared opaque-2 maize diets con-

TABLE 12.10. Effect of Protein Level and Opaque-2 Maize on Sow and Litter Performance

Criteria	*Dietary protein*					
	Maize–soybean meal diet					*Opaque-2*
	10%	*12%*	*14%*	*16%*	*18%*	*9.7%*
First litter sows						
Sow farrowing wt, kg	156.4	149.3	163.5	154.4	153.5	163.6
Sow wt loss, 0–28 days, kg	—26.9	−15.3	—13.5	—12.8	—11.0	—25.3
Sow feed consumption, 0–28 days, kg	86.6	94.6	97.0	97.8	96.0	92.2
Litter wt, 28 days, kg	45.9	49.2	51.4	54.6	53.0	49.8
Av pig wt, 28 days, kg	5.7	6.2	6.4	6.8	6.6	6.2
Second litter sows						
Sow farrowing wt, kg	188.4	195.0	187.9	197.9	186.4	184.8
Sow wt loss, kg	—25.6	—20.8	—16.5	—15.6	—13.9	—22.0
Sow feed consumption, kg	109.0	111.8	107.1	114.7	116.0	114.1
Litter wt, 28 days, kg	49.6	49.8	50.7	53.8	56.5	48.9
Av pig wt, 28 days, kg	5.0	5.0	5.1	5.4	5.7	4.9

Source: Adapted from Mahan et al. 1971A.

taining low levels of protein (9.8 and 12%) with a common maize–soybean meal diet containing 16% protein. The 9.8% protein diet contained only opaque-2 maize, vitamins, and minerals. The 12% diet contained 5.6% soybean meal in addition to vitamins and minerals. No differences in litter size or weight were observed after 35 days, but sows fed the 9.8% protein opaque-2 maize diet lost significantly more weight than the other two treatment groups (Table 12.11). It appears that a higher level of protein than that supplied by opaque-2 maize is needed for optimum lactation performance, even though its amino acid composition is superior to that of common maize.

TABLE 12.11. Production of Lactating Cows Fed Opaque-2 Maize Diets with Suboptimum Protein Levels

Criteria	*Percent of dietary protein*		
	Normal maize–soybean meal (16%)	*Opaque-2 maize–soybean meal (12%)*	*Opaque-2 maize (9.8%)*
Number of sows	23	26	25
Av wt, 24 hr postpartum, kg	175.6	184.6	174.2
Av wt, 35 days postpartum, kg	188.3	195.0	172.0
Av sow gain or loss, kg	+12.8	+10.4	—2.2
Av daily feed, 0.35 days, kg	4.63	4.85	4.57
Av pigs born alive, no.	9.9	9.6	10.0
Av pigs at 35 days, no.	7.4	7.4	8.0
Av birth wt, kg	1.3	1.3	1.3
Av weaning wt, 35 days, kg	6.1	6.2	6.1

Source: Adapted from Maner et al. 1971.

A limitation of opaque-2 maize is its 15–25% reduced yield compared to similar common maize varieties without the opaque-2 gene and its susceptibility to ear rot in the field and insect damage in storage. These poor agronomic characteristics have discouraged its production, even in areas where supplemental protein sources are scarce and expensive.

During the past 12 years breeders at the Centro Internacional de Mejoramiento de Maíz y Trigo (CIMMYT), under the direction of Dr. S. Vasal, have developed dent and flint varieties of high-lysine maize that have similar nutritional qualities to the original opaque-2 maizes but with grain characteristics similar to those of common maize. Breeding pools have been developed for both tropical and semi-tropical environments. New varieties developed from these pools have yields not different from their normal counterparts and significantly improved ear quality.

Feeding trials with growing-finishing pigs with an early selection of one of these hard-endosperm, high-lysine varieties ('VE-21') developed at CIMMYT have demonstrated that these new varieties support gains and feed conversion only slightly and nonsignificantly inferior to the soft-endosperm opaque-2 maizes (Gomez *et al.* 1975). Recently produced varieties contain 4.0–4.2% lysine and 0.9–1.0% tryptophan, and in rat trials have supported protein efficiency ratios (PER) and biological values (BV) not different from those produced with opaque-2 maize counterparts.

These agronomic improvements and stabilized nutritional value have stimulated renewed interest in these hard-endosperm, high-lysine maizes in both developed and developing countries. Guatemala leads in acreage of *quality protein maize* in commercial production, and many other countries, including Honduras, Bolivia, Haiti, Dominican Republic, Mexico, and China, are significantly increasing the area seeded to these new varieties.

FLOURY-2 MAIZE

In addition to the opaque-2 mutant gene that modifies the amino acid pattern of maize when present, a second mutant gene that affects the amino acid pattern of the endosperm is present in floury-2 maize (Nelson *et al.* 1965). According to Nelson (1969) the floury-2 gene increases lysine, tryptophan, arginine, and aspartic acid contents as well as an additional increase in methionine (Table 12.12). These data are in agreement with those reported by Cromwell *et al.* (1968). There appears also to be a slight increase in the level of total protein in the floury-2 kernel although these data are not consistent, as shown in Table 12.13 (Nelson *et al.* 1965; Nelson 1969; Maner *et al.* 1971).

Although it was early reported that floury-2 maize was equal or superior to opaque-2 maize for chicks (Cromwell *et al.* 1968), other workers have not obtained the same results. Working with many lines and double-cross hybrids containing the floury-2 gene, Colombian workers have consistently found the nutritive value of floury-2 maize to be superior to common maize but inferior to opaque-2 maize for chicks (Santos and Maner 1971), rats (Maner *et al.* 1971; Pond *et al.* 1971), and pigs (Maner *et al.* 1971).

Pigs fed floury-2 maize as the only source of protein and energy gained at a rate not different from similar groups of pigs fed common maize. Both groups were inferior in performance to groups fed opaque-2 maize (Maner *et al.* 1971).

TABLE 12.12. Percentages of Amino Acids in Grain Protein from Common and Floury-2 Maize

Amino acid	*Endosperm*[a]		*Whole grain*[b]	
	Common	*Floury-2*	*Common*	*Floury-2*
Indispensable				
Arginine	3.4	4.5	4.6	4.3
Histidine	2.9	2.2	2.9	1.1
Isoleucine	4.5	4.0	5.0	3.1
Leucine	18.8	13.3	16.1	10.5
Lysine	1.6	3.3	2.8	2.3
Methionine	2.0	3.2	3.0	0.8
Cystine	1.8	1.8	1.9	2.7
Phenylalanine	6.5	5.1	5.4	4.0
Threonine	3.5	3.3	3.5	3.1
Tryptophan	0.3	0.8	0.6	0.6
Valine	5.4	5.2	7.2	4.4
Dispensable				
Alanine	10.1	8.0	9.4	6.8
Aspartic acid	7.0	9.0	6.7	6.6
Glutamic acid	26.0	19.0	20.6	15.7
Glycine	3.0	3.4	4.1	2.9
Praline	8.6	8.3	8.9	7.5
Serine	5.6	4.8	4.6	3.0
Tyrosine	5.3	4.5	4.9	3.1

[a]Nelson 1969. [b]Maner et al. 1971.

Double mutant hybrids between opaque-2 and floury-2 have not proved promising. These hybrids in general support a level of performance that is midway between that produced by a pure floury-2 hybrid and an opaque-2 hybrid.

WAXY MAIZE

Waxy maize was first introduced into the United States from China in about 1909. It has a waxy endosperm but, more importantly, waxy maize starch consists mainly of amylopectin molecules (98.2%) whereas regular, nonwaxy maize starch contains about 25% amylose and 75% amylopectin (Bates *et al.* 1943).

TABLE 12.13. Protein Percentages in Maize Grain of Normal, Floury-2, and Opaque-2 Genotypes

Reference	*Percentage protein in maize grains*		
	Common	*Floury-2*	*Opaque-2*
Nelson et al. 1965	12.7	13.6	11.1
Nelson 1969	10.7	15.0	—
Maner et al. 1971	10.0	9.8	10.3

TABLE 12.14. Performance of Pigs Fed Regular or Waxy Maize

Performance criteria	*Regular maize*	*Waxy maize*	*Waxy maize*
Growing-finishing pigs (Hanson 1946):			
Av daily gain, kg	0.612	0.590	
Av daily feed, kg	2.66	2.48	
Feed/gain ratio	4.33	4.20	
Growing-finishing pigs (Sachtteben et al. 1975):			
Av daily gain, kg	0.74	0.69	
Av daily feed, kg	2.28	2.39	
Feed/gain ratio	3.08	3.46	
Growing pigs (Rosa et al. 1977):			
Av daily gain, kg	0.531	0.516	
Av daily feed, kg	1.30	1.19	
Feed/gain ratio	2.51	2.32	
Young starter pigs (Perez and Aumaitre 1979):			
Av daily gain, kg	0.434	0.470	
Av daily feed, kg	0.795	0.823	
Feed/gain ratio	1.84	1.76	
Young starter pigs (Wahlstrom and Libal 1975):			
Level of protein in diet, %	18	18	16
Av daily gain, kg	0.490	0.490	0.435
Av daily feed, kg	0.934	0.957	0.930
Feed/gain ratio	1.80^{a}	1.95^{a}	2.13^{b}

^{a,b}Means on the same line with different superscripts are significantly different.

Early studies (Sandstedt *et al.* 1962) indicated that waxy maize starch with less than 5% amylose and more than 95% amylopectin *in vitro* was highly susceptible to pancreatic α-amylase. More recent studies (Perez and Aumaitre 1979) using isogenic regular and waxy lines of LG11 maize showed that *in vitro* the starch in waxy maize is hydrolyzed 40% faster by pig pancreatic juice than starch in regular maize, and Gallent *et al.* (1973), using scanning electron microscopy, confirmed a faster breakdown of waxy starch granules by pig pancreatic juice.

The nutritional value of waxy maize for pigs of varying ages has been studied at several locations and contrasting results have been obtained (Table 12.14). Hanson (1946) compared the feeding value of waxy maize with that of regular nonwaxy maize and concluded that waxy maize was equal to nonwaxy maize for growing-finishing pigs but could show no advantage. When given a choice of both types of maize, the pigs consumed 2.6 times as much regular as waxy maize. However, when only one type was available, daily consumption was greater for the waxy maize. Complete mixed diets containing waxy maize were more efficiently utilized than diets containing nonwaxy maize. Similar results with growing pigs were reported by McDonald (1973), Koch *et al.* (1974), Jensen *et al.* (1973), and Rosa *et al.* (1977), but Walstrom and Libal (1975) and Robbins *et al.* (1976) observed no effect from waxy maize on early weaned pig performance with an 18% crude protein diet and a negative effect due to waxy maize on a 16% diet. Perez and Aumaitre (1979) used waxy maize in weaning diets for pigs 5–9 weeks of age and reported a clear tendency toward better pig performance.

Waxy maize types have been shown to be very similar in protein and amino acid content to isogenic nonwaxy maize (Rosa *et al.* 1977; Perez and Aumaitre 1979). (See Table 12.15). This would suggest that the waxy gene has no beneficial effect on protein synthesis and thus the protein quality of maize. Similar digestibility and performance of pigs fed either normal or waxy maize without amino acid supplementation, reported by Rosa *et al.* (1977), and similar nitrogen digestibility and utilization, reported by Perez and Aumaitre (1979), would seem to support the suggestion of equal protein quality.

Energy utilization of waxy maize has been reported to be similar (Sachtteben *et al.* 1975; Rosa *et al.* 1977) or slightly superior (Perez and Aumaitre 1979) to that of normal maize. The digestible energy, apparent metabolizable energy, and correct-metabolizable energy values of the two types of maize are presented in Table 12.16. Insufficient information is available to allow definitive conclusions about the overall relative nutritional value of normal and waxy maize; however, a modification of the molecular structure of maize starch seems to influence its feeding value, and waxy maize would be expected to be equal or slightly superior to normal maize when fed to pigs.

TABLE 12.15. Chemical Composition of Regular or Waxy Maize

Component	*Regular maize*[a]	*Waxy maize*[a]	*Regular maize*[b]	*Waxy maize*[b]
Dry matter, %	88.7	88.6	89.2	91.1
Composition (%DM)				
Organic matter	—	—	98.3	98.5
Crude protein	8.45	9.11	10.4	10.2
Crude fiber	—	—	3.2	3.0
Ether extract	4.10	4.96	—	—
Essential amino acids (g/16g N)				
Lysine	3.74	3.33	2.95	2.95
Histidine	3.34	3.33	2.65	2.75
Arginine	4.67	4.31	4.75	5.05
Threonine	3.87	3.82	3.35	3.70
Valine	4.94	4.68	4.65	5.00
Isoleucine	3.07	3.20	3.50	3.75
Leucine	12.41	11.83	11.15	12.45
Tyrosine	4.14	3.94	4.00	4.50
Phenylalanine	4.80	4.56	4.40	4.85
Cystine	3.60	3.57	2.25	2.40
Methionine	2.00	1.97	2.10	2.15
Tryptophan	0.57	0.69	—	—

[a]Adapted from Rosa et al. 1977, near isogenic lines.
[b]Adapted frm Perez and Aumaitre 1979, isogenic maize cultivars ('Funk's G 4384') differing only in waxy gene.

TABLE 12.16. Energy and Protein Utilization of Regular and Waxy Maize by Growing Pigs

	Regular maize	*Waxy maize*	*Regular maize*[a]	*Waxy maize*[a]
Nitrogen utilization				
Digestibility, %	76.0	72.0	85.6	86.3
Retention, g/day	—	—	14.2	15.2[b]
Retention, % digested N	—	—	62.8	64.2
Digestibility, %				
Organic matter	81.8	78.9	91.4	91.7
Energy	80.4	77.2	88.4	90.3[c]
Energy value (kcal/kg DM)				
Gross	—	—	4425	4406
Digestible	—	—	3914 ± 13	3980[b] ± 18
Metabolizable	—	—	3832 ± 13	3887[b] ± 27
Metabolizable corrected for N	—	—	3787 ± 13	3839 ± 27

Source: Adapted from Perez and Aumaitre 1979.
[a]Isogenic maize cultivars ('Funk's G 4384') differing only in waxy gene.
[b]Signficantly different ($P < .10$).
[c]Significantly different ($P < .05$).

GRAIN SORGHUM (MILO)

Sorghum (*Sorghum bicolor* L. Moench) is a hardy plant able to grow and produce under a wide range of environmental conditions, particularly in warmer temperatures and tropical regions of the world. In the United States, sorghums are cultivated throughout the Great Plains states from Texas to South Dakota and in the hot, irrigated valleys of southern Arizona and California. Outside of the United States, southern and eastern Asia and sub-Sahara Africa are the principal regions of cultivation. National yields for some of the major sorghum-producing countries are presented in Table 12.17 (FAO 1980).

The sorghum plant is able to withstand heat and drought better than many crops. It is well adapted to regions with as little as 17 in. (42.5 cm) of rainfall but is also highly productive on irrigated lands and in humid regions. The plant becomes practically dormant during periods of drought, resuming growth when moisture becomes available. The resistance of sorghum to drought, heat, grasshoppers, rootworm, and corn borers, its adaptation to both moderately alkaline and salty soils, and its adaptation to both heavy and sandy soils are the traditional reasons for cultivating sorghum in preference to maize in some regions of the world (Rachie 1969).

The small, round kernels, characteristic of sorghum grain, contrast with the longer flatter grains of maize and the round, elongated grains of wheat. Microscopic examination of a section through a mature grain sorghum kernel differentiates the outer seed cover or pericarp, the testa if present, the endosperm, and the embryo or germ. The pericarp surrounds the endosperm and consists of several tissue layers; the epicarp portion may be pigmented and its outer surface covered with a waxy cuticle; a central layer of large thin-walled cells containing starch constitute the mesocarp, and the inner region is composed of cross cells and tube

TABLE 12.17. Sorghum Grain Production by Regions and by Major Producing Countries

Location	*Area (millions of ha)*	*Production (millions of metric tons)*	*World production (%)*
Regions			
North America	5.148	14.936	25.56
Far East	17.761	13.440	23.00
Latin America	4.048	9.497	16.25
Africa	11.274	7.510	12.85
Near East	3.969	3.075	5.26
Oceania	0.542	0.766	1.31
Europe	0.146	0.597	1.02
Countries			
United States	5.148	14.936	25.56
India	17.000	12.800	21.90
China	3.203	7.710	13.19
Mexico	1.517	4.677	8.00
Nigeria	6.000	3.800	6.50
Argentina	1.280	2.960	5.07
Sudan	3.000	2.200	3.76
World total	46.691	58.435	—

Source: FAO Production Yearbook 1980.

cells. The testa, when present, is a tissue attached to the epicarp next to the endosperm. The testa is derived from the inner-integument and may be highly pigmented. The endosperm consists of an outer single-cell layer of aleurone tissue and an inner region of endosperm filled with starch granules which are enclosed in a protein matrix. The endosperm may be corneous (vitreous, horny, flinty) or floury or any degree between these extremes. The endosperm may be nonyellow, yellow, or hetero-yellow. Dissection indicates that the endosperm makes up 82% (range 80.0–84.6%) of the grain; the germ, 10% (range (7.8–12.1%); and the bran, which consists of the pericarp and some associated tissues, 8% (range 7.3–9.3%).

Sorghum grain is valued for its content of energy in the form of starch. Starch comprises 83% of the endosperm, 13.4% of the germ, and 34.6% of the bran or pericarp. Different types of starch are present in existing varieties and hybrids. Waxy starches are contained in varieties derived from strains introduced from China and Africa and are composed entirely of amylopectin, a branched-chain polymer of glucose units. Nonwaxy or normal sorghum starch contains about 27% of amylose, a straight-chain polymer of glucose, and 73% of amylopectin. The whole grain contains 6.5–12% protein, 72–76% starch, and about 2% fiber. Sorghums also contain intrinsic substances that occur naturally in the plant; these include tannins (polyphenols), phytates, cyanogenic glycosides, and possibly others not yet identified.

Great volumes of research have been published concerning the value of sorghum grain as a feed for swine. The composition and nutritive value of sorghum grain has been summarized by Hulse *et al.* (1980).

Early studies showed that grain sorghums were equal or superior to maize in

TABLE 12.18. Palatability and Value of Six Varieties of Sorghum for Pigs

Variety	Palability test			Feeding test	
	Protein content (%)	Percent of total sorghum consumed: First period (%)	Percent of total sorghum consumed: Second period (%)	Av daily gain (g)	Feed/gain ratio
'DeKalb F 62A'	11.94	5.4	26.9	723	3.80
'Kafir 44-14'	11.31	12.9	46.4	768	3.78
'Darset'	11.25	9.4	18.3	764	4.09
'R. S. 610'	11.56	60.7	—[a]	800	3.82
'Redlan'	12.56	6.1	4.1	782	3.87
'Amak R-12'	11.94	5.5	4.3	805	3.62

Source: Adapted from Hillier et al. 1959.
[a]Most palatable in first period; removed in second period.

supporting daily gain, but when sorghums were fed, significantly more feed was required to produce a kilogram of gain. Studies in Nebraska 30 years ago indicated that grain sorghum and maize produced approximately the same average daily gains when fed to swine, but in general 10–20% more feed was required for sorghum diets. Similar results were reported from Kansas (Aubel 1945, 1955, 1960A,B) and Oklahoma (Hillier *et al.* 1953). Peo and Hudman (1958) compared several standard varieties and hybrids of grain sorghum and demonstrated differences in palatability and nutritional value. Studies by Hillier *et al.* (1959) showed that when given a choice, pigs would select among sorghum varieties, but when each was offered separately, only small variations in feed consumption and efficiency were reported (Table 12.18). It was suggested that differences in palatability could be associated with tannin content.

Tanksley (1961) summarized a large portion of the early work from Arkansas, Colorado, Illinois, Mississippi, Nebraska, Oklahoma, and Texas. This summary shows that when ground grain sorghum is fed free-choice with a protein supplement and compared to ground maize fed in a similar manner, gains are not different for the two grain sources (Table 12.19). However, efficiency of feed

TABLE 12.19. A Comparison of the Feeding Value of Sorghum and Maize, and of Method and Form of Feeding Sorghum to Pigs

Types of feed	Number of pigs	Av initial wt (kg)	Av final wt (kg)	Av daily gain (g)	Feed/gain ratio	Percentage of control: Daily gain	Percentage of control: Efficiency
Ground sorghum + supplement	305	29.3	89.9	746	3.94	100.9	94.7
Ground maize + supplement	158	28.8	90.2	741	3.74	100.0	100.0
Ground sorghum + supplement	289	28.7	96.6	723	3.94	101.3	101.4
Whole sorghum + supplement	296	28.8	97.0	714	3.99	100.0	100.0
Ground sorghum + supplement	213	25.6	90.2	655	3.78	95.4	93.5
Ground sorghum, complete mix	157	25.2	89.9	686	3.55	100.0	100.0

Source: Adapted from Tanksley 1961.

utilization by pigs fed the grain sorghum diet is slightly less than with maize. When ground and whole grain sorghum were compared in free-choice rations, performance was very similar. Very little feed was saved by grinding, so that cost of grinding would have to be evaluated to determine the most economical method of feeding the grain when offered free-choice with a protein supplement.

These summaries also clearly demonstrate that maximum gains and feed efficiency from grain sorghums are obtained when the grain is incorporated into a complete mixed feed. Complete mixed rations based on grain sorghum produce faster and more efficient gains than when ground grain sorghum is fed free-choice with a protein supplement. In these comparisons from Oklahoma, free-choice grain sorghum and supplement gave only 95.4% of the gains and only 93.5% of the efficiency of feed conversion recorded when the supplement and ground grain sorghum were mixed before feeding.

Studies during the period 1960–1970 compared the feeding value of sorghum with that of maize, wheat, oats, and barley. These studies are summarized in Table 12.20. The results of these studies demonstrate that when good quality protein supplement is fed to either growing or finishing pigs, sorghum grain has a feeding value not different from that of maize (Jensen *et al.* 1965; Vanschoubroek

TABLE 12.20. Comparison of the Feeding Value of Sorghum Grain with That of Maize, Barley, Oats, and Wheat

Protein supplement	*Grain*	*Type of substitution*[a]	*Class of swine*[b]	*Av daily gain*		*Feed/gain ratio*	
				g	*Percent of sorghum*	*Ratio*	*Percent of sorghum*
Mixed protein[c]	Sorghum	Weight	G-F	710	100.0	3.58	100.0
Mixed protein[c]	Maize	Weight	G-F	720	101.4	3.54	101.1
Soybean meal[d]	Sorghum	Protein	F	780	100.0	3.85	100.0
Soybean meal[d]	Maize	Protein	F	810	103.8	3.85	100.0
Peanut meal (PM)[e]	Sorghum	Protein	G	450	100.0	3.03	100.0
Peanut meal[e]	Maize	Protein	G	650	144.4	2.94	103.1
PM + fish meal[e]	Sorghum	Protein	G	550	122.2	2.94	103.1
Soybean meal[f]	Sorghum	Weight	G-F	510	100.0	3.23	100.0
Soybean meal[f]	Barley	Weight	G-F	513	100.6	3.27	98.8
Meat meal, No. 1[f]	Sorghum	Weight	G-F	429	100.0	3.47	100.0
Meat meal, No. 1[f]	Barley	Weight	G-F	437	101.9	3.49	99.4
Meat meal, No. 2[f]	Sorghum	Weight	G-F	467	100.0	3.17	100.0
Meat meal, No. 2[f]	Barley	Weight	G-F	423	90.6	3.36	94.3
Meat meal, No. 3[f]	Sorghum	Weight	G-F	445	100.0	3.33	100.0
Meat meal, No. 3[f]	Barley	Weight	G-F	433	97.3	3.61	92.2
Soybean meal[f]	Sorghum	Weight	G-F	475	100.0	3.14	100.0
Soybean meal[f]	Barley	Weight	G-F	464	97.7	3.32	94.6
Meat meal[g]	Sorghum	Protein	G	363	100.0	2.76	100.0
Meat meal[g]	Barley	Protein	G	317	114.9	2.38	116.0
Meat meal[g]	Wheat	Protein	G	404	111.3	2.53	109.1
Meat meal[g]	Oats	Protein	G	431	118.7	2.71	101.8

[a]Sorghum substituted for other grains on equal weight or equal protein basis.
[b]G-F, growing-finishing; G, growing; and F, finishing periods.
[c]Vanschoueroek et al. 1964. [d]Jensen et al. 1965. [e]Ranjhan et al. 1964.
[f]Beames and Sewell 1969. [g]Batterham and Manson 1970.

et al. 1964). However, when poor quality protein such as peanut meal is fed, the value of sorghum grain is only 69.2% that of maize (Ranjhan *et al.* 1964). Replacing as much as 8% of the peanut meal with fish meal improved gains but not to a level equal to those obtained with maize. Other studies (Beames and Sewell 1969) have shown that sorghum has a feeding value equal to or superior to that of barley. Although some gains on barley were equal to those of pigs fed on sorghum, when lower quality meat meal was used as the protein supplement, gains on sorghum were superior to those obtained with barley. In all comparisons with different protein sources in these studies, feed efficiency was best when sorghum was used instead of barley. Studies by Batterham and Manson (1970) do not agree with these results and indicate that the feeding value of sorghum is inferior to that of wheat, barley, and oats when a meat meal supplement was used to supply protein.

The great variability in sorghum varieties and hybrids and the lack of accurate physical and chemical description of the materials used make it very difficult to accurately analyze the data available. More recent research has, in most cases, attempted to characterize better the materials utilized in order to evaluate better the effects of grain characteristics and nutrient and nonnutrient components on pig response. Studies have focused on (1) endosperm texture, (2) starch type, (3) endosperm color (type), (4) content of tannins and seed coat color, and (5) amino acid content, availability, and adequacy. In an attempt to evaluate the effect of these characteristics, each will be discussed separately. This is done, fully recognizing that interactions may exist among these characteristics and that different combinations of characteristics may elicit variation in nutritional response.

Endosperm Texture

The endosperm of the sorghum kernel is composed of the peripheral endosperm area, the corneous or horny endosperm, and the floury or soft endosperm. The proportion of each depends on the variety and upon environmental conditions. The peripheral endosperm area is several cell layers thick and is located beneath the aleurone layer. The starch granules are small and irregular or polygonal in shape and embedded in a dense proteinaceous matrix. The starch granules are very tenaciously held in the peripheral endosperm, and hydrolysis by digestive enzymes is difficult and slow. Newer processing methods sometimes mechanically disrupt the organization of this area which greatly increases the accessibility of the starch and protein to digestive enzymes.

The corneous endosperm (vitreous, flint, horny) is less compact than the peripheral endosperm layer. The cells of the corneous endosperm are filled with starch granules that are very angular or polyhedral in shape and enmeshed in a proteinaceous matrix. The cells are packed tightly, and light is transmitted through the endosperm so that it appears translucent.

The floury or soft endosperm is located in the innermost portion of the kernel. The cells are larger than those in the corneous endosperm, and the starch granules are larger and spherical in shape. In the floury endosperm, the protein and starch are not tightly held together and there is little if any protein matrix.

The relative proportions of floury to corneous endosperm differ among sorghum varieties and influence the properties of the kernel. Endosperm texture is now classified in three categories: floury, intermediate, and corneous.

Although studies with steers have shown a marked increased in carbohydrate digestion in the rumen as the endosperm texture changes from corneous, intermediate to floury (48.4, 68.1, and 80.1%, respectively) (Samford *et al.* 1970), these extreme differences have not been observed in swine (Cohen and Tanksley 1973A,B). When diets containing equal quantities of sorghum representing these three endosperm textures were fed to growing swine, dry matter and organic matter digestibilities of the intermediate textured sorghum were significantly greater than those with either the floury or corneous texture. Gross energy digestibility of the intermediate was also greater than either and significantly greater than the floury endosperm. There were no significant differences in protein or crude fiber digestibility among the three endosperm textures. The higher digestibility of dry matter, organic matter, and gross energy in the intermediate textured endosperm was reflected in its having the highest digestible and metabolizable energy content. There were no significant differences in nitrogen retention, gain, or feed efficiency among the three different textures.

Starch Type

As previously described, maize as well as sorghum endosperm may contain two distinct starch types. These have been described as normal or nonwaxy and waxy. Most normal sorghum grain starches contain 20–30% amylose and 70–80% amylopectin and are called the straight-chain and the branched-chain types of starch, respectively; however, "waxy" sorghum grain starches are made up of almost 100% amylopectin. *In vitro* and laboratory animal data seem to indicate that amylose starch is less digestible by pancreatic amylase than amylopectin (Sandetedt *et al.* 1962; Borchers 1962). Studies conducted with ruminants also showed that the carbohydrates in waxy endosperm sorghum were more highly digestible (Samford *et al.* 1970) and had greater net energy values (Sherrod *et al.* 1969) than normal, nonwaxy sorghum. Grain of waxy Kafir and nonwaxy Kafir 60 was fed to growing swine by Cohen and Tanksley (1973A,B). They found that gross, digestible, and metabolizable energy of waxy and nonwaxy endosperm sorghum diets were not different. However, over a short 28-day feeding period, the average daily gain and feed efficiency data, although not statistically different, did show a definite trend in favor of the waxy grain diet. Pigs fed waxy sorghum diets had 11% faster average daily gains and required 10% less feed per kilogram of gain. These data suggest that the trend for waxy sorghum to support increased gain and improved feed efficiency is not explained by inherent differences in digestibility of the endosperm starch and suggest that additional studies are needed to elucidate the reasons for these observed differences. Although total protein content of the waxy and nonwaxy sorghum kernels is essentially the same, the waxy has a more equal distribution of protein through the endosperm and less protein concentrated in the peripheral endosperm area. Although protein and starch digestion is not different over the total digestive tract, differences in digestion in the upper intestinal tract may affect not only carbohydrate but also protein and amino acid utilization.

Endosperm Color or Type

Endosperm type (color) is assigned on the amount of carotenoid pigments within the kernel endosperm; nonyellow contains the least, and yellow the heaviest pigmentation. Yellow grain sorghum is often confused with yellow endosperm sorghum. The official U.S. standards for grain sorghum recognize four classes: yellow, white, brown, and mixed. Visual appearance of the kernel is the basis for classifying sorghums, and it is influenced by a combination of pericarp color, opaqueness or translucency of the pericarp, type of endosperm, and presence or absence of a brown undercoat or pigmented testa. The yellow grain sorghum class predominates in the grain produced in the United States. It should not be confused with the yellow endosperm sorghum. The true yellow endosperm sorghum has a thin, translucent pericarp which allows the yellow color of the endosperm to impart its color to the grain. This grain falls into the yellow market class of grain sorghums. Yellow endosperm type is independent of pericarp color or presence or absence of the testa; therefore, a yellow endosperm can be found under an opaque or translucent white, red, or brown pericarp as well as a testa. Pigmented pericarps mask the endosperm color. Therefore, yellow-endosperm-type grain sorghum can be in the yellow, white, or brown market class and vary in tannin content, starch type, and endosperm texture. The hetero-yellow sorghum hybrids are produced from a cross between a yellow endosperm and a nonyellow endosperm parent.

Studies to evaluate the nutritional qualities of endosperm types for pigs and to compare them to maize have been reported by Copelin *et al.* (1974) and by Meadows *et al.* (1974). Chemical analysis showed the three endosperm types (nonyellow, hetero-yellow, and yellow) to be very similar in protein, ether extract, crude fiber, ash, and gross energy as well as in amino acid profile. When equal quantities of sorghum were combined with soybean meal to form the three test diets, nonyellow sorghum diets were consistently higher in dry matter, organic matter, and gross energy digestibilities while the hetero-yellow sorghum diets were consistently lowest among the three sorghum endosperm types. Maize was highest in protein digestibility and significantly higher than the hetero-yellow and yellow sorghum diets. The maize diet had the highest metabolizable energy value (3.56 kcal/g) which was significantly higher than the yellow, nonyellow, and hetero-yellow sorghums which were not different (3.48, 3.46, and 3.40 kcal/g, respectively).

Data for the feeding trial are presented in Table 12.21. Endosperm type had no effect on growth rate of the growing-finishing pigs but did influence efficiency of feed utilization, with the nonyellow being more efficiently utilized than the yellow and the yellow more efficiently utilized than the hetero-yellow. The efficiencies of the sorghums based on maize at 100 were nonyellow, 98.13; yellow, 95.64, and hetero-yellow, 93.77. When both the digestibility and performance data are considered, nonyellow exhibited a small but consistent advantage over yellow as did yellow over hetero-yellow.

Other data have failed to confirm the results reported by Tanksley (1975). Using young growing pigs to evaluate 27 different cultivars, Noland *et al.* (1976) reported that the digestibility of energy and protein and apparent nitrogen retention was consistently higher for yellow endosperm sorghum cultivars as compared

TABLE 12.21. Effect of Endosperm Type[a] on Performance and Diet Digestibility of Growing-Finishing Pigs

Criteria	Yellow maize	N-Y	H-Y	Y
Pig performance				
Number of animals	48	48	47	47
Av daily gain, kg	0.78	0.78	0.76	0.77
Av daily feed, kg	2.49[c]	2.57[c,d]	2.60[d]	2.59[d]
Feed/gain ratio	3.21[c]	3.27	3.41[e]	3.35[d,e]
Relative feed efficiency, %	100.00	98.13	93.77	95.64
Apparent digestibility coefficients				
Dry matter, %	83.35[b]	83.60[b]	81.08[c]	82.71[b]
Protein, %	79.70[b]	76.01[c]	74.22	75.31[c,d]
Ether extract, %	59.89	40.48	35.02	42.92
Crude fiber, %	30.19	34.49	29.80	37.40
Digestible energy, kcal/g	3.65[b]	3.55[c]	3.48[c]	3.56[c]
Metabolizable energy, kcal/g	3.56[b]	3.46[c]	3.40[c]	3.48[c]

Source: Adapted from Tanksley 1975.
[a]The endosperm types are nonyellow (N-Y), hetero-yellow (H-Y), and yellow (Y).
[b,c,d,e]Means in the same column without a common superscript differ significantly.

to white (nonyellow) cultivars. Although differences in energy digestibility between the two endosperm types were small and nonsignificant, differences in protein digestibility and nitrogen retention were of the order of 7 and 19%, respectively.

Smith and Allee (1973), in feeding trials with growing and finishing pigs, compared three sorghum types: (a) white endosperm (red seed coat), (b) hetero-yellow endosperm (bronze seed coat), and (c) yellow endosperm (yellow seed coat). Although the pigs showed a marked preference for the hetero-yellow endosperm, neither daily gain, feed efficiency, carcass characteristics, protein digestibility, nor nitrogen retention were significantly affected by endosperm type.

The variation in results of these three reports prohibits a definite conclusion related to the comparative nutritional value of different endosperm types. The results were probably confounded with varietal differences and as in the case of white and yellow endosperm maize, nutritional differences would only be expected in relation to their effect on vitamin A adequacy of the diet.

Tannins and Seed Coat Color

Although endosperm texture, color and starch type may have little effect on overall grain quality and nutritional value of different sorghum varieties and hybrids, content of tannins and protein quality have a more significant effect on the feeding value of sorghum grains.

Although some sorghum varieties and hybrids containing high levels of tannins are agronomically associated with increased bird-resistance and decreased grain deterioration in the field, they are also associated nutritionally with reduced palatability and depressed nutrient digestibility and utilization.

Plant tannins are complex phenolic polymers containing aliphatic and phenolic hydroxyl groups and in some cases carboxyl groups. The tannins are only a fraction of the polyphenols present in plants (McLeod 1974), and a major problem associated with the analysis of tannins in sorghum kernels is what precisely constitutes the tannins. In grain sorghum, the tannins are condensed tannins which yield on acid hydrolysis a flavanone and an anthocyanidin (McLeod 1974). The known reactions of this class of molecule suggest that these polymeric polyphenols are responsible for the typical tannin reactions of sorghum grain affecting color and nutritional value.

The ability of tannins to bind to proteins is probably a major factor in their observed anti-nutritional effects. The large number of phenolic groups in the tannin molecule provide many points of attachment with favorable steric opportunities for linkage by hydrogen bonding with the peptides of adjacent protein chains to form protein–tannin complexes. Tannins may undergo three types of reactions: (1) hydrogen bonding which is reversible and independent of pH, when pH is less than 8.0; (2) ionic interaction which depends on pH and is reversible; and (3) oxidative coupling which operates under aerobic conditions and is not reversible.

The evidence suggests that ingested protein may be bound by phenols present in the diet, thus rendering them insoluble and preventing their absorption from the gastrointestinal tract. Such a mechanism would be expected to reduce the protein nitrogen available to the animal. Many phenols are known also to interfere with essential biochemical processes associated with carbohydrate and fat utilization. It is possible that this type of mechanism may in part explain the apparent impaired utilization of carbohydrates by animals fed high-tannin sorghums. The activities of many enzyme systems are inhibited by tannins. Enzyme inhibition is caused by tannins complexing with enzyme protein and with the substrate.

Official standards for grain sorghum classification are based on visual appearance of the kernel; white, yellow and brown kernel classes are recognized. Pericarp color in sorghum is controlled by two independent series of genes (Shertz and Stephens 1966). One series, genes *RYI,* controls pericarp color and has no effect on the presence or absence of a pigmented testa. Pericarp color associated with these genes ranges from white to various shades of red. A second series of genes, $B_1B_2S_1$, controls presence or absence of a pigmented testa and also affects pericarp color. Materials with both dominant *B* genes have a pigmented testa. When either *B* gene is recessive, the pigmented testa is absent. When both *B* genes are dominant in combination with dominant S_1, the pericarp is brown, regardless of the pericarp color resulting from the *RYI*genes. A pigmented testa also may be present with a white or red pericarp.

McGinty (1969) used sorghum kernels of various colors produced under identical environmental conditions to determine relative digestibilities. Brown-seeded varieties were significantly lower in *in vitro* dry matter and protein digestibility than white-, red-, or yellow-seeded varieties. Removal of the pericarp of each variety eliminated differences in digestibility of the various endosperms. Addition of brown pericarps to any endosperm resulted in decreased dry matter digestibility. Harris *et al.* (1970) observed that brown-kerneled sorghum grains contained higher tannin levels than nonbrown kernels. Tannin content was directly related to decreased *in vitro* dry matter digestibility. May and Nelson

(1973) and Maxson *et al.* (1973) have also reported that brown pericarp cultivars have lower digestion coefficients when tested with rats and chicks than do the nonbrown cultivars. Noland (1976) reported that pigs fed varieties of grain sorghum with brown seed coats (pericarp) required 15% more feed and gained weight at 7% slower rate than pigs fed varieties with nonbrown seed coats. These authors also showed that with one exception all varieties with yellow seed coats were superior in energy and protein digestibility to all varieties with brown seed coat. Thrasher *et al.* (1975) found that brown-pericarp, bird-resistant hybrids supported slower and less efficient gains and had lower dry matter, crude protein, and NFE digestion coefficients than a light brown, yellow endosperm, bird-susceptible hybrid.

Sorghum samples with yellow and brown seed coats were compared in studies by Almond *et al.* (1979). The two grains were not different in proximate composition or amino acid content. The brown sorghum contained 2.5% tannins as compared to only 0.4% in the yellow sorghum. Pigs fed the yellow sorghum grew at a 10% faster rate on 9.5% less feed than did those fed the brown sorghum. The inclusion of the yellow sorghum in the diet in place of the brown-seed variety improved the apparent digestibility of dry matter, gross energy, nitrogen and fiber, and nitrogen retention. The yellow sorghum also contained higher levels of both digestible and metabolizable energy. Although the lysine contents were similar, availability of lysine in the yellow sorghum (58.6%) was higher than in the brown-seeded variety (52.5%). Overall, the yellow sorghum had a higher nutritive value than the brown-seeded variety. However, brown-kernel color as determined by visual appraisal does not appear to be a consistent indicator of tannin content or decreased digestibility. Variation in tannin content of 33 brown-seeded, bird-resistant hybrids reported by Tipton *et al.* (1975) (Table 12.22)

TABLE 12.22. Variations in Tannin Content of 33 Brown-Seeded, Bird-Resistant Grain Sorghum Hybrids

Hybrid	*Tannin, %*	*Hybrid*	*Tannin, %*
'Excel Bird Go'	7.25	'Funk BR-630'	5.10
'DeKalb BR-64'	7.22	'AKS 614'	5.06
'Funk BR-79'	7.04	'Excel Bird Go A'	5.02
'Two 75 BR'	6.51	'DeKalb BR-63'	5.00
'Pioneer brand B-815'	6.42	'AKS 618'	4.90
'NK Savanna'	6.41	'McCurdy 81'	4.86
'NK Savanna 2'	6.21	'Funk G–516 BR'	4.70
'DeKalb Br-54'	6.11	'McCurdy 83'	4.56
'Acco R-1023'	6.11	'McCurdy 13'	4.39
'KB Golden Grain BR'	5.94	'McNair 656'	4.30
'Pennington BR'	5.80	'AKS 663'	4.17
'Acco R-1093'	5.67	'RS 700'	3.97
'NK Savanna 3'	5.66	'McNair 880'	3.72
'Pioneer brand B-818'	5.22	'Acco 9417'	3.40
'Frontier 409'	5.17	'McCurdy 11'	3.19
'NcNair 546'	5.15	'DeKalb X 1602'	2.36
'FMC Shoo Bird'	5.14		

Source: Tipton et al. 1975.

partially explains these inconsistencies. Although all hybrids tested were brown seeded, the tannin content (Burns 1971) varied from 2.36 to 7.25%. Other studies (McMillian *et al.* 1972) have also demonstrated the variability in tannin content between and within pericarp colors.

The visual appearance of grain sorghum kernels is influenced by combinations of pericarp color, opaqueness or translucence of the pericarp (due to differences in mesocarp thickness), type of endosperm, presence or absence of a pigmented testa, and variations in intensity within colors. Consequently subjective visual appraisal of grain sorghum color into brown, yellow, and white is not an accurate measure of true grain color. For example, sorghums with a white or colorless pericarp but with a pigmented testa are more tan than brown in color but are classed as brown grain sorghum. Sharp *et al.* (1977) were unable to establish an index to determine grain sorghum digestibility based on an objective color rating, but they did demonstrate that tristimulus color ratings of ground blended samples and especially those related to the measurement of redness were highly negatively correlated with energy digestibility, protein digestibility, and nitrogen retention in pigs.

There seems little doubt that the tannins in grain sorghum reduce both dry matter and protein digestibility and that they inhibit the activity of various enzyme systems including amylases and probably lipases and proteases. When fed to swine, sorghum tannins reduce overall biological value, feed efficiency, and nitrogen retention. Content of tannins of different sorghum varieties and hybrids varies greatly. Although it is generally agreed that tannins impart a reddish to brown color in the pericarp and pigmented testa and brown-seeded sorghums are associated with higher tannin cultivars, seed coat color is not a reliable indicator of tannin content or nutritional quality.

Many attempts have been made to reduce the anti-nutritional effects of sorghum tannins but because of lack of definitive knowledge of the chemistry of the tannins, most methods have been somewhat empirical.

Sorghum tannins appear to be relatively reactive and easily modified to forms which do not respond to chemical tests for tannins. Price *et al.* (1978) treated a high-tannin, bird-resistant BR-54 sorghum grain with dry and steam heat, aqueous solutions of NaOH and K_2CO_3, and aqueous and anhydrous ammonia. Most of these simple treatments decreased the tannin content by 50–90% as estimated by the vanillin test. One of the treatments, moistening the grain with NH_4OH, improved the nutritional quality of the high-tannin sorghum, so that for rats and chickens it gave weight gains and feed efficiency equivalent to those obtained with low-tannin sorghum. Other treatments, including gaseous ammonia under pressure and autoclaving, also improved the nutritional quality of BR-54, but to a lesser degree. These same treatments decreased the nutritional value of a control low-tannin sorghum RS-610.

Reichert *et al.* (1980) reported that the tannin content of sorghum grain was reduced from 3.63% to 2.2, 0.6, or 0.1% by moistening with 25% by weight of water, 0.8*N* HCl, or 0.8*N* NaOH, respectively, and storing for 2 days at 25°C. With water and storage for 9 days, the tannin content was reduced to 0.3%. These treatments improved digestibility of dry matter and protein, and weight gains and feed efficiency were equivalent to those of animals given sorghum of low tannin content.

Following studies with rats and chicks, Ford (1977) and Ford and Hewitt (1977) reported that the addition of polyethylene glycol 4000 (PEG) to high-tannin sorghums improved protein digestibility and methionine availability. In studies with pigs (Savage *et al.* 1980), the addition of PEG to a diet containing ground sorghum (brown-seeded with 2.5% tannin) increased apparent digestibility of gross energy by 3.8% and improved metabolizable energy content by 5.5%. However, when added to a micronized and ground sorghum diet, PEG had no effect.

Grinding and rolling of sorghum to disrupt the intact kernel has consistently improved its nutrient digestibility (Beames 1969) and feeding value (Aubel 1945, 1955, 1960A,B; Tanksley 1961; Quisenberry and Tanksley 1970) for growing-finishing swine. Processing sorghum also makes it possible to feed complete mixed diets, which results in better pig performance than does the feeding of grain and protein supplement separately (Hillier *et al.* 1954; Hillier and Martin 1958). Quisenberry and Tanksley (1970) suggested that for swine the grain should be ground fine to moderately fine since coarsely ground grain is not as well digested. Luce *et al.* (1970) indicated that successive reduction in sorghum particle size by grinding improves feed efficiency but not daily gains. In studies by Owsley *et al.* (1981), pigs were fed a nonwaxy, intermediate endosperm, yellow sorghum with a red pericarp and no testa that had been rolled or ground in a hammer mill through a 6.4- or 3.2-mm screen, producing particles with a modulus of fineness of 3.57, 2.85, or 2.36, respectively. Each successive reduction in particle size improved the apparent ileal digestibility of dry matter, starch, gross energy, and nitrogen. Measured over the total digestive tract, digestibility was highest with the smallest particle size. Rolled or medium fine (3.57 and 2.85) were not different over the entire gastrointestinal tract because of the increased disappearance in the large intestine of dry matter, gross energy, and starch of the rolled grain. The digestibility of most amino acids at the terminal ileum was improved as the particle size decreased.

Many different methods of processing grain sorghum have been used in an attempt to improve its utilization. These include pelleting (Jensen *et al.* 1960; Tanksley 1961), reconstituting or soaking in water (Diggs 1969; Aubel 1960A; Cam Chuong *et al.* 1966), steam-flaking (Tanksley 1961), roasting (Baird 1976), and micronizing (Riggs *et al.* 1970; Shiav *et al.* 1976; Shadid *et al.* 1976; Savage *et al.* 1980). In the studies by Jensen *et al.* (1960), pelleting of the complete mixed diet containing sorghum improved gains by 47.8% over the unpelleted meal diet. Reconstituting, soaking in water, steam-flaking, or roasting had no effect on growth rate when compared to results obtained with unprocessed, ground grain. Feed efficiency was improved by steam-flaking, pelleting, reconsituting with water, and roasting (Tanksley and Osbourn 1970; Jensen *et al.* 1960; Diggs 1969; Baird 1976). Soaking in water, however, increased the quantity of feed needed to produce a kilogram of gain (Cam Chuong *et al.* 1966; Aubel 1960A,B).

Micronization [a process in which grains are subject to dry heat from a gas-fired microwave generator (Riggs *et al.* 1970, Schoke *et al.* 1972)] of sorghum grain decreases moisture content and increases *in vitro* starch availability (Harbers 1975; Noland 1976; Shiav *et al.* 1976; Savage *et al.* 1980). The micronizing process significantly improves crude protein, dry matter, and gross energy digestibility and nitrogen retention. Micronization improves the digestible and metabolizable energy content of the sorghum grain but has no effect on essential amino acid content with the exception of lysine of which 15–23% is destroyed during the process. Micronization significantly improves feed efficiency but produces only small nonsignificant improvements in growth rate (Shiav *et al.* 1976; Shadid *et al.* 1976). This process seems to have its greatest effect on bird-resistant sorghum grains with high tannin content (Shadid *et al.* 1976). Savage *et al.* (198O) showed that ground, raw, high-tannin sorghum has 3.61 kcal/g digestible energy (DE) and 3.49 kcal/g metabolizable energy (ME), while following micronization the DE and ME values are increased to 3.87 and 3.75 kcal/g, respectively.

Amino Acids

A review of the more recent research published on sorghum grain provides information that could be used to indicate that sorghum grain is inferior, equal, or superior to maize as a feed source for swine (Takahashi *et al.* 1968; Oyenuga and Fetuga 1975; Han and Ha 1977; Alcantara *et al.* 1970; Beames *et al.* 1973; Tschiderer 1966; Angelov 1965; Castaing and Leuillet 1976; Lawrence 1968; Castaing and Moal 1973; Bokorov and Srećkovic 1964; Stafijcuk and Teljatnikov 1968; Gontijo *et al.* 1976; Luce *et al.* 1972; Klett 1973; Tribble *et al.* 1972; Tonroy *et al.* 1973; Eckert and Allee 1974; Copelin *et al.* 1974, 1976A,B,C, 1978; Meadows *et al.* 1974; Orr *et al.* 1976; Cohen and Tanksley 1973A,B, 1975, 1976, 1977; Smith and Allee 1973; Hamilton *et al.* 1978; Arnot *et al.* 1978). When maize is replaced by sorghum grain in well-balanced diets supplemented with soybean meal, growth rate is generally not affected or only slightly reduced; however, the quantity of feed required per unit of weight of gain is consistently increased by 10–15%. A more marked depression in both growth rate and efficiency of feed conversion is generally noted when protein supplements such as cottonseed meal and peanut meal are included in the diet. Lysine is the most limiting and threonine is the second limiting amino acid in sorghum grain for growth (Pond *et al.* 1953; Ranjhan *et al.* 1964; Hale and Lyman 1961; Jensen *et al.* 1965; Gontijo *et al.* 1976; Cohen and Tanksley 1973A,B; Hodson *et al.* 1973; Eckert and Allee 1974; Cohen and Tanksley 1975, 1976).

Other studies (Cohen and Tanksley 1973A,B; Eckert and Allee 1974; Cohen and Tanksley 1976; Purser and Tanksley 1976) indicate that tryptophan appears to be third limiting in sorghum and that methionine and isoleucine may be fourth limiting for growing swine.

Supplementation of lysine to diets containing sorghum improves gains and feed efficiency. This response is most marked when protein supplements such as sesame, cottonseed, or peanut meal, low in lysine, are fed with grain sorghum, or when grain sorghum is used as the only protein source. A response to lysine is also sometimes obtained when soybean meal is used to provide the supplemental protein, even though studies by Tribble *et al.* (1979) failed to show any positive response from lysine supplementation to sorghum–soybean diets for either growing or finishing swine. Not only is the lysine content of sorghum grain low, but the availability of lysine in sorghum also may be low. Easter (1972) used ileocecal cannulation of finishing swine and reported a lysine availability of 69%, based on analysis of the ileal digesta. Eggum (1973) using a fecal index method with pigs, reported the lysine availability of sorghum to be 71.5%. Copelin *et al.* (1978), using a growth assay technique with growing pigs, reported an average lysine availability value of 62.7%. Even lower values based on chemical methods were reported by Almond *et al.* (1979), who indicated that lysine availability was less in a brown-seeded variety than in a yellow hybrid (49.2 vs. 56.4%, respectively). Studying 24 sorghum cultivars, Stephenson *et al.* (1971) reported lysine availability ranging from 58 to 92.5% and suggested that real differences exist in availability of lysine among different sorghum hybrids.

Easter (1972) found a lower true availability for sorghum–threonine than for maize–threonine (60.6 vs. 72.2%). Eggum (1973) and Copelin *et al.* (1978) have reported higher values (82.7 and 88.9%, respectively). However, as with lysine, varietal differences may exist as suggested by the work of Stephenson *et al.* (1971) who reported threonine availabilities of 24 different hybrids ranging from 46.8 to 91.8%.

Cohen and Tanksley (1977) formulated sorghum and soybean meal diets to contain either 16 or 13% crude protein, and supplemental threonine was added in six increments of 0.06%. When 16% protein diets containing 110% of all essential amino acids except threonine were fed to growing pigs, 0.47% threonine was required to optimize average daily gain and gain/feed. For finishing pigs fed the 14% protein diet, 0.39% threonine was required to optimize average daily gains, and 0.37% total threonine was required to optimize gain/feed.

Although data with swine are not available, studies with chicks (Stephenson *et al.* 1971) have indicated that methionine availability among 24 different sorghum hybrids was quite variable (ranging from 55.1 to 98.0%). There is some evidence that methionine may react as a methyl donor to eliminate flavinoids and other polyphenols. Therefore, methionine may be less available in sorghums high in tannins.

Grain sorghum is an excellent grain for pigs of all ages if properly fed and properly supplemented to provide adequate quantities of the essential amino acids. Tannin content can reduce the digestibility and utilization of some nutrients; although protein content and quality may not be related to tannin content, tannin in sorghums may alter amino acid availability, as has been demonstrated in the case of methionine. To be efficiently utilized, diets based on sorghum should be supplemented with protein supplements and/or amino acids to provide adequate quantities of all essential amino acids, especially lysine, threonine, tryptophan, isoleucine, and methionine which are limiting in sorghum grain.

BARLEY

Barley ranks fourth in total production among the cereal grains in the world with an annual production of 173.1 million metric tons (FAO 1980). The greatest portion of this production is concentrated in the Soviet Union and the European countries with lesser amounts being produced in Asia and in the northern part of North America.

The results of 32 experiments with swine were summarized by Morrison (1956), who concluded that barley had a feeding value of 91% that of maize. The

TABLE 12.23. Comparison of Barley with Maize, Wheat, and Sorghum for Growing-Finishing Pigs

Reference	*Class of pig*	*Grain source*	*Av daily gain, kg*	*Feed intake, kg*	*Feed/gain ratio*
Lawrence 1967	Growing (G)	Barley	0.586	1.70	2.90
	Growing	Sorghum	0.577	1.63	2.82
	Growing	Maize	0.600	1.66	2.77
	Finishing (F)	Barley	0.641	2.43	3.79
	Finishing	Sorghum	0.736	2.49	3.38
	Finishing	Maize	0.691	2.45	3.55
	G-F	Barley	0.609	2.06	3.39
	G-F	Sorghum	0.655	2.05	3.13
	G-F	Maize	0.645	2.06	3.20
Burnett and Neil 1964	G-F	Barley	0.727[a]	2.06	2.83[a]
	G-F	Maize	0.773[b]	2.06	2.67[b]
Gill et al. 1966	G-F	Barley	0.768[c]	3.05	3.97[b]
	G-F	Maize	0.918[a]	2.97	3.24[a]
	G-F	Wheat	0.836[b]	2.92	3.49[a]
Larsen and Oldfield 1960	G-F	Barley	0.786	3.08	3.92
	G-F	Maize	0.823	2.77	3.37
Bowland 1974	G	Barley	0.42	1.10	2.64
	G	Wheat	0.42	1.09	2.60
Mitchell et al. 1976	G	Barley	0.66	1.59	2.41
	G	Wheat	0.59	1.47	2.50
Hollis and Palmer 1971	G-F	Barley	0.66[b]	2.40	3.66[b]
	G-F	Wheat	0.68[b]	2.44	3.61[b]
	G-F	Maize	0.76[a]	2.40	3.17[a]
Greer et al. 1965	F	Barley	0.74[b]	3.15	4.26[b]
	F	Maize	0.82[a]	3.28	3.95[a]
McConnell et al. 1975	G	Barley	0.72	2.32	3.22[b]
	G	Wheat	0.73	2.03	2.78[a]
	G	Sorghum	0.72	2.29	3.18[b]
	G	Maize	0.75	2.09	2.79[a]
	F	Barley	0.66[b]	3.08	4.67[c]
	F	Wheat	0.81[a]	3.05	3.76[a]
	F	Sorghum	0.79[a]	3.45	4.37[b]
	F	Maize	0.79[a]	2.96	3.75[a]
	G-F	Barley	0.68[b]	2.69	3.95[b]
	G-F	Wheat	0.77[a]	2.51	3.26[a]
	G-F	Sorghum	0.75[a]	2.86	3.81[b]
	G-F	Maize	0.77[a]	2.52	3.27[a]

[a,b,c]Means in the same column and reference source without a common superscript differ significantly ($P <$ .05).

inferior feeding value of barley for growing-finishing and finishing swine has been shown by a number of researchers (Table 12.23) (Larsen and Oldfield 1960; Burnett and Neil 1964; Greer *et al.* 1965; Gill *et al.* 1966; Lawrence 1967; Hollis and Palmer 1971; McConnell *et al.* 1975). However, some studies (Lawrence 1967; Bowland 1974; McConnell *et al.* 1975; Mitchell *et al.* 1976) have clearly demonstrated that barley is equal to maize, sorghum, and wheat for young starter pigs and for growing pigs. This apparent inconsistency appears to be mitigated by a high fiber and low energy content on the one hand, and a high level of crude protein and essential amino acids on the other.

Limitation of available dietary energy is an important factor in inferior performance of growing-finishing swine fed barley rations in contrast to those fed similarly supplemented rations based on maize and wheat. This reduced feeding value of barley is due to its relatively higher crude fiber content and the apparent inability of the pig to consume enough net energy to gain at a maximum and efficient rate. Barley hull fiber acts as a diluent of available nutrients and may also contain factors which physically or chemically inhibit nutrient digestion, absorption, or utilization (Larsen and Oldfield 1960). Growth inhibitors in barley have been suggested in work with poultry (Jensen *et al.* 1957, Fry *et al.* 1958, Arscott *et al.* 1960, Morgan and McGinnis 1965). Although positive proof for their existence is lacking, it has been shown that barley hulls added to diets based on either pearled barley or maize significantly depressed growth rate and efficiency of feed conversion more than did a similar level of fiber from purified cellulose (Larsen and Oldfield 1960). Digestibility data indicated that the observed effect on growth and feed conversion is primarily caused by lowering of the digestibility of energy-supplying nutrients of the diet, a reduction that is greater with fiber from barley hulls than with cellulose. It is also suggested that the more pronounced depression with hulls could result from a physical interference with the activity of digestive enzymes due to the form of the fiber or from direct chemical inhibition of enzyme action.

A comparison of the crude protein and essential amino acid content of barley and maize (Table 12.24) indicates that barley contains more total protein and higher levels of lysine, tryptophan, and the sulfur-containing amino acids, meth-

TABLE 12.24. A Comparison of the Crude Protein and Essential Amino Acid Content of Barley and Maize

Protein and amino acid, %	*Barley*	*Maize*
Crude protein, %	11.6	8.8
Amino acids, %		
Arginine	0.6	0.5
Histidine	0.3	0.2
Isoleucine	0.5	0.4
Leucine	0.8	1.1
Lysine	0.40	0.24
Methionine	0.2	0.2
Cystine	0.3	0.2
Phenylalanine	0.6	0.5
Tyrosine	0.3	0.5
Threonine	0.42	0.39
Tryptophan	0.14	0.05
Valine	0.6	0.4

Source: National Research Council 1979.

ionine and cystine, and suggests that barley may be superior to maize in its capacity to meet the requirements of the young growing pig for these nutrients. During the early growth period, pig performance is more heavily dependent on level of dietary protein and level and balance of essential amino acids than total energy concentration. It appears, therefore, that caloric concentration of barley-based rations is adequate to meet the needs of the young growing pig and that the improved level and balance of limiting amino acids, especially lysine and tryptophan, are able to support pig performance similar to that obtained with complete mixed rations based on maize, wheat, or sorghum.

Barley contains nearly enough lysine, the first limiting amino acid in barley protein (McBeath *et al.* 1960; Soldevila and Meade, 1964), to meet the needs of growing swine. Small amounts of soybean meal alone or in combination with several other supplemental protein feeds will supply sufficient lysine and threonine to supplement adequately that supplied by barley in swine rations (Reimer and Meade 1964; Fuller *et al.* 1974). Threonine is the second limiting amino acid in barley for growing swine (Chung and Beames 1975; Fuller *et al.* 1974; Fuller and Livingston 1979; Fuller and Mennie 1979; Aw-Yong and Beames 1975). When barley is fed with only vitamins and minerals, substantial improvement in protein utilization and pig performance can be achieved by the simultaneous addition of 0.38% L-lysine and 0.18% L-threonine (Fuller and Livingstone 1979; Fuller and Mennie 1979).

The hull and fiber content of barley are major factors which cause inferior performance of swine fed barley in comparison to those fed grains low in fiber. Various methods of processing of the barley have been attempted to improve its feeding value. The removal of the hull through a process known as "pearling" greatly improves the nutritive value of barley (Dinusson *et al.* 1956). This process effectively reduces the fiber content and increases the digestible energy. Besides removal of the hulls, processing methods include treatments with enzymes, organic acids, steam pressure, and pelleting. Treatment with crude α-amylase concentrates, barley malt, and bacterial enzyme preparations has generally proven to be ineffective (Larsen and Oldfield 1960; Burnett and Neil 1964) as has steam processing (Burnett and Neil 1964; Hintz and Garrett 1967). An exception to the effect of enzyme treatment was reported by Newman *et al.* (1980) who added bacterial diastase to covered barley and noted a significant improvement in pig growth but not in efficiency of feed utilization. Treatment of barley of varying moisture contents with propionic acid did not improve pig performance over that obtained from barley stored and processed in the conventional manner (Cole *et al.* 1980). Pelleting of barley diets improves weight gains and feed intake and apparently efficiency of feed utilization (Dinusson *et al.* 1955; Thomas and Flower 1956; Dinusson and Bolin 1958; Dinusson *et al.* 1960; Larsen and Oldfield 1960; Troelsen and Bell 1962, Haugse *et al.* 1966).

Increasing the available energy content of barley diets with added fat or with whole, full-fat, roasted soybeans has been shown to improve pig performance (Heitman 1956; Oldfield and Anglemeir 1967; Anglemeir and Oldfield 1957; McConnell *et al.* 1975).

The genetic removal of hulls from barley and the production of hull-less varieties is of interest to swine producers. Even though it has been shown that differences other than crude fiber content may exist in the nutritive value of hull-

TABLE 12.25. Feeding Value of Hull-less and Hulled Barley

Treatment	Av daily gain, kg	Av daily feed, kg	Feed/gain ratio
Starter diets			
Hull-less Campana	0.42[a]	0.77[a]	1.82
Hulled Campana	0.36[b]	0.67[b]	1.97
Maize	0.40[a,b]	0.76[a]	1.89
Starter–grower diets			
Hull-less Glacier	0.67	1.83	2.74
Hulled Glacier	0.68	1.87	2.73
Starter diets			
Hull-less Nupana	0.50	1.11	2.21
Hulled Campana	0.51	1.04	2.04
Maize	0.54	1.04	1.92
Grower diets			
Hull-less Nupana	0.60	1.68	2.80
Hulled Campana	0.59	1.69	2.87
Maize	0.65	1.73	2.66
Finish			
Hull-less Nupana	0.77[a]	2.58	3.35[a]
Hulled Campana	0.69[b]	2.53	3.66[b]
Maize	0.78[a]	2.58	3.31[a]

Source: Newman et al. 1968; Newman and Eslick 1970.
[a,b]Means in the same column without a common superscript differ significantly.

less barley varieties, most hull-less varieties are superior to isogenic hulled varieties for growing-finishing and finishing swine (Joseph 1924; Gill *et al.* 1966; Newman *et al.* 1968; Newman and Eslick 1970; Bhatty *et al.* 1979; Newman *et al.* 1980). Newman *et al.* (1968) and Newman and Eslick (1970) showed not only that hull-less (Campana) barley was superior to hulled (Campana) barley for growing pigs and hull-less (Nupana) superior to hulled (Campana) for finishing pigs but also that its feeding value was equal to that of maize (Table 12.25). Their results did not, however, indicate any advantage of hull-less Glacier compared with normal hulled Glacier when fed to young pigs (8.5–10.4 kg). Hull-less barley contains more digestible energy than its hulled counterpart, 3295 vs. 3133 kcal/kg (Mitchell *et al.* 1976) and 3398 vs. 2962 kcal/kg (Bhatty *et al.* 1979).

The manner in which barley is processed before inclusion in the diet varies widely, and there is some evidence to suggest that the modulus of fineness of grinding can have a significant effect on pig response. However, the results from various workers have not always been in agreement. Haugse and co-workers (1966) indicated that cold rolling of barley was preferable to either feeding the grain whole or in a finely ground form. Lawrence (1967) showed that grinding rather than crimping of barley gave significantly better digestibility, greater energy and nitrogen retention, and improved pig performance during the finishing period. Later work by these same workers (Lawrence *et al.* 1980), however, failed to show any significant difference in pig response due to cold rolling or grinding through a 1.56- or 4.68-mm screen. Similar nonsignificant responses to differences in particle size have been reported from other stations (Simonsson and Björklund 1970). Cereal particle size has also been implicated as a cause of gastric abnormalities. The abnormalities associated with the feeding of finely ground barley vary from hyperkeratinization of varying severity to erosion and ulcers (Simonsson and Björklund 1970; Crabo *et al.* 1973; Lawrence *et al.* 1980).

When barley is included in swine diets it is best that it have a low hull (fiber) content and that it be fed in complete-mixed diets. For maximum utilization it should be pelleted and constitute not more than one-third of the diet.

OATS

Oats (*Avena sativa* L.) make up only 4.5% of the total world production of cereal grains, with most of this production of 50.5 million metric tons concentrated in Western Europe, North America, and the centrally planned countries (FAO 1978). Whole oats contain approximately one-third hull and are very high in crude fiber (12.3%) and low in digestible energy (3030 kcal/kg) and are not widely recommended as a good source of carbohydrates for young growing pigs. The gains of growing pigs as well as the efficiency of feed conversion are decreased as the quantity of ground oats is increased in the diet to as much as 30–40% (Jensen *et al.* 1959A; Meade *et al.* 1966; Owen and Ridgman 1967; Katz *et al.* 1973; Wahlstrom *et al.* 1977). This depression in both growth and feed efficiency is due to the hull portion of the oats (Jensen *et al.* 1959B) and is mediated by the dilution of utilizable energy of the ration and by lowered feed intake. Depression in performance also occurs when oat hulls are included in maize–soybean diets (Table 12.26).

It has been shown that oats vary markedly in test-weight (weight for volume) and in feeding value when replacing equal weights of maize in complete mixed rations for growing swine (Jensen *et al.* 1959A). Comparison of oats of varying

TABLE 12.26. Effect of Adding Oat Hulls and Corn Oil to Swine Diets

Treatments	*Av daily gain, kg*	*Feed intake, kg*	*Feed/gain*
Maize–soybean meal (SBM)	668	1.89	2.86
Hulled oats–SBM	557	1.94	3.36
Hulled oats–SBM + 15% hulls	473	1.89	4.06
Hulled oats–SBM + 30% hulls	373	1.50	3.99
Hulled oats–SBM + 30% hulls + 14% maize oil	550	1.65	3.01
Maize–SBM	691	2.07	3.01
Maize–SBM + 15% hulls	600	2.06	3.42
Maize–SBM + 30% hulls	441	1.84	4.15
Maize–SBM + 30% hulls + 14% maize oil	555	1.75	3.15

Source: Jensen et al. 1959B.

TABLE 12.27. Effect of Test-Weight of Oats on Composition, Pig Performance, and Metabolizable Energy Content

Test weight, kg/hl	*36*	*39*	*51*	*52*	*56*
Composition (%DM)					
Crude protein	16.1	11.2	11.7	13.8	12.6
Crude fat	6.5	6.2	6.3	6.0	5.4
Crude fiber	12.7	14.7	10.1	10.8	10.3
Ash	2.6	3.4	2.5	2.8	3.0
Nitrogen-free extract	62.1	64.5	69.4	66.6	68.7
Starch	40.6	39.7	47.7	46.6	49.7
Digestibility					
Protein	79.8	78.1	80.4	85.5	81.7
Fat	92.5	91.8	92.3	93.0	92.6
Fiber	7.6	13.5	7.1	15.0	15.0
Nitrogen-free extract	71.3[a]	70.2[a]	79.8[b]	79.8[b]	80.7[b]
Metabolizable energy, kcal/g	2.56	2.44	2.79	2.82	2.81
Performance					
Av daily gain, g	—	690[a]	775[b]	756[b]	771[b]
Feed/gain ratio	—	3.40[a]	2.95[b]	3.02[b]	2.97[b]

Source: Salo and Alaviuhkola 1980.
[a,b]Means on same line without common superscript differ significantly ($P < .05$).

test-weights has demonstrated the lower feeding value of the lighter oats (Table 12.27) (Jensen *et al.* 1959A; Salo and Alaviuhkola 1980). The feeding value of oats is principally determined by the proportion of hulls (Jensen *et al.* 1959B; Thomke 1960; Sibbald and Price 1977), which in composition are equal to rough straw and thus not utilizable by pigs. Lighter-weight oats have a higher proportion of hulls and a lower content of digestible and metabolizable energy (Table 12.27) (Salo and Alaviuhkola 1980). Low test-weight oats also have a lowered digestibility of nitrogen-free extract (NFE) because the NFE of such oats contains less starch and more hemicellulose and lignin (Salo and Alaviuhkola 1980). The pigs digest the starch of oats totally, but scarcely one-third of the hemicellulose and none of the lignin is digested (Salo 1971).

Oat groats produced mechanically, a high grade but expensive feed, tend to be used in prestarter and starter diets, since then the high digestibility and low fiber content are of most benefit. Wahlstrom *et al.* (1977) fed dehulled, high-protein oats supplemented only with lysine to young pigs and obtained the same growth and feed conversion as from pigs on 17% crude protein, maize–soybean meal starter diets. Research results (Christison and Bell 1980) with a new cultivar ('Terra') of naked oats (*Avena nuda*) indicate that this cultivar is similar to wheat and maize as an energy source for young pigs. Similar growth, feed intake, and feed efficiency were observed in young 14–17-kg pigs when this cultivar replaced all or a portion of the maize or wheat in a diet supplemented with soybean meal to provide 17–18% protein in the total diet.

New oat varieties grown today in the United States and Canada are higher in both protein and the limiting amino acid lysine than normal oats (Table 12.28). Robbins *et al.* (1971) determined the protein and amino acid content of 289

TABLE 12.28. Chemical Composition of Normal, Naked, and High-Protein Oats and Oat Groats

Item	*Normal oats*[a]	*Naked oats*		*High-protein oats*			*Oat groats*[b]
		Watson[b]	*Regina*[b]	*Hinoats*[a]	*Dal*[c]	*Spear*[d]	
Moisture, %	—	11.9	7.1	9.7	10.0	—	12.5
Crude protein, %	14.6	12.4	14.6	14.6	16.0	19.4	11.7
Crude fat	5.7	6.6	—	5.0	3.8	2.1	4.2
Crude fiber	12.3	3.1	—	12.9	11.8	8.4	3.5
Ash	3.3	1.6	—	3.5	3.2	4.3	—
Energy, kcal/kg	—	—	—	4663	—		—
Calcium, %	0.06	0.24	—	0.10	—	0.13	—
Phosphorus, %	0.36	0.39	—	0.43	—	0.42	—
Amino acids							
Arginine	0.80	0.50	0.94	1.11	1.25	1.35	0.85
Histidine	0.20	0.28	0.34	0.34	0.34	0.43	0.28
Isoleucine	0.60	0.38	0.47	0.54	0.61	0.75	0.41
Leucine	1.00	0.91	1.09	1.19	0.87	1.44	0.92
Lysine	0.40	0.57	0.62	0.65	0.66	0.74	0.52
Methionine	0.20	0.16	0.22	0.32	0.27	0.60	0.28
Cystine	0.20	0.23	0.37	0.28	0.30	—	0.17
Phenylalanine	0.70	0.96	0.79	0.86	0.75	1.05	0.67
Threonine	0.40	0.41	0.60	0.58	0.51	0.64	0.45
Tryptophan	0.20	—	—	0.21	—	—	—
Valine	0.70	0.53	0.69	0.74	0.82	1.01	0.58
Tyrosine	0.60	0.43	0.55	0.61	0.53	0.70	0.45

[a]Anderson et al. 1978 (dry-matter basis).
[b]Christison and Bell 1980 (air-dry basis).
[c]Wahlstrom and Libal 1975 (air-dry basis).
[d]Wahlstrom and Libal 1979 (dry-matter basis).

samples of oat groats and suggested that increased protein in oats is not accompanied by decreased lysine concentration in the protein as in many other grains. Selections of oat varieties having a high protein content are not available. Wahlstrom and Libal (1975) reported that high-protein oats ('Dal') containing 16% protein and 0.66% lysine did not reduce daily gains or feed-to-gain ratio when included in growing-finishing swine diets at levels up to 60% of the grain portion of the diet. Anderson *et al.* (1978) fed a high-protein oats variety ('Hinoats') containing 14.6% protein and 0.65% lysine (dry matter basis) and indicated that if this variety were used as the major cereal in swine diets, feed intake may not be adequate to promote normal growth. Similar results were obtained by Wahlstrom and Libal (1979) when more than 30% of a high-protein oat variety ('Spear') containing 19.4% protein and 0.74% lysine (dry matter basis) was included in diets for 5-week-old pigs. The depression in feed intake and performance could be overcome by pelleting (Anderson *et al.* 1978; Wahlstrom and Libal 1975). Higher levels of high-protein oats can be used in pig starter and growing-finishing diets when these diets are fed in pelleted form than when fed in meal form.

RYE

Rye is a cereal not commonly used as a grain for swine feeding although it is sometimes available at prices less than those of other grains. Early studies indicated that pigs performed poorly when fed rye. Bowland (1966) demonstrated that substituting 25% rye for an equal quantity of wheat reduced gains of finishing pigs fed the diets. Other studies have also shown the inferior growth-promoting properties of rye. Gains and feed efficiency were reduced when rye replaced all of the barley in finishing pig diets (Holzschuh and Trautmann 1962). Similar studies with finishing pigs (Friend and MacIntyre 1969) with substitution rates of 30 and 60% show that the feeding value of rye is not equal to that of barley. A 30% substitution reduced gains by 8% and a quantity of 60% rye reduced gains by 14.5% when compared to the diet containing only barley grain.

Growing and finishing pigs (31.5–90 kg) were fed diets in which 25, 50, or 100% of the maize in a maize–soybean meal diet was replaced by rye (Hale *et al.* 1967). Pigs fed the diet containing rye as the only grain, required 12 additional days to reach market weight at 90 kg. Both gains and feed efficiency were inferior to those of pigs fed the maize–soybean diet (Table 12.29). Although gains were not significantly depressed at the 25 and 50% levels of substitution, feed per unit gain was increased with each increase in level of rye. Each 25% increase in level of rye increased feed required per unit gain by approximately 4%.

In order to reduce the dustiness and to provide rye diets with gross energy levels equal to those of maize diets, 2.5% prime stabilized beef tallow was added to diets containing rye substituted for maize. Pigs fed rye diets with or without added

TABLE 12.29. Average Performance of Growing-Finishing Pigs Fed Diets Containing Different Levels of Rye

Treatment	*Criteria of performance*			
	Days on test (days)	*Av daily gain (kg)*	*Av daily feed (kg)*	*Feed/gain rato*
Trial 1[a]				
1. 100% maize—0% rye	68	0.891[x]	2.94	3.30[x]
2. 75% maize—25% rye	67	0.886[x]	3.03	3.42[x,y]
3. 50% maize—50% rye	70	0.855[x]	3.03	3.56[y]
4. 0% maize—100% rye	80	0.728[y]	2.79	3.84[z]
Trial 2[b]				
1. 100% maize—0% fat	71	0.868[x]	2.88	3.32
2. 100% maize—2.5% fat	71	0.823[x]	2.57	3.14
3. 100% rye— 0% fat	71	0.727[y]	2.68	3.68
4. 100% rye—2.5% fat	71	0.755[y]	2.62	3.48

Source: Adapted from Hale et al. 1967.
[a]Average initial weight, 31.5 kg; average final weight, 90 kg.
[b]Average initial weight, 26.8 kg.
[x,y,z]Means in the same column without a common superscript are significantly different from each other.

fat grew more slowly and required more feed to produce a kilogram of liveweight gain than did those fed maize (Table 12.29). Although the added fat improved feed efficiency in both the maize and rye diets, it was of no benefit in overcoming the growth depressing effects of the rye substitutions. As a result of these studies, it was estimated that for growing-finishing pigs, rye is worth 78–80% as much as maize when used as the only grain in a well fortified, 16% protein diet.

More recent studies (Danielson 1972; Savage *et al.* 1978) have shown rye to be equal to maize or barley when substituted on an equal weight basis. Growing-finishing pigs (22–97 kg) grew at a similar rate and efficiency when fed *ad libitum* 14% crude protein diets containing maize as the energy source or when up to 60% of the maize was replaced by ground rye (Danielson 1972) or when hand-fed diets containing either 64% barley or an equal quantity of rye.

Esnoala *et al.* (1976), however, substituted rye at 0, 12.5, 25.0, 37.5, 50, or 100% of the barley in diets for growing-finishing pigs and reported a significant linear depression in both feed intake and average daily gain. Palatability appeared to affect feed consumption which at the highest level of substitution was reduced by about 12%. The 12% reduction in feed intake resulted in a 12.6% reduction in daily gain but feed-to-gain ratio was not affected. Two trials by Honeyfield *et al.* (1977) with recently weaned 4-week-old pigs produced similar results. Feed consumption of diets containing 68.5% rye were reduced by 7–13% and gains were depressed by 10.5–10.9% when compared with similar diets with 68.5% maize. None of the criteria for pig performance were improved by the addition of lysine and methionine.

Several factors have been studied in an attempt to explain the detrimental effect of rye on pig performance. Digestibility studies comparing rye and barley (Friend and MacIntyre 1969) clearly demonstrate that lowered digestibility of rye is not the limiting factor, since the apparent digestibility of both the dry matter (88 vs. 84%) and the crude protein (84.5 vs. 82.1%) is higher for rye than for barley and nitrogen retention is similar.

The presence of ergot (*Clauiceps purpurea*) in rye grain has also been studied as a causative agent for the poor performance of pigs fed diets containing rye. The presence of ergot in rye has long been recognized as a problem in its use as a livestock feed. Ergot in rye grain fed to pregnant sows has caused reproductive problems and abortions. Nordskog and Clark (1945) reported that proportions of 0.10, 0.50, and 1.00% ergot in sows' diets produced agalactia and loss of piglets at farrowing but did not produce abortion. Reduction in body weight gains by pigs that were fed rations containing rye was attributed to the level of ergot contamination of the grain (Friend and MacIntyre 1969). Later studies by these same workers showed a severe reduction in both feed intake and growth rate when either 1.0 or 2.0% ergot was included in the ration (Friend and MacIntyre 1970). Proportions of 0.10 and 0.05% ergot allowed only moderate feed intake and weight gains which were inferior to those of pigs fed ergot-free rations. In addition to reducing weight gains and feed efficiency, 0.10% ergot also significantly reduced nigrogen retention. The ergot used in these studies contained 0.292% alkaloids with a predominance (71%) of ergocristine.

Other studies (Wieringa 1967) indicate that rye contains a fat-soluble, growth-inhibiting substance. This substance is reported to be a mixture of 5-*n*-alkyl resorcinol with odd-numbered side chains of 15–23 carbon atoms and a smaller amount of 5-*n*-alkenyl resorcinol. The growth of pigs on diets containing 50% rye or an equivalent amount of the growth-inhibiting substance contained in rye oil was 11–12% slower than on diets containing 50% barley. Studies on rats established that the harmful effects of the rye oil containing the inhibiting factors were not caused simply by the unpalatable taste of the rye oil or rye resorcinols. In studies by Honeyfield *et al.* (1977), pig performance was not improved by the addition of penicillin or by chemical or physical treatment of rye. In fact, treatment by water autoclaving, acid autoclaving, or extraction with water caused a significant reduction in pig performance.

WHEAT

Wheat and wheat products are utilized extensively for human consumption in almost every major area of the world, so that wheat is generally too expensive to be utilized for swine feeding. Only when surpluses or grain of a quality not suited for milling are available is wheat included in swine diets. Wheat has excellent palatability and is well consumed by swine of all ages. It has been suggested that there is no serious disadvantage to wheat given whole to pigs compared to that which is physically processed. However, Ivan *et al.* (1974) showed that feeding of whole wheat, as opposed to processed wheat, to both small and large pigs resulted in a reduction in digestibility of dry matter and therefore of gross energy and nitrogen. Although there is some disagreement as to the best physical form and particle size for wheat included in swine diets, palatability studies (Hale *et al.* 1979) showed that pigs consumed seven times more crushed wheat than ground wheat when both were offered simultaneously. When processed, rolled or crushed wheat has been shown to be equal or superior to ground wheat (Braude *et al.* 1961; Lawrence 1967; Ivan *et al.* 1974; Hale *et al.* 1979). Pelleting of diets containing ground wheat improves both growth rate and feed efficiency when compared with similar diets fed in meal form (Lennon *et al.* 1972; Erickson *et al.* 1980).

Wheats vary widely in chemical composition and therefore in nutritive value (Ivan and Farrell 1975; Ivan *et al.* 1975A,B). The chemical composition of wheat, as well as that of barley and other cereal grains, is affected by cereal cultivar, type of soil, and level of nitrogen fertilization (Just *et al.* 1983). However, feeding value is not greatly influenced by the changes in chemical composition induced by these variables. The protein content of wheat from the world collection was shown to vary from 6 to 22% (Johnson *et al.* 1970), but the wheat commonly used contains 13–15% crude protein. Although grains of higher protein content are of interest to swine producers, use of wheats should be based on type and amino acid analysis since it has been shown that red wheats are inferior to white wheats (Cheeke *et al.* 1980) and hard superior to soft (Ivan and Farrell 1976); and the percentage of some essential amino acids (histidine, lysine, cystine, threonine, and tryptophan) in wheat protein decreases with increasing crude protein content (Ivan and Farrell 1975). The crude protein of wheat is of high digestibility and generally high quality. Lysine has been shown to be the first most limiting amino acid in wheat protein; threonine, second; and valine or methionine, third (Ivan and Farrell 1975).

Wheat compares favorably with other grains in energy content. Expressed on a dry matter basis, wheat contains an average of 4430 kcal/kg gross energy, 3820 kcal/kg DE, 3680 kcal/kg ME, 3660 kcal/kg nitrogen-corrected metabolizable energy, and 2450 kcal/kg net energy (Wu and Ewan 1979).

The relative feeding value of wheat has been estimated to be between 91 and 106% of that of maize (Becker *et al.* 1963; Cunha 1957; Carroll *et al.* 1962; Gill *et al.* 1966; Danielson and Grabowski 1970; Hollis and Palmer 1971; Lennon *et al.* 1972; Luce *et al.* 1972; Allee 1977; Hale *et al.* 1979; Erickson *et al.* 1980), 109–114% of that of barley, and similar to that of sorghum (Evans 1960; Crampton 1962; Robinson *et al.* 1965; Gill *et al.* 1966; Allee 1977). When properly supplemented with protein and/or amino acids to supply adequate quantities and proportions of amino acids, wheat can be used as the sole cereal in growing and finishing diets for both meat and bacon pigs without reducing daily gains, feed efficiency, or carcass quality (Cole *et al.* 1969; Beames 1963; Jensen *et al.* 1969; Ivan *et al.* 1975A,B; Allee 1977; Hale *et al.* 1979; Erickson *et al.* 1980). Ivan *et al.* (1975A,B) fed wheats containing 10, 13, or 17% crude protein and showed that only supplemental lysine was needed to produce maximum gains when 17% protein wheat was fed, and both lysine and threonine were required when 13% protein wheat was fed to growing pigs. Ten percent protein wheat with supplemental lysine, threonine, valine, and methionine could not be used alone and required additional protein from fish meal, meat meal, and soybean meal to support maximum gains and feed efficiency (Table 12.30).

TABLE 12.30. The Use of 10, 13, and 17% Crude Protein Wheats as the Only Cereal for Growing Pigs

Grain	*Av daily gain, g*	*Feed/gain ratio*
Low protein series		
10% protein wheat alone	384[a]	3.76[a]
10% protein wheat + lysine	432[b]	3.35[b]
10% protein wheat + lysine, threonine, valine, and methionine	544[c]	2.69[c]
10% protein wheat + meal meal + lysine	563[a,b]	2.55[a,b]
10% protein wheat + soybean meal + lysine	584[b,c]	2.44[a]
10% protein wheat + fish meal + meat meal	600[c]	2.42[a]
Medium protein series		
13% protein wheat alone	396	3.64
13% protein wheat + lysine	496[a]	2.92[a]
13% protein wheat + lysine + threonine	562[b]	2.54[b]
13% protein wheat + lysine + threonine + methionine	592[b]	2.51[b]
High protein series		
17% protein wheat alone	433[a]	3.36[a]
17% protein wheat + lysine	582[b]	2.51[b]
17% protein wheat + fish meal + meat meal	570[b]	2.56[b]

Source: Adapted from Ivan et al. 1975A,B.
[a,b,c]Means within series in the same column without a common superscript are different from each other.

TRITICALE

Triticale is a polyploid hybrid cereal derived from an interspecific cross between durum wheat (*Triticum durum*) and rye (*Secale cereale*). Although naturally occurring wheat– rye hybrids were observed early in the nineteenth century (Leighty 1915), present day triticale hexaploid hybrids are artifically produced and most were derived from triticale developed at the University of Manitoba in Canada and improved and extended at CIMMYT in Mexico (Larter 1974).

Interest in triticale as a feed has been generated because of its higher protein and improved amino acid composition as compared to other cereal grains. Although triticale has shown promise as a nutritionally competitive feed grain with maize, wheat, sorghum, and barley in some studies with growing and finishing swine (Bowland 1968; Shimada *et al.* 1971; Noland *et al.* 1972; Wilson 1972; Allee and Hines 1972; Martynov and Shtan'ko 1979), several feeding studies with growing and finishing pigs indicated that triticale was inferior to maize, wheat, sorghum, and barley when more than 20% of the grain source was triticale (Longernecker *et al.* 1969; Stothers 1970; Harrold *et al.* 1971; Allee and Hines 1972; Stothers and Shebeski 1965; Erickson *et al.* 1979).

The differences in nutritive value of triticale reported for swine are due in part to the method of substitution of this grain in the diet. Most experiments have made comparisons based on either a substitution of grains by triticale on a weight basis or adjustment to an isonitrogenous basis (Table 12.31). Stothers and Shebeski (1965) demonstrated that triticale was equal to barley when substituted on an equal weight basis in swine starter diets; however, triticale could not be substituted on an isonitrogenous basis without having a detrimental effect on pig performance. Allee and Hines (1972) demonstrated that triticale was equal to sorghum as an energy source for growing swine, but when rations were formulated on an equal protein or equal lysine basis, a reduction in gain and feed efficiency was observed.

The inconsistencies in nutritive value reported for triticale are related to the fact that triticale is not a discrete entity. Considerable variation exists among the numerous crosses and varieties that have been developed. Screening tests for protein and lysine conducted at CIMMYT on 2400 lines of triticale showed a mean protein content of 13.4% with a range of 12.9–19.1%, and mean lysine content was 3.4% with a range of 2.6–3.9% (Villegas 1973). Erickson *et al.* (1978) also observed wide variation in amino acid content of four varieties produced during the same season at Michigan State University (Table 12.32).

A comparison of various cereal grains as energy sources for swine is shown in Table 12.33. The digestible and metabolizable energy content of triticale is similar to that of wheat, inferior to maize, and superior to barley (Cornejo *et al.* 1973). It has also been shown that varieties differ significantly in gross, digestible, and metabolizable energy (Erickson *et al.* 1978). Some of these varietal differences have been attributed to differences in fat content.

As early as 1965, Stothers and Shebeski (1965) indicated that palatability was a problem with some triticale samples used in swine starter diets. Bowland (1968) suggested that ergot may have been present in the triticale fed in these studies and proposed that contamination of triticale by ergot may be responsible for the

TABLE 12.31. Comparison of the Feeding Value of Triticale Grain with That of Maize, Wheat, Barley, and Sorghum

Source	Class of swine[a]	Grain substitution			Pig performance		
		Type	Grain	Level	ADG[b], g	Feed, kg	Feed/gain
Bowland 1968	G-F	Weight	Wheat–barley	0	750	2.64	3.53
	G-F	Weight	Wheat–barley	50	732	2.55	3.48
	G-F	Weight	Wheat–barley	100	745	2.55	3.42
Shimada et al. 1971	G	Protein	Sorghum	0	559	1.82	3.25
	G	Protein	Sorghum	30	551	1.80	3.27
	G	Protein	Sorghum	60	532	1.77	3.32
	G	Protein	Sorghum	90	554	1.78	3.21
Noland et al. 1972	G-F	Weight	Maize	0	777	2.32	2.99
	G-F	Weight	Maize	50	805	2.43	3.02
	G-F	Weight	Maize	100	755	2.46	3.26
Allee and Hines 1972	G	Weight	Sorghum	0	660	1.51	2.29
	G	Weight	Sorghum	20	650	1.50	2.33
	G	Weight	Sorghum	40	670	1.53	2.27
	G	Weight	Sorghum	60	670	1.47	2.19
	G	Weight	Sorghum	80	700	1.53	2.17
	G	Weight	Sorghum	100	650	1.49	2.30
Harrold et al. 1971	G	Weight	Barley	0	682	2.08	3.18
	G	Weight	Barley	50	650	2.16	3.32
	G	Weight	Barley	100	659	2.16	3.26
	G	Weight	Barley	100	558	1.81	3.19
Allee and Hines 1972	G	Protein	Sorghum	0	660	1.51	2.29
	G	Protein	Sorghum	100	560	1.44	2.57
	G	Lysine	Sorghum	100	550	1.48	2.66
Erickson et al. 1978				0	330	0.75	2.29
				20	370	0.76	2.03
				40	330	0.73	2.20
				60	320	0.74	2.30
				80	270	0.70	2.60
				100	220	0.54	2.47
Erickson et al. 1979				0	420	0.99	2.36
				50	380	0.89	2.33
				100	270	0.72	2.85
Erickson et al. 1979				0	590	1.68	2.85
				50	610	1.80	2.95
				100	370	1.29	3.48
Nishimuta et al. 1980	G-F	Weight	Maize	0	730	2.56	3.50
	G-F	Weight	Maize	20	790	2.76	3.49
	G-F	Weight	Maize	40	720	2.49	3.47
	G-F	Weight	Maize	60	700	2.44	3.51
	G-F	Weight	Maize	80	700	2.52	3.61

[a]G-F, growing-finishing; G, growing; and F, finishing periods.
[b]ADG is average daily gain.

TABLE 12.32. Amino Acid Composition of Triticale Varieties

Amino acid[a]	*Triticale variety*[b]			
	'MSU-1'	*'MSU-2'*	*'MSU-3'*	*'MSU-4'*
Crude protein (%)	12.93	12.22	12.46	11.70
Percent of protein				
Arginine	5.14	6.39	5.39	5.32
Histidine	2.39	2.71	2.27	2.60
Isoleucine	3.96	3.58	3.70	4.12
Leucine	6.95	7.30	5.76	3.81
Lysine	2.88	4.31	2.94	3.22
Methionine	1.04	0.94	0.90	1.01
½ Cystine	1.16	1.03	0.91	0.90
Phenylalanine	4.74	4.79	4.39	4.61
Threonine	2.76	2.74	2.73	2.70
Tyrosine	2.36	1.91	2.49	2.60
Valine	4.44	4.51	4.11	3.69

Source: Erickson et al. 1978.
[a]Values expressed on air-dry basis and crude protein calculated ($N \times 5.7$).
[b]Varieties produced during same season at Michigan State University (MSU), East Lansing.

observed decrease in acceptability of triticale. However, feed intake of growing-finishing pigs was reduced by 7.0 or 10% when triticale, cleaned to remove all but 0.02% of the ergot present, was substituted for 50 or 100% of the wheat–barley mixture in the diet. Stothers (1970) suggested that there may be inherent palatability problems with triticale. Studies by Erickson *et al.* (1979) measured the units of trypsin inhibitor present in diets containing triticale and demonstrated that

TABLE 12.33. Comparative Value of Various Grains and Triticale Varieties as Energy Sources for Pigs

Grain	*Gross energy, kcal/kg*	*Digestible energy, kcal/kg*	*Metabolizable energy, kcal/kg*
Maize[a]	4403	3796	3745
Wheat[a]	4208	3709	3625
Barley[a]	4285	3375	3332[c]
Triticale[a]	4280	3603	3522
'MSU-1'[b]	3994	3461	3413
'MSU-2'[b]	4034	3438	3403
'MSU-3'[b]	4028	3497	3454
'MSU-4'[b]	4445[c]	3893[c]	3869[c]

[a]Cornejo et al. 1973.
[b]Erickson et al. 1978. [Varieties produced at Michigan State University (MSU).]
[c]Means in the same column bearing different superscripts differ significantly ($P < .05$).

TABLE 12.34. Effect of Lysine, Methionine, and Threonine Supplementation of Triticale Diets for Growing Pigs

Amino acid supplementation, %	*Av daily gain, g*	*Feed/gain ratio*
Trial 1[a]		
None	252	3.88
0.08 L-lysine	306	3.57
0.16 L-lysine	405	3.10
Trial 2[b]		
None	239	4.88
0.1 DL-methionine	226	5.09
0.16 L-lysine	302	3.50
0.1 DL-methionine + 0.16 L-lysine	346	3.54
Trial 3[c]		
None	59[c]	8.44[c]
0.2 DL-threonine	57[c]	8.10[c]
0.3 L-lysine	130[d]	5.75[d]
0.2 DL-threonine + 0.3 L-lysine	201[e]	3.73[e]

Source: Shimada *et al.* 1971; Shimada and Cline 1974.
[a]Significant linear response to lysine ($P < .05$).
[b]Significant lysine effect ($P < .01$) with no significant effect of methionine.
[c,d,e]Means in the same column and trial bearing different superscripts differ significantly ($P < .05$).

anti-trypsin activity in triticale was negatively correlated with weight gain ($r = -0.87$) and feed efficiency ($r = -0.84$) of pigs given diets containing high levels of triticale.

Although triticale contains approximately twice the lysine content of sorghum grain and maize and a higher level of lysine than most cereal grains, Bravo *et al.* (1971) reported a linear response in growth and nitrogen retention with graded levels of L-lysine added to 85%-triticale diets fed to growing rats. It was concluded that lysine is the triticale's first limiting amino acid. These same colleagues (Shimada *et al.* 1971; Bravo *et al.* 1971) conducted several experiments with growing pigs and observed a linear response to the addition of graded levels of L-lysine to 96% triticale diets, but no significant response to DL-methionine supplementation (Table 12.34). Later growth trials (Shimada and Cline 1974) and nitrogen balance studies (Erickson *et al.* 1978) suggested that threonine is the second limiting amino acid of triticale.

The increased protein and lysine content of triticale suggests that this cereal grain might be adequate for finishing pigs with only vitamin and mineral supplementation. However, Allee and Hines (1972) demonstrated that triticale alone was inadequate for finishing pigs as measured by average daily gain. Supplementation of the triticale diet with 0.01% L-lysine increased daily gain and improved feed efficiency to a level not different from that obtained with wheat, sorghum, or triticale supplemented with soybean meal.

These data indicate that triticale can be used as a substitute for a portion or all of the grain in swine diets. However, the level of substitution must be based on the chemical composition (protein, lysine, energy, ergot, and trypsin inhibitor) of the specific grain source used, and each source must be properly supplemented to provide nutritionally adequate diets.

RICE

Rice (*Oryza sativa* L.) is the most important food crop in Asia and plays a very important role in the nutrition of millions of people in the Latin American countries. There are some 400 million metric tons produced annually (FAO 1980) and a wide variety of by-products arises from the milling of paddy rice to produce polished rice for human consumption. In particular, the bran and polishings make valuable animal feedstuffs as do the broken grains or brewers' rice.

Large-scale milling consists of many processes, which include cleaning by sieves, screens, blowers, and magnetic separators to remove extraneous material, drying to reduce moisture below 15%, hulling to remove the outer coating, and milling and polishing to remove the epicarp, mesocarp, cross-layer, testa, and aleurone layer.

After the hull has been removed from the *paddy rice,* the germ and other layers are removed and form the product known as *rice bran,* of which there may be more than one grade. After this process, the grains are nearly white but still rather rough, and they are polished until they become *polished rice.* The fine particles resulting from the polishing of the grain are called *rice polishings.* The very small pieces of broken seeds that are removed are separated and sold as *brewers' rice* or *broken rice.* The seed coat layers and embryo (germ) contain more than one-half of the mineral matter of the grain, one-quarter of the protein, practically all of the vitamins, and about three-quarters of the fat (Kik 1956). The oil is largely contained in the germ and about 85% of the entire oil content of the hulled rice goes into the by-product.

Rough Rice or Paddy Rice

Rough rice is sometimes available at a price that makes it economical to use as a feed for growing and finishing swine. Rough rice contains about 10.0% moisture, 8.2% crude protein, 9.2% crude fiber, 1.9% ether extract, 6.5% ash, and 64.2% nitrogen-free extract (Bienvenido *et al.* 1964). The hulls, which contain about 40% crude fiber and 11–19% silica, make up about 25% of the weight of rough rice. The hull portion of the rough rice greatly reduces the energy value of the product.

Noland and Scott (1963) used rough rice to substitute for 0, 20, 40, 60, 80, and 100% of the maize in growing rations for swine. For 15.5–61-kg liveweight pigs there were no significant growth rate differences due to the amounts of rough rice substitution; however, the quantity of feed required per kilogram of gain increased when rough rice was substituted for maize. Other unpublished studies by Colombian research workers have demonstrated similar results. When rough rice was substituted for 100% of the maize in a maize–soybean meal diet for growing-finishing pigs (17.9–90 kg) gains were not different but feed required per kilo-

TABLE 12.35. Chemical Composition of Rough Rice, Brown Rice, Polished Rice, Rice Bran, Rice Polishings, and Rice Hulls

Constituents (based on dry matter)	*Analysis (%)*					
	Rough rice[a]	*Brown rice*[a]	*Polished rice*[a]	*Rice bran*[b]	*Rice polishing*[b]	*Rice hulls*[c]
Crude protein, %	9.14	11.02	9.75	12.30	13.50	3.27
Crude fiber, %	10.21	1.23	0.34	11.30	5.60	44.3
Ether extract, %	2.13	2.72	0.44	12.10	16.50	0.87
Ash, %	7.18	1.86	0.56	10.20	8.70	20.77
Nitrogen-free extract, %	71.33	83.17	88.91	54.1	55.7	30.87

[a]Adapted from Bienvenido et al. 1964.
[b]Adapted from Arnott and Lim 1966.
[c]Adapted from Morrison 1956.

gram of gain was increased by 18%. Economic studies based on these trials (P. Andersen and J. H. Maner, unpublished) indicate that the price of rough rice must not be greater than 75% of that of maize for equal cost of utilization.

Brown and Polished Rice

The chemical composition and amino acid analysis of both brown and polished rice are presented in Tables 12.35 and 12.36. Brown rice contains more protein,

TABLE 12.36. Amino Acid Analysis of Brown Rice, Polished Rice, Rice Bran, Rice Polishings, Rice Bran and Polishings, and Rice Hulls

Amino acid	*Analysis (% protein)*						
	Polished rice[a,b]		*Brown rice*[a]	*Rice bran*[b]	*Rice polish*[b]	*Rice bran +polish*[b]	*Rice hulls*[b]
Arginine	8.94	8.28	7.74	8.28	8.19	7.98	4.30
Histidine	2.58	2.25	2.32	2.71	2.70	2.78	1.22
Isoleucine	4.69	4.63	3.89	3.94	3.80	3.77	3.66
Leucine	9.47	8.04	8.02	6.91	6.58	6.56	6.47
Lysine	4.00	3.51	3.68	4.81	4.66	5.06	3.82
Methionine	1.98	2.88	2.45	2.32	2.78	2.04	1.76
Cystine	1.12	2.52	1.14	2.32	2.57	2.40	1.90
Phenylalanine	6.06	5.20	5.16	4.47	4.18	4.24	4.40
Threonine	3.99	3.53	4.10	3.78	3.52	3.82	4.20
Valine	6.16	6.45	5.67	6.00	5.57	5.80	5.69
Alanine	6.28	5.51	5.99	6.15	5.87	6.25	6.13
Aspartic acid	10.88	9.05	10.43	9.09	8.83	9.05	8.60
Glutamic acid	23.19	17.74	20.36	13.58	14.60	13.10	10.42
Glycine	5.10	4.54	4.96	5.47	5.05	5.57	5.43
Proline	5.34	4.42	4.79	4.23	3.81	4.16	6.50
Serine	5.72	5.12	5.23	4.68	4.50	4.53	4.65
Tyrosine	3.65	4.86	3.33	3.13	3.39	2.90	2.16

[a]Adapted from Bienvenido et al. 1964. [b]Adapted from Houston et al. 1969.

TABLE 12.37. Comparison of Rice and Rice By-products for Growing-Finishing Swine[a]

	Diet					
	1	*2*	*3*	*4*	*5*	*6*
Performance	*Energy source*[a]					
	Maize	*Paddy rice*	*Brown rice*	*White rice*	*Bran & polishings*	*B & P +WR*[b]
Average daily gain, kg	.70^{y}	.63^{y}	.69^{y}	.70^{y}	.45^{x}	.65^{y}
Average daily feed, kg	2.16	2.30	2.17	2.10	1.33	1.97
Feed/gain ratio	3.11^{x}	3.67^{y}	3.14^{x}	2.98^{x}	2.95^{x}	3.04^{x}

Source: Maner 1971.
[a]All rice products used as complete substitute for maize; soybean meal adjusted to give 16% protein diet.
[b]B & P plus WR is rice bran and polishings in 50:50 mixture with white rice.
^{x,y}Means in the same line without a common superscript are significantly different from each other.

fiber, ether extract, and ash, but less nitrogen-free extract than polished rice. However, in diets supplied with adequate vitamins, minerals, and protein, the feeding value of each rice is similar and not different from that of maize (Bray 1943; Noland 1960). As shown in Table 12.37, daily gains, feed intake, and efficiency of feed conversion are equal to those of maize for growing-finishing pigs. Therefore, for growing-finishing pigs both white rice and brown rice have a feeding value equal to that of maize and can be used as a substitute when prices of both are equal to that of maize. Since the rice grains are very hard, they should be finely ground for most efficient utilization.

BUCKWHEAT

Buckwheat (*Fagopyrum esculentum*) is not a true cereal (Gramineae). It belongs to the family Polygonaceae. Buckwheat is grown primarily for human consumption but it has potential in animal feeding. The protein and amino acid composition of buckwheat and its fractions are summarized in Table 12.38, and the proximate analysis is in Table 12.39. The average protein content of buckwheat is about 11% and is about 74% digestible (Anderson and Bowland 1979). Research results (Sure 1975) indicate that the protein of buckwheat is of high biological value. Whereas cereals are low in lysine content, buckwheat has a high content of this amino acid. The average lysine content of buckwheat on an as-fed basis is 0.61% compared with only 0.30–0.45% for either wheat or barley and 0.18% for maize.

Anderson and Bowland (1979) reported the gross energy in buckwheat to be 68% digestible and to have an average digestible energy value of 3077 kcal/kg. This digestible energy is slightly below that of barley but higher than that of oats. The low digestible energy content of buckwheat is due to its woody hull which constitutes 18–20% of the seed. The crude fiber content of buckwheat is about 10.5% and is similar to that of oats.

TABLE 12.38. Protein and Amino Acid Content of Buckwheat and Its Fractions

Constituent	*Whole seed* (%)	*Groats* (%)	*Light flour* (%)	*Dark flour* (%)	*Germ* (%)	*Endosperm* (%)
Protein	13.8	16.4	7.4	19.0	55.9	3.3
	Percent of protein					
Lysine	6.0	5.9	5.7	5.2	5.6	6.0
Histidine	2.6	2.6	2.7	2.4	2.5	2.1
Ammonia	2.1	1.9	2.2	1.7	1.9	1.8
Arginine	9.2	10.0	7.9	8.7	11.9	5.6
Aspartic acid	11.4	11.4	10.5	11.8	10.9	10.6
Threonine	4.0	3.8	4.1	4.1	3.7	4.5
Serine	4.9	4.6	4.7	5.5	5.0	4.2
Glutamic acid	18.5	19.3	17.6	20.1	19.3	13.4
Proline	3.8	3.8	5.1	4.1	3.8	4.3
Cystine	1.6	1.8	1.8	1.5	2.2	1.9
Glycine	6.6	6.2	6.2	6.6	6.0	6.9
Alanine	4.3	4.4	4.5	4.1	3.6	6.4
Valine	5.3	4.9	5.4	5.1	4.8	6.5
Methionine	2.3	2.8	2.8	1.9	2.0	3.7
Isoleucine	4.0	3.7	4.2	3.9	3.5	5.0
Leucine	6.7	6.2	7.0	6.5	5.8	9.0
Tyrosine	2.0	2.1	2.9	2.1	2.8	3.5
Phenylalanine	4.8	4.8	4.8	4.9	4.7	4.6

Source: Pomeranz and Robbins 1972.

Feeding high levels of buckwheat to pigs housed outdoors has led to the development of fagopyrism (buckwheat poisoning). Fagopyrism occurs only in light- or white-skinned pigs when exposed to direct sunlight. This condition is a photosensitization of the skin causing itching and dermatitis. Therefore, the inclusion of buckwheat in rations for pigs housed outdoors should be limited. When pigs are housed in total confinement, fagopyrism is not a problem and the inclusion of buckwheat does not have to be limited.

TABLE 12.39. Proximate Analysis of Buckwheat

Constituent	*Buckwheat* *With hulls* (%)	*Without hulls* (%)
Dry matter	88.0	88.0
Crude protein	10.3	14.1
Fat	2.3	3.4
Crude fiber	10.7	1.8
Nitrogen-free extract	62.8	66.5
Ash	1.9	2.2

Source: Morrison 1956.

TABLE 12.40. Buckwheat for Growing Pigs (15–32 kg)

Diets	*Av daily gain, g*	*Feed/gain ratio*
Wheat (12% crude protein)	269[a]	3.9[b]
Buckwheat (12.5% crude protein)	241[a]	5.2[a]
Buckwheat + 0.5% lysine HCl	289[a]	4.3[b]
Buckwheat + 11% meat meal + 0.16% lysine HCl	434[b]	2.9[c]
Buckwheat + 15% sunflower meal + 0.17% lysine HCl	290[a]	4.3[b]

Source: Farrell 1976.
[a,b,c]Means in same column without a common superscript are significantly different.

Buckwheat has the unique property of containing a protein of higher nutritive value than that of any other known plant seed (Sure 1955), and the protein content of some varieties is approximately 14% (Pomeranz and Robbins 1972), which indicates that buckwheat supplemented with vitamins and minerals might serve as the sole feed for finishing swine if economic conditions warrant it. However, buckwheat or buckwheat supplemented with 0.5% lysine has been shown not to be adequate for growing pigs (Farrell 1976) (Table 12.40). It was necessary to supplement the diet with 11% meat meal before satisfactory performance was obtained. Although supplemented with lysine, it is possible that isoleucine is the first limiting amino acid in buckwheat, and performance of the pigs may have been improved by addition of this amino acid as was shown with rats fed buckwheat (Farrell 1976).

In studies at the University of Manitoba (Anderson and Bowland 1979), buckwheat supported similar gain to that of barley and wheat when fed to 20–60-kg pigs as the only grain in rations supplemented with soybean meal to contain 16% protein (Table 12.41). Feed-to-gain ratio was also similar for the buckwheat-fed pigs compared with those fed either wheat or barley. A combination of buckwheat and either barley or wheat substituted in the ratio 25/75, 50/50, or 75/25 for the

TABLE 12.41. Value of Buckwheat as a Substitute for Barley and Wheat in Diets for Growing Pigs (20–60 kg)

Treatments	*Grain,[a] ratio*	*Av daily gain, kg*	*Av daily feed, kg*	*Feed/gain ratio*
1. Buckwheat	100	0.70	2.01	2.91
2. Wheat	100	0.67	1.89	2.85
3. Barley	100	0.70	1.95	2.79
4. Buckwheat/wheat	75:25	0.69	2.04	2.96
5. Buckwheat/wheat	50:50	0.69	2.18	3.18
6. Buckwheat/wheat	25:75	0.72	2.02	2.82
7. Buckwheat/barley	75:25	0.66	1.94	2.93
8. Buckwheat/barley	50:50	0.72	2.27	3.16
9. Buckwheat/barley	25:75	0.70	2.12	3.02

Source: Anderson and Bowland 1979.
[a]Percentage of grain in 16% protein diets.

cereal grain, also produced satisfactory gain, feed consumption, and feed-to-gain ratio.

Nutritionally, buckwheat can be utilized as a partial or complete replacement for wheat or barley in rations for growing swine. Additional studies are needed to determine the most appropriate protein and amino acid supplement to buckwheat for optimum utilization of its protein and energy by pigs.

MILLING BY-PRODUCTS

The manufacture of flour from cereal grains and maize for human use results in several by-products available for feeding to livestock. The seed is covered with high-fiber layers which are also higher in protein and minerals than the inner substance, which is mainly starch. A diagram of a wheat kernel is shown in Fig.

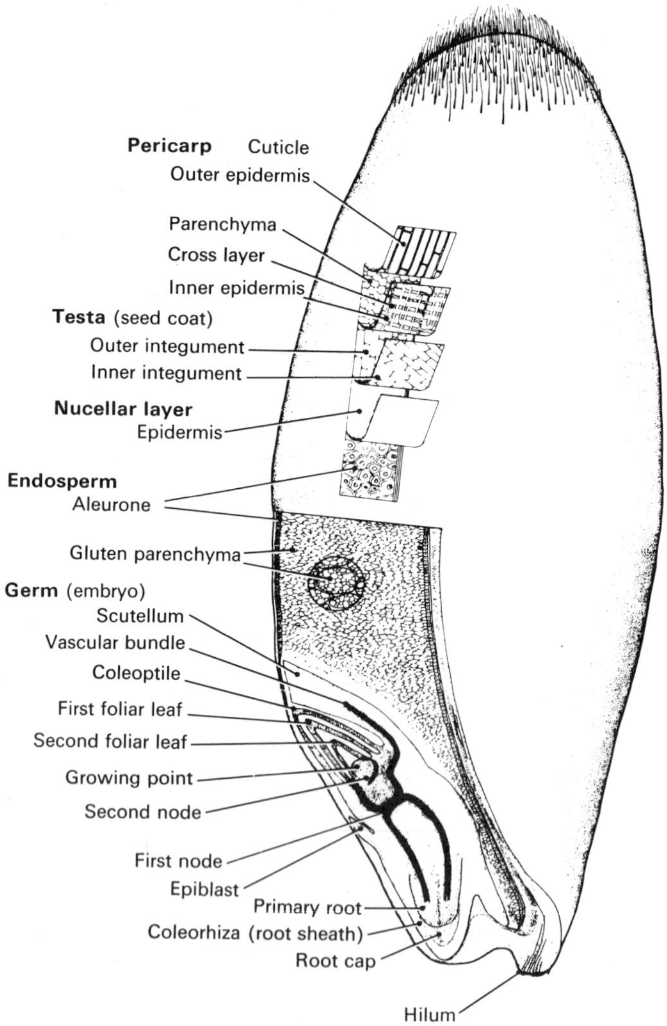

FIG. 12.2. The parts of a wheat kernel.
Courtesy USDA 1971.

12.2 to illustrate the relative locations of the various fractions. Seeds of other cereal grains have a similar general structure. The bran fraction represents the outermost layer of the seed and just under this layer is the aleurone layer. The germ, which is at the base of the seed, is high in fat, fat-soluble vitamins, and minerals. Under the aleurone layer, and constituting most of the mass of the seed, is the endosperm composed of thin-walled cells packed with starch intermixed with gluten. In milling, the objective is to remove as much starch and gluten as possible while maximizing the exclusion of the germ, aleurone, and bran fractions. Owing to variability in the degree to which this separation is made, flour by-products that contain a wide range of fiber (bran and aleurone fractions) are produced.

Wheat By-products

The currently accepted method of labeling wheat flour by-products is based on fiber content. Thus, the wheat flour by-product with more than 9.5% crude fiber is wheat bran; that with 9.5% maximum fiber is middlings or mill run; that with 7% maximum fiber is shorts, gray shorts, gray middlings, or flour middlings; that with 4% maximum fiber is red dog, white shorts, or white middlings; and that with only 1.5% maximum fiber is feed flour. Although these are the categories established by the Millers' National Federation, extreme variations exist among and between products and in nomenclature used in the literature. These variations are reflected in results obtained by different researchers and from different samples of the same product.

Wheat bran is highly palatable and has a laxative effect. Its amino acid balance is superior to that of whole wheat and improves with increasing rates of milling (Nehring *et al.* 1970). It is extremely high in phosphorus, but low in calcium, so that a nutritional imbalance occurs if wheat bran is a major component of the diet. It is high in niacin, but the niacin present is almost completely unavailable. Most of the nutrients and a major part of the potential metabolizable energy of wheat bran are contained in the cells of the aleurone layer. Contrary to what might be expected, the pig is able to completely break down the aleurone layer including the cell walls, although a portion of the carbohydrate and about 10% of the protein are not absorbed and appear in the feces (Saunders *et al.* 1969, 1974).

Feeding studies with growing and finishing pigs have demonstrated that increasing levels of wheat bran above 40–45% decreases both rate of gain and efficiency of feed conversion in both growing and finishing pigs (Ranjhan *et al.* 1972; Le Dividich *et al.* 1975; Forsythe 1979). The substitution of maize and soybean meal with wheat bran at levels below 40–45% generally has little significant effect on growth rate but tends to reduce efficiency of feed conversion (Table 12.42). Samples of bran with higher levels of crude fiber and lower density have generally been associated with poorer pig performance and increased feed required per unit of gain. The latter is related to the lower digestible [2670 kcal/kg dry matter (DM)] and metabolizable energy (2530 kcal/kg DM) (Diggs *et al.* 1965) and the pigs' observed inability to increase feed consumption of high-level bran diets to compensate for the lower digestible energy content—as has been observed when other bulky diluents have been added to swine diets.

Wheat middlings contain 3470 kcal/kg DM of digestible energy and 3340

TABLE 12.42. Effect of Level of Wheat Bran on Performance of Pigs

Source: *Level of wheat bran, %*	*Class of swine*[a]	*Av daily gain, kg*	*Av daily feed, kg*	*Feed/gain ratio*
Ranjhan et al. 1971:				
10	G	0.47	1.78	3.78
40	G	0.45	1.79	4.16
60	G	0.35	1.40	4.00
70	G	0.33	1.35	4.09
Ranjhan et al. 1972:				
30	G	0.48	1.50	3.1
60	G	0.41	1.50	3.6
Ranjhan et al. 1972:				
30	F	0.43	2.10	4.93
60	F	0.33	1.79	5.41
83	F	0.30	1.82	5.99
Forsythe 1979:				
0	G-F	0.80	1.88	2.35
16	G-F	0.78	2.02	2.60
Sharda et al. 1979:				
5	G	0.41	1.33	3.24
25	G	0.35	1.28	3.64
7.5	F	0.62	2.65	4.28
37.5	F	0.62	2.60	4.19
5–7.5	G-F	0.46	1.64	3.58
25–37.5	G-F	0.42	1.65	3.89

[a]G = growing, F = finishing, G-F = growing-finishing.

kcal/kg DM of metabolizable energy (Pals and Ewan 1978), have a high quality protein (Beames and Natoli 1969), and have a higher feeding value than wheat bran. Wheat middlings can replace a portion or all of the wheat and soybean meal or maize and soybean meal in diets for growing-finishing pigs with only small reductions in pig performance.

Evaluation studies have indicated that wheat middlings (pollards) containing 17.1% crude protein and 8.0% crude fiber could replace all the wheat and soybean meal in diets for growing-finishing pigs without altering rate of gain when fed to provide the same daily level of digestible energy intake (Beames and Natoli 1969). However, feed efficiency and dressing percentage decreased as the level of middlings (pollards) increased. Hanrahan and O'Grady (1970) observed that swine performance decreased as the level of middlings (pollards) in the diet increased. Erickson (1979) reported a small linear decrease in pig performance with increasing levels of substitution of maize and soybean meal by standard wheat middlings (Table 12.43).

Extensive studies at the University of Guelph (Patience 1977A,B; Young 1980; Young and King 1981) have shown that pig response to *wheat shorts* is variable. Patience (1977A,B) observed that wheat shorts could replace all of the maize and soybean meal with only small reductions of gain, although feed efficiency and dressing percentage were decreased. Later studies (Young 1980) showed that with

TABLE 12.43. Wheat Middlings as Partial or Total Replacement of Protein and Energy in Swine Diets

Source: *Level of wheat middlings, %*	*Class of swine*	*Av daily gain, kg*	*Av daily feed, kg*	*Feed/gain ratio*
Beames and Natali 1969:				
0	G	0.47	1.40	2.97[a]
48.4	G	0.47	1.50	3.20[b]
67.8	G	0.48	1.58	3.29[b]
96.9	G	0.46	1.65	3.58[c]
0	F	0.55	2.01	3.66[a]
48.4	F	0.55	2.22	4.04[b]
67.8	F	0.55	2.33	4.23[b]
96.9	F	0.54	2.49	4.01[c]
0	G-F	0.52	1.77	3.40[a]
48.4	G-F	0.52	1.94	3.73[b]
67.8	G-F	0.53	2.06	3.88[b]
96.9	G-F	0.51	2.15	4.22[c]
Hanrahan and O'Grady 1970:				
0	G-F	0.61	1.98	3.24[a]
20	G-F	0.58	1.94	3.37[a,b]
40	G-F	0.54	1.97	3.67[b]
Erickson 1979:				
0	G	0.77	1.94	2.52
20	G	0.74	1.74	2.35
40	G	0.70	1.83	2.61
60	G	0.70	1.79	2.56
0	F	0.89	2.81	3.16
20	F	0.84	2.78	3.31
40	F	0.84	2.87	3.42
60	F	0.81	2.79	3.45
0	G-F	0.81	2.25	2.78
20	G-F	0.77	2.14	2.78
40	G-F	0.75	2.18	2.90
60	G-F	0.75	2.17	2.89

[a,b,c]Means in same column without a common superscript are significantly different.

one sample of wheat shorts, containing 18.5% protein, 10.4% acid detergent fiber, and 0.82% lysine, pig gains were not depressed until the dietary level of shorts exceeded 64.4%; with a second sample, containing 18.6% protein and 16.5% acid detergent fiber, there was a quadratic decline in rate of gain of pigs of comparable size as the level of shorts increased. Feed efficiency followed a similar trend and was more depressed at the higher levels of substitution (Patience 1977B; Young 1980). It has been shown that lysine is the first limiting amino acid of wheat shorts for the growing pig and that threonine is the second (Patience 1977A,B).

Young and King (1981) have shown that wheat shorts can supply all of the protein and energy in diets for gestating sows. Results with sows fed wheat shorts supplemented with vitamins and minerals were not different from those of sows fed complete diets based on maize and soybean meal.

Although wheat by-products (offals) are widely used throughout the world and can supply a considerable proportion of the protein and energy in swine diets, the variation in nutritive values reported is of concern. The considerable variation in digestible energy in wheat by-products probably reflects differences in type of wheat used and different degrees of extraction of bran, middlings, and shorts during the milling process (Morgan *et al.* 1975; Henry 1976; Just *et al.* 1978; Pals and Ewan 1978; Batterham *et al.* 1980). Most of the variation is due to fiber content, which has been shown to reduce digestible energy content of feeds. Although all fiber measurements (crude fiber, acid detergent fiber, neutral detergent fiber, and hemicellulose) are negatively correlated with digestible energy content in wheat by-products (Batterham *et al.* 1980), the high apparent relationship between percentage acid detergent fiber and digestible energy (Young 1980; Batterham *et al.* 1980) plus the value of acid detergent fiber in predicting rate of gain of pigs fed wheat by-products (Young 1980) suggests that acid detergent fiber or acid detergent fiber in combination with product density may be useful parameters for estimating the value of wheat by-products for swine diets.

Maize Bran and Other By-products

A diagram of the maize kernel is shown in Fig 12.1 (p. 247). The kernel is covered by the hull which contains about 15% crude fiber and has a tip cap (where the kernel attaches to the cob) of similar composition. Just under the hull is the gluten layer which contains about 22% protein. Most of the mass of the kernel (75%) consists of the endosperm, which is composed mostly of starch and has less than 10% protein. The germ or embryo is located at the base of the kernel just above the tip cap and contains about 20% protein and 35% fat. In milling maize for producing maize meal, the germ and hull are removed. Hominy feed (hominy meal, hominy grits) is the by-product and consists of the maize bran (hulls), maize germ, and part of the endosperm. It resembles maize meal but is higher in fiber and protein and, unless part of the fat has been removed from the germ, is higher in fat than whole maize or maize meal.

Starch and glucose are obtained from maize by a wet-milling process whereby the maize is first softened in dilute acid followed by milling to free the germs, which are separated by their lower specific gravity (higher fat), resulting in their floating to the top of the liquid. The residue is finely ground and hulls (bran) are separated by filtration. The starch and gluten that remain are separated from each other by centrifugation and filtration. The evaporation of the liquid from which the germ, bran, starch, and gluten have been removed results in a residue called maize solubles. This residue includes minerals and some water soluble carbohydrates, protein, and vitamins.

Maize–gluten feed is a mixture of gluten meal and bran and may contain maize solubles and part of the germ.

Maize bran feed is too high in fiber and too low in protein quality to be used as a major component of swine feed. Maize gluten meal is high in protein (40% or more), but its protein is severely deficient in lysine and tryptophan so it cannot be used as a significant portion of the protein of the diet. Owing to its low mineral and vitamin content and its inferior protein quality, gluten meal must be used sparingly and must be considered chiefly as an energy source for swine. Since gluten feed is merely a mixture of gluten meal and bran whose proportions can be regulated at will by the processor, it is of variable feeding value and its usefulness as an energy source is governed largely by the proportion of bran (fiber percentage). Like that of wheat and other cereal by-products, the niacin content of maize by-products must be considered to be near zero, as the biological availability is very low, even though the analysis looks favorable.

Hominy feed is an excellent energy source for swine, although its crude fiber content is higher than that of whole maize. It is also higher in fat than whole maize so that it is at least equal to maize in digestible energy.

Rice Bran and Polishings

Rice bran is available in many areas of the world and especially in the tropics and subtropics. The quantity of this by-product available represents about 8% of the paddy rice produced in the world. With a reported world production of 400 million tons of paddy rice in 1980 (FAO 1980), it is estimated that 32 million tons of rice bran were produced.

Pure rice bran is very palatable and well consumed by swine of all ages when used fresh. However, because of its high oil content, it turns rancid on storage and is less palatable to the pig. As demonstrated by Tillman *et al.* (1951) and Leighton (1958), rice bran and polishings contain a lypolytic enzyme which becomes markedly active as soon as the bran is removed from the rice. In the presence of moisture, the glycerides of the rice oil hydrolyze to glycerol and a mixture of fatty acids. Rancidity may be caused by oxidation or by hydrolytic action of the oil (Arnott and Lim 1966). Rancid rice meal has an unpleasant odor and taste that is offensive to livestock.

Rice bran is of highly variable quality, depending mainly on the quantity of hulls included with the bran. The presence of hulls can be revealed by microscopic analysis if not finely ground, but if they are finely ground and added to bran and polishings it is more difficult to detect the adulteration. The deleterious effect of hulls is due to the high fiber and silica and the low digestibility of the hulls, which limit dietary energy concentration. An estimation of the percentage of adulteration can be made from the silica content of the bran. Hulls contain 11–19% silica. Therefore, calculated hull content is obtained by multiplying the silica content by 5.9 (on the basis of 17% silica in pure hulls) (Arnott and Lim 1966). A negative correlation between crude fiber and crude protein content ($r = -0.71$), a positive correlation between percent crude fiber and percent silica ($r = 0.74$), and a negative correlation between crude protein and silica ($r = -0.85$) aid in the evaluation of adulteration and bran quality (Arnott and Lim 1966).

In general, rice polishings, which come from the inner part of the grain, do not vary as widely as bran in chemical composition. In most, the protein content is slightly higher and the fiber content much lower than that of bran. Since many rice mills do not separate bran and polishings, and the characteristics of each are so poorly defined that they are difficult to distinguish, what is termed bran is usually a mixture of bran, polishings, and hulls, and this designation will be used in further discussion in this section.

In some parts of Asia where rice by-products are available and widely distributed, rice bran is commonly used as the only feed for swine of all ages.

TABLE 12.44. Effect of Level of Rice Bran on Performance of Growing-Finishing Pigs

Criterion	Level of rice bran substitution for maize				
	0	25	50	75	100
Average initial wt, kg	23.0	20.0	21.0	20.0	20.0
Average final wt, kg	89.6	88.2	87.8	79.6	70.9
Average daily gain, kg	0.81	0.80	0.75	0.66	0.57
Average daily feed, kg	2.63	2.73	2.67	2.57	2.16
Feed/gain ratio	3.23	3.41	3.58	3.87	3.77

Source: Adapted from Noland and Scott 1963.

Although under these conditions it supports economical production, pig performance and efficiency of feed utilization are reduced. Even as a substitute for 100% of the grain in well fortified growing-finishing pig diets, it supports slower and less efficient gains than if used at lower levels. Noland and Scott (1963) have shown that each increase in the level of rice bran causes a linear depression in average daily gain and an increase in feed required per kilogram of gain (Table 12.44). However, feed consumption was only affected at the 100% level of substitution. Ara *et al.* (1975) fed levels of 0, 20, 40, 60, and 80% rice bran as a substitute for sorghum to growing-finishing pigs and also noted a linear decrease in daily gains and efficiency of feed utilization. Louisiana work (Thrasher *et al.* 1966) has indicated that the substitution of 20 and 30% rice bran for maize in a maize–soybean meal diet reduces pig gains by 8 and 15%, respectively, and significantly reduces feed efficiency, back fat thickness, and carcass firmness. Other studies by Moncada and Maner (1965–1966) have demonstrated that replacing maize with rice bran in quantities of as much as 30% of the diet does not affect gains or feed efficiency. However, a 40% level of substitution depressed gains but had no effect on feed conversion. Similar results have been published by Bray (1943) and Campabadal *et al.* (1976), who reported that for growing-finishing pigs, rice bran has approximately the same feeding value as maize if the level of substitution does not exceed 30% of the total ration. Higher levels (35–45%) significantly reduce both gains and efficiency of feed utilization.

TABLE 12.45. Effect of Increasing Levels of Rice Bran in Diets for Growing Pigs

Level of rice bran, %	Av daily gain, kg	Av daily feed, kg	Feed/gain ratio, kg/kg
0	0.70[a]	1.91	2.71
15	0.71[a]	1.85	2.59
30	0.70[a]	1.93	2.78
45	0.71[a]	1.83	2.57
60	0.62[b]	1.81	2.87

Source: Chamorro and Maner 1974.
[a,b]Means in the same column without a common superscript are significantly different.

TABLE 12.46. Effect of Levels of Rice Bran With and Without Molasses on Growing-Finishing Pig Performance (15.9–78.4 kg)

Level of rice bran, %	*Av daily gain, kg*	*Av daily feed, kg*	*Feed/gain ratio, kg/kg*
0	0.74[a]	2.18	2.95
40	0.74[a]	2.32	3.09
60	0.70[a,b,c]	2.13	3.04
80	0.64[c]	2.06	3.22
40 + 32.2% molasses	0.71[a,b]	2.36	3.32
60 + 16.5% molasses	0.65[b,c]	2.20	3.38

Source: Brooks and Lumanta 1975.
[a,b,c]For a single comparable criterion, values in the same column having the same or no superscript are not significantly different ($P < .05$).

In some studies, higher levels of rice bran substitution have supported pig performance equal to that obtained with maize alone. Chamorro and Maner (1973) fed levels of 0, 15, 30, 45, and 60% rice bran containing 13.2% crude protein and reported no differences in gains, feed consumption, and feed efficiency when rice bran was substituted for up to 45% of the maize and soybean meal in the diet (Table 12.45). A significant reduction in average daily gain was observed, however, when the level of substitution was increased to 60%. Using the same levels of substitution, Martinez and Bravo (1971) reported similar results. Similar studies reported from Hawaii (Brooks and Lumanta 1975) demonstrated that although small reductions in gain were observed when 60% rice bran replaced maize in a maize–soybean meal diet, statistically significant reductions in pig gains were not recorded until all the maize was replaced with 80% rice bran (Table 12.46).

Although the exact reason for the variation in results obtained is not known, variations in rice bran quality and freshness and differences in diet formulation have been suggested. Several researchers have attempted to identify the factor or factors which limit the use of high dietary levels of rice bran for growing and finishing pigs. Among the factors studied are the level of hull adulteration, the bran and diet digestibility, the energy level, the palatability, and the supplementary protein quality and amino acid balance.

Although the degree of adulteration of rice bran with rice hulls has been suggested as an important factor responsible for limiting the use of high levels of rice bran, Chamorro and Maner (1974) fed growing-finishing pigs (20–87 kg) diets containing 60% rice bran to which had been added either 0, 10, 20, 30, or 40% pure, ground rice hulls and demonstrated that level of adulteration had no effect on rate of gain. However, the level of hull adulteration did affect feed intake and efficiency of feed utilization (Table 12.47). Each increase in the level of hulls in the bran was accompanied by a linear increase in both daily feed intake and feed required per kilogram of gain. It appears that the rice hulls were almost inert and that the pigs increased daily feed intake to compensate for the energy dilution caused by the hull additions. In fact, if daily feed is corrected by deducting the amount of rice hulls present in each diet, differences in both daily feed intake and efficiency of feed utilization are eliminated.

TABLE 12.47. Effect of Level of Rice Hulls in Rice Bran on Performance of Growing-Finishing Pigs Fed Diets Containing 60% Rice Bran Supplemented with Soybean Meal

Parameters	*Ground rice hulls in rice bran, %*				
	0	*10*	*20*	*30*	*40*
Level of hulls in diet, %	0	6.0	12.3	18.0	24.0
Level of protein in diet, %					
Growing	18.8	18.0	17.3	16.6	15.8
Finishing	15.8	15.0	14.3	13.6	12.8
Av daily gain, kg[a]	0.69	0.68	0.71	0.68	0.71
Av daily feed, kg[b]	2.07	2.12	2.29	2.50	2.68
Feed/gain ratio[c]	3.00	3.12	3.24	3.70	3.83

Source: Chamorro and Maner 1974.
[a]Initial weight 20 kg; final weight 87 kg.
[b,c]Significant linear increase in daily feed intake and feed per kilogram of gain.

Brooks and Lumanta (1975) and Campabadal *et al.* (1976) reported that the digestibility of rice bran components was lower than that of the basal soybean meal diet. In both studies, the progressive addition of rice bran to the diet decreased the apparent digestibility of dry matter, protein, nitrogen-free extract, and gross energy. It appears that most of these changes in digestibility of the diet were due to the lower digestion coefficients for rice bran and not to changes in digestibility of the basal diet. Except for dry matter, protein, and ether extract, the level of rice bran did not change digestibility of the nutrient components of the rice bran. Rice bran has approximately 82% total digestible nutrients and has a digestible energy content of 3508–3697 kcal/kg of dry matter (Robles and Ewan 1976; Ara *et al.* 1975) and a metabolizable energy content of 3338–3357 (Ara *et al.* 1975; Robles and Ewan 1976) kcal/kg of dry matter.

Although it has been shown that increasing the level of fish meal (Bistoyang *et al.* 1968) and soybean meal (Balderrama *et al.* 1968) in rice-bran-based diets caused an improvement in both gains and feed efficiency, and Gomez *et al.* (1978) showed that the replacement of soybean meal with fish meal improved gains and feed efficiency, Alvarado and Maner (1975) were unable to improve pig performance by supplying higher than normal levels of protein in rice-bran-based diets nor were the additions of lysine and methionine alone or in combination effective in overcoming the depression in performance resulting from 60% rice bran in growing-finishing pig diets.

Problems with palatability and gastrointestinal disturbances have been reported. Reduced consumption of diets containing high levels of rice bran seemed to be associated with off-flavors which developed from prolonged storage of rice bran and especially that stored in hot and humid climates. Diets containing medium to high levels of rancid rice bran are not well consumed and when eaten have, in some instances, caused gastrointestinal disturbances and loose feces. The use of fat, sugar, and molasses to improve palatability has failed to significantly increase daily feed intake. Although it has been shown that heat treatment, in the form of dry heat, steam, or extrusion, would effectively reduce or destroy the lypolytic enzyme activity (Chamorro 1978; Kopeihovskii *et al.* 1971; Shimamoto 1971; Lynn 1969; Viraktamath 1974) and thus reduce the formation of fatty acids and extend storage time, Chamorro (1978) failed to show any beneficial effects of extrusion of rice bran on either gains or efficiency of feed utilization of growing or finishing pigs fed either 40 or 60% rice bran.

Pigs slaughtered following long-term consumption of diets containing medium-to-high levels of rice bran, have displayed some degree of inflammation of the gastrointestinal mucosa (Arnott and Lim 1966). Campabadal *et al.* (1976) reported that mucosal inflammation was progressively more severe as the amount of bran was increased. Both the fine particle size of rice bran and the high level of coarse abrasive fiber (high silica) from rice hulls have been suggested as causative factors. Since Chamorro and Maner (1974) did not observe any reduction in pig gains due to increasing rates of hulls in growing-finishing pig diets, it seems unlikely that hulls per se or hull-induced inflammation is the primary factor in the reduced performance commonly observed when rice bran is used as a primary dietary constituent and the diet is fed *ad libitum*. If daily feed intake is restricted, thus preventing the pig from increasing consumption then high levels of hulls would be expected to reduce pig performance.

Mineral imbalance and especially that related to the high level of phytin phosphorus in rice bran has been implicated as a possible causative factor. Phytin phosphorus not only is poorly utilized by the pig but is also known to reduce the utilization of other minerals in the diet. Maust *et al.* (1972) reported that adding 29% rice bran did not alter digestibility of diet components of a maize–soybean meal diet when a zinc supplement was added; however, this level of rice bran without zinc depressed digestion and produced zinc deficiency.

Satisfactory gains and efficiency of feed utilization can be obtained when moderate levels of rice bran (30–45%) are used in growing-finishing pig diets. Reduced pig performance can be expected when higher levels are incorporated into the diet. For young growing pigs (5–15 kg), not more than 15–20% rice bran should be included in the diet. Economic consideration may modify these recommendations. Under some feed price conditions, slower and less efficient gains from diets containing high levels of rice bran may be more economical than fast and more efficient grains from high cost diets based on conventional feedstuffs such as maize and soybean meal.

Other Milling By-products

The milling of other cereal grains such as barley, sorghum, rye, and oats consists of generally similar procedures and comparable by-products. The usefulness of these by-products for swine feeding must be considered basically in terms of their contributions as sources of energy. Generally, the by-product feeds are higher in protein and fiber than the seeds from which they are derived, but because of the poor amino acid balance of the protein, they have their greatest usefulness as energy substitutes when economics warrant it.

DISTILLERY AND BREWERY BY-PRODUCTS

Distillery By-products

Production of distilled liquors and alcohol from cereal grain or maize includes grinding, cooking, and the addition of enzymes to hydrolyze the starch to simple sugars and yeast to cause fermentation. After fermentation, the alcohol is removed and the residue is available for livestock feeding. This residue can be fractionated into several components or used as the complete mixture, sometimes called *stillage* or *distillery slop*. The coarser particles are usually removed from the stillage and are called *wet distillers' grains*. After dehydration this product is sold as *dried distillers' grains*. The water-soluble materials and fine particles remaining are usually dehydrated to form *dried distillers' solubles*. The addition of condensed or dried distillers solubles to wet distillers grains produces distillers' grains with solubles which, after they have been dried, are sold as *distillers' dried grains with solubles*. The Association of American Feed Control Officials (AAFCO) has been effective in standardizing the nomenclature, so that each of the above products is further identified by a prefix naming the cereal grain or maize used in its manufacture.

The composition of distillers' dried grains with solubles (Table 12.48) would suggest various potential uses in swine feeds. Numerous workers have demonstrated the value of distillers' by-products as a protein substitute and as a source of known B vitamins and trace elements (Barnhart 1952; Catron *et al.* 1954; Fairbanks *et al.* 1944, 1945; Carpenter 1970). It has also been suggested that these

TABLE 12.48. Composition of Distillers' Dried Grains with Solubles

Component	Distillers' dried grains/solubles			
	1[a]	2[a]	3[b]	4[c]
Crude protein, % DM	35.8	34.3	29.6	26.1
Ether extract, % DM	6.2	5.0	7.3	—
Crude fiber, % DM	9.0	7.6	4.1	—
Ash, % DM	9.1	13.8	—	—
Nitrogen-free extract, % DM	39.9	39.3	—	—
Amino acids, % of protein				
Lysine	2.3	2.1	3.4	2.7
Methionine + cystine	1.9	1.8	2.4	—
Threonine	3.5	3.7	—	—
Tryptophan	0.8	0.7	—	—
Isoleucine	3.5	3.5	—	—
Leucine	6.3	6.5	—	—
Phenylalanine	4.2	4.5	—	—
Valine	4.4	4.5	—	—
Arginine	3.2	2.8	—	—
Histidine	1.9	1.7	—	—

Sources: [a]Salo 1978; [b]Peers et al. 1978; [c]Livingston and Livingston 1969.

products might be a source of unidentified growth factors. However, Becker *et al.* (1960), Conrad *et al.* (1959), Wallace and Combs (1965), and Speer *et al.* (1967) did not find any evidence of unidentified growth factor activity when maize distillers by-products were added to fortified maize–soybean diets for growing swine.

Livingston and Livingston (1969) found that when distiller's dried grains with solubles containing 22.8% crude protein and 0.62% lysine were included in the diets of 20–90-kg liveweight pigs at the 0 and 14.7% levels, no difference was found in the growth rate or feed conversion efficiency, but both these measures were significantly poorer when distillers dried grains with solubles were used at the 25% level. Similar feeding trials (Wahlstrom *et al.* 1970) indicated that levels of 5 and 10% distiller's dried grains with solubles can be fed in maize–soybean meal diets without adversely affecting rate of gain or feed efficiency of growing-finishing pigs. Increasing the level to 20% reduces digestibility of protein, nitrogen-free extract, and dry matter as well as the gains and the efficiency of feed conversion.

Further work with young starter pigs between 8.0 and 17.9 kg (Wahlstrom and Libal 1980) has demonstrated significant differences in average daily gain among pigs consuming diets containing 0, 10, 20, or 30% distillers' dried grains with solubles. All diets were formulated to contain 19.6% crude protein and 1.05% total lysine. Daily gains decreased in a linear fashion as the level of distillers' dried grains with solubles increased (Table 12.49). Although consumption was less for the diets containing 20 and 30% distillers' dried grains with solubles than for the 0 and 10% diets and there was a tendency for more feed to be required per unit of gain, there were no statistically significant differences among treatments in average daily feed consumed or in feed conversion.

Although it has been indicated that the amino acid pattern of the protein is too imbalanced to allow distillery by-products to be used as a major proportion of the protein supplement for swine, Harmon (1974, 1975) reported that the lysine and tryptophan in distillers' dried grains with solubles were effectively utilized by the growing pig. Furthermore, Wahlstrom *et al.* (1970) demonstrated that only 0.15% supplemental lysine was needed to overcome the depression in pig performance resulting from the addition of solubles in diets for growing-finishing pigs (Table 12.50). The crude fiber content of distillers' dried grains with solubles is

TABLE 12.49. Effect of Distillers' Dried Grains with Solubles on Performance of Young Pigs

Parameter	*Distillers' dried grains/solubles, %*			
	0	*10*	*20*	*30*
Av initial wt, kg	8.1	8.1	8.0	8.0
Av final wt, kg	17.9	17.4	17.0	16.3
Av daily gain, kg[a]	0.35	0.34	0.32	0.30
Av daily feed, kg	0.61	0.61	0.58	0.56
Feed/gain	1.75	1.82	1.84	1.90

Source: Wahlstrom and Libal 1980.
[a]Significant difference among treatments ($P < .01$).

TABLE 12.50. Effect of Lysine Supplementation on the Utilization of Distillers' Dried Grains with Solubles (DDG/S) by Growing-Finishing Pigs

Level of DDG/S, %	*0*	*20*	*20*	*20*
Level of lysine supplementation, %	*0*	*0*	*0.15*	*0.25*
Av daily gain, kg	0.79	0.61[a]	0.73	0.73
Av daily feed, kg	1.61	2.40	2.53	2.54
Feed/gain ratio	3.34	3.94[b]	3.48	3.49

Source: Wahlstrom et al. 1970.
[a,b]Means on the same line with different superscripts are significantly different.

generally higher than that of the original grain. Ewan (1976) reported a metabolizable energy value of 2940 kcal/kg for distillers' dried grains with solubles which is somewhat less than that of maize (3320) and soybean meal (3130). Thus, because of its relatively high fiber content (8.5–9.0%) and the pattern of amino acids, the dietary level of distillers' dried grains with solubles for growing-finishing swine should be limited to 10–15% or a maximum of 20% when supplemental lysine is added. Under most conditions, grains rather than their by-products will be a more economical source of energy.

Distillers' dried grains with solubles have been shown to be a useful feed for gestating swine. Thong *et al.* (1978) demonstrated that a dietary level of 44.7% distillers' dried grains with solubles could substitute for all of the soybean meal and a portion of the maize without affecting gilt gain or reproductive performance. Even at this level of substitution in which the distillers' dried grains with solubles supplied 68% of the total dietary lysine and 44% of the total metabolizable energy, average gestation gains, litter size, and weight at birth were similar to those obtained when a 11.4% crude protein maize–soybean meal control diet containing 0.42% lysine was fed. Absolute daily nitrogen retention from the 44.2% distillers' dried grains with solubles diet was equal to that from the control diet, indicating that the lysine from this distillery by-product was apparently highly available.

Brewery By-products

The brewery by-product, dried brewers' grains, is a bulky, low-energy feed containing 96.7% dry matter, 28.5% crude protein, 6.1% ether extract, 19.1% crude fiber, 3.6% ash, and 39.4% nitrogen-free extract. The protein has been shown to be 72.4% digestible and the digestible energy, metabolizable energy, and metabolizable energy corrected for nitrogen to be 2650, 2500, and 2380 kcal/kg DM, respectively, with only 52.3% of the gross energy being digestible (Kornegay 1973). The amino acid content of brewers' dried grains is shown in Table 12.51.

Young and Ingram (1968) used dried brewers' grains as a replacement for soybean meal in the diet of growing-finishing pigs. Levels of 0, 25, 50, 75, and 100% of the supplemental protein or 0, 11.4, 23.3, 34.0, or 52.4% of the diet

TABLE 12.51. Amino Acid Content of Brewers' Dried Grains

Item	*Source 1*[a]	*Source 2*[b]
Aspartic acid	1.76	1.78
Threonine	1.01	0.97
Serine	1.20	1.39
Glutamic acid	4.28	6.05
Proline	2.91	2.28
Glycine	1.06	0.94
Alanine	1.64	1.72
Valine	1.24	1.44
Cystine	0.28	—
Methionine	0.51	0.58
Isoleucine	1.71	1.95
Leucine	2.82	3.22
Tyrosine	0.95	1.22
Phenylalanine	1.57	1.44
Lysine	0.76	0.63
Histidine	0.67	0.57
Arginine	1.28	1.09
Tryptophan	—	0.25

Source: [a]Kornegay 1973; [b]Brooks 1971.

were supplied by dried brewers' grains. There were no changes in average growth rate or carcass quality of pigs given up to 50% of the supplemental protein, but when brewers' grains supplied more than 50% of the supplemental protein, there was a tendency for both gains and feed efficiency to decline. The effect of the addition of increasing levels of brewers' grains was most pronounced during the growth period, where both gains and feed efficiency were linearly depressed with each increase in level of brewer's grain. Similar studies by Baird *et al.* (1975) replaced 0, 25, 50, and 75% of the soybean meal in 15% protein maize–soybean meal diets for growing-finishing pigs. It was shown that at levels of more than 25% substitution, lysine became limiting and caused significant reductions in both daily gain and efficiency of food conversion. Nigerian studies have shown that replacement of maize by 5, 10, or 15% brewers' dried grain in diets based on maize, peanut meal, and blood meal had no effect on pig performance or efficiency of feed utilization (Babatunde *et al.* 1975).

In two trials with growing-finishing pigs, one in winter and one in summer, Crowmwell *et al.* (1979) used brewers' dried grains containing 22.4% protein and 0.74% lysine to substitute maize and soybean meal on an isolysine basis at levels of 0, 5, 10, 15, and 20%. Pooled data from the two trials (Table 12.52) indicated a quadratic reduction in growth rate and a linear increase in feed/gain with increasing level of brewers' dried grains. The authors concluded that up to 15% brewers' dried grains could be included in maize– soybean meal diets for growing-finishing swine without depressing growth; however, feed required per unit of gain increased progressively with increasing levels of brewers' dried grains.

TABLE 12.52. Effect of Dietary Level of Brewers' Dried Grains (BDG) on Performance of Growing-Finishing Swine

Level of BDG, %	Summer		Winter		Pooled	
	ADG,[a] g	Feed/gain	ADG, g	Feed/gain	ADG, g[b]	Feed/gain[c]
0	798	3.18	687	3.59	742	3.38
5	772	3.21	730	3.59	751	3.40
10	771	3.24	722	3.65	746	3.44
15	770	3.30	703	3.71	736	3.51
20	704	3.44	681	3.78	692	3.61

Source: Cromwell et al. 1979.
[a]Average daily gain.
[b]Significant quadratic reduction in growth rate.
[c]Significant linear reduction in feed/gain.

Although Young and Ingram (1968) reported that the proportion of dried brewers' grains had no effect on gains during the finishing period, Werner and Schellner (1960– 1961) demonstrated that feeding wet brewer's grains to finishing pigs (53–115 kg) reduced gains by 15.5% and increased both intake per kilogram of gain and cost of production, even when it replaced garbage in the ration.

Brewers' dried grains have been used as a nutrient source for gestating sows. Harmon *et al.* (1975) fed gestating sows 1.8 kg daily of a maize–soybean meal diet containing 0.42% lysine or similar diets in which brewers' dried grains supplied 50 or 100% of the dietary lysine. It was shown that brewers' dried grains could replace all of the soybean meal in the diet without significantly reducing reproductive performance. However, as brewers' grains increased in the diet, metabolizable energy (3.48, 3.35, and 3.06 kcal/g) and gestation weight gains were reduced. Wahlstrom and Libal (1976) used brewers' dried grains to replace 20 or 40% of a maize–soybean meal diet for pregnant sows. Diets with 0, 20, or 40% brewers' dried grains had 12.0, 13.7, or 15.4% crude protein, 0.50% lysine, and 3.12, 2.88, and 2.55 Mcal/kg of metabolizable energy (ME). They were hand fed at levels of 1.8, 2.0, and 2.2 kg daily to supply all sows with 5.6 Mcal ME daily, except during very cold weather the intakes were raised to 7.0 Mcal ME daily in the last month of pregnancy. After farrowing, all sows received a maize–sorghum meal diet containing 15% crude protein. Although there was no significant effect of diet on litter size, litter weight, or pig weight at birth or weaning, sows fed 40% brewers' dried grains gained significantly less than those fed either the control or 20% brewers' dried grains diet, and this reduction was more pronounced during extremely cold weather. Therefore, brewers' dried grains can be used to supply up to 40% of the diet for gestating sows without affecting reproductive performance. It is necessary, however, to increase daily sow intake of diets containing high levels of brewers' dried grains to compensate for lower metabolizable energy content in order to provide adequate energy for gestation weight gains.

REFERENCES

Introduction

NATIONAL RESEARCH COUNCIL. 1983. Underutilized Resources as Animal Feedstuffs, pp. 1–253. National Research Council, Natl. Acad. of Sci. Natl. Academy Press, Washington, D.C.

Common Maize

ALLEE, G. L., and BAKER, D. H. 1970. J. Anim. Sci. *30,* 748.

BAKER, D. H., BECKER, D. E., JENSEN, A. H., and HARMON, B. G. 1970. J. Anim. Sci. *30,* 364.

BAKER, D. H., KATZ, R. S., and EASTER, R. A. 1975. J. Anim. Sci. *40,* 851.

BRESSANI, R., ELIAS, L. G., SCRIMSHAW, N. S., and GUZMAN, M. A. 1962. Cereal Chem. *39,* 59.

BRESSANI, R., and MERTZ, E. T. 1958. Cereal Chem. *35,* 227.

CLAWSON, A. J., and MATRONE, G. 1963. J. Anim. Sci. *22,* 834 (Abstr.).

CROMWELL, G. L., PICKETT, R. A., CLINE, T. R., and BEESON, W. M. 1967. J. Anim. Sci. *26,* 1325.

DOBBINS, F. A., KRIDER, J. L., HAMILTON, T. S., EARLY, E. B., and TERRILL, S. W. 1950. J. Animal Sci. *9,* 625.

EASTER, R. A., and BAKER, D. H. 1980. J. Anim. Sci. *50,* 467.

FAO. 1980. Food and Agricultural Organization Production Yearbook, Vol. 34.

FREY, K. J., BRIMHALL, B., and SPRAGUE, G. F. 1949. Agron. J. *41,* 399.

GALLO, J. T., and POND, W. G. 1968. J. Anim. Sci. **27,** 73.

GALLO, J. T., POND, W. G., and LOGOMARSINO, J. V. 1968. J. Anim. Sci. *27,* 1000.

GEIGER, E., COURTNEY, G. W., and GEIGER, L. E. 1952. Arch. Biochem. Biophys. *41,* 74.

GROSBACH, D. A., LEWIS, A. J., PEO, E. R., JR., MOSER, B. D., and CRENSHAW, T. D. 1980. J. Anim. Sci. *51* (Suppl. 1), 199.

HALCK, G. E., PFANDER, W. H., and TRIBBLE, L. F. 1964. J. Anim. Sci. *23,* 879 (Abstr.).

HAMILTON, T. S., HAMILTON, B. C., JOHNSON, B. C., and MITCHELL, H. H. 1951. Cereal Chem. *28,* 163.

HAYS, V. W., ASHTON, G. C., LIV. C. H., SPEER, V. C., and CATRON, D. V. 1957. J. Anim. Sci. *16,* 44.

ILORI, J. O., and CONRAD, J. H. 1977. Nutr. Rep. Int. *15,* 19.

LLOYD, N. E., and MERTZ, E. T. 1958. Cereal Chem. *35,* 156.

MacGREGOR, J. M., TASKOWITCH, L. T., and MARTIN, W. F. 1961. Agron. J. *53,* 211.

MANER, J. H., POND, W. G., GALLO, J. T., HENAO, A., PORTELA, R. and LINARES, F. 1971. J. Anim. Sci. *33,* 791.

MERTZ, E. T. 1963. Corn proteins—A chemical and nutritional perspective. *In* Proc. 18th Annu. Hybrid Corn Industry-Research Conf., December 12 and 13, 1963, *Publ. 18,* pp. 7–12. Washington, D.C., American Seed Trade Assoc.

MERTZ, E. T., and BRESSANI, R. 1957. Cereal Chem. *34,* 63.

MERTZ, E. T., LLOYD, N. E., and BRESSANI, R. 1958. Cereal Chem. *35,* 146.

MILLER, G. M., BECKER, D. E., JENSEN, A. H., HARMON, B. G., and NORTON, H. W. 1969. J. Anim. Sci. *28,* 204.

MITCHELL, H. H., HAMILTON, T. S., and DEADLES, J. R. 1952. J. Nutr. *48,* 461.

MOSSÉ, J. 1966. Fed. Proc. *25,* 1663.

NATIONAL RESEARCH COUNCIL. 1968. Nutrient Requirements of Swine. Natl. Acad. Sci. *Publ. 1599.* Washington, D.C.

NATIONAL RESEARCH COUNCIL. 1979. Nutrient Requirements of Swine. Natl. Acad. Sci. *Publ. 2.* Washington, D.C.

NATIONAL RESEARCH COUNCIL. 1982. United States–Canadian Tables of Feed Composition, 3rd Revision, Natl. Res. Council–Natl. Acad. Sci. Natl. Academy Press, Washington, D.C.

OSBORNE, T. B., and MENDEL, L. B. 1914. J. Biol. Chem. *18,* 1.

POND, W. G., DUNN, J. A., WELLINGTON, G. H., STOUFFER, J. R., and VANVLECK, L. D. 1968A. J. Anim. Sci. *27,* 1583.

POND, W. G., WAGNER, W. C., DUNN, J. A., and WALKER, E. F., JR. 1968B. J. Nutr. *94*, 309.
POND, W. G., BARNES, R. H., and KROOK, L. 1969A. N.Y. Food Life Sci. *2*, 12.
POND W. G. STRACHAN, D. N. SINHA, Y. N., WALKER, E. F., JR., DUNN, J. A., and BARNES, R. H. 1969B, J. Nutr. *99*, 61.
RATHMAN, D. R. 1954. Zein, an Annotated Bibliography, 1891–1953. Mellon Institute, Pittsburgh, Pennsylvania.
RIPPEL, R. H., HARMON, B. G., JENSEN, A. H., NORTON, H. W., and BECKER, D. E. 1965. J. Anim. Sci. *24*, 373.
SAUBERLICH, H. E., CHANG, W. Y., and SALMON, W. D. 1953. J. Nutr. *51*, 241.
SCHNEIDER, E. O., EARLEY, E. B., and DETURK, E. E. 1952. Agron. J. *44*, 161.
TELLO, F., ALVAREZ-TOSTODO, M. A., and ALVARADO, G. 1965. Cereal Chem. *42*, 368.
USDA, 1981. Agricultural Statistics. 1981.

Opaque-2 Maize

BAKER, D. H., BECKER, D. E., JENSEN, A. H., and HARMON, B. G. 1970. J. Anim. Sci. *30*, 364.
BEESON, W. M., PICKETT, R. A., MERTZ, E. T., CROMWELL, G. L., and NELSON, O. E. 1966. *In* Proc. Distillers Feed Research Council. *21*, 70.
BRESSANI, R., and MERTZ, E. T. 1958. Cereal Chem. *35*, 227.
BRESSANI, R., ELIAS, L. G., and GOMEZ-BRENES, R. A. 1969. J. Nutr. *97*, 173.
CROMWELL, G. L., PICKETT, R. A., and BEESON, W. M. 1967. J. Anim. Sci. *26*, 1325.
EMERSON, R. A., BEADLE, G. W., and FRASER, A. C. 1935. Cornell Univ. Agric. Exp. Stn. Mem. *180*.
GALLO, J. T., MANER, J. H., and CORZO, M. 1968. J. Anim. Sci. *28*, 1152 (Abstr.).
GIPP, W. F., and CLINE, T. R. 1972. J. Anim. Sci. *34*, 963.
GOMEZ, G. G., MANER, J. H., FLORES, Z., and FRANCIS, C. A. 1975. J. Anim. Sci. *41*, 1638.
HESBY, J. H., CONRAD, J. H., PLUMLEE, M. P., and HARRINGTON, R. B. 1970A. J. Anim. Sci. *31*, 481.
HESBY, J. H., CONRAD, J. H., PLUMLEE, M. P., and MARTIN, T. G. 1970B. J. Anim. Sci. *31*, 56.
HOLDEN, P. J., LUCAS, E. W., SPEER, V. C., and HAYS, V. W. 1968. J. Anim. Sci. *27*, 1587.
LENKEIT, W. 1947 Zuchtangskunde *29*, 397.
LIVINGSTONE, R. M., MacPHERSON, R. M., and ELSLEY, F. W. H. 1966. Anim. Prod. *8*, 337.
LLOYD, N. E., and MERTZ, E. T. 1958. Cereal Chem. *35*, 156.
MAHAN, D. C., BECKER, D. E., and JENSEN, A. H. 1971A. J. Anim. Sci. *32*, 470.
MAHAN, D. C., BECKER, D. E., and JENSEN, A. H. 1971B. J. Anim. Sci. *32*, 476.
MAHAN, D. C., BECKER, D. E., HARMON, B. G., and JENSEN, A. H. 1971C. J. Anim. Sci. *32*, 482.
MANER, J. H., GALLO, J. T., OBANDO, H., and GRANT, J. 1969. Am. Soc. Agron. Detroit, Michigan (Abstr.).
MANER, J. H., POND, W. G., PORTELA, R., and MESA, J. 1971. III Reunión Latino-americana de Producción Animal, April 20–26, 1971, Bogotá, Colombia.
MANER, J. H., POND, W. G., GALLO, J. T., HENAO, A., PORTELA, R., and LINARES, F. 1971. J. Anim. Sci. *33*, 791.
MERTZ, E. T. 1963. *In* Proc. 18th Annu. Hybrid Corn Industry-Research Conf. *18*, 7–12.
MERTZ, E. T. 1968. Agric. Sci. Rev. 3rd Quarter, pp. 1–9.
MERTZ, E. T., BATES, L. S., and NELSON, O. E. 1964. Science *145*, 279.
MERTZ, E. T., VERNON, O. A., and NELSON, O. E. 1965A. Fed. Proc. *24*, 629 (Abstr.).
MERTZ, E. T., VERNON, O. A., BATES, L. S., and NELSON, O. E. 1965B. Science *148*, 1741.
MURPHY J. J., and DALBY, A. 1971. Cereal Chem. *48*, 336.
OCIO, E., and TREVIÑO, J. 1968. Actas de la VI Reunión-Cientiifica de la Sociedad Iberica de Nutrición Animal, Madrid, 1968.
OESTEMER, G. A., MEADE, R. J., STOCKLAND, W. L., and HANSON, L. E. 1970. J. Anim. Sci. *31*, 1133.
PICKETT, R. A. 1966. *In* Proc. High-Lysine Corn Conf., pp. 19–22. Corn Industry-Research Foundation, Inc., Washington, D.C.
POND, W. G., WAGNER, W. G., DUNN, J. A., and WALKER, E. F., JR. 1968. J. Nutr. *94*, 309.
POND, W. G., STRACHAN, D. N., SINHA, Y. N., WALKER, E. F., JR., DUNN, J. A., and BARNES, R. H. 1969. J. Nutr. *99*, 61.

POND, W. G., WEAVER, D., KIRTLAND, D., and WALKER, E. F., JR. 1976. Nutr. Rep. Int. *14,* 289.
RIVERA, P. H., PEO, E. R., JR., FLOWERDAY, D., CRENSHAW, T. D., MOSER, B. D., and CUNNINGHAM, P. J. 1978. J. Anim. Sci. *46,* 1024.
SANTOS, E., and MANER, J. H. 1971. III Reunión Latinoamericana de Producción Animal, April 26–30, 1971, Bogotá, Colombia.
SAUER, W. C., GIOVANETTI, P. M., and STOTHERS, S. C. 1974. Can. J. Anim. Sci. *54,* 97.
SCHELLING, G. T., LINDSEY, T. D., TUCKER, R. E., and MITCHELL, G. E., JR. 1979. J. Anim. Sci. *49* (Suppl. 1), 9.
VON ZUCHER, H., EBERSDOBLER, H. and GROPP, J. 1968. Tierernährung und Futtermittelkunde *23,* 193.
WAHLSTROM, R. C., MERRILL, R. V., REINER, L. J., and LIBAL, G. W. 1977. J. Anim. Sci. *46,* 747.

Floury-2 Maize

CROMWELL, G. L., ROGLER, J. C., FEATHERSTON, W. R., and CLINE, T. R. 1968. Poultry Sci. *47,* 840.
MANER, J. H., POND, W. G., GALLO, J. T., HENAO, A., PORTELA, R., and LINARES, F. 1971. J. Anim. Sci. *33,* 791.
NELSON, O. E. 1969. *In* New Approaches to Breeding for Improved Plant Protein. Int. Atomic Energy Agency Publ. STI/Pub./212, p. 41. Vienna, Austria.
NELSON, O. E., MERTZ, E. T., and BATES, L. S. 1965. Science *150,* 1469.
POND, W. G., MANER, J. H., and LINARES, F. A. 1971. J. Anim. Sci. *33,* 787.
SANTOS, E., and MANER, J. H. 1971. Reunión Latinoamericana de Producción Animal. April 26–30, 1971, Bogotá, Colombia.

Waxy Maize

BATES, L. F., FRENCH, D., and RUNDLE, R. E. 1943. Am. Chem. Soc. *65,* 142.
GALLENT, D., DERRIEN, A., AUMAITRE, A., and GUILBOT, A. 1973. Die Starke *25,* 56.
HANSON, L. E. 1946. J. Anim. Sci. *5,* 36.
JENSEN, A. H., BAKER, D. H., LYNCH, P. B., and HARMON, B. G. 1973. Coop. Ext. Ser. Univ. Illinois, AS665a.
KOCH, B. A., ALLEE, G. L. and HINES, R. H. 1974. Kans. Agric. Exp. Stn., Swine Industry Day Progress Rep. *220.*
McDONALD, T. A. 1973. Proc. 38th Annu. Corn Sorghum Res. Conf., December 4–6, 1973. Chicago, Illinois. 98 pp.
PEREZ, J. M., and AUMAITRE, A. 1979. Anim. Feed Sci. Tech. *4,* 109.
ROBBINS, S. L., LIBAL, W. G., and WAHLSTROM, R. C. 1976. J. Anim. Sci. *42,* 1353 (Abstr.).
ROSA, J. G., FORSYTH, D. M., GLOBER, D. V., and CLINE, T. R. J. Anim. Sci. 1977. *44,* 1004.
SACHTTEBEN, S. S., MILLER, E. R., and HENDERSON, H. F. 1975. J. Anim. Sci. *41,* 328 (Abstr.).
SANDSTEDT, R. M., STRAHAN, D., UEDA, S., and ALBOT, R. C. 1962. Cereal Chem. *39,* 123.
WAHLSTROM, R. C., and LIBAL, G. W. 1975. Coop. Ext. Serv. 19th Annu. Swine Day Proc. *SD-681.* South Dakota State Univ., Brookings.

Grain Sorghum

AUBEL, C. E. 1945. Kans. Exp. Stn. Cir. *258.*
AUBEL, C. E. 1955. Kans. Exp. Stn. Cir. *320.*
AUBEL, C. E. 1960A. Circ. Kans. Agric. Exp. Stn. *378,* 3.
AUBEL, C. E. 1960B. Circ. Kans. Agric. Exp. Stn. *378,* 5.
ALCANTARA, P. F., RIGOR, E. M., MILLER, J. C., and ARGANOSA, V. G. 1970. Philipp. Agric. *53,* 588.
ALMOND, M., SMITH, W. C., SAVAGE, G. P., and LAWRENCE, T. L. J. 1979. Anim. Prod. *29,* 143.
ANGELOV, I. 1965. Zhivotnovud. Nauki *2,* 909.
ARNOT, D. L., RICHARDSON, C. R., TRIBBLE, L. F., and ORR, D. E., JR. 1978. Agric. Sci. Tech. Rep. *T-5-138,* 126. Texas Univ. Center.
BAIRD, D. M. 1976. J. Anim. Sci. *42,* 256.
BATTERHAM, E. S., and MANSON, M. B. 1970. Aust. J. Exp. Agric. Anim. Husb. *10,* 539.

BEAMES, R. M. 1969. Aust. J. Exp. Agric. Anim. Husb. *9,* 127.
BEAMES, R.M., and SEWELL, J. O. 1969. Aust. J. Exp. Agric. and Anim. Husb. *9,* 482.
BEAMES, R. M., DANIELS, L. J., and SEWELL, J. O. 1973. Aust. J. Exp. Agric. Anim. Husb. *13,* 146.
BOKOROV, T., and SREĆKOVIC, A. 1964. Arhivza Polypriveedue Nauke *17,* 143.
BORCHERS, R. 1962. Cereal Chem. *39,* 145.
BURNS, R. E. 1971. Agron. J. *63,* 511.
CAM CHUONG, LY., MILLER, U., DEYOE, C. W., PFOST, H. B., and TRIBBLE, L. F. 1966 Kans. Agric. Exp. Stn. Bull. *493.*
CASTAING, J., AND LEUILLET, M. 1976. Journée de la Recherche Procine en France.
CASTAING, J., and MOAL, J. 1973. Ann. Zootech. *22,* 359.
COHEN, R. S., and TANKSLEY, T. D., JR. 1973A. J. Anim. Sci. *37,* 931.
COHEN, R. S., and TANKSLEY, T. D., JR. 1973B. J. Anim. Sci. *37,* 276.
COHEN, R. S., and TANKSLEY, T. D., JR. 1975. J. Anim. Sci. *41,* 309.
COHEN, R. S., and TANKSLEY, T. D., JR. 1976. J. Anim. Sci. *43,* 1028.
COHEN, R. S., and TANKSLEY, T. D., JR. 1977. J. Anim. Sci. *45,* 1079.
COPELIN, J. L., TANKSLEY, T. D., JR., and KNABE, D. A. 1974. J. Anim. Sci. *39,* 180.
COPELIN, J. L., SASSE, C. E., and TRIBBLE, L. F. 1976A. J. Anim. Sci. *42,* 245.
COPELIN, J. L., TRIBBLE, L. F., and GASKINS, C. T. 1976B. J. Anim. Sci. *43,* 251.
COPELIN, J. L., SASSE, C. E., GASKINS, C. T., and TRIBBLE, L. F. 1976C. J. Anim. Sci. *43,* 255.
COPELIN, J. L., GASKINS, C. T., and TRIBBLE, L. F. 1978. J. Anim. Sci. *46,* 133.
DIGGS, B. G. 1969. J. Anim. Sci. *29,* 133 (Abstr.).
EASTER, R. A. 1972. M.A. Thesis. Texas A. & M. Univ., College Station.
ECKERT, T. E., and ALLEE, G. L. 1974. J. Anim. Sci. *39,* 694.
EGGUM, B. O. 1973. A study of certain factors influencing protein utilization in rats and pigs. Ph.D. Thesis. 406 Beretn. Forsøgslab. Copenhagen, Denmark. 173 pp.
FAO 1980. Food and Agricultural Organization Production Yearbook.
FORD, J. E. 1977. Proc. Nutr. Soc. *36,* 124A.
FORD, J. E., and HEWITT, D. 1977. Proc. Nutr. Soc. *36,* 126A.
GONTIJO, V. de P. M., PEREIRA, J. A. A. COSTA, P. M. A., and de MELLO, H. V. 1976. Rev. Soc. Bras. Zootech. *5,* 83.
HALE, I., AND LYMAN, C. M. 1961. J. Anim. Sci. *20,* 734.
HAMILTON, C. R., ORR, D. E., JR., and TRIBBLE, L. F. 1978. Agric. Sci. Tech. Rep. *T-5-138,* 101. Texas Univ. Center.
HAN, I. K., and HA, J. K. 1977. Korean J. Anim. Sci. *19,* 180.
HARBERS, L. H. 1975. J. Anim. Sci. *41,* 1496.
HARRIS, H. B., CUMMINS, D. C., and BURNS, R. E. 1970. Agron. J. *62,* 633.
HILLIER, J. C., MacVICAR, R. W., and POND, W. 1953. Okla. Agric. Exp. Stn. Misc. Publ. *MP-31.*
HILLIER, J. C., and MARTIN, J. J. 1958. Okla. Agric. Exp. Stn. Misc. Publ. *MP-58.*
HILLIER, J. C., MARTIN, J. J., and Waller, G. R. 1959. Okla. Agric. Exp. Stn. Feeder's Day Rep. April 18, 1959, pp. 38–43.
HODSON, H. H., SNYDER, R., and KROENING, G. H. 1973. J. Anim. Sci. *37,* 283.
HULSE, J. H., LAING, E. M., and PEARSON, O. E. 1980. Sorghum and the Millets: Their Composition and Nutritive Value. Academic Press, New York.
JENSEN, A. H., BECKER, D. E., and TERRILL, S. W. 1960. J. Anim. Sci. *19,* 268 (Abstr.).
JENSEN, A. H., BECKER, D. E., and HARMON, B. G. 1965. J. Anim. Sci. *24,* 398.
KLETT, R. H. 1973. *In* 8th Bienn. Grain Sorghum Res. Util. Conf., pp. 8–14. Amarillo, Texas.
LAWRENCE, T. L. J. 1968. J. Agric. Sci. Camb. *70,* 287.
LUCE, W. G., OMTVEDT, I. T., and STEPHENS, D. G. 1970. Okla. Agric. Exp. Stn. Misc. Publ. *MP-84,* 60.
LUCE, W. G., OMTVEDT, I. T., and ROBBINS, B. S. 1972. J. Anim. Sci. *35,* 947.
MAY, M. A., and NELSON, T. S. 1973. J. Anim. Sci. *36,* 874.
Mc GINTY, D. D. 1969. Proc. 6th Bienn. Int. Grain Sorghum Res. Util. Conf., p. 20.
McLEOD, M. N. 1974. Nutr. Abstr. Rev. *44,* 803.
McMILLIAN, W. W., WISEMAN, B. R., BURNS, R. E., HARRIS, H. B., and GUENE, G. L. 1972. Agron. J. *64,* 821.
MEADOWS, D. G., TANKSLEY, T. D., JR., and HESBY, J. H., 1974. J. Anim. Sci. *39,* 185.
NOLAND, P. R. 1976. Proc. Arkansas Nutr. Conf., pp. 86–88.

NOLAND, P. R., JOHNSON, Z. B., SHARP, R. N., and CAMPBELL, D. R. 1976. Arkansas Farm Res. *25,* 11–16.
ORR, D. E., TRIBBLE, L. F., and LENNON, A. M. 1976. J. Anim. Sci. *43,* 257.
OSWALT, D. L., SHAFFERT, R. E., PICKETT, R. C., and AXTELL, J. D. 1972. Agron. Abstr. *59,* 70.
OWSLEY, W. F., KNABE, D. A., and TANKSLEY, T. D., JR. 1981. J. Anim. Sci. *52,* 557.
OYENUGA, V. A., and FETUGA, B. L. 1975. E. Afr. Agric. For. J. *40,* 388.
PEO, E. R., JR., and HUDMAN, D. B. 1958. J. Anim. Sci. *17,* 813.
POND, W. G., HILLIER, J. C., and MacVICAR, R. W. 1953. J. Anim. Sci. *12,* 928 (Abstr.).
PRICE, M. L., BULTER, G. L., FEATHERSTON, W. R., and ROGLER, J. C. 1978. Nutr. Rep. Int. *17,* 229.
PURSER, K. W., and TANKSLEY, T. D., JR. 1976. J. Anim. Sci. *43,* 257.
QUISENBERRY, J. H., and TANKSLEY, T. D., JR. 1970. *In* Sorghum Production and Utilization (J. S. Wall and W. A. Ross, Editors). AVI Publishing Co., Westport, Connecticut.
RACHIE, K. O. 1969. Cereal Sci. Today *14,* 271.
RANJHAN, S. K., JENSEN, A. H., COX, J. L., HARMON, B. G., and BECKER, D. E. 1964. J. Anim. Sci. *23,* 461.
REICHERT, R. D., FLEMING, S. E., and SCHRVOB, D. J. 1980. J. Agric. Food Chem. *28,* 824.
RIGGS, J. K., SORENSEN, J. W., JR. and HOBGOOD, P. 1970. Texas Agri. Exp. Stn. Bull. *1096.*
SAMFORD, R. A., RIGGS, J. K., ROONEY, L. W., POTTER, G. D., and COON, J. G. 1970. Proc. West Sec. Anim. Sci. *21,* 123.
SANDETEDT, R. M., STRAHAN, D., VEDA, S., and ABBOT, R. C. 1962. Cereal Chem. *39,* 123.
SAVAGE, G. P., SMITH, W. C., and BRIGGS, P. A. 1980. Anim. Prod. *30,* 157.
SCHOKE, L. M., RIGGS, J. K., and BULTER, O. D. 1972. J. Anim. Sci. *34,* 926.
SHADID, J. D., SASSE, C. E., KREIG, D. R., and TRIBBLE, L. F. 1976. J. Anim. Sci. *42,* 257.
SHARP, R. N., NOLAND, P. R., and CAMPBELL, D. R. 1977. Anim. Feed Serv. Tech. *2,* 327.
SHERROD, L. B., ALBIN, R. C., and FURR, R. D. 1969. J. Anim. Sci. *29,* 997.
SHERTZ, K. F., and STEPHENS, J. C. 1966. Texas Agric Exp. Stn. Tech. Monogr. *3.* Texas A. & M. Univ., College Station, Texas.
SHIAV, S. Y., YANG, S. P., TRIBBLE, L. F., LE-NON, A. M., and WILLIAMS, I. L. 1976. J. Anim. Sci. *43,* 258.
SMITH, K., and ALLEE, G. L. 1973. J. Anim. Sci. *37,* 291.
STAFIJCUK, A. A., and TELJATNIKOV, N. Y. 1968. Dokl. Vses. Akad. Skh. Nauk *2,* 8–10.
STEPHENSON, E. L., YORK, J. O., BRAGG, D. B., and IVY, C. A. 1971. Poult. Sci. *59,* 581.
TAKAHASHI, S., FURUYA, S., JITSUKAWA, Y., and MORIMOTO, H. 1968. Bull. Natl. Inst. Anim. Ind., Chiba *17,* 1–7.
TANKSLEY, T. D., JR. 1961. Feedstuffs *33*(20), 26–27.
TANKSLEY, T. D., JR., and OSBOURN, L. K. 1970. J. Animal Sci. *29,* 147 (Abstr.).
TANKSLEY, T. D., JR. 1976. *In* Optimizing the Utilization of Cereal Energy by Cattle and Pigs, pp. 61–75. U.S. Feed Grains Council, Washington, D.C.
TANKSLEY, T. D., JR. 1975. 9th Bienn. Grain Sorghum Res. Util. Conf., March 4–6, 1975, p. 106. Lubbock, Texas.
THRASHER, D. M., ICAZA, E. A., LADD, H. W., III, BAGLEY, C. P., and TIPTON, K. W. 1975. La. Agric. *19,* 10.
TIPTON, K. W., MABBAYAD, B. B., MARSHALL, J. G., RABB, J. L., and SLOANE, L. W. 1975. 9th Bienn. Grain Sorghum Res. Util. Conf. March 4–6, 1975, p. 113. Lubbock, Texas.
TONROY, B., PLUMLEE, M. P., CONRAD, J. H., and CLINE, T. R. 1973. J. Anim. Sci. *36,* 669.
TRIBBLE, L. F., LENNON, A. M., GASKINS, C. T., JR., CLAPP, K. L., and RAMSEY, C. B. 1972. J. Anim. Sci. *35,* 226.
TRIBBLE, L. F., INGRAM, S. H., GASKINS, C. T., and RAMSEY, C. B. 1979. J. Anim. Sci. *48,* 541.
TSCHIDERER, K. 1966. Bodenkultur. *17,* 187.
VANSCHOUBROEK, F., VAN SPOENDONK, R., and NAUWYNCK, W. 1964. Anim. Prod. *6,* 357.

Barley

ANGLEMEIR, A. F., and OLDFIELD, J. E. 1957. J. Anim. Sci. *16,* 922.
ARSCOTT, G. H., ROSE, R. J., and HARPER, J. A. 1960. Poul. Sci. *39,* 268.
AW-YOUNG, L. M., and BEAMES, R. M. 1975. Can. J. Anim. Sci. *55,* 765.
BHATTY, R. S., CHRISTISON, G. I., and ROSSNAGEL, B. G. 1979. Can. J. Anim. Sci. *59,* 585.

BOWLAND, J. P. 1974. J. Anim. Sci. *54*, 629.
BURNETT, G. S., and NEIL, E. L. 1964. Anim. Prod. *6*, 237.
CHUNG, A. S., and BEAMES, R. M. 1975. Can. J. Anim. Sci. *54*, 429.
COLE, D. J. A., WILLIAMS, I. H., ENGLISH, P. R., and LUSCOMBE, J. R. 1980. Anim. Prod. *31*, 213.
CRABO, B., BJÖRKLUND, N. E., and SIMONSSON, A. 1973. Acta Vet. Scand. *14*, 263.
DINUSSON, W. E., and BOLIN, D. W. 1958. N.D. Bimonthly Bull. *20*, 16.
DINUSSON, W. E., BOLIN, D. W., and BUCHANAN, M. L. 1955. N.D. Bimonthly Bull. *18*, 56.
DINUSSON, W. E., NYSTUEN, P. A., and BOLIN, D. W. 1956. J. Anim. Sci. *15*, 1256.
DINUSSON, W. E., NYSTUEN, P. A., BOLIN, D. W., and HAUGSE, C. N. 1960. N.D. Agr. Exp. Stn. Res. Rep. *2*.
FAO. 1980. Food and Agricultural Organization Production Yearbook, Vol. 34.
FRY, R. E., ALLRED, J. B., JENSEN, L. S., and MCGINNIS, J. 1958. Poul. Sci. *37*, 281.
FULLER, M. F. and LIVINGSTONE, R. M. 1979. Br. J. Nutr. *41*, 321.
FULLER, M. F. and MENNIE, I. 1979. Br. J. Nutr. *41*, 333.
FULLER, M. F., LIVINGSTONE, R. M., and MENNIE, I. 1974. Proc. Nutr. Soc. *33*, 91A.
GILL, D. R., OLDFIELD, J. E., and ENGLAND, D. C. 1966. J. Anim. Sci. *25*, 34.
GREER, S. A. N., HAYS, V. W., SPEER, V. C., McCALL, J. T., and HAMMOND, E. G. 1965. J. Anim. Sci. *24*, 1008.
HAUGSE, C. N., DINUSSON, W. E., ERICKSON, D. O., and BOLIN, W. E. 1966. N. D. Agric. Exp. Stn. Rep. *17*.
HEITMAN, H., JR. 1956. J. Anim. Sci. *15*, 1046.
HINTZ, H. F., and GARRETT, W. N. 1967. J. Anim. Sci. *26*, 746.
HOLLIS, G. R., and PALMER, A. Z. 1971. J. Anim. Sci. *32*, 381.
JENSEN, L. S., FRY, R. E., ALLRED, J. B., and McGINNIS, J. 1957. Poul. Sci. *36*, 919.
JOSEPH, W. E. 1924. Mont. Agric. Exp. Stn. Bull. *169*, 47.
LARSEN, L. M., and OLDFIELD, J. E. 1960. J. Anim. Sci. *19*, 601.
LAWRENCE, T. L. J. 1967. J. Agric. Sci. Camb. *69*, 271.
LAWRENCE, T. L. J., THOMLINSON, J. R., and WHITNEY, J. C. 1980. Anim. Proc. *31*, 93.
McBEATH, D. K., BENTLEY, C. E., LYNCH, D. L., and BOWLAND, J. P. 1960. Can. J. Anim. Sci. *40*, 57.
McCONNELL, J. C., SKELLY, G. C., HANDLIN, D. L., and JOHNSTON, W. E. 1975. J. Anim. Sci. *41*, 1021.
MITCHELL, K. G., BELL, J. B., and SOSULSKI, F. W. 1976. Can. J. Anim. Sci. *56*, 505.
MORGAN, E. T., JR., and McGINNIS, J. 1965. Poul. Sci. *44*, 1253.
NATIONAL RESEARCH COUNCIL. 1979. Nutrient Requirements of Swine. Natl. Acad. Sci. Publ. *2*. Washington, D.C.
NEWMAN, C. W., THOMAS, O. O., and ESLICK, R. F. 1968. J. Anim. Sci. *27*, 981.
NEWMAN, C. W., and ESLICK, R. F. 1970. Proc. West. Sec. Am. Soc. Anim. Sci. *21*, 111.
NEWMAN, C. W., ESLICK, R. F., PEPPER, J. W., and EL-NEGOUMY, A. M. 1980. Nutr. Rep. Int. *22*(6), 833.
NIELSEN, H. E., HAYS, V. W., SPEER, V. C., and CATRON, D. V. 1963. J. Anim. Sci. *22*, 1454.
OLDFIELD, J. E., and ANGLEMEIR, A. F. 1967. J. Anim. Sci. *16*, 917.
REIMER, D., and MEADE, R. J. 1964. J. Anim. Sci. *23*, 392.
SIMONSSON, A., and BJÖRKLUND, N. E. 1970. Proc. Soviet-Swedish Symp., Upsala, Sweden.
SOLDEVILA, M., and MEADE, R. J. 1964. J. Anim. Sci. *23*, 397.
THOMAS, O. O., and FLOWER, A. E. 1956. Mont. Agric. Exp. Stn. Cir. *214*.
TROELSEN, J. E., and BELL, J. M. 1962. Can. J. Anim. Sci. *42*, 63.

Oats

ANDERSON, D. M., BELL, J. M., and CHRISTISON, G. I. 1978. Can. J. Anim. Sci. *58*, 87.
CHRISTISON, G. I., and BELL, J. M. 1980. Can. J. Anim. Sci. *60*, 465.
FAO. 1978. Food and Agricultural Organization Production Yearbook.
JENSEN, A. H., BECKER, D. E., and TERRILL, S. W. 1959A. J. Anim. Sci. *18*, 701.
JENSEN, A. H., BECKER, D. E., and TERRILL, S. W. 1959B. J. Anim. Sci. *18*, 1356.
KATZ, R. S., BAKER, D. H., SASSE, C. E., JENSEN, A. H., and HARMON, B. G. 1973. J. Anim. Sci. *37*, 1165.

MEADE, R. J., DUKELOW, W. R., and GRANT, R. S. 1966. J. Anim. Sci. *25,* 58.
OWEN, J. B., and RIDGMAN, W. J. 1967. Anim. Prod. *9,* 107.
ROBBINS, G. S., POMERANZ, Y., and BRIGGLE, L. W. 1971. J. Agric. Food Chem. *19,* 536.
SALO, M. L. 1971. Soksjon. *5,* 113.
SALO, M. L., and ALAVIUHKOLA, T. 1980. Moatalovstieteelineu Aikakauskirja *52,* 435.
SIBBALD, I. R., and PRICE, K. 1977. Can. J. Anim. Sci. *57,* 365.
THOMKE, S. 1960. Kungl. Lantbrukshoegsk. Ann. *26,* 269.
WAHLSTROM, R. C., and LIBAL, G. W. 1975. J. Anim. Sci. *41,* 809.
WAHLSTROM, R. C., and LIBAL, G. W. 1979. J. Anim. Sci. *48,* 1374.
WAHLSTROM, R. C., REINER, L. J., and LIBAL, G. W. 1977. J. Anim. Sci. *45,* 948.

Rye

BOWLAND, J. P. 1966. Univ. Alberta 45th Annu. Feeder's Day Rep. *51,* 1. (Cited by Friend and MacIntyre 1969.)
DANIELSON, D. M., 1972. J. Anim. Sci. *35,* 1103 (Abstr.).
ESNOALA, M. A., ROSA, J. G., and ROMERO, O. 1976. Agric. Tec. *26,* 18.
FRIEND, D. W., and MacINTYRE, T. M. 1969. Can. J. Anim. Sci. *49,* 375.
FRIEND, D. W., and MacINTYRE, T. M. 1970. Can. J. of Comp. Med. *34,* 198.
HALE, O. M., JOHNSON, J. C. JR., and SOUTHWELL, B. L. 1967. Univ. Georgia Agric. Exp. Stn. Res. Bull. *14,* 17. September, 1967.
HÖLZSCHUL, W., and TRAUTMANN, H. 1962. Deutsch. Landwirtsch. *13,* 24. (Cited in Nutr. Abstr. Rev. *32,* 989.)
HONEYFIELD, D. C., FROSETH, J. A., and McGINNIS, J. 1977. Proc. West. Sect. Am. Soc. Anim. Sci. *28,* 91.
NORDSKOG, A. W., and CLARK, R. T. 1945. Am. J. Vet. Res. *6,* 107.
SAVAGE, G., SMITH, W. C., and PICKES, J. 1978. Anim. Prod. *26,* 398 (Abstr.).
WIERINGA, G. W. 1967. On the occurrence of growth inhibiting substances in rye. Thesis, Landboumhogesch., Wageningen, p. 68. (Cited in Nutr. Abstr. Rev. *38,* 91, No. 581.)

Wheat

ALLEE, G. L. 1977. Kansas State Univ. Swine Day Rep. 1977, p. 11.
BEAMES, R. M. 1963. Queensl. J. Agric. Sci. *20,* 231.
BECKER, D. E., JENSEN, A. J., and HARMON, B. G. 1963. Ill. Ext. Cir. *866.*
BRAUDE, R., TOWNSEND, J., HARRINGTON, G., and ROWELL, J. G. 1961. J. Agric. Sci. Cambridge *57,* 257.
CARROLL, W. E., KRIDER, J. L., and ANDREWS, N. 1962. Swine Production. McGraw-Hill, New York.
CHEEKE, P. R., DAVIDSON, T. P., and HANSELL, J. 1980. Swine Day–1980. Oregon State Univ., Dec. 1980. Special Rep. pp. 8–14.
COLE, D. J. A., CLENT, E. G., and LUSCOMBE, J. R. 1969. Anim. Prod. *11,* 325.
CRAMPTON, E. W. 1962. *In* Nutrition of Pigs and Poultry. (J. T. Morgan and D. Lewis, Editors), pp. 316–340. Butterworths, London.
CUNHA, J. T. 1957. Swine Feeding and Nutrition. Interscience Publishers, New York.
DANIELSON, D. M., and GRABOWSKI, P. H. 1970. Nebr. Agric. Exp. Stn. Rep. E. C. *70-219,* 8.
DIGGS, B. C., BECKER, D. E., TERRILL, S. W., and JENSEN, A. H. 1959. J. Animal Sci. *18,* 1492.
ERICKSON, J. P., MILLER, E. R., HILL, G. M., BLOCK, J. R., BEHIAK, D. M., and KU, P. K. 1980. J. Anim. Sci. *51,* 1065.
EVANS, R. E. 196O. Bull. Minn. Agric., Fish Fed. *48.*
GILLS, D. R., OLDFIELD, J. E., and ENGLAND, D. C. 1966. J. Animal Sci. *25,* 34.
HALE, O. M. 1979. Feedstuffs *51,* 50–51.
HALE, O. M., and MOREY, D. D. 1979. Ga. Res. Rept. *311,* April 1979, 14pp.
HALE, O. M., MOREY, D. D., and McCORMICK, M. C. 1979. Ga. Res. Rep. *311,* 1–14.
HOLLIS, G. R., and PALMER, A. Z. 1971. J. Anim. Sci. *32,* 381 (Abstr.).
IVAN, M., and FARRELL, D. J. 1975. Anim. Prod. *20,* 77.
IVAN, M., and FARRELL, D. J. 1976. Anim. Prod. *23,* 111.
IVAN, M., GILES, L. R., ALIMOW, A. R., and FARRELL, D. J. 1974. Anim. Prod. *19,* 359.
IVAN, M., FARRELL, D. J., and EDEY, T. N. 1975A. Anim. Prod. *20,* 267.

IVAN, M., FARRELL, D. J., and EDEY, T. N. 1975B. Anim. Prod. *20,* 277.
JENSEN, A. H., BAKER, D. H., BECKER, D. E., and HARMON, B. G. 1969. J. Anim. Sci. *29,* 16.
JOHNSON, V. A., MATTERN, P. J., and SCHMIDT, J. W. 1970. Proc. Nutr. Soc. *29,* 20.
JUST, A., JØRGENSEN, H., FERNÁNDEZ, J. A., AND JEPSEN, H. M. 1983. Berething fra Statens Husdyrbrugs-Forsøg. 546, Copenhagen, Denmark. pp. 1–80.
LAWRENCE, T. L. J. 1967. J. Agric. Sci. Camb. *69,* 271.
LENNON, A. M., CLAWSON, A. J., and ALSMEYER, W. L. 1972. J. Anim. Sci. *35,* 778.
LUCE, W. G., OMTVEDT, I. T., and ROBBIN, B. S. 1972. J. Anim. Sci. *35,* 947.
ROBINSON, D. W., PRESCOTT, J. H. D., and LEWIS, D. 1965. J. Agric. Sci. Cambridge *64,* 59.
WU, J. F., and EWAN, R. C. 1979. J. Anim. Sci. *49,* 1470.

Triticale

ALLEE, G. L., and HINES, R. H. 1972. J. Anim. Sci. *35,* 1101a (Abstr.).
BOWLAND, J. P. 1968. *In* 47th Annu. Feeders' Day Rep., Dep. Anim. Sci. Univ. Alberta, Edmonton, Alberta, Canada.
BRAVO, F. O., MARTÍNEZ, L., and SHIMADA, A. 1971. J. Anim. Sci. *33,* 227 (Abstr.).
CORNEJO, S., POTOCNJAK, J., HOLMES, J. H. G., and ROBINSON, D. W. 1973. J. Anim. Sci. *36,* 87.
ERICKSON, J. P., MILLER, E. R., BERGEN, W. G., and ELLIOTT, F. C. 1978. J. Anim. Sci. *46,* 417.
ERICKSON, J. P., MILLER, E. R., ELLIOTT, F. C., KU, R. K., and MILLER, D. E. 1979. J. Anim. Sci. *48,* 547.
HARROLD, R. L., DINUSSON, W. E., HAUGSE, C. N., and BUCHANNAN, M. L. 1971. N. D. Agric. Exp. Stn. Farm Res. *28,* 34.
LARTER, E. N. 1974. *In* Triticale: First Man-made Cereal. (C. C. Tsen, Editor), pp. 35–52. Am. Assoc. Cereal Chem. St. Paul, Minnesota.
LEIGHTY, C. E. 1915. J. Am. Soc. Agron. *7,* 209.
LONGERNECKER, T. C., LANGFORD, L., and STOGGS, D. 1969. High Plains Res. Found. Rep. *331.* Plainsview, Texas.
MARTYNOV, V., and SHTAN'KO, I. 1979. Svuinovodstvo *1,* 14.
NISHIMUTA, J. F., SUNKI, G. R., and RAO, D. R. 1980. Anim. Prod. *31,* 177.
NOLAND, P. R., SHARP, R. N., and McGHEE, R. W. 1972. Arkansas Farm Res. *21,* 7.
SHIMADA, A. S., and CLINE, T. R. 1974. J. Anim. Sci. *38,* 941.
SHIMADA, A. S., MARTÍNEZ R., L., and BRAVO, F. O. 1971. J. Anim. Sci. *33,* 1266.
STOTHERS, S. C. 1970. Res. Bull. A. S. *70-1,* p. 23. Dep. of Anim. Sci. Univ. of Manitoba, Winnipeg.
STOTHERS, S. C., and SHEBESKI, L. H. 1965. 14th Rep. of Livestock Res., p. 19. Dep. of Anim. Sci., Univ. of Manitoba, Winnipeg.
VILLEGAS, E. 1973. Proc. Wheat, Triticale and Barley Seminar.: (R. G. Anderson, Editor). CIMMYT, January 22–26, 1973. El Batan, Mexico.
WILSON, R. F. 1972. Ohio Agric. Res. Dev. Center Swine Res. Rep. *61-5,* 25. Wooster, Ohio.

Rice

ARNOTT, G. W., and LIM, H. K. 1966. Malaysian Agric. J. *45,* 387.
BIENVENIDO, O., BANTISTA, G. M., LUGAY, J. C., and REYS, A. C. 1964. J. Agric. Food Chem. *12,* 131.
BRAY, C. I. 1943. La. Agric. Exp. Stn. Circ. *33.*
FAO. 1968. Food and Agricultural Organization Production Yearbook.
HOUSTON, D. F., ALLIS, M. E., and KOHLER, G. O. 1969. Cereal Chem. *46,* 527.
KIK, M. C. 1956. J. Agric. Food Chem. *4,* 170.
MANER, J. H. 1971. Unpublished data. CIAT–ICA, Cali, Columbia.
MORRISON, F. B. 1956. Feeds and Feeding, 22nd Edition. Morrison, Ithaca, New York.
NOLAND, P. R. 1960. Arkansas Farm. Res. *9,* 30–38.
NOLAND, P. R., and Scott, K. W. 1963. Univ. Arkansas Agric. Exp. Stn. Bull. *668,* April.

Buckwheat

ANDERSON, D. M., and BOWLAND, J. P. 1979. 58th Annu. Feeders' Day Rep., p. 12. Dep. of Anim. Sci., Univ. Alberta, Edmonton.
FARRELL, D. J. 1976. Proc. Aust. Soc. Anim. Prod. *11,* 413.

MORRISON, F. B. 1956. Feeds and Feeding. 22nd Edition, Appendix Table I. Morrison, Ithaca, New York.
POMERANZ, Y., and ROBBINS, G. S. 1972. J. Agric. Food Chem. *20,* 270.
SURE, B. 1955. J. Agric. Food Chem. *3,* 793.

Wheat By-products

BATTERHAM, E. S., LEWIS, C. E., LOWE, R. F., and McMILLAN, C. J. 1980. Anim. Prod. *31,* 259.
BEAMES, R. M., and NATOLI, W. J. 1969. Aust. J. Exp. Agric. Anim. Husb. *9,* 594.
DIGGS, B. G., BECKER, D. E., JENSEN, A. H., and NORTON, H. W. 1965. J. Anim. Sci. *24,* 555.
ERICKSON, J. P. 1979. Mich. Res. Rep. *386,* 71–75. September, 1979.
FORSYTHE, W. A. 1979. Mich. Res. Rep. *386,* 71. September, 1979.
HANRAHAN, T. J., and O'Grady, J. F. 1970. Irish J. Agric. Res. *9,* 335.
HENRY, Y. M. 1976. Proc. 1st Int. Symp. Feed Compsition, Anim. Nutr. Req. Computerization Diets. pp. 270–281. Utah State Univ., Logan.
JUST, A., JØRGENSEN, H., and FERNÁNDEZ, J. 1978. Natl. Inst. Anim. Sci. Copenhagen, Denmark.
LE DIVIDICH, J., CANOPE, L., HEDRIVILLE, F., and DESPOIS, E. 1975. Nouvelles Agron. Antilles et Guyana. *1*(4), 284.
MORGAN, D. J., COLE, D. J. A., AND LEWIS, D. 1975. J. Agric. Sci. Cambridge *84,* 7.
NEHRING, K., KLIPPEL, W., UHLEMANN, H., and HOFFMANN, B. 1970. Archiv Tierernährung *20,* 69.
PATIENCE, J. F. 1977A. J. Anim. Sci. *45,* 1294.
PATIENCE, J. F. 1977B. J. Anim. Sci. *45,* 1302.
PALS, D. A., and EWAN, R. C. 1978. J. Anim. Sci. *46,* 402.
RAHJHAN, S. K., GUPTA, B. S., CHABBRA, S. S., and DHUDAPKER, B. S. 1971. Indian J. Anim. Sci. *41,* 373.
RAHJHAN, S. K., SHUKLO, V. P., KUMAR, I., PATHAK, N. N., and JOSKI, B. C. 1972. Indian J. Anim. Sci. *42,* 453.
SAUNDERS, R. M., WALKER, M. A., JR. and KOHLER, G. O. 1969. J. Poult. Sci. *48,* 1497.
SAUNDERS, R. M., CONNOR, M. A., KOHLER, G. O., and BALYLOCK, L. B. 1974. J. Anim. Sci. *38,* 1272.
SHARDA, D. P., YADAR, K. R., and PRADHAM, K. 1979. Indian J. Anim. Health. *18,* 13.
YOUNG, L. G. 1980. J. Anim. Sci. *51,* 1113.
YOUNG, L. G., and KING, G. L. 1981. J. Anim. Sci. *52,* 551.

Rice Bran and Polishings

ALVARADO, F., and MANER, J. H. 1975. CIAT Annu. Rep. 1975.
ARA, L., OWEN, A. A., BUITRAGO, J. and PINEDA, J. 1975. Revista ICA *10,* 127.
ARNOTT, G. W., and LIM., H. K. 1966. Malaysian Agric. J. *45,* 387.
BALDERRAMA, J. S., EUSEBIO, J. A., GONZALEZ, R. R., and ALCANTARA, P. F. 1968. Philipp. Agriculturist *52,* 146.
BISTOYANG, A. G., EUSEBIO, J. A., ALCANTARA, P. F., and GONZALEZ, R. R. 1968. Philipp. Agriculturist *52,* 287.
BROOKS, C. C., and LUMANTA, J. G. 1975. J. Anim. Sci. *41,* 308 (Abstr.) and Proc. West. Sect. Am. Soc. Anim. Sci. *26,* 115, *26,* 112.
CAMPABADAL, C., CRESWELL, D., WALLACE, H. D., and COMBS, G. E. 1976. Trop. Agric. *53,* 141.
CHAMORRO, J. T. 1978. Centro de Estudios Superiores en Nutrición y Ciencias de Alimentos, *117,* 35. Guatemala.
CHAMORRO, J. T. and MANER, J. H. 1973. CIAT Annu. Rep. 1973.
CHAMORRO, J. T. and MANER, J. H. 1974. CIAT Annu. Rep. 1974.
FAO. 1980. Food and Agricultural Organization Production Yearbook.
GOMEZ, G. A., ALVARADO, F., CHAMORRO, J. T., and MANER, J. H. 1978. CIAT, Ser. ES-29, 26 pp.
KOPEIHOVSKII, V. M., ARUTYUNVAN, N. S., and PROSKURINA, V. L. 1971. Biol. Abstr. *54,* 2463.
LEIGHTON, R. E. 1958. Feed Age, September, p. 58.

LYNN, L. 1969. *In* Protein-enriched Cereal Foods for World Needs. (Max Milner, Editor. Amer. Assoc. Cereal Chem., pp. 154–171. St. Paul, Minnesota.
MARTÍNEZ, R. L., and BRAVO, F. O. 1971. Tecnica Pecuaria *15*, 9.
MAUST, L. E., POND, W. G., and SCOTT, M. L. 1972. J. Anim. Sci. *35*, 963.
MONCADA, A., and MANER, J. H. 1965–1966. Instituto Colombiano Agropecuario. Día de campo, Bogotá, Colombia.
NOLAND, P. R., and SCOTT, K. W. 1963. Univ. Arkansas Agric. Exp. Stn. Bull. *668*, April, 1963.
ROBLES, A., and EWAN, R. C. 1976. J. Anim. Sci. *42*, 1354 (Abstr.).
SHIMAMOTO, S. 1971. Japanese patent 2:470.
THRASHER, D. M., MULLINS, A. M., and SCOTT, V. B. 1966. J. Anim. Sci. *25*, 258 (Abstr.).
TILLMAN, A. I., KIDWELL, J. F., and SINGLETARY, C. B. 1951. J. Anim. Sci. *10*, 837.
VIRAKTAMATH, C. S. 1974. J. Food Sci. Tech. India *11*, 191.

Distillery By-products

BARNHART, C. E. 1952. Proc. 7th Distillers' Feed Conf., p. 25.
BECKER, D. E., JENSEN, A. H., and TERRILL, S. W. 1960. Univ. Ill. Anim. Sci. Mimeo *AS-534*.
CARPENTER, L. E. 1970. Proc. 25th Distillers' Feed Conf., p. 54.
CATRON, D. V., DIAZ, F., SPEER, V. C., and ASHTON, G. C. 1954. Proc. 9th Distillers' Feed Conf., p. 49.
CONRAD, J. H., PLUMLEE, M. P., and BEESON, W. M. 1959. J. Anim. Sci. *8*, 1503.
EWAN, R. C. 1976. Proc. 31st Distillers' Feed Conf., p. 16.
FAIRBANKS, B. W., KRIDER, J. L., and CARROLL, W. E. 1944. J. Anim. Sci. *3*, 29.
FAIRBANKS, B. W., KRIDER, J. L., and CARROLL, W. E. 1945. J. Anim. Sci. *4*, 420.
HARMON, B. G. 1974. Distillers' Feed Res. Council *29*, 35.
HARMON, B. G. 1975. Distillers' Feed Res. Council, p. 23. Louisville, Kentucky.
PEERS, D. G., HILLYER, G. M., and MacANDREWS, A. 1978. Anim. Prod. *26*, 396 (Abstr.).
SALO, M. L. 1978. J. Sci. Agric. Soc. Finl. *50*, 291.
SPEER, V. C., BECKER, D. E., CONRAD, J. H., HOEFER, J. H., MEADE, R. J., PEO, E. R., JR., SEERLEY, R. W., and TRIBBLE, L. F. 1967. J. Anim. Sci. *26*, 1477.
THONG, L. Q., JENSEN, A. H., HARMON, B. G., and CORNELIUS, S. G. 1978. J. Anim. Sci. *46*, 674.
WALLACE, H. D., and COMBS, G. E. 1965. Univ. Fla. Anim. Sci. Mimeo *AN 66-1*.
WAHLSTROM, R. C., and LIBAL, G. W. 1980. S.D. Swine Day Res. Rep. November 20, 1980. S.D. State Univ., Brookings.
WAHLSTROM, R. C., GERMAN, C. S., and LIBAL, G. W. 1970. J. Anim. Sci. *3O*, 532.

Brewery By-products

BABATUNDE, G. M., FETUGA, B. L., OYENUGA, V. A., and AYOADE, O. A. 1975. Nigerian J. Anim. Prod. *2*, 119.
BAIRD, D. M., ALLISON, J. R., and YOUNG, C. T. 1975. J. Anim. Sci. *41*, 307 (Abstr.).
BROOKS, J. W. 1971. Proc. 26th Annu. Texas Nutr. Conf., p. 23.
CROMWELL, G. L., MENEGUE, J., and STAHLY, T. S. 1979. J. Anim. Sci. *41*, 307 (Abstr.).
HARMON, B. G., CORNELIUS, S. G., BAKER, D. H., and JENSEN, A. H. 1975. J. Anim. Sci. *41*, 315 (Abstr.).
KORNEGAY, E. T. 1973. J. Anim. Sci. *37*, 479.
WAHLSTROM, R. C., and LIBAL, G. W. 1976. J. Anim. Sci. *42*, 871.
WERNER, A. and SCHELLNER, G. 1960–1961. Jahr. Arbeitsgem. Fuetterungsberat. *3*, 27. (Cited in Nutr. Abstr. and Rev. *32*, 1388.)
YOUNG, L. G., and INGRAM, R. H. 1968. Can. J. Anim. Sci. *48*, 83.

13

Roots, Tubers, Fats, and Other Energy Sources

INTRODUCTION

The list of potential energy sources for swine is enormous; only those sources that are the most widely available (at reasonable prices) in temperate and tropical environments and those that are the most widely used have been chosen for discussion. Each generally is of importance in regions of production and most present specific problems that affect their utilization. Those sources with tremendous potential usefulness in the tropics, but whose nutritive value has not been described thoroughly heretofore, are given extensive treatment.

One of the constraints on the utilization of roots and tubers for human and animal consumption is the difficulty of processing and storage. Current knowledge on this subject has been reviewed (Plucknett 1979) and should be of special interest to those associated with developing pork production technology in locations where root crops and tubers are an appropriate potential energy source.

CASSAVA

Swine require large quantities of carbohydrate (energy) and moderate quantities of protein for growth and development. In the leading pork-producing areas, large supplies of grains available at a reasonable price provide this energy source. In other countries, production of cereal grains is insufficient for human consumption needs. Many of these countries, however, have the potential or are already producing large quantities of other feed sources that, if properly fed and supplemented, could support a large and efficient swine industry. One feed source with a great unrealized potential in many tropical areas is *Manihot esculenta* Crantz, more commonly known as cassava, yuca, manioc, tapioca, or mandioca.

Cassava is grown in the lowland tropics throughout the world, generally between latitudes 23° N and 23° S and almost always below 1800 meters altitude within these regions (Rogers 1963). These areas generally delineate regions of production because of the frost-sensitivity nature of the plant. Cassava has a wide range of adaptability, resistance to drought, tolerance of poor soils, and relative ease of cultivation. As there are varieties adapted to nearly every soil type and degree of annual rainfall, the plant offers an almost unlimited production potential. Cassava plants and roots are shown in Fig. 13.1.

The annual production of cassava for 1980 was reported to be 122,134,000 metric tons (FAO 1980). However these data may not be accurate, because cassava is not a "cash" or "plantation" crop. Most of the crop is produced on small farms in small acreages and many times as a "backyard" staple for home consumption. Of the total world production, approximately 38% or 46.8 million metric tons are produced in Africa, 36% in Asia, 25% in Latin America, and 20% in Brazil. A world collection of cassaca cultivars is maintained at the Center for International Agriculture in the Tropics (CIAT), Cali, Colombia, and an intensive research program on cassava production is underway there (CIAT 1983).

Although cassava is ninth in terms of world production of all crops and fifth in world production of tropical crops, its importance and potential as a food and feed crop in the tropical regions throughout the world has not been realized. Cassava,

a b c

FIG. 13.1. Cassava as a potential energy source for swine in the Tropics.

other root crops, and plantains are major sources of carbohydrates for humans in Latin America and Africa. The usefulness and potential of cassava as an animal feed should be given greater consideration.

Average yields of fresh cassava per hectare are generally low. The world average is about 8.3 metric tons per hectare; however, varieties that will produce 50–60 tons ha/year are available (CIAT, 1975). This would yield approximately 92 million kcal of digestible energy, an amount which is many times greater than that which could be produced by rice, corn, or other grains adapted to the tropics.

The cassava root contains an average of 65% moisture and 35% dry matter (Table 13.1). The pulp, or internal portion of the root, contains more dry matter than the peel (37.8 vs. 27.8%) and represents approximately 86.8% of the total root, with the remainder (13.2%) being made up by the peeling. The percentage composition of the root does not vary considerably with different stages of maturity.

Basically, the cassava root provides carbohydrates for the human and animal diet. An average of 30.84% of the fresh root is nitrogen-free extract, which is composed principally of starch and sugar. The nitrogen-free extract fraction of the whole root contains approximately 80% starch and 20% sugars (Vogt 1966). Variable quantities of sucrose are present and may constitute as much as 5% of the dry matter in some varieties (Brautlecht 1953). The basic unit of the starch molecule is D-glucose and this is the only monosaccharide obtained following complete hydrolysis. The two principal elements in cassava starch are amylose, which has a straight-chain structure, and amylopectin, which has a branched-chain structure. According to Kerr (1950) and Johnson and Raymond (1965), the starch of cassava contains approximately 20% amylose and 70% amylopectin.

TABLE 13.1. Proximate Analysis of 15 Colombian Cassava Varieties

Variety	*Moisture* (%)	*Protein* (%)	*Fiber* (%)	*Fat* (%)	*Ash* (%)	*Nitrogen-free extract* (%)
'Llanera'	67.90	2.33	0.97	0.18	0.95	27.67
'Santa Catalina'	64.76	2.14	1.16	0.24	1.00	30.71
'H-50'	66.71	0.56	2.03	0.35	1.71	28.58
'Tolima'	61.85	0.40	1.35	0.25	1.54	34.61
'C.M.C.-50'	60.61	1.55	1.09	0.36	1.40	35.07
'I-35 Brava'	61.21	2.06	1.18	0.31	1.03	34.14
'Blanca No. 2'	62.70	1.25	1.02	0.29	1.36	33.36
'C.M.C.-1'	62.18	1.97	2.16	0.30	2.49	30.89
'C.M.C.-3'	61.50	1.70	1.77	0.24	1.38	33.34
'C.M.C.-4'	67.59	1.71	3.46	0.35	1.59	25.27
'Seis Meses Común'	65.35	0.67	1.00	0.43	1.29	31.35
'Amarilla'	64.16	0.59	1.64	0.33	1.81	35.90
'Tempranita'	77.32	0.63	1.07	0.24	1.58	19.17
'La Respetada'	62.84	1.03	1.02	0.22	1.36	33.49
'Bartolita'	68.81	0.18	0.77	0.25	0.94	29.05
Average	65.03	1.25	1.45	0.29	1.43	30.84

Source: Nutrition Laboratory, Instituto Colombiano Agricultura (unpublished data).

TABLE 13.2. Variation of Crude Protein Contained in 87 Varieties of Cassava[a]

Protein level, % ($N \times 6.25$)	*Number of samples*
0.00 to 1.00	4
1.00 to 2.00	39
2.00 to 3.00	29
3.00 to 4.00	10
4.00 to 5.00	3
5.00 to 6.00	1
6.00 to 6.40	1
Average: 2.3	Total: 87

Source: Maner 1971.
[a]Samples are dried cassava with approximately 10% moisture.

Cassava contains only small quantities of fiber (1.45%) and only a trace of ether extract (0.20%). The whole root contains only 1.5% ash. The levels of calcium (0.12%), phosphorus (0.16%), sodium (0.06%), and magnesium (0.37%) are generally low. However, as is common in other root crops, the level of potassium is relatively high (0.86%).

The majority of the cassava varieties available contain only small quantities of crude protein (nitrogen × 6.25) (see Table 13.2). In fact, the content of the majority of the varieties analyzed does not exceed 3% when expressed on an air-dry basis (approximately 10% moisture). However, several varieties that contain much higher levels of crude protein have been identified. In Colombia, Jaramillo and Herrera (1970) reported a sample of *Manihot carthagenesis* that contained 15.4% crude protein when expressed on a moisture-free basis. Maner (1971) has studied a variety, 'Llanera,' that contains 2.33% crude protein in the fresh root or 7.25% when expressed on a dry matter or moisture-free basis (Table 13.2).

Additional unpublished studies (J. H. Maner) clearly indicate that the total nitrogen fraction of the cassava root is not present as protein. Approximately 40–60% of the total nitrogen is in the form of nonprotein nitrogen (NPN) and the percentage of total nigrogen and NPN is greater in the peel than in the pulp (Table 13.3). These data are not in agreement with those reported by Oyenuga (1955) who reported that 62% of the nitrogen and 87% of that in the peel was true protein.

A detailed chemical analysis of the nitrogen fraction indicates that nitrates, nitrites, and hydrocyanic acid represent only about 1.0% of the total nitrogen. According to these studies it appears that both aspartic and glutamic acids are present in large quantities in the free form and may, along with the glucosides, linamarin and lotaustralin, be a major portion of the NPN that does not precipitate with trichloroacetic acid (Maner and co-workers, unpublished).

The amino acid contents of four samples of cassava are presented in Table 13.4. The levels of various amino acids such as lysine and tryptophan are promising and constitute more of the protein than those of opaque-2 maize. However, the levels of both methionine and cystine are extremely deficient and appear to be

TABLE 13.3. Effect of Stage of Maturity of the Plant on the Total and Nonprotein-Nitrogen (NPN) Content of Cassava Roots of the 'Llanera' Variety[a]

Date of harvest	Whole root			Pulp			Peel		
	Total N (%)	Crude protein (N × 6.25) (%)	NPN (%)	Total N (%)	Crude protein (N × 6.25) (%)	NPN (%)	Total N (%)	Crude protein (N × 6.25) (%)	NPN (%)
1970									
January 10	0.89	5.58	0.70	0.89	5.50	0.13	1.48	9.25	0.86
February 23	0.87	5.50	0.60	0.50	3.12	0.25	1.30	8.20	0.50
March 10	0.91	5.68	0.52	0.73	4.55	0.43	1.14	7.10	0.46
March 30	0.97	5.43	0.45	0.73	4.58	0.34	0.87	5.43	0.43
April 17	1.02	6.35	0.63	0.82	5.15	0.39	1.92	12.00	1.14
May 4	0.93	5.81	0.65	0.85	5.31	0.48	2.13	13.30	—
June 3	0.74	4.63	0.31	0.56	3.50	0.26	—	—	—

Source: J. H. Maner, unpublished data.
[a]Samples contain approximately 10% moisture.

TABLE 13.4. Amino Acid Analysis of Four Cassava Samples (% Protein)

Amino acid	*1*[a]	*2*[a]	*3*[b]	*4*[c]
Arginine	17.10	12.90	3.7	7.7
Histidine	0.60	0.53	1.2	1.5
Isoleucine	0.77	1.04	2.0	5.3
Leucine	1.24	1.52	2.9	5.6
Lysine	1.54	1.56	3.5	6.2
Methionine	—	0.33	1.0	0.6
Cystine	0.51	—	0.6	—
Threonine	0.86	1.00	2.1	3.8
Phenylalanine	0.76	0.94	2.3	3.5
Valine	1.23	1.32	2.6	4.5
Tryptophan	0.50	0.50	0.5	0.5
Aspartic acid	—	—	4.4	—
Glutamic acid	—	—	12.7	—
Proline	—	—	1.6	—
Ornithine	—	—	10.4	—
Serine	—	—	1.9	—
Glycine	—	—	2.4	—
Alanine	—	—	4.6	—
Tyrosine	—	—	1.6	—

[a]Maner unpublished data.
[b]Close et al. 1953.
[c]Sreeramamurthy 1945.

first-limiting. Close (1953) and Sreeramamurthy (1945) have also reported amino acid analysis of cassava (see Table 13.4).

The presence of a toxic factor in cassava presents some problems in its utilization as a livestock feed. The toxicity of the roots and the leaves of cassava is due to the presence of hydrocyanic acid (HCN) or prussic acid (Oke 1969). Free HCN does not exist as such in the healthy, intact plant but is formed when normal plant growth is retarded or when the leaves and roots are cut or bruised. The HCN is generally formed by the chemical reaction of two substances, a glucoside and an enzyme that is contained in the plant. Neither the glucoside nor the enzyme is poisonous when considered separately (Couch 1932). The two glucosides found in the cassava plant are linamarin and lotaustralin. These glucosides consist of a chemical combination of a sugar with hydrocyanic acid and possibly with another compound, such as a ketone or an aldehyde. Hydrocyanic acid, glucose, and acetone are produced when linamarin is hydrolyzed, and only when the HCN is chemically liberated is it toxic.

The degree of toxicity of cassava roots has been amply discussed, and many differences of opinion exist. A detailed review of the role of HCN in nutrition has been prepared by Oke (1969).

In animals, the symptoms of acute HCN toxicity are increased depth and speed of respiration, accelerated pulse, failure to react to stimuli, and spasmotic muscular movements (Oke 1969). The majority of the symptoms of poisoning can be explained on the basis of the affinity of HCN with the metallic ions such as copper and iron. The cyanide combines with the hemoglobin to form cyanohemoglobin, which is not a carrier of oxygen. The HCN also forms a reversible combination with the copper of cytochrome oxidase, which inhibits its function as an oxidative enzyme in the transfer of electrons and is a classic example of histotoxic anoxia.

These chemical abnormalities cause a neuronal depression in the medullar centers that causes resporatory problems and death. Therefore, HCN is to be considered as a violent protoplasmic poison for all forms of life.

Although acute HCN toxicity is a problem, the continued ingestion of small quantities of hydrocyanic acid present a very serious toxiological, physiological, and nutritional problem. These small quantities are not sufficient to cause death, but they affect the general health and physical aspects of the individual.

It seems that many of the differences in degree of HCN toxicity can be explained by the presence or absence of substances such as methionine, cystine, sulfur, vitamin B_{12}, iodine, and other elements, such as copper and iron. At the same time, certain symptoms indicating a deficiency of methionine, vitamin B_{12}, and iodine exist even in the presence of what seem to be adequate levels of these nutrients in the diet, and these can be explained by the presence of sublethal levels of HCN that reduce the efficiency and proper metabolic utilization of these nutrients.

Various processing methods can be used to reduce the HCN toxicity of cassava. Some of these methods are more effective than others. Drying in a forced-air oven at moderate temperatures (70°–80°C), boiling, in water, and sun drying are some of the methods used.

Oven-drying with little or no moisture present, eliminates the free HCN and destroys the enzyme linamarinase that is required for the hydrolysis of glucoside to form free HCN. Dry heat has little effect in the elimination of glucoside, and this remains in the root and is consumed by the animal. If it is true, as suggested, that HCN is liberated in the stomach through the hydrolysis of linamarin by hydrochloric acid, then quantities of free HCN of metabolic importance can be absorbed from the cassava that has been processed by oven drying.

Cooking of the cassava in water destroys the HCN along with the glucosidase enzyme, linamarinase, and causes a reduction in the total quantity of glucoside that is soluble in water. If the cassava is boiled rapidly and continually for a long period of time, much of the HCN and glucoside are removed. However, if boiling is for only a short duration, some of the glucoside may remain in the root, which when ingested may cause chronic HCN toxicity, as it does in the oven-dried cassava.

The method of chopping, mixing, and sun-drying is one that is practical and also effective in reducing both the HCN and the glucoside. This process initiates the action of the enzyme on the glucoside that causes the liberation of the HCN. If drying takes place during a period of several days, the majority of the glucoside will be hydrolyzed to liberate HCN that is dissipated in the air.

Some of the greatest problems in using cassava are related to harvesting, storage, processing, and feeding.

The root system of the plant, which must be harvested, is deep as well as irregular in its placement, and it generally extends over a wide area. The large and fibrous stalks must be removed before the root can be harvested. Mechanical harvesting methods have as yet not been developed and harvesting is usually manual. In damp, lighter textured soils, hand-harvesting is not a difficult task. Drier and more compact soils present somewhat of a problem if the roots are to be removed intact. A tractor equipped with a single bottom plow or a middle-buster to expose the roots has been used to advantage, but the roots are broken and must be utilized almost immediately.

Cassava in its fresh, high-moisture state cannot be stored for more than 3–4 days without the onset of fermentation and deterioration. Refrigeration will extend the storage time considerably and freezing will help to preserve it for periods of as long as one year. If used fresh, cassava should be harvested two or three times per week to assure a satisfactory and palatable product for feeding.

Grinding, shredding, or chopping of cassava roots can be accomplished in many ways. If small quantities are needed, the use of a knife or machete is the most simple method. If larger quantities are required, a mechanical grinder or shredder is more effective.

The roots can be stored after harvest for periods of as long as 1 year if they are dried or ensiled. They can be dried either mechanically or by exposure of the chopped or shredded material to the sun. If the moisture content is reduced to 10% or less, the dried chips can be safely stored for as long as 1 year.

Since 1900, many studies to determine the value of cassava roots as an energy source have been reported (Gavin and Andovard 1914; Henke, 1919 1923; Mondoñedo and Bayan 1927; Mondoñedo 1928; Fullerton 1929; Mondoñedo and Alante 1931; Woodman et al. 1931; Alba 1937; Asico 1941; Zarate 1956; Oyenuga and Opeke 1957; Mejia 1960; Modebe 1963; Maner and Buitrago 1964; Peixoto 1965; Maner and Jiménez 1967; Maner *et al.* 1967; Aumaitre 1969; Shimada 1970; Maner 1971).

As can be seen in the summary of the results of studies reported between 1927 and 1941 to evaluate cassava as a feed for growing-finishing pigs (Table 13.5), the maximum level of cassava incorporated into the diet was 40%. In the majority of the cases, the level of production was low and the quantity of feed required per unit of gain was high. These data, however, compared favorably with the results obtained from control diets which supported only limited growth. In view of today's standards and the advances that have been made in the field of nutrition since the time these studies were conducted, these data must be considered mostly of historical interest.

Oyenuga and Opeke (1957) in Africa, utilized 40% cassava in the diet of growing pigs, and 55% in that of finishing pigs. The fresh cassava was fed to the pigs raw or cooked along with a source of supplemental feed that contained protein, energy, and minerals, but without additional vitamins. The pigs were hand-fed three times per day with the objective of equalizing daily feed intake of all diets. The authors concluded that raw cassava has a feeding value equal to that of sorghum and maize and equal or superior to that of cooked cassava.

Extensive studies have been conducted in Colombia by members of the Colombian Agricultural Institute (ICA) and in collaboration with the International Center for Tropical Agriculture (CIAT). It has been shown that raw cassava can be utilized to supply the majority of the energy in growing-finishing swine diets (Table 13.6) (Maner *et al.* 1967). Pigs fed chopped, raw cassava along with a 42% protein supplement gained as rapidly and as efficiently as pigs fed a 16%

TABLE 13.5. Summary of Early Results Obtained in Studies to Evaluate Cassava as a Feed for Swine (1927–1941)

Level of cassava (%)	*Wt of pigs (kg)*	*Av daily gain (kg)*	*Feed/gain ratio*	*Reference*
20	11–26	0.206	2.65	Mondoñedo and Bayan 1927
20	28–40	0.175	6.52	Mondoñedo and Bayan 1927
30	39–62	0.325	6.46	Mondoñedo and Bayan 1927
20	32–60	0.396	4.76	Mondoñedo 1928
10–25	—	0.559	4.10	Fullerton 1929
25	10–20	0.068	9.56	Mondoñedo and Alante 1931
24–40	35–82	0.609	4.03	Woodman et al. 1931
5	—	0.420	5.32	Alba 1937
10	—	0.420	5.38	Alba 1937
15	—	0.400	5.47	Alba 1937
20	—	0.350	6.01	Alba 1937
15	12–30	0.250	3.51	Asico 1941
20	30–50	0.280	6.35	Asico 1941
25	50–70	0.300	6.86	Asico 1941
15–25	12–70	0.270	5.68	Asico 1941

protein maize–soybean meal control diet. There was, however, a tendency for the pigs to consume more protein supplement when allowed a free-choice selection between the chopped cassava and the protein supplement consisting of soybean meal and cottonseed meal. This overconsumption of protein supplement was somewhat reduced when the pigs were fed controlled quantities to meet their daily requirement.

Similar results were also obtained if several different protein sources were used to prepare the protein supplement, as indicated in Table 13.7 (Maner *et al.* 1970). As previously shown, however, the pigs tended to consume less cassava and more protein supplement than would be considered necessary to provide a balanced

TABLE 13.6. Performance of Pigs Fed Basal Diet or Raw Cassava and Protein Supplement

Performance	*1* Basal	*2* Raw cassava + supplement ad lib	*3* Raw cassava + supplement controlled
Av daily gain, kg[a]	0.843	0.834	0.794
Av daily intake, wet cassava, kg	—	4.05	3.89
Av daily intake, dry cassava, kg[b]	—	1.63	1.57
Av daily intake, supplement, kg	—	1.17	0.73
Av daily intake, total kg[b]	2.89	2.80	2.30
Feed/gain ratio	3.43	3.36	2.90

Source: Maner et al. 1967.
[a]Five pigs per treatment; 98-day experiment; av initial wt, 17.8 kg; av final wt, 98.6 kg.
[b]Total expressed on an approximately 10% moisture basis.

TABLE 13.7. Performance of Growing-Finishing Pigs Fed Cassava with Different Protein Sources

Treatment	*Av daily gain* (g)	*Av daily consumption of fresh cassava*[a] (kg)	*Av daily consumption supplied* (kg)	*Feed/gain ratio*[b]
Fresh chopped cassava plus:				
Soybean meal	723	4.00	0.80	3.25
Meat meal	684	3.40	0.78	3.07
Cottonseed meal	592	3.13	0.79	3.38
Meat meal + blood meal	728	3.88	0.94	3.32
Meat meal + blood meal + cottonseed meal	724	4.00	0.90	3.38
Fish meal + cottonseed meal	679	4.08	0.79	3.47
Average	687	3.75	0.83	3.31

Source: Maner et al. 1970.
[a]Expressed on 65 percent moisture or 35 percent dry matter basis.
[b]Calculated on 10 percent moisture basis.

16% protein diet. In all of the trials the pigs consumed an average of only 4 kg of fresh cassava per day. This lower consumption may be related to palatability and may be due to the slightly bitter taste imparted by the low levels of HCN present in the cassava.

The consumption of fresh cassava and the voluntary consumption of protein supplement varies according to the protein content of the supplement. When fresh sweet cassava and protein supplements based on maize and soybean meal and containing 20, 30, or 40% protein were offered free-choice to growing-finishing pigs, the daily voluntary intake of cassava increased and the daily voluntary intake of protein supplement decreased as the level of protein was increased in the supplement (Table 13.8). Overconsumption of protein was not evident when 20% protein supplement was fed but there was a tendency for overconsumption of protein when the two higher levels of protein in the supplements were fed (Job 1975).

Replacement of fresh sweet cassava (low HCN) with fresh bitter cassava (high HCN) significantly modifies the consumption pattern of both protein supplement and cassava (Table 13.9). The bitter cassava is unpalatable to the pig and poorly consumed. Pigs voluntarily consume three times as much sweet cassava per day as those fed bitter cassava (2.99 vs. 0.99 kg). Underconsumption of the bitter cassava was accompanied by an overconsumption of protein supplement when offered free-choice. When the supplement offered was limited to an amount required to provide a 16% protein diet, pig performance was drastically depressed (Gomez *et al.* 1976).

Modebe (1963) used sun-dried cassava meal to replace sorghum and maize in the diets of growing and finishing pigs. Levels of 32, 37, and 40% or of 42, 47, and 50% were used in diets for 23–36-, 37–55-, and 55–75-kg pigs, respectively. The growth rates of those pigs fed the lower levels of cassava meal (32, 37, and 40%) were not significantly different from the growth rates of those that received

TABLE 13.8. Performance of Growing-Finishing Pigs Fed Fresh Cassava and a 20, 30, or 40% Protein Supplement Free-Choice[a]

Criteria	*Control diet*	*Cassava + protein supplement*		
		20%	*30%*	*40%*
Av daily gain, kg	0.63	0.70	0.67	0.65
Av daily feed, kg				
Fresh cassava, kg	—	1.78	2.74	3.32
Protein supplement, kg	—	1.39	1.00	0.75
Total feed intake, kg	2.08	2.08	2.07	2.04
Feed/gain ratio	3.30	2.97	3.09	3.14
Protein in diet consumed, %	14.3	14.6	16.6	17.3

Source: Job 1975.
[a]Mean of five individually fed pigs per treatment: average initial weight 21.1 kg; average final weight 86.1 kg; 98-day trial.
[b]Includes cassava calculated at 10% moisture.

the control diet based on sorghum and maize (0.491 vs. 0.482 kg/day) or of those that received higher levels (42, 47, and 50%) of cassava meal (0.491 vs. 0.473 kg/day). Although the efficiency of feed conversion was not optimal in any of the groups (4.38, 4.39, and 4.51 kg feed per kilogram weight gain in the control, low, and high cassava meal groups, respectively) there were no differences between treatments.

Mejia (1960) utilized levels of 20 and 40% sun-dried cassava meal to replace a similar quantity of maize in diets for growing-finishing pigs and found that the growth-rate and feed efficiency of all groups were similar.

Maner and co-workers (1967) compared graded levels of sun-dried cassava meal with and without 10% cane molasses in diets for growing-finishing pigs (Table 13.10), and demonstrated that the growth rate was reduced as the level of

TABLE 13.9. Performance of Finishing Pigs Fed Either Sweet or Bitter Cassava and a Protein Supplement Offered Free-Choice or in Controlled Quantities

Criteria	*Sweet cassava or bitter cassava + protein supplement*			
	Ad lib	*Controlled*	*Ad lib*	*Controlled*
Av daily gain, kg	0.66	0.77	0.56	−0.08
Daily feed intake, kg				
Fresh cassava, kg	2.99	3.40	0.99	0.93
Protein supplement, kg	0.81	0.82	1.21	0.22
Total feed intake, kg[a]	1.98	2.01	1.60	0.58
Feed/gain ratio	2.99	2.61	2.86	Neg
Protein in diet, %	14.1	13.3	23.5	13.3

Source: CIAT 1975.
[a]Expressed on 10% moisture basis.

TABLE 13.10. Influence of Level of Cassava Meal on Performance of Growing-Finishing Pigs

Treatment	*Av daily gain (kg)*[a]	*Daily feed intake (kg)*	*Feed/gain ratio*
Basal	0.772	2.68	3.47
25.72% cassava	0.744	2.66	3.57
48.65% cassava	0.743	2.79	3.76
69.25% cassava	0.708	2.48	3.49
Basal + 10% molasses	0.888	3.38	3.84
21.70% cassava + 10% molasses	0.827	2.95	3.56
41.04% cassava + 10% molasses	0.777	3.00	3.85
58.26% cassava + 10% molasses	0.767	2.73	3.54

Source: Maner et al. 1967.
[a]Six pigs per treatment; 111-day experiment; av initial wt, 18.5 kg; av final wt, 104.8 kg.

cassava meal increased in the diet. Although 10% molasses improved the performance of all the groups, it was not effective in overcoming the slight depression caused by substituting cassava meal for maize in the diet.

Shimada (1970) also employed graded levels (0, 22, 44, and 66%) of sun-dried cassava meal to replace maize in diets for pigs between 30 and 90 kg. It was observed that substitutions of cassava meal at levels up to 44% of the diet did not cause any reduction in performance. However, the 66% level of substitution caused a reduction in both gains and efficiency of feed conversion.

R. Portela and J. H. Maner (unpublished data) later demonstrated that the performance of pigs fed diets containing more than 50% cassava meal could be significantly improved by adding 0.1–0.2% DL-methionine (Table 13.11).

Aumaitre (1969) compared the feeding value of maize, wheat, barley, and decorticated oats with that of cassava meal. Utilizing diets containing 20% protein for 5–9-week-old pigs, he showed that the performance of pigs fed cassava meal (416 g/day) was superior to that of pigs fed barley, oats, maize, or wheat (386, 380, 354, and 360 g/day, respectively). It was also shown that the digestible

TABLE 13.11. Effect of Methionine and Fat Supplementation on the Utilization of Cassava Meal for Growing Pigs

Treatment	*Av daily gain (kg)*	*Feed/gain ratio*
Control, maize–soya	0.74	2.81
Cassava meal[a]	0.76	2.65
Cassava meal + 0.1% DL-methionine	0.83	2.46
Cassava meal + 0.2% DL-methionine	0.82	2.49
Cassava meal + 10% beef tallow	0.85	2.24
Cassava meal + tallow + 0.1% DL-methionine	0.80	2.35
Cassava meal + tallow + 0.2% DL-methionine	0.74	2.21

Source: R. Portela and J. H. Maner, unpublished data.
[a]Cassava meal, 55%.

TABLE 13.12. Digestibility, Digestible Energy, and Metabolizable Energy of Cassava

Criteria	*Coefficient*
Nutrient digestibility	
Dry matter, %	93.8
Crude protein, %	40.5
Ether extract, %	51.7
Crude fiber, %	48.8
NFE, %	98.5
TDN, %	92.5
Energy[a]	
Digestible energy, kcal/kg	3,758
Metabolizable energy, kcal/kg	3,640
Adjusted metabolizable energy, kcal/kg	3,496

Source: J. Mesa and J. H. Maner, unpublished data.
[a]Expressed on dry matter basis.

energy of cassava meal (4185 kcal/kg) is equal to or higher than that for wheat, barley, maize, and decorticated oats (3973, 3955, 4046, and 4024 kcal/kg, respectively).

J. Mesa and J. H. Maner (unpublished data) have also studied the digestibility of the various nutrient components of cassava meal and found the digestibility of the energy and carbohydrate fractions to be satisfactory (Table 13.12) and not greatly different from results reported for maize.

Buitrago and Jiménez (1968) evaluated cassava silage as an energy source for growing-finishing pigs. They found the silage to be well consumed by the pigs and to produce satisfactory performance. Buitrago and Jiménez (1968) showed that cassava silage is consumed at a rate similar to that of raw, chopped cassava and that growth rate and feed efficiency are similar (Table 13.13). Inclusion of both the leaves and stalks of the plant in the root silage, however, reduced both daily consumption and gains.

Cassava meal and fresh cassava have been used to supply energy for gestating

TABLE 13.13. Effects of Cassava Silage on Growing-Finishing Swine

Treatment	*Av daily weight gain (g)*[a]	*Av daily intake supplement (kg)*	*Av daily intake of fresh cassava (kg)*	*Feed/gain*[b] *ratio*
Protein supplement plus				
Fresh chopped cassava	750	1.00	4.04	3.43
Cassava root silage	770	1.01	3.84	3.25
Root and foliage silage	640	1.06	3.05	3.52

Source: Buitrago and Jiménez 1968.
[a]Five pigs per treatment; av initial wt, 18.34 kg; av final wt, 98.8 kg. Length of trial, 112 days.
[b]Calculated on 10 percent moisture basis.

TABLE 13.14. Effects of Fresh Cassava in Gestation in Pasture and Confinement

	Treatment		
	1 kg control diet daily on pasture	*1.7 kg fresh cassava + 0.4 kg protein supplement on pasture*	*3.1 kg fresh cassava + 0.62 kg protein supplement in confinement*
No. of sows served	10	10	10
No. of sows farrowing	9	7	7
Av sow wt at service, kg	165.8	163.6	152.8
Av sow wt at 105 days, kg	185.7	188.5	190.5
Av sow gain gestation, kg	19.9	24.9	37.7
Av no. pigs/litter	10.4	10.0	7.7
Av no. pigs at 35 days	8.3	7.3	6.9
Av birth wt, kg	1.28	1.12	1.18
Av 35-day wt, kg	6.94	6.05	6.49

and lactating sows (J. H. Maner and J. Buitrago, unpublished data). Both produce satisfactory results for gestating sows (Table 13.14) but consumption of fresh cassava is low in lactating sows (Table 13.15) even when the cassava is fed with a quantity of protein supplement calculated to supply a 16% ration.

When cassava meal was fed continually during the entire life cycle, pig performance during the growing-finishing period and reproductive performance during gestation and lactation were satisfactory, but slightly inferior to those of a similar group of gilts fed diets based on maize (Gomez *et al.* 1976).

TABLE 13.15. Performance of Lactating Sows Fed Fresh and Dried Cassava Diets

	Treatment		
	16% protein, control diet	*16 protein, dried cassava*	*40% protein supplement, fresh cassava*
No. of sows	13	15	16
Av no. of pigs at birth	10.8	10.1	9.3
Av no. of pigs at 35 days	9.0	7.9	7.6
Mortality, %	16.4	22.4	18.1
Av birth wt, kg	1.18	1.22	1.36
Av wt at 35 days, kg	6.03	6.80	7.63
Av wt of sows, 24 hrs postpartum	179.3	170.6	158.3
Av wt of sows at 35 days	190.3	183.0	165.8
Av sow gain lactation, kg	11.0	12.6	7.5
Av daily intake cassava, kg[a]	—	—	6.50
Av daily intake supplement, kg	—	—	1.21
Av total daily intake, kg	4.82	5.24	3.74

[a]Cassava contains 35 percent dry matter.

POTATOES

Potatoes (*Solanum tuberosum*) are widely used for livestock feed, and particularly in Europe, they are grown for that specific purpose. Surplus potatoes or small or cut potatoes unfit for human consumption are available in many areas for swine feeding. In addition, increasing quantities of potato waste are becoming available from potato processing industries.

The composition of the potato varies widely according to variety and environmental conditions during growth and storage, and according to processing treatment (Whittemore 1977). The water content is particularly variable, ranging from 63 to 87% (Schwimmer and Burr 1967), which has a significant effect on nutrient concentration. Expressed on a dry-matter basis, whole potatoes with peel contain 4–6% ash, 10–11% crude protein, 4% crude fiber, 0.2% ether extract, and 82–83% nitrogen-free extract (Whittemore 1977; Livingstone *et al.* 1979). Approximately 70% of the potato dry matter is starch, which is composed of about 75% amylopectin and 25% amylose.

A concentrated source of starch and energy, whole potato with peel also can make a significant contribution to the supply of both nitrogen and essential amino acids (Table 13.16). The nitrogen content of potatoes varies according to variety and location and much of the nitrogen is soluble nonprotein nitrogen composed of approximately equal parts of amide–nitrogen and amino–nitrogen. Although extremely variable, approximately 50% of the nitrogen is in the form of true protein (Whittemore 1977; Livingstone *et al.* 1979). Potato protein has been shown to have a biological value of 71–75 for the 20-kg pig (Höller 1959), which places it among the highest of the plant proteins (Westerhind 1974). The lysine content is

TABLE 13.16. Amino Acid Composition of Raw and Cooked Potatoes (% Protein)

Amino acid	*Raw*[a]	*Raw*[b]	*Cooked*[b]	*Cooked flour*[b]	*Cooked flake*[b]
Aspartic acid	17.1	22.7	18.2	18.6	20.7
Threonine	3.2	3.2	3.4	3.3	3.9
Serine	3.5	3.9	3.5	3.6	3.6
Glutamic acid	18.3	16.5	22.9	22.3	18.2
Glycine	2.7	2.8	3.0	3.0	2.9
Alanine	3.8	2.9	3.4	3.2	3.3
Valine	4.8	3.9	4.6	4.7	4.3
Methionine + cystine	3.6	2.0	1.9	2.1	2.0
Isoleucine	3.1	2.9	3.4	3.3	3.2
Leucine	4.8	4.4	5.6	5.0	4.8
Tyrosine	6.6	3.2	3.6	3.7	3.4
Phenylalanine	3.4	3.5	3.7	4.0	3.6
Lysine	5.0	5.4	5.2	5.0	5.3
Histidine	1.7	2.5	1.7	1.7	2.2
Arginine	1.7	4.5	3.5	5.0	5.6

Source: [a]Livingstone et al. 1979. [b]From Whittemore 1977; D'Mello et al. 1973; Whittemore et al. 1974.

only slightly less than that of soybean meal and almost twice that of maize and barley. Livingstone *et al.* (1979) reported that 71.3% percent of the lysine was absorbed anterior to the terminal ileum and that the lysine had a "chemical availability" of 95%. In relation to pig requirements, the sulfur amino acids, methionine, and cystine appear to be first limiting in potato protein.

Raw potatoes are unpalatable and support only very slow growth of pigs (Thompson and Hargrave 1935; Braude and Mitchell 1951). McLagan and Nasr (1950) observed that raw potatoes increased the dietary requirements for supplementary protein. With 50% of the dietary dry matter fed as raw potatoes, it was impossible to sustain pigs on a level of intake above the required for maintenance (Whittemore *et al.* 1975A,B). There is general agreement that raw (uncooked) potato causes a reduction in pig performance and that potatoes are more efficiently utilized by the pig if cooked before feeding (Whölbier and Neckermann 1951; Leroy *et al.* 1952; Martin and Buysse 1955; Zausch and Hölzschul 1960–1961; Friend *et al.* 1963; Henry and Rerat 1966; Moncada and Maner 1965–1966; Krivencou and Mosbuin 1968; Anastasijevic *et al.* 1972; Schmidt and Soroka 1972; Borgida and Laplace 1977; Popelov 1977; Saulnier and Chavuel 1978; Mirochitskii 1980).

Although raw potato starch is resistant to α-amylase attack (Sandstedt and Gates 1954), the digestibility of potato dry matter, organic matter, nitrogen-free extract, and gross energy is consistently high (85–99%) for both raw and cooked potato (Whittemore 1977). Of significance is that most of the cooked potato starch is digested and absorbed anterior to the terminal ileum while only about 30% of raw potato starch is digested and absorbed before reaching this area of the gastrointestinal tract. A large proportion of the raw starch reaches the large intestine, where it accumulates and is degraded by microorganisms (Cranwell 1968) and the end products absorbed in the lower gut. The end products of bacterial digestion may be utilized with an efficiency different from that of the end products of enzymatic digestion in the small intestine. The lower digestible energy values reported for raw potato (3.20 kcal/g DM) compared with those of cooked potato (3.83– 4.02 kcal/g DM) support this suggestion (Woodman and Evans 1943; Whittemore *et al.* 1975A,B).

While the influence of cooking on the apparent digestibility of energy of potatoes appears to be minimal, heat processing has been shown to affect nitrogen digestibility (Whittemore and Taylor 1973). The nitrogen of raw potato is digested with varying degrees of efficiency by pigs but results are generally poor relative to those obtained with cooked potato (Whittemore 1977). The digestibility of nitrogen is dependent on the degree of cooking (Whittemore and Taylor 1973).

An alkaloid complex, particularly the glycoside solanine, is a constituent of potatoes and some reports have indicated that the presence of this substance in raw potatoes is responsible for the deleterious effect on consumption and digestibility. However, the raw potato also contains a powerful protease inhibitor that is protein in nature, soluble in water, and heat labile (Sohonie and Ambe 1955). Whittemore *et al.* (1975B) showed the protease inhibitor activity to be very high in uncooked potatoes but absent in cooked potatoes, and concluded that the presence of this powerful protease inhibitor activity in raw potato adequately explains both the reduction and variability in nitrogen digestibility for uncooked as compared to cooked potatoes.

In practice, potatoes should be heat processed to convert the starch fraction into a form that can be digested by enzymatic action and to denature the protease inhibitor (Livingstone *et al.* 1980). Cooking can be accomplished by boiling the potatoes in water or by steaming. Various forms of heat drying also contribute to this process. Livingstone *et al.* (1979) demonstrated that either over- or under-processing with steam can reduce utilization of both nitrogen and energy and suggested that steaming for 20 min at 100°C is adequate to convert starch and to denature the protease inhibitor. Prolonged heating, 1 hr at 100°C, caused rupture of the cell walls with release of the gelatinized starch and resulted in potato protein of low digestibility. However, steaming for 20 min at 70°C did not cause complete gelatinization of the starch and did not effectively denature the protease inhibitor.

Cooked potatoes should be fed in combination with a protein supplement that provides adequate protein and a relatively high level of energy. Potatoes have a higher feeding value if they are not used as the sole energy source in growing-finishing diets, but are fed in combination with some grain or other concentrated energy source. Moncada and Maner (1965–1966) demonstrated that growing-finishing pigs would not consume an adequate quantity of steamed potatoes for optimal growth if the daily diet of a 43% protein concentrate were limited to a quantity calculated only to meet the pig's daily protein needs.

European workers (Braude and Mitchell 1951; Martin and Buysse 1955) have shown that a 19–20% protein concentrate can be effectively used with cooked potatoes to produce gains similar to those obtained with a 16.3% all-concentrate ration. Pigs fed a daily diet of 1.14 kg of a 19% protein concentrate and cooked potatoes to appetite performed at a rate and efficiency not different from pigs fed all-concentrate diets. Under this system of feeding, the daily consumption of cooked potatoes increased with increasing pig weights. Although 14- to 15-week-old pigs consumed only 2.18 kg of cooked potatoes per day, by 30 weeks of age they were consuming approximately 9 kg per head per day.

Pigs fed in a similar manner but fed on raw potatoes gained weight at a 31% slower rate (Braude and Mitchell 1951) and required 40 days longer to reach market weight of 100 kg. Also, those fed raw potatoes required 0.66 kg more concentrate and 2.3 kg more potatoes to produce a kilogram of gain than those fed cooked potatoes.

The best gains and utilization of both concentrates and potatoes will be obtained if cooked potatoes are fed to appetite along with daily limited quantities of a 19–20% protein concentrate. For economic calculations, 4.7 kg of cooked potatoes are approximately equal to 1 kg of a 16% concentrate. Cooked potatoes should always be allowed to cool to at least a moderate temperature before feeding.

DRIED POTATO PRODUCTS

A number of products from the potato are used in livestock feeds. Terminology used to describe these products is sometimes confusing. Potato cossettes, potato meal, potato slices, potato flakes, and potato pulp are all names given to the various products. Waste potatoes that are sliced and then dried give cossettes; when cossettes are ground, potato meal is produced. When potatoes are cooked

TABLE 13.17. Chemical Composition of Potatoes Processed in Different Ways

Product: Source	*Dry matter (%)*	*Crude protein (%)*	*Crude fiber (%)*	*Ether extract (%)*	*Ash (%)*	*Nitrogen-free extract (%)*
Raw potato:						
Woodman and Evans 1939–1943	23.8	2.1	0.9	0.1	1.0	19.7
Whittemore et al. 1973–1975	20.6	2.2	0.9	0.04	0.8	—
Cooked potato:						
Whittemore et al. 1973–1974	22.2	2.3	0.7	0.04	1.0	—
Potato silage:						
Highlands 1967	16.6	1.2	1.1	0.07	0.5	13.8
Potato meal:						
Woodward et al. 1938	88.8	9.66	2.14	0.32	4.78	71.9
Woodman and Evans 1943	89.8	8.76	2.09	0.50	3.57	74.88
Seilder et al. 1967[a]	88.5–93.0	3.68–7.44	3.15–4.98	0.71–1.08	5.72–12.84	—
Potato flake:						
Zausch and Hölzschuh 1960–1961	90.0	7.98	3.30	0.33	2.65	75.82
Hutchinson et al. 1943	87.5	8.48	—	—	—	—
Woodman and Evans 1943	89.9	8.21	2.07	0.28	3.81	75.53
Whittemore et al. 1973–1975	92.5	9.37	2.13	0.19	4.07	77.24
Potato slices:						
Woodman and Evans 1943	89.5	9.39	1.64	0.21	3.88	74.38
Zausch and Hölzschul 1960–1961	90.0	11.07	2.34	0.63	4.47	71.49
Potato pulp:						
Friend et al. 1963	88.8	7.04	7.75	0.14	2.70	82.46
Kirsch and Jantzon 1943	90.0	5.34	17.23	1.18	2.18	63.93
Whittemore et al. 1973–1975	87.0	7.83	2.26	0.52	—	—

[a]Range of values.

before drying the resultant product is potato slices; and when they are mashed after cooking and then dried on a roller, potato flakes are produced. Potato pulp is a by-product of the potato starch industry and is the residue from cold water extraction of starch from the pulped, raw potato. The composition of potato meal, flakes, slices, and pulp is summarized in Table 13.17.

Potato Meal

Dried, raw potato meal included in pig diets in more than moderate amounts may give rise to digestive disturbances and diarrhea. When potato meal makes up more than 30% of the pig's diet, the diet is unpalatable to the young growing pig (Woodman and Evans 1943) and scouring normally occurs. Sixty percent potato meal in the diet reduces daily feed intake, suppresses gains, and causes digestive disturbance when fed to pigs between 30 and 105 kg (Seidler *et al.* 1967; Glaps 1970). Levels of 10–15% potato meal are recommended for the growing pig (20–50 kg) and quantities not to exceed 30% are recommended for the finishing pig (50–100 kg).

The depression in gains and digestive disturbances of pigs fed raw potato meal is associated with the site of digestion of carbohydrates in the alimentary tract, with lower digestible energy content and with poor protein digestibility caused by the heat labile protease inhibitor activity present in raw potatoes (Whittemore *et al.* 1975B). The inclusion of potato meal in the diet of pigs gives rise to a very significant depression of the apparent digestibility of the crude protein of the potato and of the diet as a whole. Although the digestibility of the dry matter and nitrogen-free extract (Table 13.18) is relatively high, the crude protein fraction has a digestion coefficient of only 26.7–39.9% (Woodman and Evans 1943; Kellner and Newmann 1910; Witczak and Olczak 1970).

Potato Flakes

Potato flakes are prepared by steaming clean, washed potatoes for 30 min in a tank in which the pressure rises 10–15 lb/in.[2] After they are steam cooked they are mashed and then passed over drying rollers for drying, and finally removed as thin flakes. Flakes have a high digestibility of all components and have a high nutritive value (Woodman and Evans 1943). They may be included in liberal

TABLE 13.18. Apparent Digestibility Coefficients and Digestible Energy Content of Raw, Cooked, and Dried Potato Products for Pigs

Product: Source	*Crude protein (%)*	*Crude fiber (%)*	*Organic matter (%)*	*Dry matter (%)*	*Nitrogen-free extract (%)*	*Digestible energy, kcal/g DM*
Raw potato:						
Woodman and Evans 1943	76.0	55.0	94.0	—	98.0	—
Schneider 1947	51.0	79.0	91.0	—	95.0	—
Cooked potato:						
Schneider 1947	70.0	59.0	93.0	—	97.0	—
Potato silage:						
Nash et al. 1963	—	—	—	85.0	—	—
Potato meal:						
Woodman and Evans 1943	39.9	64.8	89.2	—	96.4	3.78
Kellner and Neumann 1910	26.7	85.3	94.5	—	97.9	—
Woodman and Evans 1943	40.0	65.0	89.0	89.0	96.0	—
Pfeiffer 1966	58.0	99.0	—	—	95.0	—
Potato flake:						
Woodman and Evans 1943	81.8	—	98.0	98.0	100	4.02
Whittemore et al. 1975A	86.0	—	—	—	—	3.83
Fingerling 1932	70.5	84.9	95.1	—	98.3	—
Zausch and Hölzschul 1960–1961	80.1	82.2	—	93.8	98.0	—
Potato slices:						
Woodman and Evans 1943	74.9	32.6	92.5	93.0	96.9	—
Zausch and Hölzschul 1960–1961	76.4	16.5	—	87.8	96.9	
Potato pulp:						
Kirsch and Jantzon 1943	27.0	91.0	91.2	—	91.2	—
Friend et al. 1963	25.5	76.2	—	—	80.0	3.11
Nicholson and Friend 1965	23.0	—	84.0	—	—	—

quantities in both growing-finishing diets and diets for older pigs without causing digestive disturbances or diarrhea. The most satisfactory pig performance is obtained with potato flakes if they make up not more than 30–40% of the diet for growing-finishing pigs (Ostrowski and Rys 1969). When potato flakes were used to replace all of the barley in a 16% protein diet for growing-finishing pigs between 30 and 90 kg liveweight, reduced gains, increased feed per kilogram of gain, and decreased carcass quality were recorded even in the presence of adequate protein and supplemental lysine and methionine.

With young pigs, Hillyer and Whittemore (1975) showed that 50–58% of cooked potato flakes could replace cereals in diets for suckling or early weaned pigs without affecting intake or performance. Whittemore *et al.* (1974) also found similar growth responses of individually penned pigs offered diets containing 78% of either cooked potato flakes or ground maize from 21 to 53 days of age.

Potato Slices

Potato slices are prepared by passing the slices through a hot-air rotating drier. During the process the potato slices rise to a temperature of about 150°C when they are introduced into the drier and fall to about 80°C at the outlet. The greatest part of the 2-hr processing is accomplished at 80°C. This treatment allows for both cooking and drying (Woodman and Evans 1943). Inadequate cooking of the potato slices could reduce their nutritive value.

Potato slices have been used to replace barley and maize in diets for growing-finishing pigs. A 30% level of substitution during the growing period and a 60% replacement of corn and barley during the finishing period had no effect on gains, feed efficiency, or carcass quality in the studies of Woodman and Evans (1943). Other studies (Zausch and Hölzschul 1960–1961) have shown that potato slices with a high digestibility (Table 13.18) but with 480 mg of solanin per kilogram could not be used at levels of more than 20% of the diet without causing digestive disturbances, diarrhea, and a reduction in both gains and feed efficiency. When 40% potato slices were included in the diet, diarrhea occurred after 10 days on feed and was controlled only when cereals replaced the potato slices in the ration. The variation in results of these studies may be the result of different processing of the potato slices used. Improper or inadequate cooking of the potato slices during processing might be expected to produce the poorer results reported, since it is known that raw potatoes and raw potato meal give unsatisfactory results when fed in large quantities. Care should be taken to ensure adequate cooking and proper processing of potato slices.

Potato Pulp

Potato pulp is a by-product of the starch industry and is the residue obtained after starch is extracted from the potato with cold water. Only low levels of potato pulp can be used as a grain substitute in growing and finishing swine diets (Friend *et al.* 1963; Bielinski and Bielinski 1954). When levels of 0, 15, 30, and 45% pulp were used to replace barley in growing diets and oats in finishing diets, linear depressions in both gains and efficiency of feed conversion were recorded (Table 13.19). A quantity of as much as 15% dried potato pulp allowed satisfactory

TABLE 13.19. Performance of Pigs Fed Various Levels of Dried Potato Pulp During the Growing and Finishing Periods

Level of potato pulp (%)	*Av daily gain* (*kg*)	*Feed/gain ratio*
Growing		
0	.65	2.78
15	.63	3.04
30	.59	2.96
45	.51	3.13
Finishing		
0	.83	3.63
15	.79	4.04
30	.73	4.21
45	.68	4.32
Growing-finishing		
0	.73	3.24
15	.70	3.57
30	.65	3.65
45	.59	3.75

Source: Adapted from Friend et al. 1963.

gains. Other studies (Stahl *et al.* 1932) indicated that potato pulp was not as good as potato flakes or steamed potatoes when used to replace either of these ingredients in pig diets.

The fact that potato pulp is uncooked contributes to the poor palatability and poor nutritive value of dried potato pulp. Digestibility studies with dried potato pulp (Table 13.18) yield results similar to those obtained with raw potato meal and demonstrate not only that the protein fraction is poorly digested, but also that it may contribute to lowering the digestibility of the protein of other diet components (Nicholson and Friend 1965).

SWEET POTATOES

Sweet potatoes (*Ipomaea batata* L.) are a high-moisture feed in their fresh state and, except in areas in which cull or surplus potatoes are available at economical prices, are not commonly used as a swine feed by the major swine producers of the world. However, sweet potatoes are traditionally grown for both human and animal food in many tropical and semitropical environments. In Taiwan, it is estimated that 60% of all sweet potatoes produced are fed to swine. In these areas, dry-matter yield per hectare is competitive with many of the other feed sources. The sweet potato in its fresh state contains 68–78% moisture and 22–32% dry matter. It is lower in protein than maize but as a dry meal or chip it is a concentrated source of carbohydrates (Table 13.20).

Before 1948, several stations in the southeastern United States reported results related to the use of sweet potatoes for livestock feeding. However, in view of the

TABLE 13.20. Chemical Composition of the Sweet Potato Root, Vine, and Dry Chip

Composition	*Fresh*	*Fresh*	*Moisture-free*	*Dry chip 1*[a]	*Dry chip 2*[b]	*Fresh vine*
Moisture, %	68	77.85	0	11.03	9.56	85.69
Dry matter, %	32	22.15	100	88.97	90.44	14.31
Crude protein, %	1.35	0.94	4.32	2.53	4.92	2.79
Crude fiber, %	0.53	0.89	1.70	3.14	3.09	2.76
Ether extract, %	0.20	0.78	0.64	1.16	1.31	2.97
Ash, %	0.67	0.83	2.13	4.25	3.52	1.34
Nitrogen-free extract, %	28.5	18.71	91.21	77.89	77.60	4.45
Starch, %	23.85	—	—	—	—	—
Sucrose, %	2.41	—	—	—	—	—
Invert sugar, %	0.96	—	—	—	—	—
Cartene, μg/g	40	—	130	—	—	—
Calcium, %	—	0.07	—	0.14	—	0.23
Phosphorus, %	—	0.04	—	0.12	—	0.05

Source: Adapted from Lease and Mitchell 1940; Smith et al. 1947; Bray 1933; [a]Tseng et al. 1980; [b]Wu 1980.

identification of many nutritional factors since that year, these data can be used only as an indication of the value of the sweet potato.

Sweet potatoes have been utilized in many forms. They have been fed raw, cooked, and as a silage; they have also been dehydrated to form a meal or chips. When fed wet, either raw, cooked, or as a silage, they are bulky and have a feeding value 25–33% that of maize. Early experience with sweet potatoes seemed to indicate that the sweet potato should be cooked for maximum utilization. However, it has also been stated that no beneficial effect is obtained from cooking (Calder 1960). Cooked sweet potatoes fed as 70% of the dry matter to 25-kg pigs have been shown to have only 80% of the value of maize when the two have been compared as dry matter (Springhall 1969).

The dry matter and energy digestibility of cooked sweet potatoes is slightly higher than that of raw sweet potatoes, but the protein digestibility of the cooked potato is almost double that of the raw potato (Table 13.21). Feeding trials with growing-finishing pigs (Table 13.22) indicate that pigs fed the same level of supplement grow faster and more efficiently when cooked sweet potatoes are substituted for raw potatoes. This increase is probably due to the 10–17% increase in sweet potato consumption observed when cooked potatoes were fed as well as to a small improvement in energy digestibility.

TABLE 13.21. Digestibility of Raw, Cooked, or Dehydrated Sweet Potatoes

Material	*Raw potatoes*[a]	*Cooked potatoes*[a]	*Dehydrated chips*
Dry matter, %	90.4 ± 1.57	93.5 ± 1.53	—
Crude protein, %	27.6 ± 4.36	52.8 ± 7.95	—
Energy, %	89.3 ± 2.38	93.0 ± 3.14	—
Digestible energy, kcal/kg	3373 ± 15	3463 ± 23	3497
Metabolizable energy, kcal/kg	—	—	3377
Net energy, kcal/kg	—	—	2020

[a]Mean ± standard error.

TABLE 13.22. Comparison of Raw and Cooked Sweet Potatoes for Growing-Finishing Swine

	Treatment and level of concentrate/day				
	Raw potato		*Cooked potato*		*Control*
	1.0 kg	*1.5 kg*	*1.0 kg*	*1.5 kg*	*Ad lib*
Av daily gain, kg	.434[b]	.494[c]	.513[c,d]	.544[c,e]	.597[e]
Av daily concentrate, kg	0.98	1.38	0.97	1.37	2.57
Av daily fresh potato, kg	3.01	2.61	3.50	2.74	—
Av daily air dry potato, kg[a]	1.26	1.08	1.52	1.19	—
Av daily air dry concentrate, kg[a]	0.96	1.35	0.94	1.33	2.43
Total air dry feed, kg[a]	2.22	2.43	2.46	2.52	2.43
Feed/gain ratio	5.12	4.92	4.80	4.63	4.07
Protein, % of diet	14.5	17.2	13.4	16.5	17.9
Potatoes, % of DM	56.4	44.5	61.5	47.2	

[a]Expressed on a 10% moisture, 90% dry matter basis.
[b,c,d,e]Means without a common superscript are significantly different from each other.

Dehydrated sweet potato meal, when fed as an energy source in a properly balanced diet, has a feeding value similar to that of maize (Ware 1947). Lee *et al.* (1963) showed that dehydrated sweet potato chips had approximately 87% of the feeding value of maize when supplemented with either soybean meal or a yeast mixed feed. More recent work from the same station (Lee and Yang 1979) has confirmed these early findings and has shown that the common sweet potato chip has a feeding value 88–93% that of maize when substituted for maize on a crude protein basis at levels of 25, 50, 75, and 100% in growing-finishing rations (Table 13.23). However, when high protein sweet potato chips, containing 7.3% crude protein, were used, pig performance was reduced at all levels of substitution above 25%. The results reported appear to indicate that the reduced value of the high protein sweet potato chips resulted from an overestimate of the crude protein

TABLE 13.23. Common and High Protein Sweet Potato Chips in Diets for Growing-Finishing Pigs

Diets	*Energy[a] source*	*Sweet potato level, %*	*Daily gain, g*	*Feed cons., kg*	*Feed/gain*
A	Maize	0	535	1.75	3.29
B	CSPC	25	496	1.70	3.44
C	CSPC	50	481	1.69	3.52
D	CSPC	75	470	1.59	3.39
E	CSPC	100	500	1.61	3.23
F	HPSPC	25	571	1.78	3.12
G	HPSPC	50	377	1.37	3.64
H	HPSPC	75	334	1.32	3.97
I	HPSPC	100	346	1.36	3.94

Source: Lee and Yang 1979.
[a]Common sweet potato chips (CSPC), high protein sweet potato chips (HPSPC).

TABLE 13.24. Effect of Steam Heat on the Utilization of High Protein Sweet Potato Chips and Dried Sweet Potato Vines in Diets for Growing-Finishing Pigs

Diet	*Energy*[a] *source*	*HPSPC level*[b]	*DSPV level*[b]	*Heat*[c] *treatment*	*Daily gain, g*	*Feed/gain*
A	Maize	0	0	NTP	560	3.04
B	Maize	0	0	SHP	540	3.11
C	HPSPC	50	0	NTP	470	3.58
D	HPSPC	50	0	SHP	450	3.85
E	HPSPC	100	0	NTP	440	3.86
F	HPSPC	100	0	SHP	460	3.67
G	HPSPC + DSPV	80	20	NTP	370	4.23
H	HPSPC + DSPV	80	20	SHP	380	4.16

Source: Lee and Lee 1979.
[a]High protein sweet potato chips (HPSPC), dehydrated sweet potato vines (DSPV).
[b]Level of substitution for maize.
[c]Normal temperature pelleting (NTP), steam heat at 80°C, then pelleting (SHP).

content of the chips, as analyzed crude protein values for the mixed diets containing these chips were 2–3% lower than the calculated values. This crude protein reduction increased as the level of high protein chips was increased. Lower level of dietary protein was also associated with a reduced daily feed intake for these treatment groups. Steam processing at 80°C and pelleting did not improve the feeding value of high protein sweet potato chips (Table 13.24) (Lee and Lee 1979).

It is common practice in Asia to include all or a portion of the sweet potato vine in pig rations. The vines are fed fresh, as a silage, or as a dehydrated meal. Least-cost formulation in Korea (Lee and Chung 1966) indicated that silage containing 60% roots, 30% vines, and 10% barley bran to control moisture should be fed as only 38% of the dry matter of rations from 50–70 kg of pig weight and up to 68.5% of dry matter from 70 kg to market weight. The silage was not recommended for lighter-weight pigs.

Tseng *et al.* (1980) fed sweet potato vines along with maize, fresh sweet potatoes, or sweet potato chips in rations for growing-finishing pigs and observed slower growth and reduced feed efficiency when compared to the performance of pigs fed maize–soybean meal diets. Including vines with maize as the energy source had little effect on gain but significantly reduced feed efficiency (Table 13.25). In Taiwan studies, Lee and Lee (1979) showed that replacing 20% of the sweet potato chips with dehydrated sweet potato vines caused a 16% reduction in growth rate and an 11% increase in feed requirements (Table 13.24). Although commonly used, sweet potato vines have a very low value in diets for growing-finishing pigs and may be more efficiently utilized for breeding animals.

YAMS

The "true yam," *Dioscorea* sp., is a tropical or semitropical root crop that is grown extensively in West Africa and to a lesser extent in most other tropical areas of the world. It is used as a staple human food and is generally too expensive

TABLE 13.25. Fresh Sweet Potato Vines in Diets for Growing-Finishing Pigs

Treatments[a]	*Daily gain, g*	*Feed/gain*	*Days on test*
1. Soybean meal–maize basal	472	3.52	159
2. Soybean meal–maize vines	469	5.49	160
3. Soybean meal–SPC vines	410	6.13	183
4. Soybean meal–FSP vines	346	11.43	217
5. 22% Protein suppl.–maize vines	481	5.00	156
6. 22% Protein suppl.–SPC vines	466	5.46	161
7. 22% Protein suppl.–FSP vines	510	7.82	147

Source: Tseng et al. 1980.
[a]Sweet potato chips (SPC), fresh sweet potatoes (FSP), and sweet potato vines.

to be used as a livestock feed. For details of the distribution, cultivation, composition, and utilization of *Dioscorea* sp., the reader is referred to Coursey (1967).

Of the ten species of *Dioscorea* that are of economic importance, the chemical composition varies widely, both among and within species. These variations are rendered even greater by different stages of maturity and environmental factors such as cultural practices and climate. The moisture content of the yam varies between 60 and 80%, being most frequently between 65 and 70%. The yam is basically a carbohydrate composed mainly of starch. The starch fraction appears to consist mainly of amylopectin or branched-chain starch with 10–25% amylose occurring in some species. Only a small proportion of the total carbohydrate fraction of the yam tubers is made up of mono- and disaccharides, which appear to be intermediate products in the metabolic breakdown of the starch fraction.

The protein content is very low, only 1–2% on a fresh weight basis, and very little is known about the utilization or quality of this protein. Coursey (1967) reports results that indicate that the yam protein contains substantial quantities of tryptophan (1.0%) but is very deficient in lysine (4.4%) and the sulfur-containing amino acids, methionine (1.8%) and cystine (1.1%).

Although satisfactory results with rats have been obtained (Jeffers and Haynes 1967) with the use of dried yam meal, the feasibility of this root as a swine feed is doubtful because of the alkaloid, tannin, and saponin content. It is well known that some of these substances, such as the alkaloid dioscorine, are extremely toxic, and both the tannins and the saponins have a very bitter taste.

BANANAS AND PLANTAINS

The banana belongs to the genus *Musa*, comprising 32 or more distinct species and at least 100 subspecies. Of the bananas grown in Central and South America the *Musa sapientum* ('Gros Michel') and *M. cavendishii* ('Cavendish') are the two most common, along with the plantain (*M. paradisica*) which is used for cooking (von Loesecke 1950). Because of its resistance to Panama disease, the 'Cavendish' variety is rapidly replacing the 'Gros Michel' in commercial plantations.

Bananas require a tropical climate in which the temperature does not fall below

50°F (10°C) and seldom rises above 105°F (40.5°C), and in which rainfall is frequent and scattered throughout the year. For these reasons most of the world production comes from the zone between the tropics of Cancer and Capricorn, which lie between the latitudes 23° N and 23° S. About 75% of all bananas grown in the world are produced in the Western Hemisphere. There are approximately 39.3 million metric tons of bananas and over 21 million metric tons of plantains produced annually in the world (FAO 1980).

Although bananas and plantains are grown largely for export and domestic consumption by the human population, a large quantity of this fruit is available for livestock feed. At the packing plants in the areas of commercial banana production, the bananas that are too small or too large, are slightly bruised, have spots or off color, or are not in an optimal stage of maturity for shipping are rejected for export. The "reject" bananas along with smaller quantities of farm-produced bananas and plantains constitute a good source of carbohydrates for pigs during all phases of the production cycle. The fresh whole banana with the peeling contains approximately 80% water, 20% dry matter, 1.0% protein, 1.0% fiber, 0.2% fat, 1.0% ash, and 16.8% nitrogen-free extract. The "crude fiber" of whole bananas consists of 60% lignin, 25% cellulose, and 15% hemicellulose. The ripe pulp itself contains 0.50% lignin, 0.21% cellulose, and 0.12% hemicellulose (von Loesecke 1950). Although the most common commercial varieties of bananas, 'Gros Michel' and 'Cavendish,' conform to this generalization of chemical composition, locally grown bananas and plantains vary widely in their chemical composition, as shown in Table 13.26 (Bressani *et al.* 1961).

TABLE 13.26. Chemical Composition of Several Varieties of Bananas and Banana Meal

Common name	*Water* (%)	*Protein* (%)	*Fat* (%)	*Crude fiber* (%)	*Ash* (%)	*Calcium* (mg/100 g)	*Phosphorus* (mg/100 g)	*Carotene* (mg/100g)
Guineo Oro[a]	69.4	1.50	0.20	1.20	1.20	53	21	17.10
Guineo Injerto[a]	80.5	1.13	0.20	0.70	1.20	69	12	19.75
Guineo Morado[a]	77.4	1.13	0.20	0.80	1.10	81	13	7.55
Guineo Plátano[a]	75.9	1.13	0.10	0.80	1.10	19	19	—
Guineo Dominico[a]	69.0	1.38	0.40	0.80	1.10	16	22	16.70
Plantain[a]	62.6	1.19	0.40	0.80	1.30	19	22	9.20
Average	72.5	1.24	0.25	0.85	1.17	43	18	—
Guineo Oro[b]	67.2	1.38	0.60	0.50	0.70	15	20	8.18
Guineo Injerto[b]	74.2	1.25	0.30	0.60	0.90	22	13	37.82
Guineo Morado[b]	76.1	1.25	0.40	0.50	0.80	21	12	15.09
Guineo Plátano[b]	73.2	1.06	0.40	0.50	0.90	12	16	16.86
Guineo Dominico[b]	63.0	1.50	0.20	0.40	0.80	11	24	0.84
Plantain[b]	58.8	1.00	0.20	0.40	0.90	9	21	0.20
Average	68.8	1.24	0.35	0.48	0.83	15	18	—
Banana meal								
Half-ripe without peel	9.2	4.39	0.20	3.00	3.90	—	—	—
Green with peel	15.2	3.98	1.60	3.90	5.40	—	—	—
Green without peel	11.4	4.88	0.20	2.90	3.70	—	—	—

Source: Adapted from Bressani et al. 1961.
[a]Sample analyzed with peel. [b]Sample analyzed without peel.

TABLE 13.27. Performance of Growing-Finishing Pigs Fed Ripe, Green, or Cooked Green Bananas

Criteria of performance	*Control*	*30% protein supplement + bananas*		
	Maize + supplement	*Ripe*	*Green*	*Cooked green*
Av daily gain, kg	0.68	0.56	0.46	0.50
Av daily feed, kg				
Bananas, kg	—	8.85	4.25	6.20
Supplement, kg[a]	—	0.71	1.04	0.88
Total dry feed, kg	2.31	2.48	1.89	2.13
Feed/gain ratio	3.41	4.44	4.16	4.26

Source: Hernández and Maner 1965, unpublished data. Eighteen pigs per treatment in two replications of nine pigs per group. Av initial wt, 28.3 kg. Each group removed from experiment when av wt of 92.0 kg obtained.

[a]30 percent protein supplement composed of fish meal, cottonseed meal, maize, vitamins, minerals and antibiotics.

The banana can be utilized fresh or as dried meal. Ecuadorian studies (Clavijo *et al.* 1970; Oliva *et al.* 1971) have clearly demonstrated that the pig will daily consume large quantities of bananas if they are allowed to ripen before they are fed (Table 13.27). Complementary studies (Clavijo and Maner 1971; Viteri and Maner 1971) have shown that the amount of consumption of ripe and green bananas is associated basically with differences in palatability of the two forms of presentation. Viteri and co-workers (1970) demonstrated that when growing pigs were fed equal daily quantities of either ripe or green bananas along with a controlled quantity of protein supplement, growth rate and efficiency of feed conversion were almost identical in both groups (Table 13.28).

Although Clavijo and Maner (1971) showed that the digestibility of protein, fat, and fiber of whole bananas is higher when allowed to ripen, they were unable to show any difference in digestible energy between green and ripe bananas (3114 vs. 3119 kcal/kg DM, respectively). Similar studies by Le Dividich *et al.* (1975) confirmed the lower protein digestion coefficients for green bananas and also

TABLE 13.28. Performance of Growing Pigs Fed Equal Quantities of Either Green or Ripe Bananas

Criteria of performance	*Treatment*	
	Ripe bananas	*Green bananas*
Initial wt, kg	31.9	32.1
Final wt, kg	61.1	60.3
Av daily gain, kg	.463	.449
Av daily consumption, kg	1.05	1.05
Av daily banana, kg	2.96	2.72
Feed/gain ratio	3.55	3.55

Source: Viteri et al. 1971, unpublished data.

TABLE 13.29. Changes in the Starch, Sugar, and Total Carbohydrate Content of Banana and Plantain Pulp During the Ripening Process Expressed as Percentage of Fresh Pulp

Variety	Number of ripening days							
	0	3	5	7	9	11	14	17
Gros Michel banana								
Starch	20.65	12.85	6.00	2.93	1.73	1.21	—	—
Total sugars	0.86	7.66	13.76	16.85	16.87	17.91	—	—
Total carbohydrates	21.51	20.49	19.76	19.78	18.60	19.62	—	—
Plantain								
Starch	32.20	31.68	30.90	30.48	28.52	20.17	11.69	6.12
Total sugars	0.82	0.85	1.02	0.92	3.84	9.78	18.89	21.10
Total carbohydrates	33.02	32.53	31.92	31.40	32.36	29.95	30.58	27.22

Source: Stratton and von Loesecke 1930.

reported a small improvement in digestible energy when the bananas were allowed to ripen (green, 3181, ripe, 3327 kcal/kg DM).

The difference in palatability between ripe and green bananas is immediately obvious to the consumer. The green banana is dry and bitter to the taste, whereas the ripened banana is sweeter and more moist. Palatability depends on the chemical composition of the fruit. Many chemical changes occur within the banana during the ripening process that greatly affect palatability. The water content of the pulp increases. This water is obtained from the peel and probably from the stalk. During the 10–11 days required for ripening to "eating-ripe" the moisture content of the banana pulp increases from approximately 74 to 77 or 78%; the pulp of plantains increases from 63 to 68% during the 16–18 days required for their ripening (von Loesecke 1950).

The most conspicuous change in the maturation of the banana is the conversion of starch to sugars (Table 13.29) (Stratton and von Loesecke 1930). During ripening there is a rapid decline in starch content and a corresponding increase in sugar content. Most of the sugar present is in the form of sucrose, glucose, and fructose, although maltose has been identified to be present in trace amounts. The predominant sugar in bananas is sucrose, some of which is hydrolyzed to glucose and fructose (Table 13.30) (Poland *et al.* 1938). Glucose accounts for approximately 58% of the total reducing sugars present and fructose for about 42%. Because of the needs for respiration there is a small but definite decrease in total carbohydrates in all varieties during the ripening process.

The starch and total sugar values for the peel are very small in magnitude when compared to those of the pulp, but follow a pattern of change similar to that of the pulp during the maturation process. Total sugars of the peel continue to increase as the fruit ripens and there is a corresponding decrease in total starch content.

A feature of the unripened banana is its strongly astringent taste. This taste is due to tannins present in the fruit. It has been shown that during the ripening process the banana loses its astringent taste although it has been suggested that the total amount of tannins in bananas remains practically constant during ripening. The loss of astringence is associated with a change in the state or chemical form of

TABLE 13.30. Average Distribution of Sugars in the Banana at Various Stages of Ripeness of Gros Michel Bananas

Stage of ripeness		*Percentage of fresh pulp*			*Percentage of total sugars*		
Days	*Color*	*Glucose*	*Fructose*	*Sucrose*	*Glucose*	*Fructose*	*Sucrose*
0	Green	2.24	1.45	7.65	19.24	12.46	68.30
3	Very slightly green	3.09	2.50	10.61	19.07	15.43	65.49
6	Green tip	3.99	2.75	12.00	21.29	14.67	64.03
9	Slightly speckled	4.21	3.24	12.08	21.56	16.59	61.85

Source: Poland et al. 1938; as quoted by von Loesecke 1950.

tannins. Tannins exist in bananas in two forms: (1) "free" or "active" tannin which imparts a strong, bitter taste to the fruit and (2) bound tannin or "vegatable tannate," which is insoluble and supposedly inert and which has little or no effect on palatability. During the ripening process free tannin decreases because the tannins are slowly changed to the bound state. The level of free or active tannin is much higher in the peel than in the pulp but it is significantly reduced in both by the time the fruit has matured to an "eating-ripe" stage (Barnell and Barnell 1945). The reduction in active tannin of the peel is much greater than in the pulp.

The ripe bananas, which are very palatable and readily consumed by the pig, are fed whole with the peeling. The low protein content and high moisture present in the banana require that a supplemental source of both protein and energy be supplied. Several studies have been conducted to determine the voluntary consumption pattern and performance of growing-finishing pigs fed ripe bananas and protein supplements containing varying levels of crude protein. One such study (Calles *et al.* 1970), showed that the average daily gain of growing-finishing pigs fed ripe bananas to appetite was significantly improved (770 vs. 660 g) if a 30% protein supplement was fed instead of a 40% supplement (Table 13.31). The

TABLE 13.31. Performance of Growing-Finishing Pigs Fed Ripe Bananas and Either a 30 or 40% Protein Supplement Free-Choice

Criteria of response	*Treatment*		
	Control	*Bananas + supplement*	
	16% protein	*30% protein*	*40% protein*
Av daily gain, kg[a]	0.87	0.77	0.66
Av daily fresh bananas, kg	—	8.29	8.85
Av daily bananas, kg DM[b]	—	1.84	1.97
Av daily supplement, kg	—	0.82	0.62
Total daily air-dry feed, kg	2.64	2.66	2.59
Feed/gain ratio	3.04	3.47	3.92
Protein in mixture consumed, %	16.0	12.4	18.0

Source: Calles et al. 1970.

[a]Seventy-two pigs, four replications of six pigs per treatment. Av initial wt, 28.2 kg; av final wt, 90.1 kg.

[b]Bananas expressed on 90 percent dry matter basis.

improvement in gain was assumed to be the effect of the additional energy consumed daily.

Although the total level of protein supplement consumed daily does not vary greatly during the entire period of growing and finishing, there is a marked increase in the daily voluntary consumption of ripe bananas. This increase is especially evident during the first 2–3 weeks of the feeding period. When allowed a voluntary choice of the quantity of both bananas and supplement, the pig will consume a diet during the growing-finishing period that contains 12.4–13.0% protein. From the results of a number of studies it is evident that an average of 8–8.8 kg per day of ripe bananas is all that the pig will consume during the growing-finishing period. The young pig (25–30 kg) will consume as much as 5–6 kg of bananas per day, and it appears from the results of several trials that a finishing pig approaching market weight will not consume more than 10–11 kg daily. When the 40% protein supplement is fed, energy becomes limiting. This is partially corrected when a 30% supplement replaces the 40% supplement. Although Clavijo *et al.* (1970) failed to show any further improvement in pig performance when supplemental protein levels below 30% were fed, Costa Rican workers (Musmani *et al.* 1979) reported continued linear improvement during growing and finishing when ripe bananas were supplemented with 30, 25, and 20% protein supplement (Table 13.32). As might be expected, each reduction in supplemental protein level resulted in an increase in quantity of supplement consumed and a reduction in daily banana consumption.

Under practical feeding conditions the recommended level of protein in the supplement will depend upon price relationships between protein and grain or energy sources. Locally available energy sources such as maize, milo, rice bran, and molasses can be efficiently used as an energy source for supplement dilution.

TABLE 13.32. Comparison of 20, 25, and 30% Protein Supplements for Growing and Finishing Pigs Fed Supplement and Ripe Bananas Free-Choice

Criteria of response	*Treatment*			
	Control	*Bananas + supplement*		
	16% protein	*20% protein*	*25% protein*	*30% protein*
Growing period (17–50 kg)				
Av daily gain, kg	0.68[a]	0.65[a,b]	0.63[b,c]	0.60[c]
Av daily consumption of fresh bananas, kg	—	8.92	8.61	10.61
Av daily consumption of supplement, kg	1.88	0.85	0.66	0.44
Av daily consumption of dry matter, kg	1.69[c]	2.54[a]	2.31[b]	2.51[a,b]
Feed/gain	2.49[c]	3.88[b]	3.68[b]	4.18[a]
Finishing period (50–90 kg)				
Av daily gain, kg	0.88[a]	0.81[b]	0.68[c]	0.67[c]
Av daily consumption of fresh bananas, kg	—	9.53	9.49	11.60
Av daily consumption of supplement, kg	2.91	1.49	1.12	0.87
Av daily consumption of dry matter, kg	2.62[c]	3.26[a]	2.91[b]	3.11[a,b]
Feed/gain	2.99[c]	4.00[b]	4.25[a,b]	4.61[a]

Source: Musmani et al. 1979.
[a,b,c]Means in the same line without a common superscript are different ($P < .05$).

Fresh bananas can be used efficiently for the gestating sow. Clavijo and coworkers (1971) fed gestating sows maintained on pasture a ration of ripe bananas and protein supplement and compared the performance of these sows to similar sows fed a basal 16% protein diet based on corn, wheat, wheat bran, alfalfa, and fish meal. The control sows were fed 1.5 kg of feed per day from the day of breeding until they were moved to the farrowing crates at 110 days of pregnancy. The banana-fed sows were fed a daily ration of 4.5 kg of ripe bananas and 600 g of a 40% protein supplement from day 1 to 76 of gestation. This quantity was increased to 6.0 kg of bananas and 800 g of supplement from day 77 to 110. The supplement employed was composed of 55.44% fish meal, 20% corn, 12.34% alfalfa meal, 2.82% bone meal, and 9.4% vitamin, trace mineral, and salt premix. The number and weight of the pigs farrowed was not different between the two treatments (Table 13.33). Both groups of sows entered the farrowing crates in good condition. In fact, the feeding regimen based on bananas increased sows' weights significantly above those fed the control diet. After farrowing, no differences in pig livability or performance were observed.

The lactating sow requires 5–6 kg of complete, mixed feed per day to meet the requirement for maintenance and milk secretion. If a similar quantity of dry matter were supplied from bananas and supplement, the sow would be required to consume more than 20 kg of bananas daily. This quantity appears to be greater than the sow's physical capacity. Because of the difficulty of adequate consumption and also because of the severe diarrhea that develops when the sow consumes a diet containing 14–15 kg of ripe bananas, ripe bananas are not generally recommended as the major energy source for the lactating sow. Clavijo and coworkers (1971) fed ripe bananas and protein supplement to lactating sows in the tropical banana zone of Ecuador and reported smaller and lighter litters when the sow ration was based on ripe bananas (Table 13.34). The control sows were allowed to consume a 16% grain supplement to appetite and the test sows were fed a mixture of bananas and supplement. The ratio of supplement to ripe bananas was 1.0 to 11. The sows were allowed voluntary consumption of this mixture. Under the conditions of the study pig mortality was high in both treatment groups,

TABLE 13.33. Performance of Gestating Sows Fed a Diet Based on Ripe Banana and a 40% Protein Supplement

Criteria of response	*Treatment*	
	Control (16% protein)	*Bananas + supplement (40% protein)*
Av daily concentrate, kg	1.66	0.67
Av daily bananas, kg	—	5.00
Av daily air-dried feed, kg[a]	1.66	1.67
Av daily crude protein, kg	.266	.318
Av no. of pigs per litter	8.9	8.4
Av pig wt at birth, kg	1.22	1.26
Av daily sow gain, day 1–100, kg	26.08	37.04

Source: Clavijo et al. 1971.
[a]Calculated on basis of approximately 10% moisture.

TABLE 13.34. Performance of Lactating Sows Fed Either a Complete Concentrate or Ripe Bananas Plus a Protein Supplement

Criteria of response	*Treatment*	
	Control	*Banana + supplement*
Av no. of pigs per litter	8.5	8.7
Av wt at birth, kg	1.31	1.24
Av no. of pigs weaned	6.3^{y}	5.9^{z}
Av wt at weaning, kg	10.5^{y}	9.2^{z}
Mortality, %	26.4	30.3
Av daily concentrate, kg	3.66	1.02
Av daily banana consumption, kg	—	11.22
Av daily protein consumption, kg	.586	.520
Wt loss of sows, kg	9.5^{y}	11.3^{z}

Source: Clavijo and Maner 1971.
^{y,z}Means in same line with different superscript are significantly different from each other.

and pig weights were lower than might be expected. The constant diarrhea of the sows fed bananas made it almost impossible to provide a satisfactory environment for the pigs.

The fresh plantain is similar to the banana in the appearance, but its chemical composition is somewhat different. The plantain contains a higher percentage of dry matter than the banana, and consequently it contains 10% more total carbohydrates than the banana. In diets for swine, the plantain should be fed in the same manner as previously described for bananas. Although results similar to those obtained with bananas might be expected, results (Clavijo *et al.* 1970) seem to indicate that the performance of growing-finishing pigs that were fed plantains (variety 'Dominico') was inferior to that of similar groups of pigs fed bananas.

The reason for the reduced performance is not readily apparent, but may have been associated with chemical stage of ripeness. Six to eight days more are required for the plantain to reach an ideal stage of ripeness than for the banana.

Although it is difficult to dry ripe bananas and plantains, the green fruit can be dried readily in the sun or by mechanical methods employing supplemental heat. The dried slices are then ground to form a meal. The banana meal prepared from green bananas with the peelings contains approximately 12% moisture, 4% ash, 4% protein, 3% fat, 3% fiber, and 74% nitrogen-free extract. The plantain meal from green plantains with the peeling contains about 10% moisture, 5% ash, 4% protein, 1% fat, 6% fiber, and 74% nitrogen-free extract.

Celleri and co-workers (1971B) used dried green banana meal as a substitute for grain to supply 0, 25, 50, or 75% of the total diet for growing-finishing pigs. The protein of the 16% protein diets was supplied by a combination of fish meal and cottonseed meal. Each increase in level of dried banana meal in the diet caused a small but significant linear reduction in average daily gain, and an increase in the quantity of feed required per kilogram of gain (Table 13.35). Similar studies (Oliva *et al.* 1971) confirm these findings. These workers employed levels of 0, 12, 24, 36, and 48% banana meal to replace maize in the basal ration. When the

TABLE 13.35. Performance of Growing-Finishing Pigs Fed Diets Containing Varying Levels of Green Banana Meal

Criteria of response	*Treatment*				
	1	*2*	*3*	*4*	*5*
Level of green banana meal, %[a]	0	25	50	75	—
Av days to slaughter, no.[a]	110	121	124	128	—
Av daily gain, kg[a]	.67	.65	.63	.61	—
Av daily feed, kg[a]	2.45	2.54	2.54	2.55	—
Feed/gain ratio[a]	3.66	3.88	4.04	4.19	—
Level of banana meal, %[b]	0	12	24	36	48
Days on trial, no.[b]	126	126	128	131	143
Av daily gain, kg[b]	.62	.60	.61	.59	.54
Av daily feed, kg[b]	2.62	2.59	2.78	2.78	2.82
Feed/gain ratio[b]	4.24	4.35	4.36	4.48	5.23

[a]Adapted from Celleri et al. (1971A) who utilized dehydrated commercial banana meal pellets.
[b]Adapted from Oliva et al. (1971) who utilized sun-dried, green banana slices for preparing meal.

level of meal was increased, a linear depression in growth rate was reported. This growth depression was associated with increased daily feed intake and reduced feed efficiency. It would appear from these studies that the amount of metabolizable energy is lower in banana meal than in maize.

Dried, green banana meal can supply approximately 50% of the diet of lactating sows. Studies by Ecuadorian workers (Viteri et al. 1971; Celleri et al. 1971A) indicate that the performance of sows and their offspring was not different whether they were fed a control 16% protein diet based on maize, wheat, and fish meal or whether green banana meal was used to substitute for approximately 50% of the maize (Table 13.36).

TABLE 13.36. Performance of Lactating Sows Fed Diets Containing Green Banana Meal with Peeling

Criteria of response	*Experiment 1[a]*		*Experiment 2[b]*	
	Control	*Banana meal*	*Control*	*Banana meal*
Level of banana meal, %	0	53.3	0	50
Number of sows, no.	7	7	7	8
Av pigs born per litter	8.28	8.43	9.86	10.37
Av birth wt, kg	1.20	1.32	1.08	1.11
Av pigs weaned per litter	6.00	5.86	6.71	8.12
Av weaning wt, kg	12.09	13.09	11.16	11.18
Av daily consumption per sow, kg	4.61	4.89	5.57	5.51
Av sow wt change, kg	+5.43	−7.43	−8.57	−19.62

Sources: [a]Viteri et al. 1971; [b]Celleri et al. 1971B.

CANE SUGAR

Crude and refined sugars, although not commonly used, are an excellent source of energy for swine feeding. Because both products are devoid of protein and contain only very small quantities of vitamins and minerals (Table 13.37), their efficient use is determined by the adequacy of the diet. Unlike cane molasses, sugar (sucrose), either refined or crude, is easily incorporated into the diet and does not cause digestive disturbances except in baby pigs whose digestive enzyme system is immature.

Refined sugar was used to replace maize in diets for growing-finishing pigs (Maner *et al.* 1969). Levels of 0, 15, 30, 45, and 60% were used in diets containing 16% crude protein supplied basically by soybean meal. In this study, each increase in level of sugar up to 45% of the diet was accompanied by an increase in daily gains and a decrease in feed required per kilogram of gain (Table 13.38). The performance of the pigs fed 60% sugar was not different from that of those fed 45% sugar diets. Increasing the sugar level in the diet causes an increase in digestible energy and a basic shift in the total amino acid pattern. In the control diet (maize–soya), the protein is supplied by both maize and soybean meal (approximately 50% from maize and 50% from soya). In the high-sugar diets, most of the protein comes from soybean meal.

Other studies with crude sugar (Buitrago *et al.* 1969) clearly indicate that this product can completely replace the maize in growing-finishing swine diets. Although the gains and feed conversions were not as outstanding as observed with refined sugars, proportions as large as 60% were well tolerated by the pigs and supported results not different from those observed with the maize–soybean meal control (Table 13.38). There was a tendency for improved feed efficiency when sugar composed 15–60% of the diet, in comparison with diets based on only maize.

Brooks (1972) also has demonstrated the superiority of sugar over maize as a source of digestible energy for growing-finishing pigs. Both growth and efficiency of feed conversion were improved when sugar was added to the diet (Table 13.39). The addition of 10% fat to the diet in the presence or absence of sugar (see Table 13.39) had no effect on growth, but it improved feed efficiency.

Following a series of feeding trials, Hawaiian workers (Brooks *et al.* 1970,

TABLE 13.37. Chemical Composition of Refined and Crude Sugar

Component	*Refined sugar*	*Crude sugar*
Moisture, %	0.5	2.1
Sugars, %	99.5	96.4
Protein, %	0	0
Ash, %	0	1.5
Calcium, mg/100 g.	0	85
Phosphorus, mg/100 g	0	19
Potassium, mg/100 g	3	30
Sodium, mg/100 g	1	344

Source: USDA 1963.

TABLE 13.38. Effect of Different Levels of Refined and Crude Sugar on the Performance of Growing and Finishing Pigs

	Treatment				
	1	*2*	*3*	*4*	*5*
Level of ingredients					
Sugar, %	0	15	30	45	60
Maize, %	81	62	43	24	5
Soybean meal, %	14	18	22	26	30
Performance—refined[a]					
Av daily gain, kg	0.82	0.87	0.89	0.95	0.93
Av daily feed, kg	2.68	2.78	2.81	2.83	2.75
Feed/gain ratio	3.27	3.20	3.16	3.00	2.95
Performance—crude[b]					
Av daily gain, kg	0.73	0.85	0.76	0.82	0.75
Feed/gain ratio	3.72	3.54	3.56	3.30	3.55

[a]Maner et al. 1969. Initial wt, 25 kg; final wt, 90 kg.
[b]Buitrago et al. 1969. Initial wt, 20 kg; final wt, 90 kg.

1971; Nakamura *et al.* 1971) reported the development of heart lesions of pigs fed high levels of dietary sucrose from weaning to 75–125 kg. A hemorrhagic syndrome and gross lesions of a granulomatous atrial endocarditis were observed in the left atrium of each pig fed high levels of sugar (sucrose), while no lesions could be found in the hearts of pigs fed the maize–soybean meal control diet. Later studies revealed that the hemorrhagic syndrome and atrial heart lesions were the result of a vitamin K deficiency that could be completely overcome by the addition of 11 mg of menadione per 100 kg of feed. In studies by Maner *et al.* (1969), gross and histopathological examination of hearts and arteries of pigs fed 60% sucrose with or without the addition of 10% tallow during the growing-finishing period up to 125 kg failed to show any cardiovascular lesions.

Buitrago (1972) utilized refined cane sugar to supply the energy for sows during gestation. The sows were fed constant levels (0.62kg) of protein supple-

TABLE 13.39. Effect of High Levels of Sugar and Fat on the Performance of Growing-Finishing Pigs

	Treatment			
	1 *Maize–soya*	*2* *Maize–soya (10% fat)*	*3* *64% sugar–soya*	*4* *50% sugar–soya (10% fat)*
Crude protein, %	16	18	15	17
Dig. energy, kcal/kg	3,450	3,890	3,810	4,200
Performance				
Av daily gain, kg	0.70	0.72	0.74	0.74
Feed/gain ratio	3.0	2.4	2.7	2.3
Dig. energy/g gain	10.5	10.5	10.3	9.7

Source: Brooks 1972.

TABLE 13.40. Performance of Gestating Sows Fed Refined Sugar (Sucrose) as the Major Energy Source

Criteria	*Energy level*			
	Medium		*High*	
	(kg)	*(Mcal)*[a]	*(kg)*	*(Mcal)*[a]
Diet				
Protein supplement	0.62	1.60	0.62	1.60
Refined sugar	0.95	3.54	1.73	6.44
Total diet	1.57	5.14	2.35	8.04
Performance				
No. gilts bred	27		27	
Initial wt, kg	133.7		135.8	
Final wt, kg	172.9		192.5	
Total feed intake, kg	179.0		268.0	
No. gilts farrowing	21		22	
No. pigs at birth	8.0		8.45	
Av wt birth, kg	1.18		1.21	
Av litter wt, kg	9.46		10.27	
Av weaning wt, kg	12.43		12.00	

Source: Buitrago 1972.
[a]Megacalories (therms).

ment and a medium (0.95 kg) and high (1.73 kg) level of energy provided from refined sugar. The performance of the sows maintained in confinement was satisfactory and not different from average values obtained from diets based on maize and soybean meal (Table 13.40).

Refined sugar was used in studies by Obando *et al.* (1970) to make up 60% of a lactation diet for sows. The diet contained 15.8% crude protein, the majority of which was supplied by soybean meal. When compared with the maize–soybean meal control, sows fed the sugar diet consumed more feed, gained more weight during lactation (6.8 vs. −9.1 kg), and weaned an equal number of heavier pigs. Pre-starter diets containing either 10 or 25% sugar with or without added methionine and fed to the nursing pigs had no significant effect on baby pigs performance. Independent of pre-starter diet, baby pigs from sows fed sucrose diets consumed significantly less pre-starter diet. Later studies at the same station confirmed that creep ration consumption was reduced because baby pigs preferred the 60% sugar lactation diet offered to their dam.

Only limited information on beet sugar is available; however, it is assumed that beet sugar would have properties and feeding value equal or similar to cane sugar.

CANE MOLASSES

Molasses, as a major by-product of sugar production, is a satisfactory and economical source of energy for livestock feeding in many tropical and subtropical areas of the world. An average of 25–50 kg of final molasses are produced for every 100 kg of refined sugar. Both of these quantities result from the processing of 1 ton of sugar cane (Scott 1953; Obando *et al.* 1969; Zorrilla and Merino 1970). Average yields of sugar cane vary between 60 and 100 tons per hectare under optimal agronomic and environmental conditions (FAO 1970).

Scott (1953) and Ferrando and Theodossiades (1960) have published detailed reviews on the use of sugar cane by-products in livestock feeding. More recently, Buitrago (1972) reviewed the use of sugar cane by-products for the pig.

Molasses, whose water content varies between 15 and 25%, contains a large quantity of carbohydrates. The nitrogen-free extract fraction may reach 65% and is composed mainly of water- and alcohol-soluble carbohydrates such as sucrose (32.2%), glucose (8.6%), and frustose (15.5%) (Le Dividich *et al.* 1978). The sugars present in the molasses have a high digestibility when fed to pigs of all ages except during the first 21–30 days after birth when the secretion of the enzyme, sucrase is not well developed in the baby pig. Thus, molasses is basically an energy source because of its carbohydrate composition, but because of its high level of both water and ash the gross energy level of the crude product is relatively low (2850 kcal/kg). The level of ash in molasses is high but variable (Tables 13.41 and 13.42). On the contrary, molasses is a poor source of nitrogen. It contains only about 0.50–0.56% total nitrogen of which only 38–50% is amino acid nitrogen. The remainder is made up of other nitrogen-containing compounds such as ammonia, nitric acid, amides, and albuminoids (Obando *et al.* 1969; Le Dividich *et al.* 1978). Of the amino acids present (Table 13.43), only a small percentage are essential amino acids with the main portion represented by nonessential amino acids such as aspartic and glutamic acid.

One of the major problems encountered in the utilization of cane molasses is the variability of the product. The age, type, and quality of the cane, and the system of collection and processing, are some of the factors that can drastically modify the nutrient content of cane molasses and other by-products of sugar production. For example, the percentage of ash in different types of molasses is highly variable, especially in the concentration of potassium, sodium, and magnesium, which depend in large part on the fertility of the soil and on the degree of supplemental fertilization. Other minor elements, such as iron, copper, alumi-

TABLE 13.41. Chemical Composition of Final and High-Test Molasses

Component	*Final molasses*			*High-test molasses*
	National Molasses Company 1970	*Herrera et al. 1970*	*MacLeod et al. 1968*	*Preston and Willis 1970*
Moisture, %	25.00	22.00	23.10	23.9
Total sugars, %	52.20	62.00	53.00	69.00
Crude protein, %	3.00	2.30	3.40	1.07
Ash, %	8.10	8.81	5.51	2.03
Calcium, %	0.60	0.64	0.71	0.22
Phosphorus, %	0.07	0.05	0.06	0.05
Potassium, %	2.60	—	2.00	0.58
Sodium, %	0.20	—	0.83	0.16
Magnesium, %	0.30	—	0.45	0.11

Source: Adapted from Buitrago 1972.

TABLE 13.42. Variations in Mineral Composition of Several Types of Puerto Rican Molasses

Sample	Percent of total molasses								Parts per million (mg/kg)		
	Ash	*Silicon oxide*	*Phosphorus*	*Calcium*	*Potassium*	*Magnesium*	*Sodium*	*Iron*	*Zinc*	*Manganese*	*Copper*
1	8.9	.43	.07	.56	3.55	.57	.16	.05	39	59	40
2	7.1	.95	.09	.87	1.47	.27	.09	.22	65	188	29
3	8.6	.67	.11	.71	2.39	.63	.10	.14	133	113	189
4	9.9	.28	.11	1.00	1.89	.37	.16	.29	44	278	45
5	10.8	.30	.06	.64	3.68	.58	.20	.04	46	46	36
6	6.8	.29	.04	.51	1.80	.48	.07	.02	39	32	14
7	8.1	.41	.05	1.22	2.32	.23	.35	.07	61	67	60
8	9.8	.27	.05	1.68	3.58	.33	.13	.04	11	38	64
9	8.8	.54	.08	.30	2.83	.72	.11	.07	89	43	27
10	10.0	.86	.08	1.20	3.33	.31	.19	.07	92	87	70
11	8.3	.38	.15	.73	2.46	.43	.16	.04	49	51	22

Source: Adapted from Buitrago 1972.

TABLE 13.43. Amino Acid Composition of Final Molasses

Amino acid	*g/16 g N*	*Amino acid*	*g/16 g N*
Arginine	0.5	Tyrosine	0.7
Histidine	0.3	Threonine	1.5
Isoleucine	1.1	Valine	2.3
Leucine	1.3	Alanine	4.6
Lysine	0.6	Aspartic acid	22.4
Methionine	0.5	Glutamic acid	10.6
Cystine	0.6	Glycine	1.6
Phenylalanine	0.8	Proline	1.3

Source: Christon and Le Dividich 1978.

num, zinc and manganese, can also vary according to soil fertility and processing methods (Table 13.42). Although variable, final molasses contains significant levels of some of the B vitamins (Table 13.44).

Since the initiation of research on the use of molasses in swine diets (Burns 1909; Barnett and Godell 1923; Henke 1933; Gochangeo 1933; Bray 1945), it has been recognized that large amounts of molasses (more than 25–30%) caused a laxative effect on the pig and, owing to the moisture present, tended to dilute the energy concentration of the diet, which is reflected in increased feed per unit of gain.

In the initial studies (Henke 1933; Willett *et al.* 1946), it was indicated that a maximum of 20% final molasses should be used to replace grains of the diet if satisfactory performance and weight gains were to be obtained. Iwanaga and Otagaki (1959) demonstrated that the tolerance of pigs to molasses increased with age and body weight. Maximum TDN and digestible energy of molasses were obtained when molasses made up only 10% of the diet. In these studies it was reported that 13–34-, 35–68-, or 68–90-kg pigs presented diarrhea when the diet

TABLE 13.44. Vitamin Content of Final Molasses (mg/kg)

Vitamins	*Source*				
	1[a]	*2*[b]	*3*[c]	*4*[d]	*5*[e]
Thiamine	—	0.9	0.9	0.7	1.0
Riboflavin	3.3	2.6	2.3	2.0	3.0
Pantothenic acid	17.6	37.8	39.0	50.8	39.5
Pyridoxine	—	—	—	1.5	6.7
Niacine	11.0	40.8	45.0	17.7	34.8
Folic acid	—	—	—	0.0	0.1
Choline	880.0	787.5	660.0	—	843.0
Biotin	0.77	—	0.70	1.4	0.7
Vitamin E (IU/kg)	—	—	4.4	—	

Sources: [a]Buitrago 1972; [b]Ferrando and Theodossiades 1960; [c]NRC 1979; [d]Brinkley and Walfrom 1953; [e]Harris 1974.

contained more than 10, 20, or 30% molasses, respectively. Other workers (Iwanaga *et al.* 1959; Blanco *et al.* 1964; Moncada and Maner 1964; Preston and Willis 1969; Maner *et al.* 1969A) have repeatedly confirmed the laxative effect of high levels of molasses in pig diets. However, levels lower than 30% have been used satisfactorily and the "loose-feces" resulting from this amount of molasses has not presented a major problem of either animal health or sanitation. Levels of more than 20–25% do, however, cause problems of mixing and especially in feeding management, since rations containing more than those amounts are sticky and are difficult to feed in automatic feeders.

A summary of some research findings related to the use of different quantities of molasses in diets for growing-finishing pigs is given in Table 13.45. Most of these data confirm earlier findings that the feed per gain ratio is increased with each increase in level of dietary molasses, although average daily gains may not always be significantly affected. This reduction in efficiency of feed utilization is the result of decreased energy density brought about by the level of moisture present in the molasses. Since the pig tends to consume the quantity of energy required even with diets of varying energy densities, diets high in molasses and with less total energy are generally consumed in larger quantities by the pig in order to meet daily energy needs. Although Blanco *et al.* (1964) reported a significant depression in growth rate when 30 or 45% molasses were fed to growing-finishing pigs, other workers have reported no significant effect on average daily gains (Combs and Wallace 1969, 1970; Obando *et al.* 1969) or an improvement in growth rate resulting from added molasses (Babatunde *et al.* 1975). It is, therefore, possible for the pig to gain at a rate equal to that of pigs consuming diets free of molasses if the level of molasses is not too high to prevent adequate energy intake (Iwanaga and Otagaki 1959).

The addition of molasses to the diet of pigs causes a reduction in the apparent digestibility of nitrogen and energy (Brooks and Iwanaga 1967; Combs and Wallace 1973; Le Dividich *et al.* 1974; Le Dividich and Canope 1975). The effect is accentuated at higher levels of molasses addition. From results with rats, Christon and Le Dividich (1978) suggested that the effect was the result of an increase in metabolic fecal nitrogen. However, it would appear that a general reduction in digestibility and/or absorption might result from an increased rate of passage of the diet through the gastrointestinal tract, induced by the laxative effect of the molasses. Because of the small reduction in digestibility of nitrogen, Le Dividich *et al.* (1974) studied the effect of increasing the level of crude protein in diets for young piglets as the level of molasses was increased in the diet. By maintaining equal levels of digestible crude protein they did not observe any significant differences in pig performance between dietary molasses levels of 0, 20, and 40%; however, piglets fed 40% molasses gained 7.0% more slowly than the control and consumed 13.3% less feed. Earlier studies (Obando *et al.* 1969) using growing-finishing pigs failed to show any improvement in pig performance when dietary protein level was increased from 16 to 19%. In fact, their results demonstrated that because of the improvement in amino acid balance brought about by an increase in soybean meal when maize was replaced by molasses, the level of dietary protein could be reduced to 13% without affecting either daily gains or efficiency of feed utilization.

The two major limiting factors, laxative effect and low energy density, in the

TABLE 13.45. A Summary of the Effect of Different Levels of Cane Molasses on the Performance of Growing-Finishing Swine

Level of molasses (%)	*Pig body weights (kg)*	*Av daily gain (kg)*	*Feed/gain ratio*	*Reference*
0	23–97	0.64	4.07	Blanco et al. 1964
15	23–97	0.60	4.35	Blanco et al. 1964
30	23–97	0.49	4.65	Blanco et al. 1964
45	23–97	0.38	4.82	Blanco et al. 1964
0	60–97	0.81	3.64	Blanco et al. 1964
15	60–97	0.82	3.79	Blanco et al. 1964
30	60–97	0.75	3.96	Blanco et al. 1964
45	60–97	0.67	4.34	Blanco et al. 1964
0	17–82	0.71	3.26	Brooks and Iwanaga 1967
10	17–82	0.76	3.44	Brooks and Iwanaga 1967
50	17–82	0.55	5.22	Brooks and Iwanaga 1967
5	50–85	—	3.96	Bravo and Cabello 1968
10	50–85	—	4.20	Bravo and Cabello 1968
15	50–85	—	4.62	Bravo and Cabello 1968
10	13–45	0.66	2.12	Combs and Wallace 1969, 1970
20	13–45	0.72	2.42	Combs and Wallace 1969, 1970
30	13–45	0.65	2.39	Combs and Wallace 1969, 1970
40	13–45	0.60	2.53	Combs and Wallace 1969, 1970
15	20–90	0.77	3.31	Obando et al. 1969
22.5	20–90	0.76	3.63	Obando et al. 1969
30	20–90	0.72	3.65	Obando et al. 1969
0	15–95	0.76	3.20	Obando et al. 1969
15	15–95	0.81	3.40	Obando et al. 1969
22.5	15–95	0.82	3.44	Obando et al. 1969
30	15–95	0.81	3.54	Obando et al. 1969
0	22–90	0.71	3.28	Obando et al. 1969
15	22–90	0.75	3.44	Obando et al. 1969
22.5	22–90	0.74	3.55	Obando et al. 1969
30	22–90	0.74	3.67	Obando et al. 1969
0	16–65	0.62	3.22	Babatunde et al. 1975
10	16–65	0.71	3.11	Babatunde et al. 1975
20	16–65	0.69	3.17	Babatunde et al. 1975
30	16–65	0.66	3.27	Babatunde et al. 1975
40	16–65	0.66	3.66	Babatunde et al. 1975

use of high levels of molasses have been studied by several workers (MacLeod *et al.* 1968; Corzo *et al.* 1968; Obando *et al.* 1968; Maner *et al.* 1969A,B; Velázquez *et al.* 1969; Ly and Velázquez 1969; Brooks and Iwanaga 1967; Preston and Willis 1969; Brooks 1972).

The digestive disturbances produced by high concentrations of molasses in the diet have been attributed to increased osmotic pressure caused by large quantities of potassium ions present in the molasses. However, when potassium salts of carbonate and acetate were included in the diet at levels equal to or greater than levels present when 30% molasses were fed, no changes in the consistency of the feces were observed (Obando *et al.* 1968). However, when potassium salts of sulfate and chloride were added, there was a significant increase in the moisture content of the feces, although to a lesser degree than was observed when 30% molasses were fed (Maner *et al.* 1969A). In the same study it was shown that sulfate and chloride salts of magnesium also caused some increase in fecal moisture. Although it has not been documented, it appears that the presence of not only specific salts but the total mineral concentration may play a role in the laxative effect of molasses.

It has been suggested that the laxative effect of molasses may be due to an insufficiency of intestinal sucrase to hydrolyze adequately the sucrose present in the product (Velázquez *et al.* 1969; Ly and Velázquez 1969; Ly 1971). Since sucrose is the main carbohydrate in molasses, a deficiency of this enzyme might induce digestive disturbances. The presence of an appreciable amount of soluble sugars in the cecum of pigs fed high levels of molasses (Ly 1975) supports the idea of an enzyme insufficiency related to incomplete hydrolysis of the carbohydrates present in final molasses. However, other research data would not seem to support this hypothesis. Maner *et al.* (1969B) have shown that dietary levels of sucrose as high as 60% can be efficiently utilized by the growing-finishing pig and even at this level of inclusion, neither diarrhea nor digestive disturbances were observed. Data have also been reported (Ly *et al.* 1978) which indicate no differences in sucrase and maltase specific activity between groups of pigs fed either 0, 45, 60, or 75% molasses. On the contrary, a significant increase in both total and specific activity of sucrase occurred in rats fed on 30% molasses compared with those fed sucrose or maize starch (Le Dividich *et al.* 1978). This effect could not be attributed to the sucrose, glucose, or fructose contained in molasses. The presence of other complex carbohydrates and other sugars, such as raffinose, which are poorly hydrolyzed by the pig, may become additive factors in the laxative effect (MacLeod *et al.* 1968).

One method by which the laxative effect of high levels of molasses has been reduced is the inclusion of fibrous products in the ration (Brooks and Iwanaga 1967). Thirteen percent cane bagasse, incorporated into growing-finishing swine diets containing 50% final molasses, helps to reduce and almost to eliminate the laxative effects of the molasses. However, this treatment was unsatisfactory in practice since it caused a significant decrease in both growth and feed efficiency (Table 13.46).

TABLE 13.46. Effect of Including Bagasse on Performance of Growing-Finishing Pigs Fed 50% Final Molasses

	Treatment		
Performance	*Control maize–soya*	*10% molasses*	*50% molasses 13% bagasse*
Av daily gain, kg[a]	0.72	0.76	0.55
Feed/gain ratio	3.19	3.44	5.22
Dig. energy in diet, kcal/kg	3,410	3,200	2,390
Dig. energy/gain, kcal/kg	10,900	11,000	12,500
Back fat, cm	2.7	2.8	2.5

Source: Brooks and Iwanaga 1967.
[a]Initial wt, 17 kg; final wt, 82 kg.

TABLE 13.47. Effect of Sucrose Additions on the Utilization of Final Molasses by Growing-Finishing Pigs

Performance	*Mixtures of cane sugar/final molasses*		
	20/60	*40/40*	*60/20*
Av daily gain, kg[a]	0.51	0.56	0.59
Feed/gain ratio	3.70	3.57	3.37
Moisture in feces, %	74.5	66.2	59.3

Source: Preston and Willis 1970.
[a]Initial wt, 22 kg; final wt, 92 kg.

To overcome the growth and feed efficiency depressions caused by high levels of molasses that reduce dietary energy density. Hawaiian workers (Brooks and Iwanaga 1967; Brooks 1972) have supplemented high-molasses diets with various types of fats. Combinations of molasses, wheat bran, and either 10 or 20% fat have supported pig performance equal to that obtained with molasses-free diets.

Other alternatives that have been proposed include the addition of crude or refined sugar to the final molasses diets and the utilization of high-test molasses (Preston and Willis 1969, 1970; MacLeod *et al.* 1968). Preston and Willis (1970) demonstrated that increasing the level of sugar in diets containing final molasses improved both gains and efficiency of feed conversion and showed that levels of final molasses as high as 60% of the total diet can be used satisfactorily for growing-finishing pigs if 20% crude or refined sugar is also included in the diet (Table 13.47). The replacement of final molasses with high-test molasses improves both rate of gain and efficiency of feed conversion and reduces digestive disturbances (Marrero and Ly 1976).

Cuban investigators have used both high-test molasses and whole molasses to replace all of the maize in swine diets. High-test molasses is a heavy cane syrup from which none of the sugar has been extracted, but rather partially inverted to prevent crystallization. Whole molasses is produced by concentrating cane juice that has not been submitted to the clarification process with sodium hydroxide. High-test molasses differs from final molasses in having a much lower ash content and approximately two-thirds of the total sugars in the form of monosaccharides (Table 13.48). The high-test molasses, which has been reported to have a metabolizable energy value of 3.750 kcal/kg of dry matter (Velázquez *et al.* 1969) has satisfactorily replaced all of the grain in growing-finishing swine diets (Table 13.49) (Preston and Willis 1969; MacLeod *et al.* 1968; Velázquez and Preston 1970; Velázquez and Preston 1971; Ly and Lezcano 1980). Similar studies have shown that although high-test molasses are superior to final molasses, both rate of gain and efficiency of feed utilization are inferior to those obtained with maize (Castro and Elias 1978).

Final molasses has also been used to supply a portion of the energy in gestation and lactation rations for sows and gilts. Hawaiian workers (Brooks and Iwanaga 1967) fed a control maize–soybean meal diet and a diet containing 37.36% final molasses, 13.3% bagasse pith, and 12.3% fat to gilts during a 50-day development period and throughout gestation and lactation and to mature sows during gestation and lactation. Reproductive performance in terms of number of pigs

TABLE 13.48. Chemical Composition of High-Test and Final Molasses

Item	*High-test molasses*	*Final molasses*
Dry matter, %	81.05	81.95
Sucrose, %	27.10	49.92
Reducing sugars, %	53.50	24.30
Glucose, %	0.84	10.99
Gross energy, kcal/kg DM	3.92	3.87
Nitrogen	0.28	0.60
Ash, %	3.52	11.54
Potassium	0.23	2.03
Silica, SiO_2	0.14	1.01

Source: Marrero and Ly 1976.

born and pig size at birth was not altered in gilts by feeding the molasses, fat, bagasse diet during a 50-day development period and throughout gestation. The performance of mature sows fed this diet during gestation was equal to the performance of similar sows fed maize–soybean meal diets. When these molasses diets were self-fed to gilts and sows during lactation, however, feed consumption, hence baby pig growth, was lower than for similar gilts and sows fed maize-based diets. Gilts and sows fed the molasses diets during lactation also lost more weight during lactation than the control sows and gilts.

Molasses has served as an effective laxative agent to prevent constipation of sows and gilts during late gestation and early lactation. Dietary levels of 15–25% molasses increase fecal moisture, increase rate of passage, and eliminate constipation problems often associated with the transfer of gestating animals into the farrowing facility.

TABLE 13.49. High-Test Molasses and Whole Molasses in Diets for Growing-Finishing Pigs

Criteria	*High-test molasses*		*Whole molasses*	
	1	*2*	*1*	*2*
Level of protein, %	16	12	16	12
Level of molasses, %	71	77	71	77
Av daily gain, kg	0.60	0.61	0.54	0.48
Feed/gain ratio	4.12	3.80	3.82	4.15

Source: Velázquez and Preston 1970.

DRIED BAKERY PRODUCT

Dried bakery product is produced from reclaimed bakery products that have been blended and processed to provide a product that contains 9.0–9.5% crude protein and 11.0–13.0% crude fat and a maximum of 1.0–1.5% crude fiber, 3.5% ash, and 4.5 kcal/g (Sewell 1965; Arrington *et al.* 1964).

Dried bakery product is an excellent ingredient in starter rations for young pigs. The product is highly palatable, and the high level of fat makes it a good energy source for young pigs. Sewell (1965) replaced 10, 20, or 30% of the maize meal in early weaning starter diets, containing maize, soybean meal, corn sugar (glucose), dried skim milk, animal fat, vitamins, and minerals, and reported that levels of 10 and 20% bakery product slightly increased gains and feed efficiency, but the 30% level of substitution significantly increased gains by 18.6% (Table 13.50). Dried bakery product also increased gains when 30% was used to replace corn sugar and maize; however, gains were not improved when an equal quantity replaced corn sugar, dried whey, and maize. It was further shown (Sewell 1965) that the high salt content of the dried bakery product is not deleterious to the pig.

Florida studies (Wallace *et al.* 1964; Arrington *et al.* 1964) have shown that dried bakery product has no practical value as a vitamin source for rats and for growing-finishing pigs, but that the high fat content (11–13%) makes it a high-energy ingredient that can be used to substitute for dietary maize or other grain sources when cost of the product is similar to that of the grain. Owing to the high level of fat, increasing the level of dried bakery product reduces the amount of feed required per unit of body weight gain.

Studies by Torres and Shimada (1977) have also shown the value of stale bread as a substitute for sorghum in diets for growing-finishing pigs. Complete replacement of sorghum in a sorghum–soybean meal basal diet with stale bread had no effect on average daily gain, but daily feed intake was reduced and feed efficiency was improved.

TABLE 13.50. Dried Bakery Product as Replacement for Various Ingredients in Baby Pig Starter Diets

Treatment	*Av daily gain (kg)*	*Av feed/gain ratio (kg)*
Experiment 1		
1. Basal (maize)	0.391	1.79
2. Basal with 10% DBP[a]	0.432	1.81
3. Basal with 20% DBP	0.418	1.84
4. Basal with 30% DBP	0.464	1.72
Experiment 2		
1. Basal without DBP	0.405	1.92
2. Basal with 30% DBP[b]	0.473	1.68
3. Basal with 30% DBP[c]	0.441	1.84
Experiment 3		
1. Basal without DBP	0.323	1.91
2. Basal with 30% DBP[b]	0.373	1.79
3. Basal with 30% DBP[c]	0.268	1.97
4. Basal with 50% DBP[d]	0.309	1.75
5. Basal with 50% DBP[d]	0.355	1.71

Source: Adapted from Sewell 1965.
[a]Dried bakery product (DBP) replaced equal quantity of maize.
[b]DBP replaced 10% maize sugar and 20% maize.
[c]DBP replaced 10% maize sugar, 15% dried whey and 5% maize.
[d]DBP replaced 10% maize sugar, 39% maize and 1% soybean meal.

CITRUS AND CITRUS PULP

In any fresh fruit industry, some fruit falls from the tree or is otherwise unfit for human consumption or for processing. During certain seasons of the year in citrus-producing areas of the world, significant quantities of oranges, grapefruits, and tangerines are available for livestock feeding. Cunha *et al.* (1950) have shown that tangerines can be used as an energy source for growing-finishing pigs. Pigs normally consume only the pulp of the tangerine and refuse the peel. Some problems of diarrhea or loose feces have been observed when pigs are fed a major portion of their daily ration in the form of reject or fallen tangerines. When offered tangerines, a protein supplement and a mineral mix free-choice, the pigs consumed an average of 12.64 kg of tangerines per day along with 0.97 kg of a protein supplement composed of 50% peanut meal, 25% meat scraps, and 25% dehydrated alfalfa meal (Table 13.51). Performance of these pigs was similar but slightly inferior to that in groups of pigs fed either tangerines, maize, and a protein supplement or a diet composed of protein supplement and maize. However, when properly supplemented, the tangerines provided an economical source of dietary energy.

The cannery residue ejected by citrus fruit extractors is the raw material for the production of dried citrus pulp. This residue consists of the peel, pulp (rag) and seeds of grapefruit and oranges after the juice has been extracted. This residue is shredded or ground and mixed with either calcium hydroxide or calcium oxide at a level of 0.3–0.5%. After the lime treatment to degrade and demethylate the pectins of the peel, the residue is pressed to remove a portion of the juices. After the pressing, the moisture content of the pulp is reduced to 6–8% in a rotating drier. From this dried product, three fractions—dried citrus pulp (85%), citrus meal (14%), and citrus fines (1%)—are separated (Hendrickson and Kesterson 1965).

The pulp produced contains a fairly large quantity of fiber and nitrogen-free extract (Table 13.52). The level of calcium and potassium in the final product is significant (Hendrickson and Kesterson 1965; Ammerman *et al.* 1970).

TABLE 13.51. Effect of Feeding Cull Tangerines on the Performance of Growing Swine

Performance	*Treatment*		
	1 *Tangerine +Trio Mix*	*2* *Tangerine +Trio Mix +maize*	*3* *Trio Mix +maize*
Av daily feed			
Tangerines, kg	12.64	7.82	—
Trio Mix, kg[a]	.97	.55	.65
Maize, kg	—	1.81	2.95
Minerals, kg	.008	.002	.003
Av daily gain, g	618	700	664

Source: Cunha et al. 1950.
[a]Contained 50% peanut meal, 25% meat scraps, and 25% alfalfa meal.

TABLE 13.52. Nutrient, Mineral, and Amino Acid Composition of Dried Citrus Pulp

Nutrients, %[a]		*Minerals*[b]		*Amino acids, %*[c]	
Dry matter	90.1	Calcium, %	1.47	Arginine	0.16
Crude protein	5.9	Phosphorus, %	0.11	Histidine	0.10
Crude fiber	11.5	Potassium, %	1.12	Isoleucine	0.18
Nitrogen-free extract	62.7	Sodium, %	0.10	Leucine	0.29
Ether extract	3.1	Magnesium, %	0.12	Lysine	0.11
Ash	6.9	Iron, ppm	94	Methionine	0.02
		Zinc, ppm	11.4	Cystine	0.07
		Copper, ppm	6.0	Phenylalanine	0.19
		Manganese, ppm	5.8	Threonine	0.16
				Tryptophan	0.02
				Valine	0.23
				Protein	5.87

Sources: [a]Hendrickson and Kesterson 1965; [b]Ammerman et al. 1970; and [c]Lanza and Messina 1979B.

Allen (1919) showed cooked citrus rinds to be nontoxic to hogs, but to have little feeding value. Kirk and Crown (1947) fed growing-finishing pig diets containing 0, 5, 10, and 20% dried grapefruit pulp. At the 5% level, feed required and gains were comparable with those of the control diet containing no pulp (Table 13.53). Levels higher than 5% caused a reduction in feed intake and rate of gain, increased the feed required per unit of gain, and caused frequent and serious digestive disturbances. Mineral imbalance in some citrus products may be a contributing factor in lower pig performance since hydrated lime used to degrade pectins and to remove bound water in processing citrus pulp produces citrus meal relatively high in calcium (1.5–2.9%) and low in phosphorus (0.10–0.15%) (Ammerman *et al.* 1961). Hollis *et al.* (1969) added phosphoric acid to citrus meal in swine diets to give more desirable calcium : phosphorus ratios and obtained satisfactory gains and carcasses when diets contained up to 15% citrus meal.

Citrus seed meal was fed at levels of 0, 10, and 25% to growing-finishing pigs (Glasscock *et al.* 1950). The citrus seed meal proved to be unpalatable to the pigs and caused severe depressions in pig performance (Table 13.54). It was later shown that the citrus seed meal was also toxic to chicks, causing enlarged gall

TABLE 13.53. Effect of Level of Dried Grapefruit Pulp on the Performance of Growing-Finishing Pigs

Level of pulp (%)	*Av daily gains (g)*	*Feed/gain ratio*
0	486	3.15
5	464	3.11
10	441	3.35
20	409	3.39

Source: Kirk and Crown 1947.

TABLE 13.54. Effect of Citrus Seed Meal on the Performance of Growing-Finishing Pigs

Level of citrus seed meal (%)	*Av daily gain* (g)	*Feed/gain ratio*
0	845	4.08
10	523	4.59
25	135	10.06

Source: Glasscock et al. 1950.

bladders, mottled livers, congested intestines, and high mortality. It was postulated that the toxic factor was a bitter principle in the seed known as limonin, that could be eliminated by successively extracting the meal with diethyl ether, acetone, and 95% ethyl alcohol.

Baird *et al.* (1974) fed growing-finishing pigs diets containing 0, 10, 20, and 40% dried citrus pulp. Even though by the addition of prime tallow energy levels were held constant in the diets and monosodium phosphate was added as needed to lower the calcium : phosphorus ratio, daily gains were significantly depressed with increasing levels of citrus pulp. Daily feed consumption also was depressed; however, this gave a limited feed effect and generally improved feed conversion. When estimated energy was held near constant by the addition of tallow, citrus pulp had maize replacement values of 97, 95, and 82% at the 10, 20, and 40% levels of substitution (Table 13.55). When citrus pulp replaced maize in pelleted diets without added energy, the maize replacement value was 95 and 78% at levels of 10 and 20% citrus pulp.

Higher levels of maize replacement by citrus pulp, up to 60%, have been made by Lanza and Messina (1979A,B). Levels of substitution up to 20% caused only small reductions in growth and feed consumption; however, levels of 50 and 60% significantly reduced gains and feed intake and increased the quantity of feed required to produce a kilogram of gain (Table 13.56).

Apparent digestion coefficients, digestible energy, metabolizable energy, and metabolizable energy corrected for nitrogen of dried citrus pulp were determined

TABLE 13.55. Influence of Level of Dried Citrus Pulp on Performance of 15–95-kg Pigs with Addition of Prime Tallow to Correct Energy Level

Pulp level, %	*Tallow level,* %	*Av daily gain,* kg	*Av daily feed,* kg	*Feed/gain ratio*	*Maize replacement value,* %
0	0.1	0.80[a]	2.56	3.24	100
10	1.8	0.74[b]	2.30	3.10	97.4
20	4.9	0.72[b,c]	2.19	3.06	95.5
40	11.3	0.63[c]	1.99	3.13	82.2

Source: Baird et al. 1974.
[a,b,c]Means in the same column without common superscript are significantly different from each other.

TABLE 13.56. Effect of Level of Dried Orange Pulp on Performance of 40–90 kg Pigs

Level of pulp, %	*Av daily gain, g*	*Av daily feed, kg*	*Feed/gain ratio*
Trial 1			
0	776	2.72	3.50
10	770	2.81	3.65
20	758	2.69	3.55
Trial 2			
0	765	2.56	3.34
50	708	2.50	3.53
60	585	2.28	3.90

Source: Lanza and Messina 1979A.

for 44- and 88-kg pigs by Baird *et al* (1974). The citrus pulp tested contained 92.8% dry matter, 6.0% crude protein, 17.9% crude fiber, 2.4% ether extract, 5.9% ash, 60.7% nitrogen-free extract, and 3777 kcal/kg of gross energy. All digestibility coefficients were relatively high except for crude protein which averaged 70.8% (Table 13.57). Digestible energy averaged 3357 kcal/kg with unadjusted metabolizable energy averaging 3194 kcal/kg, or about 95.1% of the digestible energy value.

Citrus pulp is not generally recommended for swine diets, but when used in growing-finishing diets, levels not greater than 10% should be used. Baird *et al.*

TABLE 13.57. Apparent Digestion Coefficients, Digestible and Metabolizable Energy of Dried Citrus Pulp

Item	*44-kg pig*	*88-kg pig*	*Av 2 trials*
Av consumption, kg			
Citrus pulp[a]	0.82	1.59	
Supplement	0.23	0.23	
Apparent digestibility, %			
Dry matter	84.4	84.6	84.5
Crude protein	69.9	71.6	70.8
Ether extract	79.0	81.3	80.2
Crude fiber	91.1	96.6	93.9
NFE	90.7	90.7	90.7
Gross energy	82.3	83.6	83.0
Total digestible nutrient, %	83.4	84.6	84.0
Digestible energy, kcal/kg	3330	3383	3357
Metabolizable energy, kcal/kg	3208	3180	3194
Metabolizable energy$_N$, kcal/kg	3098	3138	3118

Source: Baird et al. 1974.

[a]Composition of citrus pulp fed (%): dry matter, 92.8; crude protein, 6.0; crude fiber, 17.9; ether extract, 2.4; ash, 5.9; nitrogen-free extract (NFE), 60.7; and gross energy, 3777 kcal/kg.

(1974) stated that citrus pulp can safely form 20% of the swine diet when energy level is corrected in the diet by the addition of tallow.

GARBAGE

Garbage, also known as swill or kitchen refuse or waste, is a very heterogeneous product that results from food discarded from restaurants, hotels, supermarkets, institutions, military establishments, meat and fish markets, and homes. Because of the variation in origin and content, the chemical composition of garbage is also quite variable, as shown in Table 13.58 (Balazs *et al.* 1971; Barth *et al.* 1966; Kornegay *et al.* 1965, 1968). Even with this variable chemical

TABLE 13.58. Composition, Digestibility, and Energy Values of Garbage of Various Types

Components	*Analysis 1*[a]				*Analysis 2*[b]			
	Composition			*Digestibility*	*Composition*			*Digestibility*
	Max	*Min*	*Av*		*Max*	*Min*	*Av*	
Dry matter, %	16.8	8.7	13.2	87.3	25.6	16.0	18.9	84.2
Amount on dry matter basis								
Crude protein, %	27.3	15.1	19.7	83.1	17.5	14.6	15.8	82.7
Ether extract, %	54.5	22.2	34.1	94.1	24.9	14.7	23.3	91.7
Crude fiber, %	7.4	4.3	5.3	72.5	8.4	2.8	4.3	46.7
Nitrogen-free extract, %	45.4	20.1	35.0	88.6	62.6	43.8	50.3	84.9
Gross energy, kcal/g	6.38	4.72	5.53	5.70	5.67	4.81	5.2	4.56
Ash, %	8.5	5.2	6.5	—	8.6	5.2	6.3	—
Calcium, %	1.16	0.60	0.73	—	1.69	0.37	0.75	—
Phosphorus, %	0.76	0.35	0.49	—	0.39	0.29	0.35	—
Magnesium, %	0.20	0.07	0.11	—	0.10	0.05	0.07	—
Copper, mg/kg	2.9	1.4	2.4	—	50	27	39	—
Iron, mg/kg	—	—	—	—	510	346	430	—
Manganese, mg/kg	—	—	—	—	36	23	28	—
Carotene, mg/kg	—	—	—	—	4.9	3.9	4.3	—
Thiamin, mg/kg	—	—	—	—	3.9	2.2	2.8	—
Riboflavin, mg/kg	—	—	—	—	3.8	2.7	3.2	—
Niacin, mg/kg	—	—	—	—	29.9	23.0	26.5	—
Pantothenic acid, mg/kg	—	—	—	—	6.2	2.7	4.3	—
Fatty acids in fat, %								
Myristic	3.4	2.3	2.9	—	—	—	—	—
Palmitic	24.8	22.5	23.3	—	—	—	—	—
Palmitoleic	5.5	3.6	4.5	—	—	—	—	—
Stearic	11.9	8.9	10.0	—	—	—	—	—
Oleic	42.1	36.7	39.6	—	—	—	—	—
Linoleic	19.4	12.1	15.7	—	—	—	—	—

[a]Adapted from Balazs et al. 1971. [b]Adapted from Barth et al. 1966 and Kornegay et al. 1970.

composition, garbage is a significant source of feed for swine production in many areas of the world. It is used particularly in areas close to metropolitan centers. It is estimated that in 1981 in the United States approximately 1.1% of all pigs slaughtered or just over one million head were fed on garbage. Although still a significant number, this represents a reduction since 1967 when over 2.5 million head or 3% of all pigs slaughtered were fed on garbage.

One of the major problems of feeding garbage to swine is related to sanitation and disease, since garbage can be a vehicle for spreading serious infectious diseases, such as hog cholera, vesicular exanthema, foot and mouth disease and African swine fever, as well as such diseases as erysipelas and brucellosis, tuberculosis, and salmonellosis and trichinosis that not only affect swine but may also infect humans (Steel and Busch 1968).

To prevent the spread of disease through the feeding of raw garbage, the garbage should be properly processed to destroy completely all of the pathogenic agents that are present. Proper processing is usually accomplished by boiling in an open vat or steam cooker. The mass, in any case, should be thoroughly boiled (100°C) for a period of not less than 30 min in equipment designed to assure that all areas of the garbage reach a boiling temperature. The mass should be allowed to cool for a similar length of time without external cooling. This procedure assures adequate sterilization of the product. Proper handling after cooking also prevents subsequent contamination of the processed material.

Garbage contains a high percentage of water (75–85%) and variable concentrations of essential nutrients for growing-finishing swine. Although studies have demonstrated that a 2 : 1 or 2.5 : 1 mixture of water and dry meal gives optimum performance in liquid feeding, in undiluted garbage, on the average, the relationship of water to dry matter is approximately 7 : 1. Water per se is not believed to reduce efficiency of utilization of the diet components; it does, however, limit consumption when present in excessive amounts in the diet. When garbage is used as the only source of feed, pigs are forced to consume large quantities of water present in the garbage, thus limiting the total daily consumption of nutrients.

The high water content of garbage poses the most serious problem for young growing pigs between 15 and 50 kg because of the limited capacity of their gastrointestinal tract. As the pig develops in size, the greater gut capacity tends to minimize the importance of diet concentration, allowing the pig to consume quantities of high-moisture garbage almost adequate to meet its daily requirements. To increase nutrient consumption, the practice of adding water to garbage for cooking and feeding should be discouraged.

If garbage is utilized as the sole source of feed, young pigs grow very slowly and require as much as 230–245 days to reach market weight (Kornegay *et al.* 1970; Maylin and Dominguez 1978; Bermont 1979). The most efficient utilization of cooked garbage can be obtained by supplementing the garbage ration with a concentrate. Kornegay *et al.* (1970) showed that supplementing cooked garbage, fed free-choice, with a 15–17.5% protein supplement improved both daily gains and the efficiency of garbage utilization. Furthermore, it was very profitable to add supplement when garbage was fed to growing pigs between weaning and about 55 kg. The advantage, however, decreased as the pigs became larger. When levels of 0.34, 0.69, and 1.23 (*ad libitum*) kg of supplement feed were added to a free-choice garbage diet, responses of 0.47, 0.35, and 0.27 kg of additional gain per kilogram of supplement were reported. These results demonstrate a decreasing response per kilogram of feed as the supplemental level is increased. From the same data, it would appear that supplemental levels between 0.34 and 0.69 kg of a 15–17.5% protein concentrate fed with free-choice garbage will yield the highest level of profit. Higher levels of protein in the supplement were of no value in improving performance (Kornegay *et al.* 1970; Soldevila 1977). Although garbage would appear to be deficient in calcium, phosphorus, zinc, and vitamins, larger than normal quantities of these in the concentrate feed were of no value in improving growth or feed utilization (Kornegay *et al.* 1970; Shevchenko 1979).

Although some reports (Peterson 1967) suggest that there are no differences in carcass fatness or quality when garbage supplies the only source of feed, others (Modebe 1963; Lovatt 1943; Daccord 1971; Ozerova 1981) indicate that garbage, when fed alone, produces greater fat deposition and a softer fat than do diets consisting of all concentrate or garbage plus concentrate. Carcasses with higher percentages of soft fat might be expected from the large quantities of unsaturated fat present in most garbage (Table 13.58).

Morbidity and mortality is often high among garbage-fed pigs, but proper cooking and the addition of a complete supplement and, in some cases, an antibiotic to the diet appears to reduce both death losses and the level of sickness among the pigs. The greater numbers of death losses, often encountered in garbage-feeding establishments, are probably related more to poor husbandry, inadequate processing, and unsanitary conditions than to nutritional value of the garbage.

FATS AND OILS

Animal fats, including tallow, lard, lamb fat and fish oils, as well as vegatable fats from such plants as coconut, soybean, cottonseed, maize, safflower, sesame and others, are by-products of other food and industrial uses, and are available as such for swine feeding. The distinction between fats and oils is based on physical characteristics. If the product is solid at room temperature, it is generally termed fat; if liquid at room temperature, it is termed oil. The physical properties of fats and oils are determined mainly by their fatty acid composition. Oils are high in polyunsaturated fatty acids, especially linoleic acid, while fats are high in saturated fatty acids and may be nearly devoid of linoleic acid, as is, for example, coconut fat.

The American Association of Feed Control Officials has established categories and definitions of fats and oils used in animal feeds. These include the following:

Animal fats. These are obtained from the tissues of mammals and/or poultry in the commercial processes of rendering or extracting. If the product bears a name descriptive of its kind or origin (tallow, lard, grease), that name must correspond exactly to the content.

Hydrolyzed fat or oil (feed grade). This product is obtained in the fat-processing procedures commonly used in soap making. It must contain not less than 85% total fatty acids, not more than 6% unsaponifiable substances, and not more than 1% insoluble materials. It must be described according to source (hydrolyzed animal fat or hydrolyzed vegetable fat or a mixture).

Fat product (feed grade). Any fat product not meeting definitions for animal or vegetable fat, hydrolyzed fat, or fat ester belongs to this category. Minimum percentage of total fatty acid and maximum percentages of unsaponifiable and insoluble fractions must be specified. The fat product must be adequately tested for safety in animal feeding before being marketed, and its components must be listed.

Vegetable fat or oil. These products are obtained by extracting the oil from seeds or fruits commonly processed for human food. They must contain not less than 90% total fatty acids, not more than 2% unsaponifiable matter, and not more than 1% insoluble matter. The label must specify the source (soybean oil, coconut fat, and so on).

Vegetable oil refinery lipid (feed grade). This is obtained in alkaline refining of vegetable oil for human use. It may contain not more than 17% ash on a dry basis. Since it is the product of alkaline treatment, it must be neutralized with acid before use in animal feeds. This product is often called ''soapstock'' and consists primarily of the salts of fatty acids, glycerides, and phosphatides. Its feeding value for swine has not been thoroughly studied, but the rate of weight gain may be depressed (Devilat *et al.* 1971) by replacing the basal diet with approximately 7% soapstock (15% of the calories).

The main nutrient supplied by fats and oils is energy, and their caloric content is approximately 2.25 times that of carbohydrate and protein. The energy concentration of pig diets is commonly increased by inclusion of fats and oils. The increased caloric density of the diet results in less total consumption by the animal. Therefore, if fats and oils are added to the diet the percentage of other nutrients in the diets should be increased to obtain an adequate ratio of calories to other nutrients. The reason is that the pig tends to eat to satisfy energy requirements, and a deficiency of protein or other nutrients can result if the fat content of the diet is increased without otherwise adjusting the diet. The importance of the ratio of calorie to protein in swine feeding was discussed in Chapter 11.

Various responses have been reported in growth performance of pigs fed diets containing added fats or oil. Several groups (Barrick *et al.* 1953; Abernathy *et al.* 1958; Clawson *et al.* 1962; Kuryvial *et al.* 1962; Greeley *et al.* 1964A,B; Leibbrandt *et al.* 1975; Cromwell *et al.* 1978; Moser *et al.* 1975) found that the addition of fat to the diet improved daily weight gains; in contrast, others (Asplund *et al.* 1960; Kropf *et al.* 1954; Hale *et al.* 1968; Noland and Scott 1960; Newman *et al.* 1967; Seerley *et al.* 1975; McDonald and Hamilton 1976; Meade 1975) found that addition of fat to the diet had no effect on rate of gain. Addition of corn oil to the diet was found to depress the weight gains of growing pigs (Asplund *et al.* 1960; Greeley *et al.* 1964A) while rapeseed oil (Aherne *et al.* 1975; McDonald and Hamilton 1976), soybean oil (McDonald and Hamilton 1976), and palm oil (Fetuga *et al.* 1975) had no effect on average daily weight gains. In general, inclusion of at least 5% fat and oil in the diets of growing-finishing pigs reduces daily feed intake and improves the efficiency of feed utilization (Barrick *et al.* 1953; Abernathy *et al.* 1958; Clawson *et al.* 1962; Greeley *et al.* 1964A,B; Newman *et al.* 1967; Meade 1975; Moser *et al.* 1975; Seerley *et al.* 1975) but has little if any effect on weight gain.

It is well established, according to reviews by Morgan and Robinson (1962), Hays *et al.* (1963), Hays (1968), and Moser (1977) that dietary energy levels affect ratio of fat to lean in the carcass. Although low levels of added fat (2–3%) appear to have little effect on back fat thickness, 4% or more will probably have the effect of slightly increasing back fat thickness. The amount of increase appears to depend upon the level of fat added and the calorie : protein ratio (Moser 1977; Cromwell *et al.* 1978).

The fatty acid composition of some common plant and animal fats is shown in Table 13.59. The pig tends to deposit fat with a fatty acid composition similar to that of the diet. The higher the fat content of the diet and the longer the duration of feeding of a particular diet, the greater is the effect on depot fat composition. The feeding of whole cooked soybeans, which contain approximately 18% fat with a high degree of unsaturation, produces softer carcasses than are produced with a maize–soybean meal diet. Conversely, feeding hydrogenated coconut fat at a level of only 3% of the diet from weaning to market weight produces a firm carcass as a result of the deposition of highly saturated fat. Addition of corn oil, safflower oil, soybean oil, and rapeseed oil to the diet of growing-finishing pigs resulted in the deposition of unsaturated fatty acids to the depot fat (Leat *et al.* 1964; Sink *et al.* 1964; Koch *et al.* 1968; MacGrath *et al.* 1968; Walker 1972; McDonald and Hamilton 1976). The addition of animal fats to swine diets tends to have little effect on depot fat composition. Fish oils, if fed in quantities greater than 5% of the diet, impart a "fishy" flavor to pork. A process of polymerization of menhaden fish oil has been shown to reduce the flavor and odor problem of pork from pigs fed 5% menhaden fish oil, but this oil caused reduced weight gain when it was polymerized (Anglemier and Oldfield 1957). Further improvements in processing of crude fish oils may increase their usefulness for swine feeding.

It has been suggested that the addition of 5–10% fat to the diet of young pigs will improve palatability and increase feed intake and growth rate. Published

TABLE 13.59. Fatty Acid Composition of Some Common Animal and Vegetable Fats Used in Swine Feeding

Fat	Fatty acid designation										
	$C_{8:0}$	$C_{10:0}$	$C_{12:0}$	$C_{14:0}$	$C_{14:1}$	$C_{16:0}$	$C_{16:1}$	$C_{18:0}$	$C_{18:1}$	$C_{18:2}$	$C_{18:3}$
Beef fat[a]	—	—	—	6.3	1.7	32.2	4.9	11.0	41.9	2.0	—
Pork fat[b]	—	trace	trace	1.4	—	25.3	3.6	12.9	44.6	11.2	0.9
Safflower oil[c]	—	—	—	trace	—	8.8	—	1.5	9.3	80.2	trace
Hydrogenated coconut fat[c]	7.8	5.7	44.4	17.2	—	9.2	—	10.4	5.0	0.2	—

[a]Cabezas et al. 1965. The composition of the depot fat of ruminants is affected only to a small extent by diet.

[b]Pond, W. G., and R. L. Anderson (unpublished data). The composition of the depot fat of pigs is affected markedly by diet; depot fat composition tends to resemble diet fat composition. These pigs were fed maize–soybean meal diets.

[c]Babatunde et al. 1967.

TABLE 13.60. Effect of Dietary Fat Level on Performance of Early Weaned Pigs

Criteria	*Diets*			
	1	*2*	*3*	*4*
Fat level, %[c]	0	4	8	12
Av daily gain, week 1, kg	0.10[a,b]	0.11[a]	0.14[a]	0.01[b]
Av daily gain, overall, kg	0.64	0.65	0.68	0.66
Feed/gain ratio	2.05	1.84	2.02	1.96

Source: Lawrence et al. 1980.
[a,b]Values on the same line with different superscripts are significantly different.
[c]Choice white grease.

research data, although somewhat inconsistent, fail to support this claim. Lawrence *et al.* (1980) fed 0, 4, 8, and 12% choice white grease to 30-day-old pigs and over the 5-week growing period observed no effect on average daily gain or feed efficiency (Table 13.60). Pigs fed 12% added fat grew very poorly during the first week of the trial but performance subsequently improved and at the end of the 5-week period was similar to that produced by the other fat levels. Peo *et al.* (1957) and Ewan (1970) reported that the addition of fat to the diet of baby pigs (5–15 kg) had no effect on rate of gain but increased the feed required per unit of gain. Eusebio *et al.* (1965) observed a gradual decrease in growth rate of pigs fed diets in which the supplemental fat level was increased from 2 to 28%. Leibbrandt *et al.* (1975) observed that although 5% added fat tended to improve baby pig gains, the addition of 10% fat significantly depressed growth rate, and lard but not hydrogenated fat improved feed-to-gain-ratio in baby pigs started at 4.6 kg body weight. Tullis and Whittemore (1980) demonstrated that fully hydrogenated tallow is poorly digested by the pig which is consistent with the findings of Sherry *et al.* (1978) that showed that in 23-day-old neonatal pigs, the more-saturated fats reduced feed efficiency and digestibility components. In contrast, Frobish *et al.* (1970) reported improved weight gains when hydrolyzed fat was added to baby pig rations.

Piglets are weak at birth compared with animals of other species. It is well documented that one-fourth to one-third of pigs born alive in Europe and the United States die before weaning. The majority of these deaths occur during the first few days after birth, and rapid depletion of limited body stores of energy is considered as a major cause of pig mortality. If the neonatal pig is denied food, the blood sugar concentration drops and death occurs within a few days. Since the pig is dependent on the sow for its food supply, proper sow nutrition is essential for survival and growth of the pig.

Following the report of the original research of Seerley *et al.* (1974), various studies (reviewed by Pettigrew 1979; Moser and Lewis 1980; and Seerley 1981) have suggested that feeding high fat rations to sows during late pregnancy and lactation can improve piglet survival. A summary of 31 comparisons (Moser and Lewis 1980) in Canada, Denmark, and the United States is tabulated in Table 13.61. In 25 of the 31 comparisons, feeding fat or oil to sows in late pregnancy

TABLE 13.61. Summary of 31 Comparisons on the Effect of Adding Fat to Sow Diets in Late Gestation and in Lactation[a]

Criteria	*Control*	*(No. of litters)*	*Added fat*	*(No. of litters)*	*Difference*
Pigs born alive per litter	10.0	(677)	9.9	(814)	−0.1
Pigs weaned per litter	8.1	(677)	8.4	(814)	+0.3
Survival, %	82.0	(736)	84.6	(938)	+2.6
Av birth weight, kg	1.41	(677)	1.39	(814)	−0.02
Av 21-day weight, kg	5.57	(356)	5.66	(432)	+0.09
Fat in colostrum, %	7.3	(360)	9.1	(512)	+1.8
Fat in milk, %	9.1	(322)	10.1	(506)	+1.0

Source: From review by Moser and Lewis 1980.
[a]Data are weighted for the number of litters involved in each experiment.

and/or lactation increased survival rate by an average of 2.6%, which resulted in an increase in litter size at weaning of 0.3 pigs. Individual experiments have in some cases shown greater increases in the survival rate, especially of smaller pigs, owing to the addition of fat to the sow diet during late pregnancy and lactation. Reports from the University of Georgia (Seerley 1981) have shown percentage survival at 21 days of age to be 80.9, 88.3, and 90.2% among pigs in the control, 10% corn oil, and 10% animal fat groups, respectively.

Normally, neither the number of pigs born alive, the average birth weight, nor the average 21-day weight is affected by fat addition; however, total litter weight at 21 days may be higher when fat is added.

It has been postulated that increased survival rate resulting from feeding high fat diets to sows may be due to (1) increased glycogen stores at birth, (2) increased blood glucose at birth, (3) slower disappearance of blood glucose during starvation, (4) slower disappearance of liver glycogen, (5) increased fat content and milk production, and (6) change in the metabolic processes associated with fat utilization.

Almost all researchers who have analyzed milk have reported elevated fat content of milk and colostrum of sows fed added fat. Increase in milk fat content has ranged from 9 to 57% in these studies. Apparently sows metabolize lipids from adipose tissue for milk production, and consequently they lose weight and become thin. Feeding fat prior to farrowing and in early lactation reduces the need to metabolize body fat as the fat for milk synthesis is available in the blood.

Results summarized by Moser and Lewis (1980) also indicate that total milk yield is increased by 10–17% by feeding fats. Seerley (1981), citing unpublished data by M. T. Coffey at the University of Georgia, reported that on day 14 of lactation added fat increased sow milk yields by approximately 33%.

Several authors (Seerley *et al.* 1974; Cast *et al.* 1977; Boyd *et al.* 1979; Parsons 1979; Seerley 1981) have found that pigs from sows fed high fat before farrowing not only have higher glycogen stores at birth but also show a slower disappearance of liver glycogen during the first 48 hr after birth. Seerley (1981) has also reported that imposing 10% dietary fat treatments for the last trimester of pregnancy resulted in a significantly elevated blood glucose at birth and after a 48-hr fast. Evidence has been presented by Seerley (1981) to illustrate that pigs from fat-fed sows have an improved ability to metabolize lipids.

There is evidence that adding fat to sow diets during late gestation and lactation can improve pig survival, especially among small pigs and in herds with lower survival rates. If fat is added, a minimum of 7.5% fat should be added to produce a positive response, and there appears to be little advantage of levels greater than 15%. High fat diets should be fed at least 7 days before farrowing, and Seerley (1981) presented data to indicate that a greater response was elicited when rations containing fat were fed during the last trimester of pregnancy. Pettigrew (1979) suggested that at least 1 kg of fat should be fed before the sow farrows. Feeding a diet with 10% added fat for 7 days before farrowing will achieve this goal. Adding fat during both late pregnancy and lactation not only affects prenatal metabolism but also increases the fat content of both colostrum and milk, which appears to be advantageous to the suckling pig.

Fat addition to sow diets is only advantageous if it increases net return for the producer. Strict analysis must be made in each swine operation to determine if it is economical to feed fat to sows.

REFERENCES

Introduction

PLUCKNETT, D. L. 1979. Small-Scale Processing and Storage of Tropical Root Crops. Westview Press, Boulder, Colorado. 461 pp.

Cassava

ALBA, M. G. 1937. Philipp. Agric. *25,* 782.
ASICO, P. M. 1941. Philipp. Agric. *29,* 706.
AUMAITRE, A. 1969. Ann. Zootech. *18*(4), 885.
BUITRAGO, J., and I. JIMÉNEZ. 1968. Unpublished data. ICA, Cali, Colombia.
CIAT. 1975. Center for Int. Agric. in the Tropics Ann. Rep.
CIAT. 1983. Center for Int. Agric. in the Tropics, CIAT International: Report of Research and International Cooperation, vol. 2, pp 1–8, Cali, Colombia.
CLOSE, J., ADRIAENS, E. L., MOORE, S., and BIGWOOD, E. J. 1953. Bull. Soc. Chem. Biol. Brussels *35,* 985.
COUCH, J. F. 1932. USDA Leaflet 88.
FAO. 1980. Food and Agricultural Organization Production Yearbook, Vol. 34.
FULLERTON, J. 1929. J. Minist. Agric. G.B. *36,* 130.
GAVIN, A., and ANDOVARD, P. 1914. Bull. Soc. Natl. Agric. France. *74,* 481.
GOMEZ, G., CAMACHO, C., and MANER, J. H. 1976. Memoria del Seminario Internacional de Ganaderi Tropical. Acapulco, Mexico. March 8–12, 1976, pp. 91–92.
HENKE, L. A. 1919. Col. Hawaii Bull. *6,* 24.
JARAMILLO, L., and HERRERA, H. 1970. VIII Reunión de Fitotécnica, Bogota, Colombia. November 22–28, 1970.
JOB, T. A. 1975. Ph.D. Thesis, CIAT and Dep. of Anim. Sci., Univ. of Ibadan, Ibadan, Nigeria.
JOHNSON, R. M., and RAYMOND, W. D. 1965. Tropical Sci. *7*(3), 109.
KERR, R. W. 1950. Chemistry and Industry of Starch. 2nd Edition. Academic Press, New York.
MANER, J. H. 1971. Boletín Técnico *No. 9.* ICA. Colombia.
MANER, J. H., and JIMÉNEZ, I. 1967. Dia de Campo sobre Porcinos. April 1967. ICA, Palmira, Colombia.
MANER, J. H., BUITRAGO, J., and JIMÉNEZ, I. 1967. Proc. Int. Symp. Trop. Root Crops. Univ. W. I., St. Augustine. Trinidad *2*(6), 62.
MANER, J. H., BUITRAGO, J., and GALLO, J. T. 1970. J. Anim. Sci. *31,* 208 (Abstr.).
MEJIA, T. R. 1960. Rev. Fac. Nal. de Agron. Medellin, Colombia *20,* 3.
MODEBE, A. N. A. 1963. J. West Afric. Sci. Assoc. *7,* 127.
MONDOÑEDO, M. 1928. Philipp. Agric. *17*(2), 105.

MONDOÑEDO, M., and ALANTE, F. 1931. Phillipp. Agric. *2*, 113.
MONDOÑEDO, M., and BAYAN, P. V. 1927. Philipp. Agric. *15*, 523.
OKE, O. L. 1969. World Rev. Nutr. Diet. *11*, 170.
OYENUGA, V. A. 1955. Nigerian Feedstuffs. Univ. College, Ibadan. Nigeria.
OYENUGA, V. A., and OPEKE, L. K. 1957. West Afr. J. Biol. Chem. *1*, 3.
PEIXOTO, R. R. 1965. Bull. Escala de Agron. Eliseu Maciel. Univ. Rual do Sul Brasil. November, 1965.
ROGERS, D. J. 1963. Bull. Torrey Bot. Club. *90*(1), 43.
SHIMADA, A. S. 1970. IV Dia del Ganadero INIP. Veracruz, Mexico, November 21–22, 1970.
SREERAMAMURTHY, V. V. 1945. Indian J. Med. Res. *33*, 229.
VOGT, H. 1966. World's Poultry Sci. J. *22*, 113.
WOODMAN, H. E., MANZIES KITCHIN, A. W., and EVANS, R. E. 1931. J. of Agric. Sci. *21*, 526.
ZARATE, J. J. 1956. Philipp. Agric. *40*, 78.

Potatoes and Dried Potato Products

ANASTASIJEVIC, V., SLJIVOVACKI, K., RANKOVIC, P., and STANKOVIC, M. 1972. Stocarstvo *26*, 59.
BIELINSKI, K., and BIELINSKI, K. 1954. Rocz. Nauk. Roln. *68*, 297. (Cited in Nutr. Abstr. Rev. *25*, 841. 1955.)
BORGIDA, L. P., and LAPLACE, J. P. 1977. Ann. Zootech. *26*, 585.
BRAUDE, R., and MITCHELL, K. G. 1951. Agric. *57*, 501.
CRANWELL, P. D. 1968. Nutr. Abstr. Rev. *38*, 721.
D'MELLO, J. P. F., WHITTEMORE, C. T., and ELSLEY, F. W. H. 1973. J. Sci. Food Agric. *24*, 533.
FINGERLING, G. 1932. Landwirtsch Vers. Stn. *114*, 1 (Cited by Woodward *et al.* 1938).
FRIEND, D. W., CUNNINGHAM, H. M., and NICHOLSON, J. W. G. 1963. Can. J. Anim. Sci. *43*, 241.
GLAPS, J. 1970. Rocz. Nauk. Roln. *92*, 247 (Cited in Nutr. Abstr. Rev. *41*, 709.)
HENRY, Y., and RERAT, A. 1966. Ann. Zootech. *15*, 231.
HIGHLANDS, M. E. 1967. *In* Potato Processing (W. F. Talburt and O. Smith, Editors), pp. 540–549. AVI Publishing Co., Westport, Connecticut.
HILLYER, G. M., and WHITTEMORE, C. T. 1975. J. Sci. Food Agric. *26*, 215.
HÖLLER, H. 1959. Arch. Tierernähr. *9*, 236.
HUTCHINSON, J. C. D., BACON, J. S. D., MacRAE, T. F., and WORDEN, A. N. 1943. Biochem. J. *37*, 550.
KELLNER, O. M., and NEUMANN, R. 1910. Landwirtsch. Vers. Stn. *73*, 235. (Cited by Woodward *et al.* 1938.)
KIRSCH, W., and JANTZON, H. 1943. Biedermanns Zentralbl. Tierernähr. *15*, 206. (Cited in Nutr. Abstr. Rev. *14*, 169. 1944.)
KRIVENCOV, J. M., and MOSBUIN. N. A. 1968. Svinovodstvo *4*, 29.
LEROY, A. M., ZELTER, S., and FEBRIER, R. 1952. Ann. Zootech. *1*, 87.
LIVINGSTONE, R. M., BAIRD, B. A., ALKINSON, T., and CROFTS, R. M. J. 1979. Anim. Feed Sci. Tech. *4*, 295.
LIVINGSTONE, R. M., BAIRD, B. A., ALKINSON, T. and CROFTS, R. M. J. 1980. J. Sci. Food Agric. *31*(7), 695.
MARTIN, J., and BUYSSE, F. 1955. Rev. Agric. Brussels *8*, 1135. (Abstr. in Nutr. Abstr. Rev. *26*, 824).
McLAGAN, J. R., and NASR, H. 1950. Scot. J. Agric. *30*, 143.
MIROCHITSKII, F. V. 1980. Khim. Sel'sk. Khoz. *18*(6), 48.
MONCADA, A., and MANER, J. H. 1965–1966. Dia de Campo 1965–1966. Instituto Colombiana Agropecuario, Bogota, Colombia.
NICHOLSON, J. W. G., and FRIEND, D. W. 1965. Can. J. Anim. Sci. *45*, 141.
OSTROWSKI, H., and RYS, R. 1969. Rocz. Nauk. Roln. *92*, 71. (Cited in Nutr. Abstr. Rev. *40*, 1469.)
PFEIFFER, H. 1966. Dtsch. Landwirtsch *17*, 37.
POPELOV, V. V. 1977. Zhivotnovud. *9*, 53.
SANDSTEDT, R. M., and GATES, R. L. 1954. Food Res. *19*, 190.
SAULNIER, J., and CHAVUEL, J. 1978. L'elevage Porcine. *79*, 30.
SCHMIDT, F., and SOROKA, T. 1972. Rocz. Nauk. Roln. Ser. B *94*, 55.
SCHWIMMER, S., and BURR, H. K. 1967. *In* Potato Processing. (W. F. Talburt and O. Smith, Editors), pp. 12–43. AVI Publishing Co., Westport, Connecticut.

SEIDLER, S., WARCHOL, P., and WECKOWICZ, E. 1967. Zesz. Nauk. WSR. Szczecinie *26,* 35. (Cited in Nutr. Abstr. Rev. *39,* 274.)
SOHONIE, K., and AMBE, K. S. 1955. Nature. *176,* 972.
STAHL, W., GÖhner, A., and BARTH, F. 1932. Z. Schweinezucht. *39,* 229. (Cited in Nutr. Abstr. Rev. *2,* 417.)
THOMPSON, J. K., and HARGRAVE, J. 1935. Agric. *42,* 1123.
WHITTEMORE, C. T. 1977. Anim. Feed Sci. Tech. *2,* 171.
WHITTEMORE, C. T., and TAYLOR, A. G. 1973. Proc. Br. Soc. Anim. Prod. *2,* 75.
WHITTEMORE, C. T., TAYLOR, A. G., and CROOKS, P. 1974. J. Agric. Sci. Cambridge *83,* 1.
WHITTEMORE, C. T., MOFFAT, I. W., and TAYLOR, A. G. 1975A. J. Sci. Food Agric. *26,* 1567.
WHITTEMORE, C. T., TAYLOR, A. G., MOFFAT, I. W., and SCOTT, A. 1975B. J. Sci. Food Agric. *26,* 255.
WHÖLBIER, W., and NECKERMANN, L. 1951. Arch. Tierernähr. *2,* 166.
WITCZAK, F., and OLCZAK, S. 1970. Zesz. Probl. Postepow Nauk Roln. *101,* 219.
WOODMAN, H. E., and EVANS, R. E. 1943. J. Agric. Sci. *33,* 1.
WOODWARD, J. C., DAVIDSON, W. M., WATSON, C. J., ROBINSON, C. H., and MUIR, G. W. 1938. Sci. Agric. *18,* 629.
ZAUSCH, M., and HÖLZSCHUL, W. 1960–1961. Jahrb. Arbeitsgem. Fiueitturungsberat. *3,* 173. (Cited in Nutr. Abstr. Rev. *32,* 1385.)

Sweet Potatoes

BRAY, C. I. 1933. Proc. Am. Soc. Anim. Prod. *32,* 160.
CALDER, A. 1960. Rhod. Agric. J. *57,* 293.
LEASE, E. J., and MITCHELL, J. H. 1940. S.C. Agric. Exp. Stn. Bull. *329.*
LEE, J. H., and CHUNG, C. H. 1966. Res. Bull Chinju Agric. College *5,* 29–38.
LEE, P. K., and LEE, M. S. 1979. J. Taiwan Livestock Res. *12,* 49.
LEE, P. K., and YANG, Y. F. 1979. J. Taiwan Livestock Res. *12*(1), 31.
SMITH, D. M., RUSTOFF, L. L., MILLER, G. D., and BRANTON, C. 1947. La Agric. Exp. Bull. *423.*
SPRINGHALL, J. A. 1969. Papua New Guinea Agric. J. *21,* 76.
TSENG, C. M., TSAI, A. H., and CHAN, M. P. 1980. J. Agric. Assoc. China. *112,* 54.
WARE, L. M. 1947. Ala. Agric. Exp. Stn. Progress Rep. Ser. *12.*
WU, J. F. 1980. J. Anim. Sci. *51,* 1261.

Yams

COURSEY, D. G. 1967. Yams. Longmans, Green and Co., London, England.
JEFFERS, H. F., and HAYNES, P. H. 1967. Proc. Int. Symp. Tropical Root Crops. April 2–8, 1967, Trinidad.

Bananas and Plantains

BARNELL, H. R., and BARNELL, E. 1945. Ann. Botany *9,* 77. (Quoted by von Loesecke 1950.)
BRESSANI, R., AGUIRRE, A., ARROYANE, R., and JAVQUIN, R. 1961. Turrialba *11*(4), 127.
CALLES, A., CLAVIJO, H., HERVAS, E., and MANER, J. H. 1970. J. Anim. Sci. *31,* 197 (Abstr.).
CELLERI, H., OLIVA, F., and MANER, J. H. 1971A. Reunión Latinamericana de Producción Anim. April 26–30, 1971, p. 91, Bogota, Colombia.
CELLERI, H., OLIVA, F., and MANER, J. H. 1971B. Unpublished data. INIAP, Quito, Ecuador.
CLAVIJO, H., and MANER, J. H. 1971. III Reunión Latinamericana de Producción Anim. April 26–30, 1971, p. 93, Bogota, Colombia.
CLAVIJO, H., CALLES, A., OLIVA, F., and MANER, J. H. 1970. Unpublished data. INIAP. Santo Domingo, Ecuador.
CLAVIJO, H., MANER, J. H., and CALLES, A. 1971. III Reunión Latinamericana de Producción Anim. April 26–30, 1971, p. 92, Bogota, Colombia.
FAO. 1980. Food and Agricultural Organization Production Yearbook, Vol. 34.
HERNÁNDEZ, F. and MANER, J. H. 1965. Unpublished data.
LE DIVIDICH, J., GOEFFREY, F , CANOPE, I., and CHENOST, M. 1975. World Anim. Rev. *20,* 22.
MUSMANI, M., CAMPABADAL, C., and VARGAS, E. 1979. Agronomía Costarricense *3*(2), 129.
OLIVA, F., ALVARADO, F., CELLERI, H., and MANER, J. H. 1971. Unpublished data. INIAP. Santo Domingo, Ecuador.
OLIVA, F., VITERI, J., CALLES, A., and MANER, J. H. 1971. INIAP. April, 1971. Quito, Ecuador.

POLAND, G. L., MARION, J. T., BRENNER, W. O. M., and HARRIS, P. L. 1938. Ind. Eng. Chem. *30,* 340. (Quoted by von Loesecke 1950.)
STRATTON, F. O., and von LOESECKE, H. W. 1930. United Fruit Co., Research Dep. Bull. No. *32* (Quoted by von Loesecke 1950.)
VITERI, J., OLIVA, F., and MANER, J. H. 1971. Unpublished data. INIAP. Quito, Ecuador.
VITERI, J., ALVARADO, F. and MANER, J. H. 1970. Unpublished data. INIAP. Quito Ecuador.
von LOESECKE, H. W. 1950. Bananas. Chemistry, Physiology, Technology. Interscience Publishers, New York.

Cane Sugar

BROOKS, C. C. 1972. J. Anim. Sci. *34,* 217.
BROOKS, C. C., MIYAHARA, A. Y., and ISHIZAKI, S. M. 1970. J. Anim. Sci. *30,* 1031.
BROOKS, C. C., MIYAHARA, A. Y., and HUCK, D. W. 1971. J. Anim. Sci. *33,* 227 (Abstr.).
BUITRAGO, J. 1972. Reproduction of gilts and long-term effects on progeny: Effect of energy intake during gestation. Ph.D. Thesis. Cornell Univ. Ithaca, New York.
BUITRAGO, J., GARCÍA, E., MANER, J. H., and GALLO, J. T. 1969. ICA Mimeo, Palmira, Colombia.
BUITRAGO, J., GARCÍA, E., CORZO, M. A., MANER, J. H., and GALLO, J. T. 1970. Rev. ICA *5,* 419.
MANER, J. H., OBANDO, H., PORTELA, R., and GALLO, J. 1969. J. Anim. Sci. *29,* 139 (Abstr.).
NAKAMURA, R. M., BROOKS, C. C., and MIYAHARA, A. Y. 1971. J. Anim. Sci. *33,* 237 (Abstr.).
OBANDO, H., MONCADA, A., MANER, J. H., and GALLO, J. T. 1970. Rev. ICA *5,* 411.
USDA. 1963. Composition of Foods. Agriculture Handbook No. 8, Washington, D.C.

Cane Molasses

BABATUNDE, G. M., FETUGA, B. L., and OYENUGA, V. A. 1975. J. Anim. Sci. *40,* 632.
BARNETT, E., and GODELL, C. J. 1923. Miss. Agric. Exp. Stn. Bull. *218,* 32.
BLANCO, V., RAUN, N. S., and VARGAS, E. 1964. J. Anim. Sci. *23,* 868 (Abstr.).
BRAVO, F. O., and CABELLO, E. 1968. Técnica Pecuaria en México *11.*
BRAY, C. I., SNELL, M. G., MORRISON, F. L., and JACKSON, M. E. 1945. La. Agric. Exp. Stn. Bull. *394,* 1.
BRINKLEY, W. W., and WALFROM, M. L. 1953. Adv. Carbohydr. Chem. *8,* 291.
BROOKS, C. C. 1972. J. Anim. Sci. *34,* 217.
BROOKS, C. C., and IWANAGA, I. I. 1967. J. Anim. Sci. *26,* 741.
BUITRAGO, J. 1972. 1st Latin American Swine Seminar. CIAT. Sept. 18–21, 1972. Cali, Colombia.
BURNS, J. C. 1909. Texas Agric. Exp. Stn. Bull. *131,* 7.
CASTRO, M., and ELIAS, A. 1978. Cuban J. Agric. Sci. *12,* 249.
CHRISTON, R., and LE DIVIDICH, J. 1978. Ann. Zootech. *27*(2), 267–288.
COMBS, G. E., JR., and WALLACE, H. D. 1969. Fla. Agric. Exp. Stn. Mimeo Ser. No. *AN-69-13.*
COMBS, G. E., JR. and WALLACE, H. D. 1970. Fla. Agric. Exp. Stn. Mimeo Ser. No. *AN-70-10.*
COMBS, G. E., JR., and WALLACE, H. D. 1973. Anim. Sci. Res. Rep. *49,* 73-1. Gainesville, Florida.
CORZO, M., OBANDO, H., MONCADA, A., and MANER, J. H. 1968. Memoria ALPA *3,* 156 (Abstr.).
FAO. 1970. Food and Agricultural Organization Production Yearbook, 1970.
FERRANDO, R., and THEODOSSIADES, G. 1960. *In* La melasse dans l'alimentation du betail. (Vigot Freres, Editor). Paris, France.
GOCHANGEO, D. R. 1933. Philipp. Agric. *21,* 560.
HARRIS, L. E. 1974. Latin American Tables of Feed Composition. Gainesville, Florida. 509 pp.
HENKE, L. A. 1933. Hawaii Agric. Exp. Stn. Bull. No. *69.*
IWANAGA, I. I., and OTAGAKI, K. K. 1959. Proc. West. Sec. Am. Soc. Anim. Prod. *10,* 27.
IWANAGA, I. I., OTAGAKI, K. K., COBB, E., and WAYMAN, O. 1959. J. Anim. Sci. *18,* 1172 (Abstr.).
LE DIVIDICH, J., and CANOPE, I. 1975. Journ. Rech. Por. France. *7,* 145–180.
LE DIVIDICH, J., SEVE, B., CANOPE, I., and KEMP, H. 1974. Journ. Rech. Por. France *6,* 199.
LE DIVIDICH, J., CHRISTON, R., PEINIAV, J., and AUMAITRE, A. 1978. Anim. Feed Sci. Tech. *3,* 15.
LY J., and VELÁZQUEZ, M. 1969. Revista Cubana Ciencia Agricola *3,* 195.
LY, J. 1971. Cuban J. Agric. Sci. *5,* 347.
LY, J. 1975. Cuban J. Agric. Sci. *9,* 39–47.

LY, J., PERAZA, M., and DIAZ, J. 1978. Cuban J. Agric. Sci. *12*(1), 77–82.
LY, J., and LEZCANO, P. 1980. Trop. Agric. *57*(1), 91.
MACLEOD, N. A., PRESTON, T. R., LASSOTA, L. A., WILLIS, M. B., and VELÁZQUEZ, M. 1968. Revista Cubana Ciencia Agricola *2*, 205.
MANER, J. H., GALLO, J., CORZO, M., and BUITRAGO, J. 1969A. J. Anim. Sci. *29*, 139. (Abstr.).
MANER, J. H., OBANDO, H., PORTELA, R., and GALLO, J., 1969B. J. Anim. Sci. *29*, 139 (Abstr.).
MARRERO, L., and LY, J. 1976. Cuban J. Agric. Sci. *10*, 41–56. (see also 1977. *11*, 39–46.)
MONCADA, A., and MANER, J. H. 1964. ICA Publicacion Miscelanea *No. 1*. Bogota, Colombia.
NRC. 1979. Nutritional Requirements of Swine, National Research Council.
OBANDO, H., CORZO, M., MONCADA, A., and MANER, J. H. 1968. Memorias ALPA *3*, 159.
OBANDO, H., CORZO, M., MONCADA, A., and MANER, J. H. 1969. Revista ICA *4*, 3.
OBANDO, H., and MANER, J. H. 1971. ICA. Boletin Tecnico. *No. 9*. Bogota, Colombia.
PRESTON, T. R., and WILLIS, M. B. 1969. Outlook Agric. *6*, 1.
PRESTON, T. R., and WILLIS, M. B. 1970. FEEDSTUFFS *42*(13), 20.
SCOTT, M. L. 1953. Tech. Rep. Ser. *No. 9*, Sugar Research Foundation, New York.
VELÁZQUEZ, M., LY, J., and PRESTON, T. R., 1969. J. Anim. Sci. *29*, 578.
VELÁZQUEZ, M., PRESTON, T. R. 1970. Revista Cubana Ciencia Agricola *4*, 55.
VELÁZQUEZ, M., and PRESTON, T. R. 1971. Memoria ALPA *6*, 142.
WILLETT, E. L., WORK, S. H., HENKE, L. A., and MARUYAMA, C. 1946. Hawaii Agric. Exp. Stn. Bull. *No. 3*.
ZORRILLA, J. M., and MERINO, H. 1970. Técnica Pecuaria México. *No. 14*, 5.

Dried Bakery Products

ARRINGTON, L. R., MOORE, J. E., ANCALMO, R., RAULERSON, D. E., and CUNHA, T. J. 1964. J. Anim. Sci. *23*, 866 (Abstr.).
SEWELL, R. F. 1965. Georgia Agric. Res. *7*, 9.
TORRES, J. S., and SHIMADA, A. S. 1977. Tech. Pec. Mex. *33*, 101.
WALLACE, H. D., COMBS, G. E., and CONNERS, R. G. 1964. Annu. Swine Field Day Rep. Dep. of Animal Science, Univ. of Florida, Gainsville.

Citrus and Citrus Pulp

AMMERMAN, C. B., ARRINGTON, L. R., McCALL, J. T., WING, J. E., and DAVIS, G. K. 1961. J. Anim. Sci. *20*, 398 (Abstr.).
AMMERMAN, C. B., MARTIN, F. G., and ARRINGTON, L. R. 1970. J. Dairy Sci. *53*, 1514.
BAIRD, D. M., ALLISON, J. R., and HEATON, E. K. 1974. J. Anim. Sci. *38*, 545.
CUNHA, J. T., LAMAR, G. A., SHAUVER, C. B., PEARSON, A. M., FOLKS, S. J., and GLASSCOCK, R. S. 1950. Fla. Agric. Exp. Stn. Cir. *S11*, 1–7.
GLASSCOCK, R. S., CUNHA, J. T., PEARSON, A. M., POCE, J. E., and BUSCHMAN, D. M. 1950. Fla. Agric. Exp. Stn. Cir. *S12*, 1.
HENDRICKSON, R., and KESTERSON, J. W. 1965. Fla. Agric. Exp. Stn. Bull. *698*.
HOLLIS, G. R., WEST, R. L., CARPENTER, J. W., and PALMER, A. Z. 1969. Fla. Agric. Exp. Stn. Mimeo Ser. No. SUS 70-3.
KIRK, W. G., and CROWN, R. M. 1947. Fla. Agric. Exp. Stn. Bull. *428*.
LANZA, A. and G. MESSINA, 1979A. Zootecnia Nutr. Anim. *5*(102), 247.
LANZA, A. and G. MESSINA, 1979B. Zootecnia Nutr. Anim. *5*(102), 255.

Garbage

BALAZS, G. H., HUGH, W. I., and BROOKS, C. C. 1971. Hawaii Agric. Exp. Stn. Tech. Bull. *84*, 16 pp.
BARTH, K. M., VANDER NOOT, G. W., MacGRATH, W. S., and KORNEGAY, E. T. 1966. J. Anim. Sci. *25*, 52.
BERMONT, V. 1979. Suinovodstvo. *2*, 29.
DACCORD, R. 1971. Bull Información Productó Rache. Bogota, Colombia.
KORNEGAY, E. T., VANDER NOOT, G. W., BARTH, K. M., MacGRATH, W. S., WELCH, J. G., and PURKHISER, E. D. 1965. J. Anim. Sci. *24*, 319.
KORNEGAY, E. T., VANDER NOOT, G. W., MacGRATH, W. S., and BARTH, K. M. 1968. J. Anim. Sci. *27*, 1345.

KORNEGAY, E. T., VANDER NOOT, G. W., BARTH, K. M., GRABER, G., MacGRATH, W. S., GILBREATH, R. L., and BIELK, F. J. 1970. Rutgers Exp. Stn. Bull. *829*, 31 pp.
LOVATT, J., WORDEN, A. N., PICKUP, J., and BRETT, C. E. 1943. Empire J. Exp. Agric. *11*, 182.
MAYLIN, A., and DOMINGUEZ, P. L. 1978. Cienc. Tec. Agric. Ganado Porcino *1*(2), 45.
MODEBE, A. N. A. 1963. J. West Afr. Sci. Assoc. *8*, 33.
OZEROVA, V. 1981. Suinovodstvo. *1*, 16.
PETERSON, L. A. 1967. Growth and carcass comparisons of swine fed a concentrate ration, cooked garbage, and additional protein, vitamin and mineral supplements. M. S. Thesis. Univ. of Connecticut, Storrs. (Reported by Kornegay *et al.* 1970.)
SHEVCKENKO, N. 1979. Suinovodstvo. *11*, 26.
SOLDEVILA, M. 1977. J. Agric. Univ. Puerto Rico *61*(4), 513.
STEEL, J. H., and BUSCH, L. A. 1968. Proc. of U.S. Livestock Sanit. Assoc. *72*, 328.

Fats and Oils

ABERNATHY, R. P., SEWELL, R. F., and TARPLEY, R. L. 1958. J. Anim. Sci. *17*, 635.
AHERNE, F. X., BOWLAND, J. P., CHRISTIAN, R. G., VOGTMANN, H., and HARDIN, R. T. 1975. Can. J. Anim. Sci. *55*, 77.
ANGLEMIER, A. F., and OLDFIELD, J. E. 1957. J. Anim. Sci. *16*, 922.
ASPLUND, J. M., GRUMMER, R. H., and PHILLIPS, P. H. 1960. J. Anim. Sci. *19*, 709.
BABATUNDE, G. M., POND, W. G., KROOK, L., VANVLECK, L. D., WALKER, E. F., JR., and CHAPMAN, P. 1967. J. Nutr. *92*, 293.
BARRICK, E. R., BLUMER, T. N., BROWN, W. L., SMITH, F. H., TO, S. B., LUCAS, H. L., and STEWARD, H. A. 1953. J. Anim. Sci. *12*, 899. (Abstr.).
BOYD, R. D., MOSER, B. D., PEO, E. R., JR., and LEWIS, A. J. 1979. J. Anim. Sci. *49* (Suppl. 1), 236 (Abstr.).
CABEZAS, M. T., HENTGES, J. F., JR., MOORE, J. E., and OLSON, J. A. 1965. *24*, 57.
CAST, W. R., MOSER, B. D., PEO, E. R., JR. and CUNNINGHAM, P. J. 1977. J. Anim. Sci. *45*(Suppl. 1), 80 (Abstr.).
CLAWSON, A. J., BLUMER, T. N., SMART, W. W. G., JR. and BARRICK, E. R. 1962. J. Anim. Sci. *21*, 62.
CROMWELL, G. L., HAYS, V. W., TRUJILLO-FIGUEROA, V., and KEMP, J. D. 1978. J. Anim. Sci. *47*, 505.
DEVILAT, J. B., RECABARREN, I., and SKOKNIC, A. 1971. III Reunión Latinoamericana de Producción Animal, p. 83 (Abstr.).
EUSEBIO, J. A., HAYS, V. W., SPEER, V. C., and McCALL, J. T. 1965. J. Anim. Sci. *24*, 1001.
EWAN, R. C. 1970. J. Anim. Sci. *31*, 1020.
FETUGA, B. L., BABATUNDE, G. M., and OYENUGA, V. A. 1975. East Afr. Agric. For. J. *40*, 264.
FROBISH, L. T., HAYS, V. W., SPEER, V. C., and EWAN, R. C. 1970. J. Anim. Sci. *30*, 197.
GREELEY, M. G., MEADE, R. J., and HANSON, L. E. 1964A. J. Anim. Sci. *23*, 808.
GREELEY, M. G., MEADE, R. J., HANSON, L. E., and NORDSTROM, J. 1964B. J. Anim. Sci. *23*, 818.
HALE, O. M., JOHNSON, J. C., JR. and WARREN, E. P. 1968. J. Anim. Sci. *27*, 1577.
HAYS, V. W., WAGNER, G. R., and CLARK, A. J. 1963. Feedstuffs *35*(25), 28.
HAYS, V. W. 1968. The Pork Industry: Problems and Progress, p. 67. Iowa State Univ. Press, Ames.
KOCH, D. E., PEARSON, A. M., MAGEE, W. T., HAEFER, W. T., and SCHWEIGERT, B. S. 1968. J. Anim. Sci. *27*, 360.
KROPF, D. H., PEARSON, A. M., and WALLACE, H. D. 1954. J. Anim. Sci. *13*, 630.
KURYVIAL, M. S., BOMLAND, J. P., and BERG, R. T. 1962. Can. J. Anim. Sci. *42*, 23.
LAWRENCE, N. J., MAXWELL, C. V., and BELCHER, C. G. 1980. Okla. Stn. Univ. Anim. Sci. Res. Rep. *MP-107*, p. 159. April, 1980.
LEAT, W. M. F., CUTHERTSON, A., HOWARD, A. N., and G. A. GRESHAM, 1964. J. Agric. Sci. *63*, 311.
LEIBBRANDT, V. D., HAYS, V. W., EWAN, R. C., and SPEER, V. C. 1975. J. Anim. Sci. *40*, 1081.
MacGRATH, W. S., JR., VANDER HOOT, G. W., GILBREATH, R. L., and FISHER, H. 1968. J. Nutr. *96*, 461.
McDONALD, B. E., and HAMILTON, R. M. G. 1976. Can. J. Anim. Sci. *56*, 671.
MEADE, R. J. 1975. Feedstuffs, September 8, p. 22.

MORGAN, J. T., and ROBINSON, D. W. 1962. Nutrition of Pigs and Poultry, p. 255. Butterworths, London.
MOSER, B. D., PEO, E. R., JR., STAHLY, J., and NIELSEN, M. K. 1975. J. Anim. Sci. *41,* 322.
MOSER, B. D. 1977. Feedstuffs *49*(15), 20.
MOSER, B. D., and LEWIS, A. J. 1980. Feedstuffs *49*(15), 20.
NEWMAN, C. W., THRASHER, D. M., HANDSARD, S. L., MULLINS, A. M., and BOULWARE, R. F. 1967. J. Anim. Sci. *26,* 479.
NOLAND, P. R., and SCOTT, K. W. 1960. J. Anim. Sci. *19,* 67.
PARSONS, M. J. 1979. Mich. Stn. Univ. Res. Rep. *386,* 115.
PEO, E. R., JR., ASHTON, G. C., SPEER, V. C., and CATRON, D. V. 1957. J. Anim. Sci. *16,* 885.
PETTIGREW, J. E. 1979. 71st Annu. Meeting Amer. Soc. Anim. Sci. July 1979, Univ. of Arizona, Tucson.
SEERLEY, R. W. 1981. Feedstuffs *53*(10), 34.
SEERLEY, R. W., PACE, T. A., FOLEY, C. W., and SCARTH, R. D. 1974. J. Anim. Sci. *38,* 64.
SEERLEY, R. W., McCAMPBELL, H. C., and McDANIEL, M. C. 1975. J. Anim. Sci. *41,* 328.
SHERRY, M. P., SCHMIDT, M. K., and VEUM, T. L. 1978. J. Anim. Sci. *46,* 1259.
SINK, J. D., ATKINS, J. L., ZIEGLER, J. H., and MILLER, R. C. 1964. J. Anim. Sci. *23,* 121.
TULLIS, J. B., and WHITTEMORE, C. T. 1980. Anim. Feed. Sci. Tech. *5,* 87.
WALKER, B. E. 1972. Can. J. Anim. Sci. *52,* 713.

14

Animal Protein Sources

INTRODUCTION

Next to carbohydrates (energy sources), protein constitutes the nutrient class needed by swine in the largest amount. The very young pig requires more than 25% protein in the diet. The protein requirement, as a percentage of the diet, steadily declines to maturity, but even the adult requires 10% or more of protein in the diet for normal life. Since protein sources are generally more expensive than energy sources, the wise choice of feedstuffs to minimize diet costs necessitates inclusion in the diet of the smallest amount of protein compatible with normal productive function (growth, reproduction, or lactation). As indicated in Chapters 12 and 13, most energy sources contain some protein. A wide variety of protein supplements of animal origin is available. Those most often used in swine diets are discussed here. Emphasis is placed on important special characteristics, deficiencies, and special supplementary qualities for use in complementing or correcting specific amino acid inadequacies of the energy sources in the swine diet.

Processing conditions, including heat and other treatments, often have important effects on the nutritional value of plant and animal protein sources. Aspects of food and feed processing, as related to bioavailability of the proteins and amino acids of a wide variety of feed materials, are considered in detail in a two-part series (Friedman 1975) and should be consulted by those associated with swine feed formulation and plant and animal protein supplements.

MEAT MEAL AND MEAT-AND-BONE MEAL

Meat meal or meat scrap is by definition the finely ground, dried rendered residue from animal tissues, exclusive of hair, hoof, hide, trimmings, blood, and intestinal tract contents. When the dried rendered product contains more than 4.4% phosphorus it is designated as meat-and-bone meal or meat-and-bone scrap.

Although the definition of meat meal specifies the composition of both meat meal and meat-and-bone meal, large variations exist in quality, uniformity, and putrefaction. Meat meals are derived mainly from bones and associated tissues such as tendons and ligaments, gastrointestinal tract, lungs, condemned livers, and some skeletal muscle. However, hoof, hair, hides, horn, wool, and blood are often included (Sathe and McClymont 1964). Variations in the proportions of this diverse range of raw materials, with widely different chemical compositions and biological values, undoubtedly account for much of the variation in nutritive values between meals. Quality is greatly affected by the degree of dilution of the muscle protein with tendon and bone (Atkinson and Carpenter 1970). Meat meals may contain 21–61% bone (Batterham *et al.* 1970). Bone protein is 83% collagen which contains no tryptophan (Eastoe and Long 1960). Collagen also occurs in skin, connective tissue, cartilage, and tendon, so that 50–65% of the total protein in meat meals may be collagen (Eastoe and Long 1960). Quality is also affected by the amounts of fat, offal, blood, and stomach contents that are allowed into the meal (Runnels 1968). In addition, keratin protein may occur in meat meals. Keratin is a predominant protein in hoof, horn, and hair and is largely indigestible (Leibholz and Moss 1967). Keratin protein contains 3.4–12.2% cystine, while muscle protein contains 1% and collagen protein contains none. Degree of putrefaction of meat meals also affects the palatability, taste, and smell of the diets prepared from these products.

Processing conditions during the manufacture of meat meals can influence their nutritive value. Processing method and temperature are as important as muscle tissue dilution by bone and tendon in determining the final quality of meals (Atkinson and Carpenter 1970). Skurray and Cumming (1974) found that increasing the final processing temperature or extending the holding period during processing will reduce the essential amino acid content of the meat meal. High-temperature processing causes the destruction of cystine and possibly other sulphur-containing amino acids and reduces the availability of lysine through the binding of its ϵ-amino group (Bjarnason and Carpenter 1970). The degree to which the lysine is bound and thus is available to the animal is not identified by acid-hydrolysis analysis but can be determined by fluorodinitrobenzene-available lysine procedures.

The proximate and amino acid analyses of meat meals or meat-and-bone meal of varying qualities are presented in Table 14.1. The biological value and net protein utilization of commercial meat meals are low if a meal is utilized as the only source of protein in the diet (Atkinson and Carpenter 1970). The nitrogen of meat meal is 81–87% digestible and assays have shown the methionine and

TABLE 14.1. Percentage Composition of Meat Meals and Meat-and-Bone Meals of Varying Qualities

Constituent	*1*	2	3	4	5[a]	6[a]
Dry matter	93.5	96.3	92.6	93.5	92.3	95.8
Ash	35.8	31.1	26.3	24.8	10.8	7.0
Calcium	13.7	12.4	9.7	9.8	3.5	0.9
Fat	12.2	9.7	12.6	11.4	17.4	24.6
Crude protein	45.4	52.6	50.4	54.6	57.3	58.5
Amino acids						
Arginine	3.78	3.67	4.02	4.63	4.40	4.43
Histidine	1.50	1.43	1.10	1.32	1.27	1.29
Isoleucine	1.38	1.42	1.94	1.94	2.08	2.16
Leucine	2.76	3.24	3.73	3.81	3.73	3.84
Lysine	2.38	2.87	3.18	3.39	3.56	3.86
Methionine	0.43	0.56	0.65	0.76	0.90	0.85
Cystine	0.24	0.63	0.35	0.30	0.38	0.35
Phenylalanine	1.71	1.86	1.86	2.09	2.32	2.41
Tyrosine	0.88	1.05	1.23	1.36	1.65	1.66
Threonine	1.85	2.14	1.92	2.24	2.72	2.62
Tryptophan	0.22	0.38	0.43	0.49	0.68	0.69
Valine	2.03	2.86	3.08	2.69	2.73	2.78

[a]From Leibholz 1979.

tryptophan to be approximately 80% available. It has been suggested that the lysine availability may be as low as 67.5% compared to 91% for soybean meal. When used as the only protein source for pigs, meat meal supports slow growth and poor feed efficiency which are not improved by methionine, cystine, tryptophan, or lysine supplementation (Atkinson and Carpenter 1970). Therefore, special attention should be given to the diet when meat meal or meat-and-bone meal is utilized to supplement low protein energy sources such as sugar, molasses, cassava, bananas, and sweet potatoes.

Even as a supplement to cereal-based diets, meat meal is of lower quality than fish meal or soybean meal (Peo and Hudman 1962: Todd and Daniels 1965; Beames and Sewell 1969; Atkinson and Carpenter 1970; Beames and Daniels 1970; Lynch *et al.*, 1970; Springhall 1970; Evans and Leibholz 1979: Hale and McCormick 1979; Pearson and Bowland 1977; Aherne *et al.* 1979). Nebraska studies (Peo and Hudman 1962) indicated that a progressive substitution of soybean meal by meat-and-bone meal at levels of 2.5, 5.0, 7.5, and 10.0% in a maize-based diet caused a significant linear depression in rate of growth and daily feed consumption. The performance of growing-finishing pigs (20–90-kg liveweight) fed on diets containing meat meal as the protein varies from 41 to 98% of that obtained with soybean meal and probably depends on the quality of the meat meal. However, in part, the degree of depression may be related to the class of cereal used in the diet. The growth depression is greater when sorghum is used than when wheat makes up a high percentage of the diet (Beames and Daniels 1970).

The replacement of soybean meal by meat meal in the diet of young pigs (3–8 weeks of age) has been shown to reduce weight gain by as much as 88% (Puchal *et al.* 1962) or to have no effect (Evans and Leibholz 1979); again the variation in response can be largely attributed to the quality of the meat meal. Puchal *et al.* (1962) found meat meal to be very low in nutritional value when given to pigs from 22 days of age. When meat meal was the sole protein supplement to a maize-based diet, young pig (5-kg liveweight) performance was only 12% of that of pigs fed on soybean meal. In studies by Evans and Leibholz (1979) the liveweight gain and feed conversion efficiency of young pigs (24 days of age and 4–5-kg liveweight) decreased linearly as the soybean meal protein in the diet was replaced by meat meal protein. Growth of young pigs was reduced by 18% when all of the soybean meal in their wheat-based diet was replaced with meat meal containing 8.4% calcium. As meat meal protein replaced soybean meal protein there was a significant decrease in apparent dry matter digestibility and a significant linear increase in feed intake. In subsequent studies, Evans and Leibholz (1979) reported that meat meal prepared from soft offal (stomach and intestine) and containing only 3.59% calcium could replace 100% of the soybean meal in wheat-based diets for 4–8-week-old pigs without affecting performance.

Removal of the bone from meat meal improves its nutritive value (Sathe and McClymont 1964; Kennedy *et al.* 1974; Evans and Leibholz 1979). Contrary to early reports (Eastoe and Eastoe 1954; Eastoe and Long 1960) which indicated that the poor performance of pigs fed meat-and-bone meal was the result of high ash (calcium and phosphorus) content, it has been shown (Beames and Daniels 1970; Evans and Leibholz 1979) that similar depressions do not occur when the same proportion of calcium and phosphorus (either from chemical sources or bone meal) are fed in diets containing other proteins such as soybean meal. Evans and Leibholz (1979) added bone meal to diets containing soft offal meat meal or soybean meal as the only protein supplement so that the diets contained 0.80, 1.55, or 3.05% calcium. Over the 4-week feeding period, the weight gains and the efficiency of feed conversion of the pigs (4–8 weeks of age) were significantly depressed by the addition of bone meal to the meat meal-based diets, but no effect was observed in the pigs fed on the soybean meal-based diet. The effect of increased levels of ash appears to be related more to a reduction in the amount and quality of the protein in the meat meal and to the interaction of high calcium with other diet ingredients such as phosphorus and zinc.

Supplementation of cereal and meat meal diets with lysine, methionine, and threonine improved performance to levels obtained with cereal and fish meal diets (Atkinson and Carpenter 1970), and it has been suggested that the order of amino acid limitation in these diets is first lysine, followed equally by methionine and threonine. Using plasma amino acid patterns of pigs, after a 5-day test period, as a criterion of amino acid adequacy (Stockland *et al.* 1970), it was concluded that isoleucine, methionine plus cystine, threonine, tryptophan and possibly lysine were deficient in the meat-and-bone meal tested. Using young growing pigs, Evans and Leibholz (1979) suggested that tryptophan was the first limiting amino acid in wheat-and-meat meal diets. It seems clear that the exact order of limitation of amino acids in a particular sample of meat meal or meat-and-bone meal will depend on the nature of the meal with respect to carcass components included and the processing temperature and procedure used in its manufacture.

In addition to protein quality, the decrease in pig performance resulting from increasing levels of meat meal substitution for soybean meal is related to reduced dry matter and energy digestibility. Evans and Leibholz (1979) reported that the replacement of 0, 25, 50, or 100% of the soybean meal by meat meal resulted in a linear decrease in apparent digestibility of dietary dry matter (80.1–73.4%). Large variations exist in digestible energy content of different samples of meat meal and meat-and-bone meal. Mercy (1974) reported a range of 2.37 to 3.52 kcal/g in four samples of meat-and-bone meal while Batterham *et al.* (1980) reported digestible energy values for 14 samples of meat meal and meat-and-bone meal ranging from 2.25 to 3.32 kcal/g. In these studies, there was no significant relationship between digestibility of energy and any major chemical constituent of the meals.

Use of meat meal and meat-and-bone meal as a protein supplement for swine diets should take into consideration other dietary ingredients, local availability, and price. Least-cost diet formulation will include meat meal in the diet at very low levels in many commercially prepared rations. Many local conditions will justify the use of meat-and-bone meal as a sole protein source. Under these conditions, although growth rate and feed efficiency may be reduced, economic returns may justify their use.

BLOOD MEAL

A considerable amount of blood is available as a by-product of livestock and poultry slaughter. The total quantity of blood in an animal is 7–9% of the body weight. The blood contains approximately 20% solids, and since the final blood meal will contain about 88–90% dry matter it is clear that considerable dehydration is needed to produce blood meal. Losses in handling reduce the total quantity produced; therefore, cattle should produce 0.6–0.7 kg of blood meal per 100-kg liveweight, and calves sheep, goats, and pigs, 0.5–0.6 kg per 100-kg liveweight.

Dried blood meal contains over 80% crude protein and 9% total lysine but is seriously deficient in isoleucine. Its high protein and particularly high lysine content would make it a valuable supplementary protein in pig diets if the protein had a high digestibility and the lysine and other amino acids had a degree of availability equivalent to that of most other high-protein feedstuffs. However, conventional drying methods commonly employed have used a high-temperature (up to 165°C for 10–24 hr) batch-drying process that has resulted in poor availability of certain amino acids, particularly lysine, and a product lacking in palatability for swine. Commercial blood meals produced by vat- or batch-drying have been found to be quite variable in nutritional value. Waibel *et al.* (1977) reported crude protein digestibility values for vat-dried blood meal to be as low as 55%. Kratzer and Green (1957) reported lysine bioassay values equivalent to only 60% of the total lysine in vat-dried blood meals while Waibel *et al.* (1977) found biological availability values ranging from 0 to 49%.

The poor protein quality of vat-dried blood meals is not entirely an inherent property of blood but is also related to the effect of prolonged exposure to high temperatures. Hamm and Searcy (1976) subjected raw poultry blood to various methods of cooking and drying to determine their effect on the available lysine content of the final blood meal. Raw blood was (1) denatured and coagulated by

TABLE 14.2. Effect of Substituting Blood Meal for Meat-and-Bone Meal in Diets for Growing Pigs

Criteria	*Diets*		
	1	*2*	*3*
Level of meat-and-bone scraps, %	17.5	9.0	9.5
Level of blood meal, %	—	5.0	—
Level of soybean meal, %	—	—	9.5
Av daily gain, kg	0.40	0.54	0.56
Feed/gain ratio	3.91	3.01	2.72

Source: Meade and Teter 1957.

heating at 85°C in an open pan, then held for 10 min at 85°C with stirring, or (2) processed in a pressure cooker at 20 psig (125°C) for 60 min with no agitation. The cooked blood meal was then dried under various conditions at temperatures ranging from 23° to 193°C, with or without forced air movement. As the drying temperature and time of exposure increased, available lysine decreased. Air movement adversely affected lysine availability, especially at temperatures above 143°C. Waibel *et al.* (1977) heated whole, freeze-dried blood to 121°C for 1–4 hr and reported a decrease in lysine availability as the time of exposure was increased. Heating at 120°C reduced fluorodinitrobenzene-reactive lysine by 5% after 1 hr and by 13% after 4 hr.

Meade and Teter (1957) substituted one-half of the meat-and-bone scraps in the basal diet with soybean meal or vat-processed blood meal (Table 14.2) and observed a marked increase in growth rate and an improvement in feed efficiency. Hillier *et al.* (1955, 1956) utilized vat-dried blood meal to supplement diets based on cottonseed meal or soybean meal as a protein. When cottonseed meal supplied all of the supplemental protein in milo-based diets, growth rate and feed efficiency were inferior to diets supplemented with soybean meal. However, the addition of 3.3% blood meal to the diet supplemented with cottonseed meal significantly improved both gain and efficiency of feed utilization. Replacing a portion of the soybean meal in the diet with 2% blood meal was, however, without effect.

Philippine workers (Jaucian *et al.* 1964) utilized fish meal, blood meal, or meat-and-bone meal to supplement diets based on rice bran, maize, maize gluten, coconut meal, and soybean meal. No differences in growth rate or feed efficiency were observed when these low levels (fish meal, 3%; blood meal, 2.2%; or meat-and-bone meal, 3.4%) of supplements were added. Acuerdo *et al.* (1978) compared vat-dried blood meal with fish meal and meat meal as a protein supplement for growing-finishing pigs and determined that 5% blood meal supported pig performance inferior to that obtained with 6.5% fish meal but superior to 8% meat meal.

Viteri and Calles (1971) added graded levels of vat-processed blood meal, containing 80.5% crude protein, at the expense of fish meal in the diet (Table 14.3). These levels of 0, 2, 4, and 6% blood meal replaced an equal quantity of protein from fish meal. There was a quadratic response to level of blood meal on growth rate. Each level of substitution up to 4% blood meal tended to improve

TABLE 14.3. Effect of Substituting Blood Meal for Fish Meal in Diets for Growing Pigs

Criteria	Diets			
	1	*2*	*3*	*4*
Level of fish meal, %	11.37	8.30	5.12	2.08
Level of blood meal, %	0.00	2.00	4.00	6.00
Av daily gain, kg	0.61	0.62	0.64	0.57
Av daily feed, kg	2.11	2.50	2.25	2.27
Feed/gain ratio	3.42	3.99	3.49	4.01

Source: Viteri and Calles 1971.

gains, however, the highest level of substitution (6%) reduced both gains and feed efficiency.

Wahlstrom and Libal (1977) fed growing/finishing pigs (24–90 kg) 14%/12% protein diets in which blood meal was substituted for soybean meal. Pigs fed diets containing 2 and 4% of conventionally dried blood meal replacing an equal amount of soybean meal or 2% blood meal replacing an equivalent amount of soybean protein gained at similar rates but required slightly more feed per unit gain than pigs fed a maize–soybean meal diet. A diet containing 4% blood meal replacing an equivalent amount of protein from soybean meal resulted in reduced pig gains and decreased feed efficiency. Supplementation of this diet with 0.1% lysine improved performance to a level equal to that of pigs fed the maize–soybean meal basal diet. It appears from these results that up to 4% soybean meal can be replaced with an equal amount of conventionally dried blood meal, but if 4% blood meal replaces an equivalent amount of protein, a deficiency of lysine may exist. This will depend, of course, on the degree of heat damage the meal has suffered during the drying process. Since feed consumption was depressed only when 2 or 4% blood meal substituted an equal quantity of soybean protein and not when 2–4% blood meal substituted soybean meal on an equal weight basis, these data would also suggest that nutritional factors such as amino acid deficiency and/or imbalance are responsible for reduced feed intake instead of flavor factors per se.

Newer methods of drying blood, which utilize a lower drying temperature, or flash-drying procedures, including both the rotary steam drying process and the ring drying process, have improved both the palatability and lysine bioavailability of blood meal (Doty 1973). Waibel *et al.* (1974) found that lysine in "ring-dried" blood meal was 80–90% available for rats, chicks, and turkeys. Miller *et al.* (1976A) have reported that ring-dried blood meal has a lysine bioavailability for growing swine of 70%. Parsons *et al.* (1975) found that 90% of the lysine in ring-dried blood meal was available to young pigs (13–30-kg liveweight). Hamm and Searcy (1976) found that 95% of the lysine in freeze-dried blood meal was available to chicks. Waibel *et al.* (1977) studied five different commercial samples of ring-dried blood meal containing 8.6–8.9% total lysine and reported bioavailability for chicks, turkeys, and rats ranging from 80 to 97%, compared to ranges of 0–43% for four conventionally dried samples.

TABLE 14.4. Effect of Substituting Conventionally Dried or Rotary-Steam-Dried Blood Meal for Equal Quantities of Soybean Meal or for Equal Quantities of Soybean Protein for Growing-Finishing Pigs[a]

Criteria	Basal	Type of substitution: Equal weight		Equivalent protein		
	1	2	3	4	5	6
Level of blood meal, %	—	2.0	4.0	2.0	4.0	4.0
Level of soybean meal, %	14.6	12.6	10.6	11.2	8.0	8.0
Level of suppl. lysine, %	—	—	—	—	—	0.1
Conventionally dried blood meal						
Av daily gain, kg	0.70	0.72	0.70	0.60	0.64	0.72
Av daily feed, kg	2.27	2.44	2.45	2.33	2.27	2.40
Feed/gain ratio	3.26	3.38	3.47	3.37	3.56	3.34
Rotary steam-dried blood meal						
Av daily gain, kg	0.65	0.66	0.64	0.62	0.64	0.66
Av daily feed, kg	2.45	2.35	2.24	2.15	2.17	2.12
Feed/gain ratio	3.75	3.56	3.48	3.54	3.41	3.35

Source: Wahlstrom and Libal 1977.

Recent feeding experiments with meals produced from these new processes have shown that levels much higher than the previously recommended levels of 2–3% can be used for pigs without deleterious effects on feed intake or liveweight gain. Wahlstrom and Libal (1977) using maize-based diets substituted 2 or 4% rotary steam-dried blood meal for an equivalent quantity of soybean meal or an equivalent quantity of soybean meal protein and were unable to show any significant differences in pig performance (Table 14.4). At the 4% level of substitution, lysine supplementation was without effect. In a subsequent trial (Wahlstrom and Libal 1977), levels up to 6% blood meal used to replace an equivalent quantity of soybean protein supported similar feed intake and pig performance; however, a level of 8% significantly reduced feed intake during both the growing and finishing phases and pig gain during the growing phase. King and Campbell (1978) utilized five dietary levels of ring-dried blood meal (0, 3.0, 5.9, 8.9, and 11.8%) to replace soybean meal and meat-and-bone meal in diets for pigs from 21- to 70-kg liveweight. Pigs receiving either 5.9 or 8.9% blood meal grew more efficiently than pigs given either the highest or lowest level of blood meal. A reduction in feed intake at the 11.8% blood meal level retarded both growth and efficiency of feed utilization (Table 14.5).

Soluble blood flour plus methionine was used (Becker *et al.* 1963) as a sole or primary source of protein in diets for baby pigs (5 kg) and finishing pigs (45 kg). With this semisynthetic diet containing 22% protein from blood flour, the growth rate of the baby pigs was very poor (Table 14.6). Growth was improved by the addition of isoleucine to the diet. Although 0.15% supplemental isoleucine significantly improved growth, feed intake, and feed efficiency, maximum gains and feed efficiency were recorded when 0.45% isoleucine was added. For the finish-

TABLE 14.5. Effect of Ring-Dried Blood Meal as a Substitute for Soybean Meal and Meat-and-Bone Meal in Wheat-Based Diets for Growing Pigs

Treatments	*Av daily gain (g)*	*Av daily feed (kg)*	*Feed/gain ratio*
Blood meal level, %			
0	760	2.30	2.87
3.0	774	2.27	2.79
5.9	796	2.21	2.64
8.9	756	2.13	2.73
11.8	641	1.99	2.91
Blood meal and isoleucine level, %			
0	782[a]	2.24[a]	2.87
12	579[c]	1.73[c]	3.00
12 + 0.1% isoleucine	687[b]	1.97[b]	2.88

Source: King and Campbell 1978.
[a,b,c]Means in the same column without a common superscript are significantly different from each other ($P < .05$).

ing pig, the blood protein was not adequate to meet the pig's daily requirement, and 0.20% supplemental isoleucine was needed to promote maximum gains and feed conversion.

More recent studies at several stations (Rerat *et al.* 1975; Wahlstrom and Libal 1977; King and Campbell 1978) have also shown that practical pig diets containing 8–12% blood meal must be supplemented with isoleucine in order to restore

TABLE 14.6. Response of the Baby Pig and the Finishing Pig to Isoleucine Supplementation of Diets Based on Blood Protein

Level of isoleucine (% of diet)	*Level of blood meal (%)*	*Av daily gain (kg)*	*Av daily feed (kg)*	*Gain/feed ratio*
5.08-kg pigs				
0.41 (basal)	25	0.06	0.26	0.25
0.56	25	0.14	0.31	0.46
0.71	25	0.18	0.34	0.53
0.86	25	0.21	0.34	0.62
1.01	25	0.20	0.34	0.59
4.30-kg pigs				
0.47	17.3	0.05	0.26	0.19
0.62	17.3	0.12	0.34	0.36
0.77	17.3	0.17	0.38	0.44
1.07	17. 3	0.15	0.37	0.42
44.95-kg pigs				
0.15	15.0	0.07	0.96	0.07
0.25	15.0	0.39	1.78	0.22
0.35	15.0	0.59	1.78	0.33
0.55	15.0	0.60	1.78	0.34

Source: Becker et al. 1963.

the growth performance and feed intake to levels comparable to those achieved with controlled diets containing no blood meal. Supplemental isoleucine is necessary even though calculations would seem to indicate that dietary isoleucine levels are adequate to meet the pig's requirement. However, blood meal contains a very high level of leucine and there is evidence that a high level of leucine may elevate the isoleucine requirement (Harper *et al.* 1955; D'Mello and Lewis 1970; Oestemer *et al.* 1973; Taylor *et al.* 1977A,B).

Practical feeding recommendations for dietary levels of blood meal will depend on processing method used to produce the blood meal and the type of diet employed. Conventional vat-dried blood meal can be used to replace 2–4% of the protein supplement or up to 2% of the supplemental protein in grain-based diets. If higher levels are used, supplemental isoleucine and/or lysine may be needed to support satisfactory pig performance. Ring-dried blood meal is an excellent source of lysine and other amino acids and can be used at levels up to 6–8% of the diet. However, at these and higher levels, care must be taken to assure an adequate level of isoleucine.

FISH MEAL

Thirty percent of the world catch of more than 51.4 million metric tons of fish are used for industrial purposes. The most important fish product used in pig feeding is fish meal.

Raw fish is used only in very special circumstances. Fish meals are among the most valuable feed ingredients when they are of high quality, and they are among the most costly and disappointing when they are of poor quality.

Fish meals, basically, are of two types: those made from waste from the human food fishery and those made from whole fish, normally caught specifically for reduction. The second category is by far the more common and is becoming increasingly so. Fish caught for reduction also fall into two general types, those with low oil content, primarily classified as *white fish,* and those with high oil content, which are usually members of the *herring family*. Again, the second type is by far the more common. Meals processed from members of the herring family include *menhaden meals* from the Gulf and Atlantic coasts of the United States; *herring meal* from the northeastern United States, Canada, Iceland and Norway; *anchouetta meals* from Peru, Chile and Ecuador; and *pilchard meal* from South Africa (Ousterhout 1968).

Fish meal consists essentially of the dried, ground carcasses of fish; most of the oil present may be removed before drying since fish-body oil is also a valuable product. Freshly caught fish of various species may contain from 68 to 85% water, from 15 to 25% protein, and from almost 0 to 25% oil (on a fresh-wet basis). In order to make a fish meal that will be reasonably stable under normal storage conditions, the moisture content must be reduced to around 10%. If the meal contains 15% or more of water, it is possible for molds to grow. Stability requires that the fat content be reduced to about 10% and even a meal with this level of fat may become rancid in storage. With high fat levels, spontaneous heating and even combustion may occur. Fish meal leaving a typical heat dryer would normally be sterile, or nearly so, but it may easily become contaminated with bacteria during succeeding processes or storage (Cole 1968).

TABLE 14.7. Chemical Composition of Fish Meals Heated at Different Intensities and Its Effect on the Performance of Pigs

Components and effect on performance	*Fish meal treatment*		
	Not heated	*120° C for 2 hr*	*120° C for 4 hr*
Dry matter, %	91.07	92.04	89.28
Total nitrogen, %	11.15	11.15	10.94
Amino acids, g/16 g nitrogen			
Histidine	2.10	1.80	1.30
Isoleucine	4.20	4.00	3.65
Leucine	7.15	6.95	6.60
Lysine (total)	7.10	6.65	5.55
Lysine (available)	6.50	4.55	2.15
Methionine + cystine	4.65	3.50	3.05
Phenylalanine + tyrosine	6.95	6.95	6.25
Threonine	4.20	3.65	2.70
Tryptophan	0.90	0.80	0.80
Valine	5.35	4.95	5.00
Apparent dig. of dietary nitrogen, %	76	65	61
Total weight gain, kg	35.7	25.6	16.8
Total carcass nitrogen, %	2.20	1.92	1.88
Total carcass extract, %	23.70	29.50	28.50
Total nitrogen deposition, g	748	410	226
Total protein synthesized, g	7,414	5,341	4,316

Source: Madsen et al. 1965.

Since the world's major industrial fisheries in which the fish is reduced to meal are based on high oil fish species, the reduction of these fish to meal requires cooking, usually by a wet reduction with live steam, pressing to remove most of the oil, and the drying and grinding of the pressed cake. Dryer construction and operation materially affect the nutritive quality of fish meal, since overheating and overdrying damage the digestibility of both the protein and the residual oil (Ousterhout 1968). Cooling and storage are also important in fish meal quality. Fresh, hot fish meal must be lowered to a temperature of below 49° C immediately after processing since even several hours of high temperature can measurably lower nutritive value. It has been shown that heat treatment of fish meal can significantly affect its nutritive value for growing pigs (Mason and Weidner 1964; Madsen *et al.* 1965; Jones and Cadenhead 1965). Table 14.7 illustrates the effect of different heat treatments of a single lot of fish on its nutritive value for pigs. Heating significantly reduced feed intake, ration digestibility, body weight gains, nitrogen deposition, and protein synthesis (Madsen *et al.* 1965).

Since there exist innumberable types and varieties of fish that are produced by very many different methods and that are processed in many different ways, it becomes obvious that the term "fish meal" can mean many different things to different people. The detailed composition of six fish meals of different origin (Table 14.8) shows clearly that differences in specific nutrient contents may be

TABLE 14.8. Composition of Fish Meals of Different Origin

Components	Herring	Menhaden	Red fish	Sardine	Tuna	White fish
Crude protein, %	70.0	60.0	57.0	65.0	62.0	63.0
Ether extract, %	7.0	8.0	8.0	4.0	7.0	2.0
Fiber, %	1.0	1.0	1.0	1.0	1.0	1.0
Ash, %	12.0	20.0	26.0	19.0	20.0	22.0
Calcium, %	3.0	5.0	7.7	4.5	4.0	6.5
Phosphorus, %	2.0	3.0	3.8	2.4	2.5	3.5
Vitamins, mg/kg						
Choline	3,960.0	3,300.0	3,080.0	2,860.0	3,080.0	1,540.0
Nicotinic acid	88.0	55.0	42.9	66.0	66.0	66.0
Pantothenic acid	11.0	8.8	6.6	6.6	8.8	6.6
Riboflavin	8.8	4.4	6.6	5.5	8.8	8.8
Amino acids, %						
Arginine	4.0	4.0	4.0	2.7	5.3	4.2
Lysine	6.3	5.3	6.5	5.9	5.2	4.3
Methionine	2.0	1.8	1.7	2.0	1.7	1.7
Cystine	1.1	1.0	0.6	0.8	1.0	1.2[a]
Tryptophan	0.9	0.7	0.6	0.5	0.7	0.6

Source: Hubbell 1961.
[a]Braude 1961.

quite variable (Hubbell 1961). Even fish meals with a similar designation may vary widely in nutrient content, as shown in Table 14.9 (Bollmann 1959; Woodman and Evans 1940, 1945, 1951).

Differences in composition of fish meals and fish products may be attributed to many causes: origin and variation in the composition of the raw material including the proportions of whole fish and fish offals; different methods of processing including removal of fat, heat treatment, and drying; and contamination with some waste product or with sand. Some fish meals may be objectionable because they are not fresh, or because of high sand content, impurities, excessive salt content, excessive fat, or excessive moisture.

TABLE 14.9. Variation in Composition of Fish Meal (%)

Components	Fish meal[a]		White fish meal[b]	
	Average	Range	Average	Range
Moisture	8.26	5.5–11.8	11.3	5.7–14.9
Crude protein	57.82	42.3–68.8	62.9	55.0–71.2
Crude fat	6.09	1.8–11.1	3.6	1.7–5.9
Salt	4.09	0.9–15.6	—	—
$CaHPO_4$	13.93	9.0–17.9	—	—
Sand	2.71	0.3–10.5	—	—
Nitrogen-free extract	—	—	0.6	0–3.2
Ash	—	—	22.3	17.2–25.8

[a]Bollmann 1959. (Adapted from Braude 1961.) [b]Woodman and Evans 1951.

Good quality fish meal makes a valuable contribution to pig diets composed of cereals, cereal by-products, and other low protein ingredients such as bananas and cassava. Fish meal contributes protein (amino acids), vitamins (B_{12}, riboflavin, niacin, and pantothenic acid), and minerals (calcium, phosphorus, sodium, magnesium, potassium, iron, copper, zinc, and manganese). Kifer *et al.* (1968) have summarized the nutritive content of 33 samples of menhaden fish meal (Table 14.10).

The literature records a number of experiments from various countries in which the nutritive value of fish meals has been tested. Braude (1961) has made a comprehensive review of these reports. An outstanding feature of these tests is that fish meal has generally exerted a substantial growth-promoting effect and also has led to a more efficient feed conversion. In comparison to the effects of all-vegetable diets, the responses to some of the fish meals have been striking. This was especially true when the all-vegetable diet contained peanut (groundnut) meal (Woodman and Evans 1951; Carpenter *et al.* 1956) and also when soybean meal was used instead of peanut meal, though the difference in response was less. Laksesvela (1961) showed that increased levels of herring meal in diets for growing-finishing pigs (20–90 kg) produced increased growth rate and also im-

TABLE 14.10. Proximate Mineral and Amino Acid Composition of 33 Samples of Menhaden Fish Meal

Components	*Average*	*Range*
Protein, %	62.01	58.67–64.26
Ash, %	18.03	15.23–22.47
Ether extract, %	10.22	6.83–13.12
Moisture, %	8.25	4.80–11.24
Calcium, %	5.26	3.94–6.82
Phosphorus, %	2.98	2.39–3.67
Sodium, %	0.34	0.25–0.44
Magnesium, %	0.14	0.13–0.17
Potassium, %	0.72	0.60–1.10
Iron, ppm	438	250–670
Copper, ppm	11.4	6.6–17.5
Zinc, ppm	150.0	133–169
Manganese, ppm	35.6	21.1–65.0
Amino acids, % of protein		
Lysine	7.56	6.77–8.43
Methionine	2.82	2.42–3.09
Cystine	0.90	0.53–0.95
Tryptophan	1.07	0.82–1.54
Histidine	2.32	1.96–2.80
Arginine	6.04	5.26–6.76
Threonine	3.97	3.66–4.59
Valine	5.10	4.78–5.52
Isoleucine	4.40	4.00–4.86
Leucine	7.14	6.42–7.80
Phenylaline	3.95	3.53–4.39

Source: Kifer et al. 1968.

TABLE 14.11. Effect of Graded Levels of Herring Meal on the Performance of Growing-Finishing Swine (20–90 kg)

Criteria	Percentage of herring meal							
	1	*2*	*3*	*4*	*5*	*6*	*7*	*8*
Av daily gain, g[a]	631	641	669	653	691	726	713	713
Feed/gain ratio[a]	3.21	3.16	3.11	3.20	3.05	2.95	3.00	2.99

Source: Laksesvela 1961.
[a]Average calculated over all trials.

proved feed efficiency when compared to all vegetable diets based on soybean meal (Table 14.11). Most of the response was obtained with moderate levels (6–8%) of herring meal. As a result of the initial additions of 6% fish meal, the growth rate improved by an average of 8% and feed efficiency by 5% compared to the performance of pigs on all-vegetable control diets. Lower levels proved to be of little benefit.

The superior quality of fish in comparison with vegetable proteins has been reported by many researchers. Kronacher *et al.* (1932) reported that the replacement of two-thirds of the fish meal supplement with soybean meal adversely affected performance of the pigs. Frens and Ubbels (1951) showed that pigs receiving a ration supplemented with 8% herring meal grew and utilized their feed better than pigs fed a diet supplemented with 5% soybean meal and 5% sunflower meal. It was reported that white fish meal was superior to soybean meal as the only supplement to barley and fine wheat offal-based diets. Similar results were reported by Hvidsten and Husby (1955–1956) and by Comberg and Staunde (1956). Pike and Curran (1978) fed conventional diets containing 15.5% crude protein or high nutrient density diets containing 18.5% crude protein with or without herring type fish meal and reported that both weight gain and feed conversion were significantly better for the fish meal supplemented treatments. The effect of fish meal on the rate of growth was particularly marked in the period from 3 to 7 weeks of age.

In contrast, however, other workers have found no differences in performance of pigs fed either fish meal or soybean meal (Kirsch 1959; Kirsch and Fender 1960; Rabold and Schoen 1960). Schiller (1953) found that the performance of pigs on diets based on soybean meal could not be improved by the addition of fish meal.

Becker *et al.* (1954), when using menhaden meal, found that pigs up to 44-kg liveweight that were fed diets of 16% crude protein performed satisfactorily. However, equal results were obtained on diets with only 14% crude protein when soybean meal was used as the protein supplement. For pigs weighing more than 44 kg, both fish meal and soybean meal were of equal value. In diets low in crude protein the menhaden meal was less satisfactory than the soybean meal.

Smith (1976) compared fish meal with skim milk fed as a powder or in liquid form to growing pigs and concluded that partial replacement of either form of skim milk with fish meal on an equivalent protein and energy basis had no significant effect on pig performance or carcass composition.

The quality of fish meal is superior to that of most plant protein supplements (Braude, 1961).

Skim milk, either liquid or dried, is generally recognized as an excellent protein supplement for pigs. As reviewed by Braude (1961), most reports indicate that fish meal can satisfactorily replace skim milk. Danish results generally indicate that for growth rate, efficiency of feed utilization, and effect on health; fish meal and skim milk supplements are of equal value (Jaspersen and Hansen 1953). However, if the fish meal is not withdrawn several weeks before the pigs are slaughtered, the color, taste, and texture of the pork may be adversely affected.

Eggum (1973) showed that fish meal is highly suitable for feeding piglets from 2 to 3 weeks of age, particularly due to the high digestibility and utilization of its amino acids. That fish meal is beneficial in feed mixtures for young piglets was also shown by Schröder and Wehahn (1972) and Pettigrew *et al.* (1972) testing low-fat, hexane-extracted herring fish meal in creep feed for suckling piglets and in liquid diets for piglets weaned at 2–4 days of age, respectively. Using pigs one week of age, Rodriguez and Young (1980) replaced 20, 40, and 60% of the dietary protein supplied by casein and buttermilk with whole herring meal and observed a nonsignificant linear depression in both daily feed consumption and daily gains. However, when daily gains were adjusted for feed intake no differences in gain were apparent. Gjifsen *et al.* (1980) showed that, when combined with whey powder to maintain a satisfactory level of lactose in the diet, hexane-extracted, Norwegian herring-type fish meal of high quality and herring-type fish meal with normal fat content resulted in baby pig performance not different from that of pigs fed diets based on dried skim milk as a protein source.

The overall variation in response resulting from fish meal supplementation in the diets of growing-finishing swine is indicative of the variation that exists in the quality of the different fish meals. The ultimate value of fish meal will depend upon its quality and its effect on the total amino acid balance of the diet in relation to the class of swine to which it is being fed. Its comparative value will depend upon these factors and the quality of the protein to which it is compared. A better evaluation of meal quality should be based on source and final chemical analysis to determine protein level, quality, and amino acid availability.

Fish meal, fish silage, and similar materials containing fats highly decomposed by oxidation may, when mixed into pig feed, cause a strong oily or fishy off-flavor in the fresh pork and especially in the cured hams and bacon. Fishy and oily flavor may be caused by highly polyunsaturated fatty acids in feedstuffs (Asköe and Madsen 1954). If these fats originate from fish they will show a high iodine value and when fed to pigs cause an off-flavor in the fat depot tissue. As early as 1923 (Orr and Crichton 1923), this was recognized as one of the limiting factors associated with feeding fish meal or products containing fish oil to pigs especially up to market weight.

Organoleptic tests of fresh pork as well as of cured bacon were undertaken by Laksesvela (1961). Little influence of ''fish taint'' on pork quality was reported when 6–8% or less herring meal containing 7% fat was fed to pigs up to 80 kg and slaughtered at 90 kg. However, the feeding of 12% fish meal resulted in a fish off-flavor of the pork. Similarly, New Zealand workers (Shearer *et al.* 1970) reported that the presence of fish meal containing 7.1% oil in the diet did not result in subsequent ''fish taint'' in the cured bacon.

Fish meal can be fed to sows and gilts during gestion and lactation. Palmer *et al.* (1970) compared the reproductive performance of gilts and sows fed maize–soybean meal diets during gestation and lactation with the performance of animals fed similar diets containing 6% menhaden fish meal. They reported that fish meal increased gestation gain of both gilts and sows and increased the number of live pigs at birth. Total litter weight at birth and at 2 weeks was higher for litters from sows fed fish meal. In contrast, Baker *et al.* (1974) were unable to show any response during two reproductive cycles to 3% fish meal when compared to a similar diet based on maize and soybean meal. Since fish meal contributes a wide range of vitamins and minerals, part of the improved results reported by Palmer *et al.* (1970) may have been related to factors other than protein and amino acids. Fish meal would appear to be a satisfactory protein source for sows and gilts during gestation but under most circumstances probably does not provide any special beneficial effect in comparison to other protein sources that provide an adequate balance and level of essential amino acids.

FISH PROTEIN CONCENTRATE (FPC)

Milk substitutes are needed for use in feeding early-weaned pigs because of the high demand for milk for human use. Probably the most commonly used substitute is soybean protein, but the inferior utilization of soybean protein, in comparison with that of milk protein, by baby pigs (Maner *et al.* 1961; Schneider and Sarett 1969) suggests a need for alternative protein sources for use in baby pig feeding. Fish protein concentrates (FPC) have been produced by isopropyl alcohol extraction of raw whole fish or steam-dried herring meal or by enzymatic hydrolysis of fresh fish or fish waste. Each of these products contains 80–90% crude protein and favorable concentrations of amino acids (Table 14.12).

Early studies by Pond *et al.* (1971A) utilized a FPC developed by the U.S. Marine Fisheries Service by isopropyl alcohol extraction of raw whole hake (*Urophycis chuss*). Results of this study are summarized in Table 14.13. Baby pigs were fed liquid diets containing casein, isolated soybean protein, or FPC as the sole protein source. Performance was equal to that obtained with a similar diet containing milk protein (casein) as the sole source of protein and superior to that obtained with isolated soybean protein. Seve *et al.* (1975) studied the feeding value of a protein concentrate prepared by enzymatic hydrolysis of fresh fish. Baby pigs weaned at 12 days of age were fed a pelleted diet containing 95.7% milk products or diets in which 33, 66, or 100% of the milk products were replaced by fish protein concentrate and maltodextrine. Pig performance with 33% substitutes of dried skim milk was equal to that obtained with dried-skim milk alone, but higher levels of substitution reduced both gains and efficiency of feed utilization.

In studies by Newport (1979), half or total replacement of dried skim milk by fish protein concentrate and dried whey was compared with a diet containing dried skim milk (71%) as the only source of protein. All diets were prepared as spray-dried powders and reconstituted as liquids containing 200 g DM/liter and given to pigs weaned at 2 days of age during a 26-day experiment. Liveweight gain was slightly improved when half the dried skim milk was replaced but the

TABLE 14.12. Amino Acid Composition of Fish Protein Concentrate, Casein, and Isolated Soybean Protein

Amino acid	*U.S. Marine Fisheries fish protein concentrate*[a] *(g/16 g N)*	*Swedish fish protein concentrate*[b] *(g/16 g N)*	*Casein*[c] *(g/16 g N)*	*Isolated soybean protein*[c] *(g/16 g N)*
Arginine	6.4	7.1	4.1	8.3
Histidine	2.5	2.1	2.9	2.6
Isoleucine	5.5	4.6	6.4	6.5
Leucine	8.5	7.8	10.0	7.5
Lysine	9.0	8.4	8.0	6.8
Methionine	3.7	3.3	3.0	1.0
Cystine	1.0	0.8	0.4	0.6
Phenylalanine	4.7	4.2	5.3	5.0
Tyrosine	3.9	3.4	5.9	3.4
Threonine	5.1	4.5	4.4	3.9
Tryptophan	1.5	1.0	1.4	1.0
Valine	6.1	3.3	7.2	5.5

Source: Pond et al. 1971B.
[a]Bureau of Commercial Fisheries 1966.
[b]Protanimal, Astra Pharmaceutical Prod., Inc., Boston, Massachusetts.
[c]Altschul 1958.

feed/gain ratio was unaltered. Total replacement of the dried skim milk markedly reduced performance and increased the incidence of scouring and mortality (Table 14.14).

Fish protein concentrate is an excellent source of protein of high biological value for young pigs, but for optimal pig performance it should be limited to supply not more than 33–50% of the total dietary protein.

The nutritional value of fish protein concentrate has been adequately documented but the economics of commercial production still remain in doubt. Pilot or commercial plants have been established in a number of developed as well as

TABLE 14.13. Performance of Baby Pigs Fed Fish Protein Concentrate, Casein, or Isolated Soybean Protein as the Sole Source of Protein

Criteria	*Protein source*		
	Fish protein concentrate	*Casein*	*Isolated soybean protein*
No. of pigs	12	12	12
Daily gain, g	59	57	44[b]
Gain/feed ratio	0.66	0.65	0.52[b]
Final serum protein, g/100 ml	5.1	5.0	4.6[b]

Source: Pond et al. 1971A.
[a]Each diet contained 66 g protein source, 78.6 g glucose, 11.36 g minerals, 1 g vitamin premix, 40 g lard and 2 g soya lecithin per liter of liquid diet.
[b]Significantly less than in other diet groups.

TABLE 14.14. Effect of Replacing Dried Skim Milk by Fish Protein Concentrate and Dried Whey on Performance of Pigs from 2 to 28 Days of Age

Diets	*Percent replacement of dried skim milk*		
	0	*50*	*100*
Level of dried skim milk, %	71	36	0
Level of fish protein conc., %	0	12	24
Level of dried whey, %	0	23	47
Average daily gain, g	216 [b]	249 [a]	130 [c]
Feed/gain ratio	0.99[a]	0.95[a]	1.28[b]
Apparent N digestible, %	99.8[a]	98.8[b]	97.5[c]

Source: Newport 1979.
[a,b,c]Means on the same line with different superscripts are statistically different.

developing countries and although production technology seems to have been adequate, commercial production started during the 1960s and 1970s has failed in 40 countries, including the United States, Canada, Sweden, Chile, and Morocco. Therefore, the future of this product appears doubtful.

FISH SILAGE AND LIQUEFIED FISH

The material known as fish silage is a liquid product made from whole or parts of fish, to which no material has been added other than an acid and in which liquefaction of the ground fish mass is carried out by enzymes present in the fish. The principle involved in the manufacture of such silage is that enzymes already present in the fish are spread throughout the fish mass by grinding and mixing and the acidity adjusted so as to favor the rapid action of these enzymes and to inhibit bacterial action. Various acids or mixtures of acids can be used. Mineral acids such as hydrochloric and sulfuric are effective. Even though such acids are relatively inexpensive, they have the disadvantage that the acidity of the fish mass after being lowered to pH 2 in order to control bacterial growth, must be neutralized before it is fed to animals. Formic acid, although more expensive than mineral acid, is an appropriate preservative for fish silage as it reduces the fish mass to pH 4 which is suitable for enzyme activity, is sufficient to restrict bacterial growth, and has the advantage of not requiring neutralization prior to feeding to animals. Since there is no evidence of fermentation during this liquification process, the term "fish silage," often used to describe fish preserved in this manner, is a misnomer. True fish silage can be made by the addition of lactic acid bacteria such as *Lactobacillus acidophilus,* a fermenting agent, to the ground fish mass. Such studies have been described by Tibbets *et al.* (1981). To distinguish these products in this chapter, fish and fish waste preserved through enzyme action after acid addition will be termed *liquefied fish* and fish preserved by bacterial fermentation will be called *fish silage.*

Since transporting trash fish and fish waste long distances for processing in fish meal factories is costly, the waste is often dumped into the sea, causing pollution problems and loss of a valuable protein feedstuff. The preparation of products from fish and fish waste by liquefication or ensiling is not widespread. As the production of liquefied fish and fish silage requires relatively low capital expenditures, uses simple technology, and poses no pollution problems, the process is appropriate for fish and fish waste produced from fish processors at ports not possessing a drying plant for fish meal production and particularly appropriate for high quality protein production in developing countries.

Liquefied fish and fish silage are valuable mainly for their content of high quality protein. The chemical and amino acid composition of some liquefied fish products is presented in Table 14.15. Liquefied fish products contain 20–35% dry matter and have a crude protein content of 14–15%. The rate of proteolysis in liquefied fish products is high and studies of changes in nitrogen solubility indicate a rapid breakdown of the protein to low molecular weight peptides and free amino acids. The nutritional implications of high dietary levels of peptides and free amino acids are unclear, since there is no evidence that hydrolyzed protein is more or less valuable nutritionally than protein in its natural form.

TABLE 14.15. Chemical and Amino Acid Composition of Liquefied Fish and Fish Silage[a]

Item	Liquefied fish				Fish silage[d]
	Herring				
	1[b]	2[c]	De-oiled[c]	White fish[b]	
Crude protein, %	45.3	48.3	67.5	71.3	34.6
Ether extract, %	40.1	28.2	3.7	3.6	9.0
Ash, %	8.6	12.5	16.5	14.9	—
Calcium, %	2.1	3.5	4.8	3.8	5.0
Phosphorus, %	1.6	1.8	2.6	1.9	1.5
Sodium, %	—	0.6	0.8	—	—
Chlorine, %	0.5	—	—	1.1	—
Amino acids, %					
Arginine	3.65	—	5.1	5.22	1.4
Histidine	1.25	—	1.7	1.81	2.3
Isoleucine	1.92	—	2.8	2.84	0.9
Leucine	3.58	—	5.2	4.39	1.7
Lysine	4.20	—	6.2	—	1.1
Methionine	0.89	—	1.8	1.76	1.3[e]
Cystine	0.28	—	0.8	0.49	—
Phenylalanine	2.58	—	3.5	2.66	1.4[e]
Tyrosine	1.10	—	1.6	2.09	—
Threonine	2.17	—	3.0	3.08	0.7
Valine	2.46	—	3.3	3.28	1.3

[a]Expressed on dry matter basis.
[b]Smith 1977.
[c]Whittemore and Taylor 1976.
[d]Tibbets et al. 1981.
[e]Methionine plus cystine and phenylalanine plus tyrosine.

In feeding trials with growing-finishing pigs, liquefied fish prepared from white fish and fish waste low in oil (cod, haddock, and plaice) have been shown to be acceptable as the sole supplemental protein source; however, liquefied fish high in oil content (up to 40%) such as that produced from herring (*Clupea harengus*) has been shown to be unacceptable as the primary protein supplement to diets for bacon pigs and only can be used to advantage at very low levels. Luscombe (1973) used liquefied fish, containing 24% dry matter, 15% crude protein, and 2.1% oil, to replace all of the fish meal and all or a portion of the soybean meal in barley-based diets for bacon pigs. Replacement of the fish meal and soybean meal with low-oil, liquefied fish had no effect on either rate of growth or efficiency of feed utilization (Table 14.16). Organoleptic tests showed that the liquefied fish product had no effect on taste or acceptability of fat or lean meat from these carcasses. Similar results with liquefied white fish were reported by Albrecht and Gebhardt (1969) and Smith (1977). In studies by Smith (1977), a liquefied white fish product was used to replace all of the soybean meal and white fish meal in the barley-based control diet. Although a small but significant reduction in both rate of gain and feed utilization were reported when 10.3% of the dry matter was supplied by liquefied white fish, smell, taste, and fat texture and color of the uncured pork were acceptable and not different from meat obtained from the barley–white fish meal control.

Liquefied herring, because of its high oil content, is unacceptable as a primary supplemental protein source for pigs. Nutritionally, liquefied herring is equal to liquefied white fish for pigs (Smith 1977). However, high levels of liquefied herring (above 5.1%) produced pig carcasses with soft yellow fat and fish-tainted meat that is unacceptable for the human market. Feeding trials conducted in Denmark (Hansen 1979) suggested that if the oil is removed from the herring, the liquefied product may be more acceptable for supplying the protein in diets for growing-finishing pigs. Experimentally, de-oiled liquefied herring has been produced by Whittemore and Taylor (1976). Although these authors reported that de-oiled liquefied herring contained 17.6% dry matter, 13.3% crude protein, 1% available lysine, 0.65% oil, 4445 kcal of gross energy per kilogram of dry matter, 4278 kcal of digestible energy per kilogram of dry matter, and 13.1% digestible protein, neither feeding trials nor carcass evaluations were made with the product.

TABLE 14.16. Effect of Liquefied Fish as a Replacement for White-Fish Meal and Soybean Meal in Barley-Based Diets for Bacon Pigs

Treatment	*% fish in DM*	*Av daily gain, kg*	*Feed/gain ratio*	*Carcass wt, % of liveweight*
Fish meal + soybean meal + barley[a]	7.5	0.63	3.13	72.3
Liquefied fish + barley	15.3	0.60	3.07	71.1
Liquefied fish + soybean meal + barley	10.7	0.65	2.99	71.8
Barley + soybean meal + fish meal[b]	5.4	0.58	3.12	77.1
Barley + white liquefied fish	10.3	0.56	3.29	76.3
Barley + soybean meal + herring liquefied fish	2.4	0.56	3.28	76.4

Source: [a]Luscombe 1973. [b]Smith 1977.

Liquefied fish prepared from white fish with low oil content can be used as the sole protein supplement to diets for growing-finishing pigs without greatly affecting pig performance or acceptability in taste of the pork. However, because of unacceptable carcasses and tainted meat, it seems that feeding of liquefied herring will not be commercially viable.

The results of research with a true fish silage were reported by Tibbets *et al.* (1981). Using "trash" fish collected from the net of shrimpers, a fish silage was prepared with a mixture of 60% ground fish, 30% ground maize, and 5% dry molasses that had been inoculated with a fermenting agent *Lactobacillus acidophilus*. The fermented silage contained 62.6% moisture, 12.2% crude protein, 1.8% calcium, 0.5% phosphorus, and 3.2% lipids (Table 14.15).

In a digestion trial with growing pigs, digestible nutrients in a diet containing 6% silage were equal to those found in a maize–soybean meal diet. Groups of weanling and finishing pigs were fed diets containing 0, 3, 6, and 9% fish silage. Average daily gains and feed utilization for weanling pigs were not adversely affected by any level of fish silage. Average daily gains for finishing pigs fed the four diets were 0.73, 0.73, 0.74, and 0.68 kg, respectively; feed per gain ratios were 3.13, 3.37, 3.51, and 3.80. There were no significant differences between treatment groups in average daily gain; however, the inclusion of fish silage at 6 and 9% of the diet caused a decrease in feed conversion. Dietary treatments had no significant effect on carcass merit, tenderness, juiciness, detectable connective tissue, flavor, or acceptability as measured by consumer taste tests.

The silage was also fed at a dietary level of 6% to sows during breeding, gestation, and lactation. When compared to a standard maize–soybean meal diet, piglet birth weight, litter size, weaning weight, litter strength, number weaned, and average daily gain to weaning did not differ between treatments.

These results suggest that fish silage with a low level of lipids can be included successfully at a level up to 6% in pig diets from weaning through lactation without affecting reproduction, growth, or meat taste and acceptability.

SHRIMP MEAL

Shrimp meal is the dried waste of the shrimp industry, consisting of the heads and shells (hulls) and in some instances whole shrimp (Meyers and Rutledge 1971). Due to differences in both the species and condition of the raw material as well as processing and effluent recovery techniques, tremendous variability exists between different meals. Although improper storage before processing can reduce the protein level, the proportion of heads and shells significantly influences the protein level and quality. The exoskeleton contains chitin, a nitrogen-containing carbohydrate (N-acetylated glucosamine polysaccharide), that is of little nutritional value. Accurate analysis of protein levels in shrimp meal require correction of the crude protein fraction for the nonprotein nitrogen contributed by the chitin present. Shrimp hulls contain 45–46% crude protein but because of the high content of chitin (54.2%), the corrected protein level is only 22.8%. On the other hand, shrimp heads contain only about 11.1% chitin and have a corrected protein content of 53.5% (Table 14.17) and have an excellent amino acid profile (Table 14.18) (Kondos 1977; Meyers *et al.* 1973). In practice, shrimp meal seldom contains only hulls or heads but a variable mixture of both by-products. In usual

TABLE 14.17. Proximate Analysis and Protein and Chitin Nitrogen of Various Shrimp Preparations

Item	*Dehydrated meal*	*Sun-dried meal*	*Shrimp heads*	*Shrimp hulls*
Moisture, %	7.5	9.1	3.2	4.0
Fat, %	1.3	2.9	8.9	0.4
Fiber, %	21.4	10.4	11.9	27.2
Ash, %	38.2	26.9	22.6	31.7
Total crude nitrogen, % (A)	5.97	8.27	9.32	7.36
Chitin nitrogen, % (B)	1.41	0.62	0.76	3.71
Corrected nitrogen (A − B)	4.56	7.65	8.56	3.65
Crude protein, %	37.3	51.7	58.2	45.9
Corrected protein, %	28.5	47.8	53.5	22.8
Chitin, %	20.6	9.0	11.1	54.2
Calcium, %	15.0	7.0		11.10
Phosphorus, %	2.2	1.5		3.16

Source: Adapted from Meyers et al. 1973.

Kjeldahl determinations, nitrogen values obtained include both protein and chitin nitrogen. It is, therefore, essential to estimate the level of chitin to be able to calculate a corrected protein value. One method for determining the crude protein in shrimp meal, corrected for the amount of crude protein contributed by chitin nitrogen, is described by Meyers *et al.* (1973). An alternative rough estimate of the amount of chitin present can be determined by examination of the crude fiber

TABLE 14.18. Amino Acid Content of Different Samples of Shrimp Meals (g/16 g N)

Amino acid	*Sun-dried[a] meal*	*Shrimp head[b] meal*
Arginine	6.31	6.90
Histidine	1.90	1.76
Isoleucine	3.26	3.34
Leucine	7.57	5.76
Lysine	6.17	5.40
Methionine	2.84	1.72
Cystine	1.59	—
Phenylalanine	4.56	—
Threonine	4.28	3.14
Tryptophan	1.26	1.11
Valine	4.42	3.08
Alanine	5.29	—
Aspartic acid	10.74	—
Glutamic acid	15.46	—
Glycine	4.26	—
Proline	3.44	—
Serine	4.53	—
Tyrosine	3.64	—

Source: [a]Adapted from Toma and James 1975; Meyers 1976. [b]Adopted from Kondos 1977.

TABLE 14.19. Calcium Carbonate and Ash Content of Shrimp and Shrimp Waste (in Percent)

	Ash	*$CaCO_3$*
Whole shrimp	17.7	5.30
Peeler waste	20.3	9.44
Peeler waste	19.0	10.19
Shells only	27.5	24.77
Shells only	29.9	27.14
Peeler waste	—	23.01
Peeler waste	—	23.18
Range	17–30	5–27

Source: Thurston and MacMasters 1959.

content of the meal, which Lowell (1965) found to be in close agreement with the chitin level.

Both high salt (sodium chloride) and total ash levels of shrimp meal can present problems in its utilization as an animal feed. Salt is sometimes added to prevent spoilage during the sun-drying process. Salt level may vary from 0 to 7% in the final product. Definite quality standards require that the final shrimp meal must not contain more than 7% and that levels above 3% must be stated on the label. The high ash or mineral content is derived from the exoskeleton of the shrimp. Calcium carbonate is responsible for the sclerotization of the exoskeleton and constitutes most of the mineral matter which may be as high as 30% in shrimp shell waste or 17% in whole shrimp (Table 14.19) (Thurston and MacMasters 1969). Because of this high mineral level, shrimp meal is used in combination with other protein supplements.

Perez (1932) and Santiago (1926) fed diets based on maize, copra meal, and rice bran and supplemented with either tankage or shrimp meal. At the time the studies were conducted the 5% shrimp meal supplement gave satisfactory results. Angel (1935) obtained satisfactory results when 5–20% shrimp meal was fed.

Louisiana workers (Bray *et al.* 1932) used shrimp meal to replace tankage in growing-finishing swine diets. Shrimp meal was superior to tankage as a supplement to maize or to maize and rice polish regardless of whether these supplements were fed alone with maize or in combination with cottonseed meal and other protein feeds (Table 14.20). Shrimp meal was reported to be more palatable to the pigs than tankage and to contain less protein and less mineral matter.

Shrimp waste protein has a protein efficiency ratio (PER) of 2.48, as compared to 3.13 for casein, and promotes rat growth 80% as efficiently as casein (Toma and James 1975). Feeding experiments with growing pigs (initial weight 23 kg) have shown shrimp head meal used as the sole supplement to wheat–sorghum basal diets to support weight gains equal to those obtained with 80% of soybean meal, 82% of casein and blood meal, and 93% of meat meal (Table 14.21) (Kondos 1977).

When the nitrogen present in chitin is used to correct the crude protein content, shrimp meal is a valuable supplemental protein source for swine. Most efficient use is made of shrimp meal when it is combined with other supplemental protein sources such as soybean meal.

TABLE 14.20. Comparison of Shrimp Meal and Standard Tankage as a Protein Supplement to Maize or Maize and Rice Polish Diets for Growing Pigs

Performance	*Item as only supplement*		*Item with other supplements*	
	Shrimp meal	*Tankage*	*Shrimp meal*	*Tankage*
Av daily gain, kg	0.70	0.61	0.76	0.66
Feed/gain ratio	3.84	3.95	3.50	3.63

Source: Bray et al. 1932.

KRILL MEAL

Krill (*Euphausia* sp.) is a phytoplankton feeder that inhabits the waters of the Antarctic. It is a crustacean that resembles a very small shrimp in appearance. Because of their abundance, krill could play an important role in future world protein supplies, both as a human food and as a supplemental protein feedstuff for swine and poultry. Russia, Japan, Germany, Poland, Taiwan, and Chile have been active in research to improve both harvesting and processing techniques. The extremely small size of krill (20–60 mm) necessitates a special netting technology. Because of the green phytoplankton which remains in the krill's digestive tract, their high fat content and the absence of connective tissue, krill exhibit extremely poor keeping qualities and processing must take place within 3 or 4 hr after krill are caught (Martin 1979).

Raw krill contain 77–81% moisture, 11–16% crude protein, and 1–4% ether extract. Following processing, krill meal composition compares favorably with herring-type fish meal. It contains a high level of crude protein (52–59%) and is a favorable source of lysine, methionine, and tryptophan (Table 14.22). Although a small portion of krill nitrogen is contained in the chitin, protein digestibility coefficients ranging from 71.5 to 77.1% have been reported (Tronchuk *et al.* 1978).

Tests to evaluate chemical and physical treatment of krill meal to improve its storage and feeding quality indicate that meal heated to 70°–75° C and then press-dried at 95° C was equal or superior in quality and feeding value to meals to which

TABLE 14.21. Performance of Pigs Fed Shrimp Head Meal as the Sole Protein Supplement as Compared to Other Protein Supplements to Cereal-Based Diet

Protein source	*Av daily gain, kg*	*Av daily feed, kg*	*Feed/gain ratio*
Shrimp head meal	0.40[b]	1.16	2.94
Soybean meal	0.50[a]	1.30	2.59
Blood meal	0.46[a,b]	1.14	2.48
Meat meal	0.43[b]	1.14	2.68
Casein	0.46[a,b]	1.19	2.60

Source: Adapted from Kondos 1977.

TABLE 14.22. Composition of Raw Krill, Cooked Krill, Krill Protein Concentrate, and Krill Meal

Component	Krill			
	Raw	*Cooked*	*Protein concentrate*	*Meal*
Moisture, %	77–81	78–80	9.5	3–4
Crude protein, %	11–16	12–14	74.0	52–59
Ether extract, %	1–4	2–4	0.3	6–24
Ash, %	2–4	3–4	15.4	12–13
Calcium, %	—	—	—	1.6–1.8
Phosphorus, %	—	—	—	1.2–1.5
Lysine, %	—	—	—	5.6
Methionine, %	—	—	—	2.5
Tryptophan, %	—	—	—	1.6

Source: Martin 1979.

antioxidants had been added. In fact, dried untreated krill meal supported better pig performance than meals treated with antioxidants (Table 14.23) (Tronchuk *et al.* 1978). When untreated krill meal or heat-treated krill meal was substituted for equal quantities of fish meal protein, growth rates and feed utilization were reduced by 6–7% (Tronchuk *et al.* 1978). Paschma *et al.* (1979) replaced 50 or 100% of a protein concentrate with krill meal and reported a small, linear decrease in both gains and efficiency of feed conversion as the dietary level of krill meal was increased. Although replacing one-half or all of the protein concentrate in the barley-based diets with krill meal had no effect on dressing percentage, back fat was increased in response to the high level of fat in the meal.

Future use of krill meal as a protein supplement for swine diets will depend on quality, local availability, and price. Good quality krill meal can be used to supply a major portion of the supplemental protein for swine of all ages. Best results will be obtained if dietary levels are limited to 4–6% during the growing period and 2–4% during the finishing period.

TABLE 14.23. Effect of Physical and Chemical Treatment of Krill Meal on Feeding Value for Growing-Finishing Pigs

Treatment	*Av daily gain, kg*	*Feed/gain ratio*
Krill meal		
Pelleted + antioxidant	694	4.31
Meal—untreated	718	4.12
Meal—heat-treated	724	4.09
Krill meal—defatted	629	5.17
Krill meal—heat-treated	632	5.17
Fish meal	679	4.82

Source: Tronchuk et al. 1978.

KING CRAB MEAL

King crab meal is composed of shells, viscera, and unextracted meat of the crab following processing (Husby 1980). It contains 35–40% crude protein, 18% crude fiber, 20% acid detergent fiber, 1.5–3% ether extract, 37% ash, 9–12% calcium, and 1.6–1.8% phosphorus (Table 14.24). Although some of the nitrogen is present in a nonprotein nitrogen form bound to chitin, the protein contributed from viscera and unextracted meat is of excellent quality. The crude fiber content is composed mainly of chitin. The molecular structure of chitin is very similar to that of cellulose, differing only in the substitution of an acetylamine group for the hydroxyl group on carbon 2 of the glucose unit. Chitin is not digested by swine and other nonruminant animals. Because of the low level of ether extract and high level of undigestible chitin, the utilizable energy level of crab meal is low. The high ash and specially high level of calcium create an imbalance in swine rations that must be corrected by the addition of phosphorus, such as monobasic sodium phosphate, in order to maintain the preferred 1:1–2:1 ratio.

Studies at the University of Alaska have evaluated king crab meal as a source of supplemental protein for swine. Husby (1980) has shown that whole king crab meal can replace 50% of the supplemental protein from soybean meal in barley-based diets without reducing rate of growth and with only a small reduction in efficiency of feed utilization (Table 14.25). Complete replacement of the soybean meal with 12.6% crab meal significantly reduced both daily gains and efficiency of feed utilization. When maize was used as a grain source, requiring higher levels of supplemental protein to provide the total protein needs of growing swine, substitution of more than 25% of the soybean meal with whole king crab meal significantly reduced both rate of gain and efficiency of feed conversion (Table 14.25). Average daily feed intake was similar for all levels of substitution and indicated that palatability of diets containing crab meal was not a major problem.

TABLE 14.24. Chemical Composition of King Crab Meal

Item	*King crab meal*			
		Sample 2[a]		
	Sample 1	*Whole*	*<40 mesh*	*>40 mesh*
Dry matter, %	94.3	95.6	95.4	95.6
Crude protein, %	40.1	41.7	48.1	35.7
Ether extract, %	1.4	3.0	4.2	1.8
Ash, %	36.8	36.9	31.7	41.3
Crude fiber, %	17.9	17.6	—	—
Acid detergent fiber, %	20.7	20.6	15.6	27.4
Calcium, %	12.4	9.0	7.9	10.3
Phosphorus, %	1.8	1.6	1.4	1.7

Source: Husby 1980.
[a]Analysis of whole crab meal or following separation with 40 mesh screen.

TABLE 14.25. Growth Performance of Growing-Finishing Pigs Fed Several Levels of King Crab Meal with Either Barley or Maize

Item	*Basal*	*King crab meal, % of protein supplement*			
		25	*50*	*75*	*100*
Barley-based diets					
Level of crab meal, %	0	3.1	6.3	—	12.6
Av daily gains, kg	0.80[a]	0.79[a]	0.78[a]	—	0.70[b]
Av daily feed, kg	2.64	2.66	2.71	—	2.59
Feed/gain ratio	3.36[e]	3.40[c]	3.54[d]	—	3.70[e]
Maize-based diets					
Level of crab meal, %	0	5.3	10.6	16.0	21.4
Av daily gains, kg	0.85[a]	0.86[a]	0.75[b]	0.75[b]	0.67[c]
Av daily feed, kg	2.80	2.72	2.58	2.52	2.72
Feed/gain ratio	3.31[d]	3.25[d]	3.60[e]	3.81[e]	4.13[f]

Source: Husby 1980.
[a–f]Means on the same line followed by the same superscript are not significantly different ($P < .05$).

Physical separation of the whole king crab meal through a 40-mesh screen removes a portion of the larger shell waste and increases the concentration of protein from viscera and unextracted meat. Separated crab meal contains more protein and ether extract but less fiber, calcium, and ash (Table 14.24). Husby (1980) has shown that the fine crab meal resulting from this separation has a protein of higher quality and improved feeding value (Table 14.26). Whereas a maximum of 50% soybean meal could be replaced by whole king crab meal in barley-based diets without affecting pig performance, separated meal containing 48.1% crude protein can replace all of the soybean meal without significantly affecting pig performance.

In regions such as Alaska where the cost of king crab meal is 50–60% less than for soybean meal, it can be included in swine diets to replace 25% of the supplemental soybean meal in maize-based diets and up to 50% in barley-based diets. Dietary levels should not exceed 5.3–6.3%. Improved processing to reduce the

TABLE 14.26. Performance of Growing-Finishing Pigs Fed Barley-Based Diets[a]

Item	*Basal*	*50% KCM*[a]		*100% KCM*	
		Whole	*40 mesh*	*Whole*	*40 mesh*
Dietary level of crab meal, %	0	4.7	4.1	9.3	7.6
Av daily gain, kg	0.73	0.70	0.74	0.57	0.72
Av daily feed, kg	2.45	2.49	2.44	1.92	2.35
Feed/gain ratio	3.31	3.59	3.30	3.79	3.42

Source: Husby 1980.
[a]Diets are supplemented with graded levels of whole king crab meal (KCM) or fine king crab meal following separation through a 40 mesh screen.

shell waste content and to increase the protein from viscera and unextracted meat can produce a meal that can be used satisfactorily to supply 75–100% of the supplemental protein for growing-finishing and breeding animals.

DRIED MILK

Cow milk is, of course, one of the most nearly perfect foods, not only for calves and humans but also for pigs. Normally, it is in too much demand as fluid milk for human consumption to be used as swine feed. Dried whole milk and dried skim milk are also expensive, but because of their high nutritive value and palatability they are often used in early weaning diets or in creep feeds to encourage early consumption of dry feed. The composition of fresh cow milk is shown in Table 14.27, along with its composition on a dry matter basis and that of dried skim milk. The only major difference between dried whole milk and dried skim milk is that in skim milk most of the fat and fat-soluble vitamins have been removed so that all other components are increased proportionately. The only nutrients likely to be deficient in a diet containing a large proportion of dried whole milk would be iron and copper. Dried skim milk fed as a major part of the diet could cause a deficiency of fat-soluble vitamins as well, since they are removed with the milk fat. Pigs usually prefer milk products when given a choice of feedstuffs and the use of either whole or skim milk can be determined basically on economic considerations, since nutritional quality is unsurpassed. Although milk is normally used for young animals, it is tolerated by pigs of all ages. Lactase, the digestive enzyme required to split lactose into its component simple sugars, glucose and galactose, is apparently produced in adequate amounts in older animals fed whole or skim milk, although the higher concentration of lactose in whey may create problems in older pigs fed whey. There may be individual cases of lactase deficiency in older swine, but the problem is apparently not widespread as it is in the human population, in which milk intolerance is common in certain areas.

Dried whole milk and skim milk can be considered not only as good protein sources, but also as good sources of vitamins and minerals. The use of milk products for these purposes is discussed in Chapter 16.

MILK WHEY

Whey, a by-product from the manufacture of cheese and lactic-acid-precipitated casein from milk, is an abundantly available feed for pigs in certain areas of the world where cheese is produced. Many of these areas have poor or nonexistent fluid milk markets and must rely on cheese for a market for the product. Consumption of cheese products has increased sevenfold in the last decade, with a proportional increase in the amount of whey produced. This has created serious disposal problems for cheese plants. Many plants and farmers in developing countries dump excess whey, resulting in a loss of a potential feed source and in some cases pollution problems.

TABLE 14.27. Composition of Fresh Whole Cow Milk, of Its Dry Matter, and of Dried Skim Milk

Component	*Fresh milk*	*Dried whole milk*	*Dried skim milk*
Dry matter, %	12.0	96.0	94.0
Ash, %	0.8	5.4	7.9
Ether extract, %	3.6	26.6	0.8
Protein (nitrogen × 6.25)%	3.3	25.4	33.7
Digestible energy, kcal/kg	680	5500	3866
Metabolizable energy, kcal/kg	616	4522	3548
Vitamins[a]			
Choline, mg/kg	875.6	NR[b]	1390.0
Niacin, mg/kg	58.3	8.0	11.0
Pantothenic acid, mg/kg	8.1	22.8	36.3
Riboflavin, mg/kg	1.8	19.7	19.3
Thiamin, mg/kg	7.3	3.8	3.7
Vitamin B_{12}, mcg/kg	NR[b]	NR	51.0
Amino acids			
Arginine, %	NR	0.92	1.15
Histidine, %	NR	0.72	0.86
Isoleucine, %	0.32	1.33	2.18
Leucine, %	0.25	2.56	3.32
Lysine, %	0.28	2.25	2.53
Methionine, %	0.09	0.61	0.90
Cystine	NR	NR	0.45
Phenylalanine, %	0.16	1.33	1.56
Threonine, %	0.16	1.02	1.56
Tryptophan	0.05	0.41	0.43
Tyrosine, %	NR	1.33	1.14
Valine, %	0.25	1.74	2.28
Minerals			
Calcium, %	0.12	0.91	1.28
Phosphorus, %	0.09	0.71	1.02
Sodium, %	0.05	0.37	0.46
Chlorine, %	0.11	0.88	0.90
Magnesium, %	0.01	0.09	0.12
Copper, mg/kg	—	1.0	1.0
Iron, mg/kg	1.0	10.0	9.0
Manganese, mg/kg	NR	NR	2.0
Selenium, mg/kg	NR	NR	0.13
Zinc, mg/kg	3.0	22.0	38.0

Source: National Research Council 1982.
[a]Vitamins in fresh milk are from National Research Council 1964.
[b]NR = not reported.

The dry matter and moisture content of whey varies according to the processing method used to produce the cheese and the quantity of wash water that is allowed to drain into the whey storage tanks. However, generally, fresh whey contains approximately 93–96% moisture, 4–7% dry matter, 0.7–0.9% crude protein,

TABLE 14.28. Chemical Composition of Liquid, Condensed, and Dried Whey

Item	*Liquid*			*Condensed*	*Dried*			
	1[a]	*2*[b]	*3*[c]	*4*[b]	*5*[b]	*6*[d]	*7*[e]	*8*[f]
Dry matter, %	6.8	6.9	6.7	63.6	93.0	96.5	95.0	95.4
Crude protein, %	0.7	0.9	0.9	8.7	12.2	13.4	12.7	12.0
Lactose, %	—	5.0	5.0	47.7	71.5	74.4	—	—
Nitrogen-free extract, %	4.4	—	—	—	—	—	73.5	73.3
Ether extract, %	0.7	—	0.2	—	—	0.8	0.4	1.3
Ash, %	1.1	0.7	0.7	6.4	8.4	7.9	8.4	10.3
Calcium, %	0.03	0.05	—	0.38	0.91	—	—	0.7
Phosphorus, %	0.02	0.04	—	0.58	0.76	—	—	0.6

Sources: Adapted from [a]Cortez et al. 1979; [b]Schingoethe 1976; [c]Barber et al. 1978; [d]Alaviuhkola et al. 1980; [e]Pals and Ewan 1978; [f]Landblom and Nelson 1980.

5.0% lactose, and 0.7–1.0% ash. To facilitate storage, transportation, and feeding fresh, liquid whey is sometimes condensed or dried. The chemical composition of liquid, condensed, and dried whey is tabulated in Table 14.28, and each will be discussed separately. The vitamin, trace mineral, and amino acid composition is tabulated in Table 14.29.

The high moisture (approximately 94%) and lactose (approximately 75%) contents dictate the use of fresh whey in pig rations and contribute to the comparatively inferior food value of whey (Fevrier 1966). Pigs receiving 4% dry matter whey grow more slowly than pigs receiving 6% dry matter whey, even when both receive the same daily quantity of dry matter (Dunkin and Carr 1969). The low concentration of dry matter of whey limits its use as an energy source and the high lactose content reduces its use as a feed for older pigs; these have been regarded as the main reasons for the difficulties of high-whey feeding. The ability

TABLE 14.29. Vitamin, Trace Mineral, and Amino Acid Composition of Dried Whey

Vitamin	*1*[a]	*Minerals*	*2*[b]
Vitamin A, IU/kg	101	Sodium, %	3.91
Riboflavin, mg/kg	5.1	Potassium, %	1.54
Thiamin, mg/kg	0.9	Magnesium, ppm	950
Niacin, mg/kg	1.9	Iron, ppm	405
Pantothenic acid, mg/kg	9.6	Zinc, ppm	6.56
Pyridoxine, mg/kg	0.7	Copper, ppm	0.84
Biotin, mg/kg	0.09	Magnesium, ppm	1.86
Choline, mg/kg	494.4	Dry matter, %	95.4
Ascorbic acid, mg/kg	—		
Folic acid, mg/kg	0.2		
Vitamin B_{12}, mg/kg	0.004		
Dry matter, %	95.4		

Sources: Adapted from [a]Cortez et al. 1979; [b]Landblom and Nelson 1980.

of pigs to utilize lactose has been studied in many feeding trials. Satisfactory growth has been reported with 40–57% lactose in baby pig diets (McCrea and Tribe 1956: Becker *et al.* 1957). Ekström *et al.* (1975A) reported that the growing pig tolerated, without symptoms, feeds containing up to 30% lactose. In contrast, other workers have found that levels of more than 15–25% result in reduced food intake, slow growth, and a moderate diarrhea (Whittier 1935; Becker and Terrill 1954; Shearer and Dunkin 1968). Shearer and Dunkin (1968) found that 30% lactose in feed reduced the rate of growth by 10% when wheat starch was replaced by lactose in the feed of growing-finishing pigs; with a 45% lactose feed the growth was 39% slower than in the control group. The main reasons for the slow growth were the poor feed palatability and the diarrhea produced when the diet contained more than 30% lactose.

The ability of the pig to utilize lactose has been studied in many feeding trials. Ekström *et al.* (1976) found that 37–52% of the lactose entering the small intestine during the first 1.5 hr after feeding passes through to the colon intact when there is 40% lactose in the feed. Kim *et al.* (1978B) found with the same kind of diet (40% lactose) that about 30% of the lactose enters the colon during the first 5 hr after feeding. Aumaitre *et al.* (1977) found no correlation between the lactase activity of the small intestine and the amount of lactose in feed with pigs of liveweight between 25 and 100 kg. However, lactose may be hydrolyzed by microbial enzymes in the cecum–colon. Ekström *et al.* (1975B) found that feeding a diet containing 0, 10, 25, or 40% dried whey for 120 days did not change the lactase activity in the mucosa of the small intestine. It is not necessary to feed lactose continuously in order to assure tolerance of a 30% lactose diet (Ekstrom *et al.* 1975A); however, a continuous or early supply of lactose in the diet does not lead to subsequent adaptation to the use of this carbohydrate (Fevrier and Aumaitre 1978A,B).

Whey contains 13–17% protein in its dry matter, and this protein is of high biological value (Riggs *et al.* 1955). When fed free-choice as the only protein supplement to a barley–oat-based diet, pig performance was not different from that obtained with a similar diet supplemented with soybean meal (Landblom and Nelson 1980). In these studies, liquid whey fed free-choice accounted for 25% of the total dry matter and successfully replaced 100 kg of concentrate during the growing-finishing period (23–93-kg liveweight). Therefore, when whey is used, energy, not protein, is the factor limiting pig performance.

Pigs require a few days to adapt to liquid whey and should be started gradually on the product to prevent digestive disturbances and diarrhea. Feeding method and level of whey as well as type and level of concentrate supplementation have been the subject of study for many years. Whey is generally fed to appetite or in controlled daily quantities along with a constant level of concentrate. The level of concentrate feeding is important since low levels of daily concentrates (0.45–0.68 kg) along with whey support poorer performance when fed to 22–55-kg pigs than do higher levels of concentrates (0.91–1.14 kg) or an all-concentrate control diet (Dunkin 1958; Carr and Dunkin 1969; Rosa 1971; Esnaola 1971; Lerner 1973). In an attempt to determine the optimal level of concentrate, Cortez *et al.* (1979) fed growing-finishing pigs liquid whey to appetite along with graded levels of a 15.5% protein diet during the growing period and a 13.7% protein diet during the finishing period. The results of their study are summarized in Table 14.30.

TABLE 14.30. Effect of Level of Concentrate on Performance Growing-Finishing Pigs Fed Liquid Whey to Appetite

Treatment	*Level of concentrate*		*Average daily*			*Feed/gain ratio*	
	Growing	*Finishing*	*Whey, liter*	*Conc., kg*	*Gain, kg*	*Conc.*	*Conc. + whey*
1	*Ad lib*	*Ad lib*	—	2.25	0.58	3.86	—
2	0.75	1.75	13.4	1.29	0.52	2.49	4.11
3	1.00	2.00	13.3	1.42	0.52	2.72	4.34
4	1.25	2.25	13.4	1.54	0.60	2.56	4.17
5	1.50	2.50	13.3	1.70	0.59	2.86	4.43

Source: Adapted from Cortez et al. 1979.

Irrespective of level of concentrate fed, average daily voluntary consumption of whey (approximately 13.3 liters) did not vary and the two lower levels of concentrate feeding (0.75–1.75 and 1.00–2.00 kg) failed to support pig gains equal to those of the two higher levels (1.25– 2.25 and 1.50–2.50 kg/day) or the all-concentrate control. Dressing percentage was also reduced at the two lower levels of concentrate feeding, suggesting that voluntary consumption of liquid whey fed without preservatives is limited and that an adequate quantity of concentrate is required to meet the pig's needs for normal growth. In these studies, satisfactory performance was reported when whey supplied up to 35% of the dietary dry matter. Similar studies using restricted feeding (Barber *et al.* 1978) demonstrated that good pig performance could be obtained when 30% of the daily dry matter intake was supplied by liquid whey but a level of 50% dry matter from whey depressed both gains and efficiency of feed utilization.

In practice, satisfactory results have been obtained with several feeding systems. Three distinct systems that have been successfully employed are outlined in Table 14.31.

In general, under restricted feeding regimes, during the period from 20- to 90-kg body weight, an average of 1.40 kg of complete mixed diet (13–16% crude protein) and an average 8 liters of liquid whey will be consumed per pig per day (Barber *et al.* 1978). Pigs fed under full-feed systems will consume an average of 1.25 kg of 16% diet plus 11 liters of liquid whey per day during the growing phase and 2.25 kg of 13% diet plus 16 liters of liquid whey per day during the finishing period (Cortez *et al.* 1979).

Before feeding, the whey should be warmed and allowed to reach room temperature. Feeding cold whey reduces performance of the pig. Pigs fed cool whey (15°C) grow more slowly and require more dry matter to produce a kilogram of liveweight gain. This reduced performance resulting from the feeding of cool whey is more pronounced at lower (16°C) ambient temperatures than at warmer (22°C) temperatures. It is assumed that the body heat required to be used by the pig to warm the cool whey might be a major factor in causing the reduced gains because pigs fed the same whey heated to 40°C before feeding grew at a significantly faster and more efficient rate (Holmes 1971).

Acidity is a factor that affects the voluntary intake of sour whey (Barber *et al.* 1968). The titratable acidity of whey may exceed 2 when held for several days at

TABLE 14.31. Feeding Systems That Have Given Satisfactory Results with Liquid Whey Feeding

Source: liveweight of pig, kg	*Concentrate/pig/day, kg*	*Whey/pig/day*
Carr and Dunkin 1969:		
22.7	1.15	2.27 (kg)
27.3	1.15	4.54
31.8	1.15	6.80
36.4	1.15	9.08
40.9	1.15	10.90
45.5	1.15	12.27
50.0	1.15	14.09
54.5	1.15	15.90
Barber et al. 1978:		
19–20	0.70	4.25 (liter)
23–24	0.80	4.75
27–28	0.90	5.25
33–34	1.00	5.75
37–38	1.10	6.25
42–43	1.20	6.75
45–46	1.30	7.50
69–90	1.85	10.75
Cortez et al. 1979:		
20–50	1.25	*Ad lib* (11 liters)
50–90	2.25	*Ad lib* (16 liters)

medium to high ambient temperatures. During the winter months and in cooler climates the titratable acidity is lower (1–2). Storage at 4°C or the addition of 0.1% formalin to the whey when drawn off of the cheese vats will preserve the sweet whey at a low level of acidity (0.2–0.4 titratable acidity). Sweet whey is more palatable than sour whey and especially under warmer climate conditions voluntary consumption will increase when compared to sour whey. However, nutritionally, sour whey is equal to sweet whey when similar quantities are fed (Barber *et al.* 1978).

Whey is a highly fermentable substrate that contains substantial quantities of dissolved gas (Todd *et al.* 1977). Growing pigs fed on whey occasionally suffer from *colonic bloat*. The condition is characterized by sudden death and at autopsy by distension and intense reddening of the colon and lower small intestine (Dodd *et al.* 1963). Probable pathogenesis is the release of large amounts of gaseous products of whey fermentation in the cecum and colon (Todd *et al.* 1977). The rapid, fatal gas distension of colonic bloat has been directly related to the volume of whey ingested (O'Hara 1971). A small percentage of deaths due to colonic bloat is probably an inevitable consequence of high levels of free-choice consumption of whey (Todd 1977). Feeding management is important when liquid whey is fed. Morbidity and mortality can be reduced by providing dry feed prior to whey feeding and by restricting the quantity of whey fed. Adequate trough space is essential to allow all pigs access to the whey and to prevent stronger pigs from monopolizing limited trough space and overconsuming the whey.

Sows fed unrestricted quantities of whey also have been shown to suffer from gaseous distension. In one study (McClausland and Southgate 1980), a death rate of 7% was reported among sows fed 1 kg of dry pellets per day and allowed free access to whey from 10:OO A.M. to 4:00 P.M. daily (average daily whey consumption, 22 kg). All sows that died were recumbent with distended abdomens. Postmortem examination revealed acute severe peritonitis associated with rupture of the large or lower small intestine. It was suggested that the ruptures resulted from overdistension with gas produced from whey and that the likelihood of their occurrence was directly related to the volume of whey ingested.

When condensed whey (60% dry matter) is available, it can be fed as fresh whey when diluted 1:1 with water; however, fresh drinking water must be available at all times to allow the pigs adequate water consumption. If water is not available whey consumption and performance of the pigs will be reduced.

Economics of whey utilization will depend, of course, upon the relative prices of the concentrates and the whey. As a rule of thumb, 16.7 kg of 5.8% dry matter whey has a feeding value of 1 kg of concentrate for a 22–45-kg pig, 21.1 kg is equal to 1 kg of concentrate for a 45–68-kg pig, and 19.5 kg is similar in value to 1 kg of concentrate for a pig between 22- and 68-kg liveweight.

Dried whey contains 13–17% crude protein and 73–75% lactose. Pals and Ewan (1978) have shown the gross energy of dried whey to be efficiently utilized, with average gross, digestible, metabolizable, N-corrected metabolizable, and net energy values of 3.91, 3.50, 3.35, 3.34, and 2.20, respectively. However, the efficiency of dried whey utilization varies with the age of the pig and the dietary level fed. The young, early weaned pig about 2 weeks of age can make efficient use of as much as 60% dried whey in the diet without producing any diarrhea. Although early studies (Becker *et al.* 1957; Krider *et al.* 1948) reported that weanling pigs (8 weeks) do very poorly when fed diets containing more than 10% dried whey, more recent studies (Fevrier 1978) have shown that the growing pig (22–35 kg) grows very well on diets containing 30% dried whey but very poor results are obtained with 60% dietary whey powder. As the pig's weight increases, the dried whey appears to be less efficiently utilized (Table 14.32). However, even though a level of 60% dietary dried whey significantly reduced gains and efficiency of feed utilization, only feed efficiency was affected by a level of 30% whey. Landblom and Nelson (1980) also compared graded levels of dried whey (0, 15, 30, and 45% to give 0, 10, 21, and 32% dietary lactose, respectively) as a substitute for barley in diets for growing-finishing pigs. Pig gains and efficiency of feed utilization were improved at all levels of whey addition (Table 14.32).

Although others have shown a growth depression by feeding a 40% whey diet (Becker *et al.* 1957) or a 30 or 40% lactose diet (Shearer and Dunkin 1968), Ekström *et al.* (1975A,B) and Kim *et al.* (1978A,B) found no effect of feeding 40% whey to growing-finishing pigs. These results suggest that in practice, dietary levels of 30–45% dried whey can be employed successfully in the diet during the entire growing-finishing period and also that a gradual reduction in level of whey as the pigs mature might prove most efficient.

TABLE 14.32. Performance of Pigs Fed Graded Levels of Dried Whey During Three Stages of Growth

Liveweight: Dried whey, %	*Feed, kg/day*	*Gain, kg/day*	*Feed/gain ratio*
22–35 kg[a]:			
0	1.19	333	3.66
30	1.46	433	3.39
60	1.16	298	3.92
35–60 kg[a]:			
0	1.87	577	3.27
30	1.83	514	3.58
60	1.66	395	4.28
60–98 kg[a]:			
0	2.56	664	3.90
30	2.56	611	4.22
60	2.38	524	4.57
22–98 kg[a]:			
0	1.95	534	3.66
30	2.05	533	3.87
60	1.83	421	4.36
21.3–103 kg[b]:			
0	2.65	667	3.96
15	2.71	743	3.64
30	2.77	771	3.63
45	2.67	752	3.55

Sources: [a]From Fevrier 1978; [b]From Landbloom and Nelson 1980.

HATCHERY WASTE

In very localized areas adjacent to hatcheries, variable quantities of unheated incubator eggs are available. Some of these eggs are wasted. Willman (1941) reported that satisfactory gains were obtained from 23-kg pigs fed yellow maize and salt supplemented with waste hatchery eggs. The hogs had free access to the maize and salt in self-feeders and were fed all the incubator eggs they would clean up three times a day.

Willman *et al.* (1942) fed 23-kg pigs 1.3 kg of yellow maize, 2.2 kg of unhatched incubator eggs, and salt. The pigs on this diet gained 0.55 kg/day up to 90 kg. No digestive upsets or problems were observed during the trial. Digestibility trials conducted with 30-kg pigs showed the protein of the incubated eggs to be 89% digestible, with the fat having a digestion coefficient of 86%. These raw incubator eggs were utilized as efficiently as cooked eggs.

Tomhave and Hoffman (1945) fed 23-kg pigs 1.64 kg of yellow maize per 44 kg of liveweight and all the incubated eggs they would consume. The performance of these pigs was compared with that of pigs fed a control diet based on maize, alfalfa meal, tankage, and linseed meal and with that of a group fed the same amount of maize but all the dead poultry they would consume (Table 14.33). All three groups had access to a salt-mineral mixture. Pigs fed raw

TABLE 14.33. Performance of Growing-Finishing Pigs Fed Tankage, Incubated Eggs, or Dead Poultry as Protein Supplement

Performance	*Tankage*	*Incubated eggs*	*Dead poultry*
Maize consumed/hog, kg	142.3	116.0	90.5
Supplement fed, kg	29.0	130.0	66.0
Av daily gain, kg	0.40	0.39	0.35

Source: Tomhave and Hoffman 1945.

incubator eggs grew as rapidly as those fed tankage. Growth was less for pigs fed dead poultry. It was shown that 100 kg of incubator eggs replaced 20.2 kg of maize, 10.9 kg of tankage, 5.4 kg of alfalfa meal, and 5.4 kg of linseed oil meal. One hundred kilograms of dead poultry replaced 78 kg of shelled maize, 21.4 kg of tankage, 10.8 kg of linseed oil meal, and 10.8 lb of alfalfa meal.

It would appear either that incubation modifies the egg albumin in a way that makes it readily digestible by the pig or that the pig is unique in its ability to digest raw egg albumin. Whatever the reason, growing-finishing pigs can efficiently utilize raw incubator eggs and dead poultry to satisfy a major portion of their daily feed needs, thereby providing a way to salvage some value from unhatched eggs as well as a convenient method for disposal of the wastes. There is a danger of a biotin deficiency manifested by cracked hooves, dry rough skin, and loss of hair in pigs fed 30% or more of dried, raw egg whites (Cunha *et al.* 1946). A protein, avidin, in raw egg white ties up biotin in the gastrointestinal tract and decreases its absorption. Heating destroys the avidin and prevents a biotin deficiency.

FEATHER MEAL

Feather meal with a crude protein content (nitrogen × 6.25) of between 85 and 87% is one of the most protein-rich feedstuffs available for swine feeding (Table 14.34). However, because of the keratinous nature of its protein, it is of little, if any, nutritional value because of the lack of digestibility in its natural state. Feather meal is, however, the principal by-product from poultry processing and has, in addition to a very high protein content, an extremely high content of cystine, threonine, and arginine (Table 14.35). To be useful for swine feeding it must be hydrolyzed. Although it appears that steam cooking at elevated pressures (3.2 atm) produces feather meal of the highest digestibility and biological value, acid hydrolysis (HCl at pH 6 for 20 hr) can be employed to produce meals of similar value (Eggum 1970). Feather meal which has been subjected to steam hydrolyzation has a rather high digestibility (84.4%) with a true availability of single amino acids from 80 to 90%.

Feather meal quality is quite variable, with variability due largely to the raw ingredients included and the processing method employed. Protein digestibility and amino acid availability are more dependent on processing method than composition of raw ingredients. Heat and pressure treatments greatly increase di-

TABLE 14.34. Chemical Composition of Hydrolyzed Feather Meal

Component	*Content* (%)
Moisture	8.1
Crude protein	86.1
Ether extract	2.0
Fiber	0.3
Ash	3.5
Calcium	0.4
Phosphorus	0.2
Sodium chloride	0.3

Source: Barber et al. 1965.

TABLE 14.35. Amino Acid and Nitrogen Analysis of Raw and Treated Feather Meal (FM) (g/16 g N)

Amino acid	*Raw FM*[a]	*Pressure cooked FM*[b]	*HCl-treated FM*[c]
Lysine	2.22	2.08	2.23
Methionine	0.83	0.72	0.76
Cystine	9.02	6.29	6.06
Methionine + cystine	9.85	7.01	7.82
Aspartic acid	6.71	6.58	6.55
Threonine	5.21	4.84	4.87
Serine	12.52	11.81	12.15
Glutamic acid	12.11	11.91	12.12
Glycine	7.92	7.54	7.93
Alanine	4.29	4.30	4.32
Valine	7.97	7.25	7.73
Isoleucine	5.25	4.82	5.55
Leucine	8.40	8.05	8.27
Tyrosine	3.11	2.48	3.11
Phenylalanine	4.91	4.61	4.85
Histidine	0.80	0.72	0.63
Arginine	7.08	6.15	7.09
Tryptophan	0.86	0.73	0.62
Nitrogen, % of DM	15.43	15.38	15.30

Source: Eggum 1970.
[a]Unhydrolyzed feather meal.
[b]Treated with steam, where the pressure slowly increases to 3.2 atm during 45 min and falls to normal during a similar period. Dried at 60°C.
[c]Boiled under reflux for 20 h at pH 6 (HCl), freeze-dried.

gestibility of feather meal but greatly reduce sulfur–amino acid content. Samples from different plants and processing methods were tested for sulfur–amino acid availability (Kim *et al.* 1980) and shown to vary in availability from 41 to 82%. Available sulfur–amino acid content of the samples varied from 0.92 to 4.01% when expressed on a dry matter basis. Chemical analysis of methionine and cystine was shown to be a good indicator of available sulfur–amino acid content. Feather meal is deficient in both methionine and lysine but is a good source of cystine as well as zinc and iron.

Even with the relatively high level of digestibility, studies have demonstrated that feather meal cannot be used to supply all of the supplemental protein for cereal-based diets. Ten percent feather meal used as the sole supplement to a maize-based diet severely reduced daily feed intake, depressed rate of growth, and reduced efficiency of feed conversion (Combs *et al.* 1958). Even when fed at a 7.5% level as a partial replacement for soybean meal in the basal diet, growth rate and feed efficiency were depressed. Although feeding trials have shown that 5–7% hydrolyzed feather meal used to replace fish meal or soybean meal for growing-finishing pigs does not significantly reduce growth rates (Hall 1957; Combs *et al.* 1958; Barber *et al.* 1965), these levels significantly lowered nitrogen retention (Barber *et al.* 1965). Although in feeding trials lysine supplementation to diets containing 5–7% feather meal appears to support improved performance of pigs (Combs *et al.* 1958), methionine appears to be the first limiting amino acid in maize–soybean meal–feather meal diets for growing pigs.

In practice, very low levels (3–5%) of properly processed, hydrolyzed feather meal should be used in swine diets if maximum growth and efficiency of feed utilization are desired.

HOG HAIR MEAL

Significant quantities of hair are available from the annual slaughter of millions of pigs. Unprocessed keratinous proteins, such as hog hair and feathers, are of little or no nutritional value in their natural state because of poor digestibility. However, if properly processed through steam or acid hydrolysis, hydrolyzed hog hair is a potential source of protein and can be used in swine diets.

Hydrolyzed hog hair meal contains 93–95% crude protein which has been shown by Kornegay (1978) to be 74.7% digestible. Kornegay and Thomas (1973) reported that lysine and methionine are found in smaller concentrations in hydrolyzed hog hair (3.70 and 1.06% of protein) than in soybean meal (6.35 and 1.32%, respectively) and that the amino acids in hydrolyzed hog hair meal are not as available as in soybean meal and maize.

Feeding trials reported by Kornegay and Thomas (1973) indicate that hydrolyzed hog hair meal can be used in pig diets at a level of 2–5% without reducing pig performance. Pigs initially weighing 5 kg were fed a 20% protein diet, and they showed only a slight depression in gain and feed efficiency when 5% hydrolyzed hog hair meal was added on an isonitrogenous basis to replace soybean meal protein. However, there was a significant decrease in pig performance when 5% hydrolyzed hog hair meal was substituted in a 16% protein diet.

In a series of feeding trials, Kornegay and Thomas (1973) evaluated hydrolyzed hog hair meal for both growing and growing-finishing pigs. In one trial, using pigs initially weighing 9.1 kg, diets containing 18% crude protein were formulated to contain 0, 2, 4, 6, and 8% hydrolyzed hog hair meal. In each diet, hydrolyzed hog hair meal replaced an equal quantity of soybean meal protein on an isonitrogenous basis. Average daily gain and gain-to-feed ratio were linearly decreased as graded levels of hair meal were added to the diet. Feed intake was not significantly different; however, feed intake was depressed at the highest level (8%) of hog hair meal substitution (Table 14.36). In a subsequent trial, growing pigs (11.4–24 kg) were fed diets containing 0, 2, or 4% hydrolyzed hog hair meal, added on a digestible protein (70%) basis to replace the digestible protein of soybean meal (91%). Although there were no differences in feed intake among pigs fed the various diets containing 0, 2, or 4% hydrolyzed hog hair meal, pigs fed 4% hydrolyzed hog hair meal grew at a slower rate compared to pigs fed soybean meal or 2% hydrolyzed hog hair meal and gain-to-feed ratio decreased linearly as the percentage of hydrolyzed hog hair meal increased in the diet (Table 14.36).

Growing-finishing pigs were fed diets formulated by Kornegay and Thomas (1973) to contain equal quantities of digestible protein from soybean meal and maize and graded levels (0, 2, 4, and 6%) of hog hair meal. Each addition of hydrolyzed hog hair meal up to 4% increased daily gains and improved feed efficiency; however, no further increase in gain or gain-to-feed ratio was observed

TABLE 14.36. Hydrolyzed Hog Hair Meal in Diets for Growing Pigs

Level of hydrolyzed hog hair meal	*Level of protein*	*Av daily gain, kg*	*Av daily feed, kg*	*Gain/feed ratio*
Isonitrogenous				
0	18	0.49[a]	0.92	0.53[a]
2	18	0.43	0.90	0.48
4	18	0.42	0.93	0.45
6	18	0.38	0.90	0.42
8	18	0.30	0.80	0.38
Digestible protein				
0	16–14	0.49[c]	1.09	0.45[b]
2	16–14	0.50[c]	1.13	0.44
4	16–14	0.45[d]	1.07	0.42
Isonitrogenous				
0	14–12–11	0.73[c]	2.42[c]	0.30[c]
0	11.3–9.4–8.3	0.46[e]	2.00[d]	0.23[e]
2	12.6–10.7–9.7	0.60[d]	2.30[c,d]	0.26[d]
4	14–12–11	0.69[c]	2.43[c]	0.28[c]
6	15.3–13.3–12.3	0.69[c]	2.48[c]	0.28[c]

Source: Kornegay and Thomas 1973.
[a,b]Significant linear effect of diets ($P < .05$).
[c–e]Means with different superscripts are significantly different ($P < .01$).

when an additional 2% hydrolyzed hog hair meal (6% total) was added above the 4% level. Pigs fed the positive control (soybean meal– maize) had the highest gain and gain-to-feed ratio, although performance was not significantly different from that obtained with 4–6% hydrolyzed hog hair meal.

These results indicate that hydrolyzed hog hair meal is of value for swine when utilized at low dietary levels. Maximum value is realized when not more than 2–4% is included in growing-finishing swine diets.

NONPROTEIN NITROGEN (NPN)

Reports on NPN utilization by swine have been contradictory. Much of the early work utilized urea as the NPN source and among the papers reporting studies with nonruminants, there is a difference of opinion on the ability of the pig to utilize urea. Grafe and Turban (1913) and Grafe (1913) reported that the addition of urea to a ration for pigs increased nitrogen retention. However, Abderholden and Lampe (1913) repeated Grafe's experiments and were unable to confirm his conclusions. Piepenbrock (1927) fed urea at levels of 20–45 g per head daily to 45-kg pigs and concluded that 30–40% of the dietary protein can be replaced by urea without decreasing the weight gains of the pig.

Braude and Foot (1942), Liu *et al.* (1955), Kornegay *et al.* (1965), Bowland (1967), Meacham *et al.* (1967–1968), Kornegay (1969), Hintz *et al.* (1969), and Tylecek *et al.* (1971) concluded that urea is "without effect" as a nitrogen source for pigs. Even in instances where the diet was deficient in total protein, the addition of urea to the basal diet was ineffective in improving weight gains or efficiency of feed utilization (Hanson and Ferrin 1955; Kornegay 1972) or had a detrimental effect on pig performance (Hays *et al.* 1957; Grimson and Bowland 1971). Hays *et al.* (1957) reported a decreased average daily gain and feed intake when urea was used as a source of NPN to replace protein in the diet of growing pigs. Grimson and Bowland (1971) substituted 2% urea for herring meal on an isonitrogenous basis and reported depressed feed intake, daily gains, and feed conversion in pigs from 4 to 11 weeks of age when they were limit-fed, but urea substitution had no effect on pig performance when fed *ad libitum*. Growing pigs fed a urea-containing diet from 11 weeks of age to 90-kg liveweight gained less weight and required more feed per kilogram of gain than nonurea fed controls.

Other studies have shown that in some circumstances urea was utilized by young growing pigs (Burkser *et al.* 1965; Kornegay *et al.* 1965, 1970; Meacham and Thomas 1966–1967; Baia *et al.* 1967; Kornegay 1972). Kornegay *et al.* (1970) demonstrated a limited degree of urea utilization in some experiments by growing pigs fed maize– soybean meal diets in which the urea was added at the expense of either maize or soybean meal. They noted that the basal diet must supply a proper balance of essential amino acids before a positive response was obtained. This work is in agreement with results reported by Grimson and Bowland (1971) which showed that pig performance on a diet containing 2% urea could be improved to a level similar to that of pigs fed the control diet which contained an equivalent amount of nitrogen from soybean and herring meal, if the urea-containing diet was supplemented with L-lysine.

Sources of NPN other than urea have also been fed. The addition of urea, diammonium citrate, diammonium phosphate, and glutamic acid in limited amounts to diets of pigs of varying age, in general, resulted in no beneficial response (Wilczynski and Owsiejczuk 1967). Barber *et al.* (1969) attempted unsatisfactorily to improve a cereal diet supplemented with lysine by the addition of diammonium phosphate. Veláquez *et al.* (1970) supplemented diets with diammonium phosphate and reported this NPN source to be nutritionally inert. Wehrbein *et al.* (1970) substituted an equimolar mixture of diammonium citrate and diammonium phosphate for graded levels of protein in a 16% protein maize–soybean diet for growing-finishing pigs and reported feed intake, daily gains, and nitrogen retention (as percentage of intake) decreased as increasing amounts of NPN were substituted for soybean protein. Gain depression due to NPN substitution was less severe when 0.2% L-lysine, 0.05% DL-tryptophan, and 0.05% DL-methionine were added. Platter *et al.* (1973) reported that the substitution of NPN (1.5% protein equivalent supplied by monoammonium phosphate) for soybean meal protein produced no effect on body weight gains or feed consumption by pigs given either 12 or 15% protein diets. Platter *et al.* (1973) added 3.1% protein equivalent from a mixture of 6% monosodium glutamate, 28% glycine, and 66% diammonium citrate to a 10.9% protein opaque-2 maize–soybean meal diet. Various levels of amino acids, lysine, methionine, tryptophan, and threonine, were added and comparisons were made with a 14% protein maize–soybean meal diet. There was a nonsignificant trend toward increased nitrogen retention when NPN was added to the 10.9% protein diet fortified with amino acids. Clawson and Armstrong (1981) added ammonium polyphosphate or ammonium chloride to a 10% protein maize–soybean diet to increase dietary nitrogen by 7.25 or 14.5% and reported that pig performance was not influenced by either of these NPN sources with or without amino acid supplementation.

Studies by Liu *et al.* (1955) and Grimson *et al.* (1971) using urea labeled with ^{15}N have shown that small but definite amounts of administered urea were incorporated into body tissue. Liver, plasma, and intestinal scrapings had markedly higher levels of ^{15}N label in the protein than existed in skeletal muscle. Aliev and Kosharov (1971) placed a catheter in the portal vein of two pigs and infused [^{15}N]glycine and [^{15}N]urea directly by canula into the cecum and concluded that nitrogen from glycine and urea was used for synthesis of amino acids and protein. Flam and Bednarova (1971) concluded that the utilization of a part of the N of synthetic urea by the bacteria in the pig's digestive tract was possible. The incorporation of ^{15}N from dietary [^{15}N]diammonium citrate into amino acids isolated from neonatal pig livers and muscle protein was examined by Deguchi *et al.* (1980). Results indicate that N transfer from diammonium citrate into all amino acids (essential and nonessential) appears to occur in the specific-pathogen-free pig and possibly in the germ-free pig.

Although some evidence exists to indicate that nitrogen from dietary nonprotein nitrogen compounds is incorporated into amino acids and tissue protein, it has been shown that NPN cannot be used to any great extent and cannot be recommended at the present time as having practical value in swine diets even when limiting essential amino acids are supplied.

ANIMAL AND POULTRY WASTE

As the problem of disposal of animal waste continues to increase, every effort will be made to utilize these residues to the best advantage in animal feeding. Today, it appears that the ruminant animal is best equipped to utilize these fecal wastes and will probably provide the main avenue of useful utilization in the near future (Fontenot *et al.* 1983; Noland *et al.* 1955; Southwell *et al.* 1958; Belasco 1954; Hillard *et al.* 1979; Ngian and Pearce 1979; Anthony 1974; Smith and Wheeler 1979). Since the pig makes little use of nonprotein nitrogen which forms a significant proportion of animal and poultry feces, generally animal wastes are of very little value for supplying protein and energy for swine rations. In comparative studies, it has been shown that species from which manure is collected have a significant effect on concentration of digestible nutrients for swine, with poultry manure providing the greatest concentration followed in order by swine manure and cattle manure (Table 14.37). The comprehensive review by Bhattacharya and Taylor (1975) clearly showed that management conditions, diet fed, storage time and conditions, and species from which feces are collected for recycling affect composition and nutritional value. The composition of waste products from cage layers, broiler litter, and cattle manure is shown in Table 14.38 from data summarized by Bhattacharya and Taylor (1975). Therefore the value of poultry litter, poultry manure, cattle manure, swine manure, and oxidation ditch liquor and slurry will be treated separately.

Many pathogenic organisms (bacteria, viruses, and fungi) as well as high concentrations of drugs, chemicals, pesticides, and hormones that are capable of causing disease and health problems in humans, livestock, and poultry have been isolated from animal waste. These factors were reviewed comprehensively by Bhattacharya and Taylor (1975).

Poultry Litter

The nutrient composition of poultry litter, as shown in Table 14.38, has been found to vary. Among the more important factors that contribute to the wide range of variation are (1) class of poultry from which litter is collected; laying house litter is lower in crude protein (20%) than broiler litter (30%) and higher in calcium (5.7%) and phosphorus (2.2%) than broiler litter (Ca, 1.7%; P, 1.5%);

TABLE 14.37. Effect of Species from Which Feces Were Collected on Digestible Nutrients for Swine

Criteria	Dehydrated waste		
	Poultry	*Swine*	*Cattle*
Digestible crude protein, %	14.72	6.91	3.21
Total digestible nutrients, %	30.53	15.76	12.14
Digestible energy, kcal/kg	1472	850	458

Source: Yuk et al. 1978.

TABLE 14.38. Nutrient Composition of Animal Waste Products from Cage Layers, Broiler Litter, and Cattle Manure

Component	*Dehydrated cage layer excreta*	*Broiler litter*	*Cattle manure (dry basis)*
Dry matter, %	89.7	84.7	100.0
Crude protein, %	28	31.3	20.3
Digestible protein, %	14.4	23.3	4.7
Crude fiber, %	12.7	16.8	37.5
Ether extract, %	2	3.3	2.5
Ash, %	28	15	11.5
Calcium, %	8.8	2.4	0.9
Phosphorus, %	2.5	1.8	1.6
Magnesium, %	0.67	0.44	0.40
Sodium, %	0.94	0.54	—
Potassium, %	2.33	1.78	0.5
Iron, mg/kg	2000	451	1340
Manganese, mg/kg	406	225	147
Zinc, mg/kg	463	235	242
Arginine, %	0.50	0.51	0.18
Histidine, %	0.22	0.24	0.12
Isoleucine, %	0.53	0.64	0.21
Leucine, %	0.86	1.00	N.R.
Lysine, %	0.51	0.57	0.47
Methionine, %	0.10	0.13	0.09
Cystine, %	1.17	0.09	N.R.
Phenylalanine, %	0.48	0.54	—
Tyrosine, %	0.28	0.33	0.03
Threonine, %	0.51	0.57	0.29
Valine, %	0.65	0.82	0.38

Source: Bhattacharya and Taylor 1975.

(2) type of bedding employed; absorption capacity and fiber content vary among bedding materials such as citrus pulp, maize cob, wood shavings, peanut hulls, and rice hulls; (3) length of time birds have been on litter; and (4) time of storage before feeding (Bhattacharya and Taylor 1975). The average crude protein content of broiler litter is about 30% with about one-half of the crude protein made up of true protein and one-half of nonprotein nitrogen, of which 50% is uric acid. The true protein is low in arginine, lysine, methionine, and cystine (Bhattacharya 1964).

Feeding trials with poultry litter at dietary levels of 10–60% have uniformly shown that as the level of dried poultry litter is increased in the diet of growing-finishing pigs, a significant linear decrease in growth rate, feed conversion efficiency, and dressing percentage is observed (Table 14.39) (Perez-Aleman *et al.* 1971; Li and Yang 1976; Hernandez *et al.* 1978). Cost considerations will determine the economic value of poultry litter but in most situations a maximum dietary level of 10% will be most advantageous.

TABLE 14.39. Effect of Increasing Levels of Dried Poultry Litter on the Performance of Growing-Finishing Swine

Level of poultry litter, %	*Gain, kg/day*	*Feed, kg/day*	*Feed/gain ratio*	*Dressing, %*
Broiler (Perez-Aleman et al. 1971)				
0	0.58[a]	2.00	3.48[a]	75.5[a]
10	0.55	2.03	3.72	74.4
20	0.53	2.13	4.07	73.5
30	0.51	2.13	4.20	72.6
Laying hens (Hernandez et al. 1978)				
For growing pigs				
0	0.59[a]	2.02	3.41[a]	—
10	0.57	2.06	3.60	—
20	0.50	2.07	4.16	—
30	0.47	2.15	4.59	—
For finishing pigs				
0	0.75[a]	3.18[a]	4.26[a]	—
20	0.58	3.09	5.36	—
40	0.39	3.09	8.01	—
60	0.22	2.71	12.45	—

[a]Significant linear depression in performance due to the addition of dried poultry litter.

Poultry Manure

Dried poultry manure has been extensively used in feeding ruminant animals, as reviewed by Bhattacharya and Taylor (1975). The chemical composition of poultry manure is tabulated in Table 14.38. It varies widely in its composition and especially in its crude protein content, which is significantly reduced as the duration of storage is increased. The abundance of nonprotein nitrogen in poultry manure (47–64% of total nitrogen in the dry matter) limits its protein value for swine. Uric acid, which makes up 30–60% of the nonprotein nitrogen, is not utilized by swine and in chicks has been shown to depress growth even at levels of only 1% of the ration.

Although Yuk *et al.* (1978) reported that dried poultry manure contained 14.72% digestible protein, 30.53% TDN, and 1.472 kcal/kg of digestible energy, reported results of feeding trials with swine fed fresh or dried poultry manure obtained without litter from beneath cage layers have uniformly shown that when used to replace increasing levels (5–20%) of a complete basal ration, rate of gain and efficiency of feed utilization are reduced (Table 14.40) (Tsui *et al.* 1975; Li and Yang 1976; Sibalic *et al.* 1977; Coffey *et al.* 1978; Chung *et al.* 1977; Ahn *et al.* 1977; Barbosa *et al.* 1980). The only exception was reported by Babatunde *et al.* (1979), when sun-dried poultry manure obtained from cage layers and containing 28.8% protein was used to replace 0, 5, and 10% of the basal rations containing maize and peanut meal. In these studies, increasing levels of manure increased feed intake and depressed efficiency of feed utilization but had no effect on growth rate. In fact, although not significant, increasing levels of dried poultry manure slightly increased average daily gains, even though pathological examination suggested the presence of pathogenic organisms in pigs fed dried poultry manure.

Li and Yang (1978) prepared a chicken manure fermentative feed with 60% fresh chicken manure, 20% sweet potato vines, and 20% rice bran plus water to raise the mixture to 55% moisture. The feed was allowed to ferment at room temperature for a minimum of 5 days. The dried fermentative feed contained 92.8% dry matter, 10.1% crude protein, 3.6% ether extract, 22.6% crude fiber, 22.3% ash, 34.2% nitrogen-free extract, 2.89% calcium, 1.32% phosphorus, and 3.335 kcal/kg of gross energy. In feeding trials, the fermentative feed was substituted for 0, 10, 15, and 20% of the basal maize–soybean meal diet during the growing period from 20 to 50 kg and at levels of 0, 20, 30, and 40% during the finishing period from 51- to 90-kg liveweight. During the overall growing-finishing period, growth rate and efficiency of feed utilization were depressed as the level of fermentative feed was increased in the ration (Table 14.40).

Pigs fed poultry manure either fresh, dried, or as a dried fermentative feed were reported to have lower dressing percentages, lower marbling scores, and significantly lighter color of lean meat.

Under most conditions, poultry manure is not recommended as a nutrient source for swine and when fed can be expected to be of negative economic value.

Cattle Manure

Most cattle manure is lower in nutrient content than either poultry or swine manure. The exception is possible when large quantities of whole grain pass through the animal without being digested. The chemical composition, especially the crude protein, fiber, and nitrogen-free extract content of cattle manure varies.

TABLE 14.40. Effect of Substituting Fresh or Fermentative Poultry Manure Feed for Maize–Soybean Meal on Performance and Carcass Quality of Growing-Finishing Pigs

Level of substitution, %	*Gain, kg/day*	*Feed, kg/day*	*Feed/gain ratio*
Poultry manure[a]			
0	0.75[d]	2.47	3.29[d]
5	0.68	2.49	3.66
10	0.63	2.65	4.20
Poultry manure[b]			
0	0.73[d]	—	3.17[d]
20	0.62	—	4.53
40	0.31	—	7.61
Poultry manure fermentative feed[c]			
0–0	0.61	2.34	3.80
10–20	0.47	2.22	4.71
15–30	0.49	2.28	4.65
20–40	0.43	2.26	5.21

Sources: [a]Coffey et al. 1978. [b]Barbosa et al. 1980. [c]Li and Yang 1978. [d]Significant linear effect.

Cell wall content on a percentage of dry matter basis is 46–63% depending upon the level of dry matter intake, roughage level, digestibility of the ration, and type of cattle from which the manure is collected. Crude protein content of this cattle manure is about 20% on a dry matter basis (Anthony 1971; Tinnimit *et al.* 1972), while that of dairy cattle is lower (12.6%) (Smith *et al.* 1971; Hammond 1944; Tinnimit *et al.* 1972). The complete nutrient composition of cattle manure is tabulated in Table 14.38. Lipstein and Bornstein (1971) reported that cattle manure had little or no value as a source of protein or energy even when dietary levels were low. Yuk *et al.* (1978) reported that dehydrated cattle manure contained only 3.21% digestible crude protein, 12.14% TDN, and 458 kcal/kg dry matter of digestible energy for swine. Chung *et al.* (1977) fed increasing levels of dried cow manure (0, 10, 20, and 30%) and demonstrated a linear depression in growth rate and efficiency of feed utilization and a linear increase in feed intake.

Although early feeding trials by Cuban researchers (Lezcano *et al.* 1974; Diaz and Elias 1976) demonstrated fairly good average daily gains when cattle manure silage was included in diets for growing pigs, Diaz *et al.* (1979) later reported a linear depression in feed intake, average daily gain, and efficiency of feed utilization when a silage prepared with cattle manure and final molasses was substituted for maize and torula yeast at levels of 0, 25, 50, 75, and 100% in diets for growing-finishing pigs. Pathological lesions were also observed in kidneys of pigs fed diets containing over 25% manure silage. These results indicate that cattle manure is of very little value for swine.

Swine Manure

Samples of dried swine feces as analyzed by Orr *et al.* (1971) contain 22% protein, 2.5% calcium, 1.6% phosphorus, 1.0% potassium, 0.26% sodium, 800 ppm magnesium, 455 ppm iron, 509 ppm zinc, 177 ppm manganese, and 455 ppm copper. Analysis of swine feces used by Kornegay *et al.* (1977) are tabulated in Table 14.38.

Diggs *et al.* (1965A) reported that, after being dried, swine waste scraped from a concrete feeding floor could be included in a diet up to a level of 15% without depressing performance of finishing pigs. But when substituted for 30% of the diet, efficiency of feed utilization was depressed. However, Orr *et al.* (1973) reported that freshly collected feces do not support normal performance when substituted for 22% of the complete diet. Kornegay *et al.* (1977) showed that although fresh or dried swine feces from finishing pigs could be included up to 37% of the dry matter in diets for gilts before consumption was significantly reduced, feces were found to be of less nutritive value than a basal maize–soybean ration. The digestible energy of dried swine feces for growing pigs was found in one study (Diggs *et al.* 1965B) to be 1312 kcal/kg and in another study (Yuk *et al.* 1978) to be only 850 kcal/kg dry matter. Yuk *et al.* (1978) also showed dehydrated swine feces to contain 6.91% digestible crude protein and 15.76% TDN. In 120-kg gilts (Kornegay *et al.* 1977), digestibility of all proximate components decreased significantly as level of feces increased in the diet. Digestibility percentages of dry matter, energy, ash, crude fiber, ether extract, crude protein, true protein, and nitrogen-free extract were 48, 46.7, 31.6, 40.9, 54.1, 60.1, 51.4, and 45.9 respectively. With the exception of serine, glycine and cystine, in which the digestion coefficients were lower, the digestibilities of the other amino acids ranged from 51.2 to 65.1%. The apparent absorption of most minerals analyzed was reduced as the level of dietary feces substituted for the basal diet increased, but total absorption and retention were similar or greater as feces were substituted for the basal diet. Parker *et al.* (1959) had found that phosphorus in animal waste is highly available.

Hugh *et al.* (1977) evaluated swine solid waste as a nutrient source in diets for finishing swine from 41- to 84-kg liveweight. Fresh solid wastes were incorporated in diets at levels of 25 and 35% (by wet weight) with or without the addition of a fermenting agent composed of strains of *Aspergillus* and *Lactobacillus* organisms. Although the diets containing waste were readily accepted and daily feed intake was high, live animal performance as indicated by average daily gain and feed efficiency was significantly poorer for the waste-fed groups than for the basal diet groups. No differences in growth or feed efficiency existed among the groups fed waste diets (Table 14.41) nor were there any differences in carcass measurements among treatment groups. There was no performance advantage or feed cost saving from the inclusion of swine solid waste in finishing rations, nor was any benefit observed by the incorporation of a fermentation-producing agent.

Ra and Kwan (1978) prepared a fermented product from pig feces. Although the fermentation process eliminated the specific odor of pig manure, the content of nutrients before and after fermentation was not changed. When the pig manure fermentation feed (PMFF) was fed to finishing pigs at levels of 0, 20, 40, and 60%, daily gains and feed efficiency were linearly depressed as the level of PMFF was increased in the diet. Feed cost per kilogram of liveweight gain was increased as level of intake of PMFF increased.

A swine waste–maize-silage mixture was prepared by mixing 85% fresh swine feces with 15% ground corn and ensiled, with additional water to achieve a 66% moisture, in air-tight plastic bags in steel barrels for a minimum of 5 days (Marrs *et al.* 1979). Bacteriological tests showed that no living pathogens were present in the waste after the fermentation process had occurred. The ensiled waste was fed to gestating sows at a rate of 3.63 kg per head daily along with 0.68 kg of a 13% protein maize–soybean meal mixture fortified with vitamins and minerals. The total diet provided approximately 1.8-kg dry matter intake per sow daily. When compared with sows fed 1.8 kg of the basal diet, sows fed the ensiled swine waste gained an average of 27.2 kg during gestation as compared to 21.3 kg of the control group. Both groups of sows farrowed litters of equal size, although sows

TABLE 14.41. Swine Solid Waste as a Nutrient Source in Finishing Swine Diets

Treatments	*Gain, kg/day*	*Feed, kg/day*	*DM intake, kg/day*	*Feed/gain ratio*
Basal	0.82[a]	2.72	2.42	3.31[a]
25% waste	0.75[b]	3.30	2.58	4.39[b]
25% waste + fermenting agent	0.76[b]	3.37	2.69	4.46[b]
35% waste	0.73[b]	3.57	2.50	4.93[b]
35% waste + fermenting agent	0.77[b]	3.67	2.67	4.79[b]

Source: Hugh et al. 1977.
[a,b]Means within the same category with different superscripts are significantly different ($P < .01$).

fed the control diet produced more live pigs. Pigs of sows fed the waste silage diet averaged 0.14 kg heavier at birth than pigs of sows fed the control diet. Although in these studies no significant differences in sow reproductive performance were observed, Berger *et al.* (1978) reported that the apparent digestibility of dry matter, crude protein, and crude fiber was decreased with increasing amounts of swine waste silage in gilt diets. The results suggest that solid swine waste ensiled with maize can be used to supply approximately two-thirds of gestating sows' nutrient intake without significantly affecting reproductive performance.

Oxidation Ditch Liquor and Slurry

Technological advances may provide the breakthrough necessary to convert animal waste products into feed useful to the nonruminant animal. For example, important sources of minerals and amino acids and possibly B vitamins may be derived from swine waste. Research at Michigan State University (Orr *et al.* 1973) and at the University of Illinois (Harmon *et al.* 1971, 1973; Harmon 1976) demonstrated that the use of the oxidation method of storing swine waste offered potential for the biological upgrading of the waste as nutrient sources. This work has demonstrated that animal waste is of greater nutritional value after bacterial fermentation. Nonprotein nitrogen is converted into microbial cells (mainly *Staphylococci*) that have a greater concentration of protein and amino acids than the original swine waste (Holmes *et al.* 1971; Harmon 1972), while poorly digested protein such as hog hair and sloughed cellular debris are hydrolyzed and fiber is digested (Harmon 1976). Evidence of the efficiency of this biological enhancement is provided in Table 14.42 (Gouwens 1966; Orr *et al.* 1973; Harmon *et al.* 1972) which shows the increase in concentration of the different amino acids. A typical nutrient analysis of oxidation ditch mixed liquor (ODML) is shown in Table 14.43 (Harmon 1976).

Studies with swine manure subjected to anaerobic digestion and added at 24.5% to a fortified maize–soybean meal diet have shown this mixture to support satisfactory gain and feed efficiency in rats (Harmon *et al.* 1969). In studies with finishing pigs (50–100 kg), ODML was mixed with a 12% protein diet containing 0.5% lysine. Pigs fed this mixture gained weight significantly faster and more

TABLE 14.42. Amino Acid Composition of Swine Feces

Amino acid	*Fresh swine feces*[a]	*Dried swine feces*[b]	*Oxidation ditch mixed liquor*[c]	*Oxidation ditch liquor*[b]
Phenylalanine	0.81	0.87	1.48	1.66
Lysine	0.60	1.11	1.42	1.60
Arginine	0.44	0.67	1.28	1.45
Threonine	0.53	0.80	1.96	1.22
Methionine	—	0.58	0.77	0.60
Isoleucine	0.52	1.03	1.49	1.54
Leucine	0.92	1.57	2.79	2.13

Source: From Harmon 1976, who summarized work from [a]Gouwens 1966; [b]Orr et al. 1973; [c]Harmon et al. 1972.

TABLE 14.43. Nutrient Composition of Dry Matter of Swine Oxidation Ditch Mixed Liquor

Nutrients	*% of DM*	*Nutrients*	*% of DM*
Amino acids		Minerals	
Arginine	1.28	Calcium	3.33
Histidine	0.47	Phosphorus	3.83
Lysine	1.42	Magnesium	1.49
Leucine	2.79	Sodium	2.75
Isoleucine	1.49	Potassium	4.04
Methionine	0.77	Iron	0.55
Phenylalanine	1.48	Copper	0.01
Threonine	1.96	Zinc	0.11
Tryptophan	0.28		
Valine	2.06	Protein (sieved particles)[a]	
Tyrosine	1.17	20 mesh	8.4
Alanine	2.83	50 mesh	11.1
Glycine	2.29	150 mesh	20.0
Proline	1.29	200 mesh	75.6
Glutamic acid	5.06		
Aspartic acid	3.73		
Serine	2.55		

Source: Adapted from Harmon (1976) who summarized work by Harmon (1972).

[a]Microbial cells increased as sieve size decreased and contained the greatest concentration of protein and amino acids.

efficiently than those offered the 12% diet mix with fresh tap water (Harmon *et al.* 1973). In subsequent trials, an alternate feeding method was devised to simplify the feeding procedure. The same 12% protein diet was offered *ad libitum* in self-feeders and ODML was provided from the oxidation ditch by constant or intermittent pumping through a trough in each pen. Control pigs received tap water. Pigs offered ODML gained more rapidly and efficiently through the 40–100-kg weight range than those given fresh tap water (Harmon and Day 1974). Miller and Miller (1975) also showed that finishing pigs received recycled oxidation ditch liquor as the sole source of drinking water grew faster and more efficiently than those receiving fresh tap water. However, other studies by Harmon *et al.* (1973), Harmon (1976), Miller *et al.* (1974), and Miller and Miller (1975) and more recent studies by Frank *et al.* (1980) have reported that performance of growing and finishing pigs was reduced when ODML was fed. Any substitution of anaerobic products for maize or soybean meal resulted in reduced gain and efficiency. Studies by Frank *et al.* (1980) also showed that the dry, fat-free humerus weights expressed in grams per 100-kg liveweight, were reduced when ODML was fed, even when adequate calcium and phosphorus were supplied in the ration, and suggested that some component of oxidation ditch liquor interfered with utilization of the supplementary dietary calcium and phosphorus. They also showed a reduction in gain and feed intake when ODML was fed in the absence of supplementary B vitamins.

Harmon (1976) has shown that in some instances these poor results with

ODML are caused by inadequate aeration of the oxidation ditches. By operating the oxidation at less than full speed, dissolved oxygen was detectable only part way round the ditch as evidenced by odor and color of the material. Thus, the ditch did not remain aerobic and few microbial cells were produced, as was the case in studies by Harmon *et al.* (1973) and Miller *et al.* (1974). An overaerated oxidation ditch also creates problems in that nitrates can increase in the liquids to a toxic level. Overaeration may result from an excess input of oxygen or an insufficient addition of substrate (Harmon 1976). A normal range for use in recycling studies reported by Harmon and Day (1974) is 70–225 ppm nitrate in ditches. In the studies by Frank *et al.* (1980), in which pig performance was depressed when ODML was fed, the average nitrate level of the ODML was much higher, ranging from 452 to 870 ppm. When nitrate levels were highest, the difference in rate of gain between tap water and ODML treatment was greatest. Furthermore, a reduction in feed intake by ODML pigs was observed, presumably because of the decreased water intake, in the form of ODML to reduce nitrate intake.

Because most pathogenic organisms do not compete effectively in an oxidation ditch, few disease problems have been observed. However, ascarid (*Ascaris lumbricoides*) eggs are sustained in the oxidation ditch liquor and when a ditch becomes contaminated will present a problem to the pigs consuming this liquor. Inspection of slaughtered pigs that have consumed ODML has not revealed any pathological problems.

Properly aerated oxidation ditches can provide a medium for single-cell protein production that is beneficial when fed to swine; however, solutions to many technological problems are needed before oxidation ditch liquor can be widely recommended as a source of nutrients for swine feeding.

REFERENCES

Introduction

FRIEDMAN, M. 1975. Protein Nutritional Quality of Foods and Feeds, Vol. 1, Parts 1 and 2. Marcel Dekker, New York (Part 1, 626 pp., Part 2, 674 pp.).

Meat Meal and Meat-and-Bone Meal

AHERNE, F. X., SAVER, W. C., and BERG, RUTH 1979. 58th Annu. Feeders' Day Rep. (1979), p. 10. June 22, 1979, Univ. of Alberta, Edmonton.

ATKINSON, J., and CARPENTER, K. J. 1970. J. Sci. Food Agric. *21,* 360; *21,* 366; *21,* 373.

BATTERHAM, E. S., MANSON, M. B., and KIRTON, H. C. 1970. Aust. J. Exp. Agric. Anim. Husb. *10,* 384.

BATTERHAM, E. S., LEWIS, C. E., LOWE, R. F., and MCMILLAN, C. J. 1980. Anim. Prod. *31*(3), 273.

BEAMES, R. M. and SEWELL, J. D. 1969. Aust. J. Exp. Agric. Anim. Husb. *9,* 482.

BEAMES, R. M., and DANIELS, L. J. 1970. Aust. J. Exp. Agric. Amin. Husb. *10,* 249.

BJARNASON, J., and CARPENTER, K. J. 1970. Br. J. Nutr. *24,* 313.

EASTOE, J. B., and EASTOE, B. 1954. Biochem. J. *57,* 453.

EASTOE, J. B., and LONG, J. E. 1960. J. Sci. Food Agric. *11,* 87.

EVANS, D. F., and LEIBHOLZ, J. 1979. Anim. Feed Sci. and Tech. *4,* 33; *4,* 43.

HALE, O. M., and McCORMICK, W. C. 1979. Res. Rep. Ga. Exp. Stn., No. *323,* August, 1979.

KENNEDY, J. J., AHERNE, F. X., KELLEHER, D. L., and COFFREY, P. J. 1974. Irish J. Agric. Res. *13,* 1.

LEIBHOLZ, J. 1979. Anim. Feed. Sci. Tech. *4,* 53.
LEIBHOLZ, J., and MOSS, F. P. 1967. Aust. J. Agric. Res. *18,* 157.
LYNCH, P. B., O'CONNELL, W. J., and COFFREY, P. J. 1970. Irish J. Agric. Res. *9,* 1.
MERCY, A. R. 1974. The nutritive value of meat meals for growing pigs. M. Agric. Sci. Thesis, Univ. of Melbourne, Australia.
PEARSON, G., and BOWLAND, J. P. 1977. 56th Annu. Feeders' Day Rep. (1977), p. 12. Univ. of Alberta, Edmonton.
PEO, E. R., JR., and HUDMAN, D. B. 1962. J. Anim. Sci. *21,* 787.
PUCHAL, F., HAYS, V. W., SPEER, V. C., JAMES, J. D., and CATRON, D. V. 1962. J. Nutr. *76,* 11.
RUNNELS, T. D., 1968. Feedstuffs *40*(42), 27.
SATHE, B. S., and McCLYMONT, G. L. 1964. Aust. J. Agric. Res. *16,* 243.
SKURRAY, G. R., and CUMMING, R. B. 1974. J. Sci. Food Agric. *25,* 529.
SPRINGHALL, J. A. 1970. Aust. Soc. of Anim. Prod. *8,* 50.
STOCKLAND, W. L., MEADE, R. J., and NORDSTROM, J. W. 1970. J. Anim. Sci. *31,* 1142.
TODD, A. C. E., and DANIELS, L. J. 1965. Aust. J. Exp. Agric. Anim. Husb. *5,* 404.

Blood Meal

BECKER, D. E., SMITH, I. D., TERRILL, S. W., JENSEN, A. H., and NORTON, H. W. 1963. J. Anim. Sci. *22,* 1093.
D'MELLO, D. P. F., and LEWIS, D. 1970. Br. Poultry Sci. *11,* 313.
DOTY, D. M. 1973. *In* Alternate Sources of Protein for Animal Production. Natl. Acad. Sci., Washington, D.C.
HAMM, D., and SEARCY, G. K. 1976. Poultry Sci. *55,* 582.
HILLIER, J. C., MacVICAR, R., and EWING, S. A. 1955. Okla. State Misc. Publ. No. *MP-43,* June, 1955.
HILLIER, J. C., MacVICAR, R., EWING, S. A., and NICKELSON, W. 1956. Okla. State Misc. Publ. No. *MP-45* April, 1956.
JAUCIAN, A. A., SUPNET, M. G., PUYAOAN, R. B., and RIGOR, E. M. 1964. Philipp. Agic. *48,* 242.
KING, R. H., and CAMPBELL, R. G. 1978. Anim. Feed Sci. Technol. *3,* 191.
KRATZER, F. H., and GREEN, N. 1957. Poultry Sci. *36,* 562.
MEADE, R. S., and TETER, W. S. 1957. J. Anim. Sci. *16,* 892.
MILLER, E. R., PARSONS, M. J., KU, P. K., BERGEN, W. G., and ULLREY, D. E. 1976A. J. Anim. Sci. *42,* 1356 (Abstr.).
MILLER, E. R., PARSONS, M. J., RAMOS, D. R., and ULLREY, D. E. 1976B. J. Anim. Sci. *42,* 1356. (Abstr.)
OESTEMER, G. A., HANSON, L. E., and MEADE, R. J. 1973A. J. Anim. Sci. *36,* 679.
OESTEMER, G. A., HANSON, L. E., and MEADE, R. J. 1973B. J. Anim. Sci. *36,* 674.
PARSONS, M. J., MILLER, E. R., BERGEN, W. G., KU, P. K., GREEN, F. F., and ULLREY, D. E. 1975. J. Anim. Sci. *41,* 325.
RERAT, A., HENRY, V. and BOURDON, D. 1975. J. Rech. Porcine France, 1975, pp. 45–52.
TAYLOR, S. J., COLE, D. J. A., and LEWIS, D. 1977A. Anim. Prod. *24,* 137. (Abstr.).
TAYLOR, S. J., COLE, D. J. A., and LEWIS, D. 1977B. Proc. Nutr. Soc. *36,* 36A (Abstr.).
VITERI, J., and CALLES, A. 1971. Unpublished data. INIAP, April, 1971, Quito, Ecuador.
WAHLSTROM, R. C., and LIBAL, G. W. 1977. J. Anim. Sci. *44,* 778.
WAIBEL, P. E., CARPENTER, K. J., and BEHRENDS, B. R. 1974. Fed. Proc. *33,* 657.
WAIBEL, P. E., CUPERLOVIC, M., HURRELL, R. F. and CARPENTER, K. J. 1977. J. Agric. Food Chem. *25*(1), 171.

Fish Meal

ASKÖE, E., and MADSEN, J. 1954. Acta. Agric. Scand. *4,* 266.
BAKER, D. H., MOLITORIS, B. A., JENSEN, A. H., and HARMON, B. G. 1974. J. Anim. Sci. *38,* 325.
BECKER, D. E., LASSITER, J. W., TERRILL, S. W., and NORTON, H. W. 1954. J. Anim. Sci. *13,* 611.
BOLLMANN, A. 1959. Kraftfutter *42,* 412. (Cited by Braude 1961.)
BRAUDE, R. 1961. FAO Int. Conf. Fish Nutr., September 19–22, 1961, Washington, D.C.
CARPENTER, K. J., DUCKWORTH, J., LUCAS, I. A. M., SHRIMPTON, D. H., and WALKER, D. M. 1956. J. Agric. Sci. *42,* 435.

COLE, R. C. 1968. Trop. Sci. *10,* 6.
COMBERG, G., and STAUNDE, H. 1956. Tierzucht *10,* 299. (Cited by Braude 1961.)
EGGUM, B. O. 1973. Forsogslab Kobenhavn Beretn. *406,* 173 pp.
FRENS, A. M., and UBBELS, P. 1951. Versl. Landbouwk. Onderssek. *57.* (Cited by Braude 1961.)
GJEFSEN, T., Gulbrandsen, K. E., and Opstvedt, J. 1980. Acta Agric. Scand. *30,* 157.
HVIDSTEN, H., and HUSBY, M. 1955–1956. Meld. Norg. Landbruks. *35,* 23. (Cited by Braude 1961.)
HUBBELL, C. H. 1961. Feedstuffs *33*(13), 42.
JASPERSEN, J., and HANSEN, V. E. H. 1953. Forsogslab. Kobenhavn Beretn. *264,* 366.
KIFER, R. R., PAYNE, W. L., MILLER, D., and AMBROSE, M. E. 1968. Feedstuffs *40*(20), 36.
KIRSCH, W. 1959. Futter Fütterung *10*(12), 96. (Cited by Braude 1961.)
KIRSCH, W., and FENDER, M. 196O. Z. Tierphysiol. Tierernehrung *15,* 257. (Cited by Braude 1961.)
KRONACHER, C., KLIESCH, J., and BUCHHOLTZ, A. 1932. Deutsch. Landwirtsch. Tierzucht. *36,* 147. (Cited by Braude 1961.)
LAKSESVELA, B. 1961. J. Agric. Sci. *56,* 307.
MADSEN, A., MASON, V. C., and WEIDNER, K. 1965. Acta Agric. Scand. *15,* 213.
MASON, V. C., and Weidner, K. 1964. Acta. Agric. Scand. *14,* 87.
OUSTERHOUT, L. E. 1968. Feedstuffs *40*(17), 18.
PALMER, W. M., TEAGUE, H. S., and GRIFO, A. P., JR. 1970. J. Anim. Sci. *31,* 535.
PETTIGREW, J. E., HARMON, B. G., JENSEN, A. H., and BAKER, D. H. 1972. J. Anim. Sci. *35,* 11O9.
PIKE, I. H., and CURRAN, M. K. 1978. Anim. Prod. *26*(3), 394 (Abstr.).
RABOLD, K., and SCHOEN, P. 1960. Arbeiten Landwirtsch. Hochschule Hohenheim *1,* 45. (Cited by Braude 1961.)
RODRIGUEZ, J. P., and YOUNG, L. G. 1980. J. Anim. Sci. *51* (Suppl. 1), 218.
SCHILLER, K. 1953. Arch. Tierernaehr. *3*(6), 342. (Cited by Braude 1961.)
SCHRÖDER, J., and WEHAHN, E. 1972. Kraftfutter *55,* 416.
SHEARER, L. J., ADAMS, J. L., McRAE, S. E. 1970. N. Z. J. Agric. Res. *13*(2), 414.
SMITH, P. 1976. Anim. Prod. *22,* 161.
WOODMAN, H. E., and EVANS, R. E. 1940. J. Agric. Sci. *30,* 83.
WOODMAN, H. E., and EVANS, R. E. 1945. J. Agric. Sci. *35,* 133; *38,* 154; *41,* 102.
WOODMAN, H. E., and EVANS, R. E. 1951. J. Agric. Sci. *41,* 102.

Fish Protein Concentrate

ALTSCHUL, A. M. 1958. Processed Plant Protein Foodstuffs. Academic Press, New York.
BUREAU OF COMMERCIAL FISHERIES. 1966. Marine Protein Concentrate, Fishery Leaflet *584.* U.S. Dep. of Interior, Washington, D.C.
MANER, J. H., POND, W. G., and LOOSLI, J. K. 1961. J. Anim. Sci. *20,* 614.
NEWPORT, M. J. 1979. Brit. J. Nutr. *41,* 103.
POND, W. G., SNYDER, W., WALKER, E. F., JR., STILLINGS, B. R., and SIDWELL, V. 1971A. J. Anim. Sci. *33,* 587.
POND, W. G., SNYDER, W., SNOOK, J. T., WALKER, E. F., JR., McNEILL, D. A., and STILLINGS, B. R. 1971B. J. Nutr. *101,* 1193.
SCHNEIDER, D. L., and SARETT, H. P. 1969. J. Nutr. *98,* 279.
SEVE, B., AUMAITRE, A., and TORD, P. 1975. Ann. Zootech. *24,* 21.

Fish Silage and Liquefied Fish

ALBRECHT, A., and GEBHARDT, G. 1969. Arch. Tierernaehr. *19,* 345.
BACKHOFF, H. P. 1976. Some chemical changes in fish silage. J. Food Technol. *11,* 353.
HANSEN, V. 1979. Beretn. Forsøgslab. *385,* 42.
LUSCOMBE, D. K. 1973. Feeding fish silage. Pig farming. December 73, 61–3.
SMITH, P. 1976. An evaluation of liquified fish as a protein source for fattening pigs. Anim. Prod. *22,* 161.
SMITH, P. 1977. An evaluation of liquid fish as a protein source for fattening pigs. Exp. Husb. *32,* 34.
TIBBETTS, G. W., SEERLEY, R. W., McCAMPBELL, H. C., and VEZEY, S. A. 1981. J. Anim. Sci. *52,* 93.
WHITTEMORE, C. T., and TAYLOR, A. G. 1976. Nutritive value for the growing pig of deoiled liquified herring offal preserved with formic acid (fish silage). J. Sci. Food Agric. *27,* 239.

Shrimp Meal

ANGEL, M. M. 1935. Philipp. Agric. *24,* 488.

BRAY, C. I., FRANCIONI, J. B., JR. and GREGORY, E. M. 1932. La. Agric. Exp. Stn. Bull. *228* (July, 1932).

KONDOS, A. C. 1977. Aust. J. Exp. Agric. and Anim. Husb. *173,* 263.

LOWELL, J. 1965. J. Agric. Food Chem. *16,* 204.

MEYERS, S. P. 1976. Feedstuffs *48*(19), 36.

MEYERS, S. P., and RUTLEDGE, J. E. 1971. Feedstuffs *43*(49), 31.

MEYERS, S. P., RUTLEDGE, J. E., and SONU, S. C. 1973. Feedstuffs *45*(47), 34.

PEREZ, A. C. 1932. A comparative study of copra meal, shrimp meal and tankage as protein supplements to a basal ration of pigs. B. S. Thesis. Philipp. College of Agric.

SANTIAGO, M. S. 1926. Philipp. Agric. *15,* 205.

THURSTON, C. E., and MacMASTERS, P. P. 1969. J. Nutr. *98,* 279.

TOMA, R. B., and JAMES, W. H. 1975. J Agric. Food Chem. *23,* 1168.

Krill Meal

MARTIN, R. E. 1979. Food Tech. *33*(1), 46.

PASCHMA, J., PLONKA, S., and BIATAS, R. 1979. Biul. Inf., Inst. Zootech. Zoklad Inf. *17*(4), 25.

TRONCHUK, I., ZHOVNIR, B., MROCHKOV, K., and NOVIKOV, A. 1978. Svinovadstvo *3,* 28.

King Crab Meal

HUSBY, F. M. 1980. Agroborealis *12,* 4–8.

Dried Milk

NATIONAL RESEARCH COUNCIL. 1964. Joint U.S.–Canadian Tables of Feed Composition. Natl. Acad. Sci. Publ. *1232.* Washington, D.C.

NATIONAL RESEARCH COUNCIL. 1982. Joint U.S.–Canadian Tables of Feed Composition, 3rd Revision. Natl. Res. Council–Natl. Acad. Sci., Natl. Academy Press, Washington, D.C.

Milk Whey

ALAVIUHKOLA, T., HARJU, M., HEIKONEN, M., and KREULA, M. 1980. Acta Agric. Scand. *30,* 13.

AUMAITRE, A., FEVRIER, C., and DEWULF, H. 1977. C. R. Acad. Sci. Paris. *284,* 2135.

BARBER, R. S., BRAUDE, R., and MITCHELL, K. G. 1968. Anim. Prod. *10,* 238.

BARBER, R. S., BRAUDE, R., MITCHELL, K. G., and PITTMAN, R. J. 1978. Anim. Feed Sci. Tech. *3,* 163.

BECKER, D. E., and TERRILL, S. W. 1954. Arch. Biochem. Biophys. *50,* 399.

BECKER, D. E., TERRILL, S. W., JENSEN, A. H., and HANSON, L. J. 1957. J. Anim. Sci. *16,* 404.

CARR, J. R., and DUNKIN, A. C. 1969. N. Z. J. Agric. Res. *12,* 519.

CORTEZ, M. L., MONCADA, A., and OWEN, A. 1979. Revista ICA *14,* 13.

DODD, D. C., CORDES, D. O., and O'HARA, P. J. 1963. N. Z. Vet. J. *11,* 102.

DUNKIN, A. C. 1958. N. Z. J. Agric. Res. *1,* 531.

EKSTRÖM, K. E., BENEVENGA, N. J., and GRUMMER, R. H. 1975A. J. Nutr. *105,* 846.

EKSTRÖM, K. E., BENEVENGA, N. J., and GRUMMER, R. H. 1975B. J. Nutr. *105,* 851.

EKSTRÖM, K. E., GRUMMER, R. H., and BENEVENGA, N. J. 1976. J. Anim. Sci. *42,* 106.

ESNAOLA, M. A. 1971. *In* Instituto de Investigaciones Agropecuarias. Informe de Investigaciones, 1969–1970.

FEVRIER, C. 1966. Abstr. 9th Int. Congr. Animal Prod. Edinburgh, 1966.

FEVRIER, C. 1978. Ann. Zootech. *27,* 195.

FEVRIER, C., and AUMAITRE, A. 1978A. Ann. Zootech. *27,* 471.

FEVRIER, C., and AUMAITRE, A. 1978B. Ann. Zootech. *28,* 19.

HOLMES, C. W. 1971. Anim. Prod. *13,* 1.

KIM, K., BENEVENGA, N. J., and GRUMMER, R. H. 1978A. J. Anim. Sci. *46,* 1648.

KIM, K., JEWELL, D. E., BENEVENGA, N. J., and GRUMMER, R. H. 1978B. J. Anim. Sci. *46,* 1658.

KRIDER, J. L., BECKER, D. E., CURTIN, L. V., and Van Poucke, R. F. 1949. J. Anim. Sci. *8,* 112.

LANDBLOM, D. G., and NELSON, J. L. 1980. N. D. Farm Res. *37*(5), 4.

LERNER, J. T. 1973. Revista Asoc. Argentina de Criaderos de Cerdos *614*, 19.
McCAUSLAND, I. P., and SOUTHGATE, W. 1980. Aust. Vet. J. *56*, 190.
McCREA, M. R., and TRIBE, D. E. 1956. Vet Rec. *68*, 741.
O'HARA, P. J. 1971. Vet. Rec. *91*, 517.
PALS, D. A., and EWAN, R. C. 1978. J. Anim. Sci. *46*, 402.
RIGGS, L. K., BEATY, A., and MALLOU, B. 1955. J. Agric. Food. Chem. *3*, 333.
ROSA, J. G. 1971. *In* Instituto de Investigaciones Agropecuarias. Informe de Investigaciones 1969–1970, pp. 95–97.
SCHINGOETHE, D. J. 1976. J. Dairy Sci. *59*, 556.
SHEARER, I. J., and DUNKIN, A. C. 1968. N. Z. J. Agric. Res. *11*, 465.
WHITTIER, E. O., CARY, C. A., and ELLIS, N. R. 1935. J. Nutr. *9*, 521.
TODD, J. N. 1977. Pig Farming *25*, 51.
TODD, J. N., JONES, T. D., MORGAN, T. L., FRANCIS, P. G., and HEWITT, S. G. 1977. Vet. Rec. *100*, 11.

Hatchery Waste

CUNHA, T. J., LINDLEY, D. C., and ENSMINGER, M. E. 1946. J. Anim. Sci. *5*:219.
TOMHAVE, A. E., and HOFFMAN, E. 1945. J. Anim. Sci. *4*, 247.
WILLMAN, J. P. 1941. Hatchery Tribune *15*, 3, 71.
WILLMAN, J. P., McCAY, C. M., SALMAN, O. N., and KRIDER, J. L. 1942. J. Anim. Sci. *1*, 38.

Feather Meal

BARBER, R. G., BRAUDE, R., CHAMBERLAIN, A. G., HOSKING, Z. D., and MITCHELL, K. G. 1965. Anim. Prod. *7*, 103.
COMBS, G. E., ALSMEYER, W. L., and WALLACE, H. D. 1958. J. Anim. Sci. *17*, 468.
EGGUM. B. O. 1970. Acta Agric. Scand. *20*, 230.
HALL, O. G. 1957. J. Anim. Sci. *16*, 1076.
KIM, S. M., WINEGARDNER, M., and DUDLEY, W. A. 1980. Poultry Sci. *59*, 1627.

Hog Hair Meal

KORNEGAY, E. T. 1978. Anim. Feed Sci. Tech. *3*, 323.
KORNEGAY, E. T., and THOMAS, H. R. 1973. J. Anim. Sci. *36*, 279.

Nonprotein Nitrogen

ABDERHOLDEN, E., and LAMPE, A. E. 1913. Zeit. Physiol. Chem. *84*, 218.
ALIEV, A. A., and KOSHAROV, A. N. 1971. Vest. Sel'skokhoz. Nauki, Mosk. *16*, 27.
BAIA, G., ARISANU, I., VINTILA, M., and GHEORGHIU, V. 1967. Lucr. Stünt. Inst. Cerc. Zootech. *25*, 597.
BARBER, R. S., BRAUDE, R., and MITCHELL, K. G. 1969. Anim. Prod. *11*, 292 (Abstr.).
BOWLAND, J. P. 1967. 46th Annual Feeders' Day, p. 20. Dep. of Anim. Sci., Univ. Alberta, Edmonton.
BRAUDE, R., and FOOT, A. S. 1942. J. Agric. Sci. *32*, 70.
BURKSER, G. V., TIHINOVA, M. B., and IBRAEV, K. I. 1965. Suinovodstvo *19*, 28.
CLAWSON, A. J., and ARMSTRONG, W. D. 1981. J. Anim. Sci. *52*, 1.
DEGUCHI, E., NIIYAMA, M., KAGOTA, K., and NAMIOKA, S. 1980. Am. J. Vet. Res. *41*, 212.
FLAM, F., and BEDNAROVA, M. 1971. Zivocisna Vyroba *16*, 889.
GRAFE, E. 1913. Zeit Physiol. Chem. *84*, 69.
GRAFE, E., and TURBAN, K. 1913. Zeit Physiol. Chem. *83*, 25.
GRIMSON, R. E., and BOWLAND, J. P. 1971. J. Anim. Sci. *33*, 58.
HANSON, L. E., and FERRIN, E. F. 1955. J. Anim. Sci. *14*, 43.
HAYS, V. W., ASHTON, G. C., LIU, C. H., SPEER, V. C., and CATRON, D. V. 1957. J. Anim. Sci. *16*, 44.
HINTZ, H. F., POND, W. G., and VISEK, W. J. 1969. Anim. Prod. *11*, 553.
KORNEGAY, E. T. 1969. J. Anim. Sci. *29*, 138 (Abstr. 140).
KORNEGAY, E. T. 1972. J. Anim. Sci. *34*, 55.
KORNEGAY, E. T., MILLER, E. R., ULLREY, D. E., VINCENT, B. H., and HOEFER, J. A. 1965. J. Anim. Sci. *24*, 951.

KORNEGAY, E. T., MOSANIGHINI, V., and SNEE, H. D. 1970. J. Nutr. *100,* 330.
LIU, C. H., HAYS, V. W., SUEC, H. J., CATRON, D. V., ASHTON, G. C., and SPEER, V. C. 1955. J. Nutr. *57,* 241.
MEACHAM, T. N., and THOMAS, H. R. 1966–1967. Va. Polytechnic Inst. Res. Div. Livestock Rep. *122,* 36.
MEACHAM, T. N., THOMAS, H. R., and HORSLEY, W. E., JR. 1967–1968. Va. Polytechnic Inst. Res. Div. Livestock Rep. *126,* 82.
PIEPENBROCK, A. 1927. Fortschr. Landw. *2,* 650.
PLATTER, P. D., PEO, E. R., JR., VIPPERMAN, P. E., and CUNNINGHAM, P. J. 1973. J. Anim. Sci. *37,* 514.
TYLECEK, J., SKALOVA, J., and ZEDNIK, M. 1971. Biol. Chem. Vyz. Zvirat. *7,* 165.
VELÁZQUEZ, M., MARRERO, L., and McLEOD, N. A. 1970. Rev. Cubana Cienc. Agric. *4,* 105.
WEHRBEIN, G. F., VIPPERMAN, P. E., JR., PEO, E. R., JR. and CUNNINGHAM, P. J. 1970. J. Anim. Sci. *31,* 327.
WILCZYNSKI, M., and OWSIEJCZUK, W. 1967. Med. Vet. *23,* 466.

Animal and Poultry Waste

AHN, B. H., KIN, Y. H., and AHN, D. W. 1977. Korean J. Anim. Sci. *19,* 297.
ANTHONY, W. B. 1971. *In* Livestock Waste Manage. Pollution Abatement. Proc. Int. Symp. on Livestock Wastes, p. 293. Columbus, Ohio.
ANTHONY, W. G. 1974. Fed. Proc. *33,* 1939.
BARBOSA, A. S., VELOSO, J. A., DINIZ, M. L., CAVALCANTI, S., CAMPOS, E. J., and COELHO, L. S. S. 198O. Arquivos, Escola Veterinaria Univ. Federal Minas Gerais, *32,* 93.
BABATUNDE, G. M., Fetuga, B. L., ODEBUNMI, J. O., and ODUYE, O. O. 1979. Anim. Feed. Sci. Tech. *4,* 23.
BELASCO, I. J. 1954. J. Anim. Sci. *13,* 601.
BERGER, J. C. A., KORNEGAY, E. T., FONTENAT, J. P., and WEBB, K. E., JR. 1978. J. Anim. Sci. *47,* 460 (Suppl. 1) (Abstr.).
BHATTACHARYA, A. N. 1964. Ph.D. Thesis. Virginia Polytechnic Inst. and State Univ., Blacksburg.
BHATTACHARYA, A. N., and TAYLOR, J. C. 1975. J. Anim. Sci. *41,* 1438.
CHUNG, K., YOO, M. I., HAN, I. K., and CHUNG, C. Y. 1977. Korean J. Anim. Sci. *19,* 103.
COFFEY, M. T., SEERLY, R. W., and McCAMPBELL, H. C. 1978. J. Anim. Sci. *47*(Suppl. 1), 462 (Abstr. 622).
DIAZ, C. P., LEZCANO, P., and ELIAS, A. 1979. Cuban J. Agric. Sci. *13,* 33.
DIGGS, B. G., BAKER, B., JR. and JAMES, F. G. J. Anim. Sci. 1965A. *24,* 291 (Abstr.).
DIGGS, B. G., BAKER, B., JR., JAMES, F. G., and ROGERS, W. R. 1965B. Miss. Agric. Exp. Stn. Annu. Livestock Field Day Rep., pp. 47–49.
FONTENOT, J. P., SMITH, L. W., and SUTTON, A. L. 1983. J. Anim. Sci. *57*(Suppl. 2), 221.
FRANK, G. R., JENSEN, A. H., and DAY, D. L. 1980. J. Anim. Sci. *50,* 1031.
GOUWENS, D. W. 1966. Influence of dietary protein and fiber on faecal amino acid excretion. Univ. of Ill., Urbana (Mimeo).
HAMMOND, J. C. 1944. Poul. Sci. *23,* 471.
HARMON, B. G., DAY, D. L., JENSEN, A. H., and BAKER, D. H. 1971. J. Anim. Sci. *33,* 1149 (Abstr.).
HARMON, B. G., 1976. World Anim. Rev. *18,* 34.
HARMON, B. G., and DAY, D. L. 1974. Proc. 1974. Cornell Agric. Waste Manage. Conf. p. 375.
HARMON, B. G., and DAY, D. L. 1975. Proc. 3rd. Int. Symp. Livestock Waste. ASAE, St. Joseph, Michigan.
HARMON, B. G., JENSEN, A. H., and BAKER, D. H. 1969. J. Anim. Sci. *29,* 136 (Abstr.).
HARMON, B. G., DAY, D. L., JENSEN, A. H., and BAKER, D. H. 1972. J. Anim. Sci. *34,* 403.
HARMON, B. G., DAY, D. L., BAKER, D. H., and JENSEN, A. H. 1973. J. Anim. Sci. *37,* 510.
HERNANDEZ, J. F., ENRIQUEZ, F., AVILA, E., and SHIMADA, A. 1978. Veterinaria (Mexico) *9,* 149.
HILLARD, E. P., BEARD, J., and PEARA, G. R. 1979. Agric. Environ. *4,* 171.
HOLMES, L. W. J., DAY, D. L., and PFEFFER, J. T. 1971. Proc. Int. Symp. Livestock Wastes. pp. 271–351. Univ. of Ohio, Columbus, Am. Soc. Agric. Eng. Proc.
HUGH, W. I., BROOKS, C. C., OSHINO, D., and TOMA, W. 1977. Hawaii Agric. Exp. Res. Rep. *186.*

KORNEGAY, E. T., HOLLAND, M. R., WEBB, K. E., JR., BOWARD, K. P., and HEDGES, J. D. 1977. J. Anim. Sci. *44,* 608.

LI, P. K., and YANG, Y. F. 1976. J. Taiwan Liv. Res. *9,* 113.

LI, P. K., and YANG, Y. F. 1978. J. Taiwan Liv. Res. *11,* 133.

LIPSTEIN, B., and BORNSTEIN, S. 1971. Israel J. Agric. Res. *21,* 163.

MARRS, D. C., CAMPBELL, D. R., FEATHERSTONE, H. E., and NOLAND, P. R. 1979. Arkansas Farm. Res. *28,* 7.

MILLER, E. R., and MILLER, E. C. 1975. J. Anim. Sci. *41,* 321.

MILLER, E. R., MILLER, E. C., GREEN, F. F., KU, R. K., WHETTER, P. A., and ULLREY, D. E. 1974. J. Anim. Sci. *39,* 186 (Abstr.).

NGIAN, M. F., and PEARCE, G. R. 1979. Agric. Environ. *4,* 189.

NOLAND, P. R., FORD, B. F., and RAY, R. L. 1955. J. Anim. Sci. *14,* 860.

ORR, D. E., MILLER, E. R., KU, P. K., BERGEN, W. G., and ULLREY, D. E. 1971. J. Anim. Sci. *33,* 1152 (Abstr.).

ORR, D. E., MILLER, E. R., KU, P. K., BERGEN, W. G., ULLREY, D. E., and MILLER, E. C. 1973. Michigan State Univ. Rep. *232,* AH-SW-7319, East Lansing.

PARKER, M. B., PERKINS, H. F., and FULLER, H. L. 1959. Poultry Sci. *38,* 1154.

PEREZ-ALEMAN, S., DEMPSTER, D. G., ENGLISH, P. R., and TOPPS, J. H. 1971. Anim. Prod. *13,* 361.

RA, K. H., and KWAN, K. 1978. Korean J. Anim. Sci. *20,* 541.

SIBALIC, I., KIM, S. and VORNJU. E. 1977. Stocarstvo *31,* 379.

SMITH, L. W., GOERING, H. K., and GORDON, C. H. 1971. *In* Livestock Waste Manage. Pollution Abatement. Proc. Int. Symp. on Livestock Wastes, p. 314. Columbus, Ohio.

SMITH, L. W., and WHEELER, W. E. 1979. J. Anim. Sci. *48,* 144.

SOUTHWELL, B. L., HALE, O. M., and McCORMICK, W. C. 1958. Ga. Agric. Exp. Stn. Mimeo. Ser. N.S. *55.*

TINNIMIT, P., YU, Y., McGUFFY, K., and THOMAS, J. W. 1972. J. Anim. Sci. *35,* 431.

TSUI, K. C., TAM, Y. Y., and MAK, K. S. 1975. Agric. Hong Kong *1,* 328.

YUK, C. Y., HAN, I. K., CHUNG, C. S., YOO, M. I., LI, K. H., CHUNG, I. K., HA, J. K., and LI, T. W. 1978. Korean J. Anim. Sci. *20,* 430.

15

Plant Protein Sources

OILSEED MEALS

SOYBEAN MEAL

Soybean meal (SBM) is one of the most important plant proteins in livestock feeding in the United States as well as in many Latin American countries such as Brazil and Colombia. The use of soybean meal in swine rations has been under investigation for more than 40 years. Studies have related primarily to the influence of different processing methods on feeding quality of the meal and to the specific nutrient requirements of growing swine when soybean meal is used as a portion or all of the supplemental protein.

The nutritional superiority of heated soybeans over raw soybeans for monogastric animals was recognized more than 66 years ago by Osborne and Mendel

(1917) who reported that rats grew poorly when fed raw soybeans and that dry heat had no significant effect on the nutritional value of the meal. Cooking on a steam bath for about 3 hr yielded a meal that promoted normal rat growth. Robison (1930) found that the nutritive value of soybeans for the pig was improved after heating. Shrewsbury and co-workers (1932) confirmed these findings in studies with both pigs and rats, and postulated the presence of toxic factors in the raw soybean that could be removed or destroyed by heating.

The growth inhibition that results from the consumption of the raw soybean, either full-fat or as a defatted meal, has been attributed to factors such as poor palatability (Osborne and Mendel 1917), digestive enzyme inhibition (Bowman 1944), hemagglutinin (Liener and Pallansch 1952), saponins (Potter and Kummerow 1954), and decreased availability of amino acids (Borchers 1962).

Even now, more than 66 years later, the moist heat treatment used by Osborne and Mendel is employed to improve the nutritive value of soybeans and soybean meal. A number of soybean components have been isolated and suggested as anti-nutritional factors. Trypsin inhibitor was one of the first soybean components proposed as an anti-nutritional factor. Following its discovery (Bowman 1944; Ham and Sandstedt 1944), trypsin inhibitor was crystallized by Kunitz (1945). Other factors include hemagglutinins, saponins, isoflavones, and goitrogens. The role of these anti-nutritional factors in soybeans and other seed legume proteins has been reviewed by Liener (1975).

It has been shown repeatedly that both raw soybean meal and crystalline inhibitor cause pancreatic hypertrophy in both rats and chicks. An effect that is closely related to pancreatic hypertrophy, and possibly the cause of it, is stimulation of the pancreas by the trypsin inhibitor. Rats fed an amino acid mixture and added trypsin inhibitor grew less rapidly, had a larger pancreas, had higher intestinal enzyme activities, and retained more undigested protein in the small intestine than rats fed the same diet without the inhibitor (Khayambashi and Lyman 1966). The physiological response of the growing pig to raw soybean differs from that of the rat and chick in that the pancreas does not enlarge and pancreatic enzyme synthesis is not stimulated in the pig (Yen *et al.* 1977). The growth failure observed in pigs fed raw soybeans apparently results from reduced feed intake and inhibition of protein digestion.

Growth inhibition was prevented by dietary supplementation with methionine, threonine, and valine. However, amino acid supplementation did not prevent pancreatic hypertrophy or intestinal changes produced by trypsin inhibitor. These results support the theory that trypsin inhibitor stimulates the pancreas to secrete excessive amounts of digestive enzymes, which are largely undigested and pass into the cecum and colon where they are lost to the animal.

The high cystine content present in the undigested pancreatic enzymes presumably causes a depletion of methionine through its conversion to cystine, which is required for the synthesis of the secreted enzymes. In support of this mechanism, the work of Barnes and Kwong (1965) has shown that ingested crystalline trypsin inhibitor stimulated conversion of dietary methionine-S^{35} into cystine-S^{35} in the pancreas and that radioactive cystine was found in the small intestine in high concentration. The depletion of methionine and possibly other essential amino acids by this mechanism is postulated as the cause of growth inhibition. It is thought that pancreatic hypertrophy is a direct physiological response to trypsin inhibitor stimulation.

Reduction of bacterial degradation of cystine in the lower gut and subsequently coprophagy have been postulated as a possible explanation for the effect of antibiotics in improving the nutritive value of raw soybean meal. The antibiotic reduces bacterial activity and thus degradation of cystine, and coprophagy allows for the recycling of the excreted cystine.

Following the work of Liener (1951), which showed that crude trypsin inhibitor (soybean whey protein) would cause growth inhibition and death of rats when injected intraperitoneally and that injected crystalline trypsin inhibitor had no effect on growth, soybean hemagglutinins were isolated and studied (Liener and Pallansch 1952; Pallansch and Liener 1953). Chemically the soybean hemagglutinin was shown to be an albumin as contrasted to trypsin inhibitor, which is a globulin.

Biologically, soybean hemagglutinins have been shown to agglutinate red blood cells *in vitro,* to be toxic when injected intraperitoneally, and to inhibit growth when fed to rats. The ability of hemagglutinins to clump red blood cells in a test tube is a good assay procedure (Liener 1953B); however, because it is readily inactivated by pepsin during digestion, it is doubtful that ingested hemagglutinin can cause agglutination of red cells *in vivo.* The mechanism for the toxicity caused by the intraperitoneal injection of hemagglutinin is not known.

When fed, purified hemagglutinin reduces growth but only at a rate equal to about one-half of that caused by feeding raw soybean meal (Liener 1953A,B). Growth inhibition due to hemagglutinin ingestion is accompanied by a reduction in feed intake, suggesting a problem of palatability or an appetite depressant. It is suggested that only a small portion of the growth inhibition caused by feeding raw soybeans or soybean meal is due to the presence of hemagglutinins.

Defatted soybean meal contains 0.6% saponins (Gestetner *et al.* 1966A) which are glycosides of five triterpenoid alcohols (Willner *et al.* 1964; Smith *et al.* 1958) and contain galactose, glucose, mannose, xylose, arabinose, and glucuronic acid (Gestetner *et al.* 1966A). Although soybean saponins have been shown to cause hemolysis of red blood cells *in vitro* and inhibition of some enzymes, evidence is not available to indicate that they have any biological effect *in vivo.* In fact, results have shown that when fed to rats, mice, and chicks, soybean saponins pass through the entire gastrointestinal tract without being hydrolyzed or absorbed (Gestetner *et al.* 1966B). Until further evidence is available, it appears that saponins can be discounted as an anti-nutritional factor in soybeans.

Isoflavones have been considered in assessing the nutritive value of raw soybeans. Two isoflavones, genistein and daidzein (Walz 1931; Ahluwalia *et al.* 1953), are believed at present to occur in soybeans. Isoflavones are of interest biologically because of their estrogenic properties, which are characterized by their effect in increasing uterine weight (Wong and Flux 1962) and in causing vaginal cornification (Biggers and Curnow 1954) and infertility (Bennetts *et al.* 1946). However, since the estrogenic activity of both isoflavones present in soybeans is very low and no adverse effects under practical conditions are reported in the literature, it is doubtful that they are of practical significance in swine feeding.

Of the four soybean constituents listed, the trypsin inhibitors and hemagglutinins are proteins and are readily inactivated by heat treatment. The remaining two, soybean saponins and isoflavones, are heat stable, but neither has been implicated by research information as an anti-nutritional factor. Therefore, from the information presently available, it appears that these soybean constituents are of importance only when soybean or soybean meal are improperly processed.

Using polyacrylamide gel electrophoresis, workers have found an experimental soybean variant that contains a trypsin inhibitor less active than that obtained from regular commercial varieties of soybeans (Singh *et al.* 1969). Biological tests with the unheated meal from this variant clearly demonstrated that it was nutritionally superior to other unheated meals prepared from two commercial soybean varieties (Yen *et al.* 1971). Through breeding and selection, it may be soon possible to produce varieties that contain few if any of the anti-nutritional factors.

Soybean meal that is properly processed contains somewhat less utilizable energy than maize. Kachaenchai and Diggs (1979) have reported the following values for 44% protein soybean meal: gross energy, 4453; digestible energy, 3538; metabolizable energy, 3271; metabolizable energy corrected for nitrogen, 3146 cal. These values are slightly higher than those reported by the NRC (1982).

Properly processed soybean meal contains a protein of excellent quality. Depending on completeness of oil extraction and degree of hull removal, soybean meal contains between 43 and 51% crude protein (Table 15.1) and has an excellent content of amino acids (Table 15.2) that are well digested by pigs of all ages except the very young (2–21 days of age) pigs (Table 15.3). Soybean protein has

TABLE 15.1. Proximate Analysis of Soybean Meal

Component	*Expeller*	*Solvent*	*Solvent dehulled*
Dry matter, %	90	90	90
Digestible energy, kcal/kg	3610	3318	3942
Metabolizable energy, kcal/kg	2972	2817	3155
Protein, %	42.9	44.6	49.7
Ash, %	6.0	6.5	5.8
Crude fiber, %	5.9	6.2	3.4
Ether extract, %	4.8	1.4	0.9
Total digestible nutrients, %	79.0	75.0	76.0
Calcium, %	0.26	0.30	0.26
Chloride, %	0.07	0.04	0.04
Magnesium, %	0.25	0.27	0.28
Phosphorus, %	0.61	0.63	0.63
Potassium, %	1.79	1.97	2.07
Sodium, %	0.03	0.04	0.03
Sulfur, %	0.33	0.43	0.44
Cobalt, mg/kg	0.18	0.09	0.07
Copper, mg/kg	22.0	23.0	20.0
Iodine, mg/kg	—	0.13	0.11
Iron, mg/kg	157.0	119.0	133.0
Manganese, mg/kg	31.0	29.0	37.0
Selenium, mg/kg	0.10	0.30	0.10
Zinc, mg/kg	60.0	43.0	55.0

Source: National Research Council 1982.

TABLE 15.2. Vitamin and Essential Amino Acid Composition of Soybean Meal

Nutrient	*Expeller*	*Solvent*	*Solvent dehulled*
Amino acid			
Arginine	3.07	3.03	3.67
Histidine	1.14	1.07	1.22
Isoleucine	2.63	2.03	2.46
Leucine	3.62	3.27	3.73
Lysine	2.79	2.68	3.17
Methionine	0.65	0.52	0.71
Cystine	0.56	0.75	0.75
Phenylalanine	2.20	2.11	2.44
Tyrosine	1.55	1.33	1.68
Threonine	1.72	1.66	1.94
Tryptophan	0.61	0.64	0.69
Voline	2.28	2.02	2.55
Vitamin			
Vitamin E, mg/kg	7.0	3.0	2.0
Biotin, mg/kg	0.33	0.32	0.32
Choline, mg/kg	2623	2614	2753
Folic acid, mg/kg	6.4	0.7	0.7
Niacin, mg/kg	31.0	28.0	22.0
Pantothenic acid, mg/kg	14.3	16.3	14.8
Riboflavin, mg/kg	3.4	2.9	2.9
Thiamin, mg/kg	3.9	5.6	3.1

Source: National Research Council 1982.

a good balance of amino acids essential for the growing pig. Its distribution of amino acids is similar to that of animal protein. It is not severely limiting in any amino acid except methionine. Relative to the requirement of the pig, most amino acids are quantitatively in excess when soybean meal is used to supply all of the dietary protein. However, in practice, soybean meal is seldom used as the only source of protein but is used as a supplemental protein in combination with other feed ingredients, such as maize, sorghum, barley, cassava meal, banana meal, and by-product feeds. The resulting balance of amino acids is of primary importance in practical feeding systems.

A combination of maize and soybean meal provides an excellent diet for pigs of all ages if the proper proportions of maize and soybean meal are used to supply an adequate balance of amino acids. North Carolina State University workers have carried out a series of tests to determine the proper ratio of protein from maize and soybean meal. In one such study with growing-finishing pigs (19–92 kg), 10% protein maize–soybean meal diets were formulated to provide 12.5, 25, and 50% of the protein from de-hulled soybean meal, with the remainder supplied by maize. Increasing the proportion of protein supplied by soybean meal resulted in significant increases in both feed intake and daily gains (Table 15.4). The diets in which the soybean meal supplied only 12.5 and 25% of protein were inadequate for maximum gain, efficiency of feed conversion, and carcass leanness.

TABLE 15.3. Apparent Ileal and Fecal Availabilities (Digestibility) of Amino Acids in Soybean Meal

Item	*Percent apparent availability*			
	Small intestine (%)		*Total gastro-intestinal tract (%)*	
Dry matter	74.9[a]	78.3[b]	85.7[a]	90.5[b]
Protein	81.6	80.6	88.6	88.9
Essential amino acids				
Arginine	91.7	90.3	94.2	94.4
Histidine	89.2	82.1	93.3	92.5
Isoleucine	83.4	85.5	86.5	88.7
Leucine	93.3	84.2	87.9	89.6
Lysine	88.5	85.6	89.1	89.8
Methionine	77.1	86.3	81.0	88.3
Phenylalanine	87.7	86.3	88.9	90.4
Threonine	77.0	75.8	85.4	86.6
Tryptophan	81.6	—	90.8	—
Valine	80.6	74.3	86.2	85.5
Nonessential amino acids				
Alanine	80.8	80.1	84.8	85.1
Aspartic acid	84.7	83.8	90.2	91.5
Glutamic acid	87.1	85.2	92.7	93.6
Glycine	76.2	72.3	85.2	87.9
Proline	82.0	75.1	90.7	91.9
Serine	82.9	81.9	89.7	91.5
Tyrosine	85.2	83.8	88.0	86.5
Cystine	—	82.1	—	95.3

Source: [a]Tanksley et al. 1981, used 44% protein meal. [b]Sauer et al. 1982, used 48% protein meal.

TABLE 15.4. Performance of Growing-Finishing Pigs with 10% Protein Diets with Soybean Meal Supplying Varying Levels of Total Protein

Criteria of response[a]	*Percent of protein supplied by soybean meal*		
	12.5	*25.0*	*50.0*
Av daily gain, kg	.38	.47	.77
Feed/gain ratio	4.75	4.27	3.48
Back fat thickness, cm	3.73	3.61	3.71
Longissimus muscle area, cm^2	22.9	25.1	30.8

Source: Clawson et al. 1963.
[a]Twelve animals per treatment: initial weight, 19 kg; final weight, 92 kg.

TABLE 15.5. Effect of Proportion of Protein from Soybean Meal and from Maize on Performance and Carcass Quality of Growing-Finishing Pigs Fed 12.8% Protein Diets

Criteria of response[a]	*Proportion of protein from soybean meal and maize*		
	40–60	*50–50*	*60–40*
Av daily gain, kg	0.73	0.81	0.81
Feed/gain ratio	3.48	3.36	3.31
Back fat thickness, cm	5.9	6.1	5.9
Lean cuts, %	51.6	51.6	51.7

Source: Smith et al. 1967.
[a]Sixteen pigs per treatment: initial weight, 21 kg; final weight, 96 kg.

Subsequent studies by the same workers (Smith *et al.* 1967), in which soybean meal and maize supplied 40:60, 50:50, and 60:40%, respectively, of the total protein of the diet, demonstrated that pigs grew slightly more when a proportion of 50:50 or 60:40 was fed and that as the proportion of protein from soybean meal increased the feed/gain ratio was significantly improved (Table 15.5). They further showed that the level of protein is not as critical as the quality of the protein. The performance of the pigs was the same whether they were fed diets containing 11.3, 12.8, 14.3, or 17.2% protein when the balance of amino acids was held constant by providing a diet formulated to supply 60% from soybean meal and 40% from maize (Table 15.6). The only difference in performance was related to percentage of lean cuts, which was less when 11.3% protein was fed.

A calculation of the sequence in which amino acids of a protein become limiting for the animal can be made from a knowledge of the amino acid requirement of the animal and the amino acid composition of the dietary protein. However, as pointed out by Berry *et al.* (1962), this method has three principal sources of error; inadequate information about biological availability of amino acids, inaccuracies in values for the amino acid composition of proteins, and inac-

TABLE 15.6. Effect of Level of Protein with Constant Balance of Amino Acids[a] on Performance and Carcass Quality of Growing-Finishing Pigs

Criteria of response[b]	*Percent of protein*[c]			
	11.3	*12.8*	*14.3*	*17.2*
Av daily gain, kg	0.79	0.81	0.79	0.78
Feed/gain ratio	3.43	3.31	3.37	3.36
Back fat thickness, cm	6.1	5.9	6.0	5.7
Lean cuts, %	50.7	51.7	51.4	51.9

Source: Smith et al. 1967.
[a]Protein supplied at a constant proportion of 60 percent from soybean meal and 40 percent from maize.
[b]Sixteen pigs per treatment: initial weight, 21 kg; final weight, 96 kg.
[c]Lower levels of protein were obtained by diluting the 17.2 percent protein diet with sugar.

curacies in the measurements of amino acid requirements. For example, it has been shown (Harper 1959) that the sequence in which the amino acids of casein become limiting for rat growth, as determined experimentally, differs from that suggested by calculations.

For the pig, Berry *et al.* (1966) calculated that methionine, lysine, tryptophan, and threonine were limiting in soybean protein in the order indicated. However, Maner *et al.* (1961) and Berry *et al.* (1962) had previously obtained results which demonstrated that for pigs methionine was the first limiting amino acid of soybean protein and further suggested that threonine was more limiting than tryptophan and possibly even more limiting than lysine (Table 15.7). Subsequent work by Kroening *et al.* (1964) tended to confirm the importance of threonine. In experimental studies, Berry *et al.* (1966) confirmed that methionine was the first limiting amino acid in soybean meal and that both lysine and threonine were limiting, but they were unable to establish whether one or both were second limiting for the growing pig.

The sequence in which amino acids become limiting when soybean meal is used in combination with other feeds depends upon the level of protein employed and the amino acid composition of feedstuff used. Since in practice dietary protein level is reduced by increasing the level of grain and reducing the level of supplementary protein, amino acid balance and thus sequence of amino acid limitation is altered. Although variations in chemical and amino acid composition exist among feedstuffs, it has been generally shown that a 16% protein diet for growing pigs based on maize–soybean meal is not improved by either methionine or lysine supplementation, in either the presence or the absence of tryptophan (Meade 1956A,B; Meade and Teter 1956) or in the presence of added dietary fat (Tribble *et al.* 1979).

However, at lower levels of protein (10–12%) a response to lysine supplementation has been reported from several stations (Jensen 1953; Acker *et al.* 1959; Clawson *et al.* 1963). Berry *et al.* (1961) reported that when pigs were fed a 12% maize–soybean meal diet supplemented with 0.25 or 0.50% L-lysine without

TABLE 15.7. Response of Weanling Pig to Supplementary Levels of Methionine, Lysine, Threonine, and Tryptophan

Amino acid added[a]	*Calculated percent of adequacy*				*Adjusted av daily gain (kg)*
	Methionine	*Lysine*	*Threonine*	*Tryptophan*	
None	65	106	115	130	0.44
Methionine	120	106	115	130	0.55
Methionine + lysine	120	121	115	130	0.60
Methionine + lysine + threonine	120	121	130	130	0.63
Methionine + lysine + threonine + tryptophan	120	121	130	145	0.56

Source: Berry et al. 1962.

[a]The control semisynthetic diet was based on soybean meal, glucose, corn oil, Solka Floc (purified cellulose), and vitamins and minerals. The levels of DL-methionine, L-lysine HCl, DL-threonine, and DL-tryptophan added were 0.2%, 0.098%, 0.078%, and 0.021% of diet, respectively.

added methionine, the lower level of lysine improved gains slightly whereas the higher level depressed gains. When 0.10% DL-methionine was added in the presence of 0.50% lysine, gains equal to those of the pigs fed the 16% control diet were recorded. Noland *et al.* (1959) also obtained a growth response in pigs and improved feed efficiency from 0.2% added DL-methionine in the presence of 0, 0.2, 0.4, or 0.6% supplemented lysine, when added to a 12% protein maize–soybean meal diet.

Corley and Easter (1980) studied the effect of lysine and L-tryptophan supplementation to low protein maize–soybean meal diets for growing and finishing pigs. Supplementation of a 10% protein diet with 0.48% L-lysine and 0.02% L-tryptophan improved pig performance but not as much for growing pigs as a 16% protein maize–SBM diet even when the low protein diet was also supplemented with 0.14% L-isoleucine, 0.08% DL-methionine, and 0.08% L-threonine. Nitrogen balance data indicated that the low level of total nitrogen in the 10% protein diet limited amino acid response. However, the addition of 0.25% lysine and 0.02% tryptophan to a 12% protein diet based on maize–SBM produced performance similar to that of growing pigs fed a 16% protein diet. Similar studies reported by Russell *et al.* (1980) failed to show the same response to lysine and tryptophan supplementation. A 12% protein corn–SBM diet plus 0.3% lysine and 0.2% DL-methionine, calculated to be adequate in all amino acids except tryptophan, was supplemented with graded levels of tryptophan (0 to 0.045%). Performance was considerably below that obtained with the 16% protein control diet. When 0.1% each of L-threonine and L-isoleucine were added to the 12% protein diet supplemented with lysine and methionine, the addition of tryptophan resulted in a linear improvement in performance with 0.04% tryptophan producing performance similar to the 16% protein control diet. In the studies reported by Corley and Easter (1980), the performance of finishing pigs fed a 10% protein maize–soybean meal diet plus 0.34% lysine and 0.02% tryptophan was similar to that of pigs fed a 13% protein control diet.

As shown by Clawson *et al.* (1963) and Smith *et al.* (1967), the adequacy of a maize–soybean meal diet depends more upon the percentage of protein supplied by each of the two feedstuffs than upon the level of protein. However, if diets contain levels of protein that are less than those recommended and an optimum balance of amino acids is not maintained, amino acid deficiencies will occur.

Maner *et al.* (1969) utilized varying levels of refined sugar to substitute for maize in 16% protein growing-finishing swine diets and demonstrated that soybean meal supplied adequate levels of amino acids to support excellent gains and efficiency of feed conversion. Corzo *et al.* (1968) reported that reducing the level of protein from 19 to 13% in diets containing 15, 22.5, or 30% molasses had no significant effect on daily gains or feed conversion when soybean meal was used as the supplemental protein source. This is to be expected since each increase in sugar and molasses levels represents a reduction in the protein supplied by maize and an increase in that supplied by soybean meal.

Soybean proteins have been fed extensively to older pigs with excellent results but with artificially reared neonatal pigs, little success has been achieved when soybean protein totally replaced milk protein in the diet (Hays *et al.* 1959; Maner *et al.* 1961; Schneider and Sarett 1969: Pond *et al.* 1971; Schmidt *et al.* 1973; Mateo and Veum 1980; Newport 1980; Kong and Veum 1980; Wilson and

Leibholz 1981). However, when soybean meal was used as a partial replacement for milk protein, neonatal pigs were raised successfully (Sherry *et al.* 1978; Zamora *et al.* 1975; Zamora and Veum 1979; Barr *et al.* 1980; Wilson and Leibholz 1981). In studies by Wilson and Leibholz (1981), the substitution of milk protein (skim milk and casein) by soybean meal reduced liveweight gains at each level of soybean meal inclusion (Table 15.8). During the period from 7 to 14 days, the weight gains of pigs given the diets in which SBM supplied 50 and 75% of the total dietary protein were 51 and 54%, respectively, of those of the pigs given the all-milk protein diets. By 21–28 days of age, the difference in performance was only 16 and 31%, respectively. Similar studies by Zamora and Veum (1979) showed that increasing levels of milk protein replacement by soybean meal reduced both daily gains and feed conversion efficiency.

The better performance of baby pigs fed casein is due in part to the higher digestibility of casein as compared to soybean meal. Wilson and Leibholz (1981) demonstrated that each increase in dietary level of soybean meal, at the expense of casein, reduced DM and N digestion and N retention. Mateo and Veum also reported that the digestibility of dry matter, energy, crude protein, ash, and ether extract were significantly greater for casein than for soybean protein.

Although early weaned neonatal pigs are not able to utilize soybean meal efficiently, weanling pigs (5–15 kg) perform well on practical diets supplemented with soybean meal (Zimmerman 1980; Lewis *et al.* 1981). However, practical diets based on maize and soybean meal are improved by the addition of lysine. Working with 5–10-kg pigs, Zimmerman (1980) showed a significant linear decrease in feed intake and a linear increase in gain-to-feed ratio when 0, 0.10, and 0.20% lysine were added to a maize– soybean meal–whey basal diet and indicated that at least 1.05% lysine was required for optimal production. Similar studies by Lewis *et al.* (1981), who used weanling pigs between 5 and 15 kg, also showed that a basal diet composed principally of maize and soybean meal was improved when 0.10, 0.20, and 0.30% lysine were added to supply 1.05, 1.15, and 1.25% total dietary lysine. Higher levels of lysine supplementation (0.40 and 0.50%) supported gains, feed intake, and feed efficiency superior to the basal diet but somewhat lower than those obtained with a dietary level of 1.25%.

In a series of four growth trials, Corley and Easter (1980) evaluated the addition of lysine, methionine, tryptophan, and threonine to low-protein (14%) diets based

TABLE 15.8. Performance of Early-Weaned Pigs Fed Diets Containing Varying Levels of Soybean Meal as a Replacement for Milk Protein

Soybean meal level (% of protein)	*Average daily gain, g*				*F/G*[a]
	7–14[b]	*14–21*[b]	*21–28*[b]	*7–28*[b]	*7–28*[b]
0	132	189	428	265	0.80
25	127	160	366	217	0.89
50	67	124	370	188	0.96
75	71	137	327	178	0.95

Source: Wilson and Liebholz 1981.
[a]F/G is the feed-to-gain ratio.
[b]Ages of pigs in days.

on maize and soybean meal for 4–8-week old pigs. Supplementation of the diet with only lysine (0.5% L-lysine HCl) failed to improve pig performance to a level equal to that of the 18% crude protein control. However, the addition of both lysine and 0.05 percent L-tryptophan to the 14% crude protein diet supported gains and feed efficiency equal to that of the 18% corn–soybean meal control. When 0.24% DL-methionine was added in addition to lysine and tryptophan, pig performance was depressed, but the addition of 0.13% L-threonine had no effect on pig performance.

All-plant diets based on cereals and soybean meal have provided adequate nutrition and a proper balance of amino acids for satisfactory reproductive performance of sows and gilts. Although Robison (1951, 1953) and Robison *et al.* (1956) reported that the performance of pigs fed properly balanced maize–soybean meal diets was equal to that of pigs fed diets containing a combination of plant and animal proteins, Fowler and Robertson (1954) found that animal protein in the diet of gilts caused puberty to occur at an earlier age and the ovulation rate to increase.

Teague and Rutledge (1960) conducted generation studies with swine that were fed diets containing only soybean meal or soybean meal in combination with meat and bone scraps as the supplemental protein source. Satisfactory performance was maintained on both dietary treatments. After four generations, breeding performance, lactation performance, pig growth, and livability to weaning and growing-finishing did not appear to be affected by the presence or absence of supplemental animal protein. However, both birth weights and average vigor of pigs at birth were improved in the second- and third -generation females that were fed a combination of both plant and animal proteins.

Palmer *et al.* (1970) compared the reproductive performance of gilts and sows fed maize–soybean meal diets during gestation and lactation with the performance of animals fed similar diets containing 6% menhaden fish meal. They observed that fish meal increased gain of the gilts and sows and increased the number of live pigs at birth. Total litter weight was greater at birth and at 2 weeks for litters from sows fed fish meal. However, their diet was based on Ohio-grown maize that has been shown to be deficient in both vitamin E and selenium (Mahan *et al.* 1971). Since their basal diet contained neither supplemental vitamin E or selenium and it has been shown that fish meal is an excellent source of available selenium (Griffo and Moxon, 1973), as pointed out by Baker *et al.* (1974), it is conceivable that the selenium furnished by the fish meal may have accounted for the efficiency of this dietary addition. Baker *et al.* (1974) reported a similar study in which the reproductive performance of gilts and sows fed a maize–soybean meal diet or a similar diet containing 3% fish meal was compared. During two reproductive cycles, they reported that menhaden fish meal at 3% of the gestation diet did not enhance reproductive performance above that observed on a vitamin E-fortified maize–soybean meal diet.

COCONUT MEAL

Coconut meal or copra meal, the residual product after extraction of oil from the dried endosperm of the coconut (*Cocos nucifera*) is available in many parts of the world. In large areas of the tropics, coconut meal may be one of the few protein sources available for livestock feeding. Even though the meal contains only moderate levels of crude protein, it is an economically important source of protein in areas in which other sources are not readily available. Hawaii and some of the Caribbean islands are good examples of such areas.

Coconut meal contains between 20 and 26% crude protein, about 10% crude fiber, and depending on processing method and efficiency between 6 and 12% of ether extract (Creswell and Brooks 1971A; Simas and Maner 1981) (Table 15.9). The lysine content of coconut meal on a percent of protein basis is somewhat lower than soybean meal but the level of total sulfur amino acids is significantly higher.

Different levels of coconut meal have been used (Grieves *et al.* 1966; Adzic *et al.* 1966; Creswell and Brooks 1971A,B; Malynicz 1973; Lima *et al.* 1981) to supply a portion of the protein in complete mixed diets for growing-finishing pigs. Generally, each increase in level of coconut meal used to replace soybean meal caused a decrease in feed intake and daily gains and an increase in feed required per unit of gain. Dietary levels up to 20% caused only minor reduction in pig performance (Creswell and Brooks 1971B); however, 40% dietary coconut meal caused a 39.5 increase in feed required per unit of gain (Table 15.10). Loin eye measurements were also reduced at the highest level of substitution. In a series of three experiments with levels of coconut meal up to 30%, Malynicz (1973) presented data to indicate a reduced performance with increasing levels of coconut meal. More recent studies (Lima *et al.* 1981), in which growing-finishing pigs were fed levels of 0, 10, 20, and 30% coconut meal as a replacement for soybean meal protein in maize-based diets, have clearly demonstrated a linear depression in both feed intake and daily gains as the level of coconut meal was increased in the diet. Feed required per unit of gain increased linearly with increasing levels of coconut meal (Table 15.10). Although it would appear that

TABLE 15.9. Mineral, Proximate, and Amino Acid Composition of Coconut Meal (Expeller)

Energy, protein, and minerals	*Proportion*	*Amino acids*	*Percent of protein*
Dry matter, %	89.9	Arginine	9.37
Crude protein, %	20.9	Histidine	1.96
Crude fiber, %	10.5	Isoleucine	2.87
Ether extract, %	5.8	Leucine	5.78
Ash, %	6.5	Lysine	2.29
Nitrogen-free extract, %	46.2	Methionine	1.77
Gross energy, kcal/g	4.2	Cystine	1.14
DE, kcal/g	3.6	Methionine + cystine	2.91
Calcium, %	0.16	Phenylalanine	3.87
Phosphorus, %	0.55	Threonine	3.15
Magnesium, %	0.23	Valine	4.25
Potassium, %	1.75	Glycine	4.25
Zinc, mg/kg	53.0	Tyrosine	2.20
Copper, mg/kg	40.0	Aspartic acid	7.75
Manganese, mg/kg	75.0	Glutamic acid	17.41

Source: Creswell and Brooks 1971A.

TABLE 15.10. Effect of Level of Expeller-Processed Coconut Meal on the Performance of Growing-Finishing Pigs

Level of coconut meal, %	*Av daily gain,* kg	*Av daily feed,* kg	*Feed/gain*	*Loin eye area,* cm²
0[a]	0.76[c]	1.88	2.51	27.1[c]
10[a]	0.74[c]	1.97	2.68	26.9[c]
20[a]	0.65[c]	1.80	2.75	25.5[c]
40[a]	0.46[e]	1.49	3.27	22.3[d]
0[b]	0.67	2.19	3.34	—
10[b]	0.61	2.16	3.61	—
20[b]	0.56	2.08	3.77	—
30[b]	0.39	1.81	4.67	—

Source: [a]Creswell and Brooks 1971A. [b]Lima et al. 1981, who showed a significant linear treatment effect ($P < .05$).

[c,d,e]Means in the same column without a common superscript are significantly different from each other ($P < .05$).

lysine might be deficient in the diets prepared with expeller-processed (oil extracted by heat and mechanical pressure) coconut meal, supplementing lysine or increasing the protein level by adding more soybean meal had no effect on pig performance or muscle development for diets containing 20 or 40% coconut meal (Cresswell and Brooks 1971B).

The digestibility of the various constituents of coconut meal has been reported from several stations (Loosli *et al.* 1954; Butterworth and Fox 1963; Creswell and Brooks 1971A; Simas and Maner 1981) and results have shown considerable variation. Creswell and Brooks (1971A) reported apparent digestibility values for coconut meal as follows: dry matter, 83.7%; ether extract, 100%; nitrogen-free extract, 94.1%; and energy, 85.4%. Simas and Maner (1981) reported lower values for coconut meal produced in Brazil: dry matter, 73.0%; crude fiber, 70.0%; ether extract, 90.6%; nitrogen-free extract, 76.4%; and energy, 72.8%. These workers also reported digestible energy, metabolizable energy, and nitrogen-corrected metabolizable energy values of 3.54, 2.95, and 2.94 kcal/g, respectively. Reported values for protein digestibility differ quite widely. Creswell and Brooks (1971A) reported a protein digestibility of only 50.7% for pigs, while other Philippine and Brazilian workers have shown the crude protein fraction to be 72% (Loosli *et al.* 1954) and 71.5% (Simas and Maner 1981) digestible, respectively. Mee and Brooks (1973) have shown that not only is the true digestibility of all amino acids in coconut meal very low, but also that utilization at the tissue level is much below that observed for diets based on maize and soybean meal.

The variation in protein digestibility may be due to differences in processing methods and more specifically to processing temperature. Butterworth and Fox (1963) heated cold-solvent extracted coconut meal for 30 min at oven temperatures from 40° to 150°C and clearly demonstrated that each increase in treatment temperature caused a corresponding linear reduction in protein digestibility when the final material was fed to rats (Table 15.11). There was also a corresponding reduction in available lysine and net protein utilization as the temperature was

TABLE 15.11. Effect of Various Degrees of Heat Treatment on Composition and Nutritive Value of Coconut Meal

Temperature of treatment (°C)	*Protein content (%)*	*Available lysine (g/16 g N)*	*Nitrogen digestibility (%)*	*Net protein utilization (%)*
40	25.3	3.29	77.7	45.9
90	25.8	3.09	78.3	41.0
105	26.3	2.81	74.6	36.1
120	25.6	2.34	73.3	35.8
135	26.4	1.63	68.2	33.9
150	26.3	1.12	56.1	17.1

Source: Butterworth and Fox 1963.

increased. It has also been shown (Loosli *et al.* 1954) that expeller processing reduces the digestibility of the protein. In these studies, fresh coconut meals had a crude protein digestibility of 83.8% and expeller-processed meal protein was only 73.4% digestible.

Because of the low protein and amino acid digestibility, except in unusual circumstances of price relationships and availability of protein supplements, coconut meal should not be included in proportions of more than 15–20% in growing-finishing swine diets if maximum performance is to be expected.

COTTONSEED MEAL

Cottonseed meal is a protein supplement that is available in many areas of the world for swine feeding. It is commonly used as a portion of the supplemental protein in swine rations in the southeastern and southwestern areas of the United States and in many Latin American countries, such as Ecuador, Colombia, Peru, Costa Rica, Mexico, and Guatemala, where it is often the least expensive and most abundant protein supplement available.

Because of reports of poor performance, toxicity, and death of pigs fed moderate to high levels of cottonseed meal, much prejudice has developed against the use of cottonseed meal as a swine feed. Cottonseed meal, when properly used and its limiting factors understood and corrected, is an excellent source of protein for use in swine rations. Excellent pig performance can be obtained when limited amounts of cottonseed meal are fed in combination with other sources of high quality protein.

The limitations of cottonseed meal must be recognized. The seeds of the cotton plant from which cottonseed meal is produced contain a natural pigment, gossypol, that is toxic to pigs. *Gossypol* is a yellow solid, insoluble in water but soluble in most fat solvents (Adams 1938; Edwards 1958).

Gossypol, under favorable conditions, combines with the free amino groups of cottonseed protein and becomes insoluble in the usual extraction solvents, such as aqueous acetone. *Free gossypol* is that gossypol which is extracted with aqueous acetone, and *total gossypol* is that which is obtained only by acid treatment of the cottonseed product, or by some other treatment harsh enough to release all of the gossypol (Smith 1958; Pons *et al.* 1958). "Bound gossypol" is usually estimated as the difference between total and free gossypol.

The bound gossypol is of little significance to the pig (Baliga and Lyman 1957), since it is unavailable and passes through the gastrointestinal tract of the pig without being absorbed and is eliminated in the feces. The free gossypol is highly toxic to the pig. Any cottonseed meal containing more than 0.04% free gossypol should be used sparingly in swine diets. It has been determined that a standard 15–16% protein diet should not contain more than 0.01% free gossypol (Table 15.12). Percentages higher than this safe level have produced slow growth, gossypol toxicity, and death. Texas studies have indicated that higher levels of free gossypol can be tolerated by the growing pig if the level of crude protein is increased in the diet (Hale and Lyman 1957). Levels of as much as 0.03% gossypol were without detrimental effect when fed in diets containing 30% crude protein supplied principally by soybean meal. Other studies (Kornegay *et al.* 1961) have shown that the toxicity of free gossypol, as measured by feed intake, gain, and mortality of swine, was influenced by both source of protein and level of free gossypol (Table 15.13). Soybean meal diets gave superior performance to peanut or cottonseed meal rations at all levels of free gossypol. The mortality rate was lower for pigs fed diets containing soybean meal than for pigs fed peanut or cottonseed meal.

Clinically, pigs suffering from gossypol toxicity are listless, inactive, and have violently labored respirations (thumping) (Smith 1957). Progressive weakness and emaciation are accompanied by a good appetite almost until death. Patholog-

TABLE 15.12. Effect of Protein Level of Diet on Gossypol Tolerance in Growing-Finishing Pigs

Item	*Percent free gossypol in diet*			
	0	*.01*	*.02*	*.03*
Standard, 15% crude protein				
Number of animals, no.	8	8	8	8
Initial wt, kg	26.4	26.4	26.4	26.4
Final wt, kg	92.7	86.4	77.3	60.5
Av daily gain, kg	.791	.714	.595	.527
Feed/gain ratio	3.66	3.76	4.24	4.61
Number of deaths, no.	0	0	2	4
Remaining live animals with gossypol toxicity symptoms	0	0	3	2
High protein, 30% crude protein				
Number of animals, no.	8	8	8	8
Initial wt, kg	26.4	26.4	26.4	26.4
Final wt, kg	85.5	86.8	84.5	80.5
Av daily gains, kg	.704	.718	.691	.645
Feed/gain ratio	3.87	3.79	3.81	3.84
Number of deaths, no.	0	0	0	0
Remaining live animals with gossypol toxicity symptoms, no.	0	0	0	0

Source: Hale et al. 1958.

TABLE 15.13. Effect of Protein Source on Gossypol Toxicity

Free gossypol level, %	*Peanut meal*			*Soybean meal*		
	0.00	*0.02*	*0.04*	*0.00*	*0.02*	*0.04*
Av daily gains, kg						
9 weeks	.464	.464	.282	.682	.650	.423
15 weeks	.536	.423	.282	.750	.659	.345
Av daily feed, kg						
9 weeks	1.75	1.70	1.36	2.41	2.14	1.51
15 weeks	2.23	1.78	1.27	2.85	2.39	1.37
Feed/gain ratio, kg						
9 weeks	3.84	3.62	5.16	3.58	3.29	3.59
15 weeks	4.17	4.22	5.18	3.83	3.64	4.06
Number of deaths						
9 weeks	0	0	3	0	0	2
15 weeks	0	2	6	0	0	6

Source: Kornegay et al. 1961.

ical examination shows widespread congestion and edema. The lungs, kidneys, spleen, lymph nodes, and liver are congested. Large amounts of fluid are often encountered in the pleural, pericardial, and peritoneal cavities. The edema is considered to be dependent on passive congestion and venous stasis, which is attributed to a progressively failing heart (Smith 1957; West 1940; Ferguson *et al.* 1959).

Gossypol toxicity is cumulative. In contrast to an acute toxin that causes a rapid decline in body functions and rapid death, gossypol accumulates in the body and death usually does not occur until after the pig has received the gossypol-containing diet for 4–8 weeks. The exact time of the production of toxicity symptoms and death will depend upon the concentration of gossypol in the ration and the presence of certain mineral elements.

The toxicity of dietary gossypol can be prevented by the addition of iron salts to the diet (Withers and Carruth 1917; Robison 1934; Flecher 1948; Maner 1964; Jarquin *et al.* 1965; Stevenson *et al.* 1965; Phelps 1966B; Lyman 1966; Barrentine 1966; Clawson *et al.* 1967; Braham *et al.* 1967). The subject was covered thoroughly at a conference on the Inactivation of Gossypol by Mineral Salts (Smith 1966). The most effective iron salt is ferrous sulfate. It is thought that a strong complex forms between iron and gossypol in the intestinal tract, thereby preventing the absorption of gossypol. The addition of ferrous sulfate on a 1:1 weight ratio, iron to free gossypol, to diets containing more than 0.01% free gossypol is effective in reducing gossypol toxicity and improving gains and feed efficiency. Maner (1964) demonstrated the effectiveness of ferrous sulfate in reducing gossypol toxicity and of lysine in improving both gains and feed efficiency but observed that when both lysine and ferrous iron were added to the diet, the effectiveness of both were reduced. Subsequent studies (Wesmine and Maner 1967) demonstrated that ferrous iron and lysine form at least two and possibly more complexes that could reduce or completely inactivate the effectiveness of both.

Cottonseed meals manufactured under different processing conditions differ over a wide range with respect to their nutritional value as a protein supplement for pigs (Sure 1953; Phelps 1966A). The cottonseed meals with high protein quality and low free gossypol content are the most desirable. Most meals are produced by one of three processing methods; screw press (expeller), prepress-solvent, and direct solvent extraction. The screw press method involves application of heat and mechanical pressure only; the prepress–solvent method involves heat and pressure followed by extraction with a fat solvent; and the direct solvent extraction involves the use of fat solvent only.

An estimation of the nutritive value of a cottonseed meal can be made if the processing procedure used in its production is known. Meals produced by screw press extraction of the oil from the cottonseed are low in free gossypol but have poor quality protein. During processing the moist-heat generated allows the free gossypol to be chemically bound to the lysine which inactivates the gossypol and makes the lysine unavailable to the pig (Baliga *et al.* 1959; Lyman *et al.* 1959). Dry heat, even in the course of extended periods, produces relatively small amounts of bound gossypol (Danke 1962). If excessive processing temperatures are used, further reactions occur between the carbohydrates and the epsilon amino group of lysine, thereby reducing the amount of available lysine in the meal.

Prepress-solvent meals are low in free gossypol and moderate to high in protein quality. The heat generated in the initial prepressing binds much of the free gossypol and the subsequent extraction with solvents removes the remaining oil. During the prepress-solvent extraction, different prepressing methods are employed and, therefore, cottonseed meals of varying protein quality are produced. If the original pressure and temperature during prepressing is excessive, a low quality protein meal will result. If original pressures and temperatures are low, a better quality protein will be produced, but the meal will contain higher levels of free gossypol. Analysis to determine level of free gossypol, available lysine, and nitrogen solubility are employed to measure protein quality and gossypol toxicity levels (Smith 1946A; Hoffpauir and Pons 1953; Sherwood 1926; Pons *et al.* 1950; Rao *et al.* 1963).

The direct solvent meals prepared from cottonseed extracted with hexane are high in protein quality and also high in free gossypol. Both of these characteristics are the result of low temperatures during solvent extraction.

A 3-solvent extraction method has been developed by Agricultural Research Service scientists Frampton and King at the Southern Utilization Research Laboratory at New Orleans, Louisana. Hexane, acetone, and water are combined in the extraction process. Hexane is the solvent commonly used for oil extraction, but this solvent removes only a small part of the gossypol. The mixture of acetone and water removes most of the gossypol, but does not readily extract the oil. The 3-solvent mixture extracts both the oil and the gossypol and produces a meal of high protein quality and of low (0.02%) free gossypol content. Residual oils in the meals were below 0.5%. Commercial use of this 3-solvent process depends upon methods to recapture and control the solvents for recycling and for complete extraction of the gossypol from both the oil and the solvents.

Cotton breeders have developed a variety in which seed is very low in gossypol-containing glands.

Many workers have studied the nutritive value of cottonseed meal and its use as a protein supplement in swine diets and have clearly demonstrated that excellent performance can be obtained when cottonseed meal is properly used in swine rations. Although cottonseed meal is a good source of tryptophan and a fair source of methionine, it is a very poor source of lysine. The lysine content of cottonseed meal is only about two-thirds that of soybean meal. Comparing the effect of dietary lysine, crude fiber, and free gossypol in cottonseed meal on the performance of growing pigs, Knabe *et al.* (1979) concluded that the lower performance of growing pigs fed sorghum–cottonseed meal diets is due primarily to low dietary lysine. Dietary crude fiber affected only feed utilization while dietary free gossypol had no effect on pig performance when iron was included in the diets on a 1:1 weight ratio to free gossypol.

Perhaps of more importance is the fact that a portion of the lysine in cottonseed meal cannot be digested, absorbed, and utilized by swine. During processing, application of excessive heat may render 10–35% of the lysine unavailable. Tanksley *et al.* (1981) demonstrated the extreme differences in digestibility of amino acids in cottonseed meals prepared from different processes. Digestibilities of nitrogen and the ten essential amino acids at the end of the small intestine and over the total digestive tract were shown to be higher in laboratory-processed glandless (GL) cottonseed meal than in commercially-processed direct solvent (DS) and screw press (SP) cottonseed meal. Screw pressed cottonseed meal had higher digestibilities of all the essential amino acids, except lysine, methionine and tryptophan, than did direct solvent cottonseed meal. Lysine was the least digestible, in the lower small intestine, of the essential amino acids in direct solvent (61.7%) and screw press (64.0%); in contrast, it was highly digestible in glandless cottonseed meal (88.5%) which compared favorably with the lysine in soybean meal (87.1%). Lysine availability is, therefore, a major factor in determining the quality of protein in cottonseed meal, and probably no other major protein source varies as much in quality as does cottonseed meal.

It has been demonstrated that one-half of the soybean meal in a maize–soybean meal diet could be replaced by degossypolized cottonseed meal without affecting growth or efficiency of feed utilization of growing-finishing pigs. It was subsequently shown that 25, 50, or 75% of the soybean meal of the diet could be replaced with solvent-processed, degossypolized cottonseed meal without altering pig performance; however, the same levels of expeller cottonseed meal with similar levels of free gossypol but lower levels of soluble nitrogen depressed growth and feed efficiency. Moncada and Maner (1970) reported data which indicated that replacing 25, 50, or 75% of the soybean protein in maize–soybean meal diets for growing-finishing pigs allowed normal growth (Table 15.14).

The lysine deficiency of cottonseed meal–cereal grain diets can be overcome by blending the cottonseed meal with other lysine-rich protein feeds such as soybean meal or fish meal or by the addition of crystalline L-lysine. Maize–cottonseed meal diets with 0.1% supplemented L-lysine were superior to unsupplemented rations when fed to growing pigs. In studies by Knabe *et al.* (1979), increasing levels of dietary lysine (0.60, 0.70, and 0.80%) in sorghum–cottonseed meal rations were associated with a linear increase in gain, feed intake, and efficiency of feed utilization.

Noland and co-workers (1968) fed glandless cottonseed meal with a nitrogen solubility of 85.6% and only 0.009% free gossypol as a replacement for soybean

TABLE 15.14. Effect of Replacement of Soybean Meal in Maize–Soybean Meal Diets by Cottonseed Meal for Growing-Finishing Pigs

Criteria	*1*	*2*	*3*	*4*
Level of cottonseed meal, %	0	3.53	7.32	11.41
Level of soybean meal, %	13.6	10.59	7.32	3.81
Av daily gain, kg	0.76	0.84	0.83	0.83
Av daily feed, kg	2.67	3.07	3.00	2.98
Feed/gain ratio	3.54	3.65	3.59	3.50

Source: Moncada and Maner 1970.

meal in maize–soybean meal diets for 9-kg pigs, and reported a linear reduction in both growth and feed efficiency as the rate of substitution increased from 0 to 100%. L-lysine supplementation at a level of 0.1% was without effect in overcoming the depression produced by increasing the level of cottonseed meal. Subsequent studies by these same authors demonstrated that supplemental levels as high as 0.4% L-lysine were required to overcome the growth depression produced by replacing 100% of the soybean meal with glandless cottonseed meal.

Controlled studies to evaluate the use of cottonseed meal in gestation and lactation diets have been limited. Tanksley *et al.* (1981) found that sows fed sorghum–cottonseed meal diets farrowed pigs that were lighter at birth and weaning compared to sows fed sorghum–soybean meal diets. Haught *et al.* (1977) reported than neither live pigs farrowed, pigs weaned per litter, creep diet intake, lactation intake, lactation weight change, or farrowing interval were affected when cottonseed meal was substituted for one-half of the supplemental protein in a 12% protein maize–soybean meal diet. However, sows fed sorghum–soybean meal–cottonseed meal diets gained less during gestation and farrowed pigs that were lighter at birth and weaning than those fed maize–soybean meal diets.

Cottonseed meal can be used to supply up to 50% of the supplemental protein in maize– or sorghum–soybean meal diets for gestating and lactating sows. However, small reductions in farrowing and weaning weights may be observed. Maximum performance will depend on the lysine adequacy of the diet.

LINSEED MEAL

Linseed meal is a protein-rich by-product of drying oil production from linseed or flax (*Linum usitatissimum*). Flax grows best in temperate climates (Janick *et al.* 1969), and approximately 60% of the world's production of 2.3 million tons annually is from the United States, Canada, and Argentina (FAO 1980). Solvent-extracted linseed meal constitutes less than 1% of the protein supplements used for livestock feeding in the United States and only an insignificant part is used for swine. It contains approximately 35% protein, 1.7% fat and 9% crude fiber (NRC 1964). It is similar in calcium, phosphorus, and vitamin content to soybean meal but it is seriously limiting in lysine.

In feeding experiments reported (Bethke *et al.* 1928; Crampton *et al.* 1955; Wahlstrom *et al.* 1956; Taxilin 1961; Jusyczak and Zieminski 1968), diets containing up to 25% linseed meal have been fed to growing-finishing pigs without adverse effect. However, because of a serious lysine deficiency and poor amino acid balance, it should always be combined with other lysine-rich protein supplements and should not make up more than one-third of the protein supplement when used with maize or other cereal grains. It tends to have a laxative effect and is sometimes added to sow diets at parturition to prevent constipation.

A vitamin B_6 antagonist has been isolated from linseed meal (Mickelson and Yang 1966; Klostermann *et al.* 1967). The antagonist is a dipeptide linatine that, when hydrolyzed, yields 1-amino-D-proline and glutamic acid; the former was found to combine with pyridoxal phosphate *in vitro*. Bishara and Walker (1977) have reported that pigs fed diets containing 30% linseed meal grow faster if supplemental pyridoxine is added and suggest that, under some conditions, diets containing linseed meal may be marginally deficient in vitamin B_6.

PEANUT MEAL

Peanut meal (also called groundnut meal) is the by-product of the extraction of the oil from raw peanuts (*Arachis hypogaea*). The protein content of peanut meal varies from 45 to 55% and the quality is influenced by the quantity of hulls in the meal, and by the type and level of heat it has received during processing.

Early studies with peanut meals indicated that it was of little value for supplementing cereal diets when used as the only source of supplemental protein (Massey *et al.* 1942; Sewell *et al.* 1957; Brooks and Thomas 1958, 1959). Pigs fed cereal-based diets supplemented with peanut meal grow very slowly and have low feed intake and inefficient rates of feed utilization.

The poor nutritive value of the peanut meal protein is not due to reduced digestibility (Combs *et al.* 1963; Combs and Wallace 1962), as diets containing properly processed peanut meal have protein digestion coefficients (approximately 80%) that are similar to those of diets containing soybean meal or fish meal. The poor response of growing pigs fed cereal diets supplemented with peanut meal is due primarily to a lysine deficiency of the mixed diet. Peanut meal is low in methionine and lysine (Table 15.15). Performance and nitrogen balance are greatly improved by supplementing the diet with 0.13–0.3 percent L-lysine (Brooks and Thomas 1959; Ranjhan *et al.* 1964). Thomas and Kornegay (1972) and Orok *et al.* (1975) also demonstrated that a maize–peanut meal diet is deficient in lysine for growing swine. Thomas and Kornegay (1972) further demonstrated that a total dietary lysine content of about 0.85% was required to support growth and feed efficiency equal to that produced with soybean meal as a supplement instead of peanut meal. If normal maize were used, 0.35% supplemental lysine was required; however, when high-lysine maize replaced the normal maize in the maize–peanut meal diet, only 0.25% additional lysine was required for maximum performance. Although gains are improved by added lysine, performance is not improved by the addition of methionine and/or threonine.

In practice, the amino acid deficiencies of cereal–peanut meal diets are corrected by blending the peanut meal with other sources of protein such as soybean meal, fish meal, and blood meal. Sewell *et al.* (1957) have shown that 50% of the

TABLE 15.15. Amino Acid Content (Expressed in g/100 g of Nitrogen) of Different Peanut Meals

Amino acid	*Sample*			
	1	*2*	*3*	*4*
Arginine	10.65	11.59	10.70	12.44
Histidine	2.78	2.20	2.44	2.44
Isoleucine	3.29	3.60	3.45	3.84
Leucine	6.59	6.56	6.25	6.38
Lysine	3.89	3.46	3.45	3.74
Methionine	0.90	1.42	1.10	1.35
Cystine	1.62[a]	1.45	—	1.39
Phenylalanine	5.17	5.03	5.35	6.30
Tryptophan	0.96	1.03	1.15	—
Threonine	2.51	2.70	2.90	3.06
Valine	3.80	4.38	4.35	4.22

Source: Woodhamand and Dawson 1968.
[a]As cysteic acid.

soybean meal in a 16% protein maize–soybean meal diet can be replaced by peanut meal without lowering performance. Kornegay *et al.* (1968) reported that the maximum amount of soybean meal that can be replaced by peanut meal for growing pigs appeared to be between 30 and 50%. Brooks and Thomas (1959) were able to maintain optimum performance by adding either 3% fish meal or 3% blood meal to the diet.

In Canadian studies, use of peanut meal as the only supplemental protein source to starter growing pigs, given wheat–barley based diets, had a marked depression on pig performance. Increasing levels of soybean meal improved gains, feed intake, and feed efficiency. Substitution of 50% peanut meal for 50% soybean meal did not significantly reduce average daily feed or average daily gain, but it reduced efficiency of feed conversion. The addition of lysine to the diet supplemented with only peanut meal resulted in performance similar to that obtained with a 50–50 peanut meal–soybean meal mixture. Peanut meal for finishing pigs reduced gains and feed intake but had no significant effect on feed efficiency. Although rapeseed meal was superior to peanut meal as the only source of protein, it was inferior to soybean meal and not an optimal supplement to peanut meal.

In a more recent study (Hale and McCormick 1979), graded levels of either soybean meal or tankage used to supplement a maize-based diet for growing-finishing pigs were replaced with 25, 50, 75, or 100% peanut meal. One hundred percent replacement of soybean meal or tankage by peanut meal produced unsatisfactory pig performance. Each increase in level of peanut meal reduced average daily gains of pigs fed either soybean meal or tankage (Table 15.16). Twenty-five percent replacement of soybean meal had little effect on gain, feed intake, or feed efficiency. Higher levels reduced gains, increased feed intake and depressed feed efficiency. When peanut meal replaced tankage, feed intake, average daily gain, and feed efficiency were not greatly affected, except at the 100% level of substitution.

TABLE 15.16. Supplemental Effect of Soybean Meal (SBM) and Tankage on Growing-Finishing Swine Diets Based on Maize and Peanut Meal (PM)

Dietary treatments	*Av daily gain*	*Feed/gain ratio*
100% soybean meal (SBM)	0.88[a]	3.03[e]
75% SBM + 25% peanut meal (PM)	0.87[a,b]	3.00[e]
50% SBM + 50% PM	0.83[b,c]	3.17[d]
25% SBM + 75% PM	0.82[b,c]	3.23[c,d]
100% PM	0.72[e]	3.80[a]
75% PM + 25% tankage	0.74[d,e]	3.50[b]
50% PM + 50% tankage	0.75[d,e]	3.36[b,c]
25% PM + 75% tankage	0.77[d,e]	3.37[b,c]
100% tankage	0.80[c,d]	3.39[b,c]

Source: Hale and McCormick 1979.
[a,b,c,d,e]Means in same category with different superscript are significantly different ($P < .05$).

During commercial processing, after crushing, the seed is cooked by means of jacket or live steam, the temperature ranging from 80° to 115° C and the cooking time from 20 to 120 min. If the hydraulic process is used, the temperature may range from 65° to 105° C and the time from 15 to 120 min. Drying temperatures vary from 75° to 115° C. If these or similar conditions are used during processing, little damage is expected (Woodhamand and Dawson 1968) and a destruction of the trypsin inhibitor occurs. Variation from a moderate-moist heat would be expected to cause a change in nutritive value of the protein, since dry heat will not bring about a rapid removal of the trypsin inhibitor and, if prolonged, can cause a reduction in lysine availability. A similar but less drastic reduction in biological value of the protein can occur if the peanuts are subjected to moist heat and excessive duration of heating.

Like many other stored feeds, peanuts are subject to the growth of certain fungi which produce aflatoxins. *Aspergillus flavus* occurs rather commonly in peanuts and in peanut meal. The presence of these aflatoxins in diets of pigs causes loss of appetite and reduced growth rate (Barber *et al.* 1968), the severity of reduction depending on the level of aflatoxins present. Field outbreaks of peanut meal poisoning were described in 1961. The toxic factors of these peanut meals are aflatoxins. Aflatoxin B_1 is the component mainly found. In addition to producing anorexia and reduced rates of growth, aflatoxin B_1 also tends to increase serum alkaline phosphatase levels and to reduce the liver vitamin A concentration.

Differences in response to peanut meal reported by different authors do not seem to be related to the variety of peanut used to produce the meal, since it has been reported that only limited differences in protein quality exist between varieties. It appears that many of the differences in response may be due to processing or to the presence of aflatoxins in the meal.

Before the significant increase in demand for vegetable oils, raw peanuts were commonly used as a feedstuff for swine and their potential as a livestock feed has been long recognized. The occasional availability of full-fat peanuts, unsuitable for crushing or human consumption, favors their use in swine diets. Peanuts contain 31.2% crude protein and 44.8% oil. Kornegay *et al.* (1975) have indicated that although differences in lysine exist between varieties, differences in total crude protein and oil are greater between grades. Conkerton *et al.* (1978) found that certain peanut genotypes have more methionine than other similar genotypes. These reports indicate that peanut geneticists may be able to produce cultivars with high levels of lysine and methionine and a more favorable balance of amino acids for swine. Although raw peanuts have been shown to contain a trypsin inhibitor, Balogum and Koch (1979) indicated that heating at 120°C for 1 hr was of no benefit to the pig. Twenty percent raw or cooked peanuts replaced an equal quantity of maize and soybean meal in the diet without modifying pig performance. If higher dietary levels of peanuts had been used, thus increasing the total trypsin inhibitor content of the diet, proper heating might have proven beneficial.

Although soft fat has been observed commonly when full-fat peanuts constitute a significant portion of growing-finishing swine diets, local quality requirements and relative cost of peanuts, cereal grains, soybean meal, and other feed ingredients will determine the use of full-fat peanuts in swine diets.

RAPESEED MEAL (CANOLA MEAL)

Rapeseed meal is a by-product of the production of vegetable oil from rapeseed (also called canola rape, colza, or raps). In the Cruciferae or mustard family, many species of the genus *Brassica* are known; however, two of these species *B. napus* L. and *B. campestris* L. are most commonly grown. Because of its adaptation to the temperate and subtropical zones, rapeseed is produced as a winter or cool season crop in Europe, Asia, and the Western Hemisphere. Its production in many of these areas is of great importance since rapeseed and sunflower are the only edible oil crops that can be produced effectively in the colder areas of these regions. Detailed information on rapeseed meal for livestock has been reviewed elsewhere (Downey 1965; Bell and Belzile 1965; Bowland 1965; Wetter 1965; Josefsson 1972; Bowland 1975; Fenwick and Curtis 1980; Clandinin 1981).

Eighty percent of the world's annual production of some 10.6 million metric tons (FAO 1980) is produced in Asia and Europe. Within the Western Hemisphere, it is an important crop only in Canada and Chile.

Rapeseed contain about 40–50% oil by weight, which may be removed by expeller, solvent extraction, or a combination of both techniques. The oil has a history of industrial use as a special lubricant, but until recently it contained a high level of glycerides derived from erucic acid. This is nutritionally undesirable, but plant breeders have been successful in developing low glucosinolate rapeseed cultivars. Seed, oil, and meal derived from these cultivars have come to be known as canola seed, canola oil, and canola meal to distinguish them from high glucosinolate cultivars grown extensively in other parts of the world (Clandinin 1981).

Meals produced after removal of the oil may contain 32–44% crude protein with the majority containing 35–37% protein. Average analysis values of 40.5, 1.1, and 0.3% for crude protein, fat, and fiber can be expected from a typical solvent-extracted meal. The amino acid pattern (Table 15.17) of the rapeseed protein is similar to that of other vegetable protein meals, with rapeseed meal containing more methionine and less lysine than soybean meal. However, the

TABLE 15.17. Amino Acid Composition of 'Tower' and 'Candle' Rapeseed Meal and Soybean Meal

Amino acid	*Soybean meal, %*	*'Tower' meal, %*	*'Candle' meal, %*
Aspartic acid	4.44	2.65	2.91
Threonine	1.78	1.51	1.77
Serine	2.39	1.60	1.81
Glutamic acid	6.20	5.11	4.99
Proline	2.52	2.13	2.31
Glycine	1.78	1.67	1.76
Alanine	1.82	1.57	1.65
Cystine	0.74	0.51	0.64
Valine	1.74	1.53	1.62
Methionine	0.83	0.67	0.94
Methionine + cystine	1.57	1.18	1.58
Isoleucine	1.74	1.25	1.32
Leucine	3.15	2.39	2.54
Tyrosine	1.64	1.03	1.16
Phenylalanine	2.27	1.45	1.55
Lysine	2.74	1.90	2.13
Histidine	1.17	1.00	0.99
Tryptophan	0.60	0.48	0.50
Arginine	3.36	2.16	2.22
Crude protein	49.7	43.2	41.4

Source: Bell et al. 1981.

processing method can greatly influence the availability of these amino acids and thus the nutritive value of the meal. The high heat of processing by the expeller method probably is associated with poor amino acid availability; however, a combination expeller–solvent extraction, which is conducted at a lower temperature, produces a higher quality meal.

Although the amino acid composition and availability of quality rapeseed meal would indicate that it is an excellent protein supplement for swine, it has been recognized for some years that there are limits to the amount of the older varieties which can be added to swine diets without serious effects on growth and reproduction.

For young weanling or pre-weanling pigs from 3 weeks of age, levels of rapeseed meal of not more than 4–5% of the total diet should be fed. Higher levels of the meal may reduce palatability and consumption and therefore have adverse effects on the growth rate of the pig. When growth depression occurs on higher levels of rapeseed meal, it does not affect feed utilization and cannot be corrected by lysine supplementation, but usually is associated with reduced feed intake.

Similar, though less severe, adverse effects of high levels of rapeseed meal have been observed with growing-finishing pigs (Devilat 1965). Replacement of 25% of the protein supplement of the diet with rapeseed meal does not significantly affect feed intake, rate of gain, or feed efficiency, but when higher levels are utilized both rate of gain and feed efficiency are depressed.

As the result of many investigations which have been carried out on problems associated with the consumption of rapeseed meal, it is clear that the detrimental effects of high levels of conventional rapeseed meal are due to the presence of up to 8% glucosinolates in the meal depending on variety. Glucosinolate breakdown products, isothiocyanates and oxazalidinethione, have been shown to exert a goitrogenic effect on nonruminant animals, causing histological changes of the thyroid. Lower levels of these toxic substances consumed by pigs in diets containing rapeseed meal cause limited thyroid hypertrophy; higher levels that are commonly supplied by 10–12% meal cause a marked hypertrophy and an increase in cellular components.

In recent years, perhaps the greatest factor in the increased production and use of rapeseed meal for animal feed has been the development of low (or *zero*) glucosinolate cultivars (Clandinin and Robblee 1978). Two cultivars commonly grown today, 'Tower' (in Canada) and 'Erglu' (in Europe), and other recently developed varieties, are so-called *double zero* varieties, possessing low levels of both erucic acid and glucosinolates. To distinguish rapeseed meal derived from low erucic acid, low-glucosinolate cultivars of *B. campestris* or *B. napus* from conventional rapeseed meal, the term *canola meal* has been created.

One of the main factors that tends to limit the feeding value of rapeseed meal for pigs is its relatively low digestible energy (DE) or metabolizable energy (ME) values. Rapeseed meal has lower DE and ME values than soybean meal and the high fiber content of rapeseed meal has been suggested as an important reason for the lower energy values. Plant breeders have successfully developed varieties with thinner, less fibrous seed coats. The introduction of yellow seed varieties, such as 'Candle,' achieve both a 4% reduction of fiber content and a zero level of erucic acid and glucosinolates; such seeds have been termed *triple zero* varieties. To distinguish these new varieties in subsequent discussions in this chapter, the term canola meal will be used for double zero rapeseed meal and low-fiber canola meal for double zero varieties with low fiber content ('Candle').

All swine experiments that have compared canola repeseed meal ('Tower') with regular rapeseed meal have demonstrated the superiority of the double zero meal as a protein source for pigs (McCuaing and Bell 1976; Moody *et al.* 1976; Castell 1977; McKinnon and Bowland 1977; Bowland 1975A,B,C). Canola meals are finding increased acceptance as protein sources in swine diets and are being used as a replacement for soybean meal. Some studies have shown that 'Tower' canola meal may totally (Bowland 1975) or partially (Omole and Bowland 1974; Grandhi *et al.* 1976A,B, 1977; McKinnon and Bowland 1977; Narendran *et al.* 1981) replace soybean meal in starting, growing, and finishing pig diets without adverse effects on performance, while other studies have shown that complete (McKinnon and Bowland 1977; Aherne and Lewis 1978; Singam and Lawrence 1979) or partial substitution (Kennelly *et al.* 1978) of soybean meal with canola meal ('Tower' and 'Erglu') adversely affects pig performance.

Although most studies (Bowland 1975A,B; McKinnon and Bowland 1977; Aherne and Lewis 1978) have not shown any adverse effects of partial or complete replacement of soybean meal by canola meal during the finishing period, complete or partial replacement of soybean meal with canola meal during the starting or growing period has resulted in depressed pig performance (Castell 1977; Moody *et al.* 1976; Grandhi *et al.* 1976A,B; McKinnon and Bowland 1977; Aherne and Lewis 1978; Singam and Lawrence 1979). Castell (1977) added 7.5%

'Tower' canola meal to the diets of pigs in the starter period, and 12.5% during the growing period, resulting in a significant decrease in growth rate and feed conversion efficiency. Moody *et al.* (1976) and Grandhi *et al.* (1976A,B) also observed a reduction in performance of growing pigs fed on diets in which some or all of the soybean meal in the diet was replaced by 'Tower' canola meal, but the differences were not significant. In studies by McKinnon and Bowland (1977), complete substitution of soybean meal with 'Tower' canola meal resulted in reduced feed intake during the starting phase as well as significantly lower daily gain and feed conversion efficiency during the growing phase. Aherne and Lewis (1978) not only showed a reduced pig performance during the growing phase when 'Tower' canola meal completely replaced soybean meal, but also demonstrated reduced feed conversion efficiency when only 50% of the soybean meal was replaced by canola meal (Table 15.18). Although Singam and Lawrence (1979) were unable to show any difference between 'Tower' and 'Erglu' canola meal, both were shown to be inferior to soybean meal when used as the only supplemental protein source to barley-based diets.

Although studies of the amino acid levels and availability (Cho and Bagley 1972) of rapeseed meal suggest that lysine is potentially first limiting, the addition of lysine to canola meal-supplemented diets has been shown to have no influence on feed intake, daily gain, or feed efficiency. Moody *et al.* (1976) added 0.125 or 0.250% lysine to diets based on maize and canola meal for young pigs and

TABLE 15.18. Performance of Growing and Finishing Pigs Fed Diets Containing Soybean Meal (SBM) or Two Levels of 'Tower' Canola Meal (CM)

Source/Criteria	*Diets*		
	SBM	SMB/CM	CM
McKinnon and Bowland 1977			
Growing period (20–43 kg)			
Av daily feed, kg	1.67	1.68	1.63
Av daily gain, kg	0.64[a]	0.63[a]	0.55[b]
Feed/gain ratio	2.63[c]	2.67[c]	2.96[b]
Finishing period (43–85 kg)			
Av daily feed, kg	2.36	2.25	2.24
Av daily gain, kg	0.62	0.58	0.58
Feed/gain ratio	3.80	3.87	3.87
Aherne and Lewis 1978			
Growing period (20–60 kg)			
Av daily feed, kg	2.05	2.15	2.19
Av daily gain, kg	0.73[a]	0.71[a,b]	0.67[b]
Feed/gain ratio	2.80	3.05[b]	3.27[b]
Finishing period (60–100 kg)			
Av daily feed, kg	3.33	3.28	3.26
Av daily gain, kg	0.79	0.76	0.75
Feed/gain ratio	4.17	4.32	4.33

[a,b,c]Numbers in same line with different superscripts are significantly different.

observed no significant improvement in pig performance. Similar studies with growing and finishing pigs reported by Aherne and Lewis (1978) also failed to show any effect of the addition of 0.065% lysine HCl (O.05% lysine) to a barley–wheat based diet supplemented with canola meal. Experimental evidence available to date would suggest that a response is more likely to be obtained from supplemental methionine than from lysine (Bell 1975).

It is possible that the reduced performance observed when canola meal replaces soybean meal may be a reflection of its glucosinolate level. Although pigs fed low-glucosinolate cultivars (canola meal) have shown less thyroid dysfunction than those fed regular cultivars with high-glucosinolate as demonstrated by tetraiodothyronine ('T-4') levels in young pigs (Bowland 1975A), these and other studies (Bowland 1975A; Grandhi *et al.* 1976; Aherne and Lewis 1978; Ochetim *et al.* 1980A,B) have shown histological changes of the thyroid and lowered 'T-3' and 'T-4' levels when pigs were fed canola meal as a complete or partial replacement for soybean meal. These results suggest that although impaired thyroid function is less severe than with conventional, high-glucosinolate rapeseed meal, levels of glucosinolate (approximately 0.98 mg/g) in 'Tower' canola meal were sufficiently high to cause some impairment in thyroid hormone synthesis and thus in pig performance.

Canola meal ('Tower') contains a substantially higher crude fiber level than soybean meal (13 vs. 6%) and this has been suggested as contributing to the lower performance observed when canola meal is fed as the sole protein supplement in diets of young pigs (Castell 1977; Aherne *et al.* 1977A,B). The limitations imposed by the high fiber levels of rapeseed meal led to the development by plant breeders of a yellow-coated cultivar ('Candle') that is low in glucosinolate and also contains about 4% less fiber. Although Bell *et al.* (1981) reported a DE value of 3370 kcal/kg that is slightly greater than those obtained by May and Bell (1971) and Saben *et al.* (1971) of 3280 and 3210 kcal/kg, respectively, for conventional rapeseed meal. Muztar *et al.* (1978) have shown with chickens that 'Tower' and 'Candle' rapeseed meals do not differ significantly in apparent or true ME. Feeding trials with pigs have also failed to demonstrate any improvement in pig performance when canola meals prepared from 'Candle' and 'Tower' varieties are compared (Bell *et al.* 1981; Kennelly *et al.* 1978).

In studies to measure the effect of steam pelleting of the diet on the performance of growing-finishing pigs fed 'Tower' and 'Candle' rapeseed meal, or soybean meal, Narendran *et al.* (1979) reported that the performance of pigs fed soybean meal was superior to that of pigs fed either 15% 'Tower' or 15% 'Candle' rapeseed meal. Pigs fed 'Tower' rapeseed meal had a lower average daily gain and back fat thickness and pigs fed Candle rapeseed meal had a lower average daily gain and feed intake and higher feed-to-gain ratio compared to pigs fed soybean meal. Steam pelleting of the maize–soybean meal and maize–soybean meal–rapeseed meal diets improved average daily gain, feed intake, and feed-to-gain ratio of the pigs, with the effects being greatest for the rapeseed meal diets. A similar improvement due to steam pelleting was also reported by Grandhi *et al.* (1979).

Several Canadian and Chilean experiments with gilts and sows fed high glucosinolate (10–11 mg/g) rapeseed meal have demonstrated that conception and litter size may be reduced when rapeseed meal constitutes more than 6–8% of the diet of pregnant gilts and sows (Manns and Bowland 1963; Bell and Belzile 1965; Schuld and Bowland 1968; Esnaola and Ochoa 1970; Devilat and Skoknic 1971; Bowland and Hardin 1973). Gilts fed high levels (10–12%) of high glucosinolate rapeseed meal have been shown to require more estrous cycles for conception, to farrow lighter and smaller litters, and to wean fewer and lighter pigs than similar animals fed diets not containing this meal. These reports indicated that not only is maturation of the ovaries delayed but also lactation is depressed when high glucosinolate rapeseed meal supplies a major portion of the diet (Manns and Bowland 1963; Esnaola and Ochoa 1970). Devilat and Skoknic (1971) showed very severe effects on the number of pigs born alive and their survival when feeding 12% rapeseed meal and a marginal iodine level to gilts during gestation and lactation. A report by Saben and Bowland (1971) using rapeseed meal with a somewhat lower glucosinolate level (5 mg/g) suggested that a dietary rapeseed meal level of 8% may be satisfactory, but there was still a trend toward smaller litters for gilts and sows fed diets containing this level of rapeseed meal. Later studies (Bowland and Hardin 1973) indicated that up to 6% rapeseed meal with no more than 5 mg/g of glucosinolate can be fed throughout three reproductive cycles without any reduction in sow performance. As a consequence of the detrimental effects observed in some experiments with high glucosinolate rapeseed meal, it is recommended that not more than 3% conventional rapeseed meal should be used in diets of breeding swine during pregestation, gestation, and lactation.

Recently (Flipot and Dufour 1977; Lewis *et al.* 1978), studies with rapeseed meal from low glucosinolate cultivars of rapeseed have indicated that 'Tower' rapeseed meal (containing 0.98 mg/g glucosinolates) may be used as a partial or complete replacement for soybean meal in the diets of pregnant and lactating swine for at least two reproductive cycles with no apparent reduction in sow reproductive performance. Flipot and Dufour (1977) studied the reproductive traits of 64 gilts fed diets containing 0 or 10% canola meal ('Tower' rapeseed meal) when they were diagnosed as pregnant. No significant differences in litter size and weight at birth and at weaning were reported between gilts fed 0 to 10% canola meal. The first postpartum estrus observed after weaning at 28 days and the ovulation rates at first estrus were similar for the two groups. Thyroid weight was not affected by the treatment.

Lewis *et al.* (1978) fed low glucosinolate rapeseed meal (derived from the cultivar 'Tower') to replace 50 or 100% of the soybean meal in diets for 72 gilts from 20-kg liveweight through two gestation and lactation periods. The gilts were fed diets containing 17% protein from 20 to 60 kg and diets containing 15% protein from 60 kg until the end of the second lactation. All of the gilts successfully completed one reproductive cycle and 51 animals completed two. Treatments had no significant effect on litter size or weight at birth or weaning, sow weight changes in gestation or lactation, or gestation length in either the first or second parity.

Based on these studies, it appears that rapeseed meal produced from the new cultivars with low glucosinolate content (less than 1 mg/g), such as 'Tower,' 'Erglu,' and 'Candle,' can be used to supply all of the supplemental protein for gilts and sows during pregestation, gestation, and lactation without any detrimental effects on reproductive performance.

RUBBER SEED AND RUBBER SEED MEAL

The rubber tree (*Hevea brasiliensis*) produces, in addition to latex for rubber, an average of 800–1200 kg of seed per hectare per year (Siriwardene and Nugara 1972). The seeds contain approximately 50% oil and 20% protein (Table 15.19). Rubber seeds also contain a cyanogenic glucoside, linamarin, which on decomposition produces hydrocyanic acid, a substance that can cause acute poisoning in pigs and other livestock. Fresh seeds have been shown to produce 2000 mg hydrocyanic acid per kilogram of seed. However, the hydrocyanic acid content diminishes rapidly during the first week of storage and also can be reduced by some physical and chemical treatments.

Although Fetuga (1977) indicated that neither autoclaving, dilute acid, nor dilute sodium hydroxide treatment were effective in improving the nutritive value of full-fat or defatted rubber seed fed to rats, Stosic and Kaykay (1981) reported good performance of growing pigs fed diets containing 40% ground rubber seed that had either been roasted at 350°C for 15 min or soaked in an ash solution for 12 hr. The addition of treated rubber seed to the commercial diet improved both daily gains and efficiency of feed conversion (Table 15.20). Palatability of roasted rubber seed was significantly better than that of ash-treated seed.

Rubber seed meal was used by Rajaguru and Ravindran (1979) to replace 10, 20, or 30% of the coconut meal in 16% protein growing pig diets based on wheat, rice polish, meat meal, and coconut meal. Each increase in the level of rubber seed meal depressed pig performance. Replacement levels above 10% significantly reduced both gains and efficiency of feed utilization as compared to those obtained with the control diet. Since Fetuga *et al.* (1977) reported that lysine and

TABLE 15.19. Chemical Composition of Rubber Seed and Rubber Seed Meal

Constituent	*Decorticated rubber seed*		*Rubber seed meal, Expeller 1[c]*
	Full-fat 1[a]	*Defatted 2[a,b]*	
Dry matter, %	86.5	96.1	91.0
Crude protein, %	19.5	18.3	29.4
Crude fiber, %	3.3	3.8	6.6
Ether extract, %	42.8	43.4	14.7
Ash, %	3.0	3.1	5.1
Nitrogen-free extract, %	17.9	27.5	35.3
Calcium, %	0.48	—	—
Phosphorus, %	0.64	—	—
Potassium, %	0.96	—	—
Sodium, %	0.09	—	—
Magnesium, %	0.28	—	—
Chloride, %	0.07	—	—
Iron, mg/kg	92.7	—	—
Manganese, mg/kg	23.2	—	—
Zinc, mg/kg	78.5	—	—
Copper, mg/kg	25.5	—	—

Sources: [a]Fetuga et al. 1977, [b]Orok and Bowland 1981, [c]Stosic and KayKay 1981.

TABLE 15.20. Use of Ash-Treated or Heat-Treated, Full-Fat Rubber Seed in Diets for Growing Swine

Treatment	*Av daily gain, kg*	*Av daily feed, kg*	*Feed/gain ratio*
Control	406	1.42	3.50
40% ash-treated rubber seed	464	1.42	3.06
40% heat-treated rubber seed	475	1.42	2.90

Source: Stosic and KayKay 1981.

methionine were most limiting in rubber seed protein (Table 15.21), it appears that this depressing effect of rubber seed meal was due to an amino acid imbalance related to a dietary lysine and methionine deficiency instead of hydrocyanic acid toxicity, as the rubber seed meal was analyzed to contain only 0.02% hydrocyanic acid. In studies reported by Ong and Yeong (1978), rubber seed meal had no significant effect on rate of gain when levels up to 20% were incorporated into the growing-finishing diet. Six isonitrogenous diets containing 0, 5, 10, 15, 20, and 25% rubber seed meal were fed to growing-finishing (growing, 17%; finishing, 15%) pigs. Twenty-five percent rubber seed meal significantly depressed both gains and efficiency of feed utilization but was without effect on daily feed intake. Although graded levels up to 20% did not affect daily gains, each 5% increase in rubber seed meal caused a small but linear increase in feed required per unit of gain (Table 15.22).

Because of the low level and/or availability of lysine and methionine and the presence of cyanogenic glucosides in rubber seed, extreme care should be taken in its use in swine diets. Rubber seed meal low in hydrocyanic acid and fed in diets

TABLE 15.21. Amino Acid Content of Rubber Seed and Rubber Seed Meal (g/16 g N)

Amino acid	*Full-fat rubber seed*			*Rubber seed meal*
	1[a]	*2*[b]	*3*[c]	*4*[c]
Arginine	9.4	10.2	8.6	9.4
Histidine	2.3	2.0	1.6	1.8
Isoleucine	3.8	3.4	2.4	3.2
Leucine	7.1	6.2	4.9	6.1
Lysine	3.6	3.3	3.0	3.0
Methionine	1.4	1.1	2.0	1.1
Cystine	—	1.4	1.1	1.3
Phenylalanine	4.8	4.9	3.0	3.9
Tyrosine	2.6	2.7	1.9	2.7
Threonine	3.8	3.2	2.7	3.1
Tryptophan	—	1.4	—	—
Valine	8.0	6.0	5.2	6.6

Sources: [a]Orok and Bowland 1981. [b]Fetuga et al. 1977. [c]Stosic and KayKay 1981.

TABLE 15.22. Rubber Seed Meal as a Protein Source for Growing-Finishing Swine

Item	Level of rubber seed meal, %					
	0	5	10	15	20	25
Av daily gain, kg	0.50	0.50	0.49	0.48	0.47	0.41
Av daily feed, kg	1.73	1.83	1.80	1.80	1.82	1.76
Feed/gain ratio	3.46	3.66	3.67	3.75	3.87	4.29

Source: Ong and Yeong 1978.

properly balanced in amino acids can be used to supply as much as 20% of the total diet for growing-finishing swine.

SAFFLOWER MEAL

Safflower (*Carthamus tinctorius*) has long been cultivated as an oilseed crop. Although the seeds contain 36–40% oil, the hull constitutes about 40% of the seed. After the oil is removed, the product that remains contains only 18–22% crude protein and 40% crude fiber. The hulls are about 70% holocellulose, 21% lignin, and 1% ash and greatly limit the energy concentration and utilization of the meal by swine and other classes of livestock. The protein of safflower oil meal is a very poor source of lysine and somewhat deficient in the sulfur-containing amino acids, methionine, and cystine (Kratzer and Williams 1951) (Table 15.23).

Definitive research to determine the value of undecorticated safflower meal in swine diets is limited; however, Mexican workers have conducted trials that help to assess its value. Shimada and Aguilera (1966) used an 18.8% protein safflower meal, containing 30.9% crude fiber, to substitute for soybean meal in sorghum-based diets for growing pigs (17.4–56.5 kg). Safflower meal supplied 0, 12.5, or 25% of the supplemental protein, replacing soybean meal and sorghum. At the 12.5% level of substitution, no changes in pig performance were observed; however, at the 25% level both gains and efficiency of feed utilization were inferior to those obtained with either of the other two diets.

In subsequent studies, Mexican workers (Shimada and Brambila 1966) employed a 19.8% protein undecorticated safflower meal containing 40.1% crude fiber. In diets containing fish meal (25% of the supplemental protein) and cottonseed meal (50% of the supplemental protein), the addition of safflower meal (25% of the supplemental protein) reduced gains and increased feed per unit gain in comparison with the results produced when soybean meal supplied 25% of the supplemental protein. The authors suggest a maximum level of 7–8% safflower meal in the diet of growing pigs to supply not more than 12.5% of the supplemental protein.

Bravo and Ceballo (1969) working at the same station with finishing pigs, used dietary levels of 15, 20, and 25% undecorticated safflower meal in grain–molasses basal diets. Although gains were not greatly different among treatments, overall performance of all groups was poor, and feed required per kilogram of

TABLE 15.23. Amino Acid Composition of Safflower Seed Protein

Amino acid	*% of protein*
Arginine	9.3
Histidine	2.4
Isoleucine	4.0
Leucine	6.2
Lysine	3.1
Methionine + cystine	3.3
Methionine	1.7
Phenylalanine	4.4
Tyrosine	3.1
Threonine	3.3
Tryptophan	1.6
Valine	5.7

Source: Kohler et al. 1966.

gain increased with each increase of safflower meal. The authors suggested that not more than 10–12% of the diet for finishing pigs should be supplied by undecorticated safflower meal.

Because of its low protein, low lysine, and high fiber, undecorticated safflower meal is of limited value in growing-finishing swine diets. It would appear that it could best be utilized in rations for gestating sows.

Recently, attempts have been made to improve the quality of safflower meal by the physical separation of the hulls from kernels and by the development of thin-hull varieties through plant breeding. Although the complete removal of the hull results in a seed with 60–70% oil and a seed cake of about 60% protein, in practice it is very difficult to effect a complete separation because of the hardness of the seed coat and the extreme softness of the kernel. Screening of undecorticated safflower meal to partially remove the hulls produces a meal with more crude protein (42–46%) and less crude fiber (15–16%) (Williams and Daniels 1973).

In studies by Williams and Daniels (1973) and Williams and O'Rourke (1974), decorticated safflower meal containing 46.1% crude protein and 16.2% crude fiber was found to be unsatisfactory as the sole supplementary protein source for sorghum and wheat-based diets for pigs. The growth response to the safflower meal was increased by the addition of synthetic lysine but the performance of the pigs receiving the lysine-enriched safflower meal supplements was always inferior to that of pigs fed either isonitrogenous soybean meal or fish meal supplements in both sorghum and wheat-based diets. Increasing the level of safflower meal at the expense of fish meal resulted in slower rates of growth and reduced efficiency of feed utilization, particularly when fed to pigs less than 44-kg liveweight. Although safflower has been reported to contain a bitter compound (Kohler *et al.* 1966) and palatability problems have been reported (Williams and Daniels 1973), Williams and O'Rourke (1974) fed barrows and gilts diets containing up to 14.5% decorticated safflower meal without observing any limiting effect on feed intake.

Decorticated safflower meal can supply a limited portion of the supplementary protein in growing-finishing swine diets (5–10%), but when used should always be combined with other "lysine-rich" protein concentrates.

SESAME MEAL

Sesame, known botanically as *Sesamum indicum* (also called benne, til, gingerly, simsin, gergelin, and in Latin America ajonjoli), is probably one of the oldest of the cultivated oilseed crops (Cornelius and Raymond 1967). Annual world production of sesame amounts to 1.93 million tons (FAO 1980) with China and India, the chief producers, providing 41% of the total. Sesame also grows in many other countries in the Far and Near East and on the American continents. Sudan is the leading African producer and Mexico the leading American producer. Most of the countries in which sesame is grown consume the bulk of the crop domestically, so that international trade in seed and oil is small in comparison with the quantities produced.

The seeds, which vary from white through black in color, according to variety, are small, weighing usually 2–3.5 g/1000 seeds and are contained in capsules which tend to split open and shed when ripe (Johnson and Raymond 1965). Because of this tendency to "pop open," practically all the harvesting is carried out by hand. Nonshattering lines and varieties must be evolved if mechanical harvesting is to be successfully employed to allow the development of sesame as a competitive crop in areas of high labor cost.

The seed has a thin shell which can be separated from the kernel in decorticating machines or by manually soaking and rubbing. Normally however, the seed is milled without decortication to remove the 44–54% oil present in the air-dried seed. The sesame meal produced from the whole seed contains both the kernel and the thin shell.

The chemical analysis of sesame meal varies according to the processing method employed for extracting the oil (Table 15.24). Meals prepared by the screw press or hydraulic methods contain higher levels of fat and lower levels of protein than those produced by solvent extraction. Chemically, sesame meals contain 90–95% dry matter, 38–48% crude protein, 0.7–11% fat, 5–7% fiber, 9–13%

TABLE 15.24. Effect of Extraction Method on Chemical Composition of Sesame Meal (%)

Components	*Hydraulic press*[a]	*Solvent*[b]
Moisture	7.8	3.8
Fat	10.8	0.7
Crude protein	38.7	45.1
Crude fiber	6.4	5.2
Ash	9.7	13.2
Nitrogen-free extract	26.9	32.0

Source: Zaghi and Bressani 1969.
[a]Average of 5 samples. [b]Average of 2 samples.

TABLE 15.25. Amino Acid Composition of Sesame Meal (%)

Components	*1*[a]	*2*[b]	*3*[c]	*4*[d]
Protein	40.00	42.0	51.9–67.3	—
Arginine	11.91	10.95	—	—
Histidine	2.21	—	—	—
Isoleucine	4.27	—	2.61–3.52	4.21
Leucine	6.92	—	4.45–5.55	6.52
Lysine	2.76	2.85	2.71–2.83	2.71
Methionine	2.65	3.52	1.85–2.33	2.34
Phenylalanine	4.73	—	2.32–3.25	4.60
Threonine	3.64	—	2.22–3.49	3.64
Tryptophan	1.91	1.86	—	—
Valine	5.06	—	1.66–2.53	4.80

[a]Catron and Hays 1958. [b]Hubbell 1968.
[c]Villegas et al. 1968. [d]Average values in literature.

ash and 25–30% nitrogen-free extract. As the amino acid profile (Table 15.25) clearly indicates, sesame meal is a poor source of lysine but an excellent source of both methionine and tryptophan.

When used in swine diets, both the deficiencies and strengths of sesame meal should be taken into consideration. As a protein supplement for diets based on cereal grains it should not be used to supply all of the supplemental protein required but should be blended with other supplemental protein sources such as soybean meal and fish meal, which are good sources of lysine (Maner and Gallo 1963). Sesame meal can replace one-half or all of the soybean meal in a maize–soybean diet containing 4% fish meal, and it can be substituted for soybean meal in quantities as high as 10% of the diet in maize and soybean meal diets (Maner and Gallo 1963). Because of the low level of lysine in both cottonseed meal and sesame meal, blends of these two supplements are inadequate for supplementing cereal-based growing-finishing diets; however, 50–50 blends of these two protein supplements, fed with cereal grains and 5–6% fish meal, produce satisfactory gains and feed utilization (Hervas *et al.* 1965).

SUNFLOWER SEED AND MEAL

In many areas of the world, sunflower seed (*Helianthus annuus*) is the best adapted high-protein crop available. A big advantage of sunflowers is their ability to adapt to a variety of climates and soil compositions, from the heat of Southern United States to the frost of Central Asia. More than 13.1 million metric tons of the seed are produced each year (FAO 1980). The U.S.S.R. and the other centrally-planned countries produced more than 6.96 million metric tons or some 53% of the total world crop. Europe produces 2.62 million metric tons and South America 1.76 million metric tons. In the United States, hectarage has been increased from 2865 hectares in 1968 to more than 1.8 million hectares in 1980, mostly in the northern great plains region. Production during this time has in-

TABLE 15.26. Average Composition of Sunflower Meal

Components	*Type of processing*	
	Expeller	*Solvent*
Moisture, %	7.0	7.0
Crude protein, %	41.0	46.8
Crude fiber, %	13.0	11.0
Ether extract, %	7.6	2.9
Ash, %	6.8	7.7
Calcium, %	0.43	0.43
Phosphorus, %	1.08	1.08
Magnesium, %	1.00	1.00
Potassium, %	1.08	1.08
Manganese, parts per million	13	13

creased from 110 thousand tons to 1.73 million metric tons. There are two basic types of sunflower: oilseed and confectionary seed. In recent years, production of high-oil varieties of sunflowers has increased substantially.

Sunflower meal is produced from sunflower seed following oil extraction. The composition of the meal will vary according to the composition of the seed and, as with other oil-seed proteins, the nutritive value of the sunflower meal varies with the method of processing (Clandinin 1958; Smith 1968). Mitchell *et al.* (1945) found that sunflower protein produced by solvent extraction of a low-oil variety had a true digestibility of 94.3% and a biological value of 64.5%. There has been a marked improvement in meal quality during recent years due to advances in oil extraction and meal processing that has markedly lowered the crude fiber and ether extract content of the finished product. The average composition of both expeller and solvent-extracted meals is presented in Table 15.26, and an amino acid profile is given in Table 15.27. Expeller-processed meals contain more fat

TABLE 15.27. Amino Acid Content of Sunflower Meal (g/16 g N)

Amino acid	*Expeller meal*		*Solvent-extracted meal*
	Low	*High*	
Arginine	9.4	8.7	8.2
Histidine	2.1	2.1	1.7
Isoleucine	4.0	3.9	5.2
Leucine	6.1	5.9	6.2
Lysine	3.3	2.8	3.8
Methionine	1.6	1.5	3.4
Phenylalanine	4.2	4.3	5.7
Threonine	3.2	3.2	4.0
Tryptophan	1.0	1.0	1.3
Valine	4.8	4.9	5.2

and fiber and lower quantities of crude protein than do meals produced by solvent extraction. Solvent extraction and low-temperature processing improve the quality of the protein by reducing the destruction and/or loss of lysine, methionine, threonine, and tryptophan (Alexander and Hill 1952; Renner *et al.* 1953; Bandemer and Evans 1963; Amos *et al.* 1975). High-temperature processing and especially dry heating causes marked reduction in lysine content and availability. Other amino acids such as arginine and tryptophan are also reduced by high-temperature processing. Russian work (Tkacev *et al.* 1964, 1965) indicated that when meals were processed at three different temperatures (125°–130°C, severe; 115°–125°C, normal; and 105°–115°C, mild), the mild heat-treated meals produced the highest average daily gain, lowest feed per unit gain, and the highest digestibility of organic matter, ether extract, and nitrogen-free extract when fed to swine. Amos *et al.* (1975) heated sunflower seed at 0°, 75°, 100°, 115°, or 127°C for 1 hr before oil extraction and concluded from rat growth studies that any benefit obtained from heating the seed at 100°C for 1 hr was lost in seed heated to 127°C for 1 hr due either to increased amino acid destruction or the degree of availability.

Although new varieties of sunflower seed with lower percentages of hulls are now available, the energy content and protein concentration of sunflower meal vary with the quantity of hulls present. The meals with a high-protein content are produced by removing (decortication) the largest quantity of hulls possible before final processing by screening the meal after extraction. Sunflower meal used to supply all of the supplemental protein in growing-finishing swine diets is a very unsatisfactory source of protein because of its very low lysine content. It produces gains and efficiency of feed utilization inferior to those produced by soybean meal, fish meal, peanut meal, or combinations of these ingredients (Pearson *et al.* 1954; Delic *et al.* 1963; Delic *et al.* 1964; Wenk and Landis 1979). Diggs and Baker (1972) reported that replacement of one-half of the soybean meal in growing pig diets with sunflower meal did not adversely affect rate of gain or feed efficiency. Complete replacement of soybean meal with sunflower meal reduced daily gain and increased the feed required per unit of gain. Seerley *et al.* (1974) used expeller-processed sunflower meal (heated to 120°C) to replace 0, 25, 50, or 100% of the soybean meal in a 16% maize–soybean meal diet. Although there were no differences in apparent digestibility of the maize–soybean meal and maize–sunflower meal diets with respect to energy, protein, fiber, ether extract, or dry matter, daily gains of growing-finishing pigs were significantly depressed when the sunflower meal replaced more than 25% of the soybean meal in the diet (Table 15.28). Replacing 25% of the protein of soybean meal did not reduce gains but feed required per unit gain was increased. The poor pig performance at the higher levels of substitution was due to the low level of lysine (1.5%) in the sunflower meal, which may have resulted in part from overheating (Amos *et al.* 1975); the addition of 0.3% L-lysine monohydrochloride to the 50 and 100% sunflower meal diets resulted in improved gain and feed efficiency (Table 15.29). Baird (1980) used a sunflower meal containing 29.2% crude protein and 21.9% crude fiber to replace 25 or 50% of the soybean meal for growing-finishing (19–84-kg) pigs and reported no differences in either gain or feed efficiency. Lysine supplementation improved gains but not efficiency of feed utilization (Table 15.30).

TABLE 15.28. Effect of Replacing Soybean Meal with Sunflower Meal Produced by Expeller Processing of Decorticated Seed for Growing-Finishing Pigs

Criteria	Treatments 1 (%)	2 (25%)	3 (50%)	4 (100%)
Level of sunflower meal, %	—	5.0	9.8	20.0
Level of soybean meal, %	17.3	13.0	8.7	—
Av daily gain, kg	0.66[a]	0.66[a]	0.44[b]	0.38[c]
Av daily feed, kg	2.20	2.37	1.83	1.68
Feed/gain ratio	3.30	3.61	4.21	4.54

Source: Adopted from Seerley et al. 1974, who used pigs with 24-kg initial weight.

[a,b,c]Means without a common superscript are significantly different from each other.

The most efficient utilization of the meal is obtained by combining it with supplements, such as fish meal or soybean meal, which are rich in available lysine, or by supplementation with crystalline lysine (Siebrits *et al.*, 1980).

The increased production and availability of sunflower seed have, at times, made it economically competitive with maize and soybean meal. Laudert and Allee (1975) reported the use of raw sunflower seed as an energy and protein source for pigs. The seeds which were used in the study contained 19.39% crude protein and 33.96% ether extract and had an apparent digestibility of dry matter, 69; energy, 77; ether extract, 91; and crude protein, 79%. Replacement of soybean meal and maize (10 parts of sunflower meal for 8 parts maize and 2 parts soybean meal) with 0, 20, 40, or 60% ground sunflower seed in diets for growing pigs resulted in a reduced daily gain, a significant decrease in feed intake, and

TABLE 15.29. Effect of Supplemental L-Lysine–HCl in Diets with 50 or 100% of Supplemental Protein Supplied by Sunflower Meal for Finishing Pigs

Criteria	Treatments: 50% sunflower meal		100% sunflower meal	
	No lysine	0.3% lysine	No lysine	0.3% lysine
Av daily gain, kg	0.37[a]	0.77[b]	0.31[a]	0.65[b]
Feed/gain ratio	4.94	3.46	4.85	3.74
Final weight, kg	70.0	86.4	73.2	86.4

Source: Adapted from Seerley et al. 1974, who used pigs with 48-kg initial weight.

[a,b]Means without a common superscript are significantly different from each other.

TABLE 15.30. Effect of Substitution of Soybean Meal with Sunflower Meal or Maize with Sunflower Seed for Growing-Finishing Pigs

Treatments	*Av daily gain, kg*	*Av daily feed, kg*	*Feed/gain ratio*
1. Soybean meal 100%	0.67	1.94	2.86
2. Sunflower meal 25%	0.68	2.02	2.96
3. Sunflower meal 25% + 0.3% lysine	0.74	2.20	2.97
4. Sunflower meal 50%	0.68	2.02	2.99
5. Sunflower meal 50% + 0.3% lysine	0.76	2.28	2.99
6. Sunflower seed 25%	0.70	1.99	2.84
7. Sunflower seed 25% + 0.3% lysine	0.72	1.91	2.66
8. Sunflower seed 50%	0.65	1.74	2.67
9. Sunflower seed 50% + 0.3% lysine	0.77	1.92	2.51

Source: Adapted from Baird (1980), who fed pigs from 19 to 84 kg. Sunflower meal contained 29.2% crude protein and 21.9% crude fiber and sunflower seed contained 20.5% crude protein and 40.9% ether extract.

improvement in the efficiency of feed utilization. Baird (1980) used sunflower seed to replace 25 or 50% of the maize in a maize–soybean meal diet for growing-finishing pigs without significantly affecting growth rate. However, the addition of 0.3% lysine improved both gains and efficiency of feed utilization. Noland *et al.* (1980) substituted unextracted sunflower seeds for either 0, 25, 50, or 100% of the soybean meal in pig diets and reported no significant differences in digestible energy or apparent nitrogen retention. However, all diets containing unextracted sunflower seeds had significantly higher digestible nitrogen than the diets containing soybean meal as a protein source. Sunflower seeds appear to be well utilized by growing-finishing pigs, but are deficient in lysine.

Research on the usefulness of sunflower seed as a feed ingredient in gestation and lactation diets is limited. Besides supplying protein and energy for normal gestation and lactation functions, the high fat content of some sunflower seed (≈40%) lends itself as a possible source of fat in the diets of late gestating and early lactating sows. Recent research has suggested that fat addition to the diet of sows during late gestation might increase piglet survivability by increasing the amount of energy stored in the piglet body at birth, making it less vulnerable to starvation and also by increasing the fat content of colostrum milk to provide an increased energy source to the young pig.

Ground sunflower seed replaced 25 or 50% of the maize and soybean meal in diets fed at a level of 2.26 kg/head/day from day 100 of gestation and *ad libitum* during lactation (Kepler *et al.* 1980). At the higher level of substitution (50%), the sunflower seed completely replaced the soybean meal and about one-half of the maize. Some difficulties were encountered with diet acceptability when 50% ground sunflower seeds were included in the diet. Dietary treatment had no effect on number or weight of pigs farrowed and although percentage milk fat was significantly increased for sows fed diets containing 25 or 50% sunflower seed, no advantage in piglet survival or 14-day weight were observed due to sunflower addition.

GRAIN LEGUMES

The seeds of grain legumes have many advantages and offer a wide range of germ plasm of wide environmental adaptability with which to meet the protein needs of both the human and animal population. Of the 600 genera with more than 13,000 species that make up the family Leguminosae, only some 20 of the seed-producing species are utilized regularly in appreciable quantities (Aykroyd and Doughty 1964). However, with the exception of soybean and peanuts, which contribute significantly to the total supply of feed protein, the majority are low yielding and contribute only in a limited way to the world's total protein supply (Table 15.31) (FAO 1981). Because of low yields, many of these crops cannot compete economically with other higher-yielding crops, such as maize, rice, wheat, barley, and cassava. The amount of agronomic research conducted on most of the grain legumes has been limited. Production levels, therefore, remain low because of the inherent characteristics of the varieties grown, the damage from diseases and insect pests, and poor cultural practices. The majority of the food and grain legumes with the exception of soybeans (*Glycine max*), peanuts (*Arachis hypogaea*), beans (*Phaseolus vulgaris*) and field or garden peas (*Pisum* spp.), which are important in the developed world, are of importance only within the developing countries.

Summaries of the chemical composition (Table 15.32) and amino acid content (Table 15.33) of each are presented.

SOYBEANS

Soybeans (*Glycine max*) are produced in large quantities in both temperate and tropical climates. Annual world production has reached 87.9 million metric tons, with more than 49.5 million metric tons or approximately 59% being produced in the United States where they constitute by far the largest source of protein supplement for animal feeding (FAO 1980). Most soybeans are used for oil production with the by-product, soybean meal, also being produced. However, when supply

TABLE 15.31. World Production of Certain Grain Legumes in 1969–1971 and 1981

Legume seed	*Production (million metric tons)*		*Yield (kg/ha)*	
	1969–1971	*1981*	*1969–1971*	*1981*
Soybeans	43.487	87.941	1487	1751
Dry beans	11.684	14.053	520	567
Dry peas	9.399	8.215	1233	1041
Chick peas	6.508	6.292	655	623
Dry broad beans	4.399	4.178	961	1154
Lentils	0.985	1.128	613	577

Source: FAO 1981.

TABLE 15.32. Chemical Composition of Some Grain Legumes

Legume seed	*Moisture* (%)	*Protein* (%)	*Ether Extract* (%)	*Fiber* (%)	*Ash* (%)	*Nitrogen-free extract* (%)
Forage (chickpeas)[a]	11.8	17.3	3.5	8.4	2.8	56.2
Cowpeas[b]	11.0	23.4	1.8	4.3	3.5	56.0
Dry beans[b]	11.0	23.9	1.3	4.2	3.4	56.2
Field beans[b]	11.0	23.4	2.0	7.8	3.4	52.4
Field peas[c]	10.1	25.9	1.5	6.0	2.6	53.9
Pigeon peas[b]	11.0	20.9	1.7	8.0	3.5	54.9

[a]Shimada and Brambila 1967. [b]Aykrod and Doughty 1964. [c]Kroening 1968.

and demand relationships push the price of soybean meal upward in relation to the price of soybeans, the possibility exists for swine producers to utilize full-fat soybeans instead of soybean meal. This practice can be especially economical for soybean producers who are also swine producers. The presence of several inhibitors of protein utilization in raw soybeans requires that they be heated to destroy these inhibitors before the pig can utilize the soybeans efficiently. This aspect of processing of soybeans is discussed in detail in Chapter 17. Several methods are available for applying heat to soybeans, including steaming or autoclaving, infrared gas heating, and extrusion. The development of on-farm soybean cookers has further stimulated interest in on-farm processing of soybeans for swine and has overcome some of the early problems of full-fat soybean feeding.

The proximate and amino acid composition of whole soybeans is shown in Table 15.34. The unusually favorable amino acid balance of soybean protein, in comparison with most other plant proteins, makes it possible to utilize it as a major source and in some cases the only source of protein in mixed swine diets. Swine, particularly young growning pigs, fed uncooked soybeans gain slowly and

TABLE 15.33. Amino Acid Content of Some Grain Legumes

Amino acid (*% of protein*)	*Forage* (*chickpeas*)[a]	*Cowpeas*[b]	*Dry beans*[b]	*Field beans*[c]	*Field peas*[d]	*Pigeon peas*[b]
Arginine	8.93	—	—	9.60	7.15	—
Histidine	3.07	—	—	2.59	2.62	—
Isoleucine	3.87	3.00	3.60	4.34	3.59	3.80
Leucine	7.46	4.70	5.40	7.44	6.28	4.90
Lysine	4.03	4.10	4.60	6.22	6.52	4.50
Methionine	1.36	1.00	0.60	0.82	0.76	0.70
Cystine	1.04	0.80	0.60	1.28	0.55[e]	0.90
Phenylalanine	5.90	3.30	3.50	4.28	4.35	5.40
Threonine	3.12	2.50	2.70	3.68	4.04	2.40
Tryptophan	0.98	0.60	0.60	0.99	0.86[e]	0.30
Valine	4.24	3.50	3.80	4.74	3.87	3.30

[a]Abu-Shakra et al. 1970. [b]Aykroyd and Doughty 1964. [c]Henry 1970.
[d]Bell and Wilson 1970; Kroening 1968. [e]Kroening 1968. Unpublished data.

TABLE 15.34. Proximate Composition and Amino Acid Content of Full-Fat Soybeans

Proximate analysis		*Amino acid content, %*	
Dry matter, %	90.0	Arginine	2.8
Crude protein, %	37.0	Histidine	0.9
Ether extract, %	18.0	Isoleucine	2.0
Crude fiber, %	5.5	Leucine	2.8
Ash, %	5.0	Lysine	2.4
Nitrogen-free extract, %	24.0	Methionine	0.5
Calcium, %	0.25	Cystine	0.6
Phosphorus, %	0.58	Phenylalanine	1.8
Magnesium, %	0.21	Tyrosine	1.2
Copper, mg/kg	15.8	Threonine	1.5
Iron, mg/kg	80	Tryptophan	0.55
Manganese, mg/kg	29.8	Valine	1.8
Selenium, mg/kg	0.11		
Zinc, mg/kg	16		

Source: NRC 1979.

inefficiently due to inhibition of the proteolytic enzyme, trypsin (Combs *et al.* 1967; Combs and Wallace 1969; Doty 1965; Jimenez *et al.* 1963; Young 1970). Gestating swine are able to reproduce normally when fed diets containing raw whole soybeans as the only source of supplemental protein (Jensen *et al.* 1970; Crenshaw and Danielson 1983). The reason may be that protein required for growth is more critical than for reproduction because of the capacity of the dam to buffer the fetus against nutritional insults. When soybeans are properly cooked either by boiling, steam processing, autoclaving, infrared roasting, or extrusion, they can replace soybean meal in cereal-based diets for growing-finishing pigs with equal or superior results (Heyward 1970; Combs *et al.* 1967; Young 1970; Jimenez *et al.* 1963; Noland *et al.* 1969; Noland *et al.* 1970; Wahlstrom *et al.* 1971; Nickelson *et al.* 1960; Hanson *et al.* 1970; Zimmerman *et al.* 1970; Nardiello *et al.* 1980; Myer and Froseth 1980; Zamora and Veum 1979; Velloso 1964). The slightly lower feed intake and lower amount of feed required per unit of body weight gain usually encountered when whole soybeans replace soybean meal is related to the greater energy density of the whole soybeans.

The high fat content of soybean and its high proportion of unsaturated fatty acids creates a tendency for soft pork from pigs fed whole cooked soybeans as a major part of the diet. Hanson *et al.* (1970) observed less firm carcasses from pigs fed cooked soybeans as compared to pigs fed soybean meal. Pigs fed soybean meal had adipose tissue containing 13% linoleic acid as compared to 18.1% for pigs fed roasted soybeans. Work at many experiment stations in the United States (Anderson 1971, as summarized in the report by Wahlstrom *et al.* 1971) indicates that whole soybeans can be fed at 22% of a maize-based diet with no adverse effects on pig performance or on consumer acceptance of pork. Although the pork fat from pigs fed whole soybeans was noticeably softer and had a higher proportion of linoleic acid, total fat content of the carcass was not increased. The results are summarized in Table 15.35. Bergkamp and Topel (1970) studied the carcasses of pigs fed either roasted soybeans or soybean meal during growing and finishing and reported that although pigs fed roasted full-fat soybeans had softer fat, diet had no major influence on longissimus muscle characteristics. Taste panel response to flavor, juiciness, tenderness, and overall acceptability revealed no significant difference due to treatment effect. Increasing dietary protein will partially overcome increased carcass fatness which may occur when high levels of full-fat soybeans are fed (Zimmerman *et al.* 1970).

Reports from Iowa State University (Zimmerman *et al.* 1970) reveal that the protein quality, and protein and energy availability are equal for infrared roasted soybeans and commercial soybean meal for growing-finishing pigs, but not for baby pigs. In two trials, total weight gains of young pigs fed infrared roasted soybeans were lower than gains of pigs fed soybean meal, indicating better utilization of protein and/or energy from commercial soybean meal plus added soybean oil than from infrared roasted, full-fat soybeans. Feed was converted more efficiently by baby pigs fed soybean meal than those fed roasted soybeans. In similar studies, Jensen *et al.* (1970) reported that feed intake and rate of gain were reduced when roasted soybeans replaced soybean meal as a supplementary protein in a 16% crude protein diet for pigs from 7 to 11 weeks of age. When given a choice in preference tests, young pigs consistently consumed more feed when soybean meal was used as the supplementary protein source than when roasted soybeans were used.

The decision to use soybean meal or cooked whole soybeans for swine feeding must be based on economic considerations, since the nutritive value of a diet containing soybean meal is about equal to one containing cooked whole soybeans if both are properly supplemented with vitamins and minerals. The cost or inconvenience of providing the power and equipment necessary to heat the bean satisfactorily to approximately 110°C for 3 min to destroy inhibitors of protein utilization may be prohibitive in some areas or under some conditions. However, if the facilities are available, cooking offers a valuable means of utilizing home-grown soybeans as a swine feed and, if not available, raw soybeans may be used for gestating sows.

TABLE 15.35. Cooked Whole Soybeans as a Protein Supplement in Maize Diets for Growing-Finishing Swine

Pig performance and carcass traits	*Soybean meal*	*Cooked whole soybeans*
Av daily gain, kg	0.84	0.85
Av daily feed, kg	2.83	2.68
Feed/gain ratio	3.36	3.15
Carcass length, cm	76.02	76.56
Back fat, cm	3.23	3.30
Longissimus muscle area, cm^2	28.69	29.61
Ham and loin, %	37.31	38.12
Iodine no.	63.0	70.5
Carcass firmness[b]	4.0	3.2

Source: Wahlstrom et al. 1971.
[a]Firmness scores increase with hardness (1 to 5).

DRY BEANS

The Phaseolus genus covers a large number of species including *Phaseolus vulgaris* (known as the dry bean, common bean, snap bean, white kidney bean, haricot bean, or navy bean), *P. aureus* (mung bean), *P. mungo* (urd bean), *P. lunatus* (lima bean), *P. calcaratus* (rice bean), and several others of minor importance. Of these *P. vulgaris* is by far the most important in both worldwide acreage and geographic distribution and consumption (Roberts 1970).

The common bean is native to Central America but is grown on all continents as a source of protein for the human population. Major areas of production are the north central and northwestern states of the United States, medium to high elevations of Central and South America, Europe, and the Near and Middle East.

Although yield potential for the annual herbaceous plants are high (4000 kg/ha), reported annual yields range from 300–1500 kg/ha. The dry beans are susceptible to a wide range of insects and diseases, especially in the hot humid tropics. The bacteria and fungi are more important in the lowland humid tropics and subtropics, whereas viruses are generally more severe in the drier climates (Roberts 1970).

Intensive research efforts at several laboratories around the world, including the Center for International Agriculture in the Tropics (CIAT) in Cali, Colombia (CIAT, 1983), have resulted in the release during the past few years of more than 12 new bean varieties, which are in commercial production in many of the developing countries, and several other varieties with desired disease and pest resistance, which are in the testing stage for eventual release.

Although the primary activity in food and grain legume breeding research is to increase yields of grain so that the crop can become economically competitive, limited research is being directed to the improvement of protein content and quality (Rutger 1970; Kelly 1971; Leleji 1971; Ries 1971; Porter 1972). Results to date indicate that the yield is inversely correlated with protein percentage; the genotype of the maternal plant exercises control of the protein percentage of the dry bean (Lelegi 1971; Porter 1972); the protein quality as measured chemically by assaying for methionine or elemental sulfur is determined genetically; and sufficient variation exists within the species to permit improvement through hybridization and selection (Kelly 1971; Porter 1972). Porter (1972) has shown that the nutritive value of some beans follow a trend similar to that of their elemental sulfur content (Tables 15.36 and 15.37). Silbernagel (1970) states, "years of extensive and expensive research will be required, but eventually we should be able to develop a bean that is as good a source of dietary protein as meat." These efforts continue.

As early as 1920, dry beans were shown to have a marked toxic effect on rats (Johns and Finks 1920). Toxic factors identified in raw beans include a trypsin inhibitor (Bowman 1944), a factor which inhibits protein synthesis in skeletal muscles (Kakade *et al.* 1968), and a hemagglutinin (Rigas and Osgood 1955). The trypsin inhibitor caused significant pancreatic hypertrophy that was reduced by heat treatment of the raw beans (Bowman 1944; Kakade and Evans 1965; Kakade *et al.* 1968; Conner *et al.* 1971).

In addition, the presence of the complex sugars (oligosaccharides), raffinose, and stachyose (Olson *et al.* 1975) results in flatulence and gastrointestinal dis-

TABLE 15.36. Effect of Sulfur and Sulfur–Amino Acids (S-AA) in Three Varieties of Dry Beans on Protein Utilization by Rats

Variety	*Methionine* (%)	*Cystine* (%)	*Total S-AA* (%)	*Sulfur* (% *of protein*)	*Protein efficiency ratio*[a]
Casein (control)	—	—	—	—	2.81
'Calima'	1.21	1.15	2.36	.91	1.96
'Nicer 59'	1.13	1.13	2.26	.99	1.76
'Cargamanta'	1.26	0.75	2.01	.71	1.36

Source: Porter 1972.
[a]g gain/g protein consumed.

tress, due to the absence of digestive enzymes that hydrolyze these carbohydrates into their component simpler sugars, galactose and sucrose, in the alimentary tract of humans and other animals, including the pig.

Kidney beans contain a medium to high level of hydrocyanic acid. Chilean studies (Jacob 1967) indicate that four varieties of dry beans contained from 47.1 to 86.1 mg of hydrocyanic acid (HCN) per kilogram of beans. Industrial cooking reduced the level but still did not completely eliminate the HCN from the beans. Consumption of these industrially processed beans caused digestive disturbances, which were not observed if the beans were first soaked overnight in water and then boiled in new water for 2 hr.

Nutritionally, dry beans, like most legumes, are deficient in the sulfur-containing amino acids and contain only low levels of tryptophan (Bandermer and Evans 1963). Bressani (1969) has studied the variation in protein, methionine, cystine, and lysine content of selected lines of dry beans. Analysis of 268 samples of dry beans showed that the protein content varied from 16.8 to 28.2%. Analysis of 129 of these varieties for methionine, cystine, and lysine indicated a range of 0.087–0.355% methionine, 0.075–0.208% cystine, and 0.80–2.39% lysine. There was no correlation between total nitrogen content and the content of amino acids

TABLE 15.37. Effect of Crude Protein and Sulfur Content of Seven Varieties of Dry Beans on Performance of Rats

Variety	*Protein content* (%)	*Sulfur content* (% *of protein*)	*Av daily gain* (*g*)	*Protein efficiency ratio*[a]
Casein (control)	—	—	88.6	3.56
'Husasano'	19.8	.955	38.4	1.89
'Calima'	18.7	.914	52.3	2.17
'20295'	18.6	.914	42.3	1.88
'Tui'	20.0	.860	36.5	1.99
'Bunsi'	20.8	.832	59.7	2.31
'Cargamanta'	22.2	.630	19.5	1.05
'Mortino'	22.2	.604	15.7	.96

Source: Porter 1972.
[a]g gain/g protein consumed.

tested, which would support the suggestion that the protein quality of beans can be improved through selection and hybridization.

A beneficial effect was shown with methionine or antibiotic supplementation (Goatcher and McGinnis, 1972), but no additive effect was noted when the bean diet was supplemented with both methionine and antibiotics. These authors suggested that beans have a depressing effect on growth (possibly by stimulating the growth of adverse microflora in the intestinal tract). Kakade and Evans (1965, 1966) obtained similar results with rats, showing that significant growth increases were obtained by the addition of antibiotic supplementation to diets containing dry beans. They attributed the improvement to increased digestibility or absorption of nutrients resulting from the antibiotics and/or to inhibiting of an enzyme which might aid in liberating a ''bound'' growth inhibitor present in the dry beans. Raw dry beans fed at 36% of maize–dry bean diet caused steady weight loss of approximately 0.25 kg/day in weanling pigs. Heating the bean at 127° C for 3 min before grinding and mixing into the diet resulted in improved feed consumption but allowed a weight gain of only about 50 g/day (Pond *et al.* 1972). It seems clear that dry beans cannot be used as a major portion of swine diets until more is known about appropriate processing to destroy inhibitors and to hydrolyze oligosaccharides (Olson *et al.* 1975).

CHICKPEAS

The chickpea (*Cicer arietinum* L.), which is also known as garbanzo and bengal gram, consists of two main types: one is a large seeded variety used for human food in the southwestern United States, Mexico, India, Pakistan, and several European countries, and the other is a small seeded variety known as the forage chickpea, which is used in the central valleys of Mexico mostly for feeding swine. The protein and fat content of the large seed varieties are about 20% higher than those of the small seeded varieties, but contain only about one-third as much fiber (Table 15.38).

Although the chickpea contains less protein than the soybean and certain others of the pulses, it can be used for swine feeding, without cooking or heating. Mexican workers (Shimada and Brambila 1967) reported a significant improvement in the utilization of chickpeas by the chick when the seeds were autoclaved; however, it has been shown that autoclaving of the peas for 30 or 60 min at 115° C reduces the digestibility of the dry matter, energy, and nitrogen. Although autoclaving had no effect on performance of pigs receiving a diet containing 88.8% chickpeas (Shimada and Brambila 1967), Tannous and Ullah (1969) have shown that the raw chickpea contains no hemagglutinating activity but has a high antitrypsin activity. Autoclaving the chickpea for 5 min at 121° C reduced the antitrypsin activity, and the activity was further reduced but was not completely eliminated when the chickpea was autoclaved at the same temperature for 20 min.

Like the majority of the pulses, the chickpea is a fairly good source of lysine but a rather poor source of methionine and cystine. After the sulfur-containing amino acids, tryptophan is probably the most limiting one (Pennacchiotti and Schmidt-Hebbel 1966). The amino acid composition of chickpeas is shown in Table 15.39).

TABLE 15.38. Chemical Composition of the Chickpea

Component	Sample			
	1[a]	2[b]	3[c]	4[d]
Moisture, %	—	5.5	11.8	12.5
Total nitrogen, %	4.24	3.1	2.77	2.8
Crude protein, % (nitrogen × 6.25)	26.5	19.4	17.3	17.5
Ether extract, %	6.4	5.1	3.5	5.7
Crude fiber, %	2.6	9.7	8.4	3.1
Ash, %	3.1	2.7	2.8	2.8
Nitrogen-free extract, %	—	57.4	56.2	58.4
Calcium, %	—	.21	—	.14
Phosphorus, %	—	.18	—	—
Magnesium, %	—	—	—	.13
Copper, mg/kg	—	75	—	9
Iron, mg/kg	—	—	—	71
Manganese, mg/kg	—	—	—	24
Zinc, mg/kg	—	—	—	21

[a]Gonazales del Cueto et al. 1960.
[b]Verma et al. 1964.
[c]Shimada and Brambila 1967.
[d]Abu-Shakra et al. 1970.

Shimada and Brambila (1967), in a series of three experiments, have studied the chickpea as a source of both protein and energy for the growing pig. Levels of 0–88.8% chickpeas have been used to replace a mixture of maize and soybean meal in 16% protein diets or to replace portions of a sorghum–cottonseed meal–fish meal diet. At all levels of substitution, pigs performed as well as those fed the control diet. The raw chickpea supplied adequate protein, energy, and

TABLE 15.39. Amino Acid Content of Chickpea, g/16 g N

Amino acid	Sample	
	1[a]	2[b]
Arginine	8.93	—
Histidine	3.07	—
Isoleucine	3.87	—
Leucine	7.46	—
Lysine	4.03	6.78
Methionine	1.36	—
Cystine	1.04	—
Methionine + cystine	2.40	2.45
Phenylalanine	5.90	—
Threonine	3.12	3.39
Tryptophan	0.98	1.34
Valine	4.24	—

[a]Abu-Shakra et al. 1970. [b]Tannous and Ullah 1969.

amino acids to support satisfactory growth and efficiency of feed utilization, even when pea meal supplied the major portion (88.8%) of the diet. In a further series of studies, Zamora *et al.* (1975) and Casarin *et al.* (1976) have shown that chickpeas can completely replace soybean meal in practical sorghum–soybean meal diets for growing-finishing pigs without affecting pig performance or carcass quality.

Methionine supplementation of 0.2% improved both growth and feed conversion in the 88.8% chickpea diet, but higher levels of methionine supplementation tended to decrease gains (Shimada and Brambila 1967). Later studies at the same station (Casarin *et al.* 1976; Shimada and Avila 1975) have confirmed the improvement in pig performance brought about by the addition of methionine to diets based on sorghum and chickpeas; however, levels of responses do vary. Shimada and Avila (1975) reported that 0.05% added DL-methionine was adequate to promote optimal growth in growing pigs fed 14% protein diets based on sorghum and garbanzo or 0.13% when a 12% protein diet was employed. Casarin *et al.* (1976) added levels of 0, 0.14, 0.28, and 0.42% DL-methionine to a sorghum chickpea diet for growing and finishing pigs (17–90 kg) and reported data indicating an increase in daily gains and an improvement in feed efficiency as the methionine was increased to 0.28%. The higher level (0.42%) decreased gains. These same authors also reported a signficant improvement in pig performance when lysine was added to the diet even in the absence of supplemental methionine. Although this item was not tested in any of the pig studies reported, it would appear that at higher levels of methionine supplementation, tryptophan may become the limiting amino acid (Pennacchiotti and Schmidt-Hebbel 1966).

The chickpea, which has a feeding value similar to that of a mixture of maize and soybean meal, can be used to supply the major portion of both protein and energy for growing pigs and can be profitably utilized in swine rations for growing-finishing, gestating, and lactating animals when the price of the chickpea is less than that of a comparable quantity of mixed feed consisting of grain and protein supplement.

PIGEON PEAS

The pigeon pea (*Cajanus cajan* or *indicus*) is a shrub, usually cultivated as a perennial, and therefore cut back every year, and it yields for 2 or 3 years. Easily cultivated and drought resistant, the pigeon pea (also called red gram dhal, no-eye pea, and guandul) is grown in both the tropics and subtropics. The pigeon pea is used as a human food, as a livestock feed, and as a green manure crop (Krauss 1932; Aykroyd and Doughty 1964; Fike 1968; Killinger 1968). The seeds have an acrid taste, which is mainly confined to the seed coat, and they are usually eaten when mature and dry. Pigeon peas contain a trypsin inhibitor that can be precipitated with 5% trichloroacetic acid and destroyed by moist heat (Pusztai 1967).

The average total protein of the pigeon pea is 20–23%. The amino acid profile of the pea, although adequate in lysine, is very deficient in tryptophan and the sulfur–amino acids. According to Orr and Watt (1957) tryptophan is more limiting than methionine. Although pigeon peas contain 0.38% oxalic acid they do not contain any hydrocyanic acid (Oke 1967).

Although world yields are low, averaging only about 600 kg/ha, yield records from experimental plantings indicate that the potential of the crop is high (Roberts 1970). Yields in excess of 5000 kg/ha have been reported in India and in excess of 4000 kg/ha in Puerto Rico.

Dako (1966) reported that cooking of pigeon peas improved their nutritional value by destroying the substances inhibiting trypsin activity. Raw peas caused hyperactivity of the pancreas, diarrhea, and apathy. Addition of tryptophan or methionine, each alone, to raw or cooked peas had no effect on the protein efficiency ratio, but the combination of the two amino acids improved the protein efficiency ratio by about 100%.

COWPEAS

The cowpea (*Vigna sinensis* or *unguiculata*) is a herbaceous vine that most probably originated in Africa but is widely cultivated in India, China, and Africa and to a lesser extent in both North and South America. Approximately 94% of the world production of 1.18 million metric tons comes from Africa. Although reported average world yields are extremely low, averaging only about 380 kg/ha, advanced cowpea yield trials conducted at the International Institute of Tropical Agriculture, Ibadan, Nigeria, indicated that many times more kilograms per hectare per year of dry cowpea grain can be produced with yields during seasons of adequate rainfall exceeding 2220 kg/ha (Rachie *et al.* 1971).

Cowpeas have an average total protein content of 24% but varieties differ greatly in crude protein and total elemental sulfur content (Table 15.40) (Maner 1971). Varieties contain from 18 to 29% crude protein, and if elemental sulfur can

TABLE 15.40. Distribution and Range of Crude Protein and Sulfur Content of 141 Varieties of Cowpeas

Protein		*Sulfur*	
Range (%)	*No.*	*Range* (% *of protein*)	*No.*
18–19	1	.60–.70	18
19–20	1	.70–.80	54
20–21	2	.80–.90	44
21–22	11	.90–1.00	17
22–23	17	1.00–1.10	6
23–24	25	1.10–1.20	0
24–25	30	1.20–1.30	1
25–26	18	1.30–1.40	0
26–27	21	1.40–1.50	1
27–28	7	—	—
28–29	6	—	—
29–30	2	—	—
Total	141	Total	141

Source: Maner 1971.

be taken as an estimate of their sulfur amino acid content (Porter 1972), they can be assumed to vary widely in methionine and cystine content.

The essential amino acid content of the cowpea has been studied by several investigators (Jaffe 1949; Orr and Watt 1957; Bressani *et al.* 1961; Elias *et al.* 1964) and the results indicate that methionine and tryptophan are the most limiting amino acids. Biological studies (Jaffe 1949; Borchers and Ackerson 1950: Chavez *et al.* 1952; Elias *et al.* 1964) have indicated that cowpea protein is a good source of lysine but a poor source of methionine and cystine. Even though cowpeas are deficient in some of the amino acids they have a higher nutritive value than common beans (Elias *et al.* 1964).

The cowpea, like other leguminous seeds, contains a trypsin inhibitor. Studies of the relationship between digestibility of cowpeas and its trypsin inhibitor (Borchers and Ackerson 1950; Jaffe 1949) indicate that heating in the autoclave was sufficient to destroy the inhibiting factor, but that this treatment did not improve the overall digestibility.

Cowpeas contain only small quantities of oxalic acid and unlike lima beans contain only minute quantities of hydrocyanic acid (Oke 1967).

Early studies with pigs fed cowpeas were conducted by Heitman and Howarth (1960). Cleaned cull black-eyed peas containing 23.2% crude protein were ground without cooking and substituted for barley at levels of 20 and 50% in diets for growing pigs. As the percentage of raw black-eyed cowpeas increased, gain in weight decreased. Feed consumption and apparently utilization were reduced by the addition of cowpeas, even though no evidence of toxicity was observed.

More recent works with rats (Maner and Pond 1971) and pigs (Maner 1971) have indicated that the nutritive value of cowpeas can be greatly increased (Table 15.41) by cooking and by supplementing with methionine.

Rats fed diets containing cowpeas as the only source of protein grew very poorly when the cowpeas were offered raw (Maner and Pond 1971). Cooking improved gains by 70% and efficiency of feed conversion by about 17%. Methionine supplementation improved weight gains of the rats fed both raw or cooked cowpeas. Cooked cowpeas adequately supplemented with methionine have a nutritive value for rats equal to that of soybean meal.

TABLE 15.41. Utilization of Cowpeas by Rats as Affected by Processing and Methionine Supplementation

Treatment	*Av daily gain (g)*	*Feed/gain ratio*
Crude (raw)	2.26^{x}	$3.51^{x,y}$
Crude + methionine	3.25^{y}	$2.94^{x,y}$
Cooked	3.84^{z}	2.81^{z}
Cooked + methionine	4.61^{z}	2.48^{z}
Soybean meal	4.60^{z}	2.72^{z}

Source: Maner and Pond 1971. Nine rats per treatment.

^{a,b,c}Means in the same column without a common superscript are significantly different from each other ($P < .05$).

TABLE 15.42. The Value of Raw and Cooked Cowpeas as a Protein Source for Growing Pigs

Diets	*Av daily gain (g)*	*Feed/gain ratio*
Maize–soybean meal	799	2.53
Raw cowpeas	551	3.43
Raw cowpeas + methionine	483	3.09
Cooked cowpeas	816	2.50
Cooked cowpeas + methionine	815	2.43

Source: Maner 1971.

Germinated cowpeas have been utilized to supplement 12% protein diets for rats based on opaque-2 maize (Maner 1972, unpublished data). Results of this study indicate that although unsupplemented diets of cowpeas and opaque-2 maize significantly improved rat gains and feed conversion, performance equal to that obtained with diets of soybean meal and opaque-2 maize was obtained only after supplementing the diet with 0.05% DL-methionine. Other studies (Maner 1972, unpublished data) indicate that although methionine supplementation of diets of opaque-2 maize and cowpeas is effective in improving performance, no improvement in performance is obtained from methionine supplementation when common maize is substituted for opaque-2 maize. Although not tested in these studies, these data probably indicate a difference in tryptophan content of the two maizes, which becomes limiting in the common maize diet before methionine.

Studies similar to those reported with rats have also been conducted with growing pigs (Maner 1971). Pigs fed 16% protein diets, in which raw or cooked cowpeas provided the only protein source in the cowpea–sucrose diet, performed very poorly on the raw cowpea diets. Cooking of the cowpeas significantly improved gains, feed consumption, and efficiency of feed conversion. Performance of pigs fed cooked cowpeas was not different from that of pigs fed the 16% protein maize–soybean meal control diet. Methionine supplementation was without effect when added to either the raw or the cooked cowpea diets (Table 15.42).

FIELD BEANS

The field bean (*Vicia faba* L. var. minor or major), also known as faba bean, horse bean, broad bean, tick bean, or haba, has not been used extensively as a feed for pigs. However, with the increasing demand for protein for human and animal nutrition, it is gaining popularity as a partial replacement for the supplemental protein in all types of swine diets. This is true especially in Europe and Canada where soybean meal is imported. The relatively high protein content (24–30%) and the general absence of inhibitory factors make the field bean a satisfactory feed for both growing-finishing pigs and reproducing sows.

Like most legume seed, the field bean is deficient in methionine (Table 15.43) but contains high levels of cystine, which partially overcome the deficiency of the

TABLE 15.43. Chemical and Amino Acid Composition of Various Samples of Field Beans (*Vicia faba*)

Component	*Sample*				
	1[a] (*spring*)	*2*[a] (*winter*)	*3*[b]	*4*[b]	*5*[b]
Dry matter	—	—	85.0	—	—
As % of dry matter					
Crude fiber	8.0	9.0	8.4	—	—
Ether extract	1.5	1.5	1.4	—	—
Nitrogen-free extract	55.2	59.0	56.2	—	—
Ash	4.0	4.0	3.5	—	—
Crude protein	31.4	26.5	30.4	22.8	26.8
Amino acids (% of protein)					
Arginine	10.30	8.44	9.60	9.00	9.65
Histidine	2.55	2.59	2.59	2.70	2.60
Isoleucine	4.35	4.04	4.34	4.50	4.40
Leucine	7.87	7.53	7.44	7.70	7.25
Lysine	6.59	6.80	6.22	7.20	6.50
Methionine	0.73	0.80	0.82	0.90	0.80
Cystine	0.84	0.92	1.28	1.45	1.60
Methionine + cystine	1.57	1.72	2.10	2.35	2.40
Phenylalanine	4.63	4.36	4.28	4.40	4.20
Threonine	4.02	3.76	3.68	3.80	3.60
Tryptophan	—	—	0.99	0.90	0.90
Valine	4.92	4.62	4.74	5.10	4.70
Tyrosine	3.86	3.53	3.59	3.60	3.60

Sources: [a]Adapted from Eden 1968 and Clark 1970. [b]Adapted from Henry 1970.

sulfur-containing amino acids. It is a good source of lysine and a relatively satisfactory source of tryptophan. The mineral and vitamin compositions of several sources are given in Table 15.44. The field bean has been demonstrated to be a valuable complementary protein source for growing-finishing pigs. In a series of three studies, Navratil (1965) clearly demonstrated that bean meal could be satisfactorily utilized to supply one-half or all of the additional protein required to supplement diets based on cereals, beets, and potatoes. No differences in performance were observed when bean meal replaced 50% of the supplemental protein from yeast and oil meals in diets for pigs up to 50 kg nor when bean meal supplied all of the supplemental protein from 50- to 90-kg liveweight. Bean meal additions of 18.5–25.0% in all-cereal diets improved gains of growing-finishing pigs by 65% and the rate of feed conversion by 25%.

In diets for growing-finishing pigs, quantities of field bean meal as large as 30% have been used successfully to replace dietary fish meal and/or soybean meal. Spanish workers (Balboa *et al.* 1966) used 30% bean meal with or without 0.17% DL-methionine to replace 11% soybean meal in a diet consisting of maize, fish meal, and soybean meal. The control, 30% bean meal, and 30% meal-plus-methionine diets contained total methionine-plus-cystine levels of 0.55, 0.43, and

TABLE 15.44. Mineral and Vitamin Composition of Field Beans

Component (*on dry matter basis*)	Sample		
	1[a] (*spring*)	*2*[a] (*winter*)	*3*[b]
Calcium, %	0.16	0.19	0.12
Phosphorus, %	0.66	0.63	0.56
Magnesium, %	0.13	0.13	—
Potassium, %	1.17	1.22	—
Manganese, mg/kg	—	14	14
Iron, mg/kg	—	64	33
Copper, mg/kg	—	54	12
Zinc, mg/kg	—	—	55
Thiamin, mg/kg	15.0		7.2
Riboflavin, mg/kg	3.1		4.5
Niacin, mg/kg	29.0		28.6
Pantothenic acid, mg/kg	11.7		4.7
Choline, mg/kg	1,100.0		—
Pyridoxine, mg/kg	—		6.2
Biotin, mg/kg	—		0.24
α-tocopherol, mg/kg	1.0		16.5

[a]Adapted from Eden 1968 and Clarke 1970.
[b]Adapted from Henry 1970.

0.60%, respectively. When these diets were fed to appetite in two daily feedings to growing pigs (25–60 kg), average daily gains were 718, 596, and 734 g, and feed conversion ratios were 2.82, 2.86, and 2.82, respectively, for the three treatments, and no toxic effects of bean meal were observed.

It was reported (Clarke 1970) that performance of growing-finishing pigs (32–91 kg) was not affected by partial or complete replacement of the soybean meal in the diet by field beans; however, neither growth nor feed efficiency were improved by the addition of methionine.

Several European experiments have demonstrated that field beans may be used to provide part or all of the protein supplement in swine diets (Aherne and McAleese 1964; Hansen and Clausen 1969; Livingstone *et al.* 1970; Cole *et al.* 1971; Feist *et al.* 1974; Lightfoot *et al.* 1974; Fowler and Livingstone 1977; Chenuz and Jast 1979). In most of these studies, and those of Henry and Rerat (1969), Henry and Bourdon (1972), Henry *et al.* (1976), Castell (1976), Bowland *et al.* (1976), and Aherne *et al.* (1977), growth rate and/or feed utilization were reduced as the level of field beans in the diet was increased.

A series of studies at the Swine Research Station in France (Henry and Rerat 1969; Henry and Bourdon 1972; Henry *et al.* 1976) have clearly demonstrated that complete replacement of soybean meal with field beans at levels of 30–33% in cereal-based diets for growing-finishing pigs produces a decline in both rate and efficiency of gain. On the other hand, when the level of field bean incorporation was restricted to 15% of the diet, thus replacing only one-half of the soybean protein, the overall performance was practically unchanged. Despite a rather low level of sulfur–amino acids in the field beans, methionine supplementation of the diets after partial or complete replacement of soybean meal by field beans did not exert any favorable effect on pig performance. However, both gains and feed efficiency were improved when the diets were supplemented with tryptophan, especially when maize was used as the grain source.

In Canadian studies by Castell (1976), replacement of soybean meal with graded levels of field beans as the protein supplement resulted in a linear reduction in growth rate of growing-finishing pigs and a linear improvement in carcass quality. Feed-to-gain ratio was not different at the 0, 15, or 30% levels of substitution, but was improved by the addition of 7.5% field beans (Table 15.45). In agreement with results reported by Aherne and McAleese (1964) and Cole *et al.* (1971), Aherne *et al.* (1977) reported a slight but definite reduction in average daily gain and feed utilization of growing pigs when 10–30% field beans were incorporated into the diet as a replacement for soybean meal, but these effects were not significant until the level of field beans reached 25% of the diet. No differences in daily feed intake were reported but feed-to-gain ratio was significantly increased at the two highest levels of substitution.

Mateos and Puchal (1981) utilized the larger field beans (*Vicia faba* L. major) for starter and growing-finishing pigs. Starting pigs (5–9 weeks of age) fed either 10 or 20% field beans grew as well and as efficiently as pigs fed the control diet. Average daily gains were 536, 529, and 536 g and feed-to-gain ratios were 2.10, 2.07, and 2.12 for the diets containing 0, 10, and 20% field beans, respectively. Growing-finishing pigs were fed diets containing 0, 12.5, and 25% field beans and adequate supplemental methionine to provide equal dietary levels of the sulfur–amino acids. Average daily gains for the growing-finishing pigs (9–21 weeks of age) were 754, 833, and 789 g for the diets containing 0, 12.5, and 25% field beans, respectively. DL-methionine supplementation in excess of that required to provide a level equal to that in barley–maize–soybean meal control, however, failed to improve the performance of pigs fed 25% field beans.

Although field beans at levels up to 15–20% have supported satisfactory performance of growing and finishing pigs fed cereal-based diets, they are not recommended as the only supplemental protein source for low protein feed sources such as bananas and cassava. Bowland *et al.* (1976) replaced soybean meal protein with field beans in diets containing 0, 20, or 40% dried cassava meal. Although the cassava meal as an energy source at dietary levels up to 40% had no significant influence on starter pig performance, complete replacement of

TABLE 15.45. Effect of Substitution of Field Beans for Soybean Meal in Diets for Growing-Finishing Swine

Diets	*1*	*2*	*3*	*4*
Level of field beans, %	*0*	*7.5*	*15*	*30*
Av daily gain, kg[a]	784	782	751	727
Av daily feed	2.78	2.57	2.70	2.65
Feed/gain ratio	3.57	3.30[b]	3.63	3.61

Source: Castell (1976).
[a]Significant linear response.
[b]Significant improvement in feed efficiency due to 7.5% bean meal.

soybean protein by field beans (30–55%) in cassava-based diets reduced both gains and efficiency of feed utilization. This depression was only partially overcome by supplemental lysine and methionine but could be completely overcome by reducing the level of field beans to 21% and supplementing the diet with soybean meal.

The possibility of total replacement of soybean meal by field beans in diets for reproducing sows has been investigated. During a 5-week lactation period, multiparous sows were fed barley-based diets supplemented with either 11% soybean meal or 21% field beans. The addition of 21% field beans in the 14% crude protein diet had no effect on the survival or weight of piglets at any stage of lactation (Etienne *et al.* 1975). Further studies by Etienne (1977) showed that the introduction of 15–25% field beans into the diets of gestating gilts after mating had no apparent effect on mean daily gain, fertilization rate, or embryonic mortality. There was some tendency, however, for a lower mean weight of fetuses and their placenta at 105 days of pregnancy. When used throughout the reproductive cycle, there is some indication from Danish trials that field beans are not adequate as the only supplemental protein source to cereal-based diets for high-producing sows (Nielsen and Kruse 1974). These researchers fed sows during gestation and lactation on a cereal-based diet supplemented with soybean meal, fish meal, and meat-and-bone meal, or on diets containing 17 or 34% field beans as a replacement of one-half or all of this protein mixture. An increase in dietary field beans significantly reduced milk yield and litter size at birth and weaning. The dry matter and protein contents of the colostrum were also reduced by the replacement of a portion of the dietary protein with field beans (Table 15.46). In

TABLE 15.46. Effect of Field Beans on Milk Composition, Milk Yield, and Reproductive Performance of Sows

Groups	*1*	*2*	*3*
Level of field beans, %	*0*	*17*	*34*
Number of litters	*53*	*58*	*56*
Milk composition, %			
Dry matter	19.0	18.9	18.6
Protein	7.1	7.0	6.5
Fat	5.5	5.6	5.7
Milk yield, kg			
Av daily production[a]	6.3	5.5	5.2
Av daily prod. corr.[a,b]	6.0	5.6	5.4
Litter size			
At birth, total[a]	12.0	10.5	9.7
At birth, live[a]	11.0	9.5	9.2
At 8 weeks[a]	9.2	8.3	8.1
Av wt per pig, kg			
At birth	1.41	1.44	1.45
At 8 weeks	19.1	19.0	19.6

Source: Nielsen and Kruse 1974.
[a]Significant linear effect.
[b]Production corrected for litter size.

agreement with results presented by Elliott *et al.* (1971), it appears that the protein quality of the field beans was not adequate to meet the needs of the sow during the complete reproductive cycle.

Although field bean protein is known to contain low levels of sulfur–amino acids (Kadirvel and Clandinin 1974; Marquardt and Campbell 1974), supplementation of swine diets containing field beans with methionine has, with few exceptions, failed to improve animal performance (Aherne and McAleese 1964; Hansen and Clausen 1969; Cole *et al.* 1971; Henry and Bourdon 1972). However, several experiments have shown that the apparent digestibilities of the protein and amino acids in field beans are considerably lower than those of soybean meal, with most reported digestibility coefficients for field bean protein being 79–83% (Aherne and McAleese 1964; Pastuszewska *et al.* 1974). Whittemore and Taylor (1973) have reported that growing pigs fed diets containing field beans retain less nitrogen than those fed an isonitrogenous soybean meal supplemented diet, a fact that is consistent with the low biological value of 54.5% reported by Walz (1974) for field bean protein.

Factors other than protein quality that have been implicated in the reduced utilization of field beans include: trypsin inhibitors, hemagglutinin or tannin levels (Marquardt and Campbell 1973, 1974; Marquardt *et al.* 1976A,B). The improvement in performance seen in several experiments when autoclaved beans replaced unheated field beans in the diet of chickens (Marquardt and Campbell 1973, 1974; Marquardt *et al.* 1974, 1976A,B) could be due to an improvement in amino acid availability (Wilson and McNab 1972) or to destruction of condensed tannins or other growth inhibitors (Marquardt *et al.* 1976A,B). However, studies by Aherne *et al.* (1977) have shown that autoclaving of field beans for 15, 30, 45, or 60 min did not improve nitrogen or energy digestibility of diets or growth and feed conversion efficiency of the pigs. In fact, digestion coefficients for energy tended to decrease progressively as the autoclaving time increased and nitrogen digestibility was considerably lower for the diets in which the field beans had been autoclaved for 60 min. These results are in agreement with those of Ivan and Bowland (1976) who autoclaved field beans for 30 or 60 min and concluded that autoclaving had no beneficial effect on the digestion of nutrients in the small intestine of the pig. The lack of growth response or improved digestibility of nutrients by pigs fed diets containing autoclaved field beans suggests that pigs are less sensitive than chicks to chemical inhibitors present in the field beans.

Although digestive disorders have been previously indicated to occur when immature beans are fed to pigs, these disorders do not appear to be a problem when dried, mature beans are fed either immediately after harvest or following extended periods of storage. Clarke (1970) fed newly harvested and dried mature beans at an inclusion rate of 10% of the diet to pregnant sows and observed no digestive upsets nor adverse effects on the litters produced. Similarly, Luscombe (1969) observed no ill effects on the health of 23-kg pigs fed freshly harvested beans in the pods.

In practice, raw field beans can be used as a partial replacement for supplemental protein sources in diets based on cereals. Levels above 15–20% of the diet may reduce pig performance. The level of substitution and need for methionine supplementation will depend upon the type of diet used and the source and quality of the other supplemental protein sources present in the diet.

FIELD PEAS

The field pea (*Pisum sativum* or *arvense*), also called pea, garden pea, and arveja, is an annual. Different varieties are grown for both human food and livestock feed. They thrive only in temperate regions, or as cool season or hill country crops in the tropics and subtropics. Field peas are grown in many areas of the world but have been used extensively as a feed for livestock only in Russia and parts of Europe. The cull peas not useable for human consumption are now being used for livestock feed in the Pacific Northwest of the United States. Peas are palatable for growing-finishing pigs (Bell and Wilson 1970; Froseth 1976) but have been reported to be quite unpalatable to the young pig (Myer and Froseth 1980).

Peas contain 19.6–29% crude protein (Bell and Wilson 1970; Reddy *et al.* 1979) and are a good source of lysine (Table 15.47). However, they contain only about 0.36% total methionine plus cystine (Husby 1971). Peas are normally utilized to replace a portion of the protein supplement in swine diets. However, smooth peas also contain 67% total carbohydrates and 48% starch and if the price of peas and cereals is similar, peas can be used as an economical grain substitute.

TABLE 15.47. Proximate and Amino Acid Analysis of Field Peas (*Pisum sativum*)

Component (*in percent*)	*Sample*			
	1[a]	*2*[b]	*3*[c]	*4*[d]
Moisture	7.51	10.1	—	0.00
Crude protein	28.97	25.90	—	27.9
Ether extract	—	1.48	—	3.2
Crude fiber	—	6.00	—	5.9
Ash	—	2.64	—	2.84
Calcium	—	—	—	0.029
Phosphorus	—	—	—	0.026
Nitrogen-free extract		54.0	—	60.1
Gross energy[e]	4.40	—	—	—
Arginine	2.07	1.31	2.77	—
Histidine	0.76	0.80	0.66	—
Isoleucine	1.04	1.25	1.01	—
Leucine	1.82	2.34	1.65	—
Lysine	1.89	1.62	1.52	—
Methionine	0.22	—	0.19	—
Cystine	—	—	0.16	—
Methionine + cystine	—	—	0.35	—
Phenylalanine	1.26	1.10	1.07	—
Threonine	1.17	0.90	0.84	—
Tryptophan	—	—	0.25	—
Valine	1.12	1.12	1.16	—

Sources: [a]Bell and Wilson 1970. [b]Zarkadas et al. 1965. [c]Kroening 1968. [d]Singh et al. 1968.

[e]kcal/g.

Peas were demonstrated to be inferior in protein quality to either meat meal or soybean meal when fed to growing-finishing pigs in drylot as the only protein supplement to cereal-based diets (Beeson and Hickman 1945; Lehrer and Hodgson 1946). However, subsequent studies have indicated that the growth depression and poor general performance of animals fed peas was most probably related to deficiencies of nutrient factors other than energy and amino acids.

Later reports (Cunha *et al.* 1948; Warwick *et al.* 1948) have shown the value of the field pea as both a protein and energy source for pigs of all classes. Research conducted in Australia (Lobban 1963) indicated that satisfactory performance of growing-finishing swine could be obtained if 20–30% of the diet was supplied by peas.

Kroening (1968) demonstrated that peas can supply all of the supplemental protein in cereal-based diets for growing-finishing pigs and that neither gains nor feed utilization are improved by methionine supplementation as might be expected from the amino acid analysis of the peas used (Table 15.47). The level of total sulfur amino acids (methionine + cystine) is sufficient to meet the needs of the growing-finishing pigs if the quantity of peas does not exceed 60% of the 16% protein diet, based on maize, wheat, or barley (Table 15.48) (Kroening 1968).

Similar results were obtained in Canada and Chile. In Canadian studies (Bell and Wilson 1970), field peas of cull grade replaced 0, 25, 50, 75, and 100% of the soybean meal–fish meal supplement of a swine grower diet without reducing rate of gain, feed utilization or carcass quality (Table 15.49). Methionine supplementation proved unnecessary when the pea protein replaced all of the soy–fish protein. In Chilean studies (Esnaola *et al.* 1977), field peas fed at levels of 0, 10, 20, 30, and 40% of the diet to replace an equal quantity of fish meal protein had no effect on feed intake, daily gains, or efficiency of feed conversion even though the high levels completely replaced the fish meal in the diet. Fowler and Livingstone (1977) compared peas with soybean meal as protein supplements to barley-based diets for growing-finishing swine and concluded that peas may be successfully used in pig diets even when replacing 50% of the dry matter.

TABLE 15.48. Effect of Feeding Maize, Wheat, or Barley with Soybean Meal or Field Peas on the Performance of Growing Pigs

Protein source	*Maize*	*Wheat*	*Barley*	*Av*
		Av daily gain, g		
Soybean meal	846	795	764	802
Peas	818	805	795	806
Average	823	800	780	
		Feed/gain ratio		
Soybean meal	3.06	2.94	3.14	3.05
Peas	2.88	2.81	2.84	2.84
Average	2.97	2.88	2.99	

Source: Kroening 1968, unpublished data.

TABLE 15.49. Performance of Growing-Finishing Pigs Fed Various Levels of Field Pea Meal

Treatments	*Av daily gain (kg)*	*Av daily feed intake (kg)*	*Feed/gain ratio*
Control, soybean–fishmeal	.703	2.16	3.07
Pea meal	.704	2.11	3.06
Pea meal + 0.3% methionine	.694	2.16	3.11
$\frac{3}{4}$ pea meal + $\frac{1}{4}$ soy–fish	.684	2.22	3.26
$\frac{1}{2}$ pea meal + $\frac{1}{2}$ soy–fish	.696	2.09	3.35
$\frac{1}{4}$ pea meal + $\frac{3}{4}$ soy–fish	.702	2.07	2.95

Source: Bell and Wilson 1970. Pigs fed from 23 to 90 kg.

More recent studies at Washington State University (Froseth 1976; Myer and Froseth 1978, 1980) have failed to confirm these findings. Froseth (1976) substituted cull peas for one-half or all the soybean meal in growing pig diets and supplemental methionine was added to the diet in which peas provided all the supplemental protein. Pigs fed cull peas plus methionine gained faster from 18.6 to 56.6 kg than did those fed unsupplemented cull pea diets. Rate of gain and feed efficiency of pigs fed cull peas plus methionine did not differ from that of pigs fed soybean meal or a combination of soybean meal and cull peas. Although methionine supplementation was necessary to overcome depressed pig performance when cull peas supplied all of the supplemental protein during the growing period, supplemental methionine was without effect during the finisher phase from 56.6 to 100.9 kg.

Studies by Froseth (1976), Myer and Froseth (1978, 1980) and Bertrand *et al.* (1980) indicate that cull peas are unpalatable and inadequate as a complete substitute for soybean meal in barley-based diets for young pigs. Performance of 7–20-kg pigs was reduced when peas replaced all of the soybean meal in the absence of supplemental methionine (Myer and Froseth 1978). Steam pelleting and methionine supplementation improved pig performance, with the young pigs responding more to methionine supplementation when the cull peas were steam pelleted and reground before being incorporated into the diet (Myer and Froseth 1978). Further studies with 8–24-kg pigs (Myer and Froseth 1980) have confirmed these earlier findings. Raw or extruded peas were used to supply all of the supplemental protein in barley-based diets. Methionine was added at levels of 0, 0.1, or 0.25%. Feeding extruded peas resulted in increased feed consumption and performance of pigs, particularly when the methionine deficiency of the diet was corrected by the addition of 0.1% DL-methionine (Table 15.50). The addition of 0.25% DL-methionine to the raw pea diet resulted in decreased gains and feed efficiency compared to raw cull peas and 0.1% methionine. Only a slight decrease in average daily gain but a significant decrease in average daily feed was reported when extruded peas plus 0.25% methionine were compared to the lower level of methionine supplementation.

Bertrand *et al.* (1980) reported that 20% raw peas fed in diets to 5–7-kg pigs reduced feed intake by 9% and daily gains by 10% and decreased efficiency of

TABLE 15.50. Performance of Young Pigs (8–24 kg) Fed Raw or Extruded Cull Pea Diets With and Without Methionine Supplementation

Dietary protein supplement	*Av daily gain, kg*	*Av daily feed, kg*	*Feed/gain ratio*
Soybean meal	0.46^{a}	0.83^{a}	1.80^{a}
Raw peas	0.25^{c}	0.61^{b}	2.49^{c}
Raw peas + 0.1% DL-methionine	0.35^{b}	0.65^{b}	$2.00^{a,b}$
Raw peas + 0.25% DL-methionine	$0.28^{b,c}$	0.62^{b}	2.25^{b}
Extruded peas	$0.28^{b,c}$	0.62^{b}	2.22^{b}
Extruded peas + 0.1% DL-methionine	$0.42^{a,b}$	0.78^{a}	1.89^{a}
Extruded peas + 0.25% DL-methionine	$0.42^{a,b}$	0.66^{b}	1.71^{a}

Source: Myer and Froseth 1980.
[a,b,c]Means in the same column without a common superscript are significantly different ($P < .05$).

feed utilization by 32%. A 30% level of peas reduced feed intake by 15% and gains by 20%, while increasing feed requirements by 7%. Similar results were reported by Bouard *et al.* (1980) who observed poor pig performance when either 30 or 45% raw peas were included in diets for young pigs.

Henry and Rerat (1969) indicated that when 40% field peas were used to replace completely the supplemental soybean meal, growth performance and feed utilization of growing-finishing pigs were significantly depressed during the growing phase between 20- and 60-kg liveweight. Normal performance was observed only after 0.15% tryptophan was added to the diet (Bourdon and Perez 1976).

The reason for the discrepancy in results reported on the use of field peas in swine diets is not readily apparent. It has been suggested that differences in type and variety of peas may be a contributing factor. However, little work has been reported on the biological quality of legume proteins from different varieties of seeds. Bajaj *et al.* (1971) utilized rats to compare the protein quality of 28 different lines of field peas grown under similar field conditions. These different lines of peas, when given to weanling rats as a sole source of protein at a level of 10% in an otherwise adequate diet, varied from 18 to 78% of that of casein in ability to support growth and nitrogen retention. There was no correlation between protein quality and the protein content of the different pea lines as measured by nitrogen incorporation efficiency (NIE) or protein efficiency ratio (PER). Reddy *et al.* (1979) chemically determined the crude protein and methionine content of 44 breeding lines of green, yellow (*Pisum sativum* L.), and Austrian winter pea (*Pisum sativum arvense* L.) and each sample was fed raw to chicks with and without supplemental methionine. The crude protein (N $\times$ 6.25) of the peas varied from 19.6 to 25.6%; methionine values ranged from 6.2 to 28% of the protein and those for cystine from 9.9 to 18.5%. Neither protein nor methionine content was correlated with protein quality. Three lines of peas contained significantly higher levels of methionine than others in the sample, but did not support chick growth any better than the lines with lower methionine content. This would suggest that protein and amino acid availability among the pea lines differs or that some lines contain anti-nutritional factors which inhibit chick growth.

In practical diets for pigs, peas can be used to supply a portion or all of the supplemental protein required in cereal-based diets for all ages of swine except the early weaned pig. Although some reports have shown no response to supplemental methionine in diets containing as much as 60% peas, under some conditions and especially for young growing pigs, it may be advantageous to steam pellet or extrude peas before incorporation into the diet and to supplement the diets with low levels (0.1%) of methionine or to combine peas with other ingredients to provide adequate sulfur–amino acids.

SWEET LUPIN

Lupin is a legume whose use has been limited largely to green manure and green forage for ruminants because of the presence of high levels of toxic and bitter alkaloids. The main action of lupin alkaloids appears to be upon the central nervous system. Symptoms of acute poisoning are lack of appetite, fever, and difficult or labored breathing (Garner 1957). Death results from respiratory paralysis.

Genetic selections by plant breeders have resulted in a number of lupin varieties with seeds of low alkaloid content and of improved nutritional value. Although sweet varieties of low toxic alkaloid content were bred and released in 1934 (*Lupinus albus*) and in 1951 (*Lupinus luteus* cv. 'Weiko III'), their seeds were not easily distinguishable from those of the bitter varieties, thus detracting from their use as feed (Gladstones 1972). The recent development of sweet varieties of *Lupinus angustifolius* (Gladstones 1972), which are color-marked with white flowers and seeds, has stimulated considerable interest in Australia and New Zealand in the potential of lupins as an alternative protein source to traditional and increasingly expensive sources of protein for pigs, such as soybean meal, fish meal, and meat meal. *Lupinus angustifolius* cv. 'Uniwhite' was released in Western Australia in 1967. However, 'Uniwhite' is prone to shatter at harvesting and was superseded by the release of cv. 'Uniharvest' in 1971, and 'Unicrop' in 1973 (Corbin and Knight 1977). More recently, a new cv., 'Marri' has been released.

The chemical composition of different species and varieties of sweet lupin is presented in Table 15.51 and the amino acid content is presented in Table 15.52. The protein content of lupin seed ranges from 25 to 44% of the dry matter (Gardiner and Gorman 1965). Based on chemical analysis, only the sulfur–amino acids appear to be limiting. The whole seed contains 5–10% ether extract and about 15% crude fiber. Lupins contain almost no starch. The fatty acid composition of lipids in seeds is remarkably similar to that reported for soybeans. The seed coat, or hull, makes up 20–25% of the weight of the seed and is very low in protein but contains more than 50% fiber. The seed contains no lignins and the seed coats have no detectable phosphorus. Concentrations of trace elements are sufficiently high to make lupins a valuable source of these essential nutrients. *Lupinus albus* has been shown to be a manganese accumulator with levels up to 3750 ppm being reported (Gladstones and Dover 1962; Hill 1977; and Batterham 1979). At this level of concentration, it is possible that manganese toxicity could develop and the use of *L. albus* should be closely monitored.

Alkaloids are present in all of the better varieties of lupins (Leonard 1950,

TABLE 15.51. Chemical Composition of Lupin Seed (*Lupinus* species)

Component	*L. angustifolius*		*L. albus*		*L. luteus*
	'Uniwhite' [a]	*'Unicrop'* [b]	*'Ultra'* [b]	*'Astra'* [c]	*'Weiko III'* [a]
Dry matter, %	89.1	88.9	90.7	85.2	10.9
Crude protein, %	26.4	28.7	31.5	34.2	35.7
Crude fiber, %	15.0	14.5	14.6	8.4	15.8
Ether extract, %	5.8	5.6	7.4	10.1	4.6
Ash, %	3.1	3.3	3.4	3.2	3.6
Gross energy, kcal/kg	—	4376	4472	5020	—
Minerals					
Calcium, %	0.22	—	—	0.23	0.24
Phosphorus, %	0.41	—	—	0.46	0.43
Magnesium, %	0.13	—	—	—	0.19
Iron, ppm	49	51	43	—	65
Zinc, ppm	34	—	—	—	48
Manganese, ppm	36	42	3750	—	102
Copper, ppm	5.9	—	—	—	—

Source: [a]Hove 1974. [b]Batterham 1979. [c]Skoknic et al. 1979.

TABLE 15.52. Amino Acid Composition of Lupin Seed (*Lupinus* species)

Amino acid	*L. angustifolius*			*L. albus*		
	'Uniwhite' [a]	*'Uniharvest'* [a]	*'Unicrop'* [b]	*'Ultra'* [b]	*'Astra'* [c]	*'Neuland'* [a]
Lysine	1.33	1.26	1.55	1.80	1.38	1.41
Histidine	0.56	0.70	0.80	0.66	0.30	0.79
Threonine	0.89	0.91	1.00	1.32	1.18	0.89
Valine	0.97	0.94	1.06	1.35	1.16	1.08
Isoleucine	1.18	1.03	1.06	1.32	1.25	1.06
Leucine	1.84	1.77	1.78	2.17	2.24	2.02
Half cystine	0.17	—	—	—	—	—
Methionine	0.16	0.12	0.52[d]	0.72[d]	0.16	0.12
Tyrosine	0.70	0.65	—	—	—	1.06
Phenylalanine	0.79	0.72	2.10[e]	2.36[e]	1.08	0.90
Arginine	2.20	2.35	3.24	3.02	3.05	2.61
Aspartic acid	2.60	2.21	—	—	3.22	2.99
Glutamic acid	5.80	5.38	—	—	7.11	5.44
Serine	1.20	1.32	—	—	1.36	1.36
Proline	1.18	1.25	—	—	1.02	1.20
Glycine	1.04	1.07	—	—	1.20	1.06
Alanine	1.00	0.93	—	—	0.86	0.99
Tryptophan	—	—	—	—	0.32	—

Source: [a]Pearson and Carr 1979. [b]Batterham 1979. [c]Skoknic et al. 1979.
[d]Methionine plus cystine.
[e]Phenylalanine plus tyrosine.

TABLE 15.53. Alkaloid Content of Seeds of Some Lupin Cultivars

Sample	*Alkaloid, %*[a]
L. angustifolius, bitter blue lupin	0.59
L. albus cv. 'Neuland (a)'	0.09
L. albus cv. 'Neuland (c)'	0.09
Ethanol-extracted 'Neuland (a)'	0.02
L. albus cv. 'Neuland (166)'	0.04
L. albus cv. 'Neuland (167)'	0.02
L. albus cv. 'WB_2'	0.02
L. luteus cv. 'Weiko III'	0.01
L. angustifolius cv. 'Uniwhite'	Trace < 0.01[b]
L. albus cv. 'Ultra'	Trace < 0.01
L. angustifolius cv. 'Unicrop'	Trace < 0.01

Source: Ruiz et al. 1977; Batterham 1979.
[a]Values are expressed on air-dried basis.
[b]Seed kernels only.

1953; Bohlmann and Schulmann 1967) and appear to be present in varying trace amounts in many of the newer sweet varieties. Among sweet lupin varieties, the alkaloid content has been reported to vary from less than 0.01% to 0.09% which is significantly less than the 0.59% determined for bitter lupin (Table 15.53) (Ruiz *et al.* 1977; Batterham 1979).

Although the sweet lupins appear to offer great potential as an alternate source of protein in some areas of Australia and New Zealand, results from feeding trials with growing-finishing pigs have been extremely variable. Variation has been observed not only between species but also between varieties within species.

Early studies on the use of sweet yellow lupin seed meal (*Lupinus luteus*) were reported by Florence (1965). Progressive replacement of fish meal by sweet yellow lupin seed meal in maize-based diets for growing-finishing pigs resulted in a progressive decrease in average daily gain and a progressive increase in feed required per kilogram of gain (Table 15.54) (Florence 1965). Complete replacement of fish meal by lupin meal resulted in a significant decrease in daily feed consumption and daily gain and a significant increase in feed conversion ratio. However, Luchsinger (1973) replaced 0, 25, 50, and 75% of the fish meal in

TABLE 15.54. Use of Sweet Yellow Lupin Seed Meal (*Lupinus luteus*) as a Substitute for Fish Meal for Growing-Finishing Pigs Fed Maize-Based Diets

Level of lupin, % *Level of fish meal, %*	*0* *10.7*	*7.2* *7.7*	*13.8* *4.7*	*19.6* *2.3*	*25.0* *0*
Av daily gain, kg	0.54	0.52	0.51	0.48	0.43
Av daily feed, kg	1.35	1.39	1.37	1.30	1.21
Feed/gain ratio	2.54	2.67	2.69	1.80	2.95

Source: Florence 1965.

growing-finishing pig diets with equal levels of protein from sweet yellow lupin seed meal without observing any detrimental effect on pig performance. Similar results were reported with up to 50% fish meal replacement by sweet yellow lupin meal by Rosa (1974).

Under restricted feeding regimes, promising results have been obtained with *Lupinus angustifolius* cvs. 'Uniwhite,' 'Uniharvest,' and 'Unicrop' (Taverner 1975; Pearson and Carr 1976, 1977; Batterham 1979; Cowdy *et al.* 1979). As supplemental protein sources to wheat and barley diets, sweet white lupin meals ('Uniwhite,' 'Uniharvest,' and 'Unicrop') have supported pig performance equal or superior to that obtained with soybean meal, fish meal, or meat meal (Table 15.55). Taverner (1975) fed 'Uniwhite' seed meal as the only supplemental protein source in a wheat-based diet to pigs between 20 and 80 kg. Growth rate and efficiency of feed utilization of pigs fed lupin meal were better than those observed for meat meal but not significantly lower than those obtained with soybean meal. Adding lysine to the lupin-supplemented diet improved pig performance but the addition of methionine alone was without effect. The best results were obtained when both lysine (0.225%) and methionine (0.041%) were added. In pigs weighing 25–85 kg (Pearson and Carr 1976), barley-based diets containing ground Uniwhite lupin seed as the only protein concentrate supported rates of gain and efficiency of feed utilization similar to those obtained with mixtures of lupin and either fish meal or meat-and-bone meal. However, the addition of lupin seed depressed dressing percentage.

In these studies, the progressive replacement of fish meal by ground lupin seed

TABLE 15.55. Use of Sweet White Lupin Seed Meal (*Lupinus angustifolius*) as a Protein Supplement in Diets for Growing-Finishing Pigs Fed Restricted Levels of Feed Daily

Source/test diets	*Av daily gain, kg*	*Feed/gain ratio*
Taverner 1975		
Soybean meal	559[a]	3.10[a]
Meat meal	508[b]	3.37[b]
Lupin meal ('Uniwhite')	529[a,b]	3.30[a,b]
Pearson and Carr 1976[c]		
Fish meal	534	3.22
12% lupin meal ('Uniwhite')	544	3.12
24% lupin meal ('Uniwhite')	573	2.96
37% lupin meal ('Uniwhite')	600	2.79
Pearson and Carr 1977[d]		
Lupin meal ('Uniwhite')	562	3.28
Lupin meal ('Uniharvest')	576	3.18
Batterham 1979[e]		
Soybean meal	655	2.70
Lupin meal ('Unicrop')	671	2.69

[a,b]Means in the same column without a common superscript are significantly different from each other ($P < .05$).
[c]Significant linear improvement ($P < .01$).
[d,e]Differences not significantly different ($P < .05$).

resulted in a significant linear improvement in both daily gains and efficiency of feed conversion of pigs growing from 55 to 85 kg and 25 to 85 kg liveweight. With the exception of dressing percentage, which declined linearly with increasing levels of lupin seed, variations in diet had no significant effect on indices of carcass composition. In subsequent trials, Pearson and Carr (1977) fed two varieties of *L. angustifolius* 'Uniwhite' and 'Uniharvest.' Performance of pigs fed diets containing 31 or 43% of either of these varieties was similar and from a nutritional standpoint, satisfactory as a replacement of more conventional protein sources in diets given to pigs in restricted amounts. The variety 'Unicrop' was used by Batterham (1979) as a complete substitute for soybean meal. From 20 to 80 kg, pigs fed equal quantities of a wheat-based diet supplemented with sweet white lupin seed meal ('Unicrop') grew at a rate and efficiency not different from that observed when soybean meal supplied the supplemental protein. Of all the parameters measured, only dressing percentage was depressed due to soybean replacement with lupin seed meal.

Although Taverner (1975) demonstrated that under *ad libitum* feeding conditions growing pigs would consume significantly more lupin-supplemented feed and gain weight more rapidly than when fed restricted amounts of the same diet, Pearson and Carr (1977, 1979), Pérez (1980), and Louvard (1980) have reported results which indicate that diets containing sweet white lupin seed meal ('Uniwhite,' 'Kalina') are not as palatable to pigs as diets supplemented with either soybean meal or fish meal and blood meal. In these studies, the replacement of either soybean meal or a mixture of fish meal and blood meal with lupin seed meal depressed voluntary intake and at the higher levels of substitution significantly depressed the rate of gain. It appears that even these white lupin varieties, under some conditions, may contain sufficient alkaloids to reduce palatability and voluntary feed intake.

Although variable results have been observed with *L. luteus* and *L. angustifolius,* the greatest variation has been observed with *L. albus*. Romero *et al.* (1977) obtained a significant linear decrease in rate of gain as the lupin meal from cv. 'Astra' was increased up to 28% of the diet during the growing-finishing period. Skoknic *et al.* (1979) replaced 0, 25, 50, and 75% of the protein from fish meal with lupin meal prepared from *L. albus* cv. 'Astra' and reported satisfactory but reduced pig performance at the two higher levels of substitution. There was a nonsignificant reduction in daily feed intake and a significant reduction in rate of gain when 50% or more of the fish meal was replaced by lupin meal.

Batterham (1979) reported that all of the soybean meal in growing-finishing pig diets could be replaced by lupin seed meal (*L. albus* cv. 'Ultra') without any detrimental effect on growth or efficiency of feed conversion. In contrast, three samples of *L. albus* cv. 'Neuland' studied by Pearson and Carr (1977) proved to be completely unsuitable as ingredients in diets for young growing pigs. Pigs consistently refused all or part of the daily ration of diets prepared from all three samples of 'Neuland.' Performance was poorer and, in some instances, zero or negative weight gains were recorded. Chemical analysis revealed that the 'Neuland' seed contained 0.09% alkaloids consisting mainly of DL-lupanine. Following ethanol extraction which reduced the alkaloid content from 0.09 to 0.02%, a marked improvement in both feed intake and daily gains was recorded, suggesting that alkaloids were the causative factor in the depressed pig performance.

It has been indicated that the ingestion of a relatively large quantity of lupin seed over a relatively brief period of time is generally required for production of acute lupin (alkaloid) poisoning because the alkaloids are readily excreted (Kingsburg 1964). However, the pig is extremely sensitive to alkaloids (Hackbarth 1961; Gladstones 1972) and is adversely affected by levels as low as 0.03% (Pearson and Carr 1977) and voluntary feed intake is depressed at even lower levels (Pearson and Carr 1977, 1979).

It is clear that considerable differences occur between species and varieties of lupin in relation to their suitability as protein sources for pigs. Variation in their alkaloid content is one important determinant of their nutritive value. However, even in cases where the alkaloid content is very low, such as the 'Uniwhite,' 'Uniharvest,' and 'Unicrop' cultivars, the value of lupin seed meal for pigs may still be limited by factors which suppress voluntary feed consumption (Pearson and Carr 1977).

OTHER PROTEIN SOURCES

ALFALFA

Alfalfa (*Medicago sativa*) is a leafy legume plant that contains large amounts of protein. As a source of crude protein, alfalfa produces 2700 kg/ha as compared to 876 kg for maize and 786 kg for soybeans (Stahmann 1968). The results of a 10-y study (Akeson *et al.* 1966) indicate that alfalfa yielded 337 kg/ha of the essential amino acids as compared to 202 kg for soybeans and 95 kg for maize. Alfalfa hay or dehydrated alfalfa meal contains between 16 and 22% crude protein. Specially prepared alfalfa leaf meal can contain up to 30–50% crude protein (Peo *et al.* 1977). Alfalfa is a good natural source of tryptophan, with a content of 0.45% as compared to 0.65% in soybean meal and only 0.09% in maize. The alfalfa plant is an excellent source of nutrients and dehydrated alfalfa meal is commercially available.

Although the potential protein production from alfalfa is encouraging, attempts to utilize alfalfa meal as a primary source of protein in swine diets and specially during the growing-finishing phase have been discouraging. With few exceptions (Ellis and King 1952) published results generally agree that as a percentage of dietary alfalfa increases in swine diets average daily gains decrease and feed-to-gain ratios increase (Table 15.56) (Bohman *et al.* 1953, 1955; Peo *et al.* 1972, 1977; Powley *et al.* 1980; Kass *et al.* 1980A,B; Pond *et al.* 1980). Although some variations in response have been reported as might be expected on the basis of differences in alfalfa composition (Van Soest and Moore 1965), saponin content (Cheeke *et al.* 1977), methods of processing (Myer and Cheeke 1975; Peo *et al.* 1972), and differences in age and condition of animal groups, the level of feed consumption was most commonly cited as the factor responsible for depressed pig performance. Even though the alfalfa diets contained relatively high levels of fiber and lower than normal energy levels, the pigs failed to consume adequate quantities of diet to meet their energy requirements. The consistent absence of an alfalfa-induced increase in dietary feed intake in the studies reported does not agree with the generally observed tendency for feed intake to increase with

TABLE 15.56. Performance of Swine Fed Various Dietary Levels of Alfalfa

Source: alfalfa, %	*Av daily gain, kg*	*Av daily feed, kg*	*Feed/gain ratio*
Bohman et al. 1955:			
0	0.83[a]	2.55	3.06[a]
10	0.78[b]	2.59	3.57[b]
30	0.62[c]	2.45	4.04[c]
50	0.46[d]	2.44	5.40[d]
Peo et al. 1977:			
0	0.87	2.73	3.14
50 of suppl. protein	0.78	2.42	3.10
100 of suppl. protein	0.63	2.36	3.75
Kass et al. 1980A,B:			
0	0.70[a]	2.26	3.23
20	0.67[a]	2.48	3.70
40	0.52[b]	2.55	4.91
60	0.35[c]	2.18	6.23
Pond et al. 1980:			
0	0.69[a]	2.49	3.70[a]
20	0.57[b]	2.44	4.39[b]
Powley et al. 1980:			
0	0.86[a]	3.05	3.66[a]
20	0.73[b]	2.97	4.05[b]
40	0.63[c]	3.17	5.11[c]
60	0.41[d]	2.73	6.41[d]

[a,b,c,d]Within groups by the same author, means in the same column with different superscripts are statistically different.

increased dietary fiber as has been observed with various sources such as purified cellulose (Teague and Hanson 1954; Myer and Cheeke 1975), wheat bran (Hochsteller *et al.* 1959), mill feed (Cole *et al.* 1967), cottonseed hulls (Baird *et al.* 1975), rice mill feed (Baird *et al.* 1970), and rice hulls (Chamorro and Maner 1974).

Factors other than high fiber and low available energy have been studied to explain why pigs will not voluntarily consume sufficient amounts of a high alfalfa diet to allow satisfactory growth performance. Research results have shown that the saponins in alfalfa play a major role in the growth depression of pigs fed diets containing high levels of alfalfa meal. The mechanism by which alfalfa saponins caused growth depression has not been definitely determined. Possible mechanisms include depressing feed intake, complexing with nutrients to render them unavailable, inhibiting digestive enzyme activity, and inhibiting cellular metabolism.

Although saponins are capable of *in vitro* binding of protein (Birk 1969), which could inhibit protein digestion and have also been implicated as trypsin inhibitors (Mooijman 1965), Cheeke *et al.* (1978) have shown that saponin content in the diet (either from high or low saponin alfalfa or from isolated saponin) had no effect on digestibility of crude protein, ether extract, or cell wall constituents. Protein digestibility of low saponin alfalfa meal (1.46% saponin) was not different from that of unselected alfalfa meal (2.37% saponin). The digestion coefficients of 63.1 and 64.0% reported were identical to the value of 64% reported earlier (Cheeke and Myer 1975) and somewhat higher than the figure of 52% reported by Keys *et al.* (1969). Cheeke *et al.* (1978) were also unable to demonstrate any reduction in digestibility of crude protein or ether extract of semi-purified diets when 1 or 2% alfalfa saponins were added.

Method of processing and drying has been indicated as a possible factor limiting the efficient utilization of alfalfa meal by swine (Peo *et al.* 1972). Commercial dehydration of alfalfa involves temperatures in the range of 800°C at the inlet of the drying drum to 125°C at the outlet (Bolton 1962). Fulton (1959) indicated that inlet temperatures have little effect on meal quality, but a change of 25°C at the exit of the drum can alter quality and moisture content considerably. High temperature dehydration can cause caramelization and a complexing of the carbohydrates with amino acids (Maillard reaction). Bitterness may be imparted by caramelization.

To evaluate the effect of processing method on nutritive value, Peo *et al.* (1971, 1972) fed early-weaned baby pigs 16% protein maize–soybean meal diets in which freeze-dried or heat-dried alfalfa meal containing 25% crude protein replaced one-half of the supplemental protein. Pigs fed the maize–soybean meal diet gained significantly faster than those fed the alfalfa diets. Neither drying temperature (121°, 132°, 143°C) nor freeze-drying had any effect on improving pig gains; however, although pigs fed freeze-dried alfalfa diets consumed 20% less feed, freeze-drying produced a highly significant improvement in feed-to-gain ratio, suggesting that nutrient availability or utilization was enhanced by the freeze-drying process. Myer and Cheeke (1975) also showed that the avoidance of heat during the drying of alfalfa meal, through the use of freeze-drying, resulted in improved growth and feed conversion of growing rats. These results are consistent with the hypothesis that bitter compounds may be formed through caramelization reactions during conventional drying methods and that heat-drying has a negative influence on stability or availability of the nutrients in alfalfa.

The growth depressing activity of saponins is probably largely a result of effects on voluntary feed intake. In two-choice feed preference tests (Leamasters and Cheeke 1979), growing swine were offered a maize–soybean meal control diet or one containing commercial, sun-cured alfalfa meal at levels of 0.5, 1, 2.5, 5, 10, 20, and 30%. Intake of the control diet as a percentage of total feed intake was higher at all alfalfa meal levels except 0.5% (Table 15.57). This indicates that the pig can detect very low dietary alfalfa levels, and when given a choice, show a significant preference for an alfalfa-free diet. In these studies, pigs also selected diets containing low-saponin alfalfa over diets containing the same dietary level of high-saponin alfalfa, but rejected both in favor of an alfalfa-free diet. These results are consistent with the hypothesis (Cheeke *et al.* 1977) that the major effect of alfalfa saponins on swine and other nonruminants is a reduction in voluntary feed intake due to the bitterness of the saponin fraction.

If the bitter taste of saponins is responsible in part for low palatability of alfalfa, there may be an interaction in the drying process. Myer and Cheeke (1975) noted that the freeze-dried alfalfa was of much greater palatability to rats than air-dried or oven-dried material. Possibly, saponins undergo heat-stimulated reactions that increase their bitterness. Cheeke and Stangel (1972) noted a marked stimulatory effect of alfalfa saponins on the Maillard reaction, suggesting that saponin–heat

TABLE 15.57. Effect of Dietary Alfalfa Level on Diet Preference by Swine

Dietary alfalfa, %	*Control diet, % intake*	*Alfalfa diet, % intake*
0.5	49[a]	51[a]
1.0	63[a]	37[b]
2.5	66[a]	34[b]
5.0	65[a]	35[b]
10.0	74[a]	26[b]
20.0	81[a]	19[b]
30.0	97[a]	3[b]

Source: Leamasters and Cheeke 1979.
[a,b]Means on the same line with different superscripts are significantly different ($P < .01$).

interactions could be important during alfalfa meal preparation. Saponins could also affect feed intake through effects on digestive tract functions.

The recent development of alfalfa cultivars low in saponins is significant, because of their much greater feeding value for swine. Cheeke *et al.* (1977) have shown that swine fed diets containing low-saponin alfalfa meal grow at a faster and more efficient rate than similar groups fed high-saponin alfalfa meal (Table 15.58). Although the development of low-saponin alfalfa is a positive step in increasing the use of alfalfa meal as a source of protein for nonruminant animals, it appears that saponin content is not the only ''anti-nutritional'' factor in alfalfa, and that continued identification of other negative components is needed before alfalfa meal can be used with optimal effectiveness in diets for nonruminants.

Alfalfa has shown its most beneficial effect as a feedstuff for swine during the reproductive phase. Since it was shown that factors supplied in alfalfa improve reproductive performance of sows fed in dry lot (Freeman 1938; Hogan and Johnson 1941; Ross *et al.* 1944; Cunha *et al.* 1944; Fairbanks *et al.* 1945), it has become a common practice to include 15% or more of alfalfa meal in breeding and gestation rations for swine. Danielson and Noonan (1975) and Pollmann *et al.* (1978) fed 96.75% sun-cured alfalfa diets through three successive gestation cycles and maintained satisfactory reproductive performance. Allee (1976) found

TABLE 15.58. Performance of Pigs Fed Diets Containing High- and Low-Saponin Alfalfa

Treatment	*Av daily gain, kg*	*Av daily feed, kg*	*Feed/gain ratio*
Control	854[a]	2.4	2.81
Low-saponin alfalfa	754[a]	2.4	3.18
High-saponin alfalfa	617[b]	2.2	3.57

Source: Cheeke et al. 1977.
[a,b]Means in the same column with a different superscript are significantly different ($P < .05$).

that self-feeding a diet containing 96.9% dehydrated alfalfa during gestation had no significant effect on reproductive performance.

Alfalfa in brood sow diets has been reported to increase ovulation rate (Teague 1955; Moser 1978), litter size at birth (Cunha *et al.* 1944; Seerley and Wahlstrom 1963; Teague 1955), and baby pig survival at 14 days (Pollmann *et al.* 1980) and at 8 weeks (Hogan and Johnson 1941; Cunha *et al.* 1944; Fairbanks *et al.* 1945). A significant reduction in birth weight of pigs farrowed from sows fed diets based primarily on alfalfa appears to be the only negative effect on reproductive performance (Danielson and Noonan 1975; Allee 1976; Pollman *et al.* 1978, 1980).

Although sows fed diets high in alfalfa meal gained less weight during gestation (Allee 1976; Pollmann *et al.* 1980), sow longevity is significantly increased. Results reported by Pollmann *et al.* (1980) indicate that a higher percentage (89 vs. 68%) of sows fed alfalfa (50% of the diet) completed three reproductive cycles than did similar groups fed the control, maize–soybean meal diet.

Boyd *et al.* (1976) showed that the crude protein in sun-cured alfalfa was 70% digestible and that it contained 4248 kcal/kg DM of gross energy (GE) an average of 2300 kcal/kg DM of DE and 1937–2126 kcal/kg DM of ME. Pollmann *et al.* (1979) reported sun-cured alfalfa to have a ME value of 2019 kcal/kg DM.

ALFALFA PROTEIN CONCENTRATE

Despite the potential for high yields of protein per hectare, alfalfa has not yet become a major source of protein for monogastric animals. Several factors contribute to the fact that alfalfa has not been satisfactorily employed as a major protein source for swine. These include the high fiber and low digestible energy content, the presence of inhibitors such as saponins, the low digestibility of the protein, and unpalatability of high dietary alfalfa levels. These problems are largely overcome if the protein is separated from the fiber and a protein concentrate prepared (Kohler *et al.* 1973) by heat coagulation and a juice pressing process. Alfalfa protein concentrates prepared by these Pro-Xan processes (Bickoff *et al.* 1975) have been shown to contain 45–55% crude protein (Kohler *et al.* 1973; Kohler 1974), less than 3% fiber, 3741 kcal/kg of DE, 3336 kcal/kg of ME (Bourdon *et al.* 1980) and less than 0.07% saponins when heat precipitation is carried out at an alkaline pH (8.5) (Livingston and Kohler 1979).

The process by which alfalfa protein concentrate is produced consists of (1) squeezing or pressing the juice from the fresh plant material followed by (2) the precipitation of the true protein from the expressed juice through the application of steam heat which is injected to a temperature of 80°C or higher. Sudden heating minimizes proteolysis and, in leaves rich in chlorophyllase, the formation of pheophorbide. Pheophorbide in protein concentrates made from slowly heated alfalfa juice has been reported to produce photosensitivity in pigs. Formation of pheophorbide is completely avoided by quick heating to 80°C or more (Pirie 1977). The heat-precipitated protein from the alfalfa juice forms distinct curds which are separated from the alfalfa solubles by settling, decantation, filtration, or centrifugation. A further reduction of the soluble solids content, with a subsequent increase in protein and carotene–xanthophyll levels in the curd, can be achieved by pressing the curd. The pressed curd is granulated or extruded prior to drying in a rotary or other air-drying system. Yield as well as protein content are affected by the age and quality of the original green alfalfa and by the extent to which the solubles are removed from the wet curd prior to drying.

The nomenclature used to describe alfalfa leaf protein concentrates differs according to manufacturer and process used for commercial production. Henceforth in this chapter, the term alfalfa protein concentrate (APC) will be used but where important for explaining differences in results, the commercial or tradename will also be indicated. Some of these products(X-Pro, Pro-Xan) not only differ in nomenclature but also in chemical composition (Table 15.59).

In early studies, Myer and Cheeke (1975), using a commercially prepared APC (X-Pro) containing 36% crude protein, 1.98% lysine, and 0.68% methionine (Table 15.59), found that X-Pro was inferior to soybean meal and cottonseed meal as a protein supplement for maize-based rat diets. Lysine appeared to be the first limiting amino acid. In studies with swine (Myer *et al.* 1975), X-Pro gave results comparable to a mixture of soybean meal and meat-and-bone meal for growing and finishing pigs. Growing and finishing pigs fed diets supplemented with 24% X-Pro grew as rapidly and as efficiently as similar groups fed barley-

TABLE 15.59. Composition of Alfalfa Protein Concentrate (APC)

	Myer et al. 1975 (X-Pro)	*Kuzmicky et al. 1972*	*Hove et al. 1974*	*Bickoff et al. 1975 (Pro-Xan)*	
				Drained	*Pressed*
Crude protein, %	36.0	38.8	35.8	54.0	61.9
Ether extract, %	4.8	6.2	7.4	6.5	8.9
Crude fiber, %	—	2.4	—	1.2	1.7
Ash, %	19.6	19.2	23.2	14.7	11.1
Calcium, %	—	2.35	—	—	—
Phosphorus, %	1.04[a]	0.44[b]	—	—	—
Sodium, %	0.35	0.30	—	—	—
Potassium, %	1.15	1.07	—	—	—
Lysine, %	1.98	2.15	2.20	—	—
Methionine, %	0.68	0.79	0.61	—	—
Amino acid composition, g/100 g amino acid					
Lysine	5.5	5.5	6.2	6.3	6.2
Histidine	2.6	2.3	2.3	2.5	2.3
Arginine	6.7	5.6	—	6.2	6.0
Threonine	5.3	5.0	4.0	5.0	4.8
Valine	6.5	6.6	6.7	6.4	6.3
Methionine	1.9	2.0	1.7	2.0	2.1
Isoleucine	5.2	5.3	5.7	5.2	4.9
Leucine	9.7	8.9	10.3	8.9	8.2
Tyrosine	4.7	4.2	3.7	4.6	4.1
Phenylalanine	7.1	5.6	6.7	5.8	5.3

Source: Adapted from Myer et al. 1975 and Bickoff et al. 1975.
[a]Total phosphorus.
[b]Available phosphorus.

based diets supplemented with a combination of soybean meal and meat-and-bone meal. As a sole protein supplement to a maize–oat based starter diet, X-Pro was generally inferior to soybean meal as a protein supplement. The amino acid composition of Pro-Xan (Bickhoff *et al.* 1975) is shown in Table 15.59 for comparison with that of products studied by Myer and Cheeke (1975), Kuzmicky *et al.* (1972), and Hove *et al.* (1974).

Because of the possibility of heat damage to the protein in the commercial process for X-Pro preparation, Cheeke *et al.* (1977) compared X-Pro with a specially prepared freeze-dried APC sample. Young pigs (18-kg initial weight) fed maize-based diets containing 19% freeze-dried APC grew at a rate and efficiency similar to those obtained with soybean meal as a supplement, whereas gains were significantly reduced with X-Pro (Table 15.60). Follow-up studies (Table 15.60) indicated a clear trend for improved pig performance with reduced drying temperatures.

In Kansas studies (Hsu and Allee 1980), APC containing 55.8% crude protein, 3.41% lysine, and 1.67% tryptophan, on a dry matter basis, was used to replace 33, 66, or 100% of the dehulled soybean meal in diets for starter, grower, and finisher pigs. The level of APC did not affect the feed intake or feed efficiency during the starter, grower or finisher phases. Pigs fed the maize–100% APC diet gained slower than pigs fed the maize–soybean meal diet during the starter period. During the finisher period, the average daily gains for pigs fed the maize–66% APC diet were greater than those fed the control maize–soybean meal diet. For the entire trial, level of APC did not affect feed efficiency but average daily gain was improved when 66% of the soybean meal was replaced with alfalfa protein concentrate. These results suggest that APC can be used to replace soybean meal in a maize-based diet for pigs from 8.7 to 95 kg without any adverse effect on feed utilization or rate of gain.

Studies by Hsu and Allee (1980) indicated that for growing pigs methionine is the first limiting amino acid in APC and suggest that isoleucine and lysine, respectively, are the second and third limiting amino acids in APC.

TABLE 15.60. Performance of Growing Pigs Fed Two Types of Alfalfa Protein Concentrate (APC) and APC Prepared with Different Drying Temperatures

Treatment	*Av daily gain, g*	*Av daily feed, kg*	*Feed/gain ratio*
Maize–soybean meal	675[a]	1.70	2.52[a]
Maize–X-Pro	259[b]	1.40	5.40[b]
Maize–freeze-dried APC	577[a]	1.57	2.72[a]
Follow-up studies:			
Maize–soybean meal	748	1.80	2.4
Maize–freeze-dried APC	713	1.78	2.5
Maize–77°C APC	640	1.66	2.6
Maize–91°C APC	590	1.71	2.9

Source: Cheeke et al. 1977.

[a,b]Means in same column with different superscripts are significantly different ($P < .01$).

The addition of 10 or 20% APC to an all wheat diet fed to growing-finishing pigs progressively improved pig performance as the level of APC was increased. Pigs fed wheat plus 20% APC gained at a 56% faster rate than pigs fed only wheat (Bourdon *et al.* 1980).

ALFALFA AND GRASS JUICE

In recent years, interest has developed in the use of fractionation products of green-chop. Alfalfa juice and grass juice, produced by a pulper and belt-press green-chop fractionation process (Connell and Foxell 1976), have been studied as sources of protein for growing-finishing swine (Barber *et al.* 1979; Barber *et al.* 1980).

Freshly produced alfalfa juice contains 9–10% solids, 2–4% crude protein, and 1.5–3.5% true protein. However, during storage there is a loss of true protein and an increased concentration of nonprotein nitrogen. Conversion of true protein to nonprotein nitrogen can be reduced by the addition of a preservative such as 1.5 g/liter of sodium metabisulphite together with acidification with hydrochloric acid to pH 3. If, in addition to these preservatives, the juice is heated to 85°C for a few seconds, loss of true protein over a period of many weeks is negligible.

Barber *et al.* (1979) fed properly preserved alfalfa juice or grass juice to replace one-half or all of the protein provided by 7% white-fish meal included in an all-meal control diet. When the juice was given according to its true protein content and was used to replace 3.5% white-fish meal protein throughout the growing-finishing period from 20 to 90 kg liveweight, performance and carcass quality were comparable to those of control pigs fed the 7% fish meal diet. When the juice replaced 7% white-fish meal protein, pig performance was inferior to that observed with the control 7% white-fish meal diet.

Alfalfa juice produced daily and fed fresh was supplied to growing pigs (20–60 kg) in amounts required to replace one-half of the protein supplied by 7% white-fish meal in the all meal control diet. Compared with the control pigs fed the all meal diet, those given juice grew at the same rate but had a significantly poorer feed-to-gain ratio and higher daily feed intake. They also had a thicker back fat and poorer carcass grade.

From these studies it may be concluded that fresh or properly preserved alfalfa juice is readily consumed by pigs and, should it become readily available and economical, can be given to growing-finishing pigs to provide one-half of the normal protein supplement in the diet, provided the limitations of feed utilization and carcass quality are acceptable.

Italian rye grass juice was prepared by Houseman *et al.* (1978A,B) and fed to pigs to replace all or part of the soybean meal in barley-based diets. Pigs fed four parts grass juice and one part barley or two parts grass juice and one part of a mixture of barley and 7.5% soybean meal grew as rapidly and as efficiently as pigs fed a barley–soybean meal basal diet (Table 15.61).

Fresh red clover juice fed in varying proportions with barley supported gain similar to that obtained with a barley–7.5% soybean mixture, but inferior to that obtained with a barley–15% soybean meal mixture (Table 15.62). Feed utiliza-

TABLE 15.61. Performance of Pigs Fed Fresh Italian Rye Grass Juice

Dietary treatments	*Av daily gain, g*	*Feed/gain ratio*
Barley	563	3.84
Barley + 75.0% SBM[a]	718	3.21
Barley + 15.0% SBM	779	3.21
Barley (1 part) + grass juice (4 parts)	775	3.65
Barley + 7.5% SMB (1 part) + grass juice (2 parts)	830	3.21

Source: Houseman et al. 1978A.
[a]SBM, soybean meal.

TABLE 15.62. Performance of Finishing Pigs Fed Fresh Red Clover Juice

Diets	*Av daily gain, g*	*Feed/gain ratio*	*Digestion*		*Nitrogen retention, %*
			N, %	*Energy, %*	
Barley	528	3.87	72.3	78.9	9.7
Barley + 7.5% SBM	693	3.08	75.7	79.9	16.4
Barley + 15.0% SBM	781	2.92	77.3	80.8	19.5
Barley (1 part) + clover juice (2.5 parts)	637	3.69	—	—	—
Barley (1 part) + clover juice (3.0 parts)	678	3.66	74.3	79.4	19.9
Barley (1 part) + clover juice (3.5 parts)	697	3.71	—	—	—

Source: Houseman et al. 1978B.

tion of pigs fed clover juice was significantly inferior to that of either of the control groups.

SINGLE-CELL PROTEIN

Consciously or not, man has always used microorganisms such as nonpathogenic bacteria, yeast, fungi, and microalgae in his food system. However, little use has, so far, been made of the microorganism-biomass itself as a component of feedstuffs. An exception is the microalgae *Spirulina* used as food in Chad by lakeshore people and in Mexico by the Aztecs living around Lake Texcoco, and yeast from industrial production used in mixed feeds for animals. In recent years, new emphasis has been placed on microorganisms for the purpose of developing methods of food production that would be independent of the traditional soil/climate ecosystem.

Several factors have stimulated the recent research and development interest in single-cell protein. First, the growing demand for human food has created a shortage of protein concentrates such as oilseed and soybean meal, and second, there is concern over environmental pollution resulting from industrial and agricultural waste products such as sulphite waste and animal waste.

Substrates for microbial growth and protein recovery have been evaluated against such criteria as technical suitability, availability and cost. The last is important because substrates, such as molasses and paraffins, may have alternate industrial demands and consequently be competitively priced. On the other hand, the substrate may be an industrial or agricultural by-product or waste material such as sulphite waste liquor, potato starch waste, or animal fecal waste that might even raise disposal problems. At the present time, research and development work is being carried out to recover nutrients from effluents and waste through fermentation technology and to produce single-cell protein from *n*-alkanes, methane, and synthetic methanol and ethanol.

Whatever the substrate may be, the end product is a biomass composed of bacteria, yeast, fungi, or algae. The original term coined for this biomass was *single-cell protein* (SCP), which is somewhat misleading since many of the fungi are not monocellular.

Single-cell protein has been undergoing thorough investigation with regard to nutritional value and safety for human and animal consumption. Such investigation has been centered on SCP grown on such hydrocarbons as *n*-alkanes and methanol. Studies of SCP grown on animal waste have also stressed the effect of accumulated feed additives such as trace elements, antibiotics, and arsenicals. On the other hand, yeast (*Candida utilis*), grown mainly on sulphite waste liquor, molasses, whey, or other carbohydrates, has never been subjected to such scrutiny. Because of differences in chemical and physical characteristics and nutritional value, fungi, yeast, algae, and bacterial protein will be discussed separately.

Fungi

In Finland, a process for the commercial production of a micro-fungus, *Paecilomyces variottii,* has been perfected. The process involves the continuous culture of the fungi on spent sulphite liquor, a waste by-product of the wood pulping industry. In addition to producing a multicellular and filamentous fungus that is readily separated from the broth by simple filtration, the process removes acetic acid and sugars from the spent liquor, hence reducing its biological oxygen demand and pollution characteristics. Following mechanical dewatering and artificial drying, the material, which has been given the name *Pekilo protein,* is cream colored, has very little taste or odor, and is readily accepted by swine as a constituent of their diet. The chemical composition is shown in Table 15.63.

Pekilo protein has been successfully utilized as a sole protein source to supplement cereal-based diets for early weaned pigs (Bobrov *et al.* 1978) and for growing-finishing pigs (Alaviuhkola *et al.* 1975; Barber *et al.* 1977). Finnish workers (Alaviuhkola 1975) replaced 0, 33, 66, and 100% of the digestible crude protein of skim milk powder with the digestible crude protein of Pekilo in a barley-based diet for growing-finishing pigs. For 20–90-kg liveweight, all levels of Pekilo protein supported pig performance equal to that recorded when skim milk powder was used as a protein supplement (Table 15.64). Barber *et al.* (1977)

TABLE 15.63. Proximate Analysis and Mineral and Amino Acid Content of Fungal Protein (Pekilo Protein)

Proximate analysis and mineral content	*%*	*Amino acids*	*g/16 g N*
Dry matter	95.0	Alanine	5.81
Protein (N × 6.25)	55.9	Arginine	5.68
Ether extract	1.1	Aspartic acid	8.65
Crude fiber	8.8	Cystine	0.84
Ash	5.6	Glutamic acid	10.55
Nitrogen-free extract	28.6	Glycine	4.52
Digestible crude protein[b]	43.4	Histidine	1.89
Total carbohydrates	33.2	Isoleucine	4.24
Sugars	1.1	Leucine	6.74
Hemicellulose	24.5	Lysine	5.85
Cellulose	7.6	Methionine	1.48
Crude lignin	4.5	Phenylalanine	3.57
Metabolizable energy, kcal/kg DM	3205	Proline	3.75
Calcium	0.07	Serine	3.85
Phosphorus	0.20	Threonine	4.60
		Tryptophan	1.31
		Tyrosine	3.29
		Valine	4.97

Source: [a]Alaviuhkola et al. 1975. [b]Salo and Pekkarinen 1981.

were also unable to show any difference in nutritive value between Pekilo protein and white-fish meal when both were used as the sole protein supplement to cereal-based diets for pigs between 20- and 90-kg liveweight. Pig performance was not different when Pekilo protein replaced white-fish meal at both a standard and low level of protein.

Since no gross or microscopic abnormalities have been observed in pigs fed Pekilo protein, it can be fed as the only supplemental protein source for starter, growing, and finishing pigs. Although no research data are available for reproducing animals, Pekilo should be an excellent supplemental protein source for them.

TABLE 15.64. Performance of Growing-Finishing Pigs Fed Fungal Protein (Pekilo Protein) as Replacement for Skim Milk Powder

Level of fungal protein, %	*0*	*32*	*66*	*100*
Performance criteria	*1*	*2*	*3*	*4*
Av daily gain, kg	0.72	0.72	0.73	0.71
Av daily feed, kg	2.06	2.05	2.05	2.06
Feed/gain ratio	2.87	2.87	2.82	2.90

Source: Alaviuhkola et al. 1975.

Bacteria

In recent years, the intense efforts to make use of microbial fermentation for the production of feeds containing protein have shown the potential of bacterial cells as an alternative to conventional protein sources. The Imperial Chemical Industries of Great Britain have used the bacteria, *Methylophilus methylotrophus* (more recently classified as *Pseudomonas methylotropha*), to produce a bacterial protein of high biological value. The product called Pruteen® is produced by continuous fermentation of the bacteria on low concentrations of methanol as the energy substrate and ammonia as the nitrogen source. Pruteen contains 72–79% crude protein (N × 6.25) of which approximately 15% is nucleic acid. The composition of Pruteen is shown in Table 15.65.

It seems that the apparent digestibility of nitrogen and gross energy of the dried bacterial protein is high. Whittemore and Moffot (1976) reported apparent digestibilities of nitrogen and gross energy to be 90 and 85% respectively, when Pruteen was fed in cereal-based diets and much higher (N, 94%; GE, 95%) when fed in a semipurified diet (D'Mello *et al.* 1976). However, Schulz and Oslage (1976) found, in experiments with pigs, lower digestibility of 84% for nitrogen and 78% for energy. Biological value of protein has been determined in young pigs to be 68–79% percent and net protein utilization to be 67–76% (D'Mello *et al.* 1976; Schulz and Oslage 1976). Pruteen contains 4972 kcal of GE, 4450 kcal of DE, and a calculated net energy (NE) value of 2185 kcal/kg DM (Eeckout

TABLE 15.65. Chemical Composition of Bacterial Protein (Pruteen)

Constituent, %	*Eeckhout 1980*	*Schulz and Oslage 1976*	*Braude et al. 1977*
Dry matter	89.63	96.7	96.44
Crude protein	72.96	81.9	74.00
Crude fiber	—	0.5	0.38
Ether extract	7.03	7.9	2.62
Ash	7.59	9.7	11.21
Nitrogen-free extract	2.05	—	8.23
Calcium	0.06	—	.04
Phosphorus	2.21	—	2.62
Arginine	3.73	3.69	3.55
Histidine	1.32	1.39	1.48
Isoleucine	3.71	3.60	3.40
Leucine	5.65	5.70	5.40
Lysine	4.90	4.83	4.81
Methionine	1.50	2.94[a]	1.85
Cystine	0.43	—	0.59
Phenylalanine	2.74	—	2.74
Tyrosine	2.23	—	2.14
Threonine	3.24	3.44	3.18
Valine	4.12	4.26	0.37
Tryptophan	—	—	0.67

[a]Methionine + cystine.

1980). Eeckout (1980) has also shown the phosphorus in Pruteen to be about 48.8% available.

In feeding trials with growing and growing-finishing pigs, Braude and Rhodes (1977) and Braude *et al.* (1977) fed Pruteen as a replacement for white-fish meal in diets based on barley and fine wheat offal. In all three trials reported, Pruteen supported pig performance similar to that of pigs fed white-fish meal as a protein supplement. There were no significant differences in growth rate, feed efficiency, nitrogen retention, or apparent digestibility of nitrogen, energy, or most of the amino acids between diets supplemented with Pruteen and those supplemented with white-fish meal. The pigs were also shown to be able to dispose of nucleic acid at the levels of intake tested (0.49–1.13% nucleic acid in the diet). Schulz and Oslage (1976) also reported good pig performance when bacterial protein (*Pseudomonas*) was used as the sole protein supplement. However, when higher levels of dietary Pruteen (28.6%) were added to barley-based diets for growing pigs to provide 30% crude protein, pig performance was unsatisfactory (Farstad *et al.* 1980). When compared to soybean meal or hydrocarbon yeast, growth rate and feed efficiency were significantly poorer for pigs fed bacterial protein (Pruteen). Furthermore, compared with other groups receiving high levels of protein from soybean meal and yeast in the feed, the kidneys of pigs from the Pruteen group were reported to be strikingly enlarged, though no gross pathological changes were observed. Histopathological examination revealed, however, some swelling of kidney glumeruli and nephrons. Farstad (1977) had previously reported the occasional occurrence of relatively serious liver damage in chickens fed on high levels of Pruteen, and Stringer (1978) noted reduced appetite initially and increased kidney weight in pigs fed high levels of Pruteen. Although high levels of nucleic acids in Pruteen cannot be ruled out as a causative agent, endotoxins produced from gram-negative bacteria, which are higher in Pruteen than in conventional feeds (Naess 1978), also must be considered.

As a protein source for early weaned pigs, Pruteen has produced variable results. Whittemore *et al.* (1976) compared dried bacterial cells (Pruteen) with white-fish meal at 10 and 20% levels of inclusion in diets for piglets from 21 to 42 days of age. At both levels of inclusion of Pruteen in mixed cereal–whey diets, pig performance was not different from that of pigs fed white-fish meal (Table 15.66); however, in studies by Newport and Keal (1980), who replaced 50, 75, and 100% of a dried skim milk diet with a spray-dried powder containing 30.8% Pruteen, levels of replacement higher than 50% reduced both average daily gains and efficiency of feed utilization of piglets between 2 and 28 days (Table 15.66).

Pruteen also has been used as the only supplemental protein source to cereal-based diets for breeding pigs through six parities (Waterworth *et al.* 1978). Diets containing 0 or 8% Pruteen replacing all of the noncereal protein were given to sister pair pigs from 8 weeks of age to weaning of their sixth litter. There were no differences between treatments for growth rate to first service or for the number of gilts reaching sexual maturity. The mean gain to 110 days of gestation and the mean weight loss in lactation to weaning at 3 weeks over the six parities were 40.5 and 19.1 kg, respectively, for the control group and 40.6 and 18.2 kg for the Pruteen-fed group. There was no difference between treatments for sow fertility as measured by the period from weaning to effective service (control 11.7, Pruteen 11.1 days). There were no significant dietary effects on the numbers of

TABLE 15.66. Effect of Replacement of White-Fish Meal or Milk Protein by Bacterial Protein (Pruteen) in Diets for Early Weaned Pigs

Source: Diet	*Average daily*		*Feed/gain ratio*
	Feed, g/day	*Gain, g/day*	
Whittemore et al. 1976:			
10% white-fish meal	542	230	2.36
20% white-fish meal	410	241	1.70
10% bacterial protein	543	227	2.39
20% bacterial protein	410	263	1.56
Newport and Keal 1980:			
100% skim milk powder	264	307[a]	0.86[a]
50% Pruteen–whey	266	299[a]	0.89[a]
75% Pruteen–whey	240	261[b]	0.92[a,b]
100% Pruteen–whey	245	250[b]	0.98[b]

[a,b]Means in the same column with different superscripts are statistically different.

pigs born alive, average birth weight, or number and weight of pigs at weaning. Twenty-five sows killed for examination revealed no health problems or pathological changes.

These results indicate that, in general, bacterial protein (Pruteen) can be used to supply all or a major portion of the supplemental protein in cereal-based diets for both starter and growing-finishing pigs and for reproducing animals during growth, gestation, and lactation. However, additional research is needed to establish satisfactory dietary levels of bacterial protein that can be fed to pigs weaned at a very early age.

Yeast

Bakers' yeast (*Saccharomyces cerevisiae*) and torula yeast (*Candida utilis*) have been used for many years to supply a portion of the supplemental protein in creep and starter diets for early weaned pigs and in some countries, such as Cuba, as the principal supplemental protein source for swine of all ages. In recent years, increasing attention has been paid to the problem of the world shortage of protein and especially protein of high quality. A search for new sources led to the production of yeast (*Candida lypolytica, Candida boidinii,* and *Pichia aganobii*) on hydrocarbons obtained from oil. A culture of yeast is propagated by blending with petroleum hydrocarbons, a nitrogen source, such as ammonia, and an appropriate mixture of mineral salts. Production of yeast (*Candida utilis*) on sulphite spent liquor also has been increased. Under controlled conditions, the yeast culture multiplies and results in a very efficiently produced protein product which can be washed and dried for animal feed (Waldroup 1972). The typical chemical compositions of petroleum-derived yeast and yeast produced from molasses and sulphite spent liquor are shown in Table 15.67. Although the nutritive content varies somewhat with species and production method, the content of amino acids

TABLE 15.67. Chemical Composition of Yeast Grown on Different Substrates

Composition	*1*[a]	*2*[b]	*3*[c]	*4*[d]	*5*[e]
Dry matter, %	93.7	95.2	92.5	94.0	94.0
Crude protein, %	53.5	62.0	58.2	46.2	46.4
Crude fiber, %	3.2	—	1.0	5.8	6.7
Ether extract, %	2.4	1.6	0.3	2.1	1.7
Ether extract (HCl), %	8.0	—	—	6.2	7.4
Ash, %	11.6	5.7	7.9	6.7	6.4
Nitrogen-free extract, %	23.0	25.9	32.6	33.2	27.1
Calcium, %	0.03	0.07	—	—	—
Phosphorus, %	2.92	1.50	—	—	—
Magnesium, %	0.44	—	—	—	—
Sodium, %	0.05	—	—	—	—
Potassium, %	3.09	—	—	—	—
Iron, mg/kg	607	—	—	—	—
Zinc, mg/kg	657	—	—	—	—
Copper, mg/kg	46	—	—	—	—
Manganese, mg/kg	156	—	—	—	—

[a]From Russo et al. 1976, for yeast produced on *n*-paraffin, "Liquipron" produced by Liquichimica Biosintesi SpA Company, Milan, Italy.
[b]From Barber et al. 1971, for yeast (*Candida lypolytica*) produced on *n*-paraffin by British Petroleum Proteins Ltd., London.
[c]From Farstad et al. 1980, for yeast (*Pichia aganabii*) produced on methanol by Mitsubishi Gas Chemical Company, Japan.
[d]From Salo and Pekkarinen 1981, who used yeast (*Candida utilis*) produced on sulphite spent liquor from pulping industry in Finland.
[e]From Salo 1977, for bakers' yeast (*Saccharomyces cerevisiae*) produced on sugar beet molasses by Oy Alko Ab, Helsinki, Finland.

is quite similar among strains (Table 15.68). The methionine–cystine content of all yeasts appears to be a limiting factor.

The use of yeast in animal feeding is subject to certain essential conditions such as: competitive production costs in relation to those of conventional protein sources; an amino acid composition which, together with cereals, can satisfy the needs of animals; and the absence of toxic and carcinogenic effects, both on the animals being fed and on humans who consume the products derived from it. The toxic effect on humans has been fully studied in The Netherlands and U.S.S.R. by administering large quantities of yeast, or products (meat, eggs) from animals fed with yeasts, to thousands of rats and mice over several generations. The results of these tests showed no toxic or carcinogenic effects (Pokrousky 1968; Engel 1972). Other experiments with broilers fed with diets containing up to 32% yeast (Russo *et al.* 1973, 1974) and with pigs fed with diets containing up to 40.0% yeast (Farstad *et al.* 1980) also have failed to show any acute toxic effects.

In recent years many reports of feeding trials with pigs have been published. Van der Wal (1966, 1970), Fevrier (1971), Madsen *et al.* (1971), Shacklody and Gatumel (1972), Bergonzini and Fabbri (1974), Barber *et al.* (1971), Gaye and Moal (1972), Schulz and Oslage (1976), and Bergonzine *et al.* (1978) have included hydrocarbon yeast in growing-finishing diets in amounts of 5–15%, and in one test up to 30%. On the average, the performance in these trials of pigs fed

TABLE 15.68. Amino Acid Composition of Yeast Grown on Hydrocarbons or Molasses

Amino acid, %	*n-paraffins*		*Molasses*
	1[a]	*2*[b]	*3*[c]
Arginine	3.37	3.04	2.60
Histidine	1.32	1.42	1.06
Isoleucine	2.85	3.03	2.16
Leucine	4.45	4.52	3.35
Lysine	4.47	4.54	4.17
Methionine	0.86	1.09	1.60[d]
Cystine	0.73	0.59	—
Phenylalanine	2.83	2.77	—
Tyrosine	2.53	2.29	—
Threonine	3.00	2.93	2.60
Tryptophan	0.85	0.84	—
Valine	3.40	3.40	2.81
Aspartic acid	5.20	5.49	—
Glutamic acid	8.75	7.44	—
Glycine	2.66	2.90	2.32
Alanine	3.63	4.69	—
Proline	2.26	2.63	—
Serine	2.65	2.72	—

Sources: [a]Russo et al. 1976 (*Candida lypolytica*). [b]Barber et al. 1971 (*Candida lypolytica*). [c]Schulz and Oslage 1976 (Bakers' yeast—*Saccharomyces cerevisiae*). [d]Methionine and cystine.

with yeast was just as good as that of animals which were given soybean meal and/or fish meal. Russo *et al.* (1976) were equally successful with 5–15% paraffin yeast, in the diet of pigs in the range of 9–155 kg liveweight, as with soybean meal and/or animal protein. Farstad *et al.* (1980) used methanol-grown yeast (*Pichia aganobii*) at a level of 40.8% to provide up to 30% crude protein in the diet and reported pig performance equal to that obtained with equivalent amounts of protein from soybean for pigs from 21 to 80 kg.

However, some reports have indicated less success using yeast in feeds. Beck *et al.* (1974) reported a decrease in growing-finishing pig performance with a dietary level of only 6% hydrocarbon yeast. Isar *et al.* (1978) also reported depressed gains and feed consumption of pigs fed diets containing paraffin yeast as the sole supplemental protein source. Frydrych *et al.* (1980) also reported depressed pig performance when 6.1% ethanol- or ethanol–sulphite-grown yeast was used to replace a mixture of plant and animal proteins in wheat–barley-based diets for growing pigs (22–50 kg liveweight), but showed that this depression could be improved by the addition of lysine and completely overcome by the addition of lysine, methionine, and isoleucine.

The inclusion of up to 15% paraffin yeast in piglet feeds also has proved successful (Shacklody and Gatumel 1972; Russo *et al.* 1976). However, with yeast levels of more than 20% in piglet feeds, adjustment of methionine–cystine level is necessary (Nielsen *et al.* 1974).

Although information on the effect of yeast on breeding animals is limited, Shacklody and Gatumel (1972) reported studies in which the effect of 10% yeast in sow diets was studied over four generations. When compared with soybean meal, yeast produced no distinct differences in either fertility, number of stillborn, or number and weight of piglets reared per sow. In each case no mutagenic or teratogenic effects were established in the next generation.

Although care should be taken to assure adequate levels of dietary sulphur–amino acids, especially for early weaned piglets, yeast grown on hydrocarbons, sulphite spent liquor, or other products such as molasses and whey can be used at normal levels to supply all or a major portion of the supplementary protein to practical rations for all classes of swine.

Algae

There is keen interest from the theoretical and applied standpoint, in the mass production of algae. Their photosynthetic ability, which eliminates the need for applying a carbon substrate and their efficiency of solar energy utilization give algae particular appeal as a protein source for both human and animal food. Although algae are among the more attractive of the microbial protein sources, having the potential ability to produce approximately 10 times as much protein per unit land area as soybeans (Oswald 1962, 1968), the difficulty of practical production and harvest reduces their immediate potential.

Most researchers have concentrated primarily on the technology of production and nutritional toxicological evaluation. Only a few species of algae have been sufficiently studied from the viewpoint of their growth and production characteristics. Among them are the green algae (*Scenedesmus acutus*) and *Chlorella pyrenidosa*) and the blue–green algae (*Arthrospira platensis* and *Spirulina maxima*).

Although many countries, including Czechoslovakia, the United States, and Great Britain, are experimenting with freshwater algae production, commercial production is in its infancy. Under present production methods and conditions, commercial production is uneconomical. In addition to their high-quality protein, algae offer a useful means of water purification. Not only are problems stemming from animal waste and waste water treatment reduced by allowing microalgae to grow in it and decompose the organic waste material, but algal proteins derived from waste and waste-water-grown algae can be used in diets for livestock.

Considerable industrial effort has been expended on the use of the effluent from methane generation for the production of algae such as *Chlorella* or *Scenedesmus* as a source of single-cell proteins (Oswald 1962; Oswald and Gouleke 1968; Asplund and Pfander 1973). Yet a major difficulty is the harvesting of the cells because they are very small and require flocculation or centrifugation which are both expensive processes (Vincent 1971). Recent work at Cornell University (Chung *et al.* 1978) has shown that the blue–green algae (*Arthrospira platensis*) can be successfully grown on swine effluent even after it has been first utilized for methane generation. The major advantage of *Arthrospira* sp. is that, being filamentous, it can be harvested by filtration, and centrifugation or flocculation are not required. At the University of Florida, practical tests presently are being carried out to measure algae production from the effluent of 300 pigs and in

TABLE 15.69. Composition of Freshwater Green Algae, Expressed as Percent of Dry Matter

Components	*Sample*	
	1[a]	*2*[b]
Dry matter	—	90.54
Crude protein	50.93	54.85
Crude fiber	6.20	7.01
Ether extract	6.01	4.66
Nitrogen-free extract	—	14.55
Ash	6.24	—
Cellulose	3.33	—
Lignin	4.21	—
Calcium	1.93	—
Phosphorus	2.22	—
Silica	1.73	—
Magnesium	1.60	—
Potassium	0.92	—
Iron	0.23	—
Sodium	0.23	—
Zinc	0.18	—
Aluminum	0.12	—
Manganese	0.03	—
Copper	0.01	—
Lead	0.01	—
Carotene	221.4[c]	—

[a]Hintz et al. 1966. Sewage grown mixture of *Chlorella* spp., *Scendesmus obliquus* and *S. quadricauda*, expressed as percent of dry matter.
[b]Witt and Schroder 1967. Sample of *Scendesmus obliquus*.
[c]Micrograms per grams.

Kansas a commercial swine farm is feeding 2000 pigs/day on algae produced from swine waste (Glenn 1979).

Chemically, the algae contain levels of crude protein similar to or higher than those found in soybean meal and a fiber content similar to that found in barley. The ash and ether extract levels are quite variable. Calcium, phosphorus, silica, magnesium, and potassium are the principal mineral components (Hintz *et al.* 1966). A summary of chemical analysis is given in Table 15.69 and a complete amino acid profile of three species is tabulated in Table 15.70.

The green algae (*Scenedesmus* and *Chlorella* sp.) have an unpleasant, bitter taste that reduces palatability and limits the quantity that can be used in both human and animal diets. These unfavorable palatability factors, however, appeared to be overcome in studies by Hintz *et al.* (1966) who utilized green algae for swine feeding after pelleting the processed algae with steam-rolled barley. In addition to the problem of acceptability, protein availability is also limited by the durable cell wall of the algae. The cell wall cannot be broken easily by the nonruminant animal and, therefore, protein release and digestibility are reduced. Protein digestibility by swine has been reported to be between 66 and 72% (Hintz

TABLE 15.70. Amino Acid Content of Algae (g/16 g N)

Amino acid	*Arthrospira platensis*[a]	*Arthrospira platensis*[b]	*Chlorella sp.*	*Scenedesmus sp.*
Arginine	6.54	8.22	5.50	5.52
Histidine	1.71	2.48	2.01	1.60
Isoleucine	4.32	4.40	3.14	2.88
Leucine	7.08	7.88	6.29	5.65
Lysine	3.67	4.60	4.73	4.08
Methionine	1.51	1.34	1.39	1.23
Cystine	0.56	0.79	—	0.30
Phenylalanine	3.32	4.07	3.52	4.11
Tyrosine	2.94	3.36	2.40	2.29
Threonine	2.39	3.03	2.24	2.70
Valine	4.24	4.80	3.73	6.42
Alanine	5.18	6.27	4.91	4.06
Aspartic acid	5.44	6.92	4.38	4.92
Glutamic acid	8.32	10.84	7.33	6.79
Glycine	3.51	4.11	4.25	3.40
Proline	3.02	3.61	3.24	2.39
Serine	2.94	3.05	1.95	2.21
Total amino acids	66.33	79.77	61.00	60.55
Dry matter, %	92.90	97.42	98.74	90.00
Crude protein, % DM	58.9	62.9	63.1	22.2

Source: Adapted from Chung et al. 1978.
[a]Grown on chemical medium.
[b]Grown on swine waste.

and Heitman 1967). Various methods of breaking the algae cell wall include the use of a ball mill, enzymatic methods, urea soaking, steam cooking, and autoclaving. Digestion trials with swine have shown that the protein fraction of green algae is approximately 72% digestible after being subjected to pelleting and that it is a low-energy feed containing only 1571 kcal of DE/kg.

The biological value of green algae protein is low when fed uncooked as the only source of protein. Although boiling slightly improves the protein efficiency ratio (PER), the value of the protein remains much lower than that obtained with nonfat dry milk (1.85 vs. 2.64) (Cook 1963). Weng (1978) showed that autoclaving of *Chlorella* significantly improved the protein efficiency ratio when fed to rats. Although untreated *Chlorella* (PER, 2.51) had a PER value lower than either casein (PER, 2.87) or soybean meal (PER, 2.61), autoclaved *Chlorella* (PER, 2.71) had a value less than that of casein but superior to soybean meal. Because of its bitter taste and protein of low biological value, green algae should not be utilized in quantities greater than about 10% in diets for growing swine (Hintz and Heitman 1967; Rigor *et al.* 1980). It can be used to replace as much as 75% of the protein supplement in fish meal–barley (Witt *et al.* 1962) or meat-and-bone scrap– barley (Hintz and Heitman 1967) diets without decreasing growth rate or efficiency of feed utilization.

TABLE 15.71. Content of Several Elements in Dried Sewage-Grown Algae

Element, ppm	*Scenedesmus*	*Chlorella*
Mercury	0.30	0.26
Copper	45.3	24.2
Cadmium	1.6	1.4
Lead	3.8	8.1
Aluminum	33.9	0.33
Arsenic	1.1	3.6

Source: Yanni et al. 1979.

Blue–green algae has a protein content which varies from 50 to 61% (Chung *et al.* 1978; Devi *et al.* 1981) and of which nearly 9.9% is nonprotein nitrogen. Chung *et al.* (1978) have shown the protein of *Arthrospira platensis* to be 76.7 digestible and to have a PER of 2.25 and a biological value of 67.7%. Feeding trials with rats and 3-day-old pigs have shown *Arthrospira* to be inferior to casein and soybean meal as a protein source (Yap *et al.* 1978; Chung *et al.* 1978). More extensive swine feeding trials are needed before its value as a supplemental protein source can be determined.

Extensive toxicological evaluation of animals fed algae species grown on animal waste and sewage waste water have not shown any evidence of toxicity. Histopathological observations of liver, kidney, heart, lungs, spleen, stomach, intestines, testes, and lymph nodes showed no abnormalities of rats fed *Arthrospira* in studies by Chung *et al.* (1978). In studies with green algae (*Scenedesmus acutus*), rats fed 10–15% algae for 12 weeks did not reveal any harmful histological or haematological effects. Yap *et al.* (1982) also were unable to show any evidence of toxicity or changes in hemoglobin, hematocrit, serum protein, albumin, or urea when *Arthrospira plantensis, Spirulina maxima,* or *Chlorella* sp. were fed for 12 days to 3-day-old pigs.

The toxic metals content of some algae species grown on waste water is quite high (Table 15.71) but this high content has not been reflected in an increase in the content of toxic metals in the tissues of animals reared on these algae (Yanni *et al.* 1979). The only exception has been the higher arsenic content noted in the livers of chickens reared on waste water algae.

SEAWEED

Seaweed is plentiful in many areas of the world. It offers a cheap source of swine feed if it can be used to supplement the diet. There are many species of seaweed, each undergoing considerable seasonal variation in chemical composition. Seaweed contains relatively low levels of crude protein (2.5–15%) and energy (526 kcal DE/kg DM) and significant quantities of minerals (ash, 15–50%) (Table 15.72). The principal metallic ions present are sodium, po-

TABLE 15.72. Chemical Composition of Seaweed

Component	Sample			
	1[a]	*2*[b]	*3*[c]	*4*[d]
Dry matter, %	27.00	83.69	89.01	—
Crude protein, %	6.00	13.86	9.42	—
Fiber, %	4.41	5.61	20.12	—
Fat, %	0.61	0.28	0.15	—
Nitrogen-free extract, %	8.44	38.52	—	—
Ash, %	9.00	24.42	12.77	—
Calcium, %	2.81	—	1.96	—
Phosphorus, %	0.58	—	0.36	—
Iodine, %	0.20	—	—	0.50
Magnesium, %	—	—	1.64	—
Organic matter, %	—	59.27	—	—
Iron, mg	—	—	—	38–504
Manganese, mg	—	—	—	0–575
Zinc, mg	—	—	—	10–83

[a]Malajskajte 1963. [b]Sobczak and Szlichcinska 1964.
[c]Melgar 1966–67. [d]Krishna 1956.

tassium, calcium, and magnesium, mainly present as chloride or sulphate. High levels of iodine are also present. The significant carbohydrate components of seaweed consist of hexitol and mannitol; the sulfated fucan polymer, fucoilan; the structural polymer, alginic acid and its salts; and the glucan polymers, laminasan, and cellulose (Whyte 1971; Woodward 1951).

Cameron (1954) included seaweed at a level of 0, 2, 4, and 6% in a basal diet, based on barley, tankage or fish meal, and minerals, and reported that the addition of seaweed, even at the highest level, had no effect on weight gain or carcass quality when fed from 18 to 90 kg. Lombard (1962) found that 8% seaweed meal had little effect on growth but reduced the biological value of the dietary protein.

Homb (1961) and Jones *et al.* (1979) found that pig performance was reduced by seaweed in the diet, but Vosloo and Steenkamp (1958), Jensen and Minsaas (1960) and Malajskajte (1963) observed an improvement in performance. Some of these results are summarized in Table 15.73.

The effect of seaweed meal on digestibility of dietary components has been studied by a number of workers (Sobczak and Szlichcinska 1964; Homb 1961; Lombard 1962; Malajskajte 1963; Whittemore and Percival 1975; Beames *et al.* 1975, 1977). All results agree that the addition of seaweed meal lowers the digestibility of the protein, fiber, organic matter, nitrogen-free extract, and dry matter, and reduces nitrogen retention.

It is obvious from the available data that seaweed is not a good source of either protein or energy and should be used basically as a mineral and especially trace mineral supplement at low dietary levels. Although dietary levels up to 10% have been used successfully, even lower levels have caused severe diarrhea. Maximum value and utilization will probably be realized if no more than 5% is included in the diet.

TABLE 15.73. Effect of Seaweed in the Diet on the Performance of Growing-Finishing Pigs

Level of seaweed (%)	*Pig weight (kg)*	*Av daily gain (g)*	*Feed/gain ratio*	*Reference*
0	20–100	909	4.30	Vosloo and Steenkamp 1958
4.35	20–100	864	3.80	Vosloo and Steenkamp 1958
0	15–90	599	3.70	Homb 1961
5	15–90	616	3.53	Homb 1961
0	15–90	602	3.55	Homb 1961
10	15–90	589	3.83	Homb 1961
0	15–90	471	4.92	Malajskajte 1963
2	15–90	516	4.68	Malajskajte 1963

SYNTHETIC AMINO ACIDS

For many years the feed industry has had pure lysine and methionine available from microbiological production at prices competitive for use in swine diets in limited amounts. However, the cost of producing other amino acids such as tryptophan and threonine, whose concentrations in most plant proteins are below the amounts required by animals, has been prohibitive in the past. Now, gene-splicing techniques are available to produce microorganisms capable of producing large amounts of these and other amino acids. These new breakthroughs in biotechnology (United States Congress, OTA 1984) likely will provide the basis for greater use of synthetic amino acids and a reduction in conventional plant and animal protein use in the future of swine production. In 1982, worldwide sales of amino acids totalled 455,000 metric tons (United States Congress, OTA 1984), one-third of which were used in animal feeds. An annual growth rate of 7–10% is projected for the next 10 years.

REFERENCES

Soybean Meal

ACKER, D. C., CATRON, D. V., and HAYS, V. W. 1959. J. Anim. Sci. *18,* 1053.

AHLUWALIA, V. K., BHASIN, M. M., and SESHADRI, T. R. 1953. Current Sci. (India) *22,* 363.

BAKER, D. H., MOLITOUS, B. A., JENSEN, A. H., and HARMON, B. G. 1974. J. Anim. Sci. *38,* 325.

BARNES, R. H., and KWONG, E. 1965. J. Nutr. *86,* 245.

BARR, G. W., ANDREWS, N. G., PERRY, H. B., and CRANE, F. M. 1980. J. Anim. Sci. *51*(Suppl. 1), 77 (Abstr.).

BENNETTS, H. W., UNDERWOOD, E. J., and SHIER, F. L. 1946. Aust. Vet. J. *22,* 2.

BERRY, T. H., BECKER, D. E., and JENSEN, A. H. 1961. J. Anim. Sci. *20,* 295 (Abstr.).

BERRY. T. H., BECKER, D. E., RASMUSSEN, O. G., JENSEN, A. H., and NORTON, H. W. 1962. J. Anim. Sci. *21,* 558.

BERRY, T. H., COMBS, G. E., WALLACE, H. D., and ROBBINS, R. C. 1966. J. Anim. Sci. *25,* 722.

BIGGERS, J. D., and CURNOW, D. H. 1954. Biochem. J. *58,* 278.

BORCHERS, R. 1962. J. Nutr. *78,* 330.

BOWMAN, D. E. 1944. Proc. Soc. Exp. Biol. Med. *57,* 139.

CORLEY, J. R., and EASTER, R. A. 1980. J. Anim. Sci. *51*(Suppl. 1), 191.

CLAWSON, A. J., BARRICK, E. R., and SMART, W. G. 1963. J. Anim. Sci. *22,* 1027.
CORZO, M., OBANDO, H., MONCADA, A., and MANER, J. H. 1968. Memoria ALPA *3,* 156 (Abstr.).
FAO. 1981. Food and Agriculture Organization of the United Nations FAO Production Yearbook, Vol. 35, Table 22. Rome, Italy.
FOWLER, S. H., and ROBERTSON, G. L. 1954. J. Anim. Sci. *13,* 949.
GESTETNER, B., BIRK, Y., and BONDI, A. 1966A. Phytochemistry *5,* 799.
GESTETNER, B., BIRK, Y., BONDI, A., and TENCER, G. 1966B. Phytochemistry *5,* 803.
GRIFFO, A. P., and MOXON, A. L. 1973. J. Anim. Sci. *37,* 280.
HAM, W. E., and SANDSTEDT, R. M. 1944. J. Biol. Chem. *154,* 505.
HARMON, B. G., BECKER, D. E., JENSEN, A. H., and BAKER, D. H. 1969. J. Anim. Sci. *28,* 459.
HARPER, A. E. 1959. J. Nutr. *67,* 109.
HAYS, V. W., SPEER, V. C., HARTMAN, P. A., and CATRON, D. V. 1959. J. Nutr. *69,* 179.
JENSEN, A. H. 1953. Effect of antibiotic and lysine supplementation on growing-fattening pigs at different protein levels. Ph.D. Thesis. Iowa State College, Ames.
KACHAENCHAI, S., and DIGGS, B. G. 1979. J. Anim. Sci. *49*(Suppl. 1), 6 (Abstr.).
KHAYAMBASHI, H., and LYMAN, R. L. 1966. J. Nutr. *89,* 455.
KONG, P., and VEUM, T. L. 1980. J. Anim. Sci. *51*(Suppl. 1), 208 (Abstr.).
KROENING, G. H., POND, W. G., JONES, J. R., and LOOSLI, J. K. 1964. J. Anim. Sci. *23,* 881.
KUNITZ, M. 1945. Science *101,* 668.
LEWIS, A. J., PEO, E. R., JR., MOSER, B. D., and CRENSHAW, T. D. 1981. J. Anim. Sci. *51,* 361.
LIENER, I. E. 1951. J. Biol. Chem. *193,* 183.
LIENER, I. E. 1953A. J. Nutr. *49,* 527.
LIENER, I. E. 1953B. Arch. Biochem. Biophys. *54,* 223.
LIENER, I. E. 1975. *In* Protein Nutritional Quality of Foods and Feeds (M. Friedman, Editor), Vol. 1, Part 2, p. 523. Marcel Dekker, New York.
LIENER, I. E., and PALLANSCH, M. T. 1952. J. Biol. Chem. *197,* 29.
MAHAN, D. C., CLINE, J. H., JONES, J. E., CROSS, R. F., TEAGUE, H. S., and GRIFFO, A. P., JR. 1971. J. Anim. Sci. *33,* 1151.
MANER, J. H., POND, W. G., and LOOSLI, J. K. 1961. J. Anim. Sci. *20,* 614.
MANER, J. H., OBANDO, H., PORTELA, R., and GALLO, J. 1969. J. Anim. Sci. *29,* 139 (Abstr.).
MATEO, J. P., and VEUM, T. L. 1980. J. Anim. Sci. *50,* 869.
MEADE, R. J. 1956A. J. Nutr. *60,* 599.
MEADE, R. J. 1956B. J. Anim. Sci. *15,* 288.
MEADE, R. J., and TETER, W. S. 1956. J. Nutr. *60,* 609.
NRC, National Research Council. 1968. Joint U.S.–Canadian Tables of Feed Composition, Nat. Acad. Sci. Publ. *1232* Washington, D.C.
NRC, National Research Council. 1972. Joint U.S.–Canadian Tables of Feed Composition, 3rd Revision. Natl. Acad. Sci. Press, Washington, D.C.
NEWPORT, M. J. 1980. Brit. J. Nutr. *44,* 171.
NOLAND, P. R., MITCHELL, J., JR., and SCOTT, K. W. 1959. J. Anim. Sci. *18,* 1513 (Abstr.).
OSBORNE, T. B., and MENDEL, L. B. 1917. J. Biol. Chem. *32,* 369.
PALLANSCH, M. J., and LIENER, I. E. 1953. Arch. Biochem. Biophys. *45,* 366.
PALMER, W. M., TEAGUE, H. S., and GRIFFO, A. P., JR. 1970. J. Anim. Sci. *31,* 535.
POND, W. G., SNYDER, W., WALKER, E. F., JR., STILLINGS, B. R., and SIDWELL, V. 1971. J. Anim. Sci. *33,* 587.
POTTER, G. C., and KUMMEROW, F. A. 1954. Science *120,* 224.
ROBISON, W. L. 1930. Ohio Agric. Exp. Stn. Bull *452.*
ROBISON, W. L. 1951. Ohio Agric. Exp. Stn. Bull. *699.*
ROBISON, W. L. 1953. Ohio Agric. Exp. Stn. Bull. *729.*
ROBISON, W. L., KUNKLE, L. E., and CAHILL, V. R. 1956. Ohio Agric. Exp. Stn. Bull. *769.*
RUSSELL, L. E., CROMWELL, G. L., and STAHLY, T. S. 1980. J. Anim. Sci. *51*(Suppl. 1), 219 (Abstr.).
SAUER, W. C., CICHON, R., and MISIR, R. 1982. J. Anim. Sci. *54,* 292.
SCHMIDT, M. K., SHERRY, M. P., and VEUM, T. L. 1973. J. Anim. Sci. *27,* 290.
SCHNEIDER, D. L., and SARETT, H. P. 1969. J. Nutr. *98,* 279.
SHERRY, M. P., SCHMIDT, M. K., and VEUM, T. L. 1978. J. Anim. Sci. *46,* 1259.
SHREWSBURY, C. L., VESTAL, C. M., and HAUGE, S. M. 1932. J. Agric. Res. *44,* 267.
SINGH, L., WILSON, C. M., and HADLEY, H. H. 1969. Crop Sci. *9,* 489.

SMITH, H. M., SMITH, J. M., and SPRING, F. S. 1958. Tetrahedron *4,* 11.
SMITH, J., JR., CLAWSON, A. J., and BARRICK, E. R. 1967. J. Anim. Sci. *26,* 752.
TANKSLEY, T. D., JR., KNABE, D. A., PURSER, K., ZEBROWSKA, T., and CORLEY, J. R. 1981. J. Anim. Sci. *52,* 769.
TEAGUE, H. S., and RUTLEDGE, E. A. 1960. *J. Anim. Sci. 19,* 902.
TRIBBLE, L. F., INGRAM, S. H., GASKINS, C. T., and RAMSEY, C. B. 1979. J. Anim. Sci. *48,* 541.
WALZ, E. 1931. Ann. Chem. *489,* 118.
WILLNER, D., GESTETNER, B., LAVIE, D., BIRK, Y., and BONDI, A. 1964. J. Chem. Soc., Suppl. *1,* 5885.
WILSON, R. H., and LEIBHOLZ, J. 1981. Brit. J. Nutr. *45,* 301, 321, 337, 347, 351.
WONG, E., and FLUX, D. S. 1962. J. Endocrinol. *24,* 341.
YEN, J. T., HYMOWITZ, T., and JENSEN, A. H. 1971. J. Anim. Sci. *33,* 1012.
YEN, J. T., JENSEN, A. H., and SIMON, J. 1977. J. Nutr. *107,* 156.
ZAMORA, R. G., and VEUM, T. L. 1979. Nutr. Rep. Int. *19,* 49.
ZAMORA, R. G., SCHMIDT, M. K., and VEUM, T. L. 1975. J. Anim. Sci. *41,* 331 (Abstr.).
ZIMMERMAN, D. R. 1980. J. Anim. Sci. *51*(Suppl. 1), 78.

Coconut Meal

ADZIC, A., GRADINAE, D., BOZICKOVIC, S., and ZIVKOVIC, S. 1966. Stocarstvo *20,* 43.
BUTTERWORTH, M. H., and FOX, H. C. 1963. Brit. J. Nutr. *17,* 445.
CRESWELL, D. C., and BROOKS, C. C. 1971A. J. Anim. Sci. *33,* 366.
CRESWELL, D. C., and BROOKS, C. C. 1971B. J. Anim. Sci. *33,* 370.
GRIEVES, C. M., OSBOURN, D. F., and GONZALES, F. O. 1966. Trop. Agric., Trin. *43,* 257.
LIMA, A. M., MANER, J. H., and SIMAS, R. 1981. CEPED Annu. Rep., December, 1981.
LOOSLI, J. K., PENA, J. O., YNALVES, L. A., and VILLEGAS, V. 1954. Philipp. Agric. *38,* 191.
MALYNICZ, G. L. 1973. Papua New Guinea Agric. J. *24,* 142.
MEE, J. M. L., and BROOKS, C. C. 1973. Nutr. Rep. Int. *8,* 261.
SIMAS, R., and MANER, J. H. 1981. CEPED Annu. Rep., December, 1981.

Cottonseed Meal

ADAMS, R., MORRIS, R. C., GEISSMAN, T. A., BUTTERBAUGH, D. J., and KIRPATRICK, E. C. 1938. J. Am. Chem. Soc. *60,* 2193.
BALIGA, B. P., BAYLISS, M. E., and LYMAN, C. M. 1959. Arch. Biochem. Biophys. *84,* 1.
BALIGA, B. P., and LYMAN, A. E. 1957. J. Am. Oil Chem. Soc. *34,* 21.
BARRENTINE, B. F. 1966. Proc. Conf. on Inactivation of Gossypol with Mineral Salts. Nat. Cottonseed Products Assoc., Memphis, Tennessee.
BRAHAM, J. E., JARQUÍN, R., BRESSANI, R., GONZÁLEZ, J. M., and ELÍAS, L. G. 1967. J. Nutr. *93,* 241.
CLAWSON, A. J., SMITH, F. H., and ALBRECHT, J. E. 1967. J. Anim. Sci. *26,* 214 (Abstr.).
DANKE, R. J. 1962. Gossypol content of cotton seeds as affected by heat and moisture. M. S. Thesis. Oklahoma State Univ., Stillwater.
EDWARDS, J. D. 1958. J. Am. Chem. Soc. *80,* 3798.
FERGUSON, T. M., COUCH, J. R., and PIGDON, R. H. 1959. Proc. 5th Conf. on Processing as Related to Nutritive Value of Cottonseed Meal, pp. 15–18, January 19–20, 1959.
FLECHER, J. L. 1948. Inf. Sheet 409, Mississippi State College, State College.
HALE, F., and LYMAN, C. M. 1957. J. Anim. Sci. *16,* 364.
HALE, F., LYMAN, C. M., and SMITH, H. A. 1958. Texas Agric. Stn. Bull *898.*
HAUGHT, D. G., TANKSLEY, T. D., JR., HESBY, J. H., and GREGG, E. J. 1977. J. Anim. Sci. *44,* 249.
HOFFPAUIR, C. L., and PONS, W. A. 1953. J. Assoc. Off. Agric. Chem. *36,* 1108.
JARQUÍN, R. J., BRAHAM, E., BRESSANI, R., ELÍAS, L. G., TEJODA, C., and GONZÁLEZ, J. M. 1965. J. Agric. Food Chem. *14,* 275.
KNABE, D. A., TANKSLEY, T. D., and HESBY, J. H. 1979. J. Anim. Sci. *49,* 134.
KORNEGAY, E. T., CLAWSON, A. J., SMITH, F. H., and BARRICK, E. R. 1961. J. Anim. Sci. *20,* 597.
LYMAN, C. M. 1966. Proc. Conf. on Inactivation of Gossypol with Mineral Salts. Nat. Cottonseed Products Assoc. Memphis, Tennessee.
LYMAN, C. M., BALIGA, B. P., and SLAY, M. W. 1959. Arch. Biochem. Biophys. *84,* 486.

MANER, J. H. 1964. II Cong. Nacional de la Industria Porcina. June 25–29, 1964, Cali, Colombia.
MONCADA, A., and MANER, J. H. 1970. Revista ICA *5,* 91.
NOLAND, P. R., FUNDERBURG, M., ATTEBERRY, J., and SCOTT, K. W. 1968. J. Anim. Sci. *27,* 1319.
PHELPS, R. A. 1966A. World's Poultry Sci. J. *22,* 86.
PHELPS, R. A. 1966B. Feedstuffs, May 7, 1966.
PONS, W. A., HOFFPAUIR, C. L., and O'CONNER, R. T. 1950. J. Am. Oil Chemists' Soc. *27,* 390.
PONS, W. A., PITTMAN, R. A., and HOFFPAUIR, C. L. 1958. J. Am. Oil Chemists' Soc. *35,* 93.
RAO, S. R., CARTER, F. L., and FRAMPTON, V. L. 1963, Anal. Chem. *35,* 1927.
ROBISON, W. L. 1934. Ohio Agric. Exp. Stn. Bull. *534.*
SHERWOOD, F. W. 1926. J. Agric. Res. *32,* 793.
SMITH, F. H. 1946A. Ind. Eng. Chem. Anal. Ed. *18,* 43.
SMITH, F. H. 1946B. Ind. Eng. Chem. Anal. Ed. *18,* 656.
SMITH, F. H. 1958. J. Am. Oil Chem. Soc. *35,* 261.
SMITH, H. A. 1957. Am. J. Path. *33,* 353.
SMITH, K. 1966. Proc. Conf. on Inactivation of Gossypol with Mineral Salts. Natl. Cottonseed Products Assoc., Memphis, Tennessee.
STEVENSON, J. W., CABELL, C. A., and KINCAID, C. M. 1965. J. Anim. Sci. *24,* 290 (Abstr.).
SURE, B. 1953. J. Agric. Food Chem. *1,* 82.
TANKSLEY, T. D., JR., KNABE, D. A., PURSER, K., ZEBROWSKA, T., and CORLEY, J. R. 1981. J. Anim. Sci. *52,* 769.
WESMINE, G., and MANER, J. H. 1967. Unpublished data. ICA, Bogota, Colombia.
WEST, J. L. 1940. J. Am. Vet. M.A. *69,* 74.
WITHERS, W. A., and CARRUTH, F. E. 1915. J. Agric. Res. *5,* 261.
WITHERS, W. A., and CARRUTH, F. E. 1917. J. Biol. Chem. *32,* 245.

Linseed Meal

BISHARA, H. N., and WALKER, H. F. 1977. Brit. J. Nutr. *37,* 321.
BETHKE, R. M., BOHSTEDT, G., SASSAMAN, H. L., KENNARD, D. C., and EDINGTON, B. H. 1928. J. Agric. Res. *36,* 855.
CRAMPTON, E. W., MacKAY, V. G., and LLOYD, L. E. 1955. J. Anim. Sci. *14,* 688.
FAO. 1980. FAO Production Yearbook 1980.
JANICK, J., SCHERY, R. W., WOODS, F. W., and RUTTAN, V. W. 1969. Plant Science, p. 447. Freeman, San Francisco.
JUSYCZAK, J., and ZIEMINSKI, R. 1968. Nutr. Abstr. Rev. *38,* 8287 (Abstr.).
KLOSTERMANN, H. J., LAMOUREUX, G. L., and PARSONS, J. L. 1967. Biochem. (Easton) *6,* 170.
MICKELSON, O., and YANG, M. G. 1966. Fed. Proc. *25,* 118.
NRC, National Research Council. 1968. Joint U.S.–Canadian Tables of Feed Conposition. Natl. Acad. of Sci. Publ. *1232.* Washington, D.C.
TAXILIN, V. A. 1961. Svinovodstvo *3,* 39.
WAHLSTROM, R. C., KAMSTRA, L. D., and OLSON, O. E. 1956. J. Anim. Sci. *15,* 794.

Peanut Meal

BALOGUM, T. F., and KOCH, B. A. 1979. Tropical Agric. *56,* 245.
BARBER, R. S., BRAUDE, R., MITCHELL, K. G., HARDING, J. D. J., LEWIS, G., and LOOSMORE, R. M. 1968. Brit. J. Nutr. *22,* 535.
BROOKS, C. C., and THOMAS, H. R. 1958. Va. Agric. Exp. Stn. Tech. Bull. *138.*
BROOKS, C. C., and THOMAS, H. R. 1959. J. Anim. Sci. *18,* 1119.
COMBS, G. E., and WALLACE, H. D. 1962. J. Anim. Sci. *21,* 95.
COMBS, G. E., OSEQUEDA, F. L., WALLACE, H. D., and AMMERMAN, C. B. 1963. J. Anim. Sci. *22,* 396.
CONKERTON, E. J., BLANCHET, E. D., ORY, R. L., and HAMMONS, R. O. 1978. Peanut Sci. *5,* 75.
HALE, O. M., and McCORMICK, W. C. 1979. Res. Rep. Ga. Exp. Stn. *323,* August, 1979.
KORNEGAY, E. T., MEACHAM, T. N., and THOMAS, H. R. 1968. Res. Div. Rep. *32,* Virginia Polytechnic Inst. 12 pp.
KORNEGAY, E. T., THOMAS, H. R., MORJINGO, R. W., COFFELT, T. A., and HEDGES, J. D. 1975. Res. Div. Rep. *163,* 197. Virginia Polytechnic Inst.

MASSEY, Z. A., BOGGESS, T. S., JR., and HOLLY, K. T. 1942. Ga. Agric. Exp. Stn. Cir. *138*.
OROK, E. J., and BOWLAND, J. P. 1976. Can. J. Anim Sci. *55*, 347.
OROK, E. J., BOWLAND, J. P., and BRIGGS, C. W. 1975. Can. J. Anim. Sci. *55*, 133.
RANJHAN, S. K., JENSEN, A. H., COX, J. L., HARMON, B. G., and BECKER, D. E. 1964. J. Anim. Sci. *23*, 461.
SEWELL, R. F., KEEN, B. C., and CARMON, J. L. 1957. J. Anim. Sci. *16*, 357.
THOMAS, H. R., and KORNEGAY, E. T. 1972. J. Anim. Sci. *34*, 587.
WOODHAMAND, A. A., and DAWSON, R. 1968. Br. J. Nutr. *22*, 589–607.

Rapeseed Meal (Canola Meal)

AHERNE, F. X., and LEWIS, A. J. 1978. Anim. Feed Sci. Technol. *3*, 235.
AHERNE, F. X., BOWLAND, J. P., and LEWIS, A. J. 1977A. Proc. 10th Annu. Mtg. pp. 162–175. Rapeseed Assoc. Can., Vancouver, British Columbia.
AHERNE, F. X., BOWLAND, J. P., and LEWIS, A. J. 1977B. Publ. No. 45. *In* Rapeseed Oil, Meal and By-products Utilization (J. M. Bell, Editor), Rapeseed Assoc., Can., Winnipeg, Manitoba.
BELL, J. M. 1975. Can. J. Anim. Sci. *55*, 61.
BELL, J. M., and BELZILE, R. J. 1965. *In* Rapeseed Meal for Livestock and Poultry. A Review, Can. Dep. Agric. Publ. *1257*. Ottawa, Ontario.
BELL, J. M., ANDERSEN, D. M., and SHIRES, A. 1981. Can. J. Anim. Sci. *61*, 453.
BOWLAND, J. P. 1965. *In* Rapeseed Meal for Livestock and Poultry. A Review. Can. Dep. Agric. Publ. *1257*. Ottawa, Ontario.
BOWLAND, J. P. 1975A. Can. J. Anim. Sci. *55*, 409.
BOWLAND, J. P. 1975B. *In* H. Swan and D. Lewis (Ed.). Feed Energy Sources for Livestock. 9th Nutr. Conf. Feed Manuf., Nottingham, England.
BOWLAND, J. P. 1975C. 54th Ann. Feeders' Day Rep., 1975, p. 6–6, June, Univ. of Alberta, Edmonton.
BOWLAND, J. P., and HARDIN, R. T. 1973. Can. J. Anim. Sci. *53*, 355.
CASTELL, A. G. 1977. Can. J. Anim. Sci. *57*, 313.
CHO, C. Y., and BAGLEY, H. S. 1972. Can. J. Physiol. Pharmacol. *50*, 513.
CLANDININ, D. R. 1981. Canola Meal for Livestock and Poultry. Canola Council of Can. Publ. *59*, Winnipeg, Manitoba.
CLANDININ, D. R., and ROBBLEE, A. R. 1978. Proc. 5th Int. Rapeseed Conf., pp. 204–212. Malmo, Sweden.
DEVILAT, JAIME. 1965. Invest. Ganaderas de Chile *1*, 61.
DEVILAT, B. J., and SKOKNIC, K. 1971. Agric. Tec. (Chile) *31*, 155.
DOWNEY, R. K. 1965. *In* Rapeseed Meal for Livestock and Poultry. A Review. Can. Dept. Agric. Publ. *1257*. Ottawa, Ontario.
ESNAOLA, L. M. A., and OCHOA, J. A. 1970. Agric. Tecnica. *30*, 90.
FAO. 1980. Food and Agricultural Organization Production Yearbook.
FENWICK, G. R., and CURTIS, R. F. 1980. Anim. Feed Sci. Tech. *5*, 255.
FLIPOT, P. and DUFOUR, J. J. 1977. Can. J. Anim. Sci. *57*, 567.
GRANDHI, R. R., BROWN, R. G., HACKER, R. R., BOWMAN, G. H., and SLINGER, S. J. 1976A. Can. J. Anim. Sci. *56*, 842.
GRANDHI, R. R., SLINGER, S. J., BROWN, R. G., and HACKER, R. R. 1976B. J. Anim. Sci. *43*, 252.
GRANDHI, R. R., HACKER, R. R., BOWMAN, G. H., and ROUNSAVILLE, T. R. 1977. Proc. 69th Ann. Mtg. Amer. Soc. Anim. Sci. July 23–27. p. 43.
GRANDHI, R. R., NAREDRAN, R., BOWMAN, G. H., and SLINGER, S. J. 1979. Can. J. Anim. Sci. *59*, 323.
JOSEFSSON, E. 1972. *In* Rapeseed Cultivation, Composition, Processing and Utilization, (L. A. Appelquist and R. Ohlson, Editors), pp. 354–378. Elsevier, Amsterdam.
KENNELLY, J. R., AHERNE, F. X., and LEWIS, A. J. 1978. Can. J. Anim. Sci. *58*, 743.
LEWIS, A. J., AHERNE, F. X., and HARDIN, R. T. 1978. Can. J. Anim. Sci. *58*, 203.
MANNS, J. G., and BOWLAND, J. P. 1963. Can. J. Anim. Sci. *43*, 252.
MAY, R. W., and BELL, J. M. 1971. Can. J. Anim. Sci. *51*, 271.
MUZTAR, A. J., LIKUSKI, H. J., and SLINGER, S. J. 1978A. Can. J. Anim. Sci. *58*, 485.
MUZTAR, A. J., SADIO, M., and SLINGER, S. J. 1978B. Nutr. Rep. Int. *18*, 639.
McCUAING, L. W., and BELL, J. M. 1976. Proc. West. Sect. Am. Soc. Anim. Sci. *27*, 112.
McKINNON, P. J., and BOWLAND, J. P. 1977. Can. J. Anim. Sci. *57*, 663.

MOODY, D. L., SLINGER, S. J., and SUMMERS, J. D. 1976. Proc. Nutr. Conf. Feed Manuf., Univ. of Guelph, Ontario.

NARENDRAN, R., BOWMAN, G. H., and SLINGER, S. J. 1979. Can. J. Anim. Sci. *59*, 577.

NARENDRAN, R., BOWMAN, G. H., LEESON, S., and PFIEFFER, W. 1981. Can. J. Anim. Sci. *61*, 213.

OCHETIM, S., BELL, J. M., DOIGE, C. E., and YOUNG, C. G. 1980A. Can. J. Anim. Sci. *60*, 407.

OCHETIM, S., BELL, J. M., and DOIGE, C. E. 1980B. Can. J. Anim. Sci. *60*, 423.

OMOLE, T. A., and BOWLAND, J. P. 1974. Can. J. Anim.Sci. *54*, 363; *54*, 481.

SABEN, H. S., BOWLAND, J. P., and HARDIN, R. T. 1971. Can. J. Anim. Sci. *51*, 419, *51*, 427.

SCHULD, F. W., and BOWLAND, J. P. 1968. Can. J. Anim. Sci. *48*, 57.

SINGAM, A. D. R., and LAWRENCE, T. L. J. 1979. J. Sci. Food Agric. *30*, 21.

WETTER, L. R. 1965. *In* Rapeseed Meal for Livestock and Poultry. A Review. Can. Dep. Agric. Publ. *1257*. Ottawa, Ontario.

Rubber Seed and Rubber Seed Meal

FETUGA, B. L., AYENI, T. O., OLANIYAN, A., BALOGUN, M. A., BABATUNDE, G. N., and OYENUGA, V. A. 1977. Nutr. Rep. Int. *15*, 497.

ONG, H. K., and YEONG, S. W. 1978. *In* Feedstuffs for Livestock in South East Asia. (C. Devendra and S. Hutoglung, Editors.) pp. 337–344. Malaysian Scoiety Anim. Prod., Selanga, Malaysia.

OROK, E. J., and BOWLAND, J. P. 1981. Can. J. Anim. Sci. *54*, 239.

RAJAGURU, A. S. B., and RAVINDRAN, V. 1979. J. Natl. Sci. Council, Sri Lanka *7*, 101.

SIRIWARDENE, J. A. DE, and NUGARA, D. 1972. Ceylon Vet. J. *20*, 61.

STOSIC, D. D., and KAYKAY, J. M. 1981. World Anim. Rev. *39*, 29.

Safflower Meal

ANON. 1967. Nutr. Rev. *25*, 29.

BRAVO, F. O., and CEBALLO, E. 1969. Tec. Pecv. Mex. *11*, 38.

KOHLER, G. O., KUZMICKY, D. D., POTTER, R., GUGGALZ, J., and HERRING, V. V. 1966. J. Am. Oil. Chem. Soc. *43*, 413.

KRATZER, F. H., and WILLIAMS, D. E. 1951. Poultry Sci. *30*, 417.

SHIMADA, A. S., and AGUILERA, A. 1966. Tec. Pecv. Mex. *7*, 6.

SHIMADA, A. S., and BRAMBILA, S. 1966. Tec. Pecv. Mex. *8*, 30.

WILLIAMS, K. C., and DANIELS, L. J. 1973. Aust. J. Exp. Agric. Anim. Husb. *13*, 48.

WILLIAMS, K. C., and O'ROURKE, P. K. 1974. Aust. J. Exp. Agric. Anim. Husb. *14*, 12.

Sesame Meal

CATRON, D. V., and HAYS, V. W. 1958. Boletin de Soybean Council of Am. Rome, Italy, 35 pp.

COMMONWEALTH ECON. COMM. 1967. Vegetable Oils and Oilseeds—A Review, H. M. Stationary Office, London.

CORNELIUS, J. A., and RAYMOND, W. D. 1967. Trop. Sci. *9*, 75–89.

FAO. 1980. FAO Production Yearbook. 1980.

HERVAS, E., VITERI, J., and MANER, J. H. 1965. Unpublished data. INIAP. Quito, Ecuador.

HUBBELL, C. H. 1968. Feedstuffs, June 8, p. 48.

JOHNSON, R. H., and RAYMOND, W. D. 1965. Trop. Sci. *6*, 173.

MANER, J. H., and GALLO, J. T. 1963. Primer Congreso Nacional de la Industria Porcina, pp. 1–3. Bogota, Colombia.

MITRA, R. C., and MISRO, P. S. 1967. J. Agric. Food Chem. *15*, 697.

SQUIBB, R. L., and SALAZAR, E. 1951. J. Anim. Sci. *10*, 545–550.

VILLEGAS, A. M., GONZÁLEZ, A., and CALDERÓN, R. 1968. Cereal Chem. *45*, 379.

ZAGHI, S., and BRESSANI, R. 1969. Turrialba *19*, 34.

Sunflower Seed and Meal

ALEXANDER, J. C., and HILL, D. C. 1952. J. Nutr. *48*, 149.

AMOS, H. E., BURDICK, D., and SEERLEY, R. W. 1975. J. Anim. Sci. *40*, 90.

BAIRD, D. M. 1980. J. Anim. Sci. *51*(Suppl. 1), 185 (Abstr. 194).

BANDEMER, S. L., and EVANS, R. J. 1963. J. Agric. Food Chem. *11*, 134.

CLANDININ, D. R. 1958. Sunflower Oil Meal. *In* Processed Plant Protein Foodstuffs (M. Altschul, Editor). Academic Press, New York.
DELIC, I., BOKOROV, T., SRECKOVIC, A., and BIOLOSKA NIKOLIC, M. 1963. Stocarstvo *17,* 464. (Abstr. in Nutr. Abstr. Rev. 1964. *34,* 596.)
DELIC, I., NIKOLIC, M., SARGIN, S., and BOKOROV, T. 1964. Veterinaiya Sarajevo *13,* 195. (Abstr. in Nutri. Abstr. Rev. 1965. *35,* 849.)
DIGGS, B. G., and BAKER, B. 1972. Miss. Farm. Res. *35,* 7.
FAO. 1980. FAO Production Yearbook.
KEPLER, M. A., LIBAL, G. W., and WAHLSTROM, R. C. 1980. Coop. Ext. Serv. S. D. Stn. Univ. Anim. Sci. Swine Day, November 20, 1980.
LAUDERT, S. B., and ALLEE, G. L. 1975. J. Anim. Sci. *41,* 318 (Abstr. 293).
MITCHELL, H. H., HAMILTON, T. S., and BEADLES, J. R. 1945. J. Nutr. *29,* 12.
NOLAND, P. R., CAMPBELL, D. R., and JOHNSON, Z. B. 1980. Anim. Feed Sci. Tech. *5,* 51.
PEARSON, A. M., WALLACE, H. D., and HENTGES, J. F., JR. 1954. Univ. of Fla. Agric. Exp. Stn. Bull. *553.*
RENNER, R., CLANDININ, D. R., MORRISON, A. B., and RABBLEE, A. R. 1953. Poultry Sci. *32,* 922.
SEERLEY, R. W., BURDICK, D., RUSSOM, W. C., LOWREY, R. S., McCAMPBELL, H. C., and AMOS, H. E. 1974. J. Anim. Sci. *38,* 947.
SIEBRITS, F. K., KEMM, E. H., and ROS, M. N. 1980. Navorsingisinstituut, p. 31.
SMITH, K. J. 1968. Feedstuffs *40,* 20.
TKACEV, I. F., TARANENKO, G. A., ZVJAGINCEV, V. I., and VACIKALO, A. P. 1964. Zivotnovodstvo, No. 7, pp. 74–77. (Abstr. in Nutr. Abstr. Rev. 1965. *35,* 207.)
TKACEV, I. F., TARANENKO, G. A., BACIKALO, A. P., ZVJAGINCEV, V. I., MACUK, JU. P., and KOUAZMYHA, JU. T. 1965. Vestu. sel'skohoz Nauki, No. 5, pp. 119–128 (Abstr. in Nutr Abstr. Rev. 1966. *36,* 244.)
WENK, C., and LANDIS, J. 1979. Kraftfutter *62,* 84.

Grain Legumes

ABU-SHAKRA, S., MIRZA, S., and TANNONS, R. 1970. J. Sci. Food Agric. *21,* 9.
AYKROYD, W. R., and DOUGHTY, J. 1964. Legumes in Human Nutrition. FAO Publ. *19.*
BELL, J. M., and WILSON, A. G. 1970. Can. J. Anim. Sci. *50,* 15.
FAO. 1981. Food and Agricultural Production Yearbook.
HENRY, Y. 1970. Bull Tech. d'Information *253,* 1.
KROENING, G. H. 1968. Unpublished paper. Washington State Univ., Pullman.
SHIMADA, A. S., and BRAMBILA, S. 1967. Tec. Pecuar. Mex. *9,* 27.

Soybeans

ANDERSON, W. 1971. Feedstuffs, August 14, p. 13.
BERGKAMP, J. L., and TOPEL, D. G. 1970. Iowa State Univ. Ext. Ser. Bull. AS. *353F.*
COMBS, G. E., and WALLACE, H. D. 1969. Swine Day Rep. Univ. of Fla. Mimeo Series No. *AN70–2.*
COMBS, G. E., CONNESS, R. G., BERRY, T. H., and WALLACE, H. D. 1967. J. Anim. Sci. *26,* 1067.
CRENSHAW, M. A., and DANIELSON, D. M. 1983. J. Anim. Sci. *57*(Suppl. 1), 93 (Abstr.).
DOTY, H. O. 1965. USDA, Economic Res. Ser. Bull. 228.
FAO. 1980. Food and Agricultural Production Yearbook.
HANSON, L. E., ALLEN, C. E., MEADE, R. J., RUST, J. W., and MILLER, K. P. 1970. 31st Minn. Nutr. Conf., Sept. 14, 1970.
HEYWARD, J. W. 1970. Feestuffs *42,* 11.
JENSEN, A. H., BAKER, D. H., BROWN, H. W., ROMANS, J. R., and HARMON, B. G. 1970. Ill. Pork Ind. Day., March 10, 1970.
JIMENEZ, A. A., PERRY, T. W., PICKET, R. A., and BEESON, W. M. 1963. J. Anim. Sci. *22,* 471.
MYER, R. O., and FROSETH, J. A. 1980. J. Anim. Sci. *51*(Suppl. 1), 73 (Abstr.).
NARDIELLO, R., OLSEN, C., CASAL, J. J., and GARCIA, P. T. 1980. Informe Tecn. INTA, Argentina *162,* 15.
NICKELSON, W. F., BECKER, D. E., JENSEN, A. H., and TERRILL S. W. 1960. J. Anim. Sci. *19,* 1279.

NOLAND, P. R., BAUGUS, C. A., LAWRENCE, R. O., and JOHNSON, Z. 1969. Ark. Farm. Res. Rep. *18,* 6.
NOLAND, P. R., SCOTT, K. W., and BAUGUS, C. W. 1970. Ark. Farm Res. Rep. *9,* 3.
NRC. 1979. Nutr. Req. of Swine. Natl. Res. Council, Natl. Acad. Sci.
VELLOSO, L. 1964. Bull. Ind. Anim. *19,* 5.
WAHLSTROM, R. C., Libal, G. W., and BERNS, R. J. 1971. J. Anim. Sci. *32,* 891.
YOUNG, L. G. 1970. Feedstuffs. *42,* 34.
ZAMORA, R. G., and VEUM, T. L. 1979. J. Anim. Sci. *48,* 63.
ZIMMERMAN, D. R., BERGKAMP, J. L., FABER, J. L., HODSON, H. H., SPEER, V. C., and TOPEL, D. G. 1970. Iowa Farm Sci. *25,* 36.

Dry Beans

BANDERMER, S. L., and EVANS, R. J. 1963. J. Agric. and Food Chem. *11,* 134.
BOWMAN, D. E. 1944. Proc. Soc. Exp. Biol. Med. *57,* 139.
BRESSANI, R. 1969. *In* Programa Cooperativo Centroamericana para el Mijoramiento de Cultivas Alimenticios PCCMCA) Frijol XI Reunion Anual Carlos L. Arias, Editor). Publ. Misc. *68,* February 24–28, 1969. San Salvador, El Salvador.
CIAT. 1983. Center for International Agriculture in the Tropics, CIAT International: Report of Research and International Cooperation, Vol. 2, pp. 1–8, Cali, Columbia.
CONNER, J. K., NEILL, R., and BURTON, H. W. 1971. Aust. J. Exp. Agric. Anim. Husb. *11,* 387.
GOATCHER, W. D., and McGINNIS, J. 1972. Poultry Sci. *51,* 435, 440.
JACOB. E. 1967. Nutricion, Bromatol. Toxicol. *6,* 75.
JOHNS, C. D., and FINKS, A. J. 1920. J. Biol. Chem. *41,* 379.
KAKADE, M. L., and EVANS, R. J. 1965. Brit. J. Nutr. *19,* 269.
KAKADE, M. L., and EVANS, R. J. 1966. J. Nutr. *90,* 191.
KAKADE, M. L., and EVANS, R. J. 1969. J. Nutr. *99,* 34.
KAKADE, M. L., SMITH, J. E., and BARCHERS, R. 1968. Proc. Soc. Exp. Biol. Med. *128,* 811.
KELLY, J. F. 1971. J. Am. Soc. Hort. Sci. *96,* 561.
LELEJI, O. I. 1971. The genetics of crude protein and its relation to physiological and agronomic factors in dry beans. Ph.D. Thesis. Cornell Univ., Ithaca, New York.
OLSON, A. C., BECKER, R., MIERS, J. C., GUMBMANN, M. R., and WAGNER, J. R. 1975. *In* Protein Nutritional Quality of Foods and Feeds (M. Friedman, Editor), Vol. 2, Part 2, p. 551, Marcell Dekker, New York.
POND, W. G., KIRTLAND, D., and GUEST, R. W. 1972. Anim. Sci. Mimeo. Ser. *21,* Cornell Univ. Ithaca, New York.
PORTER, W. 1972. Genetic control of protein and sulfur contents in dry beans, *Phaseolus vulgaris.* Ph.D. Thesis. Purdue University, Lafayette, Indiana.
PURDOM, M. E., and BROWN, R. V. 1967. Arch. Latinoamer. Nutricion *17,* 117.
RIES, S. K. 1971. J. Amer. Soc. Hort. Sci. *96,* 557.
RIGAS, D. A., and OSGOOD, E. E. 1955. J. Biol. Chem. *212,* 607.
ROBERTS, L. M. 1970. The food legumes. Recommendations for expansion and acceleration of research to increase production of certain of these high-protein crops. Mimeo. The Rockefeller Foundation. November, 1970.
RUTGER, J. N. 1970. 10th Dry Bean Research Conf., Davis, California.
SILBERNAGEL, M. L. 1970. 10th Dry Bean Res. Conf., Davis, California.

Chickpeas

ABU-SHAKRA, S., MIRZA, S., and TANNOUS, R. 1970. J. Sci. Food Agric. *21,* 91.
CASARIN, A., BRAVO, F., de URIARTE, L. A., and SHIMADA, A. 1976. Tech. Pecuar. Mex. *31,* 27.
GONZALEZ DEL CUETO, A., MARTINEZ, W. H., and FRAMPTON, V. L. 1960. J. Agric. Food Chem. *8,* 331.
PENNACCHIOTTI, M. I., and SCHMIDT-HEBBEL, H. 1966. Arch. Latinoamer. Nutricion. *18,* 233.
SHIMADA, A. S., and BRAMBILA, S. 1967. Tec. Pecuar. Mex. *9,* 27; *10,* 5, 19.
SHIMADA, A., and AVILA, E. 1975. Arch. Latinoamer. Nutr. *25,* 363.
TANNOUS, R. I., and ULLAH, M. 1969. Trop. Agric., Trin. *46,* 123.
VERMA, S. C., LAL, B. M., and PARKASH, V. 1964. J. Sci. Food Agric. *15,* 25.

ZAMORA, J. M., AGUIRRE, A., SHIMADA, A., and MARTÍNEZ, L. 1975. Tec. Pecuar. Mex. *28,* 40.

Pigeon Peas

AKYROYD, W. R., and DOUGHTY, J. 1964. Legumes in Human Nutrition. FAO Publ. *19.*
DAKO, D. H. 1966. The protein value of African legumes in relation to pre-treatment and combination with other foods. Thesis, Landwirtsch. Fak. Justus Liebig-Univ. Giessen. (Nutr. Abstr. Rev. 1968, *38,* 469, No. 2630.)
FIKE, W. T. 1968. Pigeon Peas (*Cajanus cajan*), a New Green Manure Crop for the South. Mimeo., North Carolina State Univ., Raleigh.
KILLINGER, G. B. 1968. Soil and Crop. Sci. Fla. *28,* 161.
KRAUSS, F. G. 1932. Hawaii Agric. Exp. Stn. Bull. *64.*
OKE, O. L. 1967. *S. Afri. J. Biol. Appl. Chem. 9,* 52.
ORR, M. L., and WATT, B. K. 1957. Amino Acid Contents of Foods. USDA, Home Econ. Res. Dep. *4,* Washington, D.C.
PUSZTAI, P. 1967. Nutr. Abstr. Rev. *37,* 1.
ROBERTS, L. M. 1970. The Food Legumes. Recommendations for expansion and acceleration of research to increase production of certain of these high-protein crops. Mimeo. The Rockefeller Foundation. November, 1970.

Cowpeas

BORCHERS, R., and ACKERSON, C. W. 1950. J. Nutr. *41,* 339.
BRESSANI, R., ELIAS, L. G., and NAVARRETTE, D. A. 1961. J. Food Sci. *26,* 525.
CHAVEZ, N., TEODOSIO, N. R., GOMEZ DE MATOS, A., JR., LIMA, C. A., and DE ALMUDA, J. L. 1952. Rev. Brasil, Med. *9,* 603.
ELIAS, L. G., COLINDRES, R., and BRESSANI, R. 1964. J. Food Sci. *29,* 118.
HEITMAN, H., JR., and HOWARTH, J. A. 1960. J. Anim. Sci. *19,* 164.
JAFFE, W. G. 1949. Proc. Soc. Exp. Biol. Med. *71,* 469.
MANER, J. H. 1971. Swine Production Systems. CIAT Ann. Rep. Cali, Colombia.
MANER, J. H. 1972. Unpublished data. CIAT, Cali, Columbia
MANER, J. H., and POND, W. G. 1971. J. Anim. Sci. *33,* 233 (Abstr.).
OKE, O. L. 1967. W. Afr. J. Biol. Appl. Chem. *9,* 52.
ORR, M. L., and WATT, B. K. 1957. Amino Acid Content of Foods. USDA, Home Econ. Res. Dep. *4,* Washington, D.C.
PORTER, W. 1972. Genetic control of protein and sulfur contents in dry beans, *Phaseolus vulgaris.* Ph.D. Thesis. Purdue University, Lafayette, Indiana.
RACHIE, K. O., MEHTA, P. N., and AKINPELA, M. A. 1971. Grain Legume Improvement Annu. Rep. Int. Inst. of Trop. Agric., Ibadan, Nigeria.

Field Beans

AHERNE, F. X., and McALEESE, J. 1964. Proc. R. Dublin Soc. Biol. *1,* 113.
AHERNE, F. X., LEWIS, A. J., and HARDIN, R. T. 1977. Can. J. Anim. Sci. *57,* 321.
BALBOA, J., ZORITA, E., and RODRIGUES GUEDAS, J. 1966. Rev. Nutricion Anim. Madrid *4,* 41.
BOWLAND, J. P., ONAGHISE, G. T. U., and AHERNE, F. X. 1976. 55th Annu. Feeders' Day Rep., pp. 68–70. Univ. of Alberta, Edmonton.
CASTELL, A. G. 1976. Can. J. Anim. Sci. *56,* 425.
CHENUZ, M., and JAST, M. 1979. Rev. Suisse Agric. *11,* 249.
CLARKE, H. E. 1970. Proc. Nutr. Soc. *29,* 64.
COLE, D. J. A., BLADES, R. J., TAYLOR, R., and LUSCOMBE, J. R. 1971. Exp. Husb. *20,* 6.
EDEN, A. 1968. J. Agric. Sci. Cambridge *70,* 299.
ELLIOTT, R. F., VANDER NOOT, G. W., GILBREATH, R. L., and FISCHER, H. 1971. J. Anim. Sci. *32,* 1128.
ETIENNE, M. 1977. Journ. Rech. Porcine France, Paris. INRA-ITP Edition, pp. 199–203.
ETIENNE, M., DUĆE, P. H., and PASTUSZEWSKA, B. 1975. Jour. Rech. Porcine France, Paris. INRA-ITP Edition, pp. 115–121.
FOWLER, V. R., and LIVINGSTONE, R. M. 1977. Anim. Prod. *24,* 138.

FEIST, E. von., HOFMANN, P., KIRCHGESSNER, M., and SCHWARZ, F. J. 1974. Zurchtungskunde *46,* 50.
HANSEN, V., and CLAUSEN, H. 1969. Husdyrbougsudvalg *374,* Kohenhaven.
HENRY, Y. 1970. Bull. Tech. d'Information *253,* 1.
HENRY, Y., and Bourdon, D. 1972. Journ. Rech. Porcine France, Paris. INRA-ITP Edition, pp. 175–183.
HENRY, Y. and RERAT, A. 1969. Journ. Rech. Porcine France, Paris. INRA-ITP Edition, p. 161–168.
HENRY, Y., BOURDON, D., DUĆE, P. H., and JUNG, J. 1976. Journ. Rech. Porcine France, Paris. INRA-ITP Edition, pp. 51–60.
IVAN, M., and BOWLAND, J. P. 1976. Can. J. Anim. Sci. *56,* 451.
KADIRVEL, R., and CLANDININ, D. R. 1974. Poul. Sci. *53,* 1810.
LIGHTFOOT, A. L., YEO, M. L., and BASTIMAN, B. 1974. Exp. Husb. *27,* 120.
LIVINGSTONE, R. M., FOWLER, V. R., and WOODHAM, A. A. 1970. Proc. Nutr. Soc. *29,* 46A.
MARQUARDT, R. R., and CAMPBELL, L. D. 1973. Can. J. Anim. Sci. *53,* 741.
MARQUARDT, R. R., and CAMPBELL, L. D. 1974. Can. J. Anim. Sci. *54,* 437.
MARQUARDT, R. R., CAMPBELL, L. D., STOTHERS, S. C., and McKIRDY, J. A. 1974. Can. J. Anim. Sci. *54,* 177.
MARQUARDT, R. R., CAMPBELL, L. D., and WARD, T. 1976A. J. Nutr. *106,* 275.
MARQUARDT, R. R., WARD, T., CAMPBELL, L. D., and CANSFIELD, P. 1976B. Proc. West. Sec. Am. Soc. Anim. Sci. *27,* 138.
MATEOS, G. G., and PUCHAL, F. 1981. Nutr. Rep. Int. *23,* 347.
NAVRATIL, B. 1965. Zivoc. Vyr. *10,* 899. (Abstr. in Nutr. Abstr. Rev. *36,* 1163, No. 7117).
NIELSEN, H. E., and KRUSE, P. E. 1974. Livestock Prod. Sci. *1,* 179.
PASTUSZEWSKA, B., DUĆE, P. H., HENRY, Y., BOURDON, D., and JUNG, J. 1974. Ann. Zootech. *23,* 537.
WALZ, O. P. 1974. Z. Tierophysiol. Tierernaehr. Futtermittelkd. *33,* 183.
WHITTEMORE, C. T., and TAYLOR, A. G. 1973. J. Sci. Food Agric. *24,* 1133.
WILSON, B. J., and McNAB, J. M. 1972. Brit. Poul. Sci. *13,* 67.

Field Peas

BAJAJ, S., MICKELSEN, O., BAKER, L. R., and MARKARIAN, P. 1971. Brit. J. Nutr. *25,* 207.
BEESON, W. M., and HICKMAN, C. W. 1945. Univ. Idaho Agric. Exp. Stn. Cir. *106.*
BELL, J. M., and WILSON, A. G. 1970. Can. J. Anim. Sci. *50,* 15.
BERTRAND, G., PEREZ, J. M., and QUEMERE, P. 1980. J. Rech. Porcine France, pp. 215–225. Institut Technique du Porc., Paris.
BOUARD, J. P., CASTAING, J., FEKETE, J., LEVILLET, J., and MERLE, F. 1980. J. Rech. Porcine France. pp. 203–213.
BOURDON, D. and PEREZ, J. M. 1976. Rech. Porcine France, Paris. INRA-ITP. pp. 61–68.
CUNHA, T. J., WARWICK, E. J., ENSMINGER, M. E., and HART, N. K. 1948. J. Anim. Sci. *7,* 117.
ESNAOLA, M. A., ROMERO, O., and ROSA, J. M. 1977. Agric. Tech. (Chile) *37,* 125.
FOWLER, V. R., and LIVINGSTONE, R. M. 1977. Anim. Prod. *24,* 138.
FROSETH, J. A. 1976. Coop. Ext. Ser. Ext. Bull. 680, Washington State Univ., Pullman.
HENRY, Y., and RERAT, A. 1969. J. Rech. Porcine France, pp. 161–168. Institut Technique du Porc, Paris.
HUSBY, F. M. 1971. Swine Day Rep. Washington State Univ., Pullman.
KROENING, G. H. 1968. Protein quality studies with swine. Mimeo. Washington State University, Pullman.
LEHRER, W. P., JR., and HODGSON, C. W. 1946. Idaho Agric. Exp. Stn. Cir. *107.*
LOBBAN, J. M. 1963. J. Agr. Sci. Aust. *66,* 462.
MORAN, E. T., JR., SUMMERS, J. D., and JONES, G. E. 1968. Can. J. Anim. Sci. *48,* 42.
MYER, R. O., and FROSETH, J. A. 1978. Proc. West. Sec. Am. Soc. Anim. Sci. *29,* 182.
MYER, R. O., and FROSETH, J. A. 1980. Proc. West. Sec. Amer. Soc. Anim. Sci. *31,* 121.
REDDY, S. J., McGINNIS, J., MUEHLBAUER, F., and CAMPBELL, A. 1979. Poultry Sci. *58,* 376.
SINGH, S., SINGH, H. D., and SIKKA, K. C. 1968. Cereal Chem. *45,* 13.
TANNOUS, R. I., and ULLAH, M. 1969. Trop. Agric., Trinidad *46,* 123.
THOMKE, S., and FRÖLICH, A. 1968. Rep. of agric. Col. of Sweden. Ser. *A-NR-98,* pp. 20–26.
WARWICK, E. J., CUNHA, T. J., and ENSMINGER, M. E. 1948. Wash. Agric. Exp. Stn. Bull. *500.*
ZARKADAS, C. G., HENNEBERRY, G. D., and BAKER, B. E. 1965. J. Sci. Food Agric. *16,* 734.

Sweet Lupin

BATTERHAM, E. S. 1979. Aust. J. Agric. Res. *30,* 369.

BOHLMANN, F., and SCHULMANN, D. 1967. Lupine alkaloids. *In* The Alkaloids, (R. H. F. Manske, Editor), Vol. IX. pp. 175–221. Academic Press. New York.

CORBIN, E. J., and KNIGHT, M. E. 1977. Lupin Production. N. S. W. Dep. Agric. Southern Region Publ.

COWDY, P. E. M., EDDIE, S. M., BRIGSTOCKE, T. D. A., and WILSON, P. M. 1979. Anim. Prod. *28,* 433 (Abstr.).

FLORENCE, W. C. S. 1965. S. Afr. J. Agric. Sci. *8,* 661.

GARDINER, M. R., and GORMAN, R. C. 1965. Aust. J. Exp. Agric. Anim. Husb. *5,* 410.

GARNER, R. J. 1957. Veterinary Toxicology, pp. 316–317. Bailliere, Tindall and Cox, London.

GLADSTONES, J. S., and DOVER, D. P. 1962. Aust. J. Exp. Agric. Anim. Husb. *2,* 46.

GLADSTONES, J. S. 1972. Bull. Dep. Agric., Western Aust. Perth. *3834,* 1.

HACKBARTH, J. 1961. J. Aust. Inst. Agric. Sci. *27,* 61.

HILL, G. D. 1977. Nutr. Abstr. Rev. *47,* 511–529.

HOVE, E. 1974. J. Sci. Food Agric. *25,* 851.

KINGSBURG, J. M. 1964. Poisonous plants of the United States and Canada, pp. 333–341. Prentice-Hall, Englewood, New Jersey.

LEONARD, N. J. 1950. Lupin Alkaloids. *In* The Alkaloids, (R. H. F. Manske and H. L. Holmes, Editors), Vol. III, pp. 253–317. Academic Press, New York.

LEONARD, N. J. 1953. Lupin Alkaloids. *In* The Alkaloids, ((R. H. F. Manske and H. L. Holmes, Editors), Vol. III, pp. 119–199. Academic Press, New York.

LOUVARD, G. 1980. L'Elevage Porcine. *97,* 29.

LUCHSINGER, O. J. 1973. Thesis. Facultad de Med. Vet., Univ. Austral de Chile, Valdivia, Chile.

PEARSON, G., and CARR, J. R. 1976. Anim. Feed Sci. Technol. *1,* 631.

PEARSON, G., and CARR, J. R. 1977. Anim. Feed. Sci. Technol. *2,* 49.

PEARSON, G., and CARR, J. R. 1979. New Zealand J. Exp. Agric. *7,* 99.

PEREZ, J. M. 1980. L'Elevage Porcine. *97,* 21.

ROMERO, O., ESNAOLA, M. A., and ROSA, J. G. 1977. 2a Reunión Técnica Anual. Soc. Chilena de Prod. Animal, Santiago, Chile.

ROSA, J. G. 1974. Agric. Tech. (Chile) *34,* 158.

RUIZ, L. P. JR., WHITE, S. F., and HOVE, E. L. 1977. Anim. Feed Sci. Tech. *2,* 59.

SKOKNIC, A., DIAZ, J., and CORNEJO, S. 1979. Zbl. Vet. Med. A. *26,* 542.

TAVERNER, M. R. 1975. Anim. Prod. *20,* 413.

Alfalfa

AKESON, W. R. 1966. Econ. Bat. *20,* 244.

ALLEE, G. I. 1976. J. Anim. Sci. *42,* 1348 (Abstr.).

BAIRD, D. W., CAMPBELL, H. C., and ALLISON, J. R. 1970. J. Anim. Sci. *31,* 518.

BAIRD, D. W., CAMPBELL, H. C., and ALLISON, J. R. 1975. J. Anim. Sci. *41,* 1039.

BIRK, Y. 1969. Saponins. *In* Toxic Constituents of Plant Foodstuffs (I. E. Liener, Editor), pp. 169–210. Academic Press, New York.

BOHMANN, V. R., KIDWELL, J. F., and McCORMICK, J. A. 1953. J. Anim. Sci. *12,* 876.

BOHMANN, V. R., HUNTER, J. E., and McCORMICK, J. 1955. J. Anim. Sci. *14,* 499.

BOLTON, J. L. 1962. Alfalfa Interscience, New York.

BOYD, R. D., CRENSHAW, T. D., MOSER, B. D., PEO, E. R., JR., and DANIELSON, D. M. 1976. J. Anim. Sci. *42,* 1348.

CHAMORRO, J., and MANER, J. H. 1974. CIAT Annual Report, Cali, Colombia.

CHEEKE, P. R., and STANGEL, D. E. 1972. J. Anim. Sci. *34,* 757.

CHEEKE, P. R., and MYER, R. O. 1975. Nutr. Rep. Int. *12,* 337.

CHEEKE, P. R., KINZELL, J. H., and PEDERSEN, M. N. 1977. J. Anim. Sci. *45,* 476.

CHEEKE, P. R., PEDERSEN, M. W., and ENGLAND, D. C. 1978. Can. J. Anim. Sci. *58,* 783.

COLE, D. J. A., DUCKWORTH, J. E., and HOLMES, W. 1967. Anim. Prod. *9,* 149.

CUNHA, T. J., ROSS, O. B., PHILLIPS, P. H., and BOHSTEDT, G. 1944. J. Anim. Sci. *3,* 415.

DANIELSON, D. M., and NOONAN, J. J. 1975. J. Anim. Sci. *41,* 94.

ELLIS, J. H., and KING, J. X. 1952. J. Anim. Sci. *1,* 358.

FAIRBANKS, B. W., KRIDER, J. L., and CARROLL, W. E. 1945. J. Anim. Sci. *4,* 410.

FREEMAN, J. 1938. Mich. Agric. Exp. Stn. Quart. Bull. *21,* 91.
FULTON, R. 1959. Proc. 6th Tech. Alfalfa Conf. Berkeley, California.
HOCHSTELLER, L. N., HOEFER, J. A., PEARSON, A. M., and LUECKE, R. W. 1959. J. Anim. Sci. *18,* 1397.
HOGAN, A. G., and JOHNSON, S. R. 1941. Mo. Agric. Exp. Stn. Res. Bull. *332.*
KASS, M. L., VAN SOEST, P. J., and POND, W. G. 1980A. J. Anim. Sci. *50,* 192.
KASS, M. L., VAN SOEST, P. J., POND, W. G., LEWIS, B., and McDOWELL, R. E. 1980B. J. Anim. Sci. *50,* 175.
KEYS, J. E., VAN SOEST, P. J., and YOUNG, E. P. 1969. J. Anim. Sci. *29,* 11.
LEAMASTERS, B. R., and CHEEKE, P. R. 1979. Can. J. Anim. Sci. *59,* 467.
MOOIJMAN, J. G. J. M. 1965. Diss. Abstr. *25,* 4380.
MOSER, B. D. 1978. Ontario Forage Conf., Toronto, Ontario.
MYER, R. O., and CHEEKE, P. R. 1975. J. Anim. Sci. *40,* 500.
PEO, E. R., JR., MOSER, B. D., CUNNINGHAM, P. J., KOHLER, G., and LIVINGSTON, A. L. 1971. J. Anim. Sci. *33,* 238 (Abstr.).
PEO, E. R., JR., MOSER, B. D., and CUNNINGHAM, P. J. 1972. J. Anim. Sci. *35,* 222 (Abstr.).
PEO, E. R. JR., LIVINGSTON, A. L., KNOWLES, R. E., and KOHLER, G. 1977. Proc. Maryland Nutr. Conf. Feed Manuf., pp. 78–86, March 17–18, 1977.
POLLMANN, D. S., DANIELSON, D. M., and PEO, E. R., JR. 1979. J. Anim. Sci. *48,* 1385.
POLLMANN, D. S., DANIELSON, D. M., CRENSHAW, M. A., and PEO, E. R., JR. 1980. J. Anim. Sci. *51,* 294.
POND, W. G., YEN, J. T., LINDVALL, R. N., and HILL, D. 1980. J. Anim. Sci. *51,* 367.
POWLEY, J. S., CHEEKE, P. R., ENGLAND, D. C., and KENNICK, W. H. 1980. Proc. West. Sec. Am. Soc. Anim. Sci. *31,* 125.
ROSS, O. B., PHILLIPS, P. H., BOHSTEDT, G., and CUNHA, T. J. 1944. J. Anim. Sci. *3,* 406.
SEERLEY, R. W., and WAHLSTROM, R. C. 1963. Anim. Sci. Mimeo. Ser. *63,* 4. S. D. State College, Brookings.
TEAGUE, H. S. 1955. J. Anim. Sci. *14,* 621.
TEAGUE, H. S., and HANSON, L. E. 1954. J. Anim. Sci. *13,* 206.
VAN SOEST, P. J., and MOORE, L. A. 1965. Proc. IX Int. Grassland Congr., Sao Paulo, Brasil No. *424.*

Alfalfa Protein Concentrate

BICKOFF, E. M., BOOTH, A. N., DE FREMERY, D., EDWARDS, R. H., KNUCKLES, B. E., MILLER, R. E., SAUNDERS, R. M., and KOHLER, G. O. 1975. *In* Protein Nutritional Quality of Foods and Feeds (M. Friedman, Editor), Vol. 1, Part 2, p. 319. Marcel Dekker, New York.
BOURDON, D., PEREZ, J. M., HENRY, Y., and CALMES, R. 1980. J. Rech. Porcine France, pp. 227–244.
CHEEKE, P. R., KINZELL, J. H., de FREMERY, D., and KOHLER, G. O. 1977. J. Anim. Sci. *44,* 772.
HOVE, E. L., LOHREY, E., URS, M. K., and ALLISON, R. M. 1974. Brit. J. Nutr. *31,* 147.
HSU, A., and ALLEE, G. L. 1980. J. Anim. Sci. *51*(Suppl. 1), 204, 205 (Abstr.).
KOHLER, G. O. 1974. 12th Tech. Alfalfa Conf. Proc., pp. 65–66, Berkeley, California.
KOHLER, G. O., CHRISMAN, J., and BICKOFF, E. M. 1973. *In* Alternate Sources of Protein for Animal Production. Natl. Acad. Sci., Washington, D.C.
KUZMICKY, D. D., KOHLER, G. O., and BICKOFF, E. M. 1972. 11th Tech. Alfalfa Conf. Proc. ARS/USDA ARS *74,* 60.
LIVINGSTON, A. L., and KOHLER, G. O. 1979. Pastos *9,* 108.
MYER, R. O., and CHEEKE, P. R. 1975. J. Anim. Sci. *40,* 500.
MYER, R. O., CHEEKE, P. R., and KENNICK, W. H. 1975. J. Anim. Sci. *40,* 885.
PIRIE, N. W. 1977. World Anim. Rev. *22,* 11.

Alfalfa and Grass Juice

BARBER, R. S., BRAUDE, R., MITCHELL, K. G., PASTRIDGE, J. G., and PITTMAN, R. J. 1979. Anim. Feed. Sci. Tech. *4,* 233.
BARBER, R. S., BRAUDE, R., MITCHELL, K. G., PASTRIDGE, J. G., and PITTMAN, R. J. 1980. Anim. Feed Sci. Tech. *5,* 215.
CONNELL, J., and FOXELL, P. R. 1976. Bienn. Rev. Natl. Inst. Res. Dairy, pp. 21–41.
HOUSEMAN, R. A., TRUSCOTT, S. A., and JONES, A. S. 1978A. Anim. Prod. *26,* 378 (Abstr.).

HOUSEMAN, R. A., PIRIE, R., and TRUSCOTT, S. A. 1978B. Anim. Prod. *26,* 393 (Abstr.).

Single-Cell Protein—Fungi

ALAVIUHKOLA, T., KORHENEN, I., PARTANEN, J., and LAMPILA, M. 1975. Acta. Agric. Scand. *25,* 301.
BARBER, R. S., BRAUDE, R., and MITCHELL, K. G. 1977. Anim. Feed Sci. Tech. *2,* 161.
BOBROV, E. P., CHIKOV, A. E., and ZHERNOVOJ, I. T. 1978. Agric. Res. Centre, Finl. Inst. Anim. Husb. Rep. *12,* 50.
SALO, M. L., and PEKKARINEN, E. 1981. J. Sci. Agric. Soc. Finl. *53,* 52.

Single-Cell Protein—Bacteria

ALAVIUHKOLA, T., KORHENEN, I., PARTANEN, J., and LAMPILA, M. 1975. Acta. Agric. Scand. *25,* 301.
BRAUDE, R., and RHODES, D. N. 1977. Livestock Prod. Sci. *4,* 91.
BRAUDE, R., HOSKING, Z. D., MITCHELL, K. G., PLONKA, S., and SAMBROOK, L. E. 1977. Livestock Prod. Sci. *4,* 79.
D'MELLO, J. P. F., PEERS, D. G., and WHITTEMORE, C. T. 1976. Brit. J. Nutr. *36,* 403.
EECKHOUT, W. 1980. Revue de l'Agric. *33,* 19.
FARSTAD, L. 1977. Acta Agric. Scand. *27,* 129.
FARSTAD, L., FLATLANDSMO, K., ALMLID, T., and KOPPANG, N. 1980. Acta Agric. Scand. *30,* 3.
NAESS, B. 1978. Quoted by Farstad *et al.* 1980.
NEWPORT, M. J., and KEAL, H. D. 1980. Brit. J. Nutr. *44,* 161.
SCHULZ, E., and OSLAGE, H. J. 1976. Anim. Feed Sci. Tech. *1,* 9.
STRINGER, D. A. 1978. Quoted by Farstad *et al.* 1980.
WATERWORTH, D. G., CROSS, M. E., and SMITH, K. L. 1978. Anim. Prod. *26,* 397 (Abstr. 92).
WHITTEMORE, C. T., and MOFFOT, I. W. 1976. J. Agric. Sci. Cambridge *86,* 407.
WHITTEMORE, C. T., MOFFOT, I. W., and TAYLOR, A. G. 1976. J. Sci. Food. Agric. *27,* 1163.

Single-Cell Protein—Yeast

BARBER, R. S., BRAUDE, R., MITCHELL, K. G., and MYRES, A. W. 1971. Brit. J. Nutr. *25,* 285.
BECK, H., and GROPP, J. 1974. Z. Tierphysiol. Tierernaehr., Futtermittelkd. *33,* 148, 305.
BERGONZINI, E., and FABBRI, R. 1974. 25th Annu. Meet. E.A.A.D., Copenhagen *17,* 218.
BERGONZINI, E., FABBRI, R., BONIFORTI, L., and APPLIOLA, E. 1978. Annu. 1st Sper. Zootech. *11,* 227.
ENGEL, C. 1972. *In* Proteins from Hydrocarbons (H. G. de Pontanel, Editor), pp. 53–81. Academic Press, London.
FARSTAD, L., FLATLANDSMO, K., ALMLID, T., and KOPPANG, N. 1980. Acta. Agric. Scand. *30,* 30.
FEVRIER, C. 1971. J. Rech. Porcine France, pp. 91–98. Institut Technique du Porc, Paris.
FRYDRYCH, Z., HEGER, J., and VAVAK, J. 1980. Biol. Chem. Vet. (Praba) *16*(22), 207.
GAYE, A., and MOAL, J. 1972. J. Rech. Porcine France, pp. 169–173. Institut Technique du Porc, Paris.
ISAR, M., DAMIAN, C., ISAR, O., BUNEA, M., CURCA, C., and LIVESCU, B. 1978. Nutritia Anim. *8,* 129.
MADSEN, A., MORTENSEN, H. P., and LARSEN, E. 1971. Forsøgslab. Arbog. København., pp. 107–112.
POKROUSKY, A. A. 1968. *In* Single-Cell Protein (R. I. Mateles and S. T. Tannebaum, Editors), pp. 163–165. MIT Press, Cambridge, Massachusetts.
RUSSO, V., MARIANI, P., DEL MONTE, P., and CATALANO, A. 1973. *In* Atte Cong. Petrolito e Ambiente, Roma. April 11–14. Poligrafico Artioli Moderna, pp. 73–79.
RUSSO, V., CATALANO, A., MARIANI, P., and DEL MONTE, P. 1974. Avicoltura *43,* 73.
RUSSO, V., CATALANO, A., MARIANI, P., and DEL MONTE, P. 1976. Anim. Feed. Sci. Tech. *1,* 25.
SALO, M. L. 1977. Acta Agric. Scand. *27,* 77.
SALO, M. L., and PEKKARINEN, E. 1981. J. Sci. Agric. Sci. Finl. *53,* 52.
SCHULZ, E., and OSLAGE, H. J. 1976. Anim. Feed Sci. Technol. *1,* 9.
SHACKLODY, C. A., and GATUMEL, E. 1972. *In* Proteins from Hydrocarbons (H. G. de Pontanel, Editor), pp. 27–52. Academic Press, New York.

VAN DER WAL, P. 1966. ILOB Rep. *134,* Wageningen, Holland.
VAN DER WAL, P. 1970. ILOB Rep. *253,* Wageningen, Holland.
WALDROUP, P. W. 1972. Proc. Distillers' Feed Research Council Conf. *27,* 34, Cincinnati, Ohio.

Single-Cell Protein—Algae

ASPLUND, J. M., and PFANDER, W. H. 1973. *In* Alternate Sources of Protein for Animal Production. Natl. Acad. Sci., Washington, D.C.
CHUNG, P., POND, W. G., KINGSBURY, J. M., WALKER, E. F., JR., and KROOK, L. 1978. J. Anim. Sci. *47,* 319.
COOK, B. B., LARR, E. W., and BAILEY, B. M. 1963. J. Nutr. *81,* 23.
DEVI, M. A., SUBBULAKSHMI, G., DEVI, K. M., and VENKATARAMAN, L. V. 1981. J. Agric. Food Chem. *29,* 522.
GLEN, C. 1979. Sunshine State Agric. Res. Rep. *23,* 23.
HINTZ, H. F., and HEITMAN, H., JR. 1967. Anim. Prod. *9,* 135–140.
HINTZ, H. F., HEITMAN, H., JR., WEIN, W. C., TORELL, D. T., and MEYER, J. H. 1966. J. Anim. Sci. *25,* 675–681.
OSWALD, W. J. 1962. Am. J. Public Health *52,* 235.
OSWALD, W. J., and GLUEKE, C. G. 1968. *In* Large-Scale Protein Production of Algae in Single-Cell Protein (Richard I. Moteles and Steven R Tannenbaum, Editors). MIT Press, Cambridge, Massachussets.
RIGOR, E. M., EUSEBIO, J. A., and GARCIA, B. R. 1980. NSDB Technology J. *5,* 9.
VINCENT, W. A. 1971. Proc. Symp. Soc. General Microbiol. *8,* 47.
WENG, TZE-CHOW. 1978. Taiwan Livestock Res. *11,* 79.
WITT, M., and SCHRODER, J. 1967. Landwirtsch Forsch *20,* 148–156.
WITT, M., SCHRODER, J., and MEHDEN, H. 1962. Zuchtungskunde. *34,* 272.
YANNI, S., MOKADY, S., SACHS, K., KANTOROWITZ, B., and BERK, Z. 1979. Nutr. Rep. Int. *19,* 391.
YAP, T. N., WU, J. F., POND, W. G., and KROOK, L. 1982. Nutr. Rep. Int. *25,* 543.

Seaweed

BEAMES, R. M., TAIT, R. M., WHYTE, J. N. C., and ENGLAR, J. R. 1975. Nutrient balance of pigs receiving diets containing up to 20% kelp (*Nereocystis luetkeana*) meal. Proc. Pacific Sci. Cong. *13,* 272.
BEAMES, R. M., TAIT, R. M., WHYTE, J. N. C., and ENGLAR, J. R. 1977. Can. J. Anim. Sci. *57,* 121.
CAMERON, C. D. J. 1954. Can. J. Agric. Sci. *34,* 181.
HOMB, T. 1961. Norges Landbrukshogsk. Foringsforsok Beretn. *106,* 28.
JENSEN, A., and MINSAAS, J. 1960. Tidsskr, Norske Landbruk. *67,* 81.
JONES, R. T., BLUNDEN, G., and PROBERT, A. J. 1979. Botanica Marina *22,* 393.
KRISHNA, P. V. 1956. Proc. Indian Acad. Sci. *44,* 3.
LOMBARD, P. E. 1962. S. Afr. J. Agric. Sci. *5,* 378.
MALAJSKAJTE, B. S. 1963. Svinovodstvo *4,* 33.
MELGAR ARNAIZ, F. 1966–1967. Anim. Inst. Invest. Vet. Madrid. *17,* 167.
SOBCZAK, Z., and SZLICHCINSKA, E. 1964. Zeszyty. Nauk. WSR. Wroclau, Zootech. *13,* 143.
VOSLOO, W. A., and STEENKAMP, D. J. 1958. Farming in S. Afr. *33*(12), 49.
WHITTEMORE, C. T., and PERCIVAL, J. K. 1975. J. Sci. Food Agric. *26,* 215–217.
WHYTE, J. N. C. 1971. Proc. Int. Seaweed Symp. *7,* 496.
WOODWARD, N. 1951. J. Sci. Food Agric. *2,* 477.

Synthetic Amino Acids

UNITED STATES CONGRESS. 1984. Commercial Biotechnology—An International Analysis. Office of Technology Assessment, U.S. Congress, Washington, D.C., pp. 195–198.

16

Mineral and Vitamin Sources and Nonnutritive Feed Additives

INTRODUCTION

Although the largest proportion of the diet is made up of energy and protein, the mineral elements and vitamins are vital to normal growth and reproduction. Most energy and protein sources provide some vitamins and minerals but often it is necessary to supplement the diet with specific mineral and vitamin sources to balance the diet. In recent years, the use of nonnutritive feed additives such as antibiotics has become a common practice as a means of further increasing weight gain and efficiency of feed utilization beyond that obtained with a nutritionally balanced diet. This chapter is devoted to a description of feedstuffs and substances commonly used to supply vitamins, minerals, and nonnutritive additives so as to provide an optimum diet for all phases of the life cycle of the pig.

MINERAL SOURCES

Sodium (Na) and Chlorine (Cl)

Sodium and chlorine are essential constituents of all swine diets. Sodium, as the chief cation of the extracellular fluids, is the most important ion in the maintenance of osmotic pressure, body fluid balance, and hydration of the tissues. Sodium and potassium (K) are essential for the operation of some enzyme systems and the maintenance of blood pH. Heart action and nerve impulse conduction and transmission are also highly dependent upon proper concentrations of these cations. Chloride ion functions mainly to ensure the proper fluid-electrolyte balance (National Research Council 1980).

Almost all feedstuffs are deficient in Na and Cl; therefore, common salt is routinely added to virtually all mixed diets for swine. The level commonly added is 0.4–0.5% of the diet. It has been shown that a level of 0.20–0.25% supplemental NaCl will meet the dietary sodium and chlorine requirements of the growing-finishing pig fed grain–soybean meal diets. Hagsten and Perry (1976) recommended that under practical conditions in which the base level of salt in the feed is unknown, supplementation with 0.20% NaCl should prevent salt deficiency.

Inadequate dietary levels of NaCl will significantly affect feed and water consumption and result in unthriftiness and slow growth. Long-term salt deficiency may lead to salt craving and death. In the presence of adequate drinking water, swine can tolerate dietary levels of NaCl as high as 2–3% for several days (Done *et al.* 1959; Todd *et al.* 1964); however, when water is restricted, as little as 1.0% dietary NaCl has produced signs of toxicity which include nervousness, weakness, tremors, incoordination, convulsions, paralysis, coma, and death.

Salt is often used as a convenient carrier for trace elements. Thus, for a nominal cost, essential trace elements such as zinc, copper, iron, and manganese may be incorporated into the salt in amounts appropriate to supply the needed trace elements in the right proportions when the salt is consumed at the accepted dietary level (0.5%). The composition of a typical "trace mineralized salt" available commercially is shown in Table 16.1. It should be emphasized that the mere presence of a trace element in a particular trace mineralized salt preparation does not ensure an adequate intake. For example, in the sample salt preparation shown in Table 16.1, the amount of zinc is only 0.008% or 80 ppm in the salt. When the salt is incorporated into the diet at 0.5%, the amount of zinc contributed by the salt is only 0.4 ppm, a negligible quantity in terms of the pig's requirement of 50 ppm or more. Thus, in the use of micronutrient supplements, special attention must be given to the amount of a particular nutrient provided, for it may be present in only a token amount.

TABLE 16.1. Composition of a Typical "Trace Mineralized Salt"[a] Used at 0.5% of a Mixed Diet for All Ages of Pigs

Mineral[b]	*Percent*	*Amount in complete diet*[c] *when salt is at 0.5%(ppm)*
Sodium chloride (salt)	Not less than 97.00 nor more than 99.00	5,000
Cobalt	Not less than 0.015	0.70
Copper	Not less than 0.023	1.15
Iodine	Not less than 0.007	0.35
Iron	Not less than 0.117	5.4
Manganese	Not less than 0.225	11.25
Sulfur	Not less than 0.040	2.0
Zinc	Not less than 0.008	0.40

[a]Guaranteed analysis.

[b]The ingredients supplying the above minerals are: salt, cobalt carbonate, copper oxide, calcium iodate, iron carbonate, manganese oxide, sodium sulfate, zinc oxide; yellow prussate of soda added as an anti-caking agent.

[c]The amount of zinc, and in some cases other trace elements as well, contributed by salt at 0.5% of the diet is negligible in terms of the pig's needs in the above preparation.

Calcium (Ca)

Calcium compounds are widely distributed in feedstuffs, and Ca is the most abundant cation in the animal body, comprising 1–2% of the total body weight. Approximately 99% of the Ca in the animal body is found in the bone and teeth, with the remaining 1% distributed in the soft tissues. Calcium serves a principal function in the bone where it functions not only as a structural component but also as a readily available, dynamic source of Ca for maintenance of homeostasis. Calcium is an activator of numerous enzymes and functions in many other physiological processes such as blood coagulation, muscle contractility, myocardial function, normal neuromuscular irritability, and integrity of various intracellular as well as extracellular membranes. Signs of a calcium deficiency may include rickets, osteomalacia, and low calcium tetany.

Depending on age, size, and principal function, the National Research Council (1979) has established the calcium requirements of swine as 0.50–0.90%, younger growing animals requiring the higher levels and more mature finishing animals requiring the lower levels. Although only limited data are available, breeding animals are shown to require approximately 0.75% dietary Ca. As shown in a comprehensive review on Ca (Peo 1976), the level of Ca required for optimal growth may not necessarily coincide with the dietary level that produces optimal bone development and strength. More importantly, because of the very close interrelationship Ca has with phosphorus and vitamin D, dietary Ca requirements are not rigid and may vary between diets. Calcium and phosphorus (P) requirements are closely related because of the importance of an optimum dietary ratio (approximately 1.2 parts of Ca to 1.0 part of P) for bone formation and other metabolic functions (see Chapter 10).

Almost all feedstuffs contain some Ca; however, the calcium content varies widely and is very low in maize and the cereal grains such as sorghum, barley, wheat, and oats (0.02–0.10%) and in plant protein supplements (0.17–0.66%) such as soybean meal, rapeseed meal, and cottonseed meal that normally supply the major proportion of protein supplement swine diets. Only animal protein sources such as fish meal, meat meal, meat-and-bone meal, and dried milk products contribute significant quantities of Ca. Because most natural feedstuffs are deficient in Ca, specific supplements high in Ca are almost always required. Supplemental sources of Ca used in swine diets include ground calcitic or dolomite limestone or calcium carbonate, oyster shell, high grade calcite, gypsum, and wood ashes. More common supplements are those that contain both Ca and P from sources such as bone meal, dicalcium phosphate, and defluorinated rock phosphate. The average composition of common calcium sources is listed in Table 16.2.

Dietary calcium toxicity per se is not considered a problem with swine, as the homeostatic mechanism of the animal tends to protect against the absorption of excessive quantities of the element. Calcium excesses are important because of the interrelationship of the element with other nutrients and especially with P and zinc (Zn). Large excesses of Ca can upset the desired calcium-to-phosphorus ratio, reduce gains and feed efficiency, and induce zinc deficiency (parakeratosis).

TABLE 16.2. Calcium and Phosphorus Content of Common Mineral Supplements (in Percent)

Supplement	*Calcium*	*Phosphorus*
Bone meal, raw	21	9–10
Bone meal, steamed	24–29	12–14
Bone black, spent	27	12
Dicalcium phosphate	23–26	18–21
Defluorinated rock phosphate[a]	31–34	13–17
Raw-rock phosphate[b]	24–29	13–15
Soft phosphate[a]	18	9
Curaçao phosphate[a]	35	15
Oyster shell	38	—
Limestone	38	—
Calcite, high grade	34	—
Gypsum	22	—
Dolomite limestone	22	—
Wood ashes	21	—

Source: Sievert 1967.
[a]Usually less than 0.2% F. [b]2.0–4.0% F.

Phosphorus (P)

Phosphorus is an essential nutrient for swine and is involved in almost every aspect of metabolism. In addition to being of major importance in bone formation and skeletal growth, it is an important component of many coenzymes, adenosine di- and tri-phosphate, DNA, RNA, and nucleic acid, and plays an important role in transformations of carbohydrates, fats, amino acids, and in nerve tissue metabolism. A comprehensive review of P in swine nutrition has been prepared by Hays (1976).

Although a total phosphorus requirement of between 0.40 to 0.70% has been established for swine, many dietary factors may influence these requirements. The amount absorbed is dependent on source, intestinal pH, lactose intake, and dietary levels of Ca, P, Vitamin D, iron (Fe), aluminum (Al), manganese (Mn), potassium (K), magnesium (Mg), and fat. Signs of phosphorus deficiency include poor growth, rickets, and osteomalacia and are closely associated with those for Ca and vitamin D.

Phosphorus is present in variable amounts in almost all feedstuffs used by swine. Unfortunately, the principal form of P in most cereal grains and plant protein sources is phytate (mixed salt of *myo*-inositol hexaphosphoric acid or phytic acid), which can be utilized by swine only after the compound has been hydrolyzed to yield P. Intestinal phosphatase or phytases are ineffective in hydrolyzing phytate in swine diets and only phytases of plant or microbial origin are effective in hydrolyzing significant amounts of phytate (Nelson 1980). Although available phosphorus values for plant sources are usually less than the content of nonphytate phosphorus, the higher availability values for wheat and barley compared to other grains are probably related to the amount of phytase present in the grains. Phytate may undergo some hydrolysis during certain types of feed or

TABLE 16.3. Biological Availability and Total and Available Phosphorus in Some Commonly Used Feedstuffs

Feedstuff	*Biological availability of P, %*	*Total P, %*	*Available P, %*
Maize	12	0.25	0.03
Grain sorghum	12	0.25	0.03
Oats	23	0.35	0.08
Barley	35	0.35	0.12
Wheat	48	0.35	0.17
Wheat bran	29	1.15	0.33
Wheat middlings	34	0.90	0.31
Rice bran	18	1.50	0.27
Soybean meal, dehulled	22	0.65	0.14
Cottonseed meal	0	1.00	0

Source: Adapted from Cromwell 1980.

ingredient treatment or processing such as pelleting (Bayley *et al.* 1975), fermentation (Singsen *et al.* 1972), and acid treatment (Trotter and Allee 1979); however, the amount released is small in relation to the animal needs. Estimates of the available phosphorus in some commonly used feedstuffs are shown in Table 16.3.

Most swine diets require supplemental sources of P in addition to that present in the common feedstuffs. Unlike phytate phosphorus, most of the inorganic phosphorus sources are quite high in available P (Table 16.4) (Hays 1976). The supplements most often used in swine diets, dicalcium phosphate, defluorinated rock phosphate, and steamed bone meal, are of similar bioavailability. Cromwell *et al.* (1976) reported that P in meat-and-bone meal is as available as that in dicalcium phosphate and assumed the same to be true for fish meal, tankage, dried milk, and other animal products.

TABLE 16.4. Bioavailability of Phosphorus in Some Inorganic and Animal Product Sources for Swine[a]

Source of P	*Availability of P, %*
Monosodium phosphate	100
Monocalcium phosphate	100
Dicalcium phosphate	98
Meat-and-bone meal[b]	98
Steamed bone meal	95
Diammonium phosphate	95
Defluorinated phosphate	92
Curaçao Island phosphate	69
Soft rock phosphate	34

Source: [a]adapted from Cromwell 1980. [b]Cromwell et al. 1976.

Supplemental phosphates are not considered to be highly toxic. Single oral doses can be tolerated with little if any effects. On the other hand, long-term consumption of dietary levels two to three times the requirement level will cause severe problems due to induced changes in calcium metabolism. The effect of high phosphorus levels is very dependent on the calcium level of the diet, as well as other factors such as vitamin D. A relative excess of P in relation to Ca can result in increased bone resorption in young and adult animals and reduced growth in young swine. When tolerance is expressed in terms of a multiple of the requirement level, P has one of the lowest tolerance factors of any mineral element.

Zinc (Zn)

As a component of the hormone insulin and a large number of important metalloenzymes (Vallee and Wacker 1970; Riordan and Vallee 1974) and as an activator of many enzyme systems, zinc is required for normal growth, development, and function of all swine. A comprehensive review of Zn in swine nutrition has been prepared by Miller *et al.* (1979).

Appetite depression and growth failure are two of the early signs of zinc deficiency in baby pigs (Miller *et al.* 1968). Zinc deficiency during pregnancy and lactation affects prenatal and postnatal growth and may produce congenital abnormalities. Gilts receiving zinc-deficient diets during gestation and lactation may produce fewer and smaller pigs. Testicular development and spermatogenesis of growing boars is retarded by zinc deficiency.

The function of Zn in preventing dermatitis (parakeratosis) in swine was first recognized by Tucker and Salmon (1955). Zinc deficiency is especially common in swine fed high calcium diets because zinc absorption is apparently reduced as the calcium concentration is increased. Diets containing more than 1% Ca consistently produce parakeratosis unless a zinc supplement is included in the diet (Lewis *et al.* 1957; Newland *et al.* 1958; Stevenson and Earle 1957; Luecke *et al.* 1956; Hoefer *et al.* 1959; Heigener *et al.* 1958). Many other dietary components can also influence zinc absorption. Zinc absorption is reduced by dietary phytate (O'Dell and Savage 1960; Byrd and Matrone 1965), fiber (Reinball *et al.* 1976), copper (Cu) (Van Campen 1969), cadmium (Cd) (Parizek 1957), and chromium (Cr) (Hahn and Evans 1975). Zinc deficiency is reduced by autoclaving (Smith *et al.* 1960), certain dietary ingredients, such as casein, liver extract, and distillers dried solubles (Scott and Ziegler 1963), corn oil (Pond *et al.* 1966), blood meal (Dahmer *et al.* 1972A), and EDTA and similar chelating agents (Vahra and Kratzer 1964). Supplemental vitamin D (Becker and Hoekstra 1966), histidine (Dahmer *et al.* 1972B), and cystine (Nielsen *et al.* 1966) also appear to improve zinc utilization. Zinc absorption is affected by dietary level and chemical form used in the diet. Zinc in swine feed is supplied predominantly as zinc oxide; however, other inorganic salts of Zn such as the carbonate, sulfate, chloride, nitrate, and acetate are well absorbed. Miller *et al.* (1981) indicated that for the young growing pig, Zn from metallic dust was about 30% more available than from zinc oxide powder. Zinc from certain ores such as sphalerite and franklinite are largely unabsorbed (Edwards 1959; Poppe 1968). The percentage of Zn in

TABLE 16.5. Zinc Content of Zinc Supplements and Amounts of Each Needed to Provide Zinc at 100 ppm (0.01%)

Zinc supplement	*Zinc content*	*Amount needed to provide 100 ppm Zn (ppm)*
Zinc carbonate	54	185
Zinc oxide	80	125
Zinc sulfate	40	250
Zinc metal powder	100	100

several supplements is shown in Table 16.5. Some feedstuffs high in Zn are listed in Table 16.6.

The zinc requirement of baby pigs (1–5 weeks of age) has been studied by Miller *et al.* (1970) and Manner and McCrea (1963, 1964). Based on the chemical composition of sow-reared pigs, Manner and McCrea calculated that the Zn requirement of the baby pig was 102 ppm in an air-dry diet. Miller *et al.* (1970) indicated that the zinc requirement of baby pigs fed isolated soy–glucose diets is between 45 and 90 ppm, with the requirement of the male baby pig somewhat higher than that of the female. When plant phytates or phytic acid are eliminated by replacing plant proteins with casein and glucose, the Zn requirement of the

TABLE 16.6. Good Sources of Zinc for Swine

Source	*Zinc content, ppm—dry basis*
Nonfat dried milk	41
Blood meal	306
Brewers' dried grains	98
Dried whey, roller process	27
Defluorinated rock phosphate	38
Calcium caseinate, food grade	57
Coastal bermuda grass, dried	32[a]
Fish meal (menhaden)	147
Meal meal	103
Rapeseed meal (solvent)	66
Sesame meal (expeller)	100
Soybean meal (solvent)	27[b]
Sow milk	25[c]
Wheat bran	95
Wheat shorts	106

Source: National Research Council 1979; Miller and Miller 1963.

[a]Content varies with location where grown.

[b]The bioavailability may be low. Parakeratosis is often observed in pigs fed a maize–soybean meal diet unsupplemented with inorganic zinc salts.

[c]Pond et al. 1965.

baby pig is reduced to about 15 ppm. The National Research Council (1979) has established a requirement of 100 ppm for the 1–10-kg pig.

When dietary calcium levels are not more than 1%, 50 ppm Zn is adequate to meet the needs of growing-finishing pigs fed on semipurified diets based on isolated soybean protein or natural diets containing maize and soybean meal. Studies by Pond and Jones (1964) and Hedges *et al.* (1976) indicate that dietary zinc levels of growing-finishing pigs of 33–35 ppm may be adequate providing their dam received adequate zinc and dietary calcium levels are not excessive. Liptrap *et al.* (1970) studied the zinc requirement over the entire growing-finishing period from 15 to 90 kg and concluded that 48–53 ppm is adequate to meet the requirements of boars, gilts, and barrows.

In studies by Hedges *et al.* (1976), reproductive performance of sows through five parities was not significantly affected when up to 33 ppm Zn was fed; however, higher levels were required for satisfactory lactation performance. Pigs from sows receiving 83 ppm grew faster and more efficiently than those from sows fed only 33 ppm. In recent studies by Wegger and Palludan (1978), a level of 20 ppm Zn added to a zinc-deficient, casein-based diet containing 4–5 ppm Zn adequately met the gestation–lactation requirements.

Zinc may be increased to 1000 ppm in the diet without producing signs of toxicity. Although dietary levels up to 4000 ppm from zinc oxide did not produce a toxic reaction in studies by Hsu *et al.* (1975) and Cox and Hale (1962), definite clinical signs of zinc-toxicity (including death of some pigs) within 5 weeks on maize–soybean meal diets supplemented with 2000 ppm of Zn from zinc carbonate were observed by Brink *et al.* (1959). Signs of zinc-toxicosis may include growth depression, arthritis, hemorrhage in axillary spaces, gastritis, enteritis, lameness, and congestion of the mesentery. Hemorrhages in the ventricles of the brain, lymph nodes, and spleen have also been observed.

Iron (Fe)

Iron is essential to higher forms of life because of its central role in hemoglobin which permits oxygen and electron transport. Quantitatively, the total content of the pig's body is small and varies with weight, age, sex, and nutritional and health status. Most of the body Fe exists in complex forms bound to protein. At birth of the pig, 47% of Fe is associated with blood, 15% with liver, 1.6% with spleen, and the remainder is in other body tissues and fluids (Thoren-Tolling 1975). After the neonatal period, about 70–80% of the total is contained in hemoglobin, myoglobin, heme enzymes, cofactors, and transport Fe. The remainder is storage Fe associated with nonheme protein-bound compounds such as ferritin and transferrin and found predominantly in liver, spleen, and bone marrow. Occasionally, in certain disease states and especially during a state of Fe overload, large amounts of Fe may be present as hemosiderin. A comprehensive review of Fe in swine nutrition has been prepared by Zimmerman (1980).

Iron-deficiency anemia in pigs is the most characteristic symptom of deficiency of this nutrient and is typically described as a hypochronic–microcytic type. Hemoglobin concentration in the blood is the most commonly used indicator of iron deficiency; however, packed red cell volume, number of red cells, and hemoglobin saturation progressively decline with increasing iron deficiency. He-

TABLE 16.7. Iron Status of Pigs Relative to Hemoglobin Concentrations

Blood hemoglobin, g/100 ml	*Iron status*
10 or above	Normal level; adequate iron and optimal performance.
9	Minimum level required for average performance and the dividing line between normality and anemia.
8	Borderline anemia; iron treatment needed.
7 or below	Anemic condition that has been shown to retard growth rate.
6 or below	Severe anemia accompanied by marked reduction of performance.
4 or below	Severe anemia that can be expected to result in increased mortality rate.

Source: National Research Council 1979.

moglobin concentrations in swine blood vary with iron status and are also affected by interrelationships with other nutrients. The National Research Council (1979) characterized the iron status of pigs relative to their hemoglobin concentration (Table 16.7). Anemic pigs show signs of poor growth, listlessness, labored breathing, rough hair coat, wrinkled skin, and paleness of mucous membranes. Fast-growing anemic pigs may die suddenly of anoxia.

The amount of Fe that must be absorbed from feed in order to maintain normal body iron levels is determined by the amount excreted or lost from hemorrhage and the needs for maintenance, growth, and reproduction. Iron is most important as a supplement during the suckling period. Although the intestine of the newborn pig has fully functional absorption capacity for Fe (Furugouri and Kawabatu 1979), suckling pigs are particularly prone to iron deficiency because (1) they have limited iron stores at birth; (2) their growth rate is phenomenal during the first few weeks of life; and (3) their main source of nutrients, sow's milk, contains a very limited concentration of Fe (about 1 mg/liter). The feeding of high levels of many different iron compounds to gestating and lactating sows has not effectively increased the iron-content of milk. Thus, unless a supplemental source of Fe is provided during the first few days after birth, anemia develops. Injectable forms of Fe (iron dextran, iron dextrin) can be given intramuscularly at 2 or 3 days of age to supply 150–200 mg. Because of the higher biological availability of injectable Fe (over 90% of injectable iron is utilized), a much smaller quantity is required to maintain normal hemoglobin than when dietary Fe is the sole source. If one assumes a metabolic requirement of 15 mg of Fe per day, the total need for the first 10 days of life can be met by a single injection of 150 mg. Rapid growth and the pig's need to meet the total energy requirement by consumption of dry feed, beginning at about the third week of postnatal life, mean that attention must be given to the amount and biological availability of feed iron.

The postweaning dietary iron requirement is about 80 ppm and diminishes in later growth and maturity, since there is a smaller increase in blood volume with increasing body weight. Feed ingredients in a balanced diet usually supply enough Fe to meet postweaning requirements. Some feedstuffs high in Fe are listed in Table 16.8. Only a small part of the Fe consumed by the pig is absorbed. Efficiency of iron absorption from the diet can be influenced by many factors such as iron status of the animal, size of the dose, adequacy of major nutrients,

TABLE 16.8. Good Sources of Iron and Manganese for Swine

Feedstuffs	*Iron as-fed (ppm)*	*Manganese as-fed (ppm)*
Alfalfa meal	310	2.8
Blood meal	3000	Low
Bermuda grass, fresh	400	36.7
Bluegrass hay, all analyses	200	83.9
Bone meal	800	30.4
Brewers' dried grains	250	38
Clover hay, all analyses	800	73.5
Coconut meal	190	55.4
Cowpeas hay, all analyses	800	43.9
Fish meal, menhaden	440	33
Limestone	3800	279.6
Maize dried distillers' solubles	600	73.5
Molasses, cane	200	42.2
Rice bran	190	138
Vetch hay, all analyses	400	53.7

Source: National Research Council 1964, 1979.

presence and excess of other minerals, and biological availability of the iron source. Bioavailability values of Fe from natural feedstuffs fed to swine are not available. Relative biological value (RBV) of Fe in a few feedstuffs has been determined for the rat and chick. For maize, blood meal, oat flour, isolated soybean protein, and fish protein concentrate, RBV has been shown to be 73, 35, 21, 97, and 33–59%, respectively. Natural iron sources for growing pigs should come from animal products rather than forages because of the high fiber content of forages.

Many chemical substances are being used as sources of oral iron. The availability of Fe varies from source to source. Ferrous sulfate heptahydrate ($FeSO_4 \cdot 7H_2O$) is the standard against which other iron sources usually are evaluated. It is highly available to animals. Ferrous sulfate monohydrate appears to be equally available and has better handling and mixing properties. Often one hears that the ferrous compounds are much more available than ferric sources of Fe. Although this is frequently true, as is the case when ferric oxide and ferrous sulfate are compared, not all ferric compounds have a low availability. The relative biological values for rats, chicks, and pigs of different iron compounds are listed in Table 16.9. Many of the ferric compounds are readily available to animals. The differences in availability of the various iron compounds are related to solubility in acid; absorption of Fe occurs in the upper duodenum where the influence of stomach acid secretion is greatest. Thus iron salts that are soluble in acid are more available for absorption than those of lower solubility. Such differences among iron sources in biological availability are large enough to be of practical importance in diets for growing-finishing swine.

Characteristic signs of chronic iron toxicosis are reduced feed intake, growth rate, and efficiency of feed conversion. Clinical signs of acute toxicosis include

TABLE 16.9. Relative Biological Value of Iron from Various Sources Using Ferrous Sulfate as a Standard

Iron source	*Chick–rat, %*	*Pig, %*
$FeSO_4 \cdot 7H_2O$	100	100
$FeSO_4$, anhydrous	100	100
$FeSO_4$, feed grade	100	100
$FeSO_4 \cdot H_2O$	—	100
$FeSO_4 \cdot 2H_2O$	—	100
$Fe(NH_4)_2(SO_4)_2 \cdot 6H_2O$	99	—
$FeCO_3$	2–6	0,14,27,60
$FeCl_2 \cdot 4H_2O$	98	—
$FeNH_4$ citrate	107	100
Fe choline citrate	102	—
Fe–Cu–Co choline citrate	—	140
Fe–citrate	73	100
Fe–glycerophosphate	93	—
$Fe_2(SO_4)_3$	83	—
$FeCl_3 \cdot 6H_2O$	44	—
NaFe pyrophosphate	5–26	33
Fe_2O_3	4	12
Reduced iron	18,37,63	33
Iron powder	—	70
Ferric poly phosphate	—	97
di Na EDTA	—	90

Sources: Adapted from Miller et al. 1981; Zimmerman 1980; Fritz et al. 1970.

anorexia, diarrhea, hypothermia, disphasic shock, metabolic acidosis, and death. Vascular congestion of the gastrointestinal tract, liver, heart, kidneys, lung, brain, spleen, adrenals, and thymus are the dominant histopathologic findings. While feeding high levels of Fe to detoxify gossypol in pigs fed diets containing cottonseed meal, Buitrago *et al.* (1970) demonstrated that levels as high as 3200 ppm of Fe in the diet can be fed safely while markedly reducing gossypol levels in the liver. Furogouri (1972) reported no significant effect on body weight gains of young pigs fed either 1102 or 3103 ppm Fe; however, when the dietary Fe levels were increased to 5102 and 7102 ppm, body weight gains and feed intake were decreased. As was shown previously (O'Donovan *et al.* 1963), signs of phosphorus deficiency were noted at 5102 and 7102 ppm of Fe, even though the diets contained 0.92% phosphorus. Cornelius and Harmon (1976) administered oral doses of 200 mg Fe from either ferric ammonium citrate, ferrous sulfate, ferric oxide, and/or an iron–dextran complex to piglets within 6 hr of birth. The ferric ammonium citrate proved to be highly toxic, with only a third of the pigs surviving to 21 days. Rate of survival of pigs fed ferrous sulfate and ferric oxide ranged from 82 to 100%.

Many compounds appear to form complexes with Fe in the gastrointestinal tract and to aid absorption by keeping Fe in solution during transit through the upper part of the small intestine. Some amino acids, simple sugars, and organic acids as well as ascorbic acid have been shown to increase iron absorption. Other dietary factors such as copper and phosphorus also influence iron absorption and toxicity; acute iron toxicity may occur in vitamin E–selenium-depleted baby pigs given Fe parenterally.

Copper (Cu)

The biological requirement for Cu in animals was shown by Hart *et al.* (1928), who further demonstrated a role for Cu in iron metabolism and hemoglobin synthesis. Copper in the natural inorganic state occurs primarily as copper sulfite. In biological systems the cuprous and cupric forms predominate. Copper is absorbed from the upper gastrointestinal tract and especially from acid media in the stomach. Efficiency of copper absorption depends on age, the form of copper consumed, the copper status of the animal, and the presence of other dietary compounds such as phytates, calcium, zinc, iron, lead, silver, molybdenum, and ascorbic acid, which tend to chelate copper. Although in practice, copper sulfate is the most commonly used form of supplemental copper for swine, other soluble copper compounds such as copper carbonate, copper–methionine, and copper chloride can be utilized as well as copper sulphate in supplying Cu to swine diets. The Cu from all these compounds is more readily absorbed than that supplied by copper oxide, although this compound is commonly used as a copper supplement.

Anemia that is both microcytic and hypochromic is an early sign of copper deficiency. Ataxia, leg weakness, listlessness, reduced appetite, skeletal abnormalities, depigmentation of hair, aortic rupture, reduced serum copper, and cardiac hypertrophy have also been observed. Copper is essential for the synthesis and activation of several oxidative enzymes necessary for normal metabolism in the pig. A deficiency of Cu leads to poor iron mobilization, abnormal hematopoiesis, keratinization, and synthesis of collagen, elastin, and myelin. Most deficiency signs appear to involve a deficiency of a specific copper-dependent enzyme system such as cytochrome oxidase, ferroxidase, lysyl oxidase, tyrosinase, or superoxide dismutase. A level of 6 ppm of Cu in the diet for the very young pig seems to be adequate, and the requirement is probably no greater in later stages of growth. Information on the requirements for pregnancy and lactation is lacking (National Research Council 1979).

Supplemental levels of Cu much higher than those required for normal body functions have been shown to stimulate increased growth rate and feed efficiency in young growing swine fed barley–fish meal diets (Barber *et al.* 1955A,B,C, 1957; Bowler *et al.* 1955; Bellis 1961). Braude (1967) summarized the results of British research and reported that copper supplements providing 250 ppm Cu improved daily gain by 8% and feed efficiency by 5.5% when compared to animals receiving no supplemental copper.

Canadian and American research with high levels of Cu has produced varied results. Wallace *et al.* (1960) found that 250, 300, 500, and 750 ppm Cu supplied as copper sulfate in maize–soybean meal diets proved toxic to pigs in the growing and finishing phases. A level of 200 ppm Cu did not adversely affect growth but significantly reduced hemoglobin levels. Bunch *et al.* (1961) obtained improved gains and feed efficiency when 250 ppm Cu were added to a 22% maize–soybean meal diet, but also reported a linear decrease in hemoglobin level as the level of

Cu was increased. Miller *et al.* (1969) reported a 19% improvement in performance of growing pigs by the addition of 250 ppm Cu to a basal maize–soybean meal diet; however, growth rate was depressed during the finishing phase so there was no overall benefit from the added Cu. Average data from ten stations in the United States (NCR-42 1974) showed increased daily gains for the first 8 weeks of the trial when 125, 187.5, and 280 ppm Cu as copper sulfate were added to a maize–soybean meal diet; however, over the entire experimental period (growing and finishing) 250 ppm added Cu did not affect either growth rate or efficiency of feed utilization. Castrell *et al.* (1975) found that supplying 125 and 200 ppm Cu as copper sulfate with fish meal or rapeseed meal and soybean meal resulted in 2.2 and 1.7% improvement in growth and 2.0 and 2.2% improvement in feed efficiency. Omole and Bowland (1974) with 10, 135, and 210 ppm Cu and Myers and Bowland (1973A,B,C) with similar levels found no effect on animal performance.

As Miller *et al.* (1979) point out in their review, it has been postulated that Cu suppresses undesirable bacteria and stimulates the growth of desirable bacteria. Although Miller *et al.* (1969) observed that adding 250 ppm Cu to the diet lowered fecal bacterial count, others (Fuller *et al.* 1960; Smith and Jones 1963) have failed to confirm this finding. Studies to measure the synergistic effect of Cu with antibiotics have produced confusing results. While Wallace (1967), from his review of U.S. research, concluded that Cu and chlortetracycline induced more gain on less feed when compared to chlortetracycline alone, many other workers (Barber *et al.* 1978; Braude 1967; Lillie *et al.* 1977; Beames 1969; Barber *et al.* 1965) were unable to obtain additive effect of Cu with a variety of antibiotics including virginiamycin, tylosin, zinc bacitracin, and oxytetracycline.

A significant time period is usually required for the development of chronic copper toxicosis and swine appear more tolerant of high dietary copper than ruminants. Although British workers have commonly used 250 ppm Cu for growth stimulation in growing pigs, in the United States, the addition of 250 ppm Cu to diets of pigs 3–7 weeks old has been associated with depressed growth, hemoglobin, and liver iron levels and increased liver copper and zinc (Ritchie *et al.* 1963; Gipp *et al.* 1973A,B, 1974). Although Allcroft *et al.* (1961) found that 400 ppm Cu in swine diets were nontoxic, other workers (Combs *et al.* 1966; De Gory *et al.* 1971; Suttle and Mills 1966) have shown 425–750 ppm added Cu to cause reduced feed intake, retarded growth, anemia, and death. The toxic effects of lower levels of Cu can be eliminated by supplementing with 150 ppm Zn and 150 ppm Fe. At higher levels of copper supplementation, toxicosis can be eliminated by 500 ppm Zn and 750 ppm Fe.

Iodine (I)

The only known function of iodine in the body is as a constituent of the thyroid hormones, thyroxine, and triiodothyronine, which are important in the regulation of metabolic rate. Iodine in the body is concentrated in the thyroid but is widely distributed throughout the body. It is stored in the body in the elemental form, but when trapped in the thyroid is rapidly oxidized and converted to organic iodine by combination with tyrosine to form the thyroid hormones. These iodine-containing hormones function in thermoregulation, intermediary metabolism, reproduction, growth and development, hematopoiesis and circulation, and neuromuscular functioning.

Iodine deficiency in swine is manifested mainly in the progeny of sows fed iodine-deficient diets. Pigs born to sows fed goitrogenic diets are hairless and have thickened skin and myxedema. Although stillborn pigs are observed in iodine deficiency, most pigs are born alive, but are weak and usually die within a few hours. At necropsy, the thyroid is enlarged and hemorrhagic and decreased in iodine content. Experimental estimates of the requirement for normal growth and thyroid size range from 0.05 to 0.14 ppm; a dietary level of 0.14 ppm seems to be adequate for all stages of the life cycle (National Research Council 1979). Excessive levels of arsenic, bromide, fluoride, cobalt (Co), Ca, manganese (Mn), nitrate, and NaCl and goitrogenic substances, such as the glucosinalates, thiocyanates, and perchlorates increase the requirement.

Iodine is widely distributed in animal feeds but in very small amounts. The iodine content of plant products varies widely depending on species of plant and the iodine content of the soil where grown. Samples of oilseed meals have been shown to contain 0.11–0.2 ppm I. Products of animal origin, other than fish meal, do not contain significant levels of I. To assure adequate dietary iodine, small quantities of supplemental iodine often are added to the diet as a constituent of the salt (NaCl) supplement. Iodine sources permitted by the Food and Drug Administration as feed additives include calcium iodate, calcium iodobehenate, cuprous iodide, 3,5-diiodosalicylic acid, ethylenediamine–dihydriodide (EDDI), pentocalcium orthoperiodate, potassium iodate, potassium iodide, sodium iodate, sodium iodide, and thymal iodide. Generally, iodine from these compounds is highly available to swine; however, loss of I can occur from some of these compounds when exposed to heat, moisture, or light and when improperly protected when mixed with other minerals. Calcium and potassium iodate and pentacalcium orthoperiodate are more stable in salt and feed mixtures than sodium or potassium iodides. If salt contains 0.007 ppm I, the requirement can be met by incorporating 0.2% salt in the diet.

Although pigs can tolerate iodine levels far in excess of their requirements for this element, indiscriminate use of supplemental iodine compounds may cause iodine poisoning. Newton and Clawson (1974) fed levels of I ranging from 10 to 1600 ppm to growing-finishing pigs and found that the minimum toxic level was between 400 to 800 ppm. Growth rate, feed intake, and hemoglobin levels were depressed and thyroid weight increased at 800 and 1600 ppm I, and liver iron and serum iron and thyroid weight increased at 800 ppm.

Selenium (Se)

Selenium is an essential nutrient that serves as an integral part of the enzyme glutathione peroxidase (Rotruck *et al.* 1972), which destroys lipid peroxides and functions importantly in protecting cell membranes against peroxidative damage (Ullrey 1978). Selenium deficiency results in muscular dystrophy (paleness and dystrophy of skeletal muscles commonly known as white muscle disease); hemorrhage and necrosis of the liver; edema of the lungs, spinal colon, subcutaneous tissue, and submucosa of the stomach; impaired immune response; impaired reproduction; and, occasionally, mottling and dystrophy of the myocardium (mulberry heart disease) and sudden death.

Although the quantitative dietary requirement for Se has not been established precisely for all classes of swine nor for all dietary and environmental circumstances, it is generally agreed that the selenium requirement is between 0.1 and 0.2 mg/kg of diet (Groce *et al.* 1971; Sharp *et al.* 1972A,B; Ullrey 1974; Mahan *et al.* 1974; Piper *et al.* 1975; Young *et al.* 1977; Wilkinson *et al.* 1977A,B; Mahan *et al.* 1977; Hitchcock *et al.* 1978; Mahan and Moxon 1978; Bengtsson *et al.* 1978; Hakkarainen *et al.* 1978; Mahan 1978), and dietary adequacy is dependent on the bioavailability of the Se in the feed and upon concentrations of a number of other components, especially vitamin E. Because of their shared antiperoxidant role, the selenium requirement is inversely related to the vitamin E concentration in the diet. However, although a high dietary level of vitamin E will reduce the level of Se required, it will not completely eliminate the need for dietary selenium. Conversely, if natural diets are very low in vitamin E (0.3 IU/kg), selenium levels as high as 1.3 ppm may not prevent deficiency lesions in all pigs (Young *et al.* 1977). Excess dietary levels of arsenic (As), cadmium (Cd), Cu, Zn, Fe, and oxidized unsaturated fat, as well as low dietary protein, moldy grain, infections, cold damp environments, and physical stress have been shown to interact with Se and vitamin E metabolism and to influence the development and severity of deficiency lesions (Ammerman and Miller 1972; Miller *et al.* 1979; Underwood 1977).

There is a great geographical variability in selenium content of soils as reflected in crop content and in the natural incidence of selenium deficiency in swine in various parts of the world (Table 16.10). Supplemental selenium can be adequately supplied by sodium selenite, sodium selenate, selenomethionine, or seleniferous maize. Because of the complexities of selenium nutrition and metabo-

TABLE 16.10. Some Good Sources of Selenium of Canadian Origin for Swine

Feedstuffs	*Selenium content (ppm)*	
	Mean	*Range*
Linseed meal	1.05	0.65–1.51
Rapeseed meal	0.98	0.46–1.90
Brewers' grains	0.94	0.81–1.07
Malt sprouts	0.60	0.29–0.97
Herring meal	2.84	2.21–3.36
White-fish meal	1.83	1.65–2.18
Fish meal	1.96	1.25–2.94
Feather meal	0.72	0.60–0.88
Blood meal	0.69	0.51–1.16
Corn gluten meal	0.31	0.20–0.57
Distillers' grains	0.35	0.16–0.56

Source: Arthur 1971.

[a]The concentration of selenium in plant materials varies with the geographic location and growing conditions.

TABLE 16.11. Biological Availability for Poultry of Selenium in Feedstuffs

Feedstuffs	*Biological availability (%)*
Alfalfa meal	100
Brewers' grains	89
Brewers' yeast	81
Maize	83
Cottonseed meal	78
Herring meal	31
Menhaden meal	35
Meat-and-bone meal	36
Poultry by-product meal	33
Soybean meal	64
Tuna meal	33
Wheat	100

Source: Scott and Cantor 1972.

lism, bioavailability of various compounds has not been determined for swine. Bioavailability of selenium for poultry in some feedstuffs is presented in Table 16.11.

Of practical concern in some locations is selenium toxicity in areas where soil levels and, in turn, crop levels of selenium are excessive. Although a level of 5 mg Se per kilogram of diet supplied by sodium selenite, sodium selenate, selenomethionine, or seleniferous maize does not produce toxicity, levels from 7.5 to 10.0 mg/kg may produce toxicity. Toxicity is manifest by poor growth, anorexia, hair loss, fatty infiltration of the liver, degenerative changes in liver and kidney, swelling and occasional separation of hoof and skin at the coronary bond, poor reproduction, and death. Dietary arsenicals help to alleviate selenium toxicity, apparently by reducing the amount of selenium absorbed from the gastrointestinal tract.

Cobalt (Co)

Cobalt per se does not appear to be a dietary essential for swine except as it functions as a component of vitamin B_{12}. Microbes use cobalt to synthesize B_{12}. Some microbial synthesis of B_{12} occurs in the lower digestive tract, but only limited quantities are absorbed. The pig may ingest vitamin B_{12} via coprophagy.

No dietary requirement for Co has been established. Cobalt is found in variable quantities in plants and most feeds supply adequate cobalt for B_{12} synthesis by microorganisms in the gastrointestinal tract. Supplemental sources for ruminants include cobalt acetate, carbonate, chloride, oxide, and sulfate. Bioavailability of different cobalt compounds has not been determined for swine.

Adding Co as cobalt chloride to the diet at levels up to 200 ppm did not result in toxicosis in pigs fed a diet adequate in Fe (Huck and Clawson 1976). The addition of 400 or 600 ppm Co caused anorexia, stiff-leggedness, humpback, incoordination, and extreme muscular tremors. Decreased serum iron and anemia may occur in the presence of excess cobalt.

Manganese (Mn)

Pigs suffering from manganese deficiency have exhibited retarded growth, abnormal skeletal growth with altered fat to lean body tissue; lameness; absence of, or irregular, estral cycles; poor mammary development and lactation; resorption of fetuses; and small weak pigs at birth. The bone defects associated with manganese deficiency appear to be related to chondrogenesis rather than osteogenesis. Manganese is essential for synthesis of chondroitin sulfate, which is a component of mucopolysaccharides in the organic matrix of bone. Manganese also functions in other enzymes involved in carbohydrate, lipid, and protein metabolism.

Manganese concentration varies greatly among plant species but it is widely distributed in feedstuffs (Table 16.8). Therefore, diets containing commonly used feed ingredients should contain more than adequate quantities of Mn. Manganese is absorbed through the intestinal tract; however, there is evidence that absorption is decreased by feeding isolated soybean protein or excess levels of Ca, P, and Fe.

Definitive requirements for pigs have not been established. Continued feeding of purified diets containing only 0.5 mg/kg of diet has interfered with normal development and reproduction. Growth is normal when pigs are fed as little as 1.5 mg/kg in purified diets. For gestation and lactation a minimum of 10 mg/kg is suggested. These suggested requirements should be adequately supplied by Mn present in common feed ingredients; however, because of widely diverse estimates of requirements (Liebholz *et al.* 1962; Grummer *et al.* 1985), it is desirable to provide supplemental manganese in the diet.

Manganous oxide and manganous sulfate are the two most commonly used forms in swine feeding; however, other sources such as manganese acetate, carbonate, citrate, chloride, gluconate, orthophosphate, and dibasic phosphate can be used. Data with chicks (Ammerman and Miller 1972) showed equal bioavailability of Mn from sulfate, chloride, carbonate and oxide.

The toxic level of Mn is many times the requirement. Leibholz *et al.* (1962) were unable to demonstrate signs of toxicity until a dietary level of 675 ppm was fed. Similar signs of excess were observed by Grummer *et al.* (1950) after feeding diets containing 550 ppm Mn. Signs of toxicity included decreased food intake, growth, and hemoglobin as well as limb stiffness.

Magnesium (Mg)

Magnesium is one of the major mineral elements recognized as essential for swine. However, because it is abundant in commonly used feedstuffs and supplementation of practical diets is not normally required, its significance in swine nutrition is often overlooked. Excellent reviews on magnesium functions, use, availability, and toxicity have been published (Sell 1979; Sell and Fontenot 1980; National Research Council 1980).

Magnesium is distributed widely among body tissues. Approximately 60–70% of the total body magnesium is present in the bone, where about one-third is combined with phosphate and the remainder is adsorbed loosely on the surface of the mineral structure. Except as a reserve that can be mobilized for soft tissue use when dietary magnesium is inadequate, the function of Mg in bone is not known (Aikawa 1963; Pike and Brown 1975). In the soft tissues, Mg is found within the cells, with high concentrations in the mitochondria and lesser amounts in the microsomes, nuclei, and cytoplasm and extracellular fluids. The small amount present in extracellular fluid is exchanged easily with that adsorbed on the bone surface. Blood serum normally contains 2–4 mg of Mg per 100 ml.

Magnesium is a cofactor in many important enzyme systems and is essential as an enzyme activator for cellular oxidation. Magnesium is required for energy transformations such as oxidative phosphorylation leading to adenosine triphosphate (ATP) formation. Also, Mg activates enzymes that oxidize pyruvate, converts α-ketoglutarate to succinyl coenzyme A, and facilitates the transsketolase reaction of the pentose monophosphate shunt (Pike and Brown 1975; Sell and Fontenot 1980). It is necessary for all phosphate transfer reactions. As a component of many enzyme systems, Mg is of fundamental importance in the metabolism of carbohydrates, proteins, and lipids. Although the mechanism remains obscure, Mg also is believed to be essential for nerve excitation and muscle contraction, function of ribosomes, normal coagulation, and fibrinolytic properties of blood and genetic transcription (Aikawa 1971; Seelig and Heggtueit 1974).

Since the levels of Mg present in common feed ingredients are quite high (Table 16.12), information about magnesium nutrition is meager. Mayo *et al.* (1959) induced magnesium deficiency in growing pigs using semipurified diets. Signs of magnesium deficiency include loss of appetite, slow growth, hyperirritability, muscular twitching, reluctance to stand, stepping syndrome, weak pasterns, loss of equilibrium, and tetany followed by death. Mayo *et al.* (1959) concluded from the research that newly weaned pigs require 400–500 ppm Mg.

TABLE 16.12. Magnesium Content of Selected Feedstuffs

Feedstuff	*Mg, %*	*Feedstuff*	*Mg, %*
Barley	0.14	Alfalfa meal, dehyd. 17%	0.26
Maize	0.12	Blood meal, spray or ring dried	0.22
Oats	0.16	Cottonseed meal, solvent	0.40
Sorghum	0.20	Feather meal	0.20
Wheat	0.17	Fish meal, menhaden	0.16
Brewers' dried grains	0.16	Meat-and-bone meal, 50%	1.12
Maize gluten feed	0.29	Peanut meal, solvent	0.40
Maize distillers' solubles, dehyd.	0.64	Rapeseed meal, solvent	0.51
Molasses, cane	0.35	Safflower meal, solvent	0.37
Oat groats (dehulled)	0.09	Sesame meal, expeller	0.86
Rice bran	0.95	Soybean meal, solvent	0.27
Rice polishings	0.65	Sunflower meal	0.75
Wheat bran	0.52	Skim milk, dried	0.11
Wheat middlings	0.29	Whey, dried	0.13

Source: National Research Council 1979.

Bartley *et al.* (1961) conducted similar studies and reported that pigs from 3 to 6 weeks of age require 400 ppm Mg to maintain desirable magnesium balance, serum magnesium levels, and rates of gain. When semipurified diets containing 125 ppm or less were fed to young pigs (Miller *et al.* 1965), neuromuscular incoordination, weakened pasterns, and tetany were observed. Pigs fed diets containing 225 ppm exhibited no visual magnesium deficiency signs, but serum and bone magnesium levels were abnormally low; they estimated that 325 ppm Mg were required to support efficient growth of pigs and to maintain normal serum and bone concentrations of Mg.

Although the National Research Council (1979) recommends 400 mg of Mg per kg of feed for both growing-finishing pigs and adult swine, the minimum requirement is not known. Harmon *et al.* (1976) fed semipurified diets containing either 400 or 900 ppm Mg to sows during gestation and either 150 or 650 ppm Mg during lactation. Magnesium level had no effect on reproductive performance during gestation; however, during lactation, sows fed 150 ppm Mg were in severe negative magnesium balance, while those on 650 ppm maintained magnesium balance. Harmon *et al.* (1976) showed that lower levels of Mg had no effect on magnesium level in the milk nor on body weight gains by pigs. Magnesium toxicity due to ingestion of natural feedstuffs has not been reported and does not appear likely. Magnesium toxicity can occur when supplemental sources are added to practical diets. Ingestion of excess levels of Mg has generally resulted in decreased feed intake, depressed gains, and reduced efficiency of feed utilization and can cause lethargy, disturbance in locomotion, diarrhea, and death. Svajgr *et al.* (1969) added 100 ppm Mg to practical diets containing 1.300 ppm Mg. Magnesium supplementation reduced gains of pigs on pasture but had no effect on pigs kept in confinement. In studies by Miller (1971), supplemental magnesium (270 ppm Mg from magnesium oxide added to a corn–soybean meal-based diet containing 0.5 P and 0.5–0.7% Ca) had no effect on average daily gains or feed efficiency of growing-finishing pigs. Krider *et al.* (1972) added 1080 ppm Mg to a diet containing 1570 ppm Mg and found that both rate of gain and feed intake were depressed. However, in later studies (Krider *et al.* 1975), increasing the level of Mg in the diet from 0.16 to 0.22% lowered rate and efficiency of gain in growing pigs from 20 to 45 kg, but had no effect on pig performance during the finishing period or overall. In Nebraska studies (Fritschen and Peo 1979), the addition of 135 or 270 ppm Mg to commonly used growing-finishing pig diets reduced both gains in body weight and efficiency of feed utilization. Decreases in performance appear to be caused at least in part by depressed feed intake. Field observations have suggested that supplemental magnesium has a tranquilizing effect that modifies certain behavioral traits of pigs grown in confinement and reduces problems such as ulcers and tail biting. However, controlled studies have failed to confirm these observations (Krider *et al.* 1975; Wesoloski *et al.* 1975). In fact, Fritschen and Peo (1979) reported that the addition of 135 and 270 ppm Mg from magnesium oxide to a corn–soybean meal basal diet not only reduced pig performance but actually increased the incidence of tail biting in growing-finishing pigs.

The adequacy of the Mg in common ingredients for meeting the nutritional needs of swine depends primarily on the degree of availability of this element for

TABLE 16.13. Apparent Absorption of Magnesium in Various Chemical Compounds by Rats

Form of magnesium	*Apparent absorption, %*[b]		*Bioavailability*[b]
$MgCO_3$	68.3	64.9	75.1
$MgCl_2$	—	61.0	73.3
MgO	—	58.0	73.0
$MgPO_4$	66.5	54.1	74.2
$MgSO_4$	—	53.3	73.0
$MgSiO_3$	—	54.2	—
Mg phytate, acidic	62.4	—	—
Mg phytate, basic	68.1	—	—
Mg lactate	—	—	75.1
Mg citrate	—	—	73.8
Mg acetate	—	—	76.0

Source: Adapted from Sell and Fontenot 1980.
[a]First column data from Meintzer and Steenbock 1955; second column from Cook 1973.
[b]Data from Ranhotra et al. 1976.

absorption by the animals. Since deficiencies are not commonly seen in swine fed natural ingredients, little bioavailability data are available for swine. From balance studies with young pigs, Bartley *et al.* (1961) obtained Mg retention values ranging from 16 to 35% for magnesium carbonate and magnesium sulfate. Miller (1971) reported that growing-finishing pigs retained 10–40% of the Mg ingested when fed maize–soybean meal diets. Related data of true and apparent Mg bioavailability of some organic and inorganic sources obtained with rats and chicks are tabulated in Tables 16.13 and 16.14. Although comparative data are not available, it is assumed that swine are able to utilize Mg at rates similar to those of rats and chicks.

TABLE 16.14. True Absorption of Magnesium in Magnesium Sulfate and Selected Feedstuffs as Determined in Chickens

Magnesium source	*True absorption, %*
$MgSO_4 \cdot 7H_2O$	57.4
Maize	55.7
Wheat	56.8
Oats	82.7
Barley	54.5
Rice	60.3
Soybean meal	62.0
Dried skim milk	42.5

Source: Guenter and Sell 1974.

Potassium (K)

Although the potassium content of most feedstuffs for swine is considered to be adequate to meet metabolic requirements, recent evidence suggests that supplemental potassium may be beneficial under certain conditions. Liebholz *et al.* (1966) observed improved weight gain in pigs when K was added at levels of 0.4 or 0.8% of the diet. The data suggested that lysine utilization was improved by K. In recent work (Froseth *et al.* 1982; Austic *et al.* 1982), supplemental potassium has apparently had a sparing effect on the dietary lysine requirement. Further work is needed to clarify the conditions of this potassium–lysine interaction in promoting weight gain in young pigs, but it seems clear that the balance of dietary electrolytes affects lysine utilization in pigs and that potassium supplementation of diets marginal in lysine may be economical under certain conditions. Supplemental potassium levels of 0.55–1.44% (as potassium bicarbonate) improved weight gain of pigs fed diets containing 0.43% lysine, but not when higher lysine was present (Austic *et al.* 1982). Optimum ratios of potassium to lysine in the diet are unknown and warrant further study.

VITAMIN SOURCES

Many feedstuffs commonly used as energy and protein sources for swine contain appreciable amounts of fat-soluble vitamins (A, D, E, and K) and water-soluble vitamins (thiamin, riboflavin, pantothenic acid, vitamin B_6, vitamin B_{12}, niacin, choline, biotin, and folacin), but many supply one or more vitamins in amounts inadequate to meet metabolic needs. Inclusion in the diet of a variety of feedstuffs of plant and animal origin made it possible to avoid serious vitamin deficiencies in earlier times before the identities and metabolic functions of the vitamins were known and, indeed, before the term vitamin was even coined. In general, animal products, including milk, liver, and muscle, and green forages are rich sources of most of the vitamins while seeds (such as maize and cereal grains) and roots and tubers (such as potatoes, and cassava) are poor to fair sources. In practice, the following vitamins are most likely to be in limited supply in natural diets: riboflavin, choline, pantothenic acid, niacin, vitamin B_{12}, vitamin A, vitamin E, and vitamin D. Rich sources of these are listed in Tables 16.15–16.19. Vitamin K and biotin are synthesized in adequate quantities by intestinal microflora to overcome dietary deficiencies under most conditions. Recently, field reports have indicated the possibility of clinical signs of vitamin K and biotin deficiency among growing pigs raised on slatted floors. Under such housing conditions, coprophagy would be minimized and the ingestion of vitamins synthesized in the lower intestinal tract could be decreased. Most commercial swine feeds now contain supplemental vitamin K and biotin in addition to those routinely added. Recent evidence indicates a beneficial effect of vitamin C addition to diets of weaned pigs under certain conditions (Yen and Pond 1981). Still unknown are (a) the mode of action under which vitamin C synthesis becomes insufficient to meet the needs of the animal and (b) the exact conditions under which a positive growth response to supplemental vitamin C can be expected. As husbandry conditions change, nutritionists need to keep alert to the need for corresponding changes in vitamin fortification of diets to meet new conditions.

The increase in confinement rearing of swine, in indoor environment on concrete floors or slats, has been accompanied by a decrease in the utilization of pastures and forages. While such crops cannot be considered important sources of energy, protein, and major minerals (calcium, phosphorus, sodium, chlorine) for swine, they most certainly have provided, in the past, a vital source of fat-soluble and water-soluble vitamins as well as of trace elements. The move to confinement, in addition to removing an important source of nutrients associated with forage consumption, has deprived the pig of rooting in the soil, which is a rich source of trace elements (discussed earlier in this chapter) and has often deprived the pig of sunlight, important in vitamin D_3 synthesis by conversion from 7-dehydrocholesterol present in the pig's skin and other tissues.

Through technology it is now possible to add all of the vitamins economically to the diets of swine in crystalline or highly purified form. Thus, a typical diet for a growing-finishing pig in the United States contains maize (as the energy source), soybean meal (as the protein source), trace mineralized salt, dicalcium phosphate (or another calcium–phosphorus supplement), and a commercially prepared vitamin supplement added at 0.1–0.5% of the diet in a soybean meal carrier. However, in many situations it may be more economical to provide vitamins from natural products as components of the energy or protein fraction of the diet. Fortunately, the B vitamins, riboflavin, pantothenic acid, niacin, and choline, tend to occur together in abundant quantities in specific plant and animal products, so that selecting a source high in riboflavin, for example, generally results in a substantial supply of the others as well. Table 16.15 illustrates this tendency. The only exception among the feedstuffs listed is dried milk whey, which is a good source of riboflavin, pantothenic acid, and niacin but a rather poor source of choline. These four vitamins are found in abundant amounts in forages, distillers' and fermentation by-products, fish and animal muscle, and some of the oilseed meals. Niacin, although present in abundance in many of the cereal grains, is almost completely biologically unavailable from these grains as well as from maize, so that these seeds must be considered as zero as sources of naicin.

Vitamin B_{12}, on the other hand, is strictly a product of microbial origin and is found in large quantities in fish, milk, and meat products but not in plants. This phenomenon is of historical interest in that vitamin B_{12}, until its discovery and isolation in 1948, was called APF (animal protein factor). It occurs in animal products in smaller quantities (a few micrograms per kilogram) than other vitamins (Table 16.16) but is also required in smaller quantities by swine.

Among the fat soluble vitamins, A and E are abundant in green forages (Tables 16.17 and 16.18). Vitamin A occurs mainly as its precursor, carotene, and vitamin E is present as a mixture of tocopherols, the most biologically active of which is alpha-tocopherol. Yellow maize is a good source of carotene, but the cereal grains and most roots and tubers (except carrots, sweet potatoes, and yams) are low or devoid of carotene and vitamin A. Because vitamin E is contained in high concentrations in the germ fraction of seeds, whole seeds are fair sources while germ-rich portions are excellent sources. Fish meal and liver are also good

TABLE 16.15. Good Sources of Riboflavin, Pantothenic Acid, Niacin, and Choline for Swine (mg/kg)

Feedstuffs	*Riboflavin*	*Pantothenic acid*	*Niacin*	*Choline*
Alfalfa meal, dehydrated	13.4	27.1	19.1	1,016
Clover hay, red, all analyses	15.6	9.9	37.2	—
Cottonseed meal	5.3	10.3	33.9	2,776
Dried fermentation solubles	171.4	202.8	215.6	3,113
Dried milk whey	29.9	47.7	11.2	Low
Dried whole milk	19.6	22.7	8.4	—
Dried citrus pulp	2.4	13.0	21.6	845
Dried fish solubles	7.7	44.9	231.1	5,223
Fish meal, herring	9.0	11.4	88.9	4,004
Fish meal, menhaden	4.8	8.8	55.9	3,080
Linseed meal	3.5	17.8	35.6	1,863
Maize distillers' dried grains	3.1	5.9	42.2	1,859
Meat meal	5.3	4.8	56.9	1,955
Milo distillers' dried solubles	14.7	26.4	140.8	2,222
Peanut meal	5.3	48.2	169.0	1,683
Soybean meal	3.3	14.5	26.8	2,743
Wheat germ meal, defatted	5.1	12.3	51.3	3,375
Yeast, brewers' dried	35.0	109.8	447.5	3,885
Yeast, torula dried	44.4	82.9	500.3	2,911

Sources: National Research Council 1964, 1968; Morrison 1956.

sources. Until recent years the tocopherol content of yellow maize was considered adequate, but frequent reports of vitamin E and selenium deficiency in swine have necessitated reevaluation and to avoid a deficiency it is recommended that alpha-tocopherol be added to maize–soybean meal diets for swine.

The metabolic requirement for vitamin D can be met in two ways: by ultraviolet irradiation of the skin sterols of the pig to produce cholecalciferol (vitamin D_3) or by ingestion of plants that have undergone ultraviolet irradiation of plant sterols to produce ergocalciferol (vitamin D_2). In the first method, no dietary requirement

TABLE 16.16. Good Sources of Vitamin B_{12} for Swine

Source	*Vitamin B_{12} ($\mu g/kg$)*
Blood meal[a]	440
Condensed fish solubles[b]	660
Crab meal[b]	418
Dried skim milk[c]	20–55
Dried whey[c]	22
Fish meal[a]	172–218
Meat-and-bone scrap[c]	110–125
Poultry by-product[b]	264

Sources: [a]National Research Council 1979; [b]Sievert 1967; [c]Becker et al. 1963.

TABLE 16.17. Good Sources of Carotene[a] and/or Vitamin A[b] for Swine

Source	*Vitamin A (IU/kg)*	*Carotene (mg/kg)*
Alfalfa hay, barn cured	63,200	37.9
Alfalfa leaf meal	104,200	62.4
Clover hay	99,000	59.4
Guinea grass, fresh	105,500	63.3
Lespedeza hay	71,300	42.8
Maize, yellow dent	6,800	4.1
Soybean hay	53,000	31.8

Source: National Research Council 1964.

[a]0.6 μg β-carotene = 1 IU vitamin A activity; 1.0 mg β-carotene = 1667 IU vitamin A activity.

[b]1 IU vitamin A = 1 U.S.P. unit vitamin A = 0.300 μg crystalline vitamin A alcohol = 0.344 μg vitamin A acetate = 0.550 μg vitamin A palmitate.

exists, but when sunlight is withheld, a dietary source is required. Table 16.19 lists some rich sources of vitamin D. Sun-cured forages, irradiated yeast, and cod-liver oil meal are the best sources of vitamin D. As the trend toward life-cycle confinement in the production of swine continues, more emphasis must be placed on meeting the vitamin D needs for normal skeletal development.

TABLE 16.18. Good Sources of Vitamin E for Swine

Feedstuffs	*α-tocopherol (mg/kg)*
Alfalfa leaf meal[a]	383
Alfalfa meal, dried[b]	125
Orchard grass, dried[a]	222
Bluegrass, Kentucky, dried[a]	345
Fish meal[a]	209
Maize, yellow[a]	24
Maize distillers' solubles, dried[b]	55
Soybean seed[a]	36.5
Soybean hay[a]	27
Oats[a]	36.5
Rice bran[b]	60
Wheat[a]	34
Wheat germ oil[a]	1897

Sources: [a]Morrison 1956. [b]National Research Council 1979.

TABLE 16.19. Good Sources of Vitamin D for Swine

Source	*Vitamin D activity (IU/kg)*
Alfalfa leaves from sun-cured hay	1,043
Alfalfa hay, sun-cured	1,800
Alfalfa hay, barn-cured	500
Clover hay, field-cured	1,910
Corn distillers' dried grains with solubles	550
Cod liver oil meal	39,930
Cocoa shells, sun-dried	3,434
Timothy hay, sun-cured	1,577
Activated animal sterol, 2000 IU/g	1,995,400
Activated 7-dehydro cholesterol, 2000 IU/g	1,995,400
D-activated plant sterol, 9000 IU/g	8,976,000
Irradiated yeast, 9000 IU/g	8,976,000

Sources: National Research Council 1964; Morrison 1956.

NONNUTRITIVE FEED ADDITIVES

Chemicals are important factors in the production of animal protein. Their use assures more abundant supplies at lower prices to consumers. In the past 30 years, science has introduced many new nonnutritive feed additives that have found uses in animal agriculture. These include antibiotics, hormones, and other growth-promoting materials. The use of nonnutritive additives in the feed of pigs has become common in feed preparations because they increase growth rate, increase efficiency of feed utilization, and reduce mortality and morbidity from clinical and subclinical infections.

Antibiotics were among the first nonnutritive additives to be routinely used in swine diets. Although more than 3800 microbial metabolites that are considered to have antibiotic properties have been described in the literature and some 35,000 antibiotic derivatives and analogues have been made, the large majority are unsuitable for use in animal production, usually because of low activity, toxicity to the host animal, or resulting residues in animal tissues. Only about thirteen have been approved for livestock and poultry production. Some of those approved include bacitracin, bambermycins, chlortetracycline, erythromycin, lincomycin, neomycin, oxytetracycline, oleandomycin, penicillin, streptomycin, tylosin, flavomycin, and virginiamycin. These antibiotics have been proven effective in improving pig performance and/or effective in control of specific and nonspecific diseases.

Low, subtherapeutic levels of antibiotics are fed primarily to improve growth rate and feed efficiency as well as to control infections. The magnitude of the response to antibiotics varies with stage of life cycle or stage of production and the environmental conditions to which the animals are exposed. The response is greater in young animals than in more mature animals. It is also greater during critical stages of production such as weaning, breeding, and farrowing. Environ-

TABLE 16.20. Effect of Antibiotics in the Feed on Weight Gain and Efficiency of Feed Utilization of Growing Pigs—A Summary

Daily gain of controls (kg)	*No. of comparisons*	*Improvement over controls*[a]	
		Weight gain (%)	*Feed/gain ratio (%)*
.09–.18	2	22.0	8.2
.18–.27	3	27.0	4.5
.27–.36	4	20.4	5.6
.36–.45	7	16.1	11.1
.45–.54	9	12.3	6.4
.54–.63	9	9.4	1.9
.63–.72	20	5.6	4.7
.72 and up	7	3.8	1.8
	Total 61	Average 10.7	Average 5.1

Source: Hays 1969.
[a]Antibiotics combination of penicillin and streptomycin.

mental stresses such as inadequate nutrition, crowding, moving and mixing of animals, and poor sanitation also contribute to increased response to antibiotics (Hays 1978).

Examples of the improvement in weight gains and feed conversion to be expected from the use of antibiotics in pig feeding are shown in Tables 16.20 and 16.21. The response of the poorly performing pigs to antibiotics is much greater

TABLE 16.21. Effect of Age on Pig Response to Dietary Antibiotics

Pigs tested/Level of oleandomycin	*Average weight*		*Daily gain*		*Feed/gain*	
	Initial (kg)	*Final (kg)*	*Av (g)*	*Improve. (%)*	*Av*	*Improve. (%)*
Baby pigs[a]						
No oleandomycin	4.29	8.91	163	—	2.02	—
2.5 g/ton	4.18	10.31	213	30.7	1.73	14.4
5 g/ton	4.10	10.07	213	30.7	1.62	19.8
10 g/ton	4.34	10.70	227	39.3	1.69	16.3
20 g/ton	4.35	10.98	236	44.8	1.62	19.8
Growing pigs[b]						
No oleandomycin	13.26	34.41	527	—	2.82	—
5 g/ton	13.35	36.18	572	8.5	2.60	7.8
10 g/ton	13.21	35.96	563	6.8	2.52	10.6
20 g/ton	13.44	35.87	563	6.8	2.63	6.7

Source: Adapted from Hays 1978, who summarized data from Hawbaker et al. 1960.
[a]Initial age, 13 days.
[b]Initial age, 51 days.

than the response of those performing at a higher rate (Table 16.20) (Hays 1969). Similar responses have been reported by Sainbury (1975), who showed that antibiotic supplements to normal pigs resulted in a 10% improvement in rate of gain, whereas antibiotic supplement to a group of poor performers resulted in a 30% improvement in gains. Young pigs during the early weaning, starter phase show the greatest response to antibiotics. Iowa studies (Hawbaker *et al.* 1960) have demonstrated that antibiotic response of baby pigs weaned at about 4.2 kg liveweight is much higher (30.7–44.8% improvement) than response of similar pigs fed the same antibiotics but started at 13.3-kg liveweight (6.8–8.5%) (Table 16.21). Published results reported by Speer (1970) and Hart (1976) clearly indicate the positive effect of antibiotics on conception rate and number of pigs born alive. Feeding high levels of antibiotics (chlorotetracycline or tylosin) to female pigs at breeding time increased conception rate by 8.7% and average number of pigs born alive by 0.5 pig per litter (Table 16.22).

The mode of action of antibiotics is stimulating growth, and improved reproductive performance of pigs has been postulated to be due to a metabolic effect, a nutrient-sparing effect, a disease-control effect, or combinations of these. Though there are numerous reports to illustrate that antibiotics do have metabolic implications and also that they spare certain nutrients, the bulk of the evidence supports the conclusion that the major benefit derived from the inclusion of antibiotics in feed at subtherapeutic levels result from their suppression or control of subclinical or nonspecific diseases (Hays 1978).

The overwhelming evidence indicates a major economic advantage to the use of antibiotics and other drugs in swine feeds. The reduction in cost of production is passed on to consumers in the form of lower prices for pork and other animal products. Even so, the use of antibiotics in animal feeds is controversial and the subtherapeutic use has been questioned. There has been much concern about the possibility that the antibiotics used in animal feeds may promote the development

TABLE 16.22. The Effect of Antibiotics at Breeding Time on Farrowing Rate and Litter Size

No. of sows	*Farrowing rate (%)*		*Live pigs/litter*	
	Control	*Treated*	*Control*	*Treated*
377	68.5	82.9[a]	9.8	10.1[a]
59	—	—	7.1	9.7[b]
96	87.5	95.8[c]	9.0	10.3[c]
182	60.9	70.0[c]	9.8	10.0[c]
249	66.9	75.4[a]	9.9	10.2[a]
192	93.8	91.6[d]	10.9	11.3[d]
Weighted av	73.0	81.7	9.8	10.3

Source: From Hays 1978.
[a]Chlortetracycline, 0.5–1.0 g/sow/day.
[b]Chlortetracycline, 0.5–1.0 g/sow/day, 0.54 g/sow/day.
[c]Chlortetracycline, sulfamethazine, and penicillin at 0.5, 0.5, and 0.25 g/sow/day, respectively.
[d]Tylosin phosphate, 0.6 g/sow/day.

of human disease organisms that are resistant to treatment by the antibiotics in question and perhaps others as well and that the residues of antibiotics in human foods derived from animals may be toxic to persons who consume the foods, or they may produce or exacerbate allergies. Although the possibility of contamination of meat from residues of the additives or their derivatives is negligible, the hazard from development of antibiotic-resistant human-disease organisms remains a concern. Although there are four instances recorded which link use of antibiotics in animal agriculture with diseases due to antibiotic-resistant bacteria in humans, none of the four implicates subtherapeutic use despite the production of many billions of food animals through the years with feed containing antibiotics at subtherapeutic levels. The use of antibiotics in animal feed for a period of about three decades has not created a public health problem (CAST 1981). This conclusion was reached by a panel of scientists appointed to prepare a document for the U.S. Congress and the Food and Drug Administration (FDA) from a comprehensive review and analysis of the benefits and potential hazards of the use of antibiotics in animal feeds.

The FDA conducts a comprehensive program in a conscientious effort to guard against problems with approved drugs. Furthermore, before new products are approved for use they must be proven beyond reasonable doubt to be both *safe* and *effective* in promoting growth or in meeting whatever claims have been made for them. Each year a feed additive compendium is published that provides information on identities, levels, combinations, and conditions under which all approved drugs can be legally used in animal feeds. The list of additives that have been approved for swine in the United States as of 1983 is shown in Table 16.23. This list describes the purpose of the drug as well as the levels and conditions of its safe use. Each year many applications for approval of use of new drugs are received by FDA. Therefore, the list of approved drugs changes each year, but only after thorough documentation of the *safety* and *efficacy* of the drug under conditions approved for use. Drugs that have been previously approved for use alone, for example, must be newly approved for use in combination with another drug or under other conditions not covered in the initial approval. Further information on approved drugs and combinations can be obtained from the latest Feed Additive Compendium, monthly supplement published by the Miller Publishing Company, Minneapolis, Minnesota (Anon. 1983).

Antibiotics and other nonnutritive feed supplements have been used routinely and successfully in swine production for more than 30 years. Although resistant strains of microorganisms have emerged from the general use of antibiotics, the drugs continue to be effective (Hays 1978). Thus, the continued use of antibacterials, such as antibiotics, in pig feeds is expected. Other nonnutritive growth stimulants such as the sulfa drugs, arsenicals, and nitrofurans have been approved for use in swine feeds, and others are continually receiving attention and will undoubtedly prove to be effective in increasing the efficiency of production.

The health hazard in the United States is minimal under current methods of surveillance by the Food and Drug Administration and the judicious use of feed additives is essential to efficient production in today's intensive swine production systems.

TABLE 16.23. Feed Additives Approved by FDA for Use with Swine[a]

Drug	Use level	Indications for use
I. (a) Arsanilic acid or sodium arsanilate	0.005–0.01% (45–90 g/ton)	(a) Increase rate of gain and improve feed efficiency in growing swine (prevention of swine dysentery claim permissible with 0.01% sodium arsanilate).
(b) Arsanilic acid or sodium arsanilate	0.025–0.04% (feed for 5–6 days)	(b) For control of swine dysentery.
II. Arsanilic acid or sodium arsanilate plus	0.005–0.01% (45–90 g/ton)	Same as I(a).
(a) Oxytetracycline	25–50 g/ton	Increase rate of gain and improve feed efficiency.
(b) Oxytetracycline	7.5–10 g/ton	Increase rate of gain and improve feed efficiency.
(c) Oxytetracycline	50 g/ton	Aid in the prevention of bacterial enteritis, also known as scours, vibrio, dysentery, necro, necrotic enteritis, salmonella infections, bloody dysentery, and baby pig diarrhea.
(d) Oxytetracycline	100 g/ton	Aid in the treatment of bacterial diarrhea (scours).
(e) Oxytetracycline	50–150 g/ton	Aid in the maintenance of weight gains and feed consumption in the presence of atrophic rhinitis.
(f) Furazolidone	0.0165%	Prevention of bacterial scours of baby pigs when fed in sow's ration for 3 weeks, starting 1 week before farrowing and continuing 2 weeks thereafter.
(g) Furazolidone	0.011%	Prevention of bacterial enteritis (necrotic enteritis, necro, or black scours) and vibrionic (bloody) dysentery and stimulate growth rate and feed conversion when fed in pre-starters, starters, and growing rations to baby pigs and growing swine for 5 weeks.
(h) Furazolidone	0.0165%	Prevention of bacterial enteritis (necrotic enteritis, necro, or black scours) and vibrionic (bloody) dysentery and stimulate growth rate and feed conversion when fed for 3 weeks.

(*continued*)

TABLE 16.23. *(Continued)*

Drug	Use level	Indications for use
(i) Furazolidone	0.022%	Prevention of bacterial enteritis (necrotic enteritis, necro, or black scours) and vibrionic (bloody) dysentery and stimulate growth and feed conversion when fed for 2 weeks.
(j) Hygromycin B	12 g/ton	Aid in the control of infestations of large roundworms (*Ascaris suis*), nodular worms (*Oesophagostomum dentatum*), and whipworm (*Trichuris suis*).
(k) Hygromycin B and Oxytetracycline	12 g/ton Maximum 50 g/ton	Aid in the control of infestations of large roundworms (*Ascaris suis*), nodular worms (*Oesophagostomum dentatum*), and whipworm (*Trichuris suis*); stimulate growth and improve feed efficiency.
(l) Furazolidone and Oxytetracycline	0.011–0.022% 50–100 g/ton	For specific level and claims permitted for furazolidone, see (f)–(i); for specific level and claims permitted for oxytetracycline, see II(a)–(e).
III. Arsanilic acid or sodium arsanilate plus	0.005–0.01% (45–90 g/ton)	Same as I(a).
(a) Streptomycin and Penicillin (from procaine penicillin)	7.5–15 g/ton 1.5–3 g/ton	Growth promotion and feed efficiency in growing swine.
(b) Penicillin (from procaine penicillin) or Streptomycin or Chlortetracycline or Bacitracin (from bacitracin zinc, bacitracin methylene disalicylate or feed grade bacitracin) or permitted combination	Maximum 50 g/ton	Increase rate of gain and improve feed efficiency.
(c) Chlortetracycline or Oxytetracycline or Combination	50 g/ton	Prevention of bacterial swine enteritis.
(d) Chlortetracycline or Oxytetracycline	100 g/ton	Prevention of bacterial swine enteritis.

(e)	Streptomycin and	37.5 g/ton	Aid in the prevention of bacterial swine enteritis.
	Penicillin (from procaine penicillin)	7.5 g/ton	
(f)	Streptomycin and	Minimum 75 g	Treatment of bacterial swine enteritis.
	Penicillin (from procaine penicillin)	15 g/ton	
(g)	Bacitracin (from bacitracin zinc, bacitracin methylene disalicylate or feed grade bacitracin) or	Minimum 50 g/ton	Prevention of bacterial swine enteritis.
	Penicillin (from procaine penicillin) or		
	Combination	(Minimum 25 g bacitracin in combination)	
(h)	Bacitracin (from bacitracin zinc, bacitracin methylene disalicylate or feed grade bacitracin) or	Minimum 100 g/ton	Treatment of bacterial swine enteritis.
	Penicillin (from procaine penicillin) or		
	Combination	(Minimum 50 g bacitracin in combinations)	
(i)	Hygromycin B and	12 g/ton	Aid in the control of infestations of large roundworms (*Ascaris suis*), nodular worms (*Oesophagostomum dentatum*), and whipworm (*Trichuris suis*).
	Penicillin (from procaine penicillin) or Streptomycin or Chlortetracycline or Bacitracin (from bacitracin zinc, bacitracin methylene disalicylate or feed grade bacitracin) or Combination	Maximum 50 g/ton	Increase rate of gain and improve feed efficiency.
(j)	Hygromycin B and	12 g/ton	Aid in the control of infestations of large roundworms (*Ascaris suis*), nodular worms (*Oesophagostomum dentatum*), and whipworm (*Tricharis suis*).
	Bacitracin (from bacitracin zinc, bacitracin methylene disalicylate or feed grade bacitracin) or Penicillin (from procaine penicillin) or	Minimum 50 g/ton	Prevention of bacterial swine enteritis.

(continued)

TABLE 16.23. *(Continued)*

Drug	Use level	Indications for use
Combination of bacitracin (from feed grade bacitracin) and penicillin (from procaine penicillin)	(Minimum 25 g bacitracin in combination)	
(k) Hygromycin B and	12 g/ton	Aid in the control of infestations of large roundworms (*Ascaris suis*), nodular worms (*Oesophagostomum dentatum*), and whipworm (*Tricharis suis*).
Bacitracin (from bacitracin zinc, bacitracin methylene disalicylate or feed grade bacitracin) or	Minimum 100 g/ton	Treatment of bacterial swine enteritis.
Penicillin (from procaine penicillin) or		
Combination of bacitracin (from feed grade bacitracin) and penicillin (from procaine penicillin)	(Minimum 50 g bacitracin in combination)	
(l) Hygromycin B and	12 g/ton	Aid in the control of infestations of large roundworms (*Ascaris suis*), nodular worms (*Oesophagostomum dentatum*), and whipworm (*Tricharis suis*).
Chlortetracycline or Oxytetracycline or Combination	Minimum 50 g/ton	Prevention of bacterial swine enteritis.
IV. Bacitracin methylene disalicylate	10–30 g/ton	For increased rate of weight gain and improved feed efficiency (for growing and finishing swine).
	50–100 g/ton	Aid in the prevention of bacterial swine enteritis (scours).
	100 g/ton	Treatment of bacterial swine enteritis (scours).
Bacitracin methylene disalicylate	250 g/ton*	For control of swine dysentery associated with *Treponema hyodysenteriae* on premises with a history of swine dysentery but where signs of the disease have not yet occurred, or following an approved treatment of the disease condition.

		Caution: Not for use in swine weighing more than 250 lb. Diagnosis should be confirmed by a veterinarian when results are not satisfactory. Feed containing an approved level of bacitracin methylene disalicylate should be the sole ration. [Use for growing-finishing swine.]
V. Bacitracin zinc	10–50 g/ton	Increase rate of weight gain and improve feed efficiency.
	50–100 g/ton	Aid in the prevention of bacterial swine enteritis (scours).
	100 g/ton	For the treatment of bacterial swine enteritis (scours).
Bacitracin zinc plus (a) Penicillin (from procaine penicillin)	Combined at a total level of 100 g/ton†	For the treatment of bacterial swine enteritis (scours).
(b) One, but only one, of the following:*		
(1) Arsanilic acid	0.005–0.010%	
(2) Sodium arsanilate	0.005–0.010%	
(3) Roxarsone	0.0025–0.0075%	

**Warning Statement:* Discontinue use 5 days before treated animals are slaughtered for human consumption if any one of the arsenicals is added.

†Combination containing not less than 50% nor more than 75% of zinc bacitracin, but in no case containing more than 125 g of penicillin, as procaine penicillin.

VI. Bambermycins	2 g/ton	For increased rate of weight gain and improved feed efficiency. [Use for growing-finishing swine.]
Bambermycins	2–4 g/ton	For increased rate of weight gain.
VII. Carbadox	(a) 10–25 g/ton (0.0011–0.00275%)	For increase in rate of weight gain and improvement of feed efficiency.
	(b) 50 g/ton (0.0055%)	For control of swine dysentery (vibrionic dysentery, bloody scours, or hemorrhagic dysentery); control of bacterial swine enteritis (salmonellosis or necrotic

(*continued*)

TABLE 16.23. (*Continued*)

Drug	Use level	Indications for use
		enteritis caused by *Salmonella choleraesuis*); increase rate of weight gain and improve feed efficiency.
Carbadox plus	50 g/ton (0.0055%)	For control of swine dysentery (vibrionic dysentery, bloody scours, or hemorrhagic dysentery); control of bacterial swine enteritis (salmonellosis or necrotic enteritis caused by *Salmonella choleraesuis*).
Pyrantel tartrate	96 g/ton (0.0106%)	Aid in the prevention of migration and establishment of large roundworm (*Ascaris suum*) infections and aid in the prevention of establishment of nodular worms (*Oesophagostomum dentatum*)
VIII. Chlortetracycline[‡]	(a) 10–50 g/ton or	Promote growth and improve feed efficiency.
	(b) 50–100 g/ton or	Prevention of bacterial swine enteritis. Maintenance of weight gain in the presence of atrophic rhinitis; reduction of the incidence of cervical abscesses. Prevention of bacterial swine enteritis during time or stress.
	(c) 100–200 g/ton or	Treatment of bacterial swine enteritis.
	(d) 200 g/ton or	Aid in reducing spread of leptospirosis (feed as sole medication).
	(e) 400 g/ton (feed as sole ration for 14 days)	As an aid in reducing shedding of leptospirae; as an aid in reducing the abortion rate of swine and the mortality rate of new born pigs when leptospirosis is present.
IX. Chlortetracycline plus	10–50 g/ton	Same as VIII(a).
	50–100 g/ton	Same as VIII(b).
	100–200 g/ton	Same as VIII(c).
		[Use for growing-finishing swine.]

(a) Roxarsone	0.0025–0.00375%	For increased rate of weight gain and improved feed efficiency. *Warning:* Withdraw 5 days before slaughter; as sole source of organic arsenic; feed continuously as a complete feed throughout growing season.

‡As chlortetracycline hydrochloride equivalent.

Chlortetracycline§ plus (b) Sulfamethazine and Penicillin (from procaine penicillin or from penicillin G procaine)‖	100 g/ton 0.011% (100 g/ton) 50 g/ton	Reduction of the incidence of cervical abscesses; treatment of bacterial swine enteritis (salmonellosis or necrotic enteritis caused by *Salmonella choleraesuis* and vibrionic dysentery); prevention of these diseases during times of stress; maintenance of weight gains in the presence of atrophic rhinitis; growth promotion and increased feed efficiency in swine weighing up to 75 lb. *Warning:* Withdraw 15 days before slaughter. For swine over 75 lb.: All of the above claims with the exception of growth promotion and feed efficiency. *Warning:* Withdraw 15 days before slaughter. [Use for swine including sows and gilts.]

§As chlortetracycline hydrochloride equivalent.
‖Penicillin statement varies, check manufacturer's label.

X. Chlortetracycline plus Sulfathiazole and Penicillin (from procaine penicillin)	100 g/ton 0.011% (100 g/ton) 50 g/ton	Administer to swine in a complete feed for reduction of the incidence of cervical abscesses; treatment of bacterial enteritis (salmonellosis or necrotic enteritis caused by *Salmonella choleraesuis* and vibrionic dysentery); maintenance of weight gains in the presence of atrophic rhinitis; increased rate of weight gain and improved feed efficiency up to 6 weeks postweaning.
Chlortetracycline plus Sulfathiazole and	100 g/ton 0.011% (100 g/ton)	Administer to swine in a complete feed for reduction of the incidence of cervical abscesses; treatment of

(continued)

TABLE 16.23. (*Continued*)

	Drug	*Use level*	*Indications for use*
	Penicillin	50 g/ton	bacterial enteritis (salmonellosis or necrotic enteritis caused by *Salmonella choleraesuis* and vibrionic dysentery); maintenance of weight gains in the presence of atrophic rhinitis; increased rate of weight gain from 6 to 16 weeks postweaning.
XI.	Dichlorvos	348 g/ton (0.0384%). For swine up to 70-lb body weight, feed as sole ration for 2 consecutive days. For swine from 70 lb to market weight, feed as sole ration at the rate of 8.4 lb of feed per head until the medicated feed has been consumed. For boars, open or bred gilts, and sows, feed as sole ration at the rate of 4.2 lb per head per day for 2 consecutive days	For the removal and control of mature, immature, and/or fourth stage larvae of the whipworm (*Trichuris suis*), nodular worm (*Oesophagostomum* sp.), large roundworm (*Ascaris suum*), and the thick stomach worm (*Ascarops strongylina*) occurring in the gastrointestinal tracts of pigs, boars, open or bred gilts, and sows.
	Dichlorvos	479 g/ton (0.0528%). For boars, open or bred gilts and sows, feed as sole ration at the rate of 6 lb per head for one feeding	Same as above.
	Dichlorvos	334–500 g/ton (0.0366–0.0550%). For pregnant swine, mix into a gestation feed to provide 1000 mg per head daily during the last 30 days of gestation	An aid in improving litter production efficiency by increasing pigs born alive, birth weights, survival to market and rate of weight gain. Treatment also removes and controls mature, immature and/or fourth-stage larvae of the whipworm (*Trichuris suis*), nodular worm (*Oesophagostomum* sp.), large roundworm (*Ascaris suum*), and the thick stomach worm (*Ascarops strongylina*) occurring in the gastrointestinal tract of the sow or gilt.

XII.	Erythromycin[¶]	9.25–64.75 g/ton	For increased rate of weight gain and improved feed efficiency. [Use for starter pigs up to 13-lb body weight.]
	Erythromycin[¶]	9.25 g/ton	For increased rate of weight gain and improved feed efficiency. [Use for growing-finishing pigs.]

[¶]Source of Erythromycin is Erythromycin Medicated Premix for Swine. See Product Information in Anon. (1983).

XIII.	Ethylenediamine dihydroiodide	250–500 mg daily for 7 days	By loosening mucus and congestion, aid in control of respiratory difficulties. Aids in loosening mucus in upper respiratory tract. *Warning statement:* Treat animals with caution until tolerance is determined, because of variation in susceptibility of iodides.
XIV.	Furazolidone	0.0165% (150 g/ton)	For prevention of bacterial scours in baby pigs; growth promotion while on medication. Feed to sows 1 week before farrowing and 2 weeks after farrowing. Withdraw 5 days before slaughter.
			Prevention of bacterial enteritis (necrotic enteritis, necro) and vibrionic (bloody) dysentery; growth promotion while on medication. Feed in pre-starters, starters, and growing rations to baby pigs and growing swine.
		0.011% (100 g/ton)	. . . for 5 weeks.
		0.0165% (150 g/ton)	. . . for 3 weeks.
		0.022% (200 g/ton)	. . . for 2 weeks.
		0.033% (300 g/ton)	For treatment of bacterial enteritis (necrotic enteritis, necro) and vibrionic (bloody) dysentery in swine. Feed 10–14 days.
	Furazolidone	0.022% (200 g/ton)	Prevention of bacterial enteritis (necrotic enteritis, necro) and vibrionic (bloody) dysentery; growth pro-

(continued)

TABLE 16.23 (*Continued*)

Drug	Use level	Indications for use
		motion while on medication when fed in pre-starters, starters and growing rations to baby pigs and growing swine for 2 weeks.
XV. Furazolidone plus	0.011% (100 g/ton)	Prevention of bacterial enteritis (necrotic enteritis, necro), and vibrionic (bloody) dysentery; growth promotion while on medication when fed in pre-starters, starters and growing rations to baby pigs and growing swine for 5 weeks.
(a) Oxytetracycline	50–150 g/ton	As an aid in the maintenance of weight gains and feed consumption in presence of atrophic rhinitis.
Furazolidone plus	0.0165% (150 g/ton)	Prevention of bacterial enteritis (necrotic enteritis, necro), and vibrionic (bloody) dysentery; growth promotion while on medication when fed in pre-starters, starters and growing rations to baby pigs and growing swine for 3 weeks.
(b) Oxytetracycline plus	50–150 g/ton	As an aid in the maintenance of weight gains and feed consumption in presence of atrophic rhinitis.
(c) Oxytetracycline plus	50–150 g/ton	As an aid in the maintenance of weight gains and feed consumption in presence of atrophic rhinitis.
(d) Oxytetracycline plus	50–100 g/ton	As an aid in the maintenance of weight gains and feed consumption in presence of atrophic rhinitis.
Arsanilic acid**	0.005-0.01%	Growth promotion and feed efficiency.
Furazolidone plus	0.0165% (150 g/ton)	Prevention of bacterial enteritis (necrotic enteritis, necro), and vibrionic (bloody) dysentery; growth promotion while on medication when fed in pre-starters, starters and growing rations to baby pigs and growing swine for 3 weeks.
(e) Oxytetracycline plus	50–100 g/ton	As an aid in the maintenance of weight gains and feed consumption in presence of atrophic rhinitis.

	Arsanilic acid**	0.005–0.01%	Growth promotion and feed efficiency.
	Furazolidone plus	0.022% (200 g/ton)	Prevention of bacterial enteritis (necrotic enteritis, necro), and vibrionic (bloody) dysentery; growth promotion while on medication when fed in pre-starters, starters and growing rations to baby pigs and growing swine for 2 weeks.
	(f) Oxytetracycline plus	50–100 g/ton	As an aid in the maintenance of weight gains and feed consumption in presence of atrophic rhinitis.
	Arsanilic acid**	0.005–0.01%	Growth promotion and feed efficiency.
**Warning required:* Withdraw 5 days before slaughter; as sole source of organic arsenic.			
XVI.	Hygromycin B	12 g/ton	As an aid in the control of infestations of large roundworms (*Ascaris suis*), nodular worms (*Oesophagostenum dentatum*) and whipworms (*Trichuris suis*).
	Hygromycin B plus	12 g/ton	Same as above.
	Tylosin (as tylosin phosphate)	10–20 g/ton	Increased rate of weight gain and improved feed efficiency finisher feeds.
		20–40 g/ton	Increased rate of weight gain and improved feed efficiency grower feeds.
		20–100 g/ton	Increased rate of weight gain and improved feed efficiency starter and pre-starter feeds.
XVII.	Levamisole hydrochloride (equivalent)	0.08% (0.36 g/lb)	For treating swine infected with the following nematode infections: Large roundworms (*Ascaris suum*); Nodular worms (*Oesophagostomum* spp.); Lungworms (*Metastrongylus* spp.); intestinal threadworms (*Strongyloides ransomi*); swine kidney worms (*Stephanurus dentatus*).
XVIII.	Lincomycin	40 g/ton	For control of swine dysentery. Feed as the sole ration. For use in animals or on premises with a history of swine dysentery, but where symptoms have not yet occurred.

(continued)

TABLE 16.23. *(Continued)*

	Drug	*Use level*	*Indications for use*
	Lincomycin	100 g/ton followed by 40 g/ton	For treatment and control of swine dysentery. Feed 100 g/ton of complete feed as the sole ration for 3 weeks or until signs of disease (watery, mucoid, or bloody stools) disappear, followed by 40 g of lincomycin per ton.
	Lincomycin	100 g/ton	For treatment of swine dysentery, feed as sole ration for 3 weeks or until signs of disease (water, mucoid, or bloody stools) disappear.
XIX.	Neomycin base	70–140 g/ton of complete feed	For treatment of bacterial enteritis (scours, diarrhea, bloody dysentery, vibrionic dysentery, white scours, colibacillosis, salmonellosis; diarrheas caused by *E. coli,* vibrio, and salmonella organisms).
	Neomycin base	200–400 mg/gal of reconstituted milk replacer	For treatment of bacterial enteritis (scours, diarrhea, bloody dysentery, vibrionic dysentery, white scours, colibacillosis, salmonellosis; diarrheas caused by *E. coli,* vibrio, and salmonella organisms). [Use for pigs.]
XX.	Nitrofurazone	0.055% (500 g/ton)	For treatment of infectious bacterial swine enteritis caused by *Salmonella choleraesuis* when fed 5–7 days. Repeat if necessary.
XXI.	Oxytetracycline	25–50 g/ton	To increase rate of gain and improve feed efficiency. [Use for swine 10–30 lb.]
	Oxytetracycline	7.5–10 g/ton	To increase rate of gain and improve feed efficiency. [Use for swine 30–200 lb.]
	Oxytetracycline	50 g/ton	As an aid in the prevention of bacterial enteritis, also known as scours, baby pig diarrhea, vibrio dysentery, bloody dysentery and salmonellosis (necro or necrotic enteritis).

	100 g/ton	As an aid in the prevention of bacterial enteritis, also known as scours, baby pig diarrhea, vibrio dysentery, bloody dysentery and salmonellosis (necro or necrotic enteritis).
	50–150 g/ton	As an aid in the maintenance of weight gains and feed consumption in presence of atrophic rhinitis.
	500 g/ton	In presence of porcine leptospirosis reduces instances of abortion, gives higher survival rate, heavier and healthier newborn pigs, reduces urinary shedding of leptospirae and aids in maintenance of normal weight gains and feed consumption. Feed 7–14 days, approximately one month before farrowing.
XXII. Oxytetracycline plus	150 g/ton	Same as XXI.
(a) Nitrofurazone plus one of the following:	0.0056%	Treatment of necrotic enteritis caused by *Samonella choleraesuis*, when fed for 5–7 days or longer and repeated if necessary.
Arsanilic acid[††] or	0.005–0.01%	Growth promotion.
Sodium arsanilate[††] or	0.005–0.01%	
Roxarsone[††]	0.0025–0.0075%	

[††]*Warning:* If used, label must warn that feeding be discontinued 5 days before slaughter for food.

Oxytetracycline plus	50 g/ton	As an aid in the prevention of bacterial enteritis (scours), baby pig diarrhea (in baby pigs only), vibrionic dysentery, bloody dysentery and salmonellosis (necro or necrotic enteritis).
(f) Neomycin base[‡‡§§]	35–140 g/ton	[Use for baby pigs, growing-finshing pigs, and sows.]
Oxytetracycline plus	100 g/ton	As an aid in the treatment of bacterial enteritis (scours), baby pig diarrhea (in baby pigs only), vibrionic dysentery, bloody dysentery and salmonellosis (necro or necrotic enteritis).
(g) Neomycin base[‡‡§§]	70–140 g/ton	[Use for baby pigs, growing-finishing pigs, and sows.]

(*continued*)

TABLE 16.23. (*Continued*)

	Drug	*Use level*	*Indications for use*
	Oxytetracycline plus	50–150 g/ton	As an aid in the maintenance of weight gains and feed consumption in the presence of atrophic rhinitis. [Use for sows: to be fed during gestation and lactation periods.]
	(h) Neomycin base[‡‡§§]	70–140 g/ton	As an aid in the treatment of bacterial enteritis.

‡‡Neomycin use levels are expressed as grams of neomycin base per ton (70% neomycin sulfate level). For example, 140 g neomycin base is equivalent to 200 g neomycin sulfate.

§§*Warning:* This combination requires withdrawal from feed 10 days before slaughter when the neomycin base level is 140 g/ton and 5 days withdrawal when the neomycin base level is below 140 g/ton.

XXIII.	Penicillin (from procaine penicillin)	10–50 g/ton	Aid in stimulating growth and improving feed efficiency.
	Penicillin (from procaine penicillin) and Streptomycin	Not less than 1.5 g/ton Not less than 7.5 g/ton (maximum 50 g combination/ton)	Increase rate of gain and improve feed efficiency.
	Penicillin (from procaine penicillin) and Streptomycin	Not less than 7.5 g/ton Not less than 37.5 g/ton (maximum 90 g combination/ton)	Aid in prevention of bacterial swine enteritis.
	Penicillin (from procaine penicillin) and Streptomycin	Not less than 15 g/ton Not less than 75 g/ton or 90–270 g combination/ton)	Treatment of bacterial swine enteritis. Feed not more than 14 days.
XXIV.	Phenothiazine	5 g up to 25 lb; 8 g, 26–50 lb; 10 g, 51–100 lb; 20 g, 101–200 lb; 30 g, 210 lb and up	Removes nodular worms (*Oesophagostomum*).
XXV.	Piperazine (base)	(a) Individual 1 day treatment—50 mg per pound of body-weight	Control of infestation of large roundworms (*Ascaris*) and nodular worms (*Oesophagostomum*).
		(b) Herd treatment (as sole source of feed or water)—0.1–0.2%	Control of infestation of large roundworms (*Ascaris*) and nodular worms (*Oesophagostomum*).

		in water, 0.2–0.4% in feed	(Directions for use should assure that the amount of feed or water consumed will furnish in 1 day the required amounts of piperazine base per pound of body weight of animals eating or drinking the medicated ration.)
XXVI.	Pyrantel tartrate	96 g/ton (0.0106%) Feed continuously as the sole ration in a complete feed	Aid in the prevention of migration and establishment of large roundworm (*Ascaris suum*) infections; aid in the prevention of establishment of nodular worm (*Oesophagostomum*) infections.
	Pyrantel tartrate	96 g/ton (0.0106%) Feed for 3 days as the sole ration in a complete ration	For the removal and control of large roundworm (*Ascaris suum*) infections.
	Pyrantel tartrate	800 g/ton (0.0881%) As a single therapeutic treatment in a complete feed; feed at the rate of 1 lb of feed per 40 lb of body weight for animals up to 200 lb, and 5 lb of feed per head for animals 200 lb or over	For the removal and control of large roundworm (*Ascaris suum*) and nodular worm (*Oesophagostomum*) infections.
	Pyrantel tartrate plus	96 g/ton (0.0106%)	Aid in the prevention of migration and establishment of large roundworms (*Ascaris suum*) infections; aid in the prevention of establishment of nodular worm (*Oesophagostomum*) infections.
	Carbadox	50 g/ton (0.0055%)	For control of swine dysentery (vibrionic dysentery, bloody scours or hemorrhagic dysentery); control of bacterial swine enteritis caused by *Salmonella cholerаesuis*.
	Pyrantel tartrate plus	96 g/ton (0.0106%)	Aid in the prevention of migration and establishment of large roundworms (*Ascaris suum*) infections; aid in the prevention of establishment of nodular worm (*Oesophagostomum*) infections.

(*continued*)

TABLE 16.23. *(Continued)*

	Drug	Use level	Indications for use
	Tylosin	100 g/ton for at least 3 weeks, followed by 4 g/ton until swine reach market weight	Prevention of swine dysentery (vibrionic).
XXVII.	Roxarsone	0.0025–0.00375%	For increased rate of weight gain and improved feed efficiency.
	Roxarsone	0.02% (feed for 5–6 days)	For treatment of swine dysentery.
XXVIII.	Thiabendazole	0.005–0.1% (45.4–908 g/ton) in feed (administer continuously, feed containing 0.05–0.1% for 2 weeks followed by feed containing 0.005–0.02% thiabendazole for 8–14 weeks)	Aid in the prevention of infections of large roundworms (genus *Ascaris*).
XXIX.	Tylosin	100 g/ton for at least 3 weeks, followed by 40 g/ton until swine reach market weight	Prevention of swine dysentery (vibrionic).
	Tylosin	100 g/ton	Maintaining weight gains and feed efficiency in the presence of atrophic rhinitis.
XXX.	Tylosin	10–20 g/ton	For increased rate of weight gain and improved feed efficiency, finisher feeds.
	Tylosin	20–40 g/ton	For increased rate of weight gain and improved feed efficiency, grower feeds.
	Tylosin	20–100 g/ton	For increased rate of weight gain and improved feed efficiency, starter, and pre-starter feeds.
	Tylosin	40–100 g/ton for 2–6 weeks after treatment with tylosin in drinking water (0.25 g tylosin per gallon of drinking water for 3–10 days)	Treatment and control of swine dysentery (vibrionic).
	Tylosin plus Sulfamethazine	100 g/ton 0.011% (100 g/ton)	Maintaining weight gains and feed efficiency in the presence of atrophic rhinitis; lowering the incidence

			and severity of *Bordetella bronchiseptica* rhinitis; prevention of swine dysentery (vibrionic); control of swine pneumonias caused by bacterial pathogens (*Pasteurella multocida* and/or *Corynebacterium pyogenes*).
	Tylosin plus	100 g/ton for at least 3 weeks, followed by 40 g/ton until swine reach market weight	Prevention of swine dysentery (vibrionic).
	Pyrantel tartrate	96 g/ton (0.0106%)	Aid in the prevention of migration and establishment of large roundworm (*Ascaris suum*) infections; aid in the prevention of establishment of nodular worm (*Oesophagostomum*) infections.
		As the sole ration in a complete feed	
	Tylosin plus	Same as XXX	Same as XXX.
	Hygromycin B	12 g/ton	Control of infestation of large roundworms (*Ascaris suis*), nodular worms (*Oesophagostomum*), and whipworms (*Trichuris suis*).
XXXI.	Virginiamycin	5 g/ton (0.00055%)	For improved feed efficiency in swine 120 lb to market weight.
		5–10 g/ton (0.00055–0.0011%)	For increased rate of weight gain in swine 120 lb to market weight.
		10 g/ton (0.0011%)	For increase in rate of weight gain and improvement of feed efficiency from weaning to 120 lb (starter and grower feeds only).
		25 g/ton (0.00275%)	As an aid in the control of swine dysentery in swine up to 120 lb. For use in animals or on premises with a history of swine dysentery, but where symptoms have not yet occurred.
		100 g/ton (0.011%) for 2 weeks followed by 50 g/ton (0.0055%)	For treatment and control of swine dysentery (bloody scours or hemorrhagic dysentery) up to 120 lb.
		100 g/ton (0.011%)	For treatment of swine dysentery for two weeks in non-breeding swine over 120 lb.

Source: From Anon. 1983.
[a]Specific swine are indicated in square brackets [ ].

REFERENCES

ANON. 1983. Feed Additive Compendium, Vol. 21. Miller Publishing Co., Minneapolis, Minnesota.

AIKAWA, J. K. 1963, 1971. The Role of Magnesium in Biological Processes. Charles C. Thomas, Springfield, Illinois.

ALLCROFT, R., BURNS, K. N., and LEWIS, G. 1961. Vet. Rec. *73,* 714.

ARTHUR, D. 1971. Can. J. Anim. Sci. *51,* 71.

AUSTIC, R. E., MADUBUIKE, F. N., BOYD, R. D., KLASING, K. C., and RILEY, W. W., JR. 1982. Proc. Cornell Nutr. Conf. Cornell Univ., pp. 47–64. Ithaca, New York.

AMMERMAN, C. B., and MILLER, S. M. 1972. J. Anim. Sci. *35,* 681.

BARBER, R. S., BRAUDE, R., and MITCHELL, K. G. 1955A. Br. J. Nutr. *9,* 378.

BARBER, R. S., BRAUDE, R., and MITCHELL, K. G. 1955B. Chem. Ind. *21,* 1954.

BARBER, R. S., BRAUDE, R., MITCHELL, K. G., and CASSIDY, J. 1955C. Chem. Ind. *21,* 601.

BARBER, R. S., BRAUDE, R., and MITCHELL, K. G. 1965. Brit. J. Nutr. *19,* 575.

BARBER, R. S., BRAUDE, R., MITCHELL, K. G., and PITMAN, R. J. 1978. Anim. Prod. *26,* 151.

BARTLEY, J. C., REBER, E. F., YUSKEN, J. W., and NORTON, H. W. 1961. J. Anim. Sci. *20,* 137.

BAYLEY, H. S., POS, J., and THOMSON, R. G. 1975. J. Anim. Sci. *40,* 857.

BEAMES, R. M. 1969. J. Anim. Sci. *29,* 573.

BECKER, W. M., and HOEKSTRA, W. G. 1966. J. Nutr. *90,* 301.

BECKER, D. E., JENSEN, A. H., and HARMON, B. G. 1963. Ill. Agric. Exp. Stn. Circ. *866.*

BELLIS, D. B. 1961. Anim. Prod. *3,* 89.

BENGTSSON, G., HAKKARAINEN, J., JÖNSSON, L., LANNEK, N., and LINDBERG, P. 1978. J. Anim. Sci. *46,* 143, 153.

BOWLER, R. J., BRAUDE, R., CAMPBELL, R. C., CRADDOCK-TURNBULL, J. N., FIELDSEND, H. F., GRIFFITH, E. K., LUCAS, I. A. M., MITCHELL, K. G., NICHOLLS, N. J. D., and TAYLOR, J. H. 1955. Brit. J. Nutr. *9,* 358.

BRAUDE, R. 1967. World Rev. Anim. Prod. *3*(11), 69.

BUITRAGO, J. A., CLAWSON, A. J., and SMITH, F. H. 1970. J. Anim. Sci. *31,* 554.

BYRD, C. A., and MATRONE, G. 1965. Proc. Soc. Exp. Biol. Med. *119,* 347.

BRINK, M. F., BECKER, D. E., TERRILL, S. W., and JENSEN, A. H. 1959. J. Anim. Sci. *18,* 836.

CAST. 1981. Antibiotics in Animal Feeds. Council for Agric. Sci. and Tech. Rep. No. *88,* March, 1981.

CASTRELL, A. G., ALLEN, R. D., BEAMES, R. M., BELL, J. M., BELFILE, R., BOWLAND, J. P., ELLIOT, J. I., IHNOT, M., LARMOND, E., MELLARD, T. M., SPURR, D. T., STOTHERS, S. C., WILTON, S. B., and YOUNG, L. G. 1975. Can. J. Anim. Sci. *55,* 113.

COMBS, G. E., AMMERMAN, C. B., SHIRLEY, R. L., and WALLACE, H. D. 1966. J. Anim. Sci. *25,* 618.

COOK, D. A. 1973. J. Nutr. *103,* 1365.

CORNELIUS, J. G., and HARMON, B. G. 1976. J. Anim. Sci. *42,* 1351 (Abstr.).

COX, D. H., and HALE, O. M. 1962. J. Nutr. *77,* 225.

CROMWELL, G. L. 1980. Feedstuffs *53*(10), 38, March 3, 1980.

CROMWELL, G. L., HAYS, V. W., OVERFIELD, J. R., and KRUG, J. L. 1976. J. Anim. Sci. *42,* 1350.

DAHMER, E. J., GRUMMER, R. H., and HOEKSTRA, W. G. 1972A. J. Anim. Sci. *35,* 1176.

DAHMER, E. J., COLEMAN, B. W., GRUMMER, R. H., and HOEKSTRA, W. G. 1972B. J. Anim. Sci. *35,* 1181.

DE GORY, L. W., WAHLSTROM, R. C., and EMERICK, R. J. 1971. J. Anim. Sci. *33,* 52.

DONE, J. T., HARDING, J. D. J., and LLOYD, M. K. 1959. Vet. Rec. *71,* 92.

EDWARDS, H. M., JR. 1959. J. Nutr. *69,* 306.

FRITSCHEN, R., and PEO, E. R. 1979. Unpublished data. Dep. Anim. Sci., Univ. Nebraska, Lincoln.

FRITZ, J. C., PLA, G. W., ROBERTS, T., BOEHNE, J. W., and HOVE, E. L. 1970. J. Agric. Food Chem. *18,* 647.

FROSETH, J. A., MILLER, E. R., KU, P. K., and BERGEN, W. G. 1982. J. Anim. Sci. *55,* 271 (Abstr.).

FULLER, R. L., NEWLAND, G. M., BRIGGS, C. A. E., BRAUDE, R., and MITCHELL, K. G. 1960. J. Appl. Bacteriol. *23,* 195.

FURUGOURI, K. 1972. J. Anim. Sci. *34,* 573.

FURUGOURI, K. and KAWABATA, A. 1979. J. Anim. Sci. *49,* 715.

GIPP, W. F., POND, W. G., TASHER, J., VAN CAMPEN, D., KROOK, L. and VISEK, W. J. 1973A. J. Nutr. *103,* 713.

GIPP, W. F., POND, W. G., and WALKER, E. F. 1973B. J. Anim. Sci. *36,* 91.

GIPP, W. F., POND, W. G., KALLFELZ, F. A., TASKER, J. B., VAN CAMPEN, D. R., KROOK, L. and VISEK, W. J. 1974. J. Nutr. *104,* 532.

GROCE, A. W., MILLER, E. R., KEAHEY, K. K., ULLREY, D. E., and ELLIS, D. J. 1971. J. Anim. Sci. *32,* 905.

GRUMMER, R. H., BENTLEY, O. G., PHILLIPS, P. H., and BOHSTEDT, G. 1950. J. Anim. Sci. *9,* 170.

GUENTER, W., and SELL, J. L. 1974. J. Nutr. *104,* 1446.

HAGSTEN, I., and PERRY, T. W. 1976. J. Anim. Sci. *42,* 1187, 1191.

HAHN, C. J., and EVANS, G. W. 1975. Am. J. Physiol. *228,* 1020.

HAKKARAINEN, J., LINBERG, P., BENGTSSON, G., JÖNSSON, L., and LANNEK, N. 1978. J. Anim. Sci. *46,* 1001.

HARMON, B. G., LUI, C. T., JENSEN, A. H., and BAKER, D. H. 1976. J. Anim. Sci. *42,* 860.

HART, M. L. 1976. Quoted by Cromwell 1980.

HART, E. B., STEENBOCK, H., WADDELL, J., and ELVEHJEM, C. A. 1928. J. Biol. Chem. *77,* 797.

HAWBAKER, J. A., DIAZ, F., SPEER, V. C., HAYS, V. W., and CATRON, D. V. 1960. J. Anim. Sci. *19,* 800.

HAYS, V. W. 1969. Proc. Symp. Drug Use. Natl. Acad. Sci. Publ. *1679.* Washington, D.C.

HAYS, V. W. 1976. Phosphorus in Swine Nutrition. Natl. Feed. Ingredients Assoc., Des Moines, Iowa.

HAYS, V. W. 1978. *In* Nutrition and Drug Interrelations. (John N. Hathcock and Julius Coon, Editors), pp. 545–576, Academic Press, New York.

HEDGES, J. D., KORNEGAY, E. T., and THOMAS, H. R. 1976. J. Anim. Sci. *43,* 453.

HEIGINER, H., MICHNA, A., NEWMANN, H., BEHRENS, H., and Hamann, K. 1958. Forsch *11,* 185.

HITCHCOCK, J. P., MILLER, E. R., KEAHEY, K. K., and ULLREY, D. E. 1978. J. Anim. Sci. *46,* 425.

HOEFER, J. A., LUECKE, R. W., MILLER, E. R., ULLREY, D. E., PEARSON, A. M., and RITCHIE, H. D. 1959. J. Anim. Sci. *18,* 1503.

HSU, F. S., KROOK, L., POND, W. G., and DUNCAN, J. R. 1975. J. Nutr. *105,* 112.

HUCK, D. W., and CLAWSON, A. J. 1976. J. Anim. Sci. *43,* 1231.

KRIDER, J. L., ALBRIGHT, J. L., PLUMLEE, M. P., CONRAD, J. H., UNDERWOOD, L., ARNOLD, W. I., and JONES, R. G. 1972. J. Anim. Sci. *35,* 1092 (Abstr.).

KRIDER, J. L., ALBRIGHT, J. L., PLUMLEE, M. P., CONRAD, J. H., SINCLAIR, C. L., UNDERWOOD, L., JONES, R. G., and HARRINGTON, R. B. 1975. J. Anim. Sci. *40,* 1027.

LEWIS, P. K., JR., HOEKSTRA, W. G., and GRUMMER, R. H. 1957. J. Anim. Sci. *16,* 578.

LIEBHOLZ, J. M., McCALL, J. T., HAYS, V. W., AND SPEER, V. C. 1966. J. Anim. Sci. *25,* 37.

LILLIE, R. J., FROBISH, L. T., STEELE, N. C., and GARBER, G. 1977. J. Anim. Sci. *45,* 100.

LIPTRAP, D. O., MILLER, E. R., ULLREY, D. E., WHITENACK, D. L., SCHOEPKE, B. L., and LUECKE, R. W. 1970. J. Anim. Sci. *30,* 736.

LUECKE, R. W., HOEFER, J. A., BRANWELL, W. S., and THORP, F., JR. 1956. J. Anim. Sci. *15,* 347.

MAHAN, D. C., 1978. Proc. Dist. Feed Res. Council. *33,* 30.

MAHAN, D. C., and MOXON, A. L. 1978. J. Anim. Sci. *46,* 384.

MAHAN, D. C., PENHALE, L. H., CLINE, J. H., MOXON, A. L., FETTER, A. W., and YARRINGTON, J. T. 1974. J. Anim. Sci. *39,* 536.

MAHAN, D. C., MOXON, A. L., and HUBBARD, M. 1977. J. Anim. Sci. *46,* 738.

MANNER, M. J., and McCREA, M. R. 1963. J. Nutr. *17,* 495.

MANNER, M. J., and McCREA, M. R. 1964. Ann. Zootech. *13,* 29.

MAYO, R. H., PLUMLEE, M. P., and BEESON, W. M. 1959. J. Anim. Sci. *18,* 264.

MEINTZER, R. B., and STEENBOCK, H. 1955. J. Nutr. *56,* 285.

MILLER, E. R. 1971. Res. Rep. 148. Mich. State Univ. Exp. Stn., pp. 59–62, East Lansing.

MILLER, W. J., and MILLER, J. K. 1963. J. Dairy Sci. *XLVI,* 581.

MILLER, E. R., ULLREY, D. E., ZUTANT, C. L., BALTZER, B. U., SCHMIDT, D. A., HOEFER, J. A., and LUECKE, R. W. 1965. J. Nutr. *85,* 13.

MILLER, E. R., LUECKE, R. W., ULLREY, D. E., BALTZER, B. V., BRADLEY, B. L., and HOEFER, J. A. 1968. J. Nutr. *95,* 278.

MILLER, E. R., ULLREY, D. E., ELLIS, D. J., SHOEPKE, B. L., and HOEFER, J. A. 1969. J. Anim. Sci. *29,* 140.

MILLER, E. R., LIPTRAP, D. O., and ULLREY, D. E. 1970. *In* Trace Element Metabolism in Animals, p. 377. E & S Livingstone, Edinburgh.
MILLER, E. R., STOWE, H. D., KU, P. K., and HILL, G. M. 1979. Copper and Zinc in Swine Nutrition. NFIA, Nat. Feed Ingredients Assoc.
MILLER, E. R., KU, P. K., HITCHCOCK, J. P., and MAGEE, W. T. 1981. J. Anim. Sci. *52,* 312.
MORRISON, F. B. 1956. Feeds and Feeding, 22nd Edition. Morrison, Ithaca, New York.
MYERS, A. W., and BOWLAND, J. P. 1973A. Can. J. Anim. Sci. *53,* 115.
MYERS, A. W., and BOWLAND, J. P. 1973B. Can. J. Anim. Sci. *53,* 121.
MYERS, A. W., and BOWLAND, J. P. 1973C. Can. J. Anim. Sci. *53,* 127.
NATIONAL RESEARCH COUNCIL. 1964. Joint U.S.–Canadian Tables of Feed Composition. Natl. Acad. of Sci. Publ. *1232.* Washington, D.C.
NATIONAL RESEARCH COUNCIL. 1968. Nutrient Requirments of Swine. Natl. Acad. of Sci. Publ. *1599.* Washington, D.C.
NATIONAL RESEARCH COUNCIL. 1979. Nutrient Requirements of Swine, 7th Edition. Natl. Acad. of Sci., Washington, D.C.
NATIONAL RESEARCH COUNCIL. 1980. Mineral Tolerance of Domestic Animals. Natl. Acad. of Sci., Washington, D.C.
NCR-42. 1974. Committee on Swine Nutrition. J. Anim. Sci. *39,* 512.
NELSON, T. S. 1980. Proc. 3rd Annu. Int. Min. Conf., Mundelein, Illinois.
NEWLAND, H. W., ULLREY, D. E., HOEFER, J. A., and LUECKE, R. W. 1958. J. Anim. Sci. *17,* 886.
NEWTON, G. L., and CLAWSON, A. J. 1974. J. Anim. Sci. *39,* 879.
NIELSON, F. H., SUNDE, M. L., and HOEKSTRA, W. G. 1966. J. Nutr. *89,* 35.
O'DELL, B. L., and SAVAGE, J. E. 1960. Proc. Soc. Exp. Biol. Med. *103,* 304.
O'DONOVAN, P. B., PICKETT, R. A., PLUMLEE, M. P., and BEESON, W. M. 1963. J. Anim. Sci. *22,* 1075.
OMOLO, T. A., and BOWLAND, J. P. 1974. Can. J. Anim. Sci. *54,* 481.
PARIZCK, J. 1957. J. Endocr. *15,* 56.
PEO, E. R., JR. 1976. Calcium in Swine Nutrition. Natl. Feed Ingredients Assoc., Des Moines, Iowa.
PIKE, R. L., and BROWN, M. L. 1975. Nutrition: An Integrated Approach. 2nd Edition. Wiley, New York.
PIPER, R. C., FROSETH, J. A., McDOWELL, L. R., KROENING, G. H., and DYER, I. A. 1975. Am. J. Vet. Res. *36,* 237.
POND, W. G., VEUM, T. L., and LAZAR, V. A. 1965. J. Anim. Sci. *24,* 668.
POND, W. G., CHAPMAN, P., and WALKER, E. 1966. J. Anim. Sci. *25,* 122.
POND, W. G., and JONES, J. R. 1964. J. Anim. Sci. *23,* 1057.
POPPE, S. 1968. Arch. Geflugelzucht Kleintierk *17,* 43.
RANHOTRA, G. S., LOEWE, R. J., and PUYAT, L. V. 1976. Cereal Chem. *53,* 770.
REINBALL, J. G., FARADJI, B., ABADI, P., and ISMAIL-BEIGI, F. 1976. J. Nutr. *106,* 493.
RIORDAN, J. F., and VALLEE, B. L. 1974. Adv. Exp. Med. Biol. *48,* 33.
RITCHIE, H. D., LUECKE, R. W., BALTZER, B. V., MILLER, E. R., ULLREY, D. E., and HOEFER, J. A. 1963. J. Nutr. *79,* 117.
ROTRUCK, J. T., POPE, A. L., GANTHER, H. E., SWANSON, A. B., HAFEMAN, D. G., and HOEKSTRA, W. G. 1972. Science *179,* 588.
SAINBURY, D. 1975. Pig Farming *23, 4,* 27.
SCOTT, M. L., and ZEIGLER, T. R. 1963. J. Agric. Food Chem. *11,* 123.
SCOTT, M. L., and CANTOR, A. H. 1972. Proc. Cornell Nutr. Conf., p. 66.
SEELIG, J., and HEGGTUEIT, J. 1974. Am. J. Clin. Nutr. *27,* 59–79.
SELL, J. L. 1979. The 2nd Int. Minerals Conf. Proc., p. 33–55. January 17, 1979. St. Petersburg Beach, Florida.
SELL, J. L., and FONTENOT, J. P. 1980. *In* Magnesium in Animal Nutrition, Natl. Feed Ingredients Assoc., 1980.
SHARP, B. A., YOUNG, L. G., and VAN DREUMEL, A. A. 1972A. Can. J. Comp. Med. *36,* 371.
SHARP, B. A., YOUNG, L. G., and VAN DREUMEL, A. A. 1972B. Can. J. Comp. Med. *36,* 393.
SIEVERT, C. W. 1967. Feed ingredients analyses. Feed Bag Red Book. Editorial Service Co., Milwaukee, Wisconsin.
SINGSEN, E. P., MATTERSON, L. D., and TLUSTOHOWIEZ, J. J. 1972. Proc. Distillers' Feed Res. Conf. *27,* 46.

SMITH, I. D., GRUMMER, R. H., HOEKSTRA, W. G., and PHILLIPS, P. H. 1960. J. Anim. Sci. *19*, 568.

SMITH, H. W., and JONES, J. E. T. 1963. J. Appl. Bacteriol. *26*, 262.

SPEER, V. C. 1970. Feedstuffs 42, *46*, 27.

STEVENSON, J. W., and EARLE, I. P. 1957. J. Anim. Sci. *15*, 1036.

SUTTLE, N. F., and MILLS, C. F. 1966. Br. J. Nutr. *20*, 135.

SVAJGR, A. J., PEO, E. R., and VIPPERMAN, P. E. 1969. J. Anim. Sci. *29*, 439.

THOREN-TOLLING, K. 1975. Acta Vet. Scand. Suppl. *54*, 1.

TODD, J. R., LAWSON, G. H. K., and DOW, C. 1964. J. Comp. Pathol. Ther. *74*, 331.

TROTTER, M., and ALLEE, G. L. 1979. J. Anim. Sci. *49*, (Supp. I), 50.

ULLREY, D. E. 1974. *In* Trace Element Metabolism in Animals-2 (W. G. Hoekstra, Editor). Univ. Park. Press, Baltimore, Maryland.

ULLREY, D. E. 1978. Selenium in Swine Nutrition. Cornell Nutr. Conf., 1978, Syracuse, New York.

UNDERWOOD, E. J. 1977. Trace Elements in Human and Animal Nutrition, 4th Edition, Academic Press, New York.

VAHRA, P., and KRATZEN, F. H. 1964. J. Nutr. *82*, 64.

VAN CAMPEN, D. R. 1969. J. Nutr. *97*, 104.

VALLEE, B. L., and WACKER, W. E. C. 1970. *In* The Proteins (H. Neurath, Editor) 2nd Edition, Vol. 5. Academic Press, New York.

WALLACE, H. D. 1967. High level of copper in swine feeding. Int. Copper Res. Assoc. Inc., New York.

WALLACE, H. D., McCALL, J. T., BASS, B. and COMBS, G. E. 1960. J. Anim. Sci. *19*, 1153.

WEGGER, I., and PALLUDAN, B. 1978. *In* Trace Element Metabolism in Man and Animals-3. (M. Kirchgessner, Editor), p. 128. Freising-Weihenstephen, W. Germany.

WESOLOSKI, G. D., JENSEN, A. H., LODWIG, V. D., and GOSSER, H. 1975. Am. J. Vet. Res. *36*, 773.

WILKINSON, J. E., BELL, M. C., BACON, J. A., and MASINCUPP, F. B. 1977A. J. Anim. Sci. *44*, 224.

WILKINSON, J. E., BELL, M. C., BACON, J. A., and MELTON, C. C. 1977B. J. Anim. Sci. *44*, 229.

YEN, J. T., and POND, W. G 1981. J. Anim. Sci. *53*, 1292.

YOUNG, L. G., MILLER, R. B., EDMEADES, D. E., LUN, A., SMITH, G. C., and KING, G. J. 1977. J. Anim. Sci. *45*, 1O51.

ZIMMERMAN, D. R. 1980. Iron in Swine Nutrition. *In* NFIA Literature Review on Iron in Animal and Poultry Nutrition (H. Russell Conrad, Dean R. Zimmerman, and G. F. Combs, Jr., Editors). Natl. Feed Ingredients Assoc., West Des Moines, Iowa.

17

Processing and Its Effect on Nutritive Value

INTRODUCTION

Many potentially useful sources of nutrients for swine would remain unused if appropriate processing were not possible. The feeding value of some particular feedstuffs is low because of physical properties (such as hardness of kernel) or because of some chemical property which reduced the biological availability of one or more nutrients (such as heat-labile inhibitors in raw soybeans). Several processing methods, including grinding, pelleting, heating, drying, and oil extraction, have been developed in the feed industry to improve the usefulness of a wide variety of feedstuffs for swine. This chapter is devoted to a discussion of these methods and the beneficial effects of their use.

TYPES OF FEED PROCESSING

Preparation of swine feeds has been drastically modified and in the course of the past decade, the degree of processing has increased from slight to extensive. There has been a definite shift away from the free-choice, whole-grain-plus-supplement feeding to the self-feeding of complete mixed and processed diets. Although many of the changes have been brought about by the development of complete commercial and on-the-farm processing units and the desire to minimize labor requirements through the mechanization of feeding systems, many changes

TABLE 17.1. Reasons for Processing and Their Related Importance

Purpose of treatment	*Total treatments (%)*
Alteration of form or particle size	40
Preservation	25
Isolation of specific part	22
Improvement of palatability	8
Improvement of digestibility	4
Alteration of nutrient make-up	<1
Detoxification	<1
Total	100

Source: Harris and Crampton 1972.

in processing have been utilized in an attempt to improve the efficiency of utilization of the feedstuffs available.

Processing may be defined as a series of actions, operations, motions, and the like that are employed in the accomplishment of a definite end from a definite beginning. An understanding of the reasons for processing of feedstuffs will provide a better basis for evaluation of each process. It is beyond the scope of this chapter to describe in detail each and every processing term but a glossary of processing terms is presented in Appendix 3.

Feed processing is a physical, chemical, thermal, bacterial, or other treatment or alteration of a feed ingredient before it is fed. Feeds are processed for the following reasons: (1) to alter the physical form or particle size, (2) to preserve, (3) to isolate specific parts, (4) to improve palatability, (5) to improve digestibility, (6) to alter nutrient composition, and (7) to detoxify. Table 17.1 (Harris and Crampton 1972) clearly indicates that alteration of particle size or form is the most important consideration given to processing, and that this purpose plus preservation and isolation of specific parts constitute approximately 87% of the motivation for processing.

Modification of a feed may take place during the many processes that occur between harvest and feeding. Some of these modifications may directly alter the feeding value of the products treated; therefore, it is important that the feeder be aware of the processes to which a given feed has been subjected and the effect that each has on the nutritive value of the feed. Details of the effects of processing on the nutrition value of feeds were presented by the National Academy of Sciences (1973).

GRINDING

Of the more than 60 processing methods employed, grinding, dehydration, and fat extraction are the three principle processes utilized to prepare animal feeds. If pelleting, flaking, shredding, chopping, cracking, and milling, all related to changing of particle size, are added to grinding, the problem of particle size is the predominant factor in feed processing.

To some extent the importance of grinding depends upon the age of the pig and the method of feeding. Grinding of feed ingredients and especially of grains to reduce particle size is usually considered to be profitable when the pig fails to masticate the seed thoroughly to consume bulky feed. Pigs weighing 20–70 kg tend to chew their feed more thoroughly than older pigs. Therefore, growing pigs usually make better use of whole grain than older pigs. Creep and starter rations should always be ground. Hand-fed pigs consume their feed more rapidly and tend to chew their feed less than free-choice fed pigs which tend to spend more time in mastication. Grinding can increase digestibility since seeds that escape chewing may pass through the digestive tract without appreciable digestion. Therefore, greater benefit from grinding might be expected when feeding older pigs and baby pigs, and when a hand-feeding system of whole grain and supplement is utilized. In general, small, hard grains such as rye, barley, millet, oats and sorghum should be ground. Maize also should be ground if it is hand-fed, and especially if it is a flint type.

The degree of grinding, or fineness of grind, necessary for maximum utilization of feed grains by pigs is a subject about which there are many opinions and prejudices. It is generally recommended that grains be ground through a ¼- to ½-inch screen in order to obtain maximum response from grinding. Some suggest that finely ground feeds are unpalatable to swine; however, Maner (1972) has shown that growing-finishing pigs fed a dry diet based on 48.6% finely ground cassava meal (ground to a very fine powder) consumed as much feed daily as pigs fed diets based on flint maize that had been ground to a medium-fine particle size.

Purdue workers (Pickett *et al.* 1969) reported that each reduction in the particle size of maize caused a tendency for improvement in both rate of growth and efficiency of feed utilization (Table 17.2). However, the growth response was

TABLE 17.2. Effect of Fineness of Grind of Grains on Pig Performance

Criteria of response	*Fineness of grind—screen size*		
	0.16 cm	*1.27 cm*	*2.54 cm*
Experiment I			
Modulus of fineness	2.37	3.66	3.90
Av daily gain, kg	.65	.63	.63
Feed/gain ratio	3.19	3.56	3.67
Experiment II			
Modulus of fineness	2.07	3.74	—
Av daily gain, kg	.70	.71	—
Feed/gain ratio	3.52	3.74	—
Experiment III			
Modulus of fineness	2.08	3.23	—
Av daily gain, kg	.72	.72	—
Feed/gain ratio	3.58	3.71	—

Source: Pickett et al. 1969.

TABLE 17.3. Effect of Fineness of Grind of Maize on Feed Requirement per Kilogram of Gain by Swine

Source of data	*Feed/gain ratio*		
	Fine	*Medium*	*Coarse*
Univ. of Illinois	3.03	3.12	3.29
Univ. of Illinois	—	3.38	3.59
North Carolina State Univ.	3.04	—	3.24
Purdue Univ.	3.06	3.34	3.51
Univ. of Wisconsin	3.38	3.54	4.01
Univ. of Wisconsin	3.90	4.49	4.08
Univ. of Wisconsin	3.63	3.90	4.09
Univ. of Wisconsin	5.25	—	5.35
Univ. of Wisconsin	3.84	—	—
Allied Mills	3.39	—	3.61
Allied Mills	3.17	—	3.40
Allied Mills	3.21	—	3.44
Golden Sun Mill	3.45	—	3.73
Kent Feed	3.12	3.42	3.74
York Feed	3.13	3.16	3.25

Source: V. C. Speer, as cited by Perry 1972.

apparent only in one of the three experiments. These results are adequately supported by a summary of results from four experiments and from four feed companies (Perry 1972). More feed was required to produce a unit of gain when coarse-grind maize was included in the diet than was required with either medium or fine grind. This was also true for medium and fine grinding; more feed was required when medium-grind maize was used than when fine-grind maize was present in the diet (Table 17.3).

Although experimental results clearly indicate that the finer the grind the better the feed efficiency, the association of fineness of grinding with the development of stomach ulcer tends to modify these results (Newer *et al.* 1965; Mahan *et al.* 1966; Pickett *et al.* 1969). Analysis by Pickett *et al.* (1969) of data on stomach ulcer index scores shows that each decrease in particle size resulted in a significant increase in severity of stomach ulcers. Mahan *et al.* (1966) also reported observations that agree that the incidence of esophagogastric lesions increases as more finely ground rations are fed.

If diets are completely mixed, it is essential that grains and other diet ingredients be ground. A medium-fine grind is recommended to take advantage of the effect of particle size on improving feed efficiency and to reduce the incidence of gastric lesions.

PELLETING

Pelleting is the processing technique that has attracted more interest and has been the subject of more experimentation than any other during the past 40 years. Pelleting of diets has generally resulted in improved pig performance.

Reviews of the effect of pelleting on growth, feed consumption and feed efficiency have been made by Conrad (1958), Jensen (1966), Becker (1966), Hays (1967), North Central Regional Project 42 (1969), and National Academy of Sciences (1973). Summaries of those reviews for growing-finishing pigs are presented in Tables 17.4 and 17.5. In each summary it is evident that there is a small but consistent improvement in rate of growth and a very marked improvement in efficiency of feed conversion. Contrary to early beliefs, pelleting has not resulted in increased feed intake; on the contrary, generally there is no significant change in feed intake or a very minor reduction in total feed consumed. A summary of these data clearly indicates that a 6–7% increase in gain, a 7–10% improvement in feed efficiency, and a 1–2% reduction in feed intake might be expected when meal diets are pelleted for growing-finishing pigs. Greater improvement may be expected if barley or oats or other bulky, fibrous materials, such as wheat bran, maize germ meal, alfalfa, or rice bran, are included in the diet.

For young pigs, 14–21 days of age, workers were unable to show any improvement in pig gains due to pelleting; gain–feed ratios were improved by 13.8% when the diet was pelleted and by 10.3% when the grain was pelleted and reground prior to mixing in the diet. Pelleting had very little positive effect, or even a negative effect, on daily feed intake. These results are contrary to results reported by Braude *et al.* (1960), who reported that pelleting resulted in increased consumption of creep feed, and are different from what might be expected from the results of Lewis *et al.* (1955) and Jensen *et al.* (1955) who reported that baby pigs showed a preference for pelleted feed over meal. If given a choice, baby pigs consume more feed in the form of pellets than meal; however, if one or the other form is fed, few differences in total consumption are observed.

Although beneficial effects from pelleting diets and diet components have been reported by many workers, the mechanism or mechanisms by which the pelleting process affects the nutritional quality has not been clearly defined. Reduced fiber

TABLE 17.4. Effect of Pelleting of Diets on the Performance of Growing-Finishing Pigs

Diet	*No. of pigs*	*Grain source*	*Av daily gain*	*Av daily feed*	*Feed/gain ratio*
Meal[a]	218	Barley	.60	2.65	4.41
Pellet[a]	218	Barley	.69	2.57	3.73
Meal[b]	—	Various	.73	2.58	3.52
Pellet[b]	—	Various	.78	2.61	3.36
Meal[c]	398	Maize	.75	2.47	3.29
Pellet[c]	398	Maize	.78	2.46	3.16
Meal[d]	166	Maize	.77	2.48	3.22
Pellet[d]	166	Maize	.78	2.43	3.12
Average advantage of pellets, %			+6.3	−1.1	+9.8

Sources: [a]Conrad 1958. [b]Jensen 1966. [c]Hays 1967. [d]North Central Regional Project 42 1969.

TABLE 17.5. Effect of Pelleting of Maize-Based Diets on the Performance of Growing and of Finishing Pigs

Diet	*Av daily gain (kg)*	*Av daily feed (kg)*	*Feed/gain ratio (kg)*
Growing			
Meal[a]	.71	2.00	2.81
Pellet[a]	.75	1.87	2.49
Meal[b]	.74	2.04	2.74
Pellet[b]	.74	1.85	2.50
Av advantage of pellets, %	+2.8	−7.9	+10.1
Finishing			
Meal[a]	.84	3.19	3.80
Pellet[a]	.91	3.09	3.40
Meal[b]	.85	3.01	3.56
Pellet[b]	.90	2.91	3.24
Av advantage of pellets, %	+7.1	−2.5	+9.8

Sources: [a]Becker 1966. [b]Perry 1972.

values (Lindahl and Reynolds 1959), partial gelatinization of the starches (Smith 1957), and improved digestibility and utilization of energy (Meade *et al.* 1971; Friend and MacIntyre 1969) have been reported. It has also been shown that after pelleting there is more total nitrogen and less crude fiber in the feed ingredients as shown by proximate analysis.

Slinger *et al.* (1966) and Summers *et al.* (1967) found that the utilization of plant phosphorus in maize–soybean diets by growing chicks could be increased by steam pelleting. Bayley and Thompson (1969; Bayley *et al.* 1975), have reported results which indicate that steam pelleting increases by 19–20% the utilization of the phosphorus in plant material by pigs. This improvement took place only in diets with no added phosphorus. Steam pelleting had no effect on increasing phosphorus absorbability at the highest level of phosphorus (0.54%) provided by adding calcium phosphate. Corley *et al.* (1980) showed no improvement in bioavailability of phosphorus in rice bran and wheat bran by steam pelleting.

Since calcium from all available sources appears to be well utilized by swine, feed processing including pelleting has little influence in its utilization. Feed processing is not known to influence magnesium utilization (Loosli 1972). According to Fritz (1972), pelleting has no effect in changing the total trace mineral contents of a feed or on their bioavailability.

Scott (1972) summarized the effect of pelleting on vitamin availability and nutritive value and indicated that the effect may be positive or negative. Pelleting may increase the percentage availability of certain vitamins such as vitamin E and B_{12} if the diet contains raw legume seeds. It is also indicated that nicotinic acid and biotin may be changed from the bound form to an available form by steam pelleting; however, definitive research is not available to support such a claim.

Pelleting has been shown (Bierer and Vickers 1958; Charles and Huston 1970) to destroy vitamins E, A, and K; this is especially true if sufficient antioxidants are not present to prevent accelerated oxidation of these vitamins.

Yen *et al.* (1971) reported that subjecting maize gluten feed to the pelleting process resulted in more efficient utilization of this product by growing swine and especially when it was used as a protein source. From nitrogen balance studies it was indicated that one of the major effects of pelleting was the apparent increased availability of tryptophan. These results confirm the report of Olsen and Slinger (1968) that steam pelleting increased absorption of amino acids from wheat bran. Pelleting reduces or completely eliminates salmonella from infected feeds. Crane *et al.* (1973) summarized the studies available and presented evidence from numerous feed examinations showing that salmonella are destroyed in the pelleting process. In one survey 292 feed samples were tested. Ten percent of the 110 meal samples were shown to be positive for salmonella, whereas all of the 182 pelleted feeds were negative for salmonella. Although pelleting is helpful in reducing or eliminating salmonella from livestock feeds, it has no effect on the elimination or detoxification of molds that produce aflatoxins (Crane *et al.* 1973).

HEATING

Heating of swine feeds can be like a two-edged sword. It can improve the utilization of some feed ingredients if properly applied and reduce the efficiency of utilization if excessive. Not only does heating affect the protein and carbohydrate fractions but it can be especially destructive to the vitamins present in the feed.

For its effect on protein quality, heat is the most important processing treatment. It can be beneficial or detrimental. Loss in feeding value can result from loss of palatability, destruction of amino acids by oxidation, and delayed or inhibited release of amino acids during digestion owing to modifications of linkages between amino acids and other diet components.

The main factors controlling the extent of the damage are temperature, time of heating, and the presence of moisture and reducing substances. In general, little damage takes place in normal commercial procedures. When damage occurs it is generally because not enough attention has been paid to the control of processing conditions.

It has been well known for many years that excessive dry heating of proteins has a detrimental effect upon their nutritional value, particularly when carbohydrate is present as in plant proteins. Most plant proteins contain reducing carbohydrates such as glucose, which may react with the free amino groups of a protein, particularly the epsilon amino groups of lysine. The linkages formed by such a carbohydrate–amino acid reaction (''Maillard'' or ''Browning'' reaction) are not hydrolyzed by digestive enzymes. Since the digestive enzymes cannot release the amino acids from this strong chemical linkage, the amino acids, especially lysine, can become unavailable to the pig. Heat accelerates the formation of the protein–carbohydrate linkage. Therefore the extent of heating has an effect on the digestibility and thus the availability of the amino acids.

The chemical analysis of protein, which is preceded by acid hydrolysis, does not yield information concerning the availability of the amino acid to the living animal because acid hydrolysis can break bonds that cannot be broken by enzymatic hydrolysis. At the same time, biological tests such as net protein utilization (NPU), biological value (BV), and protein efficiency ratio (PER) measure only the limiting amino acid and give no information about the other amino acids unless multiple assays are carried out with various combinations of supplementary amino acids. Only direct measurement of the available amino acids provides the necessary information. Since these methods have been applied in recent years it has become clear that in some feeds several or all of the amino acids can be damaged at the same time.

According to Sunde (1972), heating partially denatures proteins by interrupting the orderly protein structure and makes it easier for the digestive enzymes to do their work. The presence of moisture generally tends to increase the effectiveness of the heating process. However, there appears to be little evidence to support this claim, except where heat is utilized to destroy a toxin or anti-digestive factor.

Animal by-product proteins, such as fish meal and meat-and-bone meal, are processed by heating to sterilize the products, to release fat by causing a rupture of the tissue cells, to remove moisture, and if the by-product is fish meal, to destroy the anti-thiamin present. Except for its effect in accomplishing these goals, heating has no beneficial effect on the nutritive value of these proteins and, on the contrary, may be detrimental.

It appears that the variability encountered in many commercial meat meal products is due as much to the dilution of muscle protein with tendon and ossein as to processing damage, if not more (Atkinson and Carpenter 1970; Wilder 1972). Meat meal contains very little if any carbohydrate and is thus not subject to the deterioration brought about by carbohydrate–amino acid linkages. Also, the temperatures reached in the rendering process vary between 104° and 121° C, depending upon the type of equipment used, and during all except the final drying phase moisture is present. Therefore, there appears to be very little damage from the processing heat if the processing is carried out under normal conditions. This conclusion also appears to be supported by the work of Atkinson and Carpenter (1970), who showed that freeze-dried meals had as low a value as oven-dried meals and that the resulting slow growth was not corrected by supplementation of the diets with methionine, cystine, tryptophan, and lysine.

In contrast, heating fish meal appears to have a definite detrimental effect for pigs. Mason and Weidner (1964) heated fish meal at 120° C and approximately 2 atm for periods of time as long as 100 hr and found a steady decline in available lysine with prolonged heating as well as a marked decrease in pepsin-digestibility of the meal and a decline in B vitamins and amino acid content. Madsen *et al.* (1965) also observed damage by heating fish meal. One sample was unheated and the two remaining samples were heated at 120° C. After heating, these three samples FM1, FM2, and FM3 contained 6.5, 4.6, and 2.1 g of available lysine per 16 g of nitrogen (Table 17.6). When these variously heated fish meals were included in basal diet at the rate of 5% it was observed that the pigs fed the unheated meals consumed significantly more feed than their litter mates on the heated meals and grew at a faster rate. Heating of the fish meal decreased the

TABLE 17.6. Chemical Composition of a Fish Meal Sample after Receiving Varying Heat Treatment

Components	*No heat*	*120° C heat (12 hours)*	*120° C heat (72 hours)*
Dry matter, %	91.07	92.04	89.28
Total nitrogen, %	11.15	11.15	10.94
Amino acids, g/16 g nitrogen			
Histidine	2.10	1.80	1.30
Isoleucine	4.20	4.00	3.65
Leucine	7.15	6.95	6.60
Lysine (total)	7.15	6.65	5.55
Lysine (available)	6.50	4.55	2.15
Methionine + cystine	4.65	3.50	3.05
Phenylalanine + tyrosine	6.95	6.95	6.25
Threonine	4.20	3.65	2.70
Tryptophan	0.90	0.80	0.80
Valine	5.35	4.95	5.00

Source: Madsen et al. 1965.

apparent digestibility of the nitrogen from 76% to as low as 61% for the meal receiving the greatest heat treatment. Similar results have been reported by Jones and Cadenhead (1965) who presented data indicating that when fish meal heated for 36 hr at 150° C was fed to pigs, which were then compared with groups of pigs receiving the same quantity of an unheated sample of the same meal, a reduction in nitrogen retention, growth rate, and feed conversion occurred in the first group.

Studies reported by Tarr and Biely (1972) have shown that the decreased performance of some simple-stomached animals fed heated fish meal is due to the destruction of heat-labile vitamins of the B complex. Serious impairment of nutritive value, accompanied by a marked decrease in available essential amino acids, occurred only after heat treatment at and in excess of 149° C for 3 hr.

The nutritive value of most if not all legume seeds is improved by heating. This is partly due to the destruction of trypsin inhibitors and toxic substances and partly due to improved digestibility and availability of the sulfur amino acids. Soybeans in particular have been extensively studied. As early as 1936 it was found that if soybeans were cooked at a temperature of 140°–150° C for 2.5 min the nutritive value was higher than that of the raw meal. Johnson *et al.* (1939) demonstrated that heat destroyed the toxic factor in soybeans that rendered nitrogen and sulfur unavailable. For soybeans and other legumes it has been well established that to obtain a product of the highest nutritive value the application of heat must be a balance between beneficial and destructive effects. The reduced quality associated with excessive heating is due partly to impaired protein digestibility and partly to reduced lysine bioavailability (Harrell and Carpenter 1977). Autoclaving or dry heating at 120° C for 30 min of peanuts, which when fed raw, cause pancreatic hypertrophy because of the presence of a trypsin inhibitor, reduced the inhibitor and improved the biological value of the protein. However, when the heating was continued for 4 hr the biological value was reduced.

TABLE 17.7. Effect of Cooking Time on the Protein Quality of Pigeon Peas (*Cajanus indicus*)

Time of cooking at 16 lb per sq in.	*Rat wt gain (g)*	*PER*[a]
0	11	0.46
10	41	1.37
20	40	1.52
30	44	1.39
40	35	1.34
60	38	0.94

Source: Adapted from Brahman et al. 1965.
[a]Protein efficiency ratio (g wt gain/g protein consumed).

Brahman *et al.* (1965) heated pigeon peas (*Cajanus indicus*) in water under pressure at different cooking times (Table 17.7) and measured the effect of the processing on the growth of rats. The results showed that the grain was relatively nontoxic, but also that the nutritive value, as measured by weight gain and protein efficiency ratio (PER), was significantly improved by cooking for 30 min, but prolonged cooking for 40 or 60 min decreased weight gains and PER. The decrease in performance and PER after excessive cooking is due to a decrease in availability of lysine (Bressani *et al.* 1963).

A processing procedure can be at the same time detrimental and helpful. An example of this is supplied by the results obtained from the processing of cottonseed for oil extraction and the preparation of cottonseed meal (Table 17.8). Processing procedures with heat reduce the concentration of free gossypol in the cottonseed, reducing the total toxicity of the product, but at the same time lysine availability is also reduced (Craig and Broderick 1981), thus lowering the final nutritive value of the product. Solvent processing does not reduce the available lysine that remains, but at the same time the free gossypol is also not reduced.

Although the nutritive value of potatoes (Kellner 1910) and many legume seeds (Liener 1958) can be improved by moist heat, most studies indicate that moist

TABLE 17.8. Changes in Free Gossypol and Available Lysine of Cottonseed During Processing

Processing step	*Processing temperature (°C)*	*H_2O (%)*	*Free gossypol (%)*	*Available lysine (g/16 g N)*
Raw seed	—	8.7	1.12	3.29
Cooking	83.3	10.3	1.02	3.41
Drying	83–127	10.3–5.9	0.40	2.94
Conditioning	115.5	4.1	0.36	3.04
Expeller	<135.5	4.8	0.05	2.60

Source: Bressani and Elias 1968.

heat does not improve the nutritive value of properly supplemented grain for swine (Henry 1886, Oliver 1931; Smith *et al.* 1960). Hintz and Garrett (1967) steam processed barley at 1.4 kg/cm^2 for 1.5 min and demonstrated that this had no significant effect on either pig growth rate or feed efficiency. Processing at 0.56 kg/cm^2 decreased feed intake and rate of gain. Results with pigs do not agree with those obtained with ruminants, whose capability to digest cereal grains is improved by steam processing. To be of value, steam processing should increase digestibility and absorption. Moist heat apparently improves utilization of potatoes by rupturing the outer layer of the potato starch granule in order that it can be attacked by α-amylase (Jelinck *et al.* 1952). However, it is apparently not necessary to rupture the starch granules of some grains. For example, Yoshida and Morimoto (1955) reported that raw potato starch was 57% digestible by the rat and was 100% digestible when cooked; however, raw corn starch was 99% digestible and 100% when cooked.

Proper moist processing of raw soybeans has resulted in performance of swine equal to or better than with soybean meal (Jiménez *et al.* 1963; Combs *et al.* 1967). However, the equipment used for moist processing is expensive and its operation very time consuming.

A more recently tried processing method, roasting, has produced variable results with different feedstuffs. The availability of small on-the-farm equipment for dry processing of soybeans and other grains has stimulated renewed interest in utilizing unextracted soybeans for swine feeding. In these studies (Jensen *et al.* 1970, Villegas *et al.* 1970; Lafferty and Hines 1971; Ruffin *et al.* 1971; Meade, *et al.* 1971; Wahlstrom *et al.* 1971), pigs fed toasted soybeans have gained at a rate not different than that of pigs fed soybean meal. Efficiency of feed conversion is generally improved when toasted (dry roasted) soybeans substitute for soybean meal. The addition of soybean oil to the soybean meal diet improves feed conversion to a level similar to that of full-fat, roasted soybeans but does not appear to influence gain. It appears that discharge temperature from the processing equipment has little effect on the results obtained, because Olsen (1971) showed that similar results were obtained when pigs were fed roasted soybeans toasted to a discharge temperature between 122° and 162° C.

Results of Purdue trials (Perry 1972) indicate that the feeding of roasted maize had no effect on pig gains but resulted in a 7% improvement in feed utilization. Illinois trials (Jensen *et al.* 1971) showed a slight trend toward improved efficiency of feed utilization but no marked differences in response that would justify roasting.

The recent trend to earlier harvesting of feed grains has resulted in grains with higher moisture content that require special processing for storage and handling. High-moisture maize (25–30% water) may be ensiled in air-tight silos to prevent mold and fungal spoilage or may be preserved with propionic acid or other preservatives to avoid spoilage. The nutritive value of high-moisture grains appears to be equal to that from dried grains. Young pigs may show reduced weight gains on high-moisture grains because they are unable to consume sufficient dry matter from the bulkier high-moisture feed. The alternative to storage at high moisture levels is to apply heat to reduce the moisture content. This alternative has stimulated interest in post-harvest drying temperature and its effect on the

TABLE 17.9. Effect of Drying Temperature on the Vitamin Content and the Nutritive Value of Maize Fed to Young and Growing Pigs

	60°C	*80°C*	*104°C*
Vitamin content, mg/g			
Riboflavin	1.32	1.25	1.25
Pantothenic acid	5.68	5.28	5.08
Niacin	24.55	25.37	24.64
Baby pigs—2 weeks			
Av daily gain, kg	0.26	0.27	0.27
Gain/feed ratio	0.59	0.60	0.60
Weanling pigs			
Av daily gain, kg	0.72	0.70	0.70
Gain/feed ratio	0.28	0.28	0.28

Source: Jensen et al. 1960.

nutritive value of maize. The effect of drying temperature and heat treatment on the nutritive value of maize has been reviewed by Hatfield and Wilson (1972) who cited the following work.

Davis and Cabell (1957) studied the effect of drying 22% moisture maize at 57° C and of drying 42% moisture maize at either 35° or 57° C and concluded that

TABLE 17.10. Effect of Drying Temperature of Maize on Swine Performance

Air temperature of drier (°C)	*Final maize moisture (%)*	*Method of feeding*	*Av daily gain (kg)*	*Gain/feed ratio*
Air dried	13	shelled	0.65	0.30
Air dried	13	mixed	0.70	0.30
87	13	mixed	0.70	0.30
143	12	mixed	0.67	0.28
60	12	mixed	0.71	0.29
87	9.5	mixed	0.72	0.32
87	9.5	shelled	0.64	0.31
Air dried	19	shelled	0.68	0.27
Air dried	19	mixed	0.72	0.28
87	13	shelled	0.76	0.30
87	13	mixed	0.74	0.29
87	10	shelled	0.68	0.27
87	10	mixed	0.74	0.30
116	13	shelled	0.70	0.28
116	13	mixed	0.74	0.30
143	13	shelled	0.73	0.29
143	13	mixed	0.72	0.31
Air dried	24	shelled	0.71	0.25
Air dried	24	shelled	0.68	0.25

Source: Taylor et al. 1964.

drying at these temperatures had no significant effect on the performance of pigs. Similar studies by Jensen *et al.* (1960) (Table 17.9) and Taylor *et al.* (1964) (Table 17.10) have confirmed those results and indicate that drying temperatures as high as 143° C have no significant effect upon the performance of either baby, weanling, or growing-finishing pigs. However, work at Illinois indicates that very high-drying temperatures (188°–250° C) reduce the acceptability of maize for growing pigs (22 kg) when compared to field-dried maize. When the pigs were allowed to choose the grain, they preferred field-dried maize to maize rapidly dried at either 188° or at 250° C. Cabell *et al.* (1958) reported that maize dried at 143° C showed marked discoloration and parched odor, and in biological tests it proved to have a protein nutritive value less than that of similar samples of 25% moisture maize dried at 92° C.

Although the data cited on the effects of grinding, pelleting, and heating of feedstuffs for swine are based on research completed several years ago, there is no evidence from the recent literature to indicate changes in the responses of pigs of today compared to those of pigs used in earlier studies. As new crop varieties are developed by plant geneticists it will be necessary to pursue possible changes in plant composition which could affect the nutritional value of feedstuffs processed from the plant for swine.

MICRONIZATION

Micronizing is a term used to describe a dry heat treatment of grains and other feeds by microwaves emitted from infrared burners. In the process, the grain is passed beneath a series of china-clay ceramics which have been heated to a predetermined temperature by a gas/air mixture to produce a luminous red glow known as a *bright line spectrum*. This radiates the electromagnetic spectrum in the infrared visible band and the feed passing beneath the heated ceramics absorbs the radiation produced. This radiation has a wavelength of between 2 and 6 μ (microns) and it is thought that the selective absorption of infrared rays by the grain causes it to become irradiated and that this, in turn, causes the molecules within to vibrate in accordance with the resonance of their own wavelength and frequency. As a result of this there is a rapid internal friction heating and a rise in water vapor pressure which affects, in particular, the starch granules within the grains, causing them to swell, fracture, and gelatinize (Lawrence 1973A). After micronization, the grain is usually rolled and cooled.

For the pig, the process has been shown, largely by improving starch availability, to improve the nutritive value of some cereals (Lawrence, 1973A,B; Shiau *et al.* 1976; Tardif and Levillet 1976). According to Lawrence (1973B), micronization significantly improved the nutritional value of both barley and maize for growing-finishing pigs but was without significant effect on wheat. Shiau *et al.* (1976) demonstrated that micronization of sorghum grain significantly improved the digestibility of crude protein, gross energy, and dry matter and improved feed efficiency of growing-finishing pigs. Although not significant, growth rate was improved by more than 14%.

Raw whole soybeans are poorly utilized by the pig but micronization reduces the trypsin inhibitor activity and is effective in preparing the soybeans for the digestive tract. Pigs fed micronized soybeans grow faster on less feed than do those consuming unprocessed soybeans (Lawrence *et al.* 1978). In contrast, micronized whole rapeseeds would appear to be less efficiently utilized than mixtures of rape oil and extracted rapeseed meal. The micronization process reduced the level of both isothiocyanates and oxazolidinethiones, but the levels of both were considerably higher than those found in newer varieties.

Micronization is a rapid heating process that eliminates the need for water soaking or steam treatment. Results in England have been similar to those obtained by steam-flaking.

LIQUID FEEDING

In an effort to improve the utilization of feedstuffs, research workers have reevaluated the effect of liquid feeding, a practice that was generally employed during the early development of the swine industry. Thrasher *et al.* (1968) demonstrated that a 1:1 mixture of feed and water had no effect on daily gains of pigs, but resulted in an 8% increase in the quantity of dry feed required per unit of gain (Table 17.11).

Speer (1969) summarized the research that compared liquid feeding with dry feeding of swine. Pigs self-fed dry diets have gained essentially the same as pigs fed liquid diets. Furthermore, in all trials, pigs full-fed (self-fed) were more efficient than liquid-fed pigs. However, when the feed intake was restricted to less than full-feed (*ad libitum*), pigs fed the liquid diets gained more rapidly than pigs fed dry diets and showed a trend toward more efficient gains. The results of liquid feeding are variable and probably reflect feeding method and proportion of water to feed. It is suggested that a 2:1 ratio of water to feed will produce the best results.

TABLE 17.11. Effect of Feeding Finishing Pigs the Same Diet Either Wet or Dry

Criteria	*Dry feed*	*Wet feed*[a]
Number of pigs	30	29
Av initial wt, kg	53.2	53.2
Av final wt, kg	92.3	92.3
Av daily gain, kg	0.632	0.618
Av daily feed, kg	2.35	2.49
Feed/gain ratio	3.72	4.03

Source: Thrasher et al. 1968.
[a]The ratio of dry feed to water was 1:1.

PASTE FEEDING

Paste feeding (1.3–1.5 parts of water to 1.0 part dry feed) has also been studied by Ohio workers and, although limited, the data appear to show a trend toward improved pig performance when compared to dry-meal feeding of the same diet. As shown by the summary of the Ohio results presented in Table 17.12, paste feeding increased daily feed consumption by 14.8% during the growing phase and by 6.4% during the finishing period. At the same time gains were improved by an average of 15 and 10% during the growing and finishing periods, respectively. There was no difference in efficiency of feed utilization. The physiological basis for this apparent improvement in performance by paste feeding has not been explained.

TABLE 17.12. Comparative Value of Paste Feed or Dry Meal for Growing and Finishing Pigs

Type of feed	*No. of trials*	*No. of animals*	*Daily gain (kg)*	*Daily feed (kg)*	*Feed/gain ratio*
Growing pigs	3	255	—	—	—
Dry	—	—	0.664	1.82	2.73
Paste	—	—	0.764	2.09	2.72
Finishing pigs	3	203	—	—	—
Dry	—	—	0.791	2.83	3.56
Paste	—	—	0.873	3.01	3.45

Source: Perry 1972.

OIL EXTRACTION

Oil is normally extracted by one of three methods: (1) mechanical extraction (old process, hydraulic process, expeller process); (2) prepress-solvent extraction; and (3) solvent extraction. The mechanical process utilizes both heat and mechanical pressure to extract the fat or oil from the original material. It is less efficient than the other two processes listed, and therefore the meal or cake resulting from this process contains a higher level of oil. In the solvent process, the fats and oils are removed by organic solvents, usually hexane, but sometimes may include acetone and water. This process is carried out at a very low temperature and little heat is generated during the extraction. The level of fat remaining after solvent extraction is extremely low. The prepress-solvent system employs both of the other two methods of processing. The fats or oils are first partially extracted by heat and mechanical pressure and after mechanical extraction are subjected to extraction by organic solvents.

TABLE 17.13. Effect of Method of Oil Extraction on the Protein Quality and Free-Gossypol Content of Cottonseed

Processing method	*Free-gossypol content (%)*	*Available lysine (g/16 g N)*	*Av wt gain (g)*	*PER*[a]
Mechanical	0.05	2.79	62	1.47
Prepress-solvent	0.05	3.33	105	1.96
Solvent	0.13	3.75	85	1.79

Source: Elias et al. 1969.
[a]Protein efficiency ratio (g body weight gain per g protein consumed).

During extraction involving heat there is a reduction in the availability of lysine in some products but at the same time, as in soybeans and cottonseed, there is also a reduction in the anti-metabolite factors present. It has been stated that "processing must be a happy medium between the beneficial and the destructive." An example of the effect of these three methods of processing on the protein quality and free-gossypol content of cottonseed meal is presented in Table 17.13.

A cottonseed meal with higher available lysine was produced by solvent extraction, but at the same time more free gossypol was present. The mechanical process reduced the free-gossypol level, but also caused a reduction in the level of available lysine. The best biological results were obtained by feeding the prepress-solvent-extracted meal. Although the available lysine was slightly less than in the solvent-extracted meal, it was much higher than in the mechanically extracted meal. At the same time the effect of the heat during the prepressing of the seed was adequate to reduce the free-gossypol content, permitting the animals to utilize the available lysine without being affected by too high a free-gossypol level.

SUMMARY OF THE EFFECTS OF PROCESSING FEEDSTUFFS ON PIG PERFORMANCE

Hogberg *et al.* (1980) summarized the current knowledge of the effects of feed processing on the growth rate, feed intake, and efficiency of feed utilization of growing swine. This information is reproduced in Table 17.14. It must be recognized that the direction and magnitude of response to a particular set of processing conditions is never completely predictable, due to differences in pigs, plant growing conditions, crop harvesting times and methods, and plant variety. Standard weights and measures of some common grains used unprocessed or processed in swine diets are shown in Table 17.15.

TABLE 17.14. Effects of Processing Feedstuffs on Pig Performance

Processing method	*Type of grain*	*Growth rate*	*Change in feed intake*	*Feed conversion*	*Comments*
Grinding	Corn	Improved 3–5%	No effect	Improved 3–5%	Medium screen (¼–⅜ in.). Generally most acceptable.
	Grain sorghum	Improved	No effect	Improved	Too fine of grind can reduce palatability.
Pelleting	Corn, grain sorghum	Improved 3–6%	Reduced 1–3%	Improved 5–8%	
	Barley, oats, alfalfa, wheat, bran	Improved 3–6%	Reduced 1–3%	Improved 7–10%	Greater improvement with higher fibrous materials.
Paste	Corn	Improved 10–15%	Increased 10–15%	No change	Feed:water ratio of 1.2–1.5:1. Advantage with G-F pigs but not weanlings.
Liquid	Corn	Improved	Increased	Improved	Water:feed ratio of 2.1. Advantages with limit-fed systems. Full-fed levels fail to improve performance.
Roasting	Corn	No change	No change	Slightly improved	Processing cost greater than improvement return.
	Soybeans	No change	Decreased	Improved 4–6%	Necessary for adequate performance. Comparisons made with SBM.
Steam flaking	Corn, grain sorghum	No change	No change	No change	
Micronizing	Corn, grain sorghum	No change	No change	Variable	Availability of amino acids may be reduced.
Extruding	Wheat, grain sorghum	No change	No change	No change	
	Soybeans	No change	Decreased	Improved 4–6%	Comparison made with SBM—necessary for adequate performance.

Source: Hogberg et al. 1980.

TABLE 17.15. Weights and Measures of Common Feeds

Feeds	*Pounds per bushel*	*Kilograms per bushel*
Barley	48	21.8
Cowpeas	60	27.2
Maize, husked ear	70	31.8
Maize, cracked	50	22.7
Maize, shelled	56	25.4
Maize meal	50	22.7
Maize-and-cob meal	45	20.4
Oats	32	14.5
Rye	56	25.4
Soybeans	60	27.2
Wheat	60	27.2

REFERENCES

ATKINSON, J., and CARPENTER, K. J. 1970. J. Sci. Food Agric. *21,* 360; *21,* 366.

BAYLEY, H. S., POS, J., and THOMSON, R. G. 1975. J. Anim. Sci. *40,* 857.

BAYLEY, H. S., and THOMSON, R. G. 1969. J. Anim. Sci. *28,* 484.

BECKER, D. E. 1966. Merck Agric. Memo, Vol. II, No. 1, Merck and Co., Rahway, New York.

BIERER, B. W., and VICKERS, C. L. 1958. J. Am. Vet. Med. Assoc. *133,* 228.

BRAHMAN, J. E., VELA, R. M., BRESSANI, R., and JARQUIN, R. 1965. Arch. Venez. Nutr. *15,* 19.

BRESSANI, R., and ELIAS, L. G. 1968. Arch. Lat. Am. Nutr. *18,* 319.

BRESSANI, R., ELIAS, L. G., and VALIENTE, A. T. 1963. Br. J. Nutr. *17,* 69.

CABELL, C. A., DAVIS, R. E., and SAND, R. A. 1958. J. Anim. Sci. *17,* 1204.

CAMERON, C. D. T. 1960. Can. J. Anim. Sci. *40,* 126.

CHARLES, O. W., and HUSTON, T. W. 1970. World's Poultry Congress. Sept. 1970.

COMBS, G. F., CONVERS, R. G., BERRY, T. G., and WALLACE, H. D. 1967. J. Anim. Sci. *26,* 1067.

CONRAD, J. H. 1958. Feedstuffs *30*(35), 38.

CORLEY, J. R., BAKER, D. H., and EASTER, R. A. 1980. J. Anim. Sci. *50,* 286.

CRAIG, W. M., and BRODERICK, G. A. 1981. J. Anim. Sci. *52,* 292.

CRANE, F. M., HANSEN, M., YODER, R., LEPLEY, K., and COX, P. 1973. Effects of processing feeds on molds, salmonella, and other harmful substances in feeds. *In* Effects of Processing on the Nutritional Value of Feeds, pp. 72–90. Natl. Acad. Sci., Washington, D.C.

DAVIS, R. E., and CABELL, C. A. 1957. Feed Age *7,* 49.

ELIAS, L. G., SANCHEZ LOARCA, S., and BRESSANI, R. 1969. Arch. Lat. Am. Nutr. *19,* 279.

FORGACS, J., and CARLL, W. T. 1962. Adv. Vet. Sci. *7,* 273.

FRIEND, D. W., and MacINTYRE, T. M. 1969. Can. J. Anim. Sci. *9,* 375.

FRITZ, J. C. 1972. Symp. on the Effect of Processing on the Nutritional Value of Feeds. NRC–NAS and Univ. of Florida Center for Tropical Agriculture, Gainesville.

HARRIS, L. E., and CRAMPTON, E. W. 1972. Symp. on the Effect of Processing on the Nutritional Value of Feeds. NRC–NAS and Univ. of Florida Center for Tropical Agriculture, Gainesville.

HARRIS, L. E., ASPLUND, J. M., and CRAMPTON, E. W. 1968. Utah Agric. Exp. Stn. Bull. *479.*

HATFIELD, E. E., and WILSON, W. M. D. 1972. Symp. on the Effect of Processing on the Nutritional Value of Feeds. NRC–NAS and Univ. of Florida Center for Tropical Agriculture, Gainesville.

HAYS, V. W. 1967. Proc. 22nd Annu. Texas Nutr. Conf., College Station, Texas.

HENRY, W. A. 1886. 4th Annu. Rep. Wis. Agric. Exp. Stn., p. 64.

HOGBERG, M., MAHAN, D., and SEERLEY, R. 1980. Physical Forms of Feed—Feed Processing for Swine. Natl. Pork Handbook *PIH-71.* Cooperative Extension Service, Purdue Univ., West Lafayette, Indiana.

HURRELL, R. F., and CARPENTER, K. J. 1977. Nutritional significance of cross-link formation during

food processing, pp. 225–238. *In* Protein Crosslinking: Biochemical, Medical, and Nutritional Consequences (M. Friedman, Editor). Plenum Press, New York.
JELINCK, B., KATAYAMA, M. C., and HARPER, A. G. 1952. Can. J. Med. Sci. *30,* 447.
JENSEN, A. H. 1966. Feedstuffs. *38*(31), 24.
JENSEN, A. H., BECKER, D. E., and TERRILL, S. W. 1959. Ill. Agric. Exp. Stn. Publ. *AS–502,* Urbana.
JENSEN, A. H., BROWN, H. W., HARMON, B. G., and BAKER, D. H. 1970. J. Anim. Sci. *31* 1022 (Abstr.).
JENSEN, A. H., BAKER, D. H., BROWN, H. W., COSTA, P. M. A., and HARMON, B. G. 1971. Univ. of Ill. Park. Ind. Day. *1971,* p. 3.
JENSEN, A. H., LAUNER, J. E., TERRILL, S. W., and BECKER, D. E. 1955. Ill. Agric. Exp. Stn. Publ. *AS–418,* Urbana.
JENSEN, A. H., TERRILL, S. W., and BECKER, D. E. 1960. J. Anim. Sci. *19,* 629.
JIMÉNEZ, A. A., PERRY, T. W., PICKETT, L. A., and BEESON, W. M. 1963. J. Anim. Sci. *22,* 471.
JONES, A. S., and CADENHEAD, A. 1965. J. Sci. Food Agric. *16,* 38.
KELLNER, O. 1910. The Scientific Feeding of Animals. Macmillan Co., New York.
LAFFERTY, D. T., and HINES, R. H. 1971. J. Anim. Sci. *33,* 233 (Abstr.).
LAWRENCE, T. L. J. 1973A. Anim. Feed Sci. and Tech. *16,* 99.
LAWRENCE, T. L. J. 1973B. Anim. Feed Sci. and Tech. *16,* 109.
LAWRENCE, T. L. J. 1978. Anim. Feed Sci. and Tech. *3,* 179.
LEWIS, C. J., CATRON, D. V., COMBS, G. C., JR., ASHTON, G. C., and CULBERTSON, C. C. 1955. J. Anim. Sci. *14,* 1103.
LIENER, I. E. 1958. *In* Processed Plant Protein Foodstuffs (A. M. Altschul, Editor), Academic Press, New York.
LINDAHL, I. L., and REYNOLDS, P. J. 1959. J. Anim. Sci. *18,* 1074.
LOOSLI, J. K. 1972. Symp. on the Effect of Processing on the Nutritional Value of Feeds. NRC–NAS and Univ. of Florida Center for Tropical Agriculture, Gainesville.
MADSEN, A., MASON, V. C., and WEIDNER, K. 1965. Acta Agric. Scand. *15,* 213.
MAHAN, D. C., PICKETT, R. A., PERRY. T. W., CURTIN, T. W., BEESON, W. M., and FEATHERSTON, W. R. 1966. J. Anim. Sci. *25,* 1O19.
MANER, J. H. 1972. Seminario sobre Sistemas de Producción Porcina en América Latina. CIAT, Cali, Colombia, September 18–21, 1972.
MASON, V. C., and WEIDNER, K. 1964. Acta Agric. Scand. *14,* 87.
MEADE, R. J., RUST, J. W., and HANSON, L. E. 1971. J. Anim. Sci. *33,* 236 (Abstr.).
NATIONAL ACADEMY OF SCIENCES. 1973. Effect of Processing on the Nutritional Value of Feeds. Natl. Acad. Sci., Washington, D.C. 494 pp.
NORTH CENTRAL REGIONAL PROJECT 42, COMMITTEE ON SWINE NUTRITION. 1969. J. Anim. Sci. *29,* 927.
NEWER, A. J., PERRY, T. W., PICKETT, R. A., CURTIN, T. M., FEATHERSTON, W. R., and BEESON, W. M. 1965. J. Anim. Sci. *24,* 113.
OLIVER, A. W. 1931. Oregon State Agric. Exp. Stn. Cir. *104.*
OLSEN, E. M. 1971. Effect of toasting, particle size and dietary protein level on the utilization of soybeans by pigs. Ph.D. Thesis, University of Guelph, Guelph, Ontario.
OLSEN, E. M., and SLINGER, S. J. 1968. Can. J. Anim. Sci. *48,* 35.
PERRY, T. W. 1972. Symp. on the Effect of Processing on the Nutritional Value of Feeds. NRC–NAS and Univ. of Florida, Center for Tropical Agriculture, Gainesville.
PICKETT, R. A., FUGATE, W. H., HARRINGTON, R. B., PERRY, T. W., and CURTIN, T. M. 1969. J. Anim. Sci. *28,* 837.
ROBLES, A., and EWAN, R. C. 1982. J. Anim. Sci. *55,* 572.
RUFFIN, B. G., POWELL, W. E., and BROWN, V. L. 1971. J. Anim. Sci. *32,* 391 (Abstr.).
SCOTT, M. L. 1972. Symp. on the Effect of Processing on the Nutritional Value of Feeds. NRC–NAS and Univ. of Florida, Center for Tropical Agriculture, Gainesville.
SHIAU, S. Y., YANG, S. P., TRIBBLE, L. F., and WILLIAMS, I. L. 1976. J. Anim. Sci. *43,* 258 (Abstr.).
SLINGER, S. J., SUMMERS, J. D., PEPPER, W. F., BENTLEY, H. U., and CISNEROS, G. 1966. Proc. 13th World's Poultry Congr. p. 162. Kiev., USSR.
SMITH, O. B. 1957. Feedstuffs *29,* 22.

SMITH, I. D., GRUMMER, R. H., HOEKSTRA, W. G., and PHILLIPS, P. H. 1960. J. Anim. Sci. *19*, 568.
SPEER, V. C. 1969. Feedstuffs *41*(8), 30.
STANLEY, D. L., and EWAN, R. C. 1982. J. Anim. Sci. *54*, 1175.
SUMMERS, J. D., SLINGER, S. J., and CISNEROS, G. 1967. Cereal Chem. *44*, 318.
SUNDE, M. L. 1972. Symp. on the Effect of Processing on the Nutritional Value of Feeds. NRC–NAS and Univ. of Florida, Center for Tropical Agriculture, Gainesville.
TARDIF, H., and LEVILLET, M. 1976. J. Rech. Porcine, France. pp. 17–24.
TARR, H. L. A., and BIELY, J. 1972. Symp. on the Effect of Processing on the Nutritional Value of Feeds. NRC–NAS and Univ. of Florida, Center for Tropical Agriculture, Gainesville.
TAYLOR, M. E., PICKETT, R. A., ISAACS, C. W., and FOSTER, C. 1964. J. Anim. Sci. *23*, 894.
THRASHER, D. M., ROBERTS, H. P., and MULLINS, A. M. 1968. La. State Univ. Livestock Prod. Day. January 21, 1968. Baton Rouge.
VILLEGAS, F. J., VEUM, T. L., HEDRICK, H. B., and McFATE, K. L. 1970. J. Anim. Sci. *31*, 213 (Abstr.).
WAHLSTROM, R. C., LIBAL, G. L., and BERNS, R. J. 1971. J. Anim. Sci. *32*, 891.
WILDER, O. H. M. 1972. Symp. on the Effect of Processing on the Nutritional Value of Feeds. NRC–NAS and Univ. of Florida Center for Tropical Agriculture, Gainesville.
YEN, J. T., BAKER, D. H., HARMON, B. G., and JENSEN, A. H. 1971. J. Anim. Sci. *33*, 987.
YOSHIDA, M., and MORIMATO, H. 1955. J. Nutr. *57*, 565.

Part IV

Management and Marketing

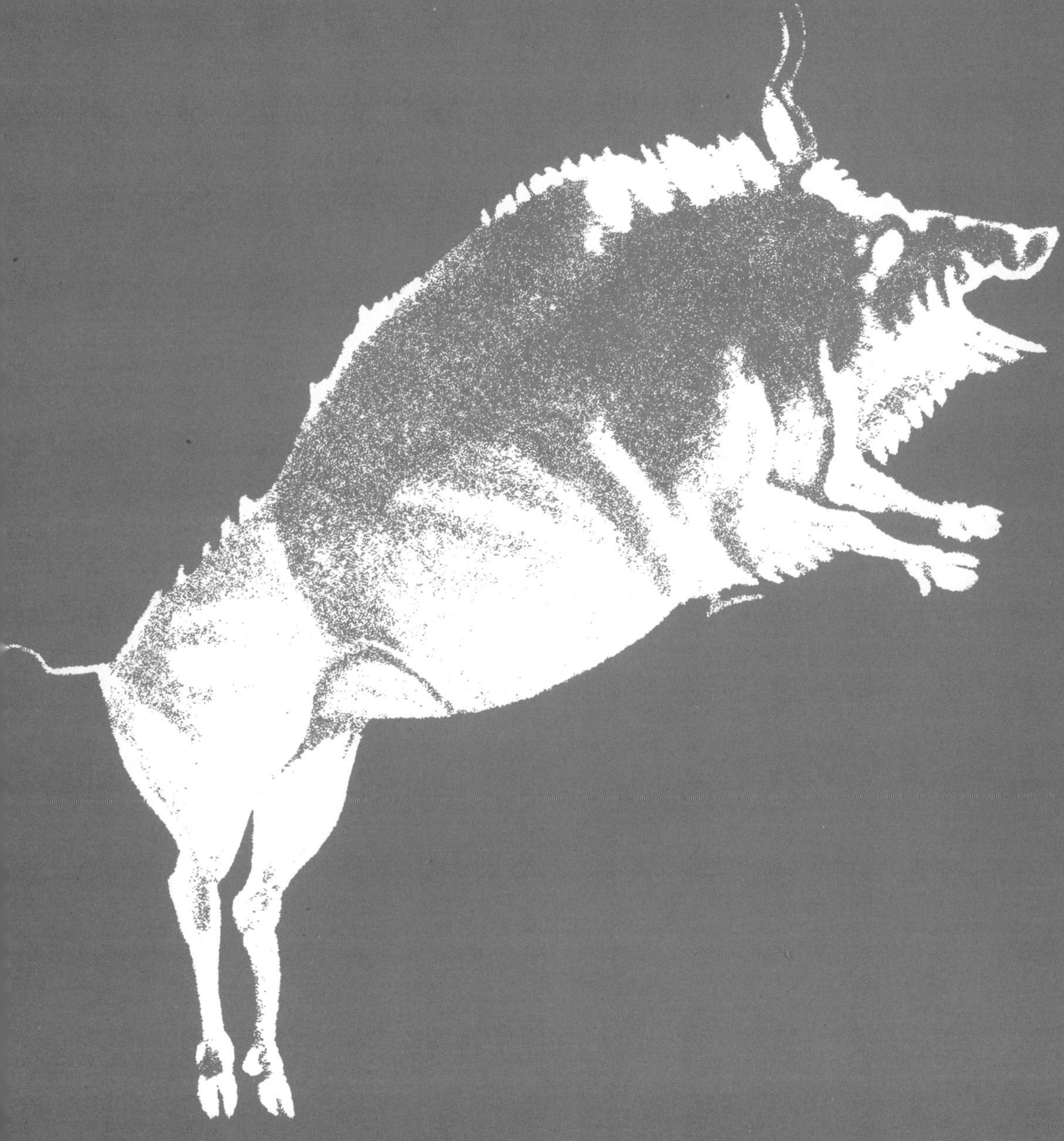

18

Environmental Physiology

INTRODUCTION

Swine production methods are rapidly changing from extensive to intensive, whereby highly mechanized, low-labor confinement facilities geared to volume production are becoming common, especially in the temperate zones. This move toward confinement systems of production has brought with it much research into the effect of the environment on the performance of the pig during all phases of the life cycle. No attempt is made here to describe exhaustively the work done on environmental physiology of pigs, but selected references are reviewed to provide the basis on which decisions and judgments on swine housing and management can be made. Curtis (1981) has reviewed and integrated the current knowledge of environmental physiology in food animals, including swine. For detailed information, this reference textbook (Curtis 1981) and the *Proceedings of a Symposium on Management of Food-Producing Animals* (Woods 1982) are highly recommended. In addition, a report by the National Research Council (National Research Council 1981) provides highly useful information on the effects of environment on nutrient requirements of animals.

EFFECT OF ENVIRONMENTAL TEMPERATURE

Young Pigs

Mount (1968) described in detail the changes that occur in metabolic rate, body temperature control and thermal insulation of the pig from birth to maturity. Some of these considerations were discussed in Chapter 6. In brief, the new born pig is prone to chilling and hypoglycemia, which may lead to death because of its relative lack of hair and subcutaneous fat cover, its high surface area in relation to body weight, its immature body temperature homeostatic control, and its low body fat and glycogen storage for available energy (Curtis 1970). Thus, supplemental heat must be provided during the first few days of postnatal life to avoid

TABLE 18.1. Oxygen Consumption Rate in Piglets on a Concrete Floor as Affected by Adding a Layer of Straw (ml/min/kg)

Environmental temperature (°C)	*Concrete floor*	*10-cm-thick layer of straw added*
10	35	29
20	31	23

Source: Stephens 1971.

losses to chilling. About 25% of live piglets born in the United States die before weaning and the relative physiological immaturity is probably a major cause. The newborn wild pig is notably more resistant to cold than the newborn domestic pig. Comparative studies have shown (Foley *et al.* 1971) that this is due partly to a greater hair covering and partly to a more mature metabolic response of the wild pig to cold. For all animals there is a range of environmental temperatures within which no extra body heat production is needed to keep the animal warm or to cool the animal. This is referred to as the *zone of thermoneutrality*. This zone corresponds to the environmental temperature range within which the metabolic rate is at a minimum. Body temperature is maintained constant within this range by vasomotor control (peripheral vasoconstriction or vasodilation, hair erection, postural changes) and by sweating (physical control of heat balance). The point at which the environmental temperature falls below this range is called *critical temperature* and an increase in metabolic rate is necessary to maintain body temperature (chemical control of heat balance). Oxygen consumption rate of cold-stressed piglets is increased; the protective effect of adding straw bedding to concrete-floor pens on piglets kept at environmental temperatures of 10° and 20° C is illustrated in Table 18.1 (Stephens 1971). The point at which the environmental temperature rises above this range is called point of *hyperthermal rise* and an increase in metabolic rate is necessary to attempt to expell excess heat production (chemical control of heat balance). A recent report of the National Research Council (1981) gives useful information about the relationship between environmental temperature and nutrition of swine and other animals. Environmental temperatures outside the zone of thermoneutrality (comfort zone) are frequent in the tropics and subtropics and have profound effects on feed intake and animal performance. Nigerian research (Babatunde *et al.* 1974), for example, has shown that protein concentration in the diet must be increased in the tropics to offset the reduced feed consumption of growing pigs in a hot environment. Ames and Ray (1983) reviewed the effect of environmental temperature on animal performance and cited Danish work that showed dramatically the relationship between environmental temperature and feed intake and efficiency of energy conversion (Table 18.2).

In the very young pig the critical temperature is of most concern. It is high in the newborn pig (near 30°C) and gradually declines as the pig grows (20°–23° C for the 50-kg pig, and 17°–22° C for the 100-kg pig) (Mount 1968). Critical temperature increases at all ages when level of feed intake is reduced from *ad libitum* to maintenance (Holmes and Close 1977).

TABLE 18.2. Effect of Temperature on Intake, Growth Rate, and Efficiency of Energy Conversion for Swine (70–100 kg)

Temp., °C	*Caloric intake, kcal DE/day*	*Feed/gain*	*Growth rate, kg/day*	*Product[a] kcal GE*	*% Caloric[b] efficiency*
0	15,377	9.5	0.54	2991	19.4
5	11,404	7.1	0.53	2936	25.7
10	10,616	4.4	0.80	4432	41.7
15	9,554	4.0	0.79	4376	45.8
20	9,766	3.8	0.85	4709	48.2
25	7,976	3.7	0.72	3988	50.1
30	6,703	4.9	0.45	2493	37.1
35	4,579	4.9	0.31	1717	37.4

Source: Thorbeck 1975.
[a]Estimated caloric value of gain for an 80-kg pig is 5.54 kcal GE/g (Thorbeck 1975). DE is digestible energy; GE, gross energy.
[b]Calculated: (kcal GE in product ÷ kcal DE intake) × 100.

The newborn pig is quickly attracted to artificial heat sources such as heat lamps and seems to prefer an ambient temperature of about 32° C (Mount 1963A). After the first week and until the sixth week, the preferred temperature is somewhat less, about 29° or 30° C (Mount 1963A) (Table 18.3). Curtis (1981) devised an operant supplemental heat system whereby young growing pigs could turn on or off a heater switch to control pen temperature on demand. Even though pigs with operant heat preferred about 27° C each afternoon, they let the temperature fall to 15° C at night, resulting in about a 50% reduction in total fuel use. Despite the widely fluctuating temperature from day to night, pigs in this environment had performance comparable to that of pigs kept in a constant 27° C environment. Such observations suggest that, in the future, more use will be made of techniques to provide to the pig an opportunity to select its preferred environment on both humane and economic grounds.

The environmental temperature during the first few weeks of life has a marked effect on performance, body leanness, and thyroid and adrenal weight (Sugahara 1970). Pigs kept at 7° C from 5 to 11 weeks of age had an average daily gain of 0.64 kg; in comparison, those kept at 23° C had a gain of 0.61 kg, and those kept at 33° C, a gain of 0.40 kg. The sharp depression in growth rate at the highest

TABLE 18.3. Ambient Temperatures Preferred by Young Pigs

Age, days[a]	*No. of animals by ambient temperature, °C*				*Mean ± standard error, °C*
	23–28	*29–31*	*32–34*	*35–37*	
Less than 1	1	2	5	2	32.3 ± 0.9
1–7	8	16	4	0	29.3 ± 0.4
8–41	16	33	11	6	29.8 ± 0.6

Source: Mount 1963A.
[a]Housed singly.

TABLE 18.4. Effect of Temperature on Gain, Feed Intake, Muscle Mass, and Organ Weight of Growing Swine

Criteria	Environmental temperature, °C		
	7	23	33
Av daily gain, kg[a]	0.64	0.61	0.40
Av daily feed, kg	1.61	1.33	0.91
Gain/feed ratio	0.40	0.46	0.45
Muscle mass, %[b]	39.0	34.6	35.9
Thyroid wt, mg/kg	112.8	76.0	57.6
Adrenal wt, mg/kg	98.2	84.5	74.1

Source: Sugahara et al. 1970.
[a]Four pigs per group, av initial wt, 9.3 kg; 35 days of age; experimental period, 41 days.
[b]Determined by ^{40}K whole body counting.

temperature was a reflection of depressed feed intake. Body weight gain per unit of feed consumed was not affected, but muscle mass was drastically reduced at the highest temperatures. The changes in weight gain and muscle mass were accompanied by reductions in the weights of the thyroid and adrenal glands. These relationships are shown in Table 18.4.

Growing-Finishing Pigs

Jensen (1971) and Curtis (1981) reviewed the literature on environmental temperature requirements for growing-finishing pigs. Several reports have demonstrated the important effects of ambient temperature on performance under laboratory conditions (Heitman *et al.* 1958; Bond *et al.* 1967; Mount 1963B; Sorenson 1962) and practical conditions (Seymour *et al.* 1964; Holme and Coey 1967; Jensen *et al.* 1969). As the pig begins to accumulate subcutaneous fat, the fat cover provides good thermal insulation against cold, but inability to dissipate large amounts of heat by perspiration makes the older pig sensitive to heat. This effect of extremes in temperature and the increased sensitivity of the pig to increased temperature with increased body weight is illustrated in Table 18.5 (Heitman *et al.* 1958).

The effect of building type (closed heated, closed unheated, or open front) on pig performance in winter is illustrated in Table 18.6 (Jensen *et al.* 1969). The average temperature in the open-front building was −7° C (range −23°–15° C), whereas that of the heated closed building was 22° C and that of the unheated closed building was 10° C. Daily gain was similar for all groups but daily feed increased as temperature decreased and the gain–feed ratio decreased accordingly during both the growing and finishing periods. Curtis (1981) reviewed a large body of data on growing-finishing swine raised in a variety of housing types and degrees of environmental control and concluded that pig performance is generally as high on pasture or in open-front housing as in enclosed houses, but labor is lower in enclosed houses. In winter studies, labor and straw costs partially offset capital investments and fuel costs, but total cost of production was often higher

TABLE 18.5. Effect of Ambient Temperature on Body Weight Gain at Various Liveweights

Mean livewt (kg)	Body weight gain (kg) by ambient temperature, °C[a]							
	4	10	16	21	27	32	38	43
45	—	0.62	0.72	0.91	0.89	0.64	0.18	—0.60
70	0.58	0.67	0.79	0.98	0.03	0.52	—0.09	—1.18
90	0.54	0.71	0.87	1.01	0.76	0.40	—0.35	—
115	0.50	0.76	0.94	0.97	0.68	9.28	—0.62	—
135	0.46	0.80	1.02	0.93	0.62	0.16	—0.88	—
160	0.43	0.85	1.09	0.90	0.55	0.15	—1.15	—

Source: Heitman et al. 1958.
[a]Relative humidity held constant at 50%.

in enclosed, heated houses. Carcass traits appear not to be greatly affected by type of accommodation in which pigs are raised.

The effect on performance that results from protecting growing-finishing pigs, kept in outdoor lots, from heat during the summer has been studied extensively (Hale *et al.* 1968; Andrews *et al.* 1960; Garrett *et al.* 1960; Heitman *et al.* 1959; Culver *et al.* 1960). In a 3-year study (Garrett *et al.* 1960), the use of shade over a concrete wallow during 32.8° C temperature (range 23.8°–41.8° C) improved weight gain and feed intake but tended to increase feed per unit gain. At a mean temperature of 23.9° C (maximum 34.4° C), provision of a shaded wallow or air-conditioned house resulted in a similar improvement in performance of growing pigs over that of the controls (Heitman *et al.* 1959). The type of shade (plywood, straw, corrugated steel and aluminum, or laminated polyethylene plastic) appears to be of no importance regardless of pig color (white, red, or black) in protecting the pig from the sun (Andrews *et al.* 1960). A similar response in daily gain, superior to that of the control group, was obtained with all shaded groups of

TABLE 18.6. Effect of Building Type (Environmental Temperature) on Performance of Growing-Finishing Pigs

Performance	Building			Probability
	Heated	Unheated	Open front	
Growing (14–48 kg)				
Temperature, °C	22(16–24)[a]	10(1–5)	—7(—23–15)	—
Daily gain, kg	0.80	0.80	0.75	—
Daily feed, kg	1.47	1.64	1.80	<.01
Gain/feed ratio	0.54	0.49	0.42	<.01
Finishing (48–102 kg)				
Temperature, °C	21(16–25)	15(2–24)	1(—14–18)	—
Daily gain, kg	0.82	0.82	0.92	<.01
Daily feed, kg	2.89	2.87	3.35	<.01
Gain/feed ratio	0.29	0.29	0.27	<.05

Source: Jensen et al. 1969.
[a]Mean with range in parentheses.

TABLE 18.7. Changes in Body Temperature and Respiration Rate Induced in Growing Pigs by Exposure to Direct Summer Sun for 30 Min

Day[a]	*Ambient temperature* (°F)	(°C)	*Relative humidity* (%)	*Rectal temperature* *Initial* (°F)	(°C)	*Change* (°F)	(°C)	*Respiration rate* *Initial* (per min)	*Change* (per min)
1	90	32.2	58	102.7	39.3	1.8	1.68	78.8	60.5
2	88	31.1	67	102.1	39.0	1.8	1.68	56.6	66.5
3	91	32.8	62	101.8	38.7	2.6	1.63	60.8	100.0
4	88	31.1	64	101.9	38.8	2.1	1.66	42.3	99.0
5	86	30.0	76	101.7	38.6	1.5	1.69	30.6	78.4
Mean	88.6	31.5	65.4	102.03	39.0	2.0	1.67	53.7	80.8

Source: Tidwell and Fletcher 1951.
[a]Eight Poland China and Duroc pigs (four each) weighing 53–83 kg were exposed to direct sunlight for 30 minutes one day per week for five weeks at the temperature and humidity indicated.

growing-finishing pigs. The use of a fine mist spray (1 gal/hour at 40-lb pressure) from a nozzle mounted beneath a shade 1.5 m above a concrete-floor pen produced a significant response over controls in daily gain, but was not significantly better than a portable steel wallow in promoting growth (Culver *et al.* 1960). The use of wallows carries with it problems of sanitation and odors so that an overhead spray may be more desirable as a means of cooling hogs in warm weather under most circumstances. Sprinkler systems can also be used under portable shade for pigs kept on pasture and have been shown to improve weight gain consistently when used in a warm, humid environment (average temperature 30°–33° C: minimum temperature, 18°–21° C) (Hale *et al.* 1966).

The detrimental effect of high summer temperatures on performance of growing-finishing pigs is related to their inefficient dissipation of body heat. The increase in rectal temperature and respiration rate induced by a short exposure (30 min) to direct sunlight in 31.5° C average environmental temperature is shown in Table 18.7 (Tidwell and Fletcher 1951). Effects of longer-term heat stress (7 days) in growing pigs are shown in Table 18.8 (Heitman and Hughes 1949). Pigs were kept at air temperatures ranging from 4.4° to 37.8° C in a psychrometric room for periods of about 7 days with constant humidity and air flow. As air temperature increased above 15.6° C, rectal temperature and respiration rate

TABLE 18.8. Effect of Air Temperature on Rectal Temperature, Pulse Rate, and Respiration Rate of Growing-Finishing Pigs Exposed for 7 Days

Room temperature (°F)	(°C)	*Rectal temperature* (°F)	(°C)	*Pulse rate* (beats/min)	*Respirations rate* (inspirations/min)
40	4.4	102.3	39.1	125	20
60	15.6	102.2	39.0	120	30
80	26.7	102.6	39.2	120	60
100	37.8	104.0	40.0	100	100

Source: Heitman and Hughes 1949.

TABLE 18.9. Effect of Ambient Temperature on Thyroid Secretion Rate in Growing Pigs

Ambient temperature[a] (°C)	*No. pigs*[b]	*Thyroid secretion rate (mg thyroxin/day)*
3	5	3.81 ± 0.67
8	12	1.72 ± 0.18
15	5	0.51 ± 0.08
24	12	0.38 ± 0.03

Source: Sorensen 1962.
[a]Relative humidity, 70%, except 90% at 24° C.
[b]40 to 90 kg body wt.

increased while pulse rate decreased. Pigs at 70 kg could be kept at temperatures of 46° C but it was not possible to keep heavier pigs at temperatures of more than 37.8° C. When relative humidity was increased from 30 to 94% during an 8-hour period, with temperature held constant at 35.6° C, body temperature increased 1° C, and respiratory rate doubled. Sorensen (1962) showed that long-term exposure to high environmental temperature leads to depressed thyroid activity. This work is illustrated in Table 18.9. Another means of estimating stress in pigs has been reported (Stefanovic *et al.*, 1970). Because urinary excretion of a breakdown product of adrenal medulla hormones has been shown to be increased sharply by cold (5° C) or heat 33° C) and decreased by starvation, it has been suggested that the urinary output of this compound (vanilmandelic acid) offers a convenient way to measure physiological stress in the pig.

Pregnant Sows

Heat stress (35° C) at day 1 to 5 of pregnancy appears to reduce the number of live embryos to a greater degree than heat stress imposed at day 20 to 25 (Tompkins *et al.* 1967). Exposure of sows to heat stress at later stages of pregnancy has been studied by several research groups. From a review of research work such as that just described, the Livestock and Poultry Management Committee of the American Feed Manufacturers Association (A.F.M.A.) Nutrition Council (1970) prepared a table of recommended optimum temperatures for growing-finishing pigs. These recommendations are in Table 18.10A.

Phillips and Bickert (1982) presented optimum temperatures and allowable ranges for all classes of swine (Table 18.10B). Ventilation rates compatible with these temperatures in closed buildings are suggested in the National Pig Industry Handbook 60, one of a series of pamphlets on swine production authored by scientists at land grant universities in the United States (Phillips and Bickert 1982).

Marple *et al.* (1972) showed that plasma corticoids and growth hormone are increased in growing swine in response to low temperature (4.4° C) and suggested that stress may change the turnover rate of corticoid hormones in the body.

Exposure to high temperatures causes fetal death in some animals, but apparently the sow is less sensitive than some other species to high-temperature stress, at least to stress causing fetal mortality. Heitman *et al.* (1951) exposed pregnant sows to temperatures up to 37.2° C for 1–3 days between the 85th day of

TABLE 18.10A. Optimum Temperatures Suggested for Growing-Finishing Pigs

Body weight (kg)	*Temperature (°F)*	*(°C)*
11–23	70	21
23–45	70	21
45–68	64	18
68–91	64	18

Source: A.F.M.A. Nutrition Council 1970.

TABLE 18.10B. Optimum Temperatures for Housed Swine of All Ages

Animal	*Temperature optimum*		*Desirable limits (°F)*
	(°F)	*(°C)*	
Lactating sow	60	16	50–70
Litter—newborn	95	35	90–100
Litter—3-week old	80	27	75–85
Pre-nursery	80	27	75–85
Nursery (13–23 kg)	75	24	70–80
Nursery (23–34 kg)	65	18	60–70
Growing-finishing	60	16	50–70
Gestating sows	60	16	50–70
Boars	60	16	50–70

Source: Phillips and Bickert 1982.

pregnancy and term. They observed decreased feed consumption and increased respiration rate and rectal temperature among the 13 sows tested, but 11 of the sows farrowed normally following return to a normal environment. One sow died of heat prostration and one aborted, but fetuses had apparently died before exposure to high temperature. These observations provided no evidence to indicate that a short-term rise in body temperature induced by high ambient temperature causes abortion of a normal litter. Exposure of sows to heat stress during the first few weeks of pregnancy has a more adverse effect on prenatal piglet survival than exposure in mid-pregnancy (Table 18.11) (Omtvedt *et al.* 1971; Kattesh, *et al.* 1980). Heat stress during late gestation (days 102–110) increases the incidence of stillbirths (Omtvedt *et al.* 1971). Kattesh *et al.* (1980) provided evidence that blood-plasma-corticosteroid-binding globulin concentration is reduced during exposure to high temperatures and may serve as an indicator of physiological stress.

Teague *et al.* (1968) and Warwick *et al.* (1965) showed reduced ovulation rate in gilts kept in a 32°–33.3° C environment. Warwick *et al.* (1965) observed further reduction in embryo survival in heat-stressed gilts, suggesting that cooling sows during high-temperature periods may improve reproductive performance. Experimental and field observations have shown that conception rate is lowest in the months of hot weather (Curtis 1981; Christenson 1981).

TABLE 18.11. Farrowing Performance of Gilts as Affected by Daily Heat Stress on Days 102–110 of Gestation

Trait	*Heat stress daily*	*Control*
No. of live pigs per litter	6.0	10.4
No. of dead pigs per litter	5.2	0.4
Mean birth weight (kg)	1.2	1.4
Mean litter weight (kg)	8.6	13.6

Source: Omtvedt et al. 1971.

In boars, as in males of other species, scrotal and testicular temperatures are increased in high environmental temperatures, and heat stress (33°C), even for 3 days, reduces semen output and morphological integrity of the spermatozoa, with maximum effects about 1 month after the stress. Heat stress for longer periods would be expected to have an even more detrimental effect on sperm production and fertility. Cold stress, on the other hand, has not been shown to affect boar fertility adversely (Swierstra 1970).

The stress of high temperature on sows in farrowing crates can be relieved by providing a forced cool air duct near the sow's snout to reduce local temperature and encourage postpartum feed intake.

SPACE AND GROUP CONSIDERATIONS

Space Allowances

Optimum floor-space allowances for growing-finishing swine were suggested by Gehlbach *et al.* (1966) (Table 18.12). Space allowances suggested for pigs under other conditions are in Table 18.13 (Jensen 1972) and Table 18.14. The number of pigs per group affects performance similarly in solid, partially slotted, or completely slotted floors (Midwest Plan Service 1964). During warm seasons 8 and 12 pigs per group gained faster than 16 pigs per group; during cool seasons the differences disappeared. Groups of four, six, and eight finishing pigs gained similarly in enclosed buildings but in open-front buildings eight pigs per group gained more slowly than four- and six-pig groups. Jensen (1971) emphasized that factors such as ambient temperature, number of pigs per pen, method of feeding, ventilation, and floor design (solid vs. slotted) have an effect on optimum space allowances. Although, in general, pregnant gilts are not adversely affected by tethering or individual stall-housing in comfortable environments, increased incidence of infantile genital tract at 1 year of age has been reported in gilts closely confined beginning before puberty (Midwest Plan Service 1964: Jensen *et al.* 1970). Difficult estrus detection is also sometimes a problem in gilts kept in close confinement.

TABLE 18.12. Recommended Minimum Floor Space Allowances per Pig for Growing-Finishing Swine in Confinement

Weight of pig (kg)	*Slotted floor*		*Solid floor*	
	(sq m)	*(sq ft)*	*(sq m)*	*(sq ft)*
11.5–18.0	0.27	3	0.36	4
18.0–45.5	0.36	4	0.54	5¾
45.5–68.0	0.54	5¾	0.81	8⅜
68.0–95.5	0.72 (0.81)[a]	7¾ (8¾)	1.08	11½

Source: Gehlbach et al. 1966.

[a]An allotment of 0.81 sq m (8¾ sq ft) is preferable in high environmental temperature.

TABLE 18.13. Recommended Floor Space per Animal Unit and Number per Pen

Animal unit	*Body weight (kg)*	*Floor design and minimum space per animal*			*Number per pen*
		Solid (sq ft)	*Partially slotted (sq ft)*	*Slotted (sq ft)*	
Sow and litter[a]	—	35[b]	35	35	—
Weaned pig	4–11	4[c]	$2\frac{3}{4}$	$2\frac{3}{4}$	20–30
Growing pig	11–18	6	3	3	20–30
	18–45	8	4	4	20–30
Finishing pig	45–68	10	6	6	10–15
	68–95	12	8(9)	8(9)	10–15
Breeding females					
Gilt	113–136	15	12	12	12–15
Sow	136–227	18	15	15	10–12
Gestating animals[d]					
Gilt	—	17	14	14	12–15
Sow	—	18	15	15	12–15

Source: These values are based on research results and management observations of intensive swine rearing programs. It is assumed that excellent management is practiced (Jensen 1972).

[a]Commonly used farrowing stall area is 5 × 7 ft. The stall itself is 22–24-in. wide, $66\frac{1}{2}$–84-in. long.

[b]Floor should be sloped both to front and rear of the farrowing stall to facilitate drainage away from the pigs.

[c]In solid floor pens the floor should slope $\frac{1}{2}$ in. /ft. Concrete floor, whether solid or slats, should be steel troweled smooth when poured.

[d]Individual stalls should be about 22-in. wide and 6-ft long for gilts; 24-in. wide and 7-ft long for sows. Tethered animals would require essentially the same amount of space.

Number per Group

Small groups of growing pigs tolerate high environmental temperatures better than do larger groups (more than 12 pigs per group during warm weather may reduce performance), because of more efficient heat dissipation.

Cannibalism

This usually takes the form of tail-biting among groups of growing-finishing pigs. Level of performance may be decreased where cannibalism is prevalent probably because of infections introduced by the open wound. The cause of the practice is not known but it appears to be related to close confinement. Usually, the problem starts with one or a few offenders in a group of pigs, so that early removal of the offenders may solve the problem. Toys such as suspended chains or rubber tires in the pen area sometimes appear to alleviate the problem by encouraging the pigs to bite and play with these objects rather than with their neighbor's tail. However, the practice of cutting the tail to a short stump at birth has become common as a routine management practice in commercial herds.

TABLE 18.14. Space Requirements for Pigs under Various Systems of Management

Conditions of management	*Requirement by pig weight*	
	Growing pigs	
	Less than 45 kg	*More than 45 kg*
	Sq ft/animal	
Total confinement		
Solid floor	4	8
Slotted or partially slotted	4	8
Partial confinement		
Shelter space	4	6
Outside pen, paved	4	6
Total	8	12
Shade	4	6
	Pigs/acre	
Pasture, permanent	20–25	20–25
Pasture, legume	35–40	35–40
	Pigs/cup	
Watering space	25–25	20–25
	Pigs/linear ft	
Self-feeder space		
Complete diet	4	4
Grain	4	4
Protein supplement	15	15
	Sows	
	Less than 182 kg	*More than 182 kg*
Farrowing stall	22 in. × 7 ft	24–26 in. × 7–8 ft
	(plus creep areas, 18 in. × 7 ft)	
	Pen dimensions	
Conventional pen or movable house	7 × 8	7 × 9 to 8 × 9
Long pen	$4\frac{1}{2}$ to 5 × 12 or longer	$4\frac{1}{2}$ to 5 × 12 or longer
	Sq ft/sow	
Shade shelter	15–18	18–20
Drylot	50–100	50–100
Feeding space	1.5	1
	Sows/acre	
Pasture	15–18 (10–12 with litters)	15–18 (10–12 with litters)

Source: Midwest Plan Service 1964.

Mixing

Strange pigs almost always fight as a means of establishing a social order. The stress that results from mixing groups of pigs or from adding strange pigs to an established group is least with small pigs and progressively increases with larger pigs. Sedation and application of odor-disguising agents, for reduction of the stress that results from mixing groups of pigs, appear to be only sometimes effective. Groups formed should be as uniform as possible in size and age.

Crowding

The stress of crowding has been implicated in increased incidence of stomach ulcers and fatalities associated with animals genetically susceptible to the porcine stress syndrome (PSS) (Bicknell 1968). This syndrome (PSS) occurs in pigs from 20 kg to market weight and causes sudden death when they are moved from confinement housing and transported to a new environment or to slaughter (Bicknell 1968). Porcine stress syndrome is discussed in more detail in Chapter 19. Lindvall (1981) found that increasing the number of nursery pigs from 8 to 12 or 16 pigs per pen (pen size 9 × 2.4 m) decreased average daily gain by 8 and 25%, respectively during a 5-week period. Feed required per unit body weight gain was unaffected.

EFFECT OF OTHER ENVIRONMENTAL CONDITIONS

Humidity

Relative humidity is apparently not an important factor in pig performance except when above 85% at high temperature. Addis *et al.* (1967) reported no effect on growth or carcass composition with humidity constant at 85%, constant at 30%, or alternating between 85 and 30% at an environmental temperature of 29° C.

Light

As reviewed by Jensen (1971) variations in light intensity or exposure appear to have little effect on performance of growing-finishing pigs, except that a tendency for improved feed utilization has been observed when the period of darkness is extended. Ovulation rate is not affected by continuous exposure to light, but continuous darkness has been reported to hasten the onset of puberty (Dufour and Bernard 1968). Exposure to 16 hr of continuous light compared with 8 hr resulted in greater milk production and an additional pig weaned per litter in sows kept in a temperature-controlled farrowing facility (Mabry *et al.* 1982), without affecting the number of days from weaning to rebreeding. Such a favorable response should encourage additional research on the effects of light on sow productivity.

Noise

Performance is apparently not affected by sounds of high intensity or long duration (Bond *et al.* 1963), but sudden noises provoke short-term excitement and distress in pigs. This kind of provocative noise or sudden entrance of personnel into a confined housing unit should be avoided to minimize excitement and sudden attempts at escape that could result in injury to the pigs or damage to equipment.

Atmospheric Gases and Contaminants

As the practice of raising swine in confinement in closed buildings increases in the temperate zones, greater attention must be paid to ventilation to reduce the build-up of noxious gases, dust, and unpleasant odors. This is especially true in slotted-floor buildings where excreta is stored in a water suspension for periods of as long as several weeks before removal to fields or sewage lagoons. Muehling (1970) described the problems with manure gases in terms of their potential danger to both pigs and man. The major gases of concern are carbon dioxide, ammonia, hydrogen sulfide, and methane. Only under special conditions of poor ventilation, such as in power failure, do concentrations usually reach harmful levels. The properties of these gases and their physiological effects have been summarized by A.F.M.A. Nutrition Council (1970) and this information is reproduced in Table 18.15.

Curtis (1982) summarized the role of carbon monoxide poisoning as a cause of abortion in sows 12–36 hr after entering farrowing crates as follows: Use of fuel fired radiant heaters in barns with inadequate ventilation may result in aerial carbon monoxide concentration of 150 ppm or higher and cause abortions; carboxyhemoglobin in sow blood, which normally is less than 2%, may rise to 15% in sows that abort.

Drummond *et al.* (1981) showed that exposure of pigs to high levels of atmospheric ammonia resulted in inflammation of the turbinate epithelium in the nasal passages of pigs with atrophic rhinitis.

Muehling (1970) summarized methods for removing odors from confinement swine buildings. Ammonia and hydrogen sulfide are water soluble and can be controlled by dilution and solution in water. Activated charcoal can be placed at exhaust fans to adsorb odors from the above gases, but this procedure is expensive, owing to the large amount of charcoal required. Masking of unpleasant odors by superimposing pleasant odors is possible, but it is expensive, as is counter-action by combining two odors to produce an odorless mixture. This counter-action method may be effective but it depends on a trial and error approach. Burning of excreta is possible, but the high water content of swine excreta, especially if collected from under a slotted-floor confinement building where water is added, makes this practice infeasible. A temperature of 815° C is required to burn swine excreta, and this is too expensive under present conditions.

TABLE 18.15. Properties of Noxious Gases and Their Physiological Effects

Gas	*Specific gravity*[a]	*Odor*	*Minimum identifiable odor (ppm)*[b]	*Maximum allowable concentration (ppm)*[c]	*Concentration (ppm)*[d]	*Physiological effects*[e]
Ammonia (NH_3)	0.6	Sharp, pungent	5.3	100	4	Irritant
					400	Irritation of throat
					700	Irritation of eyes
					1,700	Coughing and frothing
					3,000	Asphyxiating
					5,000	Could be fatal
Carbon dioxide (CO_2)	1.5	None	—	5,500	—	Asphyxiant
					20,000	Safe
					30,000	Increased breathing
					40,000	Drowsiness, headaches
					60,000	Heavy, asphyxiating breathing
					300,000	Could be fatal
Hydrogen sulfide (H_2S)	1.2	Rotten egg smell, nauseating	0.7	20	—	Poison
					100	Irritation of eyes and nose
					200	Headaches, dizziness
					500	Nausea, excitement, insomnia
					1,000	Unconsciousness, death
Methane (CH_4)	0.5	None	—	—		Asphyxiant
					500,000	Headache, nontoxic

Source: From Ohio State University Research Foundation Bulletin entitled Origin, Identification, Concentration and Control of Noxious Gases in Animal Confinement Production Units, by E. Taiganides and R. K. White, Grant No. SW-00015; cited by A.F.M.A. Nutrition Council 1970.

[a]Specific gravity: the ratio of the weight of pure gas to standard atmospheric air. If number is less than 1 the gas is lighter than air; if greater than 1 it is heavier than air.

[b]Minimum identifiable odor: the threshold odor; that is, the lowest concentration (highest dilution) from which an odor is detected.

[c]Maximum allowable concentration: the concentration set by health agencies as the maximum in an atmosphere where men work more than an 8–10-hr period. These levels must be lower in confinement units because animals stay in such environment continuously for 24 hr.

[d]Concentration, in parts of the pure gas in millions of atmospheric air; to change concentration to % by volume, divide the listed numbers by 10,000.

[e]Physiological effects: those found to occur in adult humans; similar effects would be felt by animals weighing 68 kg; lighter animals will be affected sooner and at lower levels; heavier animals at later times and higher concentrations.

Ventilation

Any closed confinement housing system for swine requires a good forced ventilation system to minimize accumulation of noxious gases and unpleasant odors as described above. The A.F.M.A. Nutrition Council (1970) has described the important features of a good ventilation system. Any ventilation system works most efficiently in insulated buildings. There are two broad types of ventilation: (1) pressure ventilation in which fresh air is forced in and stagnant air is allowed to flow out of the building through vents and (2) exhaust ventilation in which stagnant air is forced out by the building by fans and fresh air is drawn into the building through appropriately spaced openings. The adequacy of the ventilation is affected by the amount of insulation, the number, size, and placement of fans and vents, and the adequacy of the heating source. The population density in the building and the outside environment dictate the amount of ventilation needed.

Although it is important to prevent drafts, some air movement is desirable at all times. This can be accomplished by having a fan continuously operating and additional fans thermostatically controlled. If ventilation fans fail to distribute uniform humidity and temperature in all corners or other areas of the building, circulation fans should be used to accomplish it. There should be 900 cm^2 of vent or inlet area per 500 cfm fan capacity. Inlets should be installed that allow incoming fresh air to be tempered during cold weather. Inlets and vents should be adjustable to allow varying of the air flow. Air inlets and vents may be placed at opposite ends of the building to provide for good circulation. Since warm air rises and carries more moisture than cold air, it is advisable to place the inlets near the ceiling during winter to allow fresh air to mix with warm air near the ceiling. In winter stagnant air should be exhausted near the floor to conserve heat. Provision for exhausting stagnant air from either floor or ceiling is advantageous in that it allows for conservation of heat depending on weather conditions.

Recommended ventilation rates (cubic feet per minute) for various environmental temperatures and for growing pigs and sows, boars, and sows and litters are summarized in Table 18.16. Table 18.17 gives recommended rates for 58-kg

TABLE 18.16. Total Per-Head Ventilation Rates for Confinement Swine Building during Various Times of the Year (cfm)

	Cold weather	*Mild weather*	*Hot weather*
Sow and litter	20	80	500
Pre-nursery pig (12–30 lb)	2	10	25
Nursery pig (30–75 lb)	3	15	35
Growing pig (75–150 lb)	7	24	75
Finishing hog (150–220 lb)	10	35	120
Gestation sow (325 lb)	12	40	150[a]
Boar (400 lb)	14	50	180[a]

Source: Jones et al. 1980.
[a]Use 300 per sow or boar in breeding facilities due to low animal density.

TABLE 18.17. Recommended Ventilation Rates for Farrowing Houses and Growing-Finishing Houses at Various Environmental Temperatures

Outside temperatures		*Ventilation rate (cfm per pig)*	
°C	*°F*	*Sow and litter*	*Finishing pigs (58 kg body wt)*
—29	—20	12	—
—23	—10	13	7
—18	0	14	7.5
—12	10	16	8
—7	20	18	9
—1	30	22	12
4	40	36	20
10	50	50	25
10–21	50–70	50	30
Above 21	Above 70	100	75

Source: Iowa State University 1965A,B.

growing pigs and for sows with litters with outside temperatures ranging up to summer conditions (Iowa State University 1965A,B).

The amount of auxiliary heat needed to supplement that produced by the animals occupying a closed swine building will depend on the population density and the size of animals along with physical characteristics of the building (insulation, ventilation openings, open areas for drafts, and so on). The rates of heat production by animals of various sizes are shown in Table 18.18 (Mitchell and Kelley 1938).

SOLID FLOORS AND SLOTTED FLOORS

The move to confinement systems of swine production has brought with it the problem of manure removal. Until very recently this was accomplished by means of concrete-floor pen arrangements, which allowed frequent scraping or hosing of

TABLE 18.18. Heat Production of Swine of Various Sizes

Weight (kg)	*Heat production (BTU/hour)*
1	33
5	200
23	375
45	520
68	675
91	800
136 (slaughter pig)	870
136 (pregnant gilt)	1,025

Source: Mitchell and Kelley 1938.

FIG. 18.1. Sewage lagoon.

the manure to a drain which carried it to a manure lagoon or to an area at which it could be loaded into vehicles for transport to be spread on the land as fertilizer. The amount of labor required to clean and remove excreta in this way encouraged the development of slotted-floor housing systems in which the excreta drop through slots in the floor to a holding tank from which the mixture of urine and feces can be removed periodically for spreading on the land. Problems with odor control in slotted-floor confinement houses where under-floor storage is used have led to the development of agitation systems whereby a wheel or other mechanical agitator keeps the liquid aerated by constant movement. This encourages aerobic microbial populations rather than anaerobic ones, and thereby markedly reduces the offensive odor problem. A confinement feeding building from which manure is drained to a sewage lagoon is shown in Figure 18.1. Lagoons such as this can be used for long-term storage and decomposition of waste. In this system, all fertilizer is lost and odor from the lagoon may become a problem if a proper ratio of excreta volume to surface area of the lagoon and total volume of the lagoon is not attained. The lagoon can be a part of either a solid concrete-floor or a slotted-floor operation. A modern enclosed swine building with completely slotted floors under the growing-finishing pigs is shown in Figure 18.2. Besides the large agitation wheel that aerates the liquid manure under the floor, another method for the control of odor is a flushing system that moves manure to an underground storage tank or to a lagoon for decomposition or for later land-spreading as a fertilizer. Flush tanks designed to empty several times daily into shallow concrete gutters under slotted-floored pens have found favor in recent years as a means of removing wastes with minimum human labor, and with reduced odor levels in confined swine units.

Many types of floor material have been used in slotted-floor buildings and different ratios of slat-to-slot area have been compared. Wooden slats can be used, but they have a short life-span owing to wear and breakage (hardwood should be used) and they are difficult to clean. Metal slats, including expanded metal and other variations have been used, but tend to cause feet and leg soreness

FIG. 18.2. Confinement-feeding building with slotted floors for growing-finishing pigs. Note outside feed storage bins from which feed is dispensed automatically by auger.
Courtesy USDA.

and they may corrode and rust (Fig. 18.3). Aluminum or stainless steel slats appear to have a long life span and minimize problems with feet and legs. Of materials used thus far, concrete slats appear to be most satisfactory in terms of durability.

The degree of traction, roughness, and abrasiveness of slatted floors appears to be more important than slat width in influencing hoof and foot lesions in growing-finishing swine (Newton *et al.* 1980). Comparisons of porcelainized steel, plastic, concrete, and aluminum slats (Table 18.19) suggest only minor effects of slat type

FIG. 18.3. Metal slats for growing pigs. Concrete slats are in greater favor for durability and comfort.
Courtesy USDA.

TABLE 18.19. Mean Performance of Pigs Housed on Four Types of Slotted Floors[a]

Item	*Floor*			
	Steel	*Plastic*	*Concrete*	*Aluminum*
Initial wt, kg	23.2	23.3	23.2	23.2
Daily gain, kg	0.78[b]	0.72[c]	0.75[bc]	0.76[bc]
Daily feed, kg	2.59[b]	2.46[c]	2.56[bc]	2.65[b]
Feed-to-gain ratio	3.32[b]	3.41[bc]	3.43[bc]	3.50[c]

[a]Two pens of five pigs each per floor type during each of four trials averaging 100 days duration.
[bc]Means on the same line with different superscripts are different, $P < .05$.

of swine performance. The lower daily weight gain of swine on plastic slats in the experiment cited (Newton *et al.* 1980) was suggested as due to lower feed consumption associated with insecure footing and reduced activity. Plastic and concrete slats produced more hoof lesions than steel or aluminum slats. Some hoof and leg lesions sustained early in the growing period may negatively affect subsequent performance.

Recommended widths of slats and slots for pigs of various sizes are given in Table 18.20. If the recommendations for number of pigs per pen and population density are followed (see Tables 18.11, 18.12, and 18.13), there is little or no manure accumulation on the floor and pigs of the various given sizes remain clean when the corresponding ratios of slot to slat width are used. Slats should run parallel with the long axis of the pen or stall; pigs are more sure-footed with this configuration and fewer foot and leg injuries can be expected.

In swine buildings in which floors are solid, some form of bedding is usually needed to absorb moisture. All pen floors should be sloped at least 3.5–4.0 cm/m to allow adequate drainage. Storage space for bedding must be provided. Table 18.21 shows approximate storage space required for various kinds of bedding materials as well as for feed supplies (USDA, 1958). Bedding materials vary considerably in absorptive capacity; whole oat straw is perhaps the most absorptive of common materials (250 kg of water absorbed per 100 kg of straw), while pine shavings, mixed shavings, and corn cobs are about 75, 50, and 50% as effective, respectively (Curtis 1981). Under tropical conditions, solid concrete floors with open construction as illustrated in Fig. 18.4 provide, with the addition

TABLE 18.20. Recommended Widths of Slats and Slots on Floors for Pigs of Various Ages

Pig weight (kg)	*Slat width (in.)*	*Slot width (in.)*
11–22.7	3	0.75
22.7–45	4–6	1
45–68	4–6	1
68–91	4–6	1

Source: A.F.M.A. Nutrition Council 1970.

FIG. 18.4. Concrete floors with open construction in the Tropics.

of bedding materials, adequate housing, except for farrowing pens. For pigs kept on pasture in the tropics, simple sun shades made of bamboo poles and a palm-thatched roof provide adequate environmental protection (Fig. 18.5).

The chemical composition and amounts of manure produced by swine are summarized in Table 18.22 (University of Illinois 1953; University of Kentucky 1964; Sutton *et al.* 1980). Dollar value for fertilizer based on current prices of nitrogen, phosphorus, and potassium can be derived for use in making decisions about disposal systems. The use of anaerobic digesters to produce methane for energy (Table 18.23) is receiving considerable attention in the United States (Sweeten *et al.* 1981) and has been a common practice in Taiwan for many years.

Estimated production of methane and fertilizer elements from swine manure is shown in Table 18.23. A species of blue–green algae studied by Chung *et al.* (1978) grows rapidly on the effluent from fermented swine wastes. It is used

FIG. 18.5. Pigs on pasture with simple shades in the Tropics.

TABLE 18.21. Feed and Bedding Storage Space (Cubic Feet per Ton)

Feed	*Storage space (cubic ft/ton)*
Shelled maize	45
Ear maize	72
Oats	70–80
Barley, grain sorghum, wheat	40–55
Bagged concentrate	44–50
Straw, loose	500–550
Straw, baled	250–300
Wood shavings	100

Source: USDA 1958.

TABLE 18.22. Production and Composition of Swine Manure

Components	*Composition (%)*	*Amount produced per year[a] per kg liveweight (kg)*
Total fresh wt	—	32
Water	75	24
Nitrogen	0.5	.16
Phosphorus	0.14	.05
Potassium	0.38	.12

Source: University of Kentucky 1964.
[a]Includes urine and feces combined.

TABLE 18.23. Estimated Production of Methane and Fertilizer Elements from Swine Manure

Size operation[a]	*Gross yield, ft³/day[b]*		*Net methane yield, ft³/day[c]*	*Fertilizer elements lb/day[d]*		
	Biogas	*Methane*		*N*	*P_2O_5*	*K_2O*
50	240	144	100	4	3	3
100	480	290	200	7	5	5
200	960	580	400	14	10	11
500	2,400	1,440	1,000	35	25	25
1,000	4,800	2,900	2,000	70	50	50
2,000	9,600	5,800	4,000	140	100	100
5,000	24,000	14,400	10,000	350	250	250
10,000	48,000	29,000	20,000	700	500	500
15,000	96,000	43,000	30,000	1,100	750	750

Source: Sweeten et al. 1981.
[a]150-lb hog or equivalent liveweight, one-time capacity.
[b]Assumptions: Volatile solids destruction = 50%. Gross methane yield is 8.0 $ft^3 CH_4$/lb volatile solids destroyed. Biogas contains 60% methane and 40% carbon dioxide (volume basis).
[c]Energy requirement for heating and mixing the digester is 30% of gross biogas output.
[d]Recoverable fertilizer amounts are 0.07 lb N, 0.05 lb P_2O_5, and 0.05 lb K_2O per head per day, not including losses during slurry storage.

experimentally on small farms in Taiwan in conjunction with fish production. Swine wastes are collected in closed tanks to produce methane gas which is piped to the household for cooking and heating; the effluent from the methane generation is pumped into an open reservoir where algae use the inorganic nutrients (N,P,K) and sunlight for growth. Algae can be harvested by filtration or surface skimming and fed to pigs or can provide food for fish and other marine life to produce another salable product.

REFERENCES

ADDIS, P. B., JOHNSON, H. R., HEIDENREICH, C. J., JONES, H. W., and JUDGE, M.D. 1967. J. Anim. Sci. *26,* 705.

A.F.M.A. NUTRITION COUNCIL. 1970. Management Recommendations for Growing-finishing Swine in Environmental Controlled Confinement. A.F.M.A., Livestock and Poultry Management Committee. Chicago, Illinois.

AMES, D. R., and RAY, D. E. 1983. J. Anim. Sci. *57*(Suppl. 2), 209.

ANDREWS, F. N., FONTAINE, W. E., CULVER, A. A., NOFFSINGER, T. L., and GARWOOD, V. A. 1960. J. Anim. Sci. *19,* 429.

BABATUNDE, G. M., FETUGA, B. L., and OYENUGA, V. A. 1974. Niger. J. Anim. Prod. 1:169.

BICKNELL, E. J. 1968. *In* The Pork Industry: Problems and Progress (D. G. Topel, Editor). Iowa State Univ. Press, Ames.

BOND, J., WINCHESTER, C. F., CAMPBELL, L. E., and WEBB, J. C. 1963. USDA Technical Bull. 1280.

BOND, T. E., KELLY, C. F., and HEITMAN, H., JR. 1967. Anim. Prod. *9,* 453.

CHRISTENSON, R. C. 1981. J. Anim. Sci. *52,* 821.

CHUNG, P., POND, W. G., WALKER, E. F., KINGSBURY, J. F., and KROOK, L. 1978. J. Anim. Sci. *47,* 319.

CULVER, A. A., ANDREWS, F. N., CONRAD, J. H., and NOFFSINGER, T. L. 1960. J. Anim. Sci. *19,* 421.

CURTIS, S. E. 1970. J. Anim. Sci. *31,* 576.

CURTIS, S. E. 1981. Environmental Management. Animal Environment Services, Box 507, Mahomet, Illinois.

CURTIS, S. E. 1982. Natl. Hog Farmer, January, p. 29.

DICKMAN, M. A. 1982. Influence of supplemental lighting on reproduction in swine. *In* Proc. Symp. Manage. of Food Producing Anim. (W. R. Woods, Editor). Purdue Univ., West Lafayette, Indiana.

DRUMMOND, J. G., CURTIS, S. E., MEYER, R. C., SIMON, J., and NORTON, H. W. 1981. Am. J. Vet. Res. *42,* 963.

DUFOUR, J., and BERNARD, C. 1968. Can. J. Anim. Sci. *48,* 425.

FOLEY, C. W., SEERLEY, R. W., HANSEN, W. J., and CURTIS, S. E. 1971. J. Anim. Sci. *32,* 926.

GARRETT, W. N., BOND, T. E., and KELLY, C. F. 1960. J. Anim. Sci. *19,* 921.

GEHLBACH, G. D., BECKER, D. E., COX, J. L., HARMON, B. G., and JENSEN, A. H. 1966. J. Anim. Sci. *25,* 386.

HALE, O. M., GIVENS, R. L., JOHNSON, J. C., JR., and SOUTHWELL, B. L. 1966. J. Anim. Sci. *25,* 756.

HALE, O. M., JOHNSON, J. C., JR., and WARREN, E. P. 1968. J. Anim. Sci. *27,* 1577.

HEITMAN, H., JR., and HUGHES, E. H. 1949. J. Anim. Sci. *8,* 171.

HEITMAN, H., JR., HUGHES, E. H., and KELLY, C. F. 1951. J. Anim. Sci. *10,* 907.

HEITMAN, H., JR., KELLY, C. F., and BOND, T. E. 1958. J. Anim. Sci. *17,* 62.

HEITMAN, H., JR., BOND, T. E., KELLY, C. F., and HAHN, L. 1959. J. Anim. Sci. *18,* 367.

HOLME, D. W., and COEY, W. E. 1967. Anim. Prod. *9,* 209.

HOLMES, C. W., and CLOSE, W. H. 1977. *In* Nutrition and the Climatic Environment (W. Haresign, H. Swan, and D. Lewis, Editors). Butterworths, London.

IOWA STATE UNIV. 1965A. AE-993 p. 4. Iowa State Univ., Ames.

IOWA STATE UNIV. 1965B. AE-992 p. 3. Iowa State Univ., Ames.

JENSEN, A. H. 1971. J. Anim. Sci. *32,* 560.

JENSEN, A. H. 1972. Concepts and trends in modern swine housing, facilities and waste disposal. Proc. 5th Biennial Swine Producers' Short Course. Cornell Univ., Ithaca, New York.

JENSEN, A. H., KUHLMAN, D. E., BECKER, D. E., and HARMON, B. G. 1969. J. Anim. Sci. *29,* 451.

JENSEN, A. H., YEN, J. T., GEHRING, M. M., BAKER, D. H., BECKER, D. E., and HARMON, B. G. 1970. J. Anim. Sci. *31,* 745.

JOHNSON, D., and HALL, M. 1981. Maintenance and operation of ventilation fans for hog barns. Natl. Pork Industry Handbook Fact Sheet *PIH-41.* Cooperative Extension Service, Purdue Univ., West Lafayette, Indiana.

JONES, D. D., DRIGGERS, L. B., FEHR, R. L., and STEWART, B. R. 1980. Natl. Pork Industry Handbook Fact Sheet *PIH-87.* Cooperative Extension Service, Purdue Univ., West Lafayette, Indiana.

KATTESH, H. G., KORNEGAY, E. T., KNIGHT, J. W., GWAGDAUSKAS, F. N., THOMAS, H. R., and NOTTER, D. R. 1980. J. Anim. Sci. *50,* 897.

LINDVALL, R. N. 1981. J. Anim. Sci. *53,* 863.

MABRY, J. W., CUNNINGHAM, F. L., KRAELING, R. R., and RAMPACEK, G. B. 1982. J. Anim. Sci. *54,* 918.

MARPLE, D. N., ABERLE, E. D., FORREST, J. C., BLAKE, W. H., and JUDGE, M. O. 1972. J. Anim. Sci. *34,* 809.

MIDWEST PLAN SERVICE. 1964. Swine Equipment Plans and Housing Needs, Publ. No. 8. Iowa State Univ., Ames.

MITCHELL, H. H., and KELLEY, M. A. R. 1938. J. Agric. Res. *56,* 811.

MOUNT, L. E. 1963A. Nature *199,* 122.

MOUNT, L. E. 1963B. J. Physiol. *168,* 698.

MOUNT, L. E. 1968. The Climatic Physiology of the Pig. Williams and Wilkins, Baltimore.

MUEHLING, A. J. 1970. J. Anim. Sci. *30,* 526.

NATIONAL RESEARCH COUNCIL. 1981. Effect of Environment on Nutrient Requirements of Domestic Animals. Natl. Acad. Sci., Washington, D.C.

NEWTON, G. L., BOORAM, C. V., HALE, O. M., and MULLINIX, B. G., JR. 1980. J. Anim. Sci. *50,* 7.

OMTVEDT, I. T., NELSON, R. E., EDWARDS, R. L., STEPHENS, D. F., and TURMAN, E. J. 1971. J. Anim. Sci. *32,* 312.

PHILLIPS, R. E., and BICKERT, W. G. 1982. Mechanical ventilation of swine buildings. Natl. Pork Industry Handbook Fact Sheet *PIH-60.* Cooperative Extension Service, Purdue Univ., West Lafayette, Indiana.

SEYMOUR, E. W., SPEER, V. C., HAYS, V. W., MANGOLD, D. W., and HAZEN, T. E. 1964. J. Anim. Sci. *23,* 375.

SORENSEN, P. H. 1962. *In* Nutrition of Pigs and Poultry, (J. T. Morgan and D. Lewis, Editors). Proc. Univ. Nottingham 8th Easter School Agric. Sci., London, England.

STEFANOVIC, M. P., BAYLEY, H. S., and SLINGER, S. J. 1970. J. Anim. Sci. *30,* 378.

STEPHENS, D. B. 1971. Anim. Prod. *13,* 303.

SUGAHARA, M., BAKER, D. H., HARMON, B. G., and JENSEN, A. H. 1970. J. Anim. Sci. *31,* 59.

SUTTON, A. L., VANDERHOLM, D. H., and MELVIN, S. W. 1980. Natl. Pork Industry Handbook *PIH-25,* Purdue Univ., W. Lafayette, Indiana.

SWEETEN, J. M., FULHAGE, C., and HUMENIK, F. J. 1981. Natl. Pork Industry Handbook *PIH-76,* Purdue Univ., W. Lafayette, Indiana.

SWIERSTRA, E. E. 1970. Biol. Reprod. *2,* 23.

TEAGUE, H. S., ROLLER, W. L., and GRIFO, A. P. 1968. J. Anim. Sci. *27,* 408.

THORBECK, G. 1975. Studies on energy metabolism of growing pigs. No. 343 Beretning fra Statens Husdyrbrugs Forsog, Natl. Agric. Univ., Copenhagen, Denmark.

TOMPKINS, E. C., HEIDENREICH, C. J., and STOB, M. 1967. J. Anim. Sci. *26,* 377.

TIDWELL, A. L., and FLETCHER, J. L. 1951. J. Anim. Sci. *10,* 523.

UNIV. OF ILLINOIS. 1953. Univ. Ill. Circ. *595.*

UNIV. OF KENTUCKY. 1964. Univ. Ky. Circ. *593.*

USDA 1958. Misc. Publ., *744,* U.S. Gov. Printing Office, Washington, D.C.

WARNICK, A. C., WALLACE, H. D., PALMER, A. Z., SOSA, E., DUERRE, D. J., and CALDWELL, V. E. 1965. J. Anim. Sci. *24,* 89.

WOODS, W. R. (Editor). 1982. Proc. Symp. Manage. of Food Producing Anim. Purdue Univ., West Lafayette, Indiana.

19

Swine Management and Marketing

INTRODUCTION

The management of swine for profitable production is perhaps as varied as are the environments in which swine are raised, and yet there are many factors over which the successful swine manager has control and which must be considered if the enterprise is to be profitable. These factors include the applications of economic, agronomic, biologic, and engineering principles to management practices. Specific management practices related to production and marketing are described here. Although local conditions will have a bearing on specific decisions, the principles that apply to temperate environments also apply to tropical environments and therefore the same general guidelines can be used for all environmental conditions. A book by Curtis (1981) and a report by the National Research Council (1981) deal exhaustively with environmental management and address the principles of environmental effects on the biological responses of animals under a broad array of environmental conditions. Most of the data and examples presented here were obtained under conditions in the United States, but inferences can be drawn and applied to all areas of the world.

MANAGEMENT PRACTICES

Suckling Pigs

The baby pig begins to explore his environment within a few minutes after birth and soon finds his way to a nipple and begins to nurse. Because of his curiosity, it is essential that the pen environment be clean to minimize chances of exposure to disease and parasites. In addition, the following tasks should be performed.

1. The navel (umbilical cord) should be tied off to prevent loss of blood, and it should be cut so that 3–5 cm remain and this portion dipped into a solution of iodine. This section will soon dry up (Fig. 19.1a) and drop off, leaving a clean, noninfected navel. The umbilical cord is a crucial area for the entrance of infection. A common malady, "navel ill," which can cause lameness or death, may result from failure to disinfect this cord at birth.

2. The "needle" teeth should be clipped. Baby pigs have four pairs of sharp teeth (see dental pattern, Fig. 6.4), two on each jaw called needle or "wolf" teeth. They are of no practical value to the pigs themselves and they may irritate the sow's udder during nursing or cause injury to other pigs when they fight or play among themselves. Therefore, it is advisable to clip these teeth shortly after birth. Side-cutting pliers are suitable for this purpose (Fig. 19.1b). It is important to avoid loosening the base of the tooth or leaving jagged edges or causing injury to the gums.

3. Ear-notching should be done during the first day to establish identity. This is a simple and effective method of identifying individual pigs and litters and is essential for good record keeping. Tools especially designed for the purpose should be used and are available commercially (Fig. 19.1c). Any systematic notching scheme is suitable, but the one recommended by most swine breed associations in the United States has been found effective and is shown in Fig. 19.2.

4. Iron should be administered to prevent anemia. The newborn pig has only a limited reserve of iron for hemoglobin synthesis owing to inefficient placental transfer of iron to the fetus. The newborn pig contains only about 50 mg of iron, stored mainly in the liver, and the daily need is about 5 or 10 mg. Since sow milk is also very low in iron, the suckling pig must have supplemental iron during the first few days to prevent anemia. Pigs depend mainly on milk for nutrients until they begin to nibble on sow feed or creep feed at 2–3 weeks of age. Thus, the amount of supplemental iron should be 150–200 mg to meet the needs until iron-containing dry feed is eaten in significant amounts at about 3 weeks. In the past, this iron need was normally met by the pig rooting in the soil and otherwise picking up iron from the environment. When pigs are kept indoors in concrete-floor pens this source of iron is eliminated. Anemia can be prevented and cured by supplying iron either orally or by injection. Oral administration consists of spraying or swabbing the sow's udder with a solution of ferrous sulfate (1 lb ferrous sulfate in 1 gal. of water) so that some iron will be ingested when the pigs nurse. To be effective this solution must be applied daily from birth until the pigs begin to eat creep feed. Commercial sources of iron in the form of pills or pastes, to be administered to the pigs individually, are also available. Also, iron-rich sweetened feeds or pellets may be provided on the pen floor. A more dependable method is the intramuscular injection of iron–dextran compounds (Fig. 19.1d). These compounds are available commercially and a single injection given at a few days of age provides protection until dry feed is consumed. A concentration of 100–150 mg of iron/cc is customary, so that a 1- or 2-cc injection is adequate. Iron toxicity can be produced by exceeding the recommended dosage level, but there is about a fivefold safety factor.

5. Cutting the tails is advisable whenever pigs are to be raised in total confinement. This is conveniently done at the time the needle teeth are clipped. The

a

b

c

d

e

FIG. 19.2. Suggested ear-notching system for baby pigs. The right ear is used for the little mark and all pigs in the same litter must have the same ear notches in this ear. The right ear is on the pig's own right. The left ear is used for notches to show the individual pig number in the litter. Each pig will have different notches in this ear. One can mark up to 181 litters with this system. Litter number and notches for that number are shown in each square.
From Yorkshire Journal, August 1971.

FIG. 19.1. (a) Litter of newborn pigs just before ear notching, needle teeth clipping, tail cutting, and iron injection. Note navel cords before sloughing. (b) Clipping the needle teeth to prevent injury to other pigs and to udder of the dam. (c) Ear-notching for permanent individual identification. Note table to right which provides convenient work area for recording data and for holding needle teeth clippers, syringes, and bottles containing iodine solutions and injectable iron, and a pan with disinfectant. (d) Injecting iron–dextran intramuscularly to prevent anemia. (e) Cutting the tail.
Photos courtesy of USDA.

tail should be cut to as short a stub as possible, with the use of side cutting pliers or other blunt pinchers to crush the bone and reduce bleeding (Fig. 19.1e).

6. Creep feed should be supplied during the second week to encourage dry-feed consumption. Baby pigs begin to develop an appetite for dry feed at about this time as their feed requirement begins to exceed the amount supplied by the sow milk. Palatability is an important factor in creep diets and for this reason sweeteners, such as sugar, and components that improve texture, such as lard, are common constituents. Examples of suitable creep feeds are given in Appendix 3.

The creep area may be partitioned from the other feeders so that only the young pigs have access to it, or it may be a separate feeder built or positioned so that only the young pigs can eat from it.

7. Castration of male pigs can be done at any age but a good time is at 1–3 weeks of age when the problem of restraining the animal is minimum and the stress on the animal is not great.

Castration is a simple operation and, provided sanitary precautions are taken, infection or other complications are uncommon. A *clean* scalpel or sharp knife is the only equipment needed in addition to a disinfectant. Either a single incision or two incisions (one over each testis) produces satisfactory results. The incision should be made low enough on the scrotum to allow good drainage from the wound. Figure 19.3 shows the steps in castration of young pigs. It is essential that hands be thoroughly washed and utensils sterilized before use to avoid infection. To reduce bleeding an emasculator can be used to crush and break the cord and tissues as in Fig. 19.3e. Normal castration is the removal of the testis and epididymis on both sides. Cryptorchids cannot be castrated without surgery into the body cavity. To avoid loss of identity of pigs with only one testis descended, it is preferable not to remove the descended one until both can be removed.

8. The final task to be performed on the suckling pig is weaning. With the availability of highly fortified diets for young pigs, it is possible to remove them from the sow at birth and feed sow milk substitutes. However, although this is energetically the most efficient method of producing pork, the technology of devising diets and sanitation practices that can compete with sow milk and the nursing intervals and similar factors that the innate maternal–offspring relationship provides has not yet been developed. There are now experimental as well as commercially available baby pig brooder units that utilize this concept, but the industry as a whole is not ready for it yet, although certain producers with superior managerial ability and technical knowledge may find it a profitable approach.

It is customary to wean pigs at 4 to 6 weeks of age, although some producers are routinely weaning pigs at an earlier age. Some advantages of early weaning are these: sows require less feed than during lactation; sows can be rebred earlier if desired; there is a possibility of heavier, more uniform pigs at 6 weeks of age. However, it requires superior management (the sow cannot cover up poor management with a constant balanced and sterile milk supply) and the highly fortified diets required to replace sow milk are expensive.

If pigs are weaned at 5 weeks or younger they should weigh a minimum of 9.5 kg at 5 weeks, 6.8 kg at 4 weeks, 5.4 kg at 3 weeks, 4.1 kg at 2 weeks, and 2.4 kg at 1 week of age.

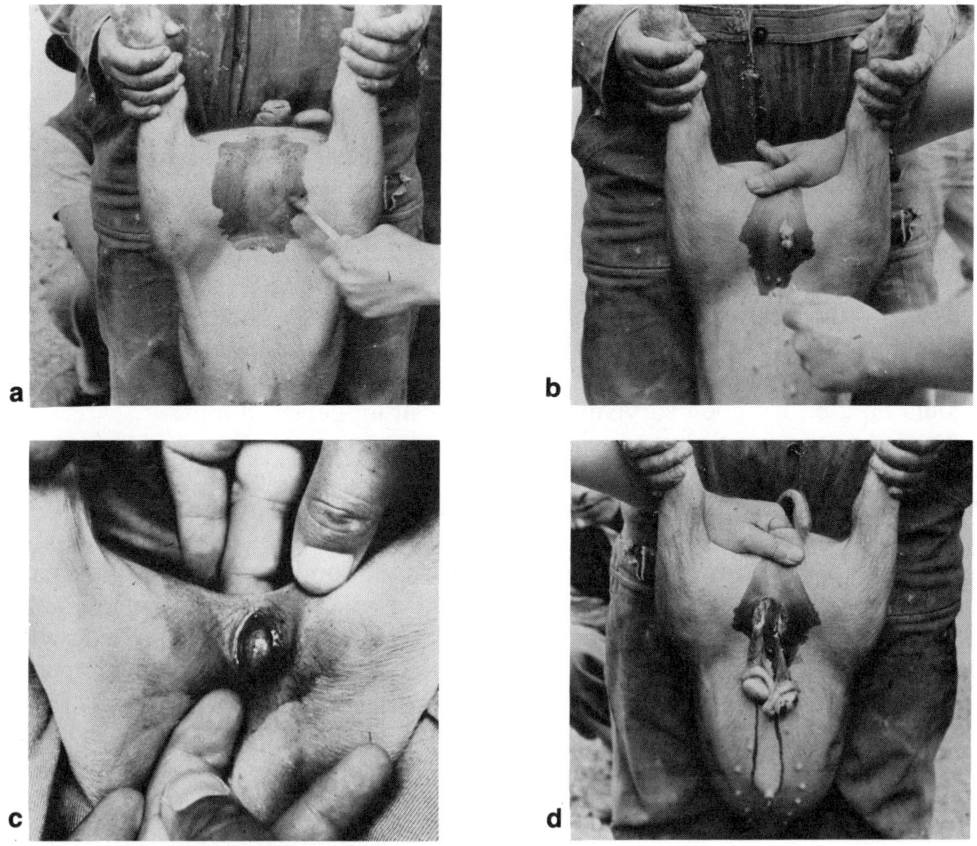

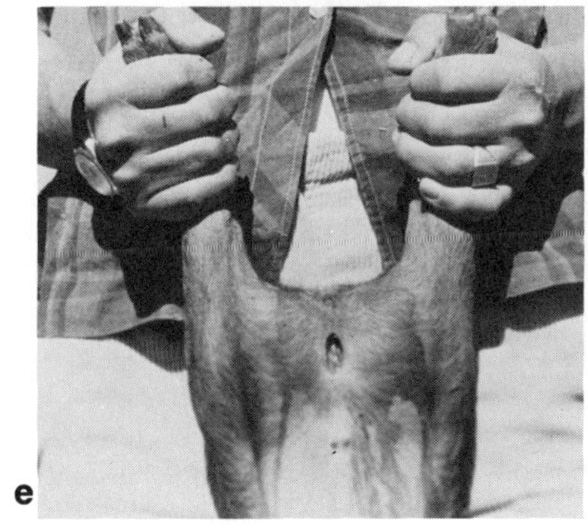

FIG. 19.3. Castration series: (a) Pig held in position for incision; area being swabbed with disinfectant. (b) Incision made. (c) Testes being squeezed out through incision. (d) Testes (with epididymis) externalized, still attached to spermatic cord. (e) Spermatic cord crushed with emasculator to minimize bleeding and trauma; incision disinfected and castrated pig ready to be returned to its littermates.

The stress of weaning can be minimized by removing the sow and leaving the pigs in the farrowing pen for the early part of the growing period. However, most facilities call for moving the pigs to a nursery or growing unit and this transition can be made with minimum setback if pen size, feeder, and waterer space and environmental temperature are adequate. Recommendations concerning these factors for swine of all ages are given in Chapter 18.

Orphan Pigs

If a sow dies after farrowing, or if udder trouble, lactation failure, or litters larger than the sow is able to raise occur, the result will be "orphan pigs." Two possibilities exist for the successful raising of such pigs: adoption by a foster sow, or the use of either cow milk or a sow milk replacer. In either solution, it is important that the newborn pig receive colostrum as the first feed. The phenomenon of "gut closure" (see Chapter 11), in which the absorption of large molecules such as the gamma globulins of colostrum ceases following ingestion of feed, dictates that colostrum be the first food to ensure immune globulin transfer to the blood. Cow colostrum apparently provides some protection and, though not as satisfactory as sow colostrum, is greatly superior to no colostrum at all. Injectable gamma globulin preparations are also available commercially and should be used for good survival rates in colostrum-deprived pigs.

If another sow has farrowed within a short time interval before or after birth of the orphaned litter, the pigs may be transferred to her. This transfer must be made within a few days after farrowing because those glands that are not used soon cease their milk production. To ensure acceptance of new pigs, the sow should be separated from her own litter while the new pigs are combined, and a disinfectant or other odorous material should be sprinkled on all the pigs to disguise odors.

Pigs raised on cow or synthetic milk have a strong suckling instinct and will suckle the ears and navels of other pigs when kept together, causing commotion and unrest that drains their strength. For this reason, they should be housed individually during the first 2 weeks. Feeding should be as frequent as feasible to stimulate normal nursing pattern (see Chapter 9). Automatic feeding devices are available commercially that allow time- and volume-controlled dispensing at 90-min intervals or less.

The newborn pig can survive for about 24 hr without food if kept warm, but the first feeding should be given during the first 12 hr if possible. It is possible to train baby pigs to drink from a shallow bowl by pushing their snout into the milk. After one or two feedings they learn to consume milk quickly from a bowl, and bowl-feeding is less time-consuming than bottle feeding.

Scouring (diarrhea) is a major problem in artificially raised pigs. It is easily induced by overfeeding. Therefore, the amount of milk fed at each feeding should be reduced by one-half for one or two feedings if diarrhea appears. It is not necessary to warm refrigerated milk before feeding except as it may encourage consumption in very weak pigs for the first few feedings.

Growing-Finishing Pigs

The management of pigs during the post-weaning period to slaughter weight is less demanding than that of suckling pigs.

Growing-finishing pigs thrive well in pasture lots or on concrete floor or slotted-floor confinement housing, provided good sanitation, disease control, and parasite control are practiced, and a balanced diet is offered. Pigs perform well in pasture lots only if a complete feed is offered. The amount of nutrients obtained from pasture is usually negligible, and the alternative uses of land usually preclude its use for pasture for swine. Figure 19.4 shows growing-finishing pigs

FIG. 19.4. Pigs can be housed on pasture with equipment shown above in either temperate or tropical environments: pasture can not be considered as a significant source of nutrients for growing pigs.
Courtesy E. A. Pierce, Cornell University.

being raised successfully in pasture systems used in both temperate and tropical environments.

In general, most rapid and economical gains are made when pigs are full-fed a complete balanced diet. However, under some marketing systems a sufficiently large premium is paid for lean carcasses to warrant limiting daily feed intake to 75–85% of *ad libitum* intake. This results in slower weight gain and leaner carcasses but may increase feed required per unit gain, owing to the higher proportion of feed needed for body maintenance.

Some feeding systems, especially in the tropics, require hand-feeding the energy portion of the diet on a fresh basis. This is true for cassava and bananas. The same type of trough feeding of garbage is practiced in many areas both in the temperate and tropical zones. When energy is provided in this way, the other diet components (protein, vitamins, and minerals) must be provided in limited quantities daily to avoid overconsumption of this expensive portion of the diet. These are usually supplied as a complete supplement.

Many swine producers raise their own grain or maize as the energy source, and purchase complete protein–vitamin–mineral supplement. Where mixing costs are high or mixing facilities are not easily accessible, it may be preferable to feed the grain (or maize) and supplement free-choice. When this is done the number of feeding holes in the self-feeder should favor the grain portion by at least four to one to avoid overconsumption of the supplement. The pig does a remarkably good job of selecting the correct proportion of grain and supplement to balance nutrient needs under such a feeding system.

As growing-finishing pigs approach market weight their daily gain tends to plateau, but their daily feed intake continues to increase while efficiency of feed utilization declines. This relationship between liveweight and gain and feed data is shown in Table 19.1 (Becker *et al.* 1966; Ensminger 1970) and Tables 19.2 and 19.3 (Headley *et al.* 1961). These values were derived from data on pigs typical of genetic populations and husbandry conditions of more than 20 yr ago and therefore, represent only approximations of present expectations. A rather wide

TABLE 19.1. Expected Daily Gain, Feed/Gain, Cumulative Feed, and Water Needs of Pigs of Various Liveweights and Ages[a]

Liveweight		*Age*	*Daily gain*		*Feed/gain*[b]	*Total feed to attain wt*[c]		*Water needed*[d]
(lb)	*(kg)*	*days*	*(lb)*	*(kg)*	*(kg/kg)*	*(lb)*	*(kg)*	*(gal/day)*
2.2	1.0	0	—	—	—	110	50	milk
10	4.5	14	0.50	0.23	1.5	130	59	$\frac{1}{2}$–1
30	13.6	48	1.00	0.45	1.9	165	75	$\frac{1}{2}$–$1\frac{1}{2}$
40	18.2	56	1.20	0.54	2.2	185	84	$\frac{1}{2}$–$1\frac{1}{2}$
80	36.3	85	1.65	0.75	2.8	285	129	$\frac{1}{2}$–$1\frac{1}{2}$
100	45.4	96	1.80	0.82	3.1	345	157	$\frac{1}{2}$–$1\frac{1}{2}$
120	54.5	108	1.90	0.86	3.3	410	186	$1\frac{1}{2}$–2
160	72.6	128	2.00	0.91	3.5	545	247	$1\frac{1}{2}$–2
200	90.8	145	2.10	0.95	4.1	700	318	2 –3
220	100.0	155	2.15	0.98	4.4	785	356	2 –3

Source: Becker et al. 1966.
[a]Includes feed for sows starting 30 days before mating and for boars.
[b]Assumes diets with energy density of approximately 3300 kcal/kg.
[c]Amount consumed by the dam per fetus during pregnancy. From Ensminger 1970.
[d]Gestating swine and boards need $4\frac{1}{2}$ gal/day, lactating sows 6 gal/day.

range above and below the mean can be expected, depending on genetic differences and environmental factors. Voluntary feed intake of pigs declines as environmental temperature increases from −20° to 35° C. As temperature increases from 20° to 35° C, feed intake declines rapidly in pigs of all body weights, but most drastically in heavy animals (National Research Council 1981; Verstegen *et al.* 1978). Feed requirements and conversion rates to carry pigs from various initial weights to various final weights are shown in Table 19.4, based on

TABLE 19.2. Averages of Observed Liveweight and Feed Consumption Data on 1200 Growing-Finishing Pigs

Age	*Liveweight*		*Feed consumed in 2-week period*		*Cumulative feed*		*Observed efficiency in 2-week period*	*Observed cumulative efficiency*
(weeks)	*(lb)*	*(kg)*	*(lb)*	*(kg)*	*(lb)*	*(kg)*	(%)	(%)
6	26.6	12.1	—	—	—	—	—	—
8	38.6	17.5	29.9	13.6	29.9	13.6	40.1	40.1
10	55.0	25.0	42.3	19.2	72.2	32.8	38.7	39.3
12	75.9	34.5	63.6	28.9	135.8	61.7	32.9	36.3
14	98.5	44.7	71.9	32.6	207.7	94.3	31.4	34.6
16	122.4	55.6	84.5	38.4	292.2	132.7	28.3	32.8
18	145.8	66.2	95.2	43.2	387.4	175.9	24.6	30.8
20	169.6	77.0	100.7	45.7	488.1	221.6	23.6	29.3
22	191.8	87.1	100.8	45.8	588.9	267.4	22.0	28.0
24	213.7	97.0	104.9	47.6	693.8	315.0	21.0	27.0

Source: Headley et al. 1961.

TABLE 19.3. Calculated Daily Feed Consumption and Maintenance Requirement

Livewt		*NRC daily feed requirements*[a]		*Calculated daily feed consumption*		*Instantaneous maintenance*	*Calculated daily maintenance requirement*		*Daily feed available for growth*	
(lb)	*(kg)*	*(lb)*	*(kg)*	*(lb)*	*(kg)*	*(%)*	*(lb)*	*(kg)*	*(lb)*	*(kg)*
50	22.7	3.2	1.45	3.4	1.54	15.1	0.5	0.23	2.9	1.32
75	34.1	—	—	4.7	2.13	22.6	1.1	0.50	3.6	1.63
100	45.4	5.3	2.40	5.3	2.41	30.1	1.6	0.73	3.7	1.68
125	56.8	—	—	6.3	2.86	37.7	2.4	1.09	3.8	1.73
150	68.1	6.8	3.09	6.8	3.09	45.2	3.1	1.41	3.7	1.68
175	79.5	—	—	7.2	3.27	52.8	3.8	1.73	3.4	1.54
200	90.8	8.0	3.63	8.5	3.86	60.3	5.1	2.32	3.4	1.54

Source: Headley et al. 1961.
[a]NRC is National Research Council.

TABLE 19.4. Feed Requirements and Conversion Rates to Carry Hogs from Various Purchased Weights to Various Market Weights[a]

Purchase wt, lb (kg)	Feed requirements	Units	Selling weight 200 lb (91 kg)	210 lb (95 kg)	220 lb (100 kg)	230 lb (104 kg)	240 lb (109 kg)	250 lb (114 kg)
30 (16.3)	Corn	bu	8.5	9.2	9.9	10.6	11.4	12.1
		or lb (kg)	478 (217)	515 (234)	554 (252)	595 (270)	636 (289)	680 (309)
	Supplement	lb (kg)	102 (46)	109 (49.5)	116 (52.6)	123 (55.8)	131 (59.5)	138 (62.7)
	Total	lb (kg)	580 (263)	624 (283)	670 (304)	718 (326)	767 (357)	818 (371)
	Conversion		3.41	3.47	3.53	3.59	3.65	3.72
50 (22.7)	Corn	bu	7.9	8.6	9.3	10.0	10.8	11.5
		or lb (kg)	443 (201)	480 (218)	520 (236)	560 (254)	602 (273)	645 (293)
	Supplement	lb (kg)	91 (41.3)	98 (44.5)	104 (47.2)	112 (50.8)	119 (54)	127 (57.7)
	Total	lb (kg)	534 (242)	578 (262)	624 (283)	672 (305)	721 (327)	772 (350)
	Conversion		3.56	3.61	3.67	3.73	3.79	3.86
70 (31.8)	Corn	bu	7.1	7.8	8.5	9.2	10.0	10.8
		or lb (kg)	400 (182)	437 (198)	476 (216)	517 (235)	558 (253)	602 (273)
	Supplement	lb (kg)	80 (36.3)	87 (39.4)	94 (42.7)	101 (45.9)	109 (49.5)	116 (52.7)
	Total	lb (kg)	480 (218)	524 (238)	570 (259)	618 (281)	667 (303)	718 (326)
	Conversion		3.69	3.74	3.80	3.86	3.92	3.99
90 (40.9)	Corn	bu	6.2	6.9	7.6	8.3	9.1	9.9
		or lb (kg)	350 (159)	388 (176)	427 (194)	467 (212)	509 (231)	552 (251)
	Supplement	lb (kg)	68 (30.8)	74 (33.6)	81 (36.8)	89 (40.4)	96 (43.6)	104 (47.2)
	Total	lb (kg)	418 (190)	462 (210)	508 (231)	556 (252)	605 (275)	656 (298)
	Conversion		3.80	3.85	3.91	3.97	4.03	4.10

Source: From Bache and Foster 1980.

[a]Feed efficiency varies considerably depending on environmental temperatures, disease level, ration fed, quality of management, and death loss. The feed requirements here are for hogs with good performance under excellent management. These figures assume zero mortality.

data for modern pigs handled under excellent management (Bache and Foster 1980).

To market animals at desired weights, it is advisable to have accurate scales in close proximity to the pen area so that body weight can be recorded easily. Portable scales that can be moved from pen to pen are also available. Pigs that have been unaccustomed to leaving the pens in which they have been grown will resist movement out of the pen and onto a truck. A narrow alley (no more than 1 m) to the loading area is helpful in controlling movement. Electric shocking devices or canvas or leather slappers can be used to move pigs, but solid clubs or sticks should never be used, first because inhumane treatment obviously should always be avoided, and also because losses in carcass value due to bruises and injuries result from harsh treatment.

Marketing of hogs during hot weather is always hazardous because of the poor ability of slaughter weight and mature swine to dissipate body heat in warm environmental temperatures. Pigs should be watered down or hosed down with water after loading, and wet sand or wet straw placed on the floor of the truck for cooling.

Fighting is inevitable among strange pigs grouped together and can result in injuries and bruising that lower carcass value.

A loading dock, either portable or stationary, is a necessary part of any swine finishing operation to make loading easier on both personnel and pigs.

Breeding Swine

Sows and boars require less care and attention than growing pigs, and yet they are the key to a successful swine enterprise.

Sows and boars are often kept in outdoor drylot or pasture lots with minimum shelter during both winter and summer in temperate and tropical environments, and they thrive very well as long as shade is provided in summer and a dry, well-bedded shelter, which protects from wind and snow, in winter. Outdoor houses of this kind for pregnant sows can be used to accommodate groups of 15–20 if adequate pen area and feeder and waterer space are provided. Boars that have not been penned together should be kept individually, as they fight ferociously when placed together. Such skirmishes often lead to the death of one of the boars. Strong woven wire or other durable fences are required for both boars and sows.

In some commercial practices boars are allowed to run with a group of sows during breeding season (pen breeding). In purebred enterprises it is often necessary to hand mate to establish ancestry and breeding date. General recommendations for number of sows per boar and frequency of use in natural mating systems are given in Table 19.5. In commercial breeding it is desirable to rotate boars frequently between groups of sows to avoid problems of low conception in case of temporary infertility of one boar. Sample sow and litter records are given in Appendix 1.

Feeding the pregnant sow and boar involves limited intake in each case whereas feeding the sow during lactation involves *ad libitum* feeding. Suggested feeding levels and performance criteria for swine in confinement are summarized in Table 19.6 (Jensen 1982). Mature swine are able to utilize higher quantities of fibrous feeds such as pasture and silage, apparently owing to their large cecum and colon

TABLE 19.5. General Recommendations for Number of Services and Number of Sows per Boar

<table>
<tr><th rowspan="2">Age of boar</th><th colspan="4">Number of services</th></tr>
<tr><th>Per day</th><th>Per week</th><th colspan="2">Per month</th></tr>
<tr><td>8–15 months</td><td>2</td><td>8</td><td colspan="2">24</td></tr>
<tr><td>15 months and older</td><td>3</td><td>12</td><td colspan="2">36</td></tr>
<tr><th rowspan="3">Length of breeding season</th><th colspan="4">Number of sows per boar by age of boar</th></tr>
<tr><th colspan="2">Hand-mating</th><th colspan="2">Pasture-mating (or pen-mating)</th></tr>
<tr><th>8–15 months</th><th>15 months and older</th><th>8–15 months</th><th>15 months and older</th></tr>
<tr><td>Three weeks</td><td>20</td><td>30</td><td>10</td><td>15</td></tr>
<tr><td>Six weeks</td><td>30</td><td>40</td><td>20</td><td>30</td></tr>
</table>

Source: Pierce 1969.

and the microbial activity contained therein. Therefore, the amounts of feed recommended in Table 19.6 would need to be decreased for swine that were provided succulent pasture. Judgments on increasing or decreasing feed intake of sows and boars on pasture must be made on the basis of general appearance of the animal and changes in body weight. Sows fed the levels of feed recommended during gestation should gain 10–20 kg or more during pregnancy, depending on

TABLE 19.6. Suggested Feeding Levels and Performance Criteria for Swine in Confinement

Animal weight or production phase	*No. of days*	*Av daily feed, lb*[a]	*Av daily gain, lb*	*Feed per lb of gain, lb*
10–30 lb	29	1.20	0.70	1.70
30–45 lb	14	2.00	1.10	1.82
45–120 lb	50	4.00	1.50	2.67
120–220 lb	55	6.50	1.80	3.60
Gestation				
Gilts	114	4.0[b]	0.90	—
Sows	114	4.0[b]	0.60	—
Lactation	21–42	7.0–10.0[c]	—	—
Boars[d]				
Inactive	—	4.0–5.0[e]	—	—
Breeding	—	5.0–7.0[e]	—	—

Source: From Jensen 1982.

[a]Corn–soybean meal, or equivalent, diets of about 90% dry matter and 1500 kcal digestible energy/lb are assumed.

[b]This level is usually adequate when animals are in a comfortable environment. Levels would need to be suitably adapted for different environments to ensure desired body weight gain.

[c]Lactating sows need sufficient energy and nutrients to satisfy need for body maintenance and milk production. Body size, length of lactation period, and number of nursing pigs are primary factors affecting total need.

[d]Diets nutritionally adequate for gestating gilts should be adequate for mature boars.

[e]These levels minimize excess body weight and size. Environment and intensity of use affect total needs.

their original condition. Boars should maintain constant weight after skeletal growth has stopped at 24–36 months of age.

Care of the Sow at Farrowing

Near term the udder congests and the underline becomes prominent. When parturition is imminent, the sow becomes restless and begins arranging bedding (if present) in preparation for farrowing. This may persist for several hours just prior to birth of the first pig. Because of the danger of crushing of newborn pigs, it is important either to provide guard rails 20–30 cm above the floor and extending 20–30 cm from the pen wall on all sides (Fig. 19.5), or, preferably, to place the sow in a farrowing crate (Fig. 19.6), which restrains movement of the sow. Both of these precautions reduce markedly the death losses of young pigs due to crushing. Some swine producers prefer to put sows into farrowing crates a week or two before the end of the 114-day gestation period to allow adjustment to it before the pigs are born.

It is advisable to be present during farrowing if possible to save pigs from suffocating in the placental membranes. Often pigs are born (either front feet first or hind feet first) with the surrounding membranes having been left in the birth canal, but many times the membranes remain surrounding the newborn pig, and suffocation results when the umbilical cord is broken without someone being present to remove the membranes.

The sow will usually farrow uneventfully with no need for special attention. Occasionally, if the process is slow (extending more than 8 hr) it may be advisable to inject oxytocin intramuscularly into the ham or intravenously into an ear vein to induce uterine contractions. Veterinary help should be sought promptly if this is not successful in expelling pigs and placental attachments.

Facilities

Recommendations for space allotment, environmental temperature, and other important physical features for growing-finishing and breeding swine have been given in Chapter 18. Facilities that will be in accord with the standards of sanitation, disease and parasite control, and environmental temperature, set forth

FIG. 19.5. Guard rails in a farrowing pen.
Courtesy E. A. Pierce, Cornell University.

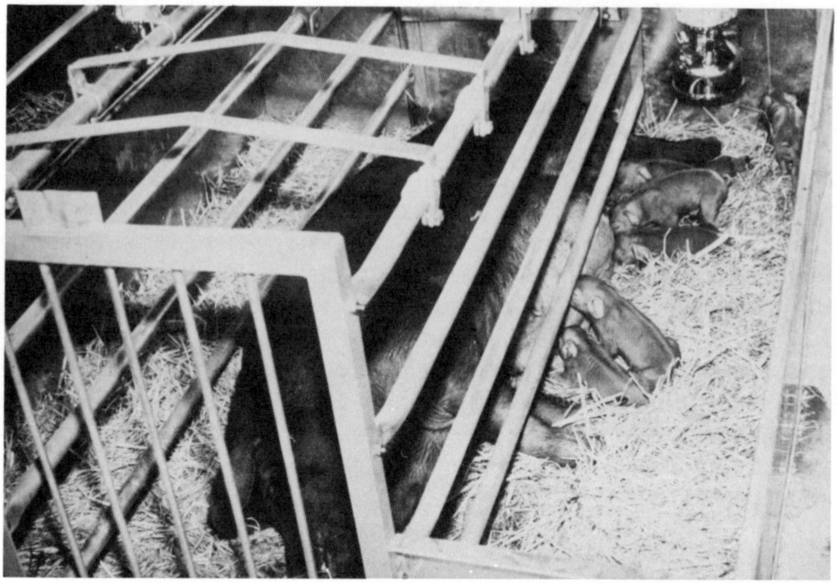

FIG. 19.6. Farrowing crate

earlier in this chapter, may be developed by remodeling older buildings and structures at a much lower cost than possible with package-arrangements such as many units available commercially. The decision on which approach to take in setting up a profitable swine operation will depend on the judgment of the individual producer under his special economic and marketing conditions.

GENERAL TYPES OF SWINE ENTERPRISES

There are two general types of swine production—purebred and commercial. Both of these have been discussed previously. Basically, they differ only in a few respects. Accurate records of ancestry, breeding and farrowing dates, and an active and continuous advertising and promotion program are essentials of a purebred enterprise. These activities require more attention to details, more labor, and better management. Though opportunities for selling breeding stock at better than slaughter prices are afforded purebred producers, the increased expenses and greater efficiency required of this type of production discourage many producers from engaging in it.

Commercial swine production can be more specialized in the sense that a producer may choose only one or all of the common methods of swine production whereas purebred operations require more facilities and devotion to both a breeding program and a feeding program for growing pigs. Commercial producers can specialize as follows: (1) the production of feeder pigs (sold at weaning); (2) the growing and finishing of purchased feeder pigs; (3) a complete sow and litter (farrow-to-finish) program, including both production of pigs to weaning and the growing and finishing of these pigs for market (as in purebred production). The type of enterprise chosen will depend on the interest and experience of the producer as well as on the equipment, facilities, labor, feed supply, and markets he has available.

Feeder pig production is perhaps best suited for those who have a surplus of labor available and a limited feed supply. Feeder pig producers have the disadvantage that they must depend on operators of growing-finishing enterprises for the market and are therefore somewhat limited in choice of the time of marketing and the volume of their sales. However, those engaging in growing-finishing enterprises must depend for their supply on those producing feeder pigs and therefore have no control over the time when pigs will be available. Because of these uncertainties, complete sow and litter programs with facilities for growing the pigs to slaughter weight are increasingly popular in many areas. Percentages of each of the above three swine production types in 1980 listed by size of enterprise are shown in Table 19.7 (Rhodes and Grimes 1982).

For those who have large grain supplies but limited labor and facilities, the growing of purchased feeder pigs to slaughter weight offers a good means of marketing the grain. The profit will depend on the price of feed, the price of feeder pigs, the efficiency of feed conversion, and the price of slaughter pigs as well as the labor and capital investment costs. Costs of capital investments, labor, and feed for various types of pork production enterprises have been reviewed by Bache and Foster (1979). Rate of earnings on the investment in feeder pig enterprises was estimated at 6.6–9.5%; farrow to finish, 12– 19.5%; and feeder pig finishing, 10.4–11.3%. Obviously, the actual rate of return on investment may be less or even negative, but under good management, the above values are expected.

A series of 4–8-page fact sheets, collectively called the *National Pork Industry Handbook,* authored by scientists at land grant universities in the United States covers a wide range of topics of practical value to pork producers. Management calendars for producers of feeder pigs, market hogs, and large and small life cycle pork production enterprises under extensive and intensive management are available from University Agricultural Extension Offices in most states. Some of these calendars are shown in Tables 19.8, 19.9, and 19.10 (Bache and Foster 1980).

TABLE 19.7. Percentages of Hog Production Types by Size, 1980

Size operation	*Production type*			
	Pig finishers	*Farrow-to-finish*	*Pig producers*	*Total*
(5000+)	20%	56%	24%	100%
(2500–4999)	16	65	19	100%
(1000–2499)	12	71	17	100%

Source: Rhodes and Grimes 1982.

TABLE 19.8. Calendar of Management Activities for a Two-Litter Pasture System (Example: 25 Sows Farrowing in March and September)[a]

Activity	Example & your plan	J	F	M	A	M	J	J	A	S	O	N	D	Comments
1. Select replacement gilts	Aug. 1 ______								●					To provide greatest selection, choose gilts before selling any slaughter hogs.
2. Buy boars	Sept. 1 ______									●				Boars should be on the farm at least 2 months before the breeding season and be at least 9 months old when mating begins.
3. Vaccinate gilts and sows	May 1 & Nov. 1 ______ & ______					●						●		Vaccinate for erysipelas and leptospirosis.
4. Breed	May 12-June 8 & Nov. 9-Dec. 6 ______ & ______					●—	—●					●—	—●	Expose sows to the boars for a 4-week period.
5. Prepare for farrowing	Feb. & Aug. ______ & ______		●						●					Repair and position fencing, water supply, shelters, feeders, etc. Confine sows to farrowing huts when they have milk.
6. Farrow	Mar. 1-28 & Sept. 1-28 ______ & ______			●						●				Although this example schedules farrowings in March and September, your farrowings might be during two other months, provided there is a 6-month interval. Castrate when pigs are 7-10 days old.
7. Wean	Apr. 15-May 12 & Oct. 15-Nov. 11 ______ & ______				●—	—●					●—	—●		Plan for 6- to 7-week weaning. Vaccinate for erysipelas at this time. To keep on the 6-month farrowing schedule, you will need to skip one heat period before rebreeding those sows which farrowed early.
8. Sell market hogs	Mar.-Apr. & Sept.-Oct. ______ & ______			●—	—●					●—	—●			Animals should reach slaughter weight at 6-7 months of age.
9. Percentage distribution of labor	100% ______	6	10	12	9	7	6	6	10	12	9	7	6	Labor needs vary widely. Large volume producers using such technology as electric fencing and water piped in the field, report a labor requirement as low as 25 hours per sow. Most producers would use about 36 hours per sow. However, both figures ignore the indirect labor spent in planning, keeping records, maintaining the farmstead, etc. For a relatively small enterprise, "indirect" labor may add another 25-35 percent, resulting in a total time commitment of 48 hours per sow.
10. Number of sows	25 ______													
11. Hours of labor per sow	48 ______													
12. Hours of labor per month	1200 ______	72	120	144	108	84	72	72	120	144	108	84	72	

Source: Bache and Foster 1980.

[a]The "sow" is the unit around which the discussion in this publication is built. A sow unit denotes a mature female in production and includes a "supporting cast" of boars, replacement gilts and progeny in various stages of growth—all of which must be provided for.

TABLE 19.9. Management Calendar for a 150-Sow Operation

Stage	*Days from breeding*	*Management practices*
Pre-breeding	−30	Co-mingle gilts with sows or provide for fence-line contact; also provide new boars fence-line contact with sows.
	−28–−14	Vaccinate sows for lepto; also vaccinate for erysipelas, if a problem.
	−14–−7	Increase feed intake for gilts to 6–7 lb per gilt per day for 1–2 weeks prior to breeding.
	−2	Spray for lice and mange.
Gestation	0	Breed, rotating boars daily.
	21–24	Remove boars.
	35–60	Pregnancy check, and sell any open females.
	90	Vaccinate sows for erysipelas, if a problem.
	105	Worm sows with a broad-spectrum wormer.
		Clean the farrowing house.
	108–110	Wash sows with warm water and soap, spray for lice and mange, and move to farrowing house; isolate farrowing house from all visitors.
		Start hand-feeding 3–4 lb per sow per day of a farrowing ration containing 20% wheat bran (or other bulky ingredients) and continue until 2–3 days after farrowing.
	112	Prepare auxiliary heat for pigs.
	Days after farrowing	
Farrowing-nursery	0–1	Dip navel cord in mild iodine solution.
		Clip needle teeth.
		Cut off tails.
		Ear notch gilts from good litters of 9 or more pigs.
		Complete the farrowing records, including date, number pigs farrowed, etc.
		Observe sows for MMA (mastitis, metritis, agalactia).
	1–3	Transfer pigs to equalize litters.
		Give iron shots.
	3–14	Castrate males.
	7–10	Provide creep feed for pigs.
	14–21	Give second iron shot, if needed.
	28–42	Wean pigs, group by size and move to nursery.
Growing-finishing	42	Worm pigs with a broad-spectrum wormer, and spray for lice and mange.
	56	Vaccinate pigs for erysipelas, if a problem.
	72	Worm pigs with a wormer other than the one used at 42 days.
	150	Select replacement gilts from those ear-notched at birth.
	150–200	Observe withdrawal regulations for feed additives, and market the hogs at 215–235 lb.

Source: Bache and Foster 1980.

TABLE 19.10. Calendar of Management Activities for a Feeder Pig Production Operation

Stage	*Days from breeding*	*Management practices*
Pre-breeding	−30	Co-mingle gilts with sows or provide for fence-line contact, also provide new boars fence-line contact with sows.
	−14–−28	Vaccinate sows for lepto; also, vaccinate for erysipelas if a problem.
	−7–−14	Increase feed intake for gilts to 6–7 lb per gilt per day for 1–2 weeks prior to breeding.
	−2	Spray for lice and mange.
Gestation	0	Breed, rotating boars daily; record known breeding dates.
	21–24	Remove boars.
	35–60	Pregnancy check, and sell any open females.
	90	Vaccinate sows for erysipelas, if a problem.
	105	Worm sows with broad spectrum wormer.
		Clean the farrowing house.
	108–110	Wash sows with warm water and soap, spray for lice and mange, and move to farrowing house; isolate farrowing house from all visitors.
		Start hand-feeding 3–4 lb per sow per day; if constipation is a problem, add 20% wheat bran (or other laxative ingredients), and continue until 2 or 3 days after farrowing.
	112	Prepare auxiliary heat for pigs.
	Days after farrowing	
Farrowing-nursery	0–1	Dip navel cord in mild iodine solution.
		Clip needle teeth.
		Cut off tails.
		Ear notch gilts from good litters of 9 or more pigs.
		Complete farrowing record, including date, number of pigs farrowed, etc.
		Observe sows for MMA (mastitis, metritis, agalactia).
	1–3	Transfer pigs to equalize litters.
		Give iron shots.
	1–14	Castrate males.
	7–10	Provide creep feed for pigs.
	14–21	Give second iron shot, if needed.
		Move sows and pigs to nursery.
	28–42	Wean pigs and group by size.
	42	Worm pigs with broad spectrum wormer.
		Spray for lice and mange.
	56	Select replacement gilts from those ear-notched at birth.
	56–70	Sell pigs.

Source: From Bache and Foster 1980.

MARKETING

Effect of Environmental Stress on Carcass Characteristics

The movement of pigs into confinement has resulted in problems related to carcass quality in some animals that did not exist or were not recognized previously. These problems have been discussed by Wismer-Pederson (1968); Briskey and Lister (1968); Judge (1968); and Cassens *et al.* (1972). Several years ago a condition was described in the carcasses of some pigs that lent a pale, soft, watery appearance to the meat. This was termed PSE (pale, soft, exudative) pork. It appears to be more prevalent in some breeds than in others, and occurs in breeds representing different countries.

The characteristics of muscle that impart the PSE syndrome have been studied and reviewed by Briskey and Lister (1968) and by Sybesma (1972). A high degree of leanness and muscularity and freedom from the PSE condition of pork (PSS, porcine stress syndrome) are not mutually exclusive (Kukoyi *et al.* 1981). This conclusion has been reached by comparing the response of different breeds and crosses of similar leanness and muscularity to standard halothane anesthesia and blood enzyme tests for stress susceptibility (PSE-prone pigs are sensitive to halothane anesthesia and have elevated blood levels of the enzyme creatine phosphokinase) (Kukoyi *et al.* 1981; Hwang *et al.* 1977).

The PSE condition is associated with a high production of lactic acid while the muscle temperature remains high. The lactic acid accumulation results in a reduced pH of the muscle and this is associated with a reduced water-binding capacity, resulting in the exudative appearance of PSE pork. There are breed differences in muscle structure and composition. The amounts of glycogen and ether extract within the muscle, and the pH, color, myoglobin content, and fiber diameter all showed breed differences (Allen *et al.* 1966). Danish workers have studied the relationship of meat color in Danish Landrace pig carcasses to muscle pH, and observed that the color score (low scores reflecting pale color) is lower when muscle pH is lower (Wismer-Pederson 1968; Jonsson, 1968). This paleness is probably similar to or the same as PSE pork. Jonsson (1968) found that Danish Landrace pigs slaughtered in their home pens where they had been raised had higher meat color scores and higher muscle pH than pigs hauled to the slaughter house for slaughter. Thus, the stress of movement to slaughter clearly has an effect on carcass characteristics. Pre-slaughter stress imposed when pigs are moved to slaughter is probably greater in pigs kept in close confinement than in pigs allowed free exercise. This was demonstrated experimentally (Lannek *et al.* 1967) by subjecting growing-finishing pigs twice per week to a treadmill which provided 10 min of forced exercise on a belt moving at 2 m/sec. They were compared at slaughter with a group kept in similar confinement pens but not given exercise. The expressible juice (taken as a measure of water-holding capacity and muscle pH) from the ham was greater in nontrained than in trained pigs, and in those subjected to the treadmill just before slaughter, the difference between the two groups still persisted. These changes in muscle have been suggested by Briskey and Lister (1968) to be related to adrenaline gland release of norepinephrine and epinephrine, which causes constriction of blood flow in skeletal muscle and thus leads to anoxia, and a build-up of lactic acid, lowering the pH. Thus, the chain of events that leads to PSE pork is well established and it is possible to alter pre-slaughter management conditions so as to minimize the incidence of PSE in genetically susceptible animals. A condition observed in live animals, referred to as the porcine stress syndrome (PSS), has been observed with increasing frequency, especially among lean strains of hogs. Pigs producing PSE pork frequently also suffer from the porcine stress syndrome. Judge (1972) has summarized the methods of detecting stress-susceptible swine. Animals showing anxiety, extreme muscularity, tail tremors, leg weakness, blanching of the skin, high rectal temperature (malignant hyperthermia syndrome, MHS), high blood creatine phosphokinase, high blood ACTH or low corticoid-ACTH ratio, high muscle glucose-6-phosphate, and low mitochondrial oxygen consumption are practically certain to suffer the PSS. More recently, intolerance to halothane anesthesia (Eikelenboom and Minkema 1974) has been noted in PSS pigs. Those with some but not all of the above traits have a higher than even chance of developing PSS. Combinations of traits denoting various degrees of certainty of an animal developing PSS, outlined by Judge (1972), are shown as follows: (1) suspect when (a) extreme muscularity and (b) anxiety; (2) high probability when (a) extreme muscularity, (b) anxiety, (c) muscle tremors (tail tremor), (d) chronic stress signs (leg weakness, etc.), (e) erythema (expressed as reddening and blanching of skin), and (f) hyperthermia (relatively high rectal temperature); (3) practically certain when (a) extreme muscularity, (b) anxiety, (c) muscle tremors (tail tremor), (d) chronic stress signs (leg weakness, etc.), (e) erythema (expressed as reddening and blanching of skin), (f) hyperthermia (relatively high rectal temperature), and (g) blood test showing (i) high creatine phosphokinase or other nonplasma-specific enzyme and (ii) high cortisol binding capacity, high ACTH, or low cortocoid–ACTH ratio, and (h) muscle biopsy showing (i) high glucose-6-phosphate or other indicator or glycolytic activity and (ii) low mitochrondrial oxygen consumption.

Christian (1972) has suggested that the PSS and malignant hyperthermia syndrome (MHS) are identical. Evidence from the Netherlands (Eikelenboom and Minkema 1974) and more recent data from the United States (Kukoyi *et al.* 1981) indicate that halothane anesthesia is a good screening tool to identify PSS-prone pigs at 5–9 weeks of age. The PSS-prone pigs are considerably less able to tolerate halothane than are normal pigs. Such a relatively simple technique appears to be very useful in selection against this trait.

Shrinking Losses on Marketing

Shipment of hogs to slaughter often requires several hours without feed and/or water for the animals. The ''shrink'' that takes place during this process often amounts to 2–10% of original off-farm weight. Thus a 90-kg hog would be expected to lose as much as 9 kg during a 24-hr trip to slaughter. Much of this weight loss is due to loss of feces and urine, but some is also due to carcass losses. The changes in carcass measurements that occur, owing to withholding of feed and water before slaughter, have been studied and are shown in Table 19.11 (Bowland and Standish 1966). Carcass length and area of the cross section of the loin eye were not affected by withholding feed, water, or both, for as long as 48

TABLE 19.11. Effect of Withholding Feed or Water or Both before Slaughter on Liveweight and Carcass Shrink

	Treatment					*Difference owing to withholding feed and/or water*
Feed removed, hr before slaughter	*no*	*24*	*no*	*24*	*48*	
Water removed, hr before slaughter	*no*	*no*	*24*	*24*	*no*	
Final wt, kg	93.2	90.8	90.9	92.1	91.6	—
Weight after feed and water withheld, kg	—	86.3	86.0	86.9	84.3	yes
Liveweight shrink, %	—	5.0	5.5	5.7	7.9	yes
Apparent carcass shrink, kg	—	0.9	1.9	1.5	2.8	yes
Carcass length, cm	77.2	76.2	77.7	77.2	77.2	no
Back fat, cm	3.7	3.7	3.6	3.7	3.3	no, except at 48 hours
Area of loin eye, sq cm	25.9	25.0	25.1	24.3	25.9	no

Source: Bowland and Standish 1966.

hr before slaughter, but carcass shrink increased progressively to 2.8 kg (approximately 3% of original liveweight) after 48 hr away from feed and water. Back fat of the carcass was not affected except after 48 hr when it was reduced by about 10%. The effects of temperature, distance, and time on the shrinkage of hogs shipped from Indiana concentration points to slaughter were studied (Stout and Armstrong 1960; Stout and Cox 1959). The findings, which are summarized in Tables 19.12 and 19.13, indicate that temperature extremes below −7°C or above 27°C increase the percentage weight loss. Minimizing transit time is especially important on very hot or very cold days. Wet sand may be used for bedding to reduce losses in shipment during hot weather. Increasing the distance from 160 to 960 km increased the shrinkage from 1.7 to 3.8% in fed hogs and from 1.4 to 3.5% in hogs not fed prior to shipment (Table 19.13).

Market Grades

The grading system upon which price of market hogs is based varies widely from country to country. In Denmark, Canada, and many other countries the price received by the swine producer is based on the carcass value as determined by

TABLE 19.12. Effect of Variation Above or Below 50° F over Time on In-Transit Shrinkage of Slaughter-Weight Market Hogs; 232 Observations

Hours in transit	*Percentage weight loss by temperature range*		
	1–20° F	*69–80° F*	*More than 80° F*
Up to 1	0.96	0.54	0.87
1–2	1.17	0.84	1.00
2–4	1.24	1.26	1.54
More than 4	3.27	1.75	1.51

Source: Stout and Cox 1959.

TABLE 19.13. Percent In-Transit Shrinkage of Hogs Shipped from Indiana Concentration Points

Distance shipped (miles)	Percent in-transit shrinkage	
	Fed	Unfed
100	1.7	1.4
200	2.0	2.1
600	3.8	3.5

Source: Stout and Armstrong 1960.

objective carcass measurements. These measurements usually include back fat thickness, length, some measure of proportion of lean cuts, such as actual cut-out data or cross-sectional area of the loin eye muscle, and often an estimate of consumer acceptability such as a color or marbling score. Except in areas of the world in which a premium is paid for fat carcasses, the conventional carcass measurements are used as indices of carcass leanness and value is increased as leanness traits increase. The payment to the producer on the basis of carcass measurements requires individual identification of pigs before slaughter so that identity of the producer can be preserved.

In the United States, the traditional method of payment to the farmer has been on the basis of live animal value as determined by its probable carcass yield. This method of pricing hogs had been almost exclusively used in the United States until within the last few years in which several individual meat packing companies have initiated grade and yield pricing systems based on the actual carcass value rather than on predicted carcass value. However, the majority of pigs marketed are still sold on a liveweight basis. The USDA started a grading system more than 70 yr ago. This system has been revised several times since then, most recently in 1968. There are separate standards for feeder pigs and slaughter pigs and yet another standard for pork carcasses. Each of these standards is related in the sense that feeder pig standards are intended to be directly related to the market hogs they produce and, in turn, the grades of market hogs are intended to be directly related to the grades of carcasses they produce.

The USDA standards include classes of slaughter and feeder pigs. Definitions of the respective classes are as follows (USDA 1970A).

> *Barrow*—A barrow is a male swine castrated when young and before development of the secondary physical characteristics of a boar. *Gilt*— A gilt is a young female swine that has not produced young and has not reached an advanced stage of pregnancy. *Sow*—A sow is a mature female swine that shows evidence of having reproduced or has reached an advanced stage of pregnancy. *Stag*—A stag is a male swine castrated after development or beginning of development of the secondary physical characteristic of a boar. Typical stags are somewhat coarse and lack balance—the head and shoulders are more fully developed than the hindquarter parts; bones and joints are large, the skin is thick and rough, and the hair is coarse. *Boar*—A boar is an uncastrated male swine.
>
> USDA grades of slaughter barrows and gilts are predicted on the same two general considerations that provide the basis for the grades of barrow and gilt carcasses: (2) quality of the lean and (2) expected combined carcass yields of the four lean cuts (ham, loin, picnic shoulder, and Boston butt). [In the determination of quality, the amount and distribution of external fat, firmness of fat and firmness of lean are used as quality-indicating factors.]

Barrows and gilts with characteristics which indicate that the carcass lean will have an unacceptable level of quality and/or the belly will be of unacceptable thickness are graded U.S. Utility. Also graded U.S. Utility—regardless of their development of other quality-indicating characteristics—are slaughter barrows and gilts with indications that their carcasses will be soft and oily.

The USDA system of grading the second general consideration, yield, is as follows.

> Four grades—U.S. No. 1, U.S. No. 2, U.S. No. 3, and U.S. No. 4—are provided for barrows and gilts with characteristics which indicate that the carcass lean will have an acceptable level of quality. These grades are based entirely on the expected combined carcass yields of the four lean cuts.
>
> Average back fat thickness in relation to carcass length or liveweight is used as a guide to expected yields of the four lean cuts in these standards. In grading barrows and gilts, these factors normally are appraised visually; liveweight is the only one that can be readily determined in the live animal. The average back fat thickness appraisal includes consideration of the distribution of fat on other parts of the animal in addition to those points on the back where it is measured on the carcass.
>
> Barrows and gilts will produce carcasses that vary in yields of four lean cuts because of variations in their degree of fatness (expressed as average thickness of back fat), and in their degree of muscling (thickness of muscling in relation to skeletal size). Since many slaughter swine have a normal development of muscling for their degree of fatness, in determining their grade the average thickness of back fat and carcass length or liveweight are the only factors considered. These relationships are illustrated (in Fig. 19.7) for barrows and gilts weighing 170–350 lb that will produce carcasses 27–36-in. long. For slaughter swine of other lengths or weights, average back fat thickness requirements for the various grades can be determined by an extension of the lines in this figure.
>
> The degree of muscling specified for each of the four grades decreases progressively from the U.S. No. 1 grade through the U.S. No. 4 grade. This reflects the fact that among barrows and gilts of the same weight, the fatter animals normally have a lesser degree of muscling. For purposes of these standards six degrees of muscling are recognized: very thick, moderately thick, slightly thick, slightly thin, thin, and very thin. These degrees are intended to cover the entire range of muscling present among slaughter swine. The degrees specified as typical for barrows and gilts at the minimum of the U.S. No. 1, U.S. No. 2, U.S. No. 3, and U.S. No. 4 grades are, respectively, thick, moderately thick, slightly thin, and thin.
>
> For animals having a development of muscling which is different from that normally associated with their degree of fatness, the average back fat thickness–carcass length or average back fat thickness–liveweight relationships for the various grades are different from

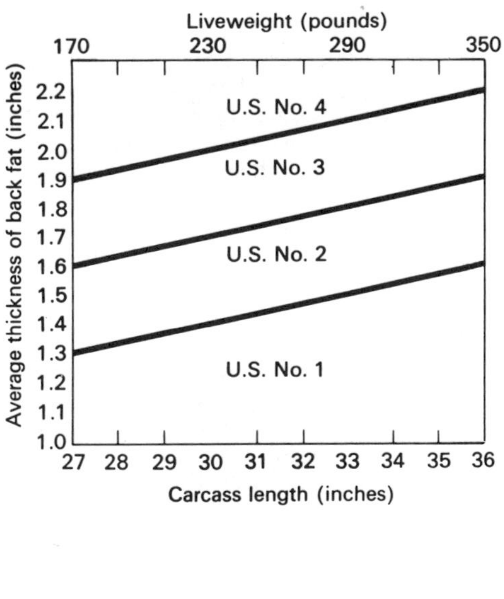

FIG. 19.7. The relationship between carcass traits (average thickness of back fat, weight or carcass length) and grade for barrows and gilts with muscling typical of their degree of fatness.
From USDA Marketing Bull., No. 49.

those shown (in Fig. 19.7). Consideration is given such unusual developments of muscling as follows: In each grade, superior muscling is permitted to compensate for greater fatness at the rate of one full degree of superior muscling for a 0.1-in. increase in average back fat. Except for the U.S. No. 1 grade, the reverse type of compensation is also permitted and at the same rate; 0.1-in. less fat compensates for a full degree of inferior muscling. In the U.S. No. 1 grade, this type of compensation is limited to one full degree of inferior muscling; barrows and gilts which have less than moderately thick muscling but which otherwise qualify for the U.S. No. 1 grade are graded U.S. No. 2. In no case, however, may variations from normal muscling alter the final grade more than one full grade.

The following specifications for official U.S. standards for grades of slaughter barrows and gilts are taken directly from the USDA bulletin on the subject (USDA 1970A):

U.S. No. 1. Barrows and gilts in this grade will produce carcasses with acceptable lean quality and acceptable belly thickness and a high percentage of lean cuts. Swine near the borderline between the U.S. No. 1 and U.S. No. 2 grades are thickly muscled in the hams, loins, and shoulders. The width through the ham is nearly equal to the width through the shoulders and both are wider than the back. The back is slightly full, is well rounded, and blends smoothly into the sides. The sides are moderately long and smooth. The rear flank is slightly full and its depth is less than the depth of the fore flank. The jowls are firm and slightly thick and full.

A development of muscling superior to that specified as minimum for the U.S. No. 1 grade may compensate for a development of fatness which is greater than that indicated (in Fig. 19.7) as maximum for the U.S. No. 1 grade at the rate of one full degree of muscling for 0.1 in. greater thickness of back fat. For example, a barrow or gilt with very thick muscling may have 0.1-in. more back fat than that indicated as maximum for this grade and remain eligible for the U.S. No. 1 grade. The reverse type of compensation is also permitted at the same rate—except that in no case may a barrow or gilt be graded U.S. No. 1 with less than moderately thick muscling.

U.S. No. 2. Slaughter barrows and gilts in this grade will produce carcasses with acceptable lean quality, acceptable belly thickness, and a slightly high percentage of lean cuts. Swine near the borderline between the U.S. No. 2 and U.S. No. 3 grades are moderately thickly muscled in the hams, loins, and shoulders. The width through the shoulders is slightly more than the width through the hams. The back is moderately full and is slightly rounded but with a slight break into the sides. The sides are slightly short and smooth. The rear flank is moderately full and its depth is slightly less than the depth of the fore flank. The jowls are moderately thick and full.

A development of muscling superior to that specified as minimum for the U.S. No. 3 grade may compensate for a development of fatness which is greater than that indicated (in Fig. 19.7) as maximum for the U.S. No. 3 grade at the rate of one full degree of muscling for 0.1-in. greater thickness of back fat. For example, a barrow or gilt with moderately thick muscling may have 0.1-in. more back fat than that indicated (in Fig. 19.7) as maximum for this grade and remain eligible for the U.S. No. 3 grade. The reverse type of compensation is also permitted at the same rate. For example, a barrow or gilt with thin muscling may have 0.1-in. less back fat than that indicated (in Fig. 19.7) as maximum for this grade and remain eligible for the U.S. No. 2 grade.

U.S. No. 3. Slaughter barrows and gilts in this grade will produce carcasses with acceptable lean quality and acceptable belly thickness and a slightly low percentage of four lean cuts. Swine near the borderline between the U.S. No. 3 and U.S. No. 4 grades are slightly thinly muscled in the hams, loins, and shoulders. The width through the shoulders is definitely greater than the width through the hams. The back is full and nearly flat with a pronounced break into the sides. The sides are short and smooth. The rear flank is full and its depth is equal to the depth of the fore flank. The jowls are thick and full.

A development of muscling superior to that specified as minimum for the U.S. No. 3 grade may compensate for a development of fatness which is greater than that indicated (in Fig. 19.7) as maximum for the U.S. No. 3 grade at the rate of one full degree of muscling for 0.1-

in. greater thickness of back fat. For example, a barrow or gilt with moderately thick muscling may have 0.1-in. more back fat than that indicated (in Fig. 19.7) as maximum for this grade and remain eligible for the U.S. No. 3 grade. The reverse type of compensation is also permitted at the same rate.

U.S. No. 4. Barrows and gilts in this grade will produce carcasses with acceptable lean quality and acceptable belly thickness. However, they are fatter and less muscular and will have a lower carcass yield of the four lean cuts than those in the U.S. No. 3 grade.

U.S. Utility. Barrows and gilts typical of this grade will have a thin covering of fat. The sides are wrinkled and the flanks are shallow and thin. Barrows and gilts in this grade will produce carcasses with unacceptable lean quality and/or unacceptable belly thickness.

Drawings of pigs representing each of the live slaughter grades and photographs of carcasses of the corresponding carcass grades are shown in Chapter 2 (Fig. 2.4). These are from the USDA bulletins on USDA Grades for Slaughter Swine and Feeder Pigs (USDA 1970A) and USDA grades for pork carcasses (USDA 1970B), respectively. Drawings of feeder pigs representing each USDA grade can be seen in the former bulletin (USDA 1970A). The official grades for feeder pigs, slaughter hogs and carcasses as they correspond to each other are given in Table 19.14.

The grade distribution of barrows and gilts slaughtered in the United States during 1968 (USDA 1968) was 8% U.S. No. 1, 42% U.S. No. 2, 36% U.S. No. 3, 12% U.S. No. 4, and 3% U.S. Utility. The proportion of U.S. No. 1 plus U.S. No. 2 slaughter hogs is currently more than 95%, reflecting the emphasis placed by the producer on carcass leanness in response to consumer demands for lean pork. The expected yields of the four lean cuts (Boston butt, picnic, loin, ham) based on the chilled carcass weight by grade (Williams and Clower 1978) are in the right-hand column of Table 19.15.

It is readily apparent that the higher yield of lean cuts from U.S. No. 1 slaughter hogs provides adequate incentive for production of this kind of pig for a market that pays a premium for lean meat. The price differential between grades is usually fifty cents to one dollar or more on most U.S. markets. This, along with the generally greater efficiency of feed utilization of lean pigs, has resulted in a steady increase in the proportion of pigs that grade U.S. No. 1 and 2.

Carcass measurements taken on more than 69,000 hogs slaughtered in carcass contests during the period 1976–1981 in the United States show the high degree

TABLE 19.14. Official USDA Grades for Feeder Pigs, Slaughter Hogs, and Pork Carcasses

Feeder pigs	*Slaughter hogs*	*Pork carcasses*
U.S. No. 1	U.S. No. 1	U.S. No. 1
U.S. No. 2	U.S. No. 2	U.S. No. 2
U.S. No. 3	U.S. No. 3	U.S. No. 3
U.S. No. 4	U.S. No. 4	U.S. No. 4
U.S. Utility[a]	U.S. Utility	U.S. Utility
U.S. Cull[a]	—	—

Source: USDA 1970A,B.

[a]Includes only pigs which lack thriftiness; differentiation between U.S. Utility and U.S. Cull grades is based entirely on differences in thriftiness.

TABLE 19.15. Carcass Grade Distribution and Expected Yield of Barrows and Gilts Slaughtered in the United States in 1978

Grade	*Yield*
U.S. No. 1	60% and over
U.S. No. 2	57.5 to 60.0%
U.S. No. 3	55.0 to 57.5%
U.S. No. 4	52.5 to 55.0%
U.S. No. 5	Less than 52.5%

Source: Williams and Clower 1978.

of leanness in modern U.S. swine (Anon. 1982). Average carcass length was 78.6 cm (31.4 in.), back fat was 3.0 cm (1.2 in.), and cross-sectional area of the longissimus muscle was 31.7 cm^2 (5.07 $in.^2$).

Genetic and Environmental Factors Affecting Carcass Yield

The carcass composition of slaughter pigs is affected by both heredity and environment. The heritability of carcass traits is generally high. Table 4.1 provides estimates of heritability of length, loin eye area, and lean cut weights as percentages of carcass weight (Craft 1958). The highest estimate of heritability of these traits was for length ($H = 0.59$) and the lowest was for the four lean cuts as percentage of carcass weight ($H = 0.31$). These heritability estimates are even higher than that of body weight gain which is also responsive to selection. Therefore, changes in carcass composition of a *permanent* nature can be accomplished by genetic selection.

Differences that are due to sex also exist in carcass composition. Boars are leaner than castrated males (barrows) and gilts, and gilts are leaner than barrows at a constant slaughter weight. This is illustrated in Table 19.16 (Cahill *et al.* 1960). There appears to be no large difference in tenderness due to sex, but boar

TABLE 19.16. Differences between Boar, Barrow, and Gilt Littermates in Carcass Measurements

Criteria	*45-kg animals*			*95-kg animals*		
	Boar	*Barrow*	*Gilt*	*Boar*	*Barrow*	*Gilt*
No. animals	10	10	10	4	5	5
Carcass length, cm	61.0	60.5	60.8	73.5	70.8	73.0
Back fat, cm	2.3	2.5	2.8	3.8	4.5	4.0
Lean cuts, %	40.6	39.8	40.3	40.2	36.2	39.5
Tenderness[a]	7.2	7.5	7.0	7.4	7.4	7.5
Flavor[b]	6.1	6.5	6.3	5.6	6.7	6.5

Source: Cahill et al. 1960.
[a]Scoring system: 10 = very tender, 1 = very tough.
[b]Scoring system: 9 = like extremely, 1 = dislike extremely.

meat has a significantly less acceptable flavor. This well-known flavor (or odor) of boar meat has been the reason for the common practice of castration of young males before they reach puberty. The onset of the off-flavor coincides with the androgen level of the blood (Echternkamp *et al.* 1969). One of the compounds that may be at least partially responsible for the odor and flavor of boar meat has been identified (Patterson 1968) as 5-α-androst-16-ene-3-one, a fat-soluble substance that is volatile and therefore is released on cooking of boar meat. Attempts have been made to reduce the development of offensive odors in boar meat by administration of estrogenic hormones to boars during the growing period to inhibit testosterone production and thereby reduce the development of secondary sex characteristics, including tainted meat. This method has been studied by implantation of 96 mg of diethylstilbestrol behind the ear at various ages ranging from 20 to 127 days (Teague *et al.* 1964; Plimpton *et al.* 1967; Palmer *et al.* 1971). Diethylstilbestrol causes some delay in sexual maturity, increases daily weight gain, and suppresses the intensity of the boar odor in the meat. Because of dangers of tissue residues, the FDA has not approved the use of this method of suppressing boar odor in meat. However, the advantages of the boar over the barrow in growth rate and carcass leanness offer incentive for continued study of the possibility of developing management practices such as described above to reduce boar odor in intact boars and thereby eliminate the practice of castration. In many countries it is already common practice to raise intact boars to market weight, without a price deduction.

Given a particular sex and genotype, there are environmental methods by which carcass characteristics of pigs can be affected. The effect of temperature was discussed earlier in this section. The effects of nutrition and management on carcass composition have been reviewed by Elsley (1968), Hays (1968), and Wallace (1968). No differences in chemical composition of newborn pigs due to protein or energy intake of the sow during gestation have been noted with the use of diets within the practical range of energy and protein intake (1.6–2.7 kg feed and 11–18% protein). Elsley (1968) suggested that restriction of feed intake of growing pigs causes a uniform retardation of development except for fat tissue. Thus, he suggested, there may be no advantages to be gained in attempts to alter by nutritional means the other components of the carcass. Restriction of the total energy intake results in reduction in rate of body weight gain, but also increases carcass leanness. This is illustrated in Table 19.17 (Greer *et al.* 1965) and can be accomplished either by restricting total diet offered or by feeding a high-fiber (low-available energy) diet *ad libitum*. Dietary protein level also affects carcass leanness as illustrated in Table 19.18 (Seymour *et al.* 1964). Carcass leanness is increased with higher levels of protein and the effect is similar at either 15.6° or 32.2° C. It should be emphasized that, although no pig can produce maximum lean meat unless provided adequate protein (and amino acids), the pig cannot be forced to increase deposition of lean meat beyond his genetic capacity to do so (Clausen 1959). In this context, it would be of value, in the interest of saving on the world protein supply to locate or develop a strain of pigs able to grow normally and produce acceptable carcasses when fed diets lower in protein than presently recommended. There is limited experimental evidence that selection for pigs with a low dietary protein requirement may be feasible (Maner *et al.* 1977), but data with larger populations of breeding stock will be needed to quantify the

TABLE 19.17. Effect of Restricting Feed Intake of Growing Swine on Carcass Measurements

Criteria	*Level of daily intake*			
	Ad libitum	*Hand fed*		
		100[a]	*85*	*70*
Daily feed, kg	3.28	2.61	2.11	1.80
Daily gain, kg	0.83	0.70	0.55	0.45
Feed/gain ratio	3.95	3.73	3.84	4.00
Back fat, cm	3.71	3.73	3.61	3.28
Ham and loin, % of carcass	38.0	37.1	38.6	39.4

Source: Greer et al. 1965.
[a]Full-fed twice daily all they would consume in a 30-min period. Eighty-five and 70% groups fed on the basis of that consumed by the 100% group.

rate of progress to be expected. It is reasonable to assume that pigs with lean carcasses have a higher dietary protein requirement. Wallace (1968) suggested a difference in dietary protein requirement between barrows and gilts on the basis of the leaner carcass of gilts, and presented data to support such a concept.

Water intake also affects carcass composition. Restriction of water intake to 2 kg of water per kilogram of dry feed was shown to increase protein and to decrease fat in the carcass without affecting weight gain (Cunningham and Friend 1966). More severe restriction (1.25 kg water per kilogram of dry feed) depressed growth while improving lean. This means of altering body composition cannot be used generally, because of wide variations in environmental temperature and humidity which probably have a large effect on water requirement.

The weight at which pigs are slaughtered has an important influence on the ratio of lean to fat in the carcass. Provision has been made for the greater fat deposition of heavier hogs in the U.S. grading standards for pork carcasses (Fig. 19.7). Wallace (1968) presented a bar graph to illustrate this relationship. It is reproduced in Fig. 19.8.

The effects of environmental temperature and dietary protein level on performance and carcass measurements of pigs are shown in Table 19.18 (Seymore *et*

TABLE 19.18. Effect of Dietary Protein and Temperature on Performance and Carcass Measurements of Growing-Finishing Pigs

Environmental temperature, °C	*15.6*		*32.2*	
Protein, %	*20–17–14*[a]	*16–13–10*[b]	*20–17–14*	*16–13–10*
Daily gain, kg	0.74	0.59	0.69	0.57
Feed/gain ratio	3.15	3.76	3.05	3.68
Lean cuts, % of carcass	51.4	49.8	51.9	49.2
Back fat, cm	3.8	4.1	3.8	4.0

Source: Seymour et al. 1964.
[a]20% protein, 3–7 weeks of age; 17%, 7 weeks to 57 kg; 14%, 57–91 kg.
[b]16% protein, 3–7 weeks of age; 13%, 7 weeks to 57 kg; 10%, 57–91 kg.

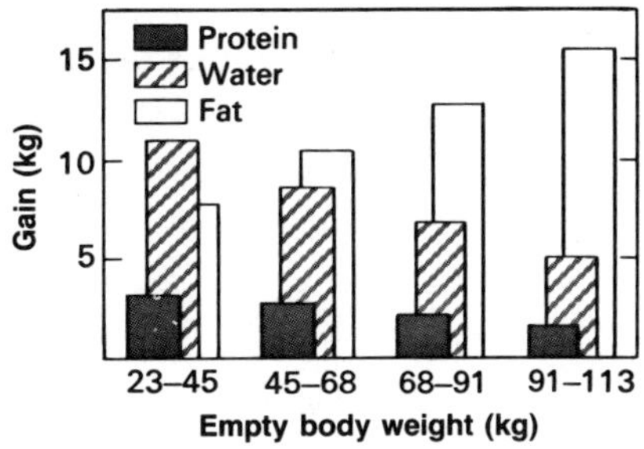

FIG. 19.8. Changes in the body composition of pigs during the growing-finishing period.
From Wallace 1968.

al. 1964). Weights and percentages of fat trim and percentages of bone and lean in barrows and gilts slaughtered at different body weights are shown in Table 19.19 (Martin *et al.* 1980). The weights of lean cuts are greater from the heavier pigs, but the percentage of lean cuts declines because of greater fatness.

Contemporary market pigs selected genetically for leanness may be fed to considerably heavier weights than pigs of earlier generations without unduly increasing carcass fatness. This phenomenon is illustrated in Tables 19.19 and 19.20.

Typical packing house operations are geared to a given number of head slaughtered per day, so that the volume of pork can be increased only by increasing liveweight of pigs slaughtered.

In earlier work, Field *et al.* (1961) evaluated growth, feed conversion, and consumer, packer, and retailer opinions of hogs slaughtered at 73 kg in comparison to those slaughtered at 100 kg. Lightweight pigs had higher gain per unit feed consumed, but heavyweight pigs had faster gains. Processing costs per pig were estimated by packers at 20% greater for lightweight than for heavyweight pigs, but total wholesale value per kilogram of live pig was greater for the lightweight pig, partially offsetting the greater processing cost.

There were no differences between the two weight groups in meat palatability based on taste panel scores or on tenderness. Table 19.21 compares the preference

TABLE 19.19. Effect of Slaughter Weight on Carcass Measurements of Barrows and Gilts Bred for Lean Carcasses on *Ad Libitum* Feed Intake[a]

Criteria	*Body weight at slaughter, kg*					
	73	*85*	*100*	*112*	*126*	*137*
Carcass length, cm	72.1	74.4	77.3	81.6	83.0	86.0
Longissimus muscle area, cm^2	27.1	29.8	32.8	35.9	39.0	40.7
Shoulder, % of carcass wt	29.9	29.0	28.2	28.6	27.4	26.6
Belly, % of carcass wt	15.9	16.7	17.2	17.7	18.5	19.3
Ham, % of carcass wt	25.5	25.4	25.1	24.4	24.6	24.4
Loin, % of carcass wt	21.6	21.6	22.5	22.6	23.3	23.4

Source: Martin et al. 1980.
[a]Data from 126 gilt and 132 barrow carcasses from Lacombe pigs full fed from 42 days of age to slaughter.

TABLE 19.20. Weights of Fat Trim and Percentages of Fat Trim, Bone, and Lean in Barrows and Gilts Slaughtered at Different Body Weights

Criteria	Sex[a]	Body weight at slaughter, kg					
		73	85	100	112	126	137
Fat trim, kg	G	3.43	4.21	4.98	5.43	6.99	8.46
	B	3.38	4.43	5.85	6.39	7.80	8.42
Fat trim, %	G	28.7	29.9	29.4	28.9	31.0	34.6
	B	28.2	30.8	34.6	33.7	36.5	36.0
Bone, %	G	14.6	13.5	12.9	12.7	11.8	10.9
	B	14.7	13.4	12.0	12.0	11.3	11.4
Lean, %	G	56.7	56.6	57.5	58.4	57.2	54.5
	B	57.1	55.8	53.4	54.3	52.2	52.6

Source: Martin et al. 1980.
[a]G = gilt (126 carcasses), B = barrow (132 carcasses).

TABLE 19.21. Preferences by Packers and Retailers for Different Weights of Wholesale Pork Cuts

Cuts	Weight		Percent preferring different weights of each cut	
	(lb)	(kg)	Packers	Retailers
Hams	Less than 11	Less than 5.0	6.8	13.3
	11–13	5.0–5.9	31.9	46.7
	13–15	5.9–6.8	37.9	30.7
	15–17	6.8–7.7	20.1	9.3
	No preference		3.1	1.3
Loins	Less than 10	Less than 4.5	13.2	4.0
	10–12	4.5–5.4	47.4	69.3
	12–14	5.4–6.4	29.8	26.7
	14–16	6.4–7.3	6.5	2.7
	More than 16	More than 7.3	2.9	0
	No preference		0.2	0
Picnics	Less than 4	Less than 1.8	0.7	0
	4–6	1.8–2.7	42.7	90.4
	6–8	2.7–3.6	51.0	12.3
	More than 8	More than 3.6	3.5	0
	No preference		2.1	0
Butts	Less than 3	Less than 1.4	2.5	5.5
	3–5	1.4–2.3	48.0	67.1
	5–7	2.3–3.2	37.0	35.6
	More than 7	More than 3.2	5.2	0
	No preference		2.6	0

Source: Field et al. 1961.

of packers and retailers for various weights of wholesale cuts. Lightweight pigs produced smaller wholesale cuts, but percentage of carcass weight of these cuts was similar for both weight groups except that weight of lard and percentage of lard in the carcass were considerably greater in the heavy-weight group (7.8 vs. 13.4 kg). In general, retailers preferred wholesale cuts similar in weight to those preferred by packers.

Marketing Avenues for Slaughter Swine in the United States

Perhaps no other single factor is more important to the overall success of a commercial swine enterprise than having a suitable and dependable market for the slaughter pigs, for no matter how superior the production methods or the quality of the pig, the profit to the producer depends on obtaining a fair price for the product. The per capita demand for pork in the United States has been maintained at a high and rather stable level for many years. The price paid to the producer is subject to the same supply and demand relationships as any other commodity in a free economy. Thus, the price fluctuates in inverse relationship to the supply, as shown in Fig. 19.9 (USDA 1982) illustrating the changes in hog prices and production in the United States for the years 1960–1976. Demand for pork is inelastic. This means that, for example, although a small increase in supply will cause a decrease in price, the demand will not change greatly. Factors that affect demand for pork include consumer preferences, consumer income, and production and prices of competing products such as beef, poultry, and fish. The corn–

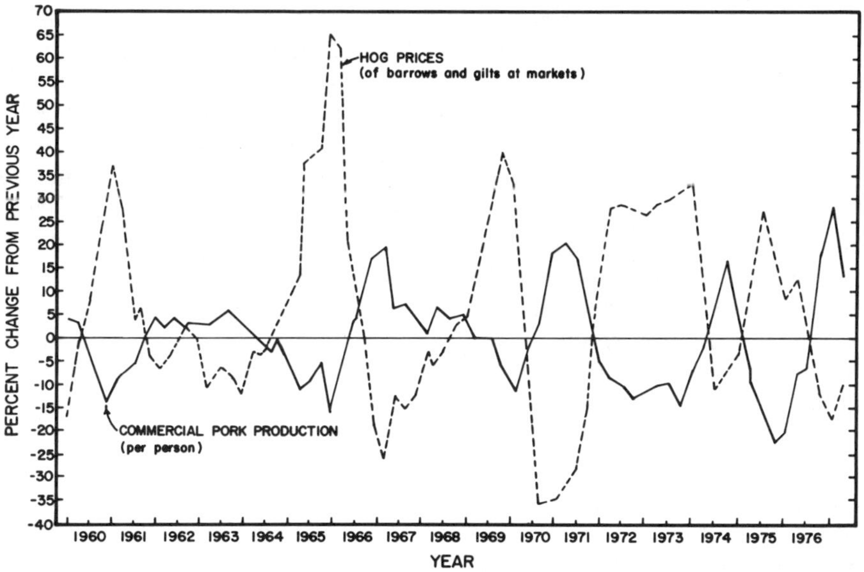

FIG. 19.9. Fluctuation in hog prices and pork production in the United States 1960–1977.
Courtesy USDA 1972 and Williams and Clower 1978.

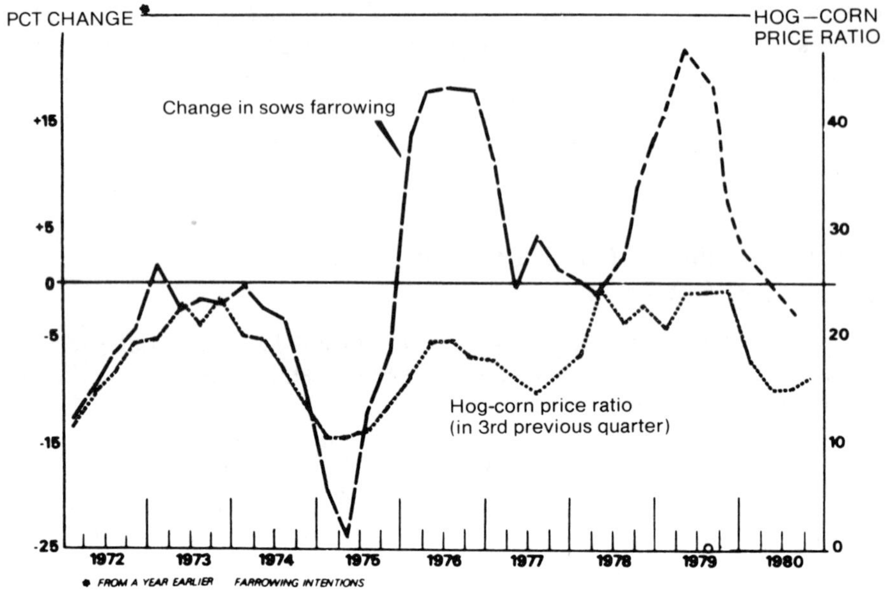

FIG. 19.10. Hog–corn price ratio versus farrowings in the United States.
Courtesy USDA.

hog (or hog–corn) price ratio[1] is an economic yardstick that helps determine whether maize will be sold for cash or be fed to livestock. Changes in hog production generally follow the changes in maize production. When the ratio increases, hog producers generally increase production. This increased supply tends to drive slaughter hog prices down, and these low prices in turn affect the ratio. Thus, the price cycles for slaughter hogs (Fig. 19.9) (USDA 1982) and the number of sows kept for breeding and the hog–corn ratio (Fig. 19.10) (USDA 1980) tend to show a cyclical pattern. Slaughter hog and pork prices in the United States are characterized by huge fluctuations as illustrated in Figs. 19.11A and B. Figure 19.11A shows prices ranging from 28 to 60 dollars during the period 1978–1983 (USDA 1983A). In late summer of 1982, an all-time high price in the United States of about 62 dollars per hundredweight was reached. The complicated set of physical and economic factors that influence the determination of hog prices and supplies has been thoroughly treated by Harlow (1962) and is diagrammed in Fig. 19.12 reproduced from USDA data. No further attempt is made here to describe these delicate interrelationships, but the factors set forth in the figure would be expected to operate in any marketing system.

In the United States, the proportion of slaughter hogs procured directly from the producer by the packer has steadily increased. The terminal markets (public stockyards), at which animals were formerly assembled in large numbers at a central location often many miles from the farm where produced, have gradually given way to more direct methods of sale. Thus, the Chicago stockyards, which were at one time the hub of slaughter livestock movement in the Midwestern

[1]The number of bushels of No. 2 yellow shelled maize that is equal in value to 100 lb (45 kg) of live hog.

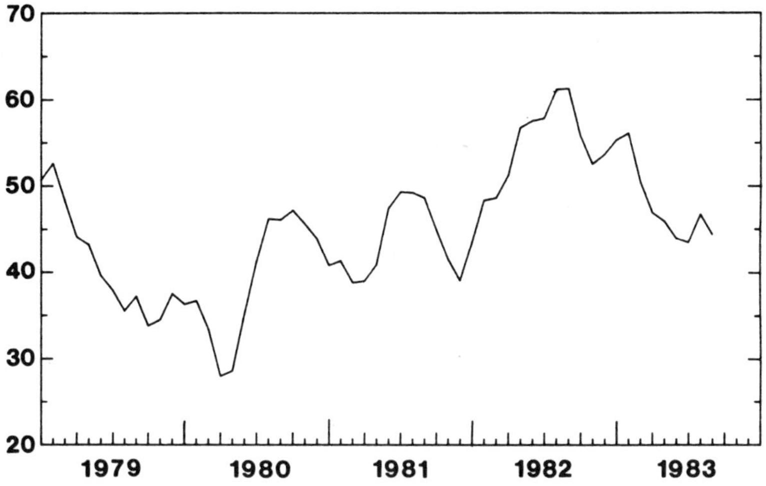

FIG. 19.11A. Slaughter hog price fluctuations (dollars per cwt).
Courtesy of USDA 1983A.

United States, are now closed and the meat packing plants which had been located in Chicago have become replaced by smaller plants in the areas of production. Thus, most swine in the United States are now marketed by truck rather than by rail and reach the slaughter plant much sooner. In 1980, the market avenues for slaughter hogs in the United States were as shown in Table 19.22.

The total number of hogs that are slaughtered commercially in the United States

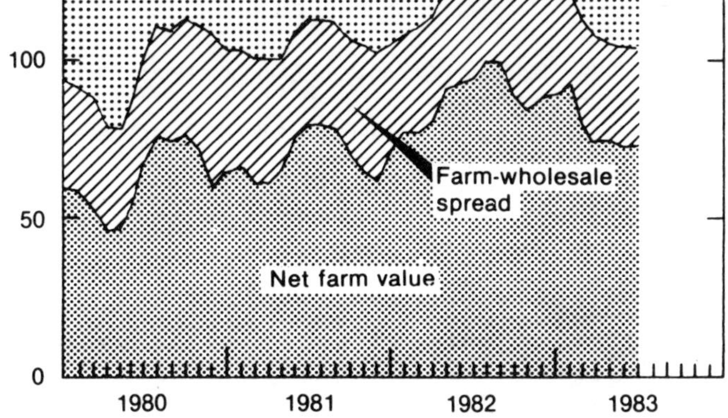

FIG. 19.11B. Price spreads for pork.
Courtesy of USDA 1983B.

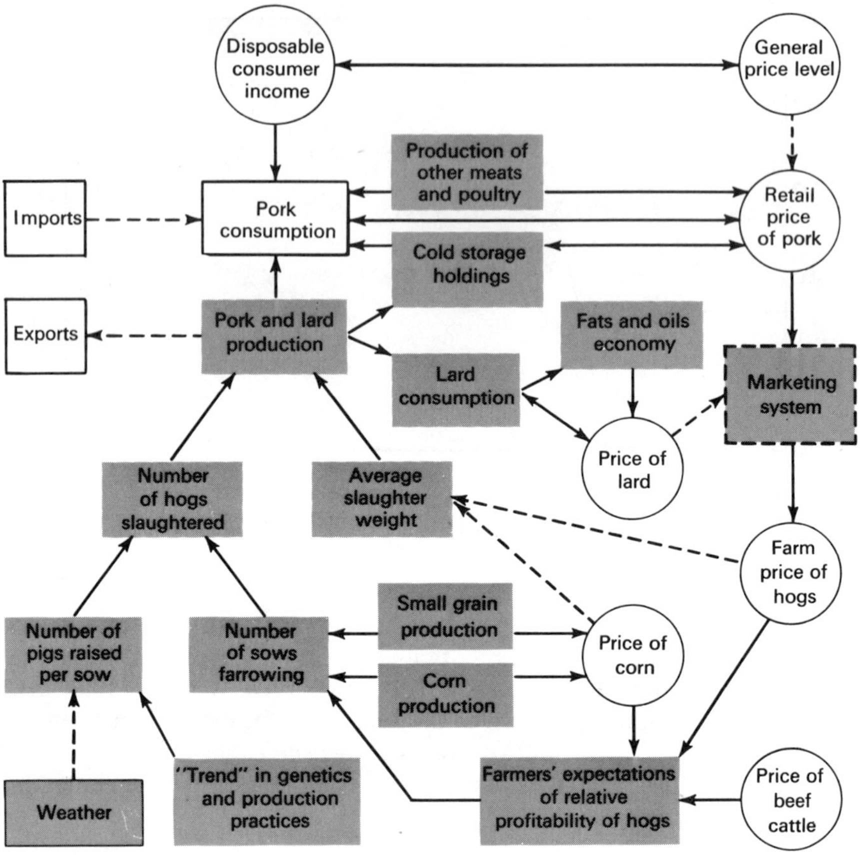

FIG. 19.12. Supply–demand–price relationship in swine production. The arrows show direction in influence. Solid lines indicate major paths of influence; broken lines show minor paths of influence.
Courtesy USDA.

TABLE 19.22. Marketing Channels of Slaughter Hogs, 1980

Size of operation	*Percentage by volume of marketings*			
	Packing plant	*Buying point*	*Terminal and other*[a]	*Total*
5,000– 6,999	57.9	31.8	11.2	100.0
7,000–14,999	65.3	31.3	3.0	100.0
≥15,000	94.1	3.3	2.6	100.0
All operations	76.3	19.1	4.6	100.0

Source: Rhodes et al. 1981.
[a]The percentage of "other" was 1% or less in each category.

accounts for nearly all hog slaughter—more than 99%. The Livestock Division, Consumer and Marketing Service, USDA, publishes a weekly market news summary of livestock, meat, and wool prices. This weekly summary is a useful guide to market trends.

Modern communication and computer technology has enhanced the attractiveness of marketing slaughter hogs on future markets. A live hog futures market is one in which prices are established for live hogs not to be delivered until some future date. The two basic incentives for the producer to use future markets are that current futures market prices may be higher than cash prices when his or her hogs are ready for delivery or that the probability of later prices lower than the current futures prices appears very high. Hedging (the simultaneous buying and selling of futures contracts in an effort to reduce risk) is a marketing tool that is effective in maximizing profit if done wisely, but may result in a less than maximum profit if poor judgement or incomplete information is used in the decision-making process. About 2.5% of slaughter hogs sold in the United States in 1980 were sold by futures contract (Rhodes *et al.* 1981); this is a substantial reduction from the figure for 1978, but similar to that of 1975, and may reflect a relationship between the prevalence of contracting and the hog price cycle (Rhodes *et al.* 1981).

SIZE OF U.S. COMMERCIAL SWINE FARMS

The average number of pigs produced per farm is increasing, and the number of farms producing pigs is declining. The trend toward larger farms is illustrated in Table 19.23 (National Pork Producers' Council 1971), which shows that in 1970 about 30% of all hogs marketed by members of the National Pork Producers' Council (NPPC) were produced on farms producing more than 1000 pigs/yr as compared to only 13% in 1965. This group of farmers represents approximately 10% of all U.S. swine producers so that the estimate may be biased upward, as NPPC members are probably larger than average producers. However, the direc-

TABLE 19.23. Pork Production Trends in the United States

Farm size (No. of pigs/yr)	*Percent of total producers*[a]	
	1965	*1970*
To 199	28	9
200–499	34	26
500–999	25	35
1,000–3,000	12	26
More than 3,000	1	4

Source: National Pork Producers' Council 1971.

[a]Represents approximately 10% of all hog producers in United States and is probably biased in that members of NPPC are probably larger producers than the overall U.S. average.

TABLE 19.24. Distribution of Marketings by Size Group, 1980

Size group	Marketings, %			
	All hogs	Slaughter hogs	Feeder hogs	Breeding stock
15,000 or more	40.4	43.6	29.4	49.7
7,000–14,999	41.8	36.6	58.6	36.7
5,000– 6,999	17.8	19.8	12.0	13.6
	100.0	100.0	100.0	100.0

Source: Rhodes et al. 1981.

tion of the trend is clear and it seems probable that now more than 50% of hogs slaughtered are from farms producing more than 1000 pigs per year. Contrary to what might be expected, individual and family farms still dominate the swine industry. In 1980, less than 1% of farms producing 1000–2500 market pigs annually were non-family-owned, and more than 70% of "factory pork farms" (farms producing more than 5000 market pigs per year) were individually or family owned (Table 19.24). Completely confined highly automated swine production units capable of producing several thousand pigs per year are now quite numerous. A survey of distribution of swine marketing from herds producing more than 15,000 pigs yearly (Rhodes *et al.* 1981) indicated that 40.4% of all of these hogs and 43.5% of slaughter hogs marketed in 1980 came from operations producing more than 15,000 pigs. The results of this survey are summarized in Table 19.25.

TABLE 19.25. Facilities Principally Used in Finishing Slaughter Hogs, 1980

Principal finishing facilities	% of reporting units by size of unit			
	U.S.	5,000–6,999	7,000–14,999	15,000+
Total confinement, environmental control	35.6	30.9	38.7	37.8
Total confinement, natural ventilation	23.0	25.4	20.9	24.4
Total confinement, environmental control in winter, natural in summer	13.5	15.9	10.4	17.8
Open front, concrete floors	20.1	18.3	22.7	15.6
Dirt lots, some shelter	5.7	7.1	5.5	2.2
Other	2.1	2.4	1.8	2.2
Total	100.0	100.0	100.0	100.0

Source: From Rhodes et al. 1981.

Marketing Avenues for Purebred Swine in the United States

Although the number of purebred swine in the United States is less than 10% of the total hog population, the purebred industry is of utmost importance to the commercial hog producer for seedstock to use in crossbreeding and upgrading programs within the commercial herd.

Sale of purebred animals for breeding is generally either by private treaty or by auction. Auctions may be private farm auctions arranged by large individual breeders, local or state breed association-sponsored auctions to which animals are consigned by several breeders, or test station sales which often market tested animals to which a ranking based on a selection index (see Chapter 4) has been assigned. Normally such auctions are held either annually or more frequently at the state or local testing station.

The National Association of Swine Records has suggested a series of procedures for use in purebred swine transactions at private treaty or in public auction. These terms are as follows:

Standard purebred swine guarantees. Note: It is understood when purebred swine are offered for sale as breeding stock they are guaranteed to be breeders by the seller. At the same time, it is further recognized that care and treatment of the animals subsequent to the date of sale may, after a reasonable length of time, affect the animal's reproductive ability, and therefore any guarantee must be confined to a reasonable time limit.

General provisions. (1) *Registration.* In all purebred transactions, the certificate of registration is an integral part of the transaction, and shall be delivered to the purchaser, properly transferred on the Association records at the seller's expense. (2) *Health.* Sellers are obligated by law to provide a health certificate specifying the types and kinds of vaccinations; blood tests showing diseases to which animal is negative; and a statement of good health. See your veterinarian for these regulations. (3) *Transportation.* When an adjustment involves the exchange of an animal by public carrier such expense shall be shared equally by buyer and seller. (4) *Animals under 5 months.* Because subsequent feed and care can also greatly affect the future usefulness for breeding purposes of animals under 5 months of age, at time of sale, the seller cannot be reasonably expected to make any adjustments.

Standard guarantee for boars. All boars after they have reached 7 months of age are guaranteed breeders *except* when let run with the sow herd. Should any boar fail to serve or settle sows, the seller shall make a replacement satisfactory to the buyer or refund the purchase price when the boar is returned, retested, and negative to same blood tests as when sold and in satisfactory state of flesh; or the seller may direct the marketing of the boar and refund the difference between the purchase price and market value as shown by sales receipts.

No adjustment will be completed until certificate of registry is re-transferred to seller, where animal is returned.

However, the seller shall always have the right, if he so desires, to try a boar for as much as 30 days before making final adjustment, to determine his usefulness as a breeder.

All requests for adjustment under this guarantee must be made through the mails within 6 weeks of date of sale or within 6 weeks after boar becomes 7 months of age. No charge for feed or maintenance shall be made by either buyer or seller.

Standard guarantee for open gilts. All open females over 7 months of age are guaranteed open and breeders. Should any prove to be bred, she may be returned at the seller's expense and the purchase price refunded. Should any prove to be nonbreeders, the seller shall make a satisfactory replacement or refund the purchase price when the animal is returned, retested, and negative to same blood tests as when sold and in a satisfactory state of flesh. Or in either case the seller may direct the marketing of such animals and refund the difference between the purchase price and the market value as shown by sales receipt.

No adjustment will be completed until certificate of registry is re-transferred to seller.

However, the seller shall always have a right, if he so desires, to hold a claimed nonbreeder 45 days to determine whether she is with pig or settles immediately, before making final adjustment.

All claims for adjustment shall be made through the mails within 3 calendar months of the date of sale.

Standard bred sow guarantee. All bred sows are guaranteed to be with pig to the indicated boar and date of service. Any proving otherwise may be returned and exchanged for another sow satisfactory to the purchaser, or refund of the purchase price, if the sow is returned, retested, and negative to same blood tests as when sold and in satisfactory state of flesh; and upon re-transfer of certificate of registry to seller.

If a buyer wishes to retain the sow, the seller shall return one-half the difference between the purchase price and the market value of the sow at time of purchase.

The seller may direct the marketing of the sow and refund the difference between the purchase price and the market value of the sow as shown by sales receipt and upon return of certificate of registry.

All claims for adjustment must be made within 114 days of the indicated date of service. However, purchasers are warned that a sow may farrow as much as a week before, or a week later than this estimated farrowing date. No guarantee is made as to the number or quality of pigs a sow will farrow or raise, unless seller specifies same in writing.

ECONOMICS OF LARGE SWINE UNITS IN THE UNITED STATES

Cost comparisons for 1970 between three different types of confinement systems were made by University of Illinois economists (Mueller and Kesler 1970). A 32-sow slotted-floor farrowing house was included in all three systems, and the farrowing schedule for each was six times per year (192 litters at 7.5 pigs weaned per litter = 1440 pigs).

System A included remodeled existing buildings for use as solid-floor nursery and finishing units.

System B included a slotted-floor nursery building and a partially slotted-floor open-front finishing unit.

System C included an enclosed slotted-floor nursery, enclosed slotted-floor finishing building, and an enclosed slotted-floor sow-confinement building. The investments in buildings and equipment were calculated from current building contractor's and engineering estimates. The estimated costs per 45.5 kg of live hog produced were greatest for system C and least for system A. The figures obtained in that study cannot, of course, be applied to all types of enterprises and conditions, but they do indicate the relative proportions of hog production costs assigned to various items.The feed cost is only 57.5–61% of the total cost of production under the conditions of high building, equipment, and labor cost that are typical of modern U.S. swine operations. This is a considerable change from the past, when 75–85% of total production costs were for feed. In many parts of the world where labor costs are less and in the tropical areas where building costs are considerably less, this higher figure of 75–85% or more for feed would still prevail.

Building and equipment cost represent 10–15% of total production costs in most U.S. swine enterprises. Production units of all sizes tend to rely heavily on confinement facilities, more than 80% of which utilize some form of liquid manure system for waste handling. The kinds of facilities used for finishing slaughter hogs in the United States in 1980, categorized by size of production unit, are shown in Table 19.25 (Rhodes *et al.* 1981).

Increasing land values and labor costs provide economic incentive to move toward confinement swine production in both temperate and tropical environments.

REFERENCES

ANON. 1982. Natl. Hog Farmer, August 15, 1982, p. 62. Webb Publishing Co. St. Paul, Minnesota.

ALLEN, E., FORREST, J. C., CHAPMAN, A. B., FIRST, N. L., BRAY, R. W., and BRISKEY, E. J. 1966. J. Anim. Sci. *25,* 1962.

BACHE, D. H., and FOSTER, J. R. 1979. Natl. Pork Industry Handbook *PIH-48*. Cooperative Extension Service, Purdue Univ., West Lafayette.

BACHE, D. H. and FOSTER, J. R. 1980. Natl. Pork Industry Handbook *PIH-13, -14, -15, -17*. Cooperative Extension Service, Purdue Univ., West Lafayette, Indiana.

BECKER, D. E., JENSEN, A. H., and HARMON, B. G. 1966. Univ. Ill. Circ. 866.

BOWLAND, J. P., and STANDISH, J. F. 1966. J. Anim. Sci. *25,* 377.

BRAY, R. W. 1968. *In* The Pork Industry: Problems and Progress. (D. G. Topel, Editor). Iowa State Univ. Press, Ames.

BRISKEY, E. J., and LISTER, D. 1968. *In* The Pork Industry: Problems and Progress (D. G. Topel, Editor). Iowa State Univ. Press, Ames.

CAHILL, V. R., TEAGUE, H. S., KUNKLE, L. E., MOXON, A. L., and RUTLEDGE, E. A. 1960. J. Anim. Sci. *19,* 1036.

CASSENS, R., GIESLER, F., and KOEB, Q. 1972. *In* Proc. Pork Quality Symp. Natl. Pork Producers Council, Des Moines, Iowa 50312.

CHRISTIAN, L. L. 1972. *In* Proc. Pork Quality Symp., Natl. Pork Producers Council, Des Moines, Iowa 50312.

CLAUSEN, H. 1959. Proc. Cornell Nutr. Conf. Cornell Univ., Ithaca, New York.

CRAFT, W. A. 1958. J. Anim. Sci. *17,* 960.

CUNNINGHAM, H. M., and FRIEND, D. W. 1966. J. Anim. Sci. *25,* 663.

CURTIS, S. E. 1981. Environmental Management. Animal Environment Services, Mahomet, Illinois.

ECHTERNKAMP, S. E., TEAGUE, H. S., PLIMPTON, R. F., JR., and GRIFO, A. P., JR. 1969. J. Anim. Sci. *28,* 653.

EIKELENBOOM, G. and MINKEMA, D. 1974. Neth. J. Vet. Sci. *99,* 421.

ELSLEY, F. W. 1968. *In* The Pork Industry: Problems and Progress (D. G. Topel, Editor). Iowa State Univ. Press, Ames.

ENSMINGER, M. E. 1970. Swine Science, 4th Edition. The Interstate Printers and Publishers, Danville, Illinois.

FIELD, R. A., VARNEY, W. Y., and KEMP, J. D. 1961. J. Anim. Sci. *20,* 742.

GREER, S. A. N., HAYS, V. W., SPEER, V. C., McCALL, J. T., and HAMMOND, E. G. 1965. J. Anim. Sci. *24,* 1008.

HARLOW, A. A. 1962. U.S.D.A. Tech. Bull. No. 1274.

HAYS, V. W. 1968. *In* The Pork Industry: Problems and Progress (D. G. Topel, Editor). Iowa State Univ. Press, Ames.

HEADLEY, V. E., MILLER, E. R., ULLREY, D. E., and HOEFER, J. A. 1961. J. Anim. Sci. *20,* 311.

HWANG, P. T., ADDIS, P. B., REMPEL, W. E., and ANTONEK, A. 1977. J. Anim. Sci. *45,* 1015.

JENSEN, A. H. 1982. Feedstuffs Redbook Issue, p. 32.

JONSSON, P. 1968. *In* The Pork Industry: Problems and Progress. (D. G. Topel, Editor). Iowa State Univ. Press, Ames.

JUDGE, M. D. 1968. *In* The Pork Industry: Problems and Progress. (D. G. Topel, Editor). Iowa State Univ. Press, Ames.

JUDGE, M. D. 1972. *In* Proc. Pork Quality Symp. Natl. Pork Producers Council, Des Moines, Iowa 50312.

KUKOYI, E. A., ADDIS, P. B., McGRATH, C. J., REMPEL, W. E., and MARTIN, F. B. 1981. J. Anim. Sci. *52,* 278.

LANNEK, N., LINDBERG, P., and RULCHER, C. 1967. Svensk Veterinar Tidning. (Cited by Wismer-Pederson, 1968).

MANER, J. H., ROUNSAVILLE, T. R., GALLO, J. T., POND, W. G., and VANVLECK, L. D. 1977. J. Anim. Sci. *45,* 1243.

MARTIN, A. H., SATHER, A. P., FREDEEN, H. T., and JOLLY, K. W. 1980. J. Anim. Sci. *50,* 699.

MUELLER, A. G., and KESLER, R. P. 1970. Natl. Hog Farmer Swine Info. Serv. Bull. *A18*.

NATIONAL PORK PRODUCERS' COUNCIL. 1971. Feedstuffs *43,* 20.

NATIONAL RESEARCH COUNCIL. 1981. Effects of Environment on Nutrient Requirements of Domestic Animals. Natl. Res. Council. Natl. Acad. Sci., Washington, D.C. 152 pp.
PALMER, W. M., TEAGUE, H. S., and GRIFO, A. P., JR. 1971. J. Anim. Sci. *32,* 525.
PATTERSON, R. L. S. 1968. J. Sci. Food Agric. *19,* 31.
PIERCE, E. A. 1969. Cornell Univ. Ext. Bull. *1210.*
PLIMPTON, R. F., JR., CAHILL, V. R., TEAGUE, H. S., GRIFO, A. P., JR., and KUNKLE, L. E. 1967. J. Anim. Sci. *26,* 1319.
POND, W. G., and PIERCE, E. A. 1971. Cornell Univ. Ext. Bull. *1045.*
RHODES, V. J., and GRIMES, G. 1982. The structure of the U.S. pork industry: National Pork Industry Handbook *PIH-85,* Cooperative Extension Service, Purdue Univ., West Lafayette, Indiana.
RHODES, V. J., GONZALES, J., and GRIMES, G. 1981. Hog Farm Management, December, p. 18.
SEYMOUR, E. W., SPEER, V. C., HAYS, V. W., MANGOLD, D. W., and HAZEN, T. E. 1964. J. Anim. Sci. *23,* 375.
STOUT, T. T., and ARMSTRONG, J. H. 1960. Purdue Univ. Res. Bull. *710.*
STOUT, T. T., and COX, C. B. 1959. Purdue Univ. Res. Bull. *685.*
SYBESMA, W. 1972. *In* Proc. Pork Quality Symp. Natl. Pork Producers Council, Des Moines, Iowa.
TEAGUE, H. S., PLIMPTON, R. F., JR., CAHILL, V. R., GRIFO, A. P., JR., and KUNKLE, L. E. 1964. J. Anim. Sci. *23,* 332.
USDA 1967. Livestock and meat statistics. Suppl. to Stat. Bull. *No. 33,* Washington, D.C.
USDA 1968. Marketing Research Rep. *No. 849,* Economic Research Service, Washington, D.C.
USDA 1970A. Grades for pork carcasses. Marketing Bulletin *No. 49.* USDA Consumer and Marketing Service, U.S. Govt. Printing Office, Washington, D.C.
USDA 1970B. Grades for slaughter swine and feeder pigs. Marketing Bulletin *No. 51.* USDA Consumer and Marketing Service, U.S. Govt. Printing Office, Washington, D.C.
USDA 1971. Livestock and Meat Situation, *LMS-182.* Economic Research Service, Washington, D.C.
USDA 1972. Market News Weekly Summary and Statistics, Livestock and Wool. Livestock Division, Consumer and Marketing Service, Washington, D.C.
USDA 1980. Livestock and Meat Situation. Economics, Statistics and Cooperatives Services *LMS-234.* Washington, D.C.
USDA 1982. Agricultural Prices. Statistical Reporting Service, June. Washington, D.C.
USDA 1983A. Agric. Prices, September. Crop Reporting Board, SRS-USDA, Washington, D.C.
USDA 1983B. Livestock and Poultry Outlook and Situation Rep. *LPS-7,* October. ERS-USDA, Washington, D.C.
VERSTEGEN, M. W. A., BRASCAMP, E. W., and VANDERHEL, W. 1978. Can. J. Anim. Sci. *58,* 1.
WALLACE, H. D. 1968. *In* The Pork Industry: Problems and Progress (D. G. Topel, Editor). Iowa State Univ. Press, Ames.
WILLIAMS, W. F., and CLOWER, J. A. 1978. Natl. Pork Industry Handbook *PIH-51.* Purdue Univ., W. Lafayette, Indiana.
WISMER-PEDERSON, J. 1968. *In* The Pork Industry: Problems and Progress (D. G. Topel, Editor). Iowa State Univ. Press, Ames.

Appendixes

Appendix 1

Sample Sow and Litter Records

Swine Breeding Record 19 __

Sow number		Breeding dates									Farrowing date	
		First service			Second service			Third service			Date due	Date farrowed
Herd	Assoc.	Nat.	Art.	Sire	Nat.	Art.		Nat.	Art.			

Sow Permanent Performance Record

Sow

Herd no. ____________ Birth date ____________ 21-day wt ____________

Assoc. no. ____________ Sire ____________ 200-lb age ____________

Breed ____________ Dam ____________ Feed/gain ____________

Disposal ________________________

Littermates

Litter size ____________ Av. back fat ____________

Av. 21-day wt ________ Av. loin eye area ________

Av. age at 200 lb ________

Litter no.	Boar	Farrowed		Birth wt	No. weaned	21-day wt	Age at 200 lb	Feed effic.	Back fat	Loin eye area	Carc. length	Remarks
		Date	No.									

Sow and Individual Litter Record

Sow record

	Birth date ______	21-day wt, lb ______	Back fat ______
Assoc. no. ______	Sire ______	200 lb age, day ______	Index ______
Herd no. ______	Dam ______	Feed/gain ______	Litter size ______

Sow's littermates (No. observed in parentheses)

Av. 21-day wt ______ (   ) Av. back fat ______ (   )
Av. 200 lb age ______ (   ) Av. loin eye area ______ (   )

Litter record Date bred ______ Boar ______ nat. or art.
Date farrowed ______ No. of services ______

Herd no.	Sex	Weight		200 lb age		Back fat	Feed/lb gain	Loin eye area	Index	Remarks
		Birth	21 day	Age	Weight					
Total										
Average										

Appendix 2

Swine Industry Directory; Pork Cookbooks and Facts about Pork

SWINE INDUSTRY DIRECTORY

National Associations

American Association of Swine Practitioners
Fred D. Wertman
5921 Fleur Drive
Des Moines, Iowa 50309

American Feed Manufacturers Association, Inc.
1701 North Fort Meyer Drive
Arlington, Virginia 22209

American Meat Institute
59 East Van Buren Street
Chicago, Illinois 60605

Livestock Conservation, Inc.
239 Livestock Exchange Building
South St. Paul, Minnesota 55075

Midwest Plan Service
122 Davidson Hall
Iowa State University
Ames, Iowa 50011

National Association of Swine Records
P.O. Box 647
Lebanon, Indiana 46052

National Feeder Pig Marketing Association
Peter Fish, Executive Secretary
Livestock Marketing Association
4900 Oak Street
Kansas City, Missouri 64122

National Live Stock and Meat Board
444 North Michigan Avenue
Chicago, Illinois 60611

National Pork Producers Council
P.O. Box 10383
Des Moines, Iowa 50306

National SPF Swine Accrediting Agency, Inc.
Conrad, Iowa 50621

National Swine Improvement Federation
C. Christians, Executive Secretary
Department of Animal Science
University of Minnesota
St. Paul, Minnesota 55108

Pork Producer Buying Guide
National Hog Farmer (November 1983)
The Webb Company
1999 Shepard Road
St. Paul, Minnesota 55116

U.S. Meat Export Federation
Suite 7200, 3333 Quebec Street
Denver, Colorado 80207

U.S. Animal Health Association
Suite 6, Monument Avenue
Richmond, Virginia 23220

Swine Industry Periodicals

Hog Farm Management
Miller Publishing Company, P.O. Box 67
2501 Wayzata Blvd.
Minneapolis, Minnesota 55440

National Hog Farmer
The Webb Company
1999 Shepard Road
St. Paul, Minnesota 55116

Pig American
Watt Publishing Company
Mount Morris, Illinois 61054

Pig Farming
Wharfedale Road
Ipswich IPI 4LG
England

Pork '83
Livestock Division
Vance Publishing Corporation
7950 College Blvd.
Shawnee Mission, Kansas 66201

Swine Breed Associations

Berkshire

American Berkshire Association
601 West Monroe Street
Springfield, Illinois 62704
Phone: (217) 523–2123
Jack Wall, Secretary
Monthly Publication: *Berkshire News*

Chester White

Chester White Swine Record Association
817 East Ninth Street, Box 228
Rochester, Indiana 46975
Phone: (219) 223-3512
Daniel Parrish, Exec. Secretary
Monthly publication: *Chester White Journal*

Duroc

United Duroc Swine Registry
1803 West Detweiller Drive
Peoria, Illinois 61615
Phone: (309) 691-8094
Gary Huffington, Secretary
Monthly publication: *Duroc News*

Hampshire

Hampshire Swine Registry
1111 Main Street
Peoria, Illinois 61606
Phone: (309) 674-9134
Robert Naylor, Field Service Director
Monthly publication: *Hampshire Herdsman*

Hereford

National Hereford Hog Record Association
Route 3
Shelbyville, Illinois 62565
Mrs. Mary Chapman, Secretary

Landrace

American Landrace Association
P.O. Box 647
Lebanon, Indiana 46052–0647
Phone: (317) 482-3042
Ernie Barnes, Secretary and Editor
Bimonthly publication: *American Landrace*

Poland China

Poland China Record Association, Inc.
P.O. Box B
Knoxville, Illinois 61448
Phone: (309) 289-6786
Mike Dunphy, Secretary
Monthly publication: *Poland China World*

Spotted Swine

Spotted Swine Record, Inc.
110 West Main Street
Bainbridge, Indiana 46105
Phone: (317) 522–6272
Harry Miller, Executive Secretary
Monthly publication: *Spotted News*

Tamworth

The Tamworth Swine Association
Rural Route 2, Box 36
West Manchester, Ohio 45697
Phone: (513) 695-0114
Tom Fenton, Jr., Secretary

Yorkshire

American Yorkshire Club
Box 2417, 1769 U.S. 52 North
West Lafayette, Indiana 47906
Phone: (317) 463-3593
Glenn Conatser, Executive Secretary
Monthly publication: *Yorkshire Journal*

Private Companies Selling Hybrid Breeds (Based on Combinations of the Above Breeds)

Babcock Swine, Inc.
Jim McPeak, President
P.O. Box 759
Rochester, Minnesota 55903

Dekalb Swine Breeders, Inc.
Dekalb, Illinois 60115

Farmers Hybrid Company, Inc.
P.O. Box 4528
Des Moines, Iowa 50306

Kleen Leen
Cedar Rapids, Iowa 52401

Pig Improvement Company
Box 128
Spring Green, Wisconsin 53588

PORK COOKBOOKS AND FACTS ABOUT PORK

Lean Facts about Pork emphasizes the nutritional benefits of pork including a graph giving dietary contributions of one 3-oz serving of pork. For a free copy send self-addressed, stamped envelope to National Live Stock and Meat Board, Dep. LFAP, 444 North Michigan Avenue, Chicago, Illinois 60611.

Lean Facts about Pork and Calories gives the latest information on pork's caloric value. For a free copy send self-addressed, stamped envelope to National Live Stock and Meat Board, Dep. LFCAL, 444 North Michigan Avenue, Chicago, Illinois 60611.

Lean Facts about Pork and Cholesterol explains that pork is not a high-cholesterol food. For a free copy send self-addressed, stamped envelope to National Live Stock and Meat Board, Dep. LFCHO, 444 North Michigan Avenue, Chicago, Illinois 60611.

Super Pork is a cookbook by Ron Lieske and "Ed" the pig. It is spiral-bound for kitchen counter ease. The cost is $8.95 plus $1.20 postage/handling, from NHF Super Pork, 1999 Shepard Road, St. Paul, Minnesota 55116. Make checks payable to National Hog Farmer; allow 4–6 weeks for delivery.

Calorie-Conscious Pork Recipes is a guide to making pork an important part of a health-promoting dietary fitness plan. It features ten calorie-calculated recipes and information on nutrition, dieting, and exercise. For a free copy, send self-addressed, stamped envelope to National Live Stock and Meat Board, Dep. CCPR, 444 North Michigan Avenue, Chicago, Illinois 60611.

Pork on a Budget features 18 economical recipes to help consumers meet the challenge of stretching their food dollar. Includes smart shopping tips and nutrition information. For a free copy, send self-addressed, stamped envelope to National Live Stock and Meat Board, Dep. POB, 444 North Michigan Avenue, Chicago, Illinois 60611.

Pork, a 65-page chapter in "Another Revolution in U.S. Farming?", explains the forces shaping the future in the industry. No cost. AER 441, ESS Publications, Room 0054-S, USDA, Washington, D.C. 20250.

Meat Handbook, 1979. 4th Edition, by Albert Levie, an in-depth look at those portions of the meat industry involved in slaughter, processing, buying, and selling, including a review of the Wholesome Meat Act and its implications, information on meat quality, sections on procurement, purchase specifications, merchandising by-products, cooking, and palatibility. 18 chapters, 354 pages, ISBN-0-87055-315-1, U.S. $24.50. AVI Publishing Co., 250 Post Rd. East, P.O. Box 831, Westport, Connecticut 06881.

Agricultural Outlook is published 11 times a year to give an overview of the economics of agriculture, situation analysis, and outlook. Commodity highlights is one of the regular features. ESS Publications, Room 0054-S, USDA, Washington, D.C. 20250.

Livestock Slaughter provides monthly numbers and liveweights of hogs and other red meat animals slaughtered. Crop Reporting Board, Room 0054-S, ESS/USDA, Washington, D.C. 20250.

Appendix 3

Glossaries of Terms Used in Swine Production and Feed Processing

Many of the terms defined under Physiological Terms were taken directly from the glossary on pp. 499–510 of "Principles of Nutrition" by E. D. Wilson, K. H. Fisher, and M. E. Fuqua, 3rd Edition, 1975, John Wiley and Sons, New York, and from the glossary on pp. 387–396 of "Basic Animal Nutrition" by D. C. Church and W. G. Pond, 2nd Edition, 1982, John Wiley and Sons, New York. Terms defined under International Feed Nomenclature were taken from Appendix B of "Names for Feed Processes and Their Use in Evaluating the Nutrient Content of Feeds" by L. E. Harris and E. W. Crampton, *in* "Effect of Processing on the Nutritional Value of Feeds" (1973), National Research Council, National Academy of Sciences, Washington, D.C. Terms defined under Feed Manufacturing and Feed Ingredient Terms were taken from the list of Official Feed Terms, Association of American Feed Control Officials, State Capitol Building, Charleston, West Virginia. Terms defined under Animal Terms and Husbandry Terms are those in common use in the swine industry.

ANIMAL TERMS

barrow. Male castrated before sexual maturity.
boar. Male swine of any age.
breeding stock. Female or male young or mature swine kept for breeding to produce piglets.
feeder pig. Young swine after weaning and before reaching slaughter weight.
finishing pig. Young swine generally weighing more than 70 kg but not yet heavy enough for slaughter.
gilt. Female swine of any age to second pregnancy.
growing pig. Young swine after weaning, generally weighing less than 70 kg.

hog. Swine of either sex, generally referring to immature gilts, barrows or boars.

market pig. *See* Slaughter pig.

mummified fetus. Dead partially resorbed fetus, characterized by shrunken, brownish appearance, expelled at the time of birth of normal live littermates.

mummy. *See* Mummified fetus.

pig. In United States, refers to young swine of either sex; in Europe, refers to all ages and either sex and is used in preference to swine.

piglet. Baby pig.

runt. Piglet or growing pig considerably smaller than littermates or contemporaries and often unhealthy in appearance.

shoat. Young swine of either sex, generally weighing less than 50 kg.

slaughter pig. Young swine ready for slaughter, usually weighing 90–120 kg.

sow. Female swine having produced one or more litters.

stag. Male castrated after reaching sexual maturity.

suckling pig. Young swine before weaning.

swine. Broad term applying to all animals belonging to *Sus suis.*

weanling pig. Young swine at or shortly after weaning.

HUSBANDRY TERMS

back fat probe. Measure the depth of back fat over the shoulder, loin, and rump on market pigs or breeding stock using a scalpel and metal rule or an ultrasonic instrument.

castrate. Remove the gonads (testes in males, ovaries in females).

creep. An area near the lactating sow accessible only to the suckling pigs.

creep feed. baby pig feed provided in the creep area beginning at about 2 weeks.

dock. Remove the tail by cutting or pinching.

farrow. To give birth to a litter of pigs.

farrowing crate. Enclosure for the sow before farrowing and during early lactation to minimize accidental newborn pig deaths and injuries overlaying by the dam.

hand breeding. *See* Hand-mating.

hand mating. Supervised mating of selected males and females for purposes of recording pedigree.

pen breeding. *See* Hand breeding.

pen mating. Mating of groups of breeding females by one or more boars penned with the females through one or more estrous cycles.

wean. Remove the piglets from their dam.

PHYSIOLOGICAL TERMS

acetylcholine. An acetic acid ester of choline; normally present in many parts of the body; functions in the transmission of nerve impulses.

active transport. The movement of substances (particularly electrolyte ions) across cell membranes, usually against a concentration gradient. Unlike diffusion or osmosis, active transport requires the expenditure of metabolic energy.

adenine. One of four nitrogenous bases found in DNA.

adenosine triphosphatase (ATPase). An enzyme in muscle that catalyzes the hydrolysis of the terminal phosphate group of adenine triphosphate.

adenosine triphosphate (ATP). A compound that consists of one molecule each of adenine (a purine) and ribose (a 5-carbon atom sugar) and three molecules of phosphoric acid; it is required for energy transfer and the phosphorylation of compounds.

ADF. Acid detergent fiber, the fraction of a feedstuff analyzed by the Van Soest scheme of detergent analysis used to divide carbohydrates in plant constituents into those highly available and poorly available to animals; ADF content of a feed reflects the amounts of carbohydrates not solubilized by acid detergent.

adrenaline. *See* Epinephrine.

albumin. Group of globular proteins; a major constituent of blood serum protein.

amine. A chemical compound formed from ammonia (NH_3) by replacement of one or more of the hydrogen atoms with hydrocarbon groups ($—CH_3$ or $—C_2H_5$ and others).

amino acids. Simplest organic structure of which proteins are formed; many different amino acids occur in nature of which about 10 (for most animals) are required in the diet; all have the common property of containing a carboxyl group and an amino group on the adjacent C atom.

amino group. A chemical structure, $—NH_2$; a constituent of all amino acids, attached to the C atom adjacent to the carboxy group.

anemia. A deficiency in the blood, either in quality or quantity.

hyperchromic anemia. A decrease in hemoglobin that is proportionately much less than the decrease in the number of erythrocytes.

hypochromic anemia. A decrease in hemoglobin that is proportionately much greater than the decrease in the number of erythrocytes.

macrocytic anemia. A condition in which the erythrocytes are much larger than normal.

megaloblastic anemia. A condition in which there are megaloblasts in the bone marrow.

microcytic anemia. An anemia in which the erythrocytes are smaller than normal.

nutritional anemia. Hypochromic and microcytic anemia due to insufficient iron.

pernicious anemia. Macrocytic and hypochromic anemia due to lack of secretion of the intrinsic factor by the gastric mucous membrane (this factor is essential for the absorption of vitamin B_{12}).

antagonist. A substance that exerts a nullifying or opposing action to another substance.

antibody. A protein produced by the body in response to the presence of a foreign agent (antigen). Antibodies are part of the body's natural defense against invasion by foreign substances.

antigen. Any substance not normally present in the body that produces an immune reaction upon introduction into the body.

antioxidant. A substance that inhibits the oxidation of other compounds.

antivitamin. A substance that interferes with the synthesis or metabolism of a vitamin.

anus. Distal opening of the gastrointestinal tract through which undigested feed residues are excreted as feces.

apoferritin. A protein in the mucosal cells of the small intestine; together with iron it forms the compound ferritin.

appetite. A desire for food or water; generally a long-term phenomenon, in contrast to short-term satiety.

arachidonic acid. A 20-carbon-atom fatty acid with four double bonds; in the body it is synthesized from the essential fatty acid linoleic acid.

ascorbic acid (vitamin C). Water soluble vitamin required in oxidation–reduction reactions (electron transport) and for normal tyrosine oxidation and collagen metabolism and in hydroxylation reactions (i.e., hydroxyproline from proline).

ash. The residue remaining after complete combustion at 500° to 600°C of a feed or animal tissue or excreta during proximate analysis.

basal metabolic rate. The basal metabolism expressed as kilocalories per unit of body size [square meter of body surface; weight to the three-fourths power ($W^{3/4}$)].

basal metabolism. Energy expenditure of the body at rest, under comfortable environmental conditions, and in the postaborsorptive state (12 hr after the ingestion of food).

bile. Secretion from the liver, containing metabolites such as cholesterol and important bile acid conjugates which aid in emulsification of fats for digestion and absorption.

bile duct. Tube carrying bile from liver to duodenal lumen.

biopsy. The removal and examination of tissue or other material from the living body, usually for diagnosis.

biotin. Water soluble vitamin required as a constituent of several enzymes related to energy and amino acid metabolism.

BMR. Basal metabolic rate.

calcitonin (thyrocalcitonin). A polypeptide secreted by the thyroid glands that lowers the level of calcium and phosphorus in the plasma.

calcium (Ca). Mineral required as a structural component of the sekelton and for controlling excitability of nerve and muscle.

calorimeter. The equipment used to measure the heat generated in a system. In nutrition it is an instrument for measuring the amount of heat produced by a food on oxidation or by an individual.

carboxyl group. A chemical structure $-C\begin{smallmatrix}=O\\ -OH\end{smallmatrix}$; a constituent of all amino acids and the characteristic group of all organic acids.

carboxypeptidase. A proteolytic digestive enzyme secreted by the pancreas; hydrolyzes peptides to amino acids.

cartilage. A connective tissue characterized by nonvascularity (without blood vessels) and firm texture; consists of cells (chondrocytes), interstitial substance (matrix), and a ground substance (chondromucoid).

cassava. A tropical plant of the spurge family with edible starchy roots.

catalyst. A substance that either increases or decreases the rate of a particular chemical reaction without itself being consumed or permanently altered; enzymes are special chemical catalysts of biological origin.

cecum. A blind pouch just distal to the small intestine and containing a large population of anerobic bacteria that ferment complex carbohydrates; appendix in humans corresponds loosely with a rudimentary cecum.

cellulose. Polymer of glucose in linkage resistant to hydrolysis by digestive enzymes.

ceruplasmin. A copper-containing protein (an alphaglobulin) in blood plasma.

chlorine (Cl). Mineral required in regulation of extracellular osmotic pressure and in maintaining acid-base balance.

cholecystokinin. Hormone secreted by duodenal cells and perhaps by certain brain cells; release stimulated by fat in lumen of duodenum and perhaps by concentrations of metabolites in blood circulating in brain; causes gall bladder contraction and may exert short term effect on food intake.

cholic acid. A family of steroids comprising the bile acids; they are derived from cholesterol.

choline. Water soluble vitamin required as a structural component of tissues (i.e., lecithin, sphingomyelin) and of acetylcholine and as a donor of labile methyl groups; there is no evidence for a role as a cofactor in enzymatic reactions.

chromium (Cr). Trace mineral required as trivalent ion for normal glucose metabolism.

chylomicron. A particle of emulsified fat present in lymph; they are especially numerous after a meal high in fat content.

chyme. A semiliquid material produced by the action of gastric juice on ingested food; it is discharged from the stomach into the duodenum.

chymotrypsin. A proteolytic digestive enzyme secreted by the pancreas.

cirrhosis of the liver. A progressive destruction of the liver cells and an abnormal increase of connective tissue.

citric acid cycle. The major series of pathways through which carbohydrates are oxidized to produce energy in animal cells.

cobalt (Co). Trace mineral required as a constituent of vitamin B_{12}.

coenzyme. An organic molecule that is required for the activation of an apoenzyme to an enzyme. The vitamin coenzymes are niacin, pyridoxine, thiamin, riboflavin, pantothenic acid, and folic acid.

coenzyme A. A complex molecule containing pantothenic acid; it is required for fatty acid oxidation and synthesis and for the synthesis of cholesterol and phospholipids; it combines with acetate to form acetyl coenzyme A (active acetate), which in turn combines with oxaloacetate to form citrate and enter the tricarboxylic acid cycle.

collagen. Group of fibrous proteins: resistant to digestive enzymes, prevalent in muscle from aged animals; become gelatin on acid or alkaline hydrolysis.

colon. Large intestine, distal to cecum and small intestine; contains a large population of anaerobic bacteria that ferment complex carbohydrates which may represent a significant energy contribution in some animals.

convulsion. An involuntary spasm or contraction of muscles.

copper (Cu). Trace mineral required for normal red blood cell formation, apparently by allowing normal iron metabolism, and as a component of several oxidases involved in normal wool and hair growth.

cornea. The clear, transparent membrane covering of the anterior part of the eye.

corpus luteum. A yellow mass formed in the ovary at the site of the ruptured ovarian follicle; if impregnation takes place, the corpus luteum grows and persists for several months; if impregnation does not take place, the corpus luteum shrinks and degenerates.

creatine. A nonprotein nitrogenous substance in muscle; it combines with phosphate to form phosphocreatine, which serves as a storage form of high energy phosphate required for muscle contraction.

creatinine. A nitrogenous compound that is formed as a metabolic end product of creatine; creatinine is produced in the muscle, passes into the blood, and is excreted in the urine.

crude fibre. The insoluble carbohydrates remaining in a feed after boiling in acid and alkali during proximate analysis; this fraction represents the poorly digested part of a feed by animals.

crude protein. The content of nitrogen in a feed or animal tissue or excreta, multiplied by a factor (usually 6.25, since most proteins contain about 16% N) to provide an estimate of protein content; both nonprotein N (amino acids, amines, ammonia, etc.) and true protein may be present.

cyanocabalamin (B_{12}). Water soluble vitamin required as a coenzyme in several enzymes including isomerases, dehydrases and enzymes involved in methionine biosynthesis.

cystine. A sulfur-containing, nonessential amino acid that occurs notably in keratin and insulin; in the diet, it exerts a sparing effect on methionine.

cytochromes. Hemoproteins that contain iron in their structure; their principal biological function is electron and/or hydrogen transport.

cytosine. One of four nitrogenous bases found in DNA.

deamination. Removal of the amino group from a compound.

degradation. Conversion of a chemical compound to one that is less complex.

dehydrogenase. Any of a class of enzymes found in plant and animal tissues that catalyze oxidation by the transference of hydrogen ions to hydrogen acceptors.

deoxyribonucleic acid (DNA). A large complex molecule in the cell nucleus that carries the code of genetic information; it is composed of the (nitrogen-containing) bases: adenine, thymine, cytosine, and guanine, each of which is attached to a pentose (deoxyribose) and to a phosphate.

diet. A regulated selection or mixture of feedstuffs or foodstuffs provided on a continuous or prescribed schedule; a balanced diet supplies all nutrients needed for normal health and productive functions.

digestibility, apparent. That percentage of a feed nutrient that is apparently digested and absorbed from the gastrointestinal tract as indicated by intake minus fecal output of the nutrient; differs from true digestibility in that feces contain endogenously derived as well as feed-derived nutrients so that true digestibility must be estimated indirectly by feeding a diet devoid of the nutrient of concern.

digestibility, true. *See* Digestibility, apparent.

disaccharide. Dimer of simple sugars: sucrose (common cane or beet sugar) yields glucose and fructose.

diuresis. Excretion of urine; commonly denotes production of unusually large volumes of urine.

diuretic. A substance that promotes the excretion of urine; diuretic drugs are used chiefly to rid the body of excess fluid.

dry matter. That portion of a feed or tissue remaining after water is removed, usually expressed as a percent.

duodenum. Upper part of small intestine into which empty the secretions of the exocrine pancreas and liver bile.

edema. An abnormal accumulation of fluid in the intercellular spaces of the body.

elastin. Group of fibrous proteins; similar to collagen but not converted to gelatin; a major constituent of tendons and cartilage.

electrolyte. Any substance that, when in solution dissociates into charged particles (ions) capable of conducting an electrical current.

emaciation. Excessive leanness; a wasted condition of the body.

endemic. A disease of low morbidity that persists over a long period of time in a certain region.
endocrine. Pertains to internal secretions; endocrine glands are those that produce one or more internal secretions (hormones) that enter directly into the blood and affect metabolic processes.
endogenous. Produced within or caused by factors within the organism.
enterocrinin. Hormone secreted by jejunal cells; release perhaps stimulated by food digestion products; stimulates secretion of intestinal fluids and enzymes.
enterogastrone. Hormone secreted by duodenal cells; release stimulated by fat and fatty acids and bile in lumen of duodenum; inhibits gastric secretion and motility.
enzyme. A protein formed in plant or animal cells that acts as an organic catalyst in initiating or speeding up specific chemical reactions.
epinephrine. Adrenal medulla hormone that stimulates autonomic nerve action; increases heart rate, perspiration.
ergosterol. A sterol found chiefly in plant tissues; on exposure to ultraviolet irradiation, it becomes vitamin D.
erythrocyte. Red blood cell.
erythropoiesis. Formation of erythrocytes.
estrogens. Ovarian steroid hormones responsible for promoting estrus and the development and maintenance of secondary sex characteristics; estriol $C_{18}H_{24}O_3$ and estrone $C_{18}H_{22}O_2$ are two important estrogens.
ether extract. The fraction of a feed or animal tissue that is soluble in a fat solvent such as ethyl ether and removed by extraction during proximate analysis.
exogenous. Originating outside or caused by factors outside the body.

fatty acids. Organic acids, composed of carbon, hydrogen, and oxygen, which combine with glycerol to form fat.
feces. Excretory product of undigested feed residues, plus endogenously produced digestive secretions, sloughed cells of the intestinal lining and metabolites reexcreted into the intestinal lumen in the bile from the liver.
feed. Food for animals.
feedstuff. Any substance suitable for animal feed; several feedstuffs are normally combined to provide a balanced diet.
ferritin. Form in which iron is stored in the body; it is an iron–protein complex made up of iron and the protein apoferritin.
fistula. An abnormal passage from an abscess, cavity, or hollow organ to the surface or from one abscess, cavity, or organ to another.
fluorine (F). Trace mineral required for protection against dental caries.
fluorosis (dental). A mottled discoloration of the enamel of the teeth due to chronic ingestion of excessive amounts of fluorine.
folacin (folic acid). Water soluble vitamin required as a cofactor in a variety of metabolic reactions involving incorporation of single C units into large molecules.
folic acid. *See* folacin.
food. Any material, usually of plant or animal origin, containing essential nutrients.
foodstuff. Any substance suitable for food; several foodstuffs are normally combined to provide a balanced diet.
fortification. The addition of one or more nutrients to a food in amounts so that the total amount will be larger than that contained in any natural (unprocessed) food of its class, for example, the fortification of fruit juices with vitamin C; the FDA has not established standards for fortification.

galactosemia. The accumulation of galactose in the blood due to a genetic lack of the enzyme galactose-1-phosphate uridyl transferase, which is necessary for the conversion of galactose to glucose; it is characterized by vomiting and diarrhea, abdominal distension, enlargement of the liver, and mental retardation.
gastrin. Hormone secreted by intestinal cells near pyloris; release stimulated by distension and movement of stomach; stimulates acid secretion by gastric glands in stomach.
globulin. Group of globular proteins; alpha, beta and gamma globulins are major constituents of blood serum protein.
glossitis. Inflammation of the tongue.

glucagon. A compound secreted by the alpha cells of the islets of Langerhans that is hyperglycemic (produces a rise in the blood glucose concentration), glycogenolytic (hydrolyzes glycogen to glucose), and stimulates gluconeogenesis.
gluconeogenesis (glyconeogenesis). The formation of glucose from noncarbohydrate sources, chiefly certain amino acids and the glycerol portion of the fat molecule.
glucose tolerance test. A test indicating the efficiency of the body in its use of glucose; changes are noted in the concentration of glucose in the blood at determined intervals after ingestion of a standard amount of sugar.
glutelins. Group of globular proteins.
glyceride. A compound (ester) formed by the combination of glycerol and fatty acids and the loss of water (the ester linkage); according to the number of ester linkages, the compound is a mono-, di-, or triglyceride.
glycerol (glycerin). The 3-carbon atom alcohol derived from the hydrolysis of fat.
glycogenesis. The formation of glycogen from glucose.
glycolysis. The conversion of glucose to lactic acid in various tissues, notably muscle; since molecular oxygen is not consumed in the process, it is referred to as "anaerobic glycolysis."
goitrin. The antithyroid, or goitrogenic, compound obtained from turnips and the seeds of cruciferous plants.
gram (g). A metric unit of weight that is equivalent to about 1/28 of an ounce (1 oz = 28.4 g), 1/1000 of a kilogram, or 1000 milligrams.
growth hormone. *See* Somatrotropin.
guanine. One of four nitrogenous bases found in DNA.

hematocrit. The volume percentage of erythrocytes in whole blood; it is determined by centrifuging a blood sample to separate the cellular elements from the plasma; the results of the test indicate the ratio of cell volume to plasma volume and are expressed as cubic millimeters of packed cells per 100 ml of blood.
hematopoiesis. The production of various types of blood cells and blood platelets.
heme. The nonprotein, insoluble pigment portion of the hemoglobin molecule; the prosthetic group of the hemoglobin molecule.
hemodilution. An increase in the fluid content of the blood, resulting in the diminution of the proportion of formed elements (red and white blood cells and platelets).
histones. Group of globular proteins.
homeostasis. The tendency of the body to maintain uniformity or stability in its internal environment or fluid matrix.
homeothermal (homothermal). Warm-blooded animals.
hormone. A secretion produced in the body (chiefly by the endocrine glands) that is carried in the bloodstream to other parts of the body; each hormone has a specific effect on cells, tissues, and organs.
hunger. Desire for food; the antithesis of satiety.
hydrogenation. The addition of hydrogen to a compound, especially to an unsaturated fat or fatty acid; adding hydrogen at the double bond will solidify soft fats or oils.
hydrolysis. A chemical process whereby a compound is broken down into simpler units with the uptake of water.
hypoglycemia. A condition characterized by a lower than normal level of glucose in the blood.
hypomagnesemia. An abnormally low magnesium content of the blood plasma.
hypothalamus. A portion of the brain, lying beneath the thalamus at the base of the cerebrum, and forming the floor and part of the walls of the third ventricle; it contains centers for temperature regulation, appetite control, and others.

ileum. Lower part of small intestine, distal to the jejunem and proximal to the large intestine; absorption of most nutrients occurs in jejunem and ileum.
immunoglobulin (Ig). Serum globulin having antibody activity; most of the antibody activity appears to be in the gamma fraction of globulin.
inorganic. Compounds that do not contain carbon.
insulin. A hormone secreted by the beta cells of the islets of Langerhans of the pancreas; it facilitates glucose oxidation, as well as the synthesis of glycogen, fat, and protein.
interstitial. Relates to spaces in any structure, such as tissue.

intestinal peptidase. A group of poly-, tri- and dipeptide digestive enzymes secreted by epithial cells in the small intestine.

intolerance. An allergy or sensitivity, for example, to certain foods or medicines.

intrinsic factor. A transferase enzyme (mucoprotein) secreted by the mucosal cells of the stomach which is required for the absorption of vitamin B_{12} through the intestinal wall; a lack of this factor will produce pernicious anemia.

iodine (I). Trace mineral required as a constituent of thyroxine.

iron (Fe). Trace mineral required for hemoglobin and myoglobin formation as a constituent of several heme enzymes and of several nonheme metalloenzymes.

isomer. A compound having the same percentage composition and molecular weight as another compound but differing in chemical or physical properties; for example, L (levorotary) amino acids are biologically active but D (dextrorotary) amino acids generally are not (see also racemic mixture).

isotope. One of two or more atoms, the nuclei of which have the same number of protons but different numbers of neutrons; for example, ^{14}N and ^{15}N are both stable isotopes of nitrogen.

jejunem. Middle segment of small intestine, distal to the duodenum and proximal to the ileum; absorption of most nutrients occurs in jejunem and ileum.

joule (J). A unit of energy that is equal to 0.0002 kcal.

kcal. Kilocalorie.

keratin. Group of fibrous proteins; resistant to digestive enzyme; major constituent of skin, hooves, horns.

Krebs cycle. See Citric acid cycle.

ketogenic. That which is conducive to the formation of ketone bodies, such as a high-fat, low-carbohydrate diet.

α-*ketoglutaric acid.* A compound that is common to the metabolic pathways of carbohydrates, fats, and certain amino acids.

ketone. A chemical compound that contains the carbonyl group (CO); ketone bodies include betahydroxybutyric acid, acetoacetic acid, and acetone.

kilocalorie (kcal). The quantity of heat required to raise the temperature of 1 kg of water 1°C (or, more precisely, from 15° to 16°C).

kilogram (kg). A metric unit of weight that is equivalent to 2.2 pounds, or 1000 g.

lactic acid. A compound formed in the body during anaerobic glycolysis; it is also produced in milk by the bacterial fermentation of lactose.

lecithin. The traditional term for phosphatidyl choline.

leucine. An essential amino acid.

ligament. A band or sheet of fibrous tissue connecting two or more bones, cartilages, or other structures, or serving as a support for muscles.

lignin. Biologically unavailable mixture of polymers of phenolic acids; it is a major structural component of mature plants and of trees; its presence affects the bioavailability of cellulose and hemicellulose in plants.

linoleic acid. The essential fatty acid; it is unsaturated; and occurs widely in plant glycerides.

lipids. Substances that are diverse in chemical nature but soluble in fat solvents (such as ethanol, either, chloroform, benzene); lipids include fats and oils; phospholipids, glycolipids, and lipoproteins; fatty acids, alcohols [glycerol, sterols (including vitamin D and vitamin A), and carotenoids].

liver. A large internal organ of the body, which lies in the upper right section of the abdomen directly beneath the diaphragm. Among other things, it produces bile, glycogen, antibodies; interconverts proteins, carbohydrates, and fats; stores iron, copper, vitamins A and D; and detoxifies harmful substances.

lymph. A fluid that circulates within the lymphatic vessels and is eventually added to the venous blood circulation; it arises from tissue fluid and from intestinal absorption of fatty acids; it is colorless, odorless, slightly alkaline, and slightly opalescent.

lymphatic system. All of the vessels and structures that carry lymph from the tissues to the blood.

lysine. An essential amino acid.

malnutrition. An overall term for poor nourishment; it may be due to an inadequate diet or to some defect in metabolism that prevents the body from using the nutrients properly.

magnesium (Mg). Mineral required as a constituent of bone, for oxidative phosphorylation of mitochondria of cells, and for activation of numerous energy-related enzymes.
manganese (Mn). Trace mineral required for formation of chondroitin sulfate and as a component of metalloenzymes and as a cofactor in several enzymes.
megaloblast. A large, nucleated, embryonic type of cell; it is found in the blood in cases of pernicious anemia, vitamin B_{12} deficiency, and folacin deficiency.
metabolic pool The total amount of a specific substance in the body that is in a state of active turnover, such as the amino acid pool or vitamin D pool; subtractions from and additions to the pool are constantly being made.
metabolic size. The body weight raised to the three-fourths power ($W^{3/4}$).
metabolite. Any compound produced during metabolism.
metalloenzyme. An enzyme containing a metal (ion) as an integral part of its active structure.
methionine. A sulfur-containing essential amino acid.
microgram (μg). A metric unit of weight that is equivalent to one-millionth of a gram or 1/1000 milligram.
micelle. An aggregate of molecules of lipids and bile acids formed in the lumen of the gastrointestinal tract during preparation of dietary lipids for absorption.
milligram (mg). A metric unit of weight that is equivalent to 1/1000 gram or 1000 micrograms.
milliliter (ml). A metric unit of liquid measure that is equivalent to 1/1000 liter.
millimeter (mm). A metric unit of length that is equivalent to 1/10 centimeter.
mitochondria. The largest organelles in the cytoplasm; the cell mitochondria contain important enzymatic, oxidative, and respiratory systems.
molecular weight. The weight of a molecule of a chemical compound as compared with the weight of an atom of hydrogen; it is equal to the sum of the weights of the constituent atoms.
molybdenum (Mo). Trace mineral required as a constituent of the metalloenzyme, xanthine oxidase.
monogastric. Simple stomach; the term often used for simple stomached (nonruminant) animals; technically a misnomer, since ruminants have only one stomach with four compartments.
monosaccharide. Simple sugar of 5 (pentoses) or 6 (hexoses) C atoms; glucose is a prime example.
mucin. Secretions containing mucopolysaccharides.
mucopolysaccharide. A complex of protein and polysaccharides.
mucoprotein. A complex of protein and oligosaccharides (yield on hydrolysis: two to ten monosaccharides).

NDF. Neutral detergent fiber, the fraction of a feedstuff analyzed by the Van Soest scheme of detergent analysis, containing mostly cell wall constituents of low biological availability.
necrosis. Death of a cell or group of cells due to irreversible damage.
nephritis. Inflammation of the glomeruli of the kidney causing impairment of the filtering process so that blood and albumin are excreted.
neuritis. Inflammation of the peripheral nerves (which link the brain and spinal cord with the muscles, skin, organs, and other parts of the body).
NFE (nitrogen-free extract). Consists primarily of readily available carbohydrates such as sugars and starches, but may contain hemicellulose and lignin; NFE is calculated by difference during proximate analysis by subtracting all measured components from 100.
niacin (nicotinic acid). Water soluble vitamin required as a constituent of coenzymes, nicotinamide adenine dinucleotide (NAD), and nicotinimide adenine dinucleotide phosphate (NADP); (another form of the vitamin is nicotinamide).
nicotinamide adenine dinucleotide (NAD) [formerly called diphosphopyridine nucleotide (DPN)]. Attached as a prosthetic group to a protein, it serves as a respiratory enzyme (part of an oxidative-reduction system, converting substrates to CO_2 and H_2O and the transfer of electrons removed to oxygen).
nicotinamide adenine dinucleotide phosphate (NADP) [formerly triphosphopyridine nucleotide (TPN)]. A coenzyme that participates in oxidation–reduction processes.
nicotinic acid. *See* Niacin.
norepinephrine. Hormone secreted by sympathetic nerve endings: it is a vasoconstrictor.
nucleic acid. Complex, high molecular weight molecules that contain phosphate, ribose, and four

bases—adenine, guanine, cytosine, and thymine; DNA and RNA, which are nucleic acids, are responsible for inherited characteristics.

nucleoprotein. A complex of protein and nucleic acid.

nucleotide. A combination of purine and pyrimidine bases, 5-carbon atom sugar, and phosphoric acid; a hydrolytic product of nucleic acid; a constituent of the coenzymes NAD and NADP.

nucleus. Typically the largest structure within cells; it contains DNA, RNA, and usually a distinct body—the nucleolus.

nutrient. A chemical substance that nourishes; such as protein, carbohydrate, mineral or vitamin.

obesity. The accumulation of body fat beyond the amount needed for good health.

oleic acid. An 18-carbon fatty acid that contains one double bond; it is found in animal and vegetable fat.

ornithine cycle. *See* Urea cycle.

osmosis. The passage of a solvent through a membrane separating two solutions from the solution of lesser concentration to that of greater concentration.

osteomalacia. Adult rickets; a softening of the bones caused by a vitamin D deficiency in adults.

osteoporosis. A reduction in the normal quantity of bone.

oxidation. The increase of positive charges on an atom or loss of negative charges; oxidation is one of the changes that take place when fats become rancid.

oxygen (O). A gaseous element that combines with other elements to form oxides.

palmitic acid. A saturated fatty acid with 16-carbon atoms; it is common in fats and oils.

pancreas. Organ located in the abdominal cavity whose functions are both endocrine and exocrine; that is, its specialized cells produce the hormones insulin and glucagon for control of energy and protein metabolism and other cells produce several digestive enzymes for food digestion.

pancreatic duct. Tube carrying pancreatic secretions, including lipases, amylase, chymotrypsin and trypsin, from pancreas to duodenal lumen.

pancreozymin. Hormone secreted by duodenal cells; release stimulated by acid and nutrients in lumen of duodenum; stimulates pancreatic secretion of enzymes.

pantothenic acid. Water soluble vitamin required as a constituent of coenzyme A, needed for acetylation of numerous compounds in energy metabolism.

parturition. The birth of offspring.

phosphorus (P). Mineral required as a structural component of the skeleton, as a component of phospholipids which are important in lipid metabolism and cell membrane structure and as a component of ATP, creatine phosphate, and other essential compounds in energy metabolism.

phosphocreatine. A creatine–phosphoric acid compound that occurs in muscle; the energy source in muscle contraction.

phospholipids. Fat-like substances consisting of glycerol, two fatty acids, a phosphate group, and a nitrogen-containing compound, such as choline (found in the phospholipid lechithin).

photosynthesis. Process by which chlorophyll-containing cells in green plants convert sunlight to chemical energy and synthesize glucose from carbon dioxide and water and release oxygen.

phytic acid (inositolhexaphosphoric acid). A phosphorous-containing organic acid; phytin is the mixed salt of phytic acid with calcium and magnesium.

placenta. Vascular specialized tissue associated with the uterus during pregnancy and through which the fetus is nourished.

plasma. The fluid portion of the blood in which corpuscles are suspended; serum is plasma from which the fibrinogen has been removed.

polyneuritis. An inflammation encompassing many peripheral nerves.

polyunsaturated fatty acids. Fatty acids containing two or more double bonds, such as linoleic, linolenic, and arachidonic acids.

polyuria. Excessive excretion of urine.

potassium (K). Mineral required for maintenance of acid–base balance, for activation of several enzymes, for normal tissue protein synthesis, for integrity of heart and kidney muscle, and for a normal electrocardiogram; it is located mostly within cells.

progesterone. Ovarian steroid hormone produced by corpus luteum; required for maintenance of pregnancy.

prolamins. A group of globular proteins.

protamines. Group of globular proteins.

protein-bound iodine (PBI). The binding of almost all thyroxine in the blood to protein (PBI); the measure of the amount of PBI is useful as an indicator of the quantity of circulating iodine.

proximate analysis. A combination of analytical procedures used to quantify the protein, lipid, dry matter, ash and carbohydrate (nitrogen-free extract) content of feed, animal tissues, or excreta.

PTH (parathyroid hormone). Hormone secreted by the parathyroid gland; maintains blood plasma calcium homeostasis by promoting calcium resorption from bone in hypocalcemia.

purine. The parent substance of the purine bases; adenine and guanine are the major purine bases of nucleic acids; other important purines are xanthine and uric acid.

phyloris. Muscular sphincter separating stomach from duodenum, which controls rate of movement of ingesta out of stomach.

pyridoxine (B_6). Water soluble vitamin required as a coenzyme in a large array of enzymes associated with protein and nitrogen metabolism (other forms of the vitamin are pyridoxamine and pyridoxal).

pyrimidine. The parent substance of several nitrogenous compounds found in nucleic acids—uracil, thymine, and cytosine.

pyruvic acid. A keto acid of 3-carbon atoms; it is formed of carbohydrate in aerobic metabolism; pyruvate is the salt or ester of pyruvic acid.

radioactive. The emission of particles during the disintegration of the nuclei of radioactive elements; the emissions include alpha particles, beta particles, and gamma rays.

radioisotope. A radioactive form of an element; the nucleus of a stable atom when charged by bombarding particles (in a nuclear reactor) becomes radioactive and is called "labeled" or "tagged."

reduction. The gain of one or more electrons by an ion or compound; for example, ferric iron (Fe^{3+}), which is the common form that is found in food, is reduced to ferrous iron (Fe^{2+}) in the acid medium of the stomach.

racemic mixture. Pertaining to a chemical compound containing equal amounts of dextrorotary (D) and levorotary (L) isomers so that it does not rotate the plane of polarized light.

ration. A fixed portion of feed, usually expressed as the amount of a diet allowed daily.

rectum. Distal portion of gastrointestinal tract, joined proximally to the colon and opening to the exterior via the anus.

riboflavin (B_2). Water soluble vitamin required as a constituent of coenzymes, flavin adenine dinucleotide (FAD), and falvin mononucleotide (FMN).

ribonucleic acid (RNA). A nucleic acid found in all living cells; on hydrolysis it yields adenine, guanine, cytosine, uracil, ribose, and phosphoric acid; it takes part in protein synthesis.

ribosome. A ribonucleic acid-containing particle in the cytoplasm of a cell; the site of protein synthesis.

satiety. The condition of being fully satisfied with food; the antithesis of hunger.

secretion. Hormone secreted by duodenal cells; release stimulated by acid and nutrients in lumen of duodenum; stimulates pancreatic secretion of water and electrolytes.

sedimentation. The settling out of a precipitate.

erythrocyte sedimentation rate (ESR). the rate at which erythrocytes settle out of unclotted blood; inflammatory processes cause an aggregation (clumping together) of the red blood cells which makes them heavier and more likely to fall.

selenium (Se). Trace mineral required as a constituent of the enzyme glutathione peroxidase which destroys peroxides arising from tissue lipid oxidation.

serum. The clear portion of any fluid from an animal's body that remains after the solid elements have been separated out; blood serum is the clear, straw-colored liquid that remains after blood has clotted; the plasma from which fibrinogen has been removed.

silicon (Si). Trace mineral required for initiation of the mineralization process in bones.

sodium (Na). Mineral required for maintenance of acid–base balance, transfer of nerve impulses and as an extracellular component of an energy dependent sodium "pump"; it is located mostly in extracellular fluids.

somatotropin. Anterior pituitary hormone active in promoting protein and mineral accretion in animals and stimulating animal growth.

sphingomyelin. A group of phospholipids found in the brain, spinal cord, and kidney; on hydrolysis it yields phosphoric acid, choline, sphingosine, and a fatty acid.

spleen. A large organ situated under the ribs in the upper left side of the abdomen; it functions in the normal destruction of old red blood cells.
starch. Polymer of glucose readily hydrolyzed by digestive enzymes.
stearic acid. A saturated fatty acid composed of 18-carbon atoms.
sterol. An alcohol of high molecular weight, such as cholesterol and ergosterol.
stillbirth. Birth of a dead fetus.
stomach. The first portion of the gastrointestinal tract on which chemical action occurs on ingested feeds in most animals species; in humans, pigs, and many other species it is a single compartment (simple stomach); in ruminants such as cattle, sheep, goats, deer it contains several compartments.
sulfur (S). Mineral required as a constituent of several organic metabolites, including methionine, cystine, thiamin, biotin, coenzyme A and mucopolysaccharides.

taste. To distinguish flavors between or among feed or water components.
testosterone. Testicular steroid hormone; stimulates sperm production and controls secondary sex characteristics and accessory reproductive organs; increases muscle protein accretion.
tetany. A condition characterized by sharp bending (flexion) of the wrist and ankle joints, muscle twitchings, cramps, and convulsions; inadequate calcium in the blood causes irritability of the nerves and muscles so that they respond to a stimulus with greater sensitivity and force than normally.
tetraiodothyronine. Thyroid hormone active in regulation of basal metabolic rate (BMR); chemical formula is $C_{15}H_{11}I_4NO_4$.
thiamin (B_1). Water soluble vitamin required as a constituent of the enzymes cocarboxylase and lipothiamide.
thymine. One of four nitrogenous bases found in DNA.
threonine. An essential amino acid.
thymine. One of the four nitrogenous bases in nucleic acid.
thymus gland. A two-lobed ductless gland located behind the upper part of the sternum and extending into the neck; it is fairly large in young animals but usually shrinks in adulthood; it is structured like a lymph node and contains lymphatic follicles; it may play a role in immune reactions.
thyroxine. An iodine-containing hormone that is produced by the thyroid gland; it is a derivative of the amino acid tyrosine and has the chemical name tetraiodothyronine.
transferrin (siderophilin). An iron-binding protein in the blood that transports iron.
tricarboxylic acid cycle (Krebs cycle and citric acid cycle). A series of biochemical reactions by which carbon chains of sugars, fatty acids, and amino acids are metabolized to yield carbon dioxide, water, and energy.
triglyceride (fat). An ester composed of glycerol and three fatty acids.
tristearin. A triglyceride of stearic acid.
tryptophan. An essential amino acid; a precursor of niacin.
trypsin. A proteolytic digestive enzyme secreted by the pancreas; its precursor trypsonogen is activated to trypsin by enterokinase.
tyrosine. A nonessential amino acid; it spares the essential amino acid phenylalanine.

undernutrition. A condition resulting from insufficient food.
urea. The chief end product of protein metabolsim in mammals and one of the chief nitrogenous constituents in the urine.
urea cycle. Major pathway of nitrogen excretion in mammals.
uric acid. A nitrogenous end product of purine metabolism; it is present in the blood and excreted in the urine.

valine. An essential amino acid.
vasoconstrictor. An agent (motor nerve or chemical compound) that acts to decrease the caliber of blood vessels.
vitamin A. Fat soluble vitamin required for normal night vision, normal epithelial cells, and normal bone growth and remodeling.
vitamin B_1. *See* Thiamin.
vitamin B_2. *See* Riboflavin.
vitamin B_6. *See* Pyridoxine.

vitamin B_{12}. *See* Cyanocobalamin.
vitamin D. Fat soluble vitamin required for normal absorption of calcium from the intestinal tract as a precursor of 1,25 dihydroxycholecaliferol which acts in the control of mobilization of calcium from bone and stimulation of calcium absorption from the gastrointestinal tract.
vitamin E. Fat soluble vitamin required as an antioxidant in preventing a wide variety of diseases relating to maintaining integrity of cellular membranes.
vitamin K. Fat soluble vitamin required for normal blood clotting, specifically for synthesis of prothrombin in liver.

xanthurenic acid. A metabolite of tryptophan; it is found in normal urine, but appears in increased amounts in cases of vitamin B_6 deficiency.
xerosis. Abnormal dryness, as of the conjunctiva (xerophthalmia) or skin (xeroderma).

zinc (Zn). Trace mineral required as a constituent of an array of enzyme systems in several body tissues, as an activator of several metalloenzymes and in binding reactants to the active site of enzymes.

INTERNATIONAL FEED NOMENCLATURE

acid-hydrolyzed. *See* Hydrolyzed.
acidified. Treated with an acid to provide a pH lower than pH 7. Sometimes a precipitate forms that may be removed by mechanical means (e.g., curd from whey).
air-ashed. Reduced by combustion in air to a mineral residue.
alcohol-extracted. Treated with alcohol to remove all alcohol-soluble substances.
ammoniated. Combined or impregnated with ammonia or an ammonium compound.
artificially dried. Moisture removed by other than natural means. *See* fan air-dried with heat.
as fed. *See* Dry-matter content of feed samples.
aspirated. Removal of light materials from heavier materials by use of air. Refers to chaff, dust, or other light materials.

barn-cured. Dried with forced ventilation in an enclosure. Refers to forage. *See* Fan air-dried without heat (international term).
bitterness-extracted. Treated to remove bitter taste.
blended. Mingled or combined. Refers to ingredients of a mixed feed; does not imply uniformity of dispersion.
blocked. Compressed into a large, solid mass.
bolted. Separated from parent material by means of a bolting cloth. Refers to two ingredients (e.g., bran separated from flour).

calcined. Heated to high temperature in the presence of air.
canned. Processed, packaged, sealed, and sterilized in cans or similar containers.
centrifuged. Separated by a force moving away from a center.
chipped. Cut or broken into fragments or cut into small, thin slices.
chopped. Reduced in particle size by cutting.
cleaned. Subjected to any process (e.g., scalping, screening, aspiration, or magnetic separation) by which unwanted material is removed.
clipped. Refers to removal of ends of whole grain.
close-planted. Planted with less than normal distance between rows.
coagulated. Curdled, clotted, or congealed, usually by the action of a coagulant.
coarse-bolted. Separated from parent material by means of a coarsely woven bolting cloth.
coarse-sifted. Separated according to particle size by passage through coarsely woven wire sieves.
condensed. Reduced in volume by removal of moisture.
conditioned. Brought to predetermined moisture characteristics or temperature (or both) prior to further processing. *See also* Tempered (international term).
cooked. Heated in the presence of moisture to alter chemical or physical characteristics (or both) or to sterilize. *See also* Pressure cooked.
cracked. Reduced in size by a combined breaking and crushing action. Refers to particles of grain.

crimped. Rolled with corrugated rollers. The grain to which this term refers may be tempered or conditioned before it is crimped, and may be cooled afterward.
crumbled. Broken with corrugated rollers. Refers to pellets.
crushed. *See* Rolled.
cultured. Produced in a culture. Refers to biological material.
cured. Prepared for keeping or use (e.g., by drying, smoking, or salting or by using a chemical preservative).

D-activated. Activated with vitamin D (e.g., by ultraviolet light). Refers to plant or animal sterol fractions.
debittered. Having had bitter substances removed.
defluorinated. Having had fluorine partially removed.
degermed. Having had the embryos wholly or partially separated from the starch endosperms. Refers to seeds.
dehulled. Having had the hulls removed.
dehydrated. Having had most of the moisture removed by heat.
deribbed. Having had the primary veins removed. Refers to leaves.
digested. Subjected to prolonged heat and moisture, or to chemicals or enzymes, with a resultant change or decomposition of the physical or chemical nature.
dressed. Made uniform in texture by breaking or screening lumps or by applying liquid(s). Refers to feed.
dried. *See* Dehydrated (international term).
dry. *See* Dry-matter content of feed samples.
dry-matter content of feed samples. The dry matter content of feed samples and other materials as expressed on three dry-matter bases: as fed, partially dry, and dry. Definitions of these terms follow:

as fed. As fed refers to the feed as it is consumed by the animal; the term "as collected" is used for materials which are not usually fed to the animal, i.e., urine, feces, etc. If the analyses on a sample are affected by partial drying, the analyses are made on the "as fed" or "as collected" sample. Similar terms: air dry, i.e., hay; as received; fresh; green; wet.

partially dry. Partially dry refers to a sample of "as-fed" or "as-collected" material that has been dried in an oven (usually with forced air) at a temperature usually at 60°C or freeze-dried and has been equilibrated with the air; the sample after these processes would usually contain more than 88% dry matter (12% moisture); some materials are prepared in this way so they may be sampled, chemically analyzed, and stored. This analysis is referred to as "partial dry matter % of 'as-fed' or 'as-collected' sample." The partially dry sample must be analyzed for dry matter (determined in an oven at 105°C) to correct subsequent chemical analyses of the sample to a "dry" basis. This analysis is referred to as "dry matter % of partial dry sample." Similar term: air dry (sometimes air dry is used for "as fed"; *see* As fed).

dry. Dry refers to a sample of material that has been dried at 105°C until all the moisture has been removed. Similar terms: 100% dry matter; moisture free. If dry matter (in an oven at 105°C) is determined on an "as-fed" sample it is referred to as "dry matter of as-fed sample." If dry matter is determined on a partial dry sample it is referred to as "dry matter of partial dry sample." It is recommended that analyses be reported on the "dry" basis (100% dry matter or moisture free), and in addition the "as fed-dry matter" should be reported.

dry-milled. Milled by tempering with a small amount of water or steam to facilitate separation into component parts. Refers to kernels of grain.
dry-rendered. Having undergone (1) cooking in open steam-jacketed vessels until the water has evaporated and (2) removal of fat by draining and pressing. Refers to residues of animal tissues.

ensiled. Preserved by ensiling, a process in which finely cut parts of plants; packed in an air-tight chamber (e.g., a silo), undergo an acid fermentation that retards spoilage.
etiolated. Grown in reduced light. Refers to plants.
eviscerated. Subjected to removal of all organs. Refers to the great cavity of an animal's body.
expanded. Increased in volume as a result of abrupt reduction in pressure. Refers to a feed or feed mixture that is extruded after being subjected to moisture, pressure, and temperature to gelatinize the starchy part.
expeller extracted. *See* Mechanically extracted.

extracted, mechanical. *See* Mechanically extracted.
extracted, solvent. *See* Solvent-extracted.
extruded. Pushed through orifices of a die under pressure. Refers to feed.
extruded with steam. Pushed through orifices of a die under pressure and after preconditioning with steam. Refers to feed.
extruded without steam. Pushed through orifices of a die under pressure and in the absence of steam. Refers to feed.

fan air-dried. Dried with a device producing a current of air.
fan air-dried with heat. Dried with a device producing a current of heated air. Syn: artifically dried.
fermented. Acted upon by yeasts, filamentous fungi, or bacteria in a controlled aerobic or anaerobic process. Refers to products (e.g., grains and molasses) used in the manufacture of alcohols, acids, vitamins of the B-complex group, and antibiotics.
finely ground. Reduced to very small particles by impact, shearing, or attrition.
finely screened. Separated according to particle size by passage through a finely woven meshed material.
finely sifted. Separated according to particle size by passage through a finely woven meshed material.
flaked. (1) Prepared by a method involving the use of high heat, tempering, and rollers set close together. (2) Cut into flat pieces (e.g., potato flakes). Syn: steam flaked.
fused. Blended by melting.

gelantinized. Ruptured by a combination of moisture, heat, and pressure. Refers to starch granules of a feed.
ground. Reduced in particle size by impact shearing or attrition.

heat- and acid-precipitated. Separated from a suspension or solution by action of heat and acid.
heat-hydrolyzed. *See* Hydrolyzed.
heat-processed. Prepared by a method involving the use of elevated temperatures, with or without pressure.
heat-processed flaked. *See* Flaked.
heat-rendered. Melted, extracted, or clarified by heating. (Water and fat are usually removed.)
homogenized. Broken down into evenly distributed globules small enough to remain as an emulsion for long periods. Refers to particles of fat.
hydraulically extracted. *See* Mechanically extracted.
hydrolyzed. Subjected to hydrolysis, a process by which complex molecules (e.g., those in proteins) are split into simpler units by chemical reaction with water molecules. (The reaction may be produced by an enzyme, catalyst, or acid or by heat and pressure.)

iodinated. Treated with iodine.
irradiated. Treated, prepared, or altered by exposure to radiant energy.

kibbled. Cracked or crushed. Refers to baked dough or to extruded feed that was cooked before or during the extrusion process.

lactase hydrolyzed. *See* Hydrolyzed.
leached. Affected by the action of percolating water or other liquid.

magnetic separation. Removal of ferrous material by magnets (e.g., removal of iron objects from mixed feed).
malted. Converted into malt or treated with malt or malt extract.
maltase processed. Treated with the enzyme maltase.
mechanically extracted. Extracted by heat and mechanical pressure. Refers to removal of fat or oil from seeds. Syn: expeller-extracted, hydraulically extracted, old process.
mechanically extracted caked. Extracted from seeds by heat and mechanical pressure in such a way that the remaining product (e.g., cottonseed meal) is caked. Refers to fat or oil.
micronized. Heated by gas-fired infrared generators to 150°C (300°F). Microwaves are emitted in the heating process. Refers to grain.
mixed. Two or more materials combined by agitation to a specific degree of dispersion.
molded. Overgrown or otherwise acted upon by fungi.

new process. *See* Solvent-extracted.

old process. *See* Mechanically extracted.

partially dry. *See* Dry-matter content of feed samples.
partially extracted. Partially removed from a feed by a chemical or mechanical process. Refers to fat or oil.
pearled. Reduced by machine brushing to smaller, smooth particles. Refers to dehulled grains (e.g., pearled barley).
pelleted. Made into pellets.
polished. Smoothed by a mechanical process. Refers to grain (e.g., polished rice).
popped. Heated with accompanying expansion until explosion or eruption occurs. Refers to grain [e.g., popped corn (maize)].
precipitated. Separated from suspension or solution as a result of a chemical or physical change.
premixed. Mixed with a diluent or carrier (or both) preliminary to final mixing with other ingredients.
prepressed solvent-extracted. Removed from materials partly by heat and mechanical pressure and (later) partly by organic solvents. Refers to fat or oil.
pressed. Compacted or molded by pressure, or extracted under pressure. Refers to fat, oil, or juice.
pressure-cooked. Heated either wet or dry and under pressure greater than atmospheric pressure.
pulverized. *See* Ground.

retort-charred. Partly burned in a closed retort. Refers to bone black.
roasted. Cooked in an oven by dry heat. For grains, cooked or heated with dry heat with an exit temperature of 150°C (300°F). Usually accompanied by expansion and toasted appearance.
rolled. Compressed between rollers. Rolling may entail tempering or conditioning. *See* Steam-rolled.

scalped. Removed by screening. Refers to large particles.
scoured. Cleansed by impact or friction. Refers to removal of the beard from the wheat kernel.
screened. Separated into different sizes by being passed over or through screens.
separation. Classification of particles by size, shape, or density.
separation, magnetic. *See* Magnetic separation.
shredded. Cut into long, thin pieces.
sifted. Separated into different sizes by being passed through wire or nylon sieves.
sizing. *See* Screened.
skimmed. Removed by settling, flotation, or centrifuging. Refers to removal of the lighter part of a liquid from the heavier part (e.g., removal of cream from milk).
solvent extracted. Removed from materials (e.g., soybean seeds) by organic solvents. Refers to fat or oil. Syn: new process.
solvent-extracted caked. Removed from materials by organic solvents and pressed to make a cake. Refers to fat or oil.
spent. Exhausted of absorbing properties (e.g., spent bone black).
stabilized. Made more resistant to chemical change by an added substance.
stack-ensiled. Ensiled while in a pile above ground.
steamed. Treated with steam, as in steam cooking. Syn: steam cooked, steam rendered, tanked.
steam-rolled. Preconditioned in steam under pressure for a short period and then compressed between rollers.
steep-extracted. Soaked in water or other liquid to remove soluble materials. Refers to grain (e.g., corn that is being wet milled).
Steffen process. A process for treating beet molasses to recover additional sugar through precipitation of calcium sucrate.
sulfur-fertilized. Supplied with a form of elemental sulfur as a nutrient. Refers to plants.
sun cured. Dried by exposure to the sun.
sun-cured brown. Partially dried by exposure to the direct rays of the sun, then put in a stack or bale, where heat from microbial action causes browning.
sun cured on riders. Stacked on tripods made of poles, then sun-cured. Refers to forage.

tanked. *See* Steamed.
tempered. Brought to predetermined moisture characteristics or temperature (or both) before further processing. Syn: conditioned.

threshed. Separated from straw by impaction and subsequent screening. Refers to grain, peas, or beans.
toasted. Browned, dried, or parched by exposure to a wood fire or to gas or electric heat.
toxicity extracted. Treated to remove a poisonous substance.

vacuum dehydrated. Dehydrated under vacuum. *See* Dehydrated.

wafered. Agglomerated by compressing into a form that usually measures more in diameter or cross section than in length. Refers to fibrous feeds (e.g., wafered alfalfa hay).
water extracted. Removed with water. Refers to a product from which soluble substances have been removed.
weathered. Exposed to air, sunlight, and precipitation.
wet milled. Steeped in water, which may contain sulfur dioxide, to facilitate separation of the parts. Refers to kernels of corn.
whole pressed. Pressed to remove oil. Refers to seeds with hulls (e.g., cotton seeds).

FEED MANUFACTURING TERMS

additive. An ingredient or combination of ingredients added to the basic feed mix or parts thereof to fulfill a specific need. Usually used in micro-quantities and requires careful handling and mixing.
(*Note:* A *food additive* is defined by federal law as any substance which becomes a component of or affects the characteristics of a feed or food if such substance is not generally recognized among experts qualified by scientific training and experience to evaluate its safety as having been adequately shown through scientific procedures to be safe under the conditions of its intended use. Excepted are substances having ''prior sanction'' and pesticide chemicals under certain conditions. See Public Law 65-929 for details.)
antibiotics. A class of drug. They are usually synthesized by a living microorganism and in proper concentration inhibit the growth of other microorganisms.
artifically dried. (Process.) Moisture having been removed by other than natural means.
aspirated, aspirating. Having removed chaff, dust, or other light materials by use of air.

balanced. A term that may be applied to a diet, ration, or feed having all known required nturients in proper amount and proportion based upon recommendations of recognized authorities in the field of animal nutrition, such as the National Research Council, for a given set of physiological animal requirements. The species for which it is intended and the functions such as maintenance or maintenance plus production (growth, fetus, fat, milk, eggs, wool, feathers or work) shall be specified.
blending. (Process.) To mingle or combine two or more ingredients of feed. It does not imply a uniformity of dispersion.
blocked, blocking. (Process.) Having agglomerated individual ingredients or mixtures into a large mass.
blocks. (Physical form.) Agglomerated feed compressed into a solid mass cohesive enough to hold its form and weighing over 2 pounds, and generally weighing 30–50 pounds.
brand name. Any word, name, symbol or device or any combination thereof identifying the commercial feed of a distributor and distinguishing it from that of others.
bricks. (Physical form.) Agglomerated feed, other than pellets, compressed into a solid mass cohesive enough to hold its form and weighing less than two pounds. (*See also* Blocks).
by-product. (Part.) Secondary products produced in addition to the principal product.

cake. (Physical form.) The mass resulting from the pressing of seeds, meat, or fish in order to remove oils, fats, or other liquids.
carcass residue, mammals. (Part.) Residues from animal tissues including bones and excluding hair, hoofs, horns and contents of the digestive tract.
carriers. An edible material to which ingredients are added to facilitate uniform incorporation of the latter into feeds. The active particles are absorbed, impregnated, or coated into or onto the edible material in such a way as to physically carry the active ingredient.

chaff. (Part.) Glumes, husks, or other seed coverings together with other plant parts separated from seed in threshing or processing.

cleaned, cleaning. (Process.) Removal of material by such methods as scalping, aspirating, magnetic separation, or by any other method.

cleanings. (Part.) Chaff, weed seeds, dust, and other foreign matter removed from cereal grains.

clipped, clipping. (Process.) Removal of the ends of whole grain.

commercial feed. The term commercial feed (as defined in the Uniform Feed Bill) means all materials which are distributed for use as feed or for mixing in feed, for animals other than man except: (1) Option A—Unmixed seed, whole or processed, made directly from the entire seed. Option B—Unmixed or unprocessed whole seeds. (2) Hay, straw, stover, silage, cobs and hulls (a) when unground, and (b) when unmixed with other materials. (3) Individual chemical compounds when not mixed with other materials.

complete feed. A nutritionally adequate feed for animals other than man; by specific formula is compounded to be fed as the sole ration and is capable of maintaining life and/or promoting production without any additional substance being consumed except water.

concentrate. A feed used with another to improve the nutritive balance of the total and intended to be further diluted and mixed to produce a supplement or a complete feed.

condensed, condensing. (Process.) Reduced to denser form by removal of moisture.

conditioned, conditioning. (Process.) Having achieved predetermined moisture characteristics and/or temperature of ingriedents or a mixture of ingredients prior to further processing.

cooked, cooking. (Process.) Heated in the presence of moisture to alter chemical and/or physical characteristics or to sterilize.

cooled, cooling. (Process.) Temperature reduced by air movement, usually accompanied by a simultaneous drying action.

cracked, cracking. (Process.) Particle size reduced by a combined breaking and crushing action.

cracklings. (Part.) Residue after removal of fat from adipose tissue or skin of animals by dry heat.

crumbled, crumbling. (Process.) Pellets reduced to granular form.

crumbles. (Physical form.) Pelleted feed reduced to granular form.

cubes. (Physical form.) See pellets.

cubes, range. (Physical form.) See pellets and range cubes.

degermed. (Process.) Having had the embryo of seeds wholly or partially separated from the starch endosperm.

dehulled, dehulling. (Process.) Having removed the outer covering from grains or other seeds.

dehydrated, dehydrating. (Process.) Having been freed of moisture by thermal means.

diet. Feed ingredient or mixture of ingredients including water, which is consumed by animals.

diluent. (Physical form.) An edible substance used to mix with and reduce the concentration of nutrients and/or additives to make them more acceptable to animals, safer to use, and more capable to being mixed uniformly in a feed. (It may also be a carrier.)

dressed, dressing. (Process.) Made uniform in texture by breaking or screening of lumps from feed and/or the application of liquid(s).

dried, drying. (Process.) Materials from which water or other liquid has been removed.

drug. (As defined by FDA as applied to feed.) A substance (a) intended for use in the diagnosis, cure, mitigation, treatment, or prevention of disease in man or other animals or a substance (b) other than food intended to affect the structure or any function of the body of man or other animals.

dry-milled. (Process.) Tempered with a small amount of water or steam to facilitate the separation of the various component parts of the kernel in the absence of any significant amount of free water.

dust. (Part.) Fine, dry pulverized particles of matter usually resulting from the cleaning or grinding of grain.

emulsifier. A material capable of causing fat or oils to remain in liquid suspension.

ensiled. (Process.) Aerial parts of plants that have been preserved by ensiling. Normally the original material is finely cut and blown into an air-tight chamber as a silo, where it is pressed to exclude air and where it undergoes an acid fermentation that retards spoilage.

evaporated, evaporating. (Process.) Reduced to a denser form; concentrated as by evaporation or distillation.

expanded, expanding. (Process.) Subjected to moisture, pressure, and temperature to gelatinize the starch portion. When extruded, its volume is increased, due to abrupt reduction in pressure.

extracted, mechanical. (Process.) Having removed fat or oil from materials by heat and mechanical pressure. Similar terms: expeller extracted, hydraulic extracted, "old process."

extracted, solvent. (Process.) Having removed fat or oil from materials by organic solvents. Similar term: "new process."

extruded. (Process.) A process by which feed has been pressed, pushed, or protruded through orifices under pressure.

fat. (Part.) A substance composed chiefly of triglycerides of fatty acids, and solid or plastic at room temperature.

fatty acids. (Part.) Aliphatic monobasic acids containing only the elements of carbon, hydrogen, and oxygen.

feed(s). Edible material(s) that are consumed by animals and contribute energy and/or nutrients to the animal's diet. (Usually refers to animals rather than man.)

feed additive concentrate. (Part.) (As defined by FDA) An article that is intended to be further diluted to produce a complete feed or a feed additive supplement and is not suitable for offering as a supplement or for offering free choice without dilution. It contains, among other things, one or more additives in amounts in a suitable feed base such that from 100 to 1000 pounds of concentrate must be diluted to produce 1 ton of a complete feed. A feed additive concentrate is unsafe if fed free choice or as a supplement because of danger to the health of the animal or because of the production of residues in the edible products from food producing animals in excess of the safe levels established.

feed additive premix. (As defined by FDA.) An article that must be diluted for safe use in feed additive concentrate, a feed additive supplement or a complete feed. It contains, among other things, one or more additives in high concentration in a suitable feed base such that up to 100 pounds must be diluted to produce 1 ton of complete feed. A feed additive premix contains additives at levels for which safety to the animal has not been demonstrated and/or which may result when fed undiluted in residues in the edible products from food producing animals in excess of the safe levels established.

feed additive supplement. (As defined by FDA.) An article for the diet of an animal which contains one or more food additives and is intended to be the following: (1) Further diluted and mixed to produce a complete feed; or (2) Fed undiluted as a supplement to other feeds; or (3) Offered free choice with other parts of the rations separated available.

(*Note:* A feed additive supplement is safe for the animal and will not produce unsafe residues in the edible products from food producing animals if fed according to directions.)

feed grade. Suitable for animal consumption.

feed mixture. *See* Formula feed.

feedstuff. *See* Feed(s).

fermentation aid. A substance added to assist in providing proper conditions that result in action by yeasts, molds, or bacteria in a controlled aerobic or anaerobic process used for the manufacture of certain products.

fermented, fermenting. (Process.) Acted upon by yeasts, molds, or bacteria in a controlled aerobic or anaerobic process in the manufacture of such products as alcohols, acids, vitamins of the B-complex group, or antibiotics.

fines. (Physical form.) Any material which will pass through a screen whose openings are immediately smaller than the specified minimum crumble size or pellet diameter.

flakes. (Physical form.) An ingredient rolled or cut into flat pieces with or without prior steam conditioning.

flaked, flaking. (Process.) See rolled.

flour. (Part.) Soft, finely ground and bolted meal obtained from the milling of cereal grains, other seeds, or products. It consists essentially of the starch and gluten of the endosperm.

food(s). When used in reference to animals, is synonymous with feed(s). *See* Feed(s).

formula feed. Two or more ingredients proportioned, mixed, and processed according to specifications.

free choice. A feeding system by which animals are given unlimited access to the separate components or groups of components constituting the diet.

gelatinized, gelatinizing. (Process.) Having had the starch granules completely ruptured by a combination of moisture, heat, and pressure, and in some instances by mechanical shear.
germ. (Part.) The embryo found in seeds and frequently separated from the bran and starch endosperm during the milling.
gluten. (Part.) The tough, viscid nitrogen substance remaining when the flour or wheat or other grain is washed to remove the starch.
gossypol. (Part.) A phenolic pigment in cottonseed that is toxic to some animals.
grain. (Part.) Seed from cereal plants.
GRAS. Acronym for the phrase "generally recognized as safe." A substance that is generally recognized as safe by experts qualified to evaluate the safety of the substance for its intended use.
grease. Animal fats with a titer below 40° C.
grits. (Part.) Coarsely ground grain from which the bran and germ have been removed, usually screened to uniform particle size.
groats. (Part.) Grain from which the hulls have been removed.
ground, grinding. (Process.) Reduced in particle size by impact, shearing, or attrition.

heat-processed, heat-processing. (Process.) Subjected to a method of preparation involving the use of elevated temperatures with or without pressure.
hulls. (Part.) Outer covering of grain or other seed.
hydrolized, hydrolizing. (Process.) Complex molecules having been split to simpler units by chemical reaction with water, usually by catalysis.

ingredient, feed ingredient. A component part or constituent of any combination or mixture making up a commercial feed.

kibbled, kibbling. (Process.) Cracked or crushed baked dough, or extruded feed that has been cooked prior to or during the extrusion process.

lard. (Part.) Rendered fat of swine.
lecithin. (Part.) A specific phospholipid. The principal constituent of crude phosphatides derived from oil-bearing seeds.

malt. (Part.) Sprouted and steamed whole grain from which the radicle has been removed.
malted, malting. (Process.) Converted into malt or treated with malt or malt extract.
mash. (Physical form.) A mixture of ingredients in meal form. Similar term: mash feed.
meal. (Physical form.) An ingredient which has been ground or otherwise reduced in particle size.
medicated feed. Any feed that contains drug ingredients intended or represented for the cure, mitigation, treatment, or prevention of diseases of animals other than man or that contains drug ingredients intended to affect the structure or function of the body of animals other than man.
Antibiotics included in a feed at growth promotion and/or feed efficiency levels are drug additives and feeds containing such antibiotics are included in the definition of medicated feed.
microingredients. Vitamins, minerals, antibiotics, drugs, and other materials normally required in small amounts and measured in milligrams, micrograms, or parts per million (ppm).
middlings. (Part.) A by-product of flour milling comprising several grades of granular particles containing different proportions of endosperm, bran, germ, each of which contains different levels of crude fiber.
mill by-product. (Part.) A secondary product obtained in addition to the principal product in milling practice.
mill dust. (Part.) Fine feed particles of undetermined origin resulting from handling and processing feed and feed ingredients.
mill run. (Part.) The state in which a material comes from the mill, ungraded and usually uninspected.
mineralize, mineralized. (Process.) To supply, impregnate, or add inorganic mineral compounds to a feed ingredient or mixture.
mixing. (Process.) To combine by agitation two or more materials to a specific degree of dispersion.
molasses. (Part.) The thick, viscous by-product resulting from refined sugar production or the concentrated, partially dehydrated juices from fruits.

pearled, pearling. (Process.) Dehulled grains reduced by machine brushing into smaller smooth particles.

pellets. (Physical form.) Agglomerated feed formed by compacting and forcing through die openings by a mechanical process. Similar terms: pelleted feed, hard pellet.

pellets, soft. (Physical form.) Pellets containing sufficient liquid to require immediate dusting and cooling. Similar term: high molasses pellets.

pelleted, pelleting. (Process.) Having agglomerated feed by compaction and forced through die openings.

pomace. (Part.) Pulp from fruit. *See* Pulp.

premix. A uniform mixture of one or more microingredients with diluent and/or carrier. Premixes are used to facilitate uniform dispersion of the microingredients in a larger mix.

premixing. (Process.) The preliminary mixing of ingredients with diluents and/or carriers.

product. (Part.) A substance produced from one or more other substances as a result of chemical or physical change.

protein. (Part.) Any of a large class of naturally occurring complex combinations of amino acids.

pulp. (Part.) The solid residue remaining after extraction of juices from fruits or stems. Similar terms: bagasse and pomace.

pulverized, pulverizing. (Process.) *See* Ground, grinding.

range cake. (Physical form.) *See* Cake.

range cubes. (Physical form.) Large pellets designed to be fed on the ground. Similar term: range wafer.

ration. The amount of the total feed that is provided to one animal over a 24-hr period.

rolled, rolling. (Process.) Having changed the shape and/or size of particles by compressing between rollers. It may entail tempering or conditioning.

scalped, scalping. (Process.) Having removed larger material by screening.

scratch. (Physical form.) Whole, cracked, or coarsely cut grain. Similar terms: scratch grain, scratch feed.

screened, screening. (Process.) Having separated various sized particles by passing over and/or through screens.

self-fed. A feeding system where animals have continuous free access to some or all component(s) of a ration, either individually or as mixtures.

separating. (Process.) Classification of particle by size, shape, and/or density.

separating, magnetic. (Process.) Removing ferrous material by magnetic attraction.

shorts. (Part.) Fine particles of bran, germ, flour or offal from the tail of the mill from commercial flour milling.

sizing. (Process.) *See* Screened, screening.

skimmed. (Process.) Material from which floating solid material has been removed. It is also applied to milk from which fat has been removed by centrifuging.

solubles. Liquid containing dissolved substances obtained from processing animal or plant materials. It may contain some fine suspended solids.

solvent extracted. (Process.) A product from which oil has been removed by solvents.

spray dehydrated. (Process.) Material that has been dried by spraying on the surface of a heated drum. It is recovered by scraping from the drum.

starch. (Part.) A white, granular polymer of plant origin. The principal part of seed endosperm.

steamed, steaming. (Process.) Having treated ingredients with steam to alter physical and/or chemical properties. Similar terms: steam cooked, steam rendered, tanked.

stillage. (Part.) The mash from fermentation of grains after removal of alcohol by distillation.

supplement. A feed used with another to improve the nutritive balance or performance of the total and intended to be (1) fed undiluted as a supplement to other feeds; or (2) offered free choice with other parts of the ration separately available; or, (3) further diluted and mixed to produce a complete feed.

tallow. (Part.) Animal fats with titer above 40° C.

tankage. (Part.) *See* Carcass residue.

tempered, tempering. (Process.) *See* Conditioned, conditioning.

titer. A property of fat determined by the solidification point of the fatty acids liberated by hydrolysis.

toasted. (Process.) Browned, dried, or parched by exposure to a fire, or to gas, or electric heat.

trace minerals. Mineral nutrients required by animals in microamounts only (measured in milligrams per pound or smaller units).

vitamins. Organic compounds that function as parts of enzyme systems essential for the transmission of energy and the regulation of metabolisms of the body.

wafered, wafering. (Process.) Having agglomerated a feed of a fibrous nature by compressing into a form usually having a diameter or cross-section measurement greater than its length.

wafers. (Physical form.) A form of agglomerated feed based on fibrous ingredients in which the finished form usually has a diameter or cross-section measurement greater than its length.

whey. (Part.) The watery part of milk separated from the curd.

whey solids. (Part.) The solids of whey (proteins, fats, lactose, ash, and lactic acid).

FEED INGREDIENT TERMS

alfalfa meal or pellets, dehydrated. The aerial portion of the alfalfa plant, reasonably free of other crop plants, weeds, and mold, which has been finely ground and dried by thermal means.

alfalfa meal or pellets, sun-cured, or ground alfalfa hay. The aerial portion of the alfalfa plant, reasonably free of other crop plants, weeds, and mold, which has been sun cured and finely ground. If it is chopped instead of finely ground, it must be designated as ''sun-cured chopped alfalfa'' or ''chopped alfalfa hay.''

animal fat. Obtained from the tissues of mammals and/or poultry in the commercial processes of rendering or extracting. It consists predominantly of glyceride esters of fatty acids and contains no additions of free fatty acids or other materials obtained from fats. It must contain not less than 90% total fatty acids, not more than 2.5% unsaponifiable matter, and not more than 1% insoluble matter. If the product bears a name descriptive of its kind or origin—i.e., tallow, lard, grease—it must correspond thereto. If antioxidant(s) is(are) used, the common name(s) must be indicated, followed by the word ''preservative(s).''

blood meal, conventional cooker dried. Produced from clean, fresh animal blood, exclusive of all extraneous material such as hair, stomach belchings, and urine except in such traces as might occur unavoidably in good manufacturing processes. The product usually has a dark blacklike color and is rather insoluble in water.

blood meal, flash dried. Produced from clean, fresh animal blood, exclusive of all extraneous material such as hair, stomach belchings, and urine except in such traces as might occur unavoidably in good manufacturing processes. A large portion of the moisture (water) is usually removed by a mechanical dewatering process or by condensing by cooking to a semisolid state. The semisolid blood mass is then transferred to a rapid drying facility where the more tightly bound water is rapidly removed. The minimum biological activity of lysine shall be 80%.

brewers' dried grains. The dried extracted residue of barley malt alone or in mixture with other cereal grain or grain products resulting from the manufacture of wort or beer and may contain pulverized dried spent hops in an amount not to exceed 3%, evenly distributed.

coconut meal. The ground residue that remains after removal of most of the oil from dried meat of coconuts, by mechanical or solvent extraction process. May also be called ''copra meal.''

corn gluten feed. That part of the commercial shelled corn that remains after extraction of the larger portion of the starch, gluten, and germ by the processes employed in the wet milling manufacture of corn starch or syrup. It may or may not contain fermented corn extractives and/or corn germ meal.

corn gluten meal. The dried residue from corn after the removal of the larger part of the starch and germ, and the separation of the bran by the process employed in the wet milling manufacture of corn starch or syrup, or by the enzymatic treatment of the endosperm. It may contain fermented corn extractives and/or corn germ meal.

cottonseed hulls. Consists primarily of the outer covering of the cottonseed.

cottonseed meal. Cottonseed meal, mechanical or solvent extracted, is the product obtained by finely grinding the cake that remains after removal of most of the oil from cottonseed by a mechanical or solvent extraction process. It must contain not less than 36% crude protein.

Low gossypol cottonseed meal is a meal in which the gossypol is not more than 0.4% free gossypol.

distillers' dried grains. Obtained after removal of ethyl alcohol by distillation from the yeast fermentation of a grain or a grain mixture by separating the resultant coarse grain fraction of the whole stillage and drying it by methods employed in the grain distilling industry. The predominating grain shall be declared as the first word in the name.

distillers' dried solubles. Obtained after the removal of ethyl alcohol by distillation from the yeast fermentation of a grain or a grain mixture by condensing the thin stillage fraction and drying it by methods employed in the grain distilling industry. The predominating grain shall be declared as the first word in the name.

distillers' dried grains with solubles. The product obtained after the removal of ethyl alcohol by distillation from the yeast fermentation of a grain or a grain mixture by condensing and drying at least three-fourths of the solids of the resultant whole stillage by methods employed in the grain distilling industry. The predominating grain shall be declared as the first word in the name.

dried bakery product. Mixture of bread, cookies, cake, crackers, flours and doughs that has been mechanically separated from nonedible material, artificially dried, and ground. If the product contains more than 3.5% salt, the maximum percentage of salt must be part of the name.

dried beet pulp, plain. The dried residue from sugar beets that has been cleaned and freed from crowns, leaves and sand, and that has been extracted in the process of manufacturing sugar.

dried citrus meal. The finer particles obtained by screening dried citrus pulp.

dried citrus pulp. The ground peel, residue of the inside portions, and occasional cull fruits of the citrus family that have been dried, producing a coarse, flaky product. It may contain dried citrus meal or pellets and whole citrus seeds. If calcium oxide or calcium hydroxide is added as an aid in processing, the maximum percentage present, expressed as calcium, must be shown. If it bears a name descriptive of its kind or origin, it must correspond thereto.

dried skimmed milk, feed grade. The residue obtained by drying defatted milk. It contains 8% maximum moisture. The words ''feed grade'' are not required when listing as an ingredient in manufactured feed.

dried whey. The product obtained by removing water from whey. It contains not less than 11% protein nor less than 61% lactose.

dried whey product. The residue obtained by drying whey from which a portion of the lactose has been removed. The minimum percentage lactose must be prominently declared on the label.

fat, hydrolyzed, or oil, feed grade. Obtained in the fat processing procedures commonly used in edible fat processing or soap making. It consists predominantly of fatty acids and must contain not less than 85% total fatty acids, not more than 6% unsaponifiable matter, and not more than 1% insoluble matter. Its source must be stated in the product name—i.e., hydrolyzed animal fat, hydrolyzed vegetable oil, hydrolyzed animal and vegetable fat. If antioxidant(s) is used, the common name(s) must be indicated, followed by the word preservative(s).

feeding oat meal. Obtained in the manufacture of rolled oat groats or rolled oats and consists of broken oat groats, oat groat chips, and floury portions of the oat groats, with only such quantity of finely ground oat hulls as is unavoidable in the usual process of commercial milling. It must not contain more than 4% crude fiber.

fish meal. Clean, dried, ground tissues of undecomposed whole fish or fish cuttings, either or both, with or without the extraction of part of the oil. It it contains more than 3% salt, the amount of salt must constitute a part of the brand name, provided that in no case must the salt content of this product exceed 7%.

fish solubles, condensed. Obtained by condensing the stickwater. It must contain not less than 30% crude protein.

fish solubles, dried. Obtained by dehydrating the stickwater. It must contain not less than 60% crude protein.

guar meal. Obtained from whole guar beans after removal of most of the endosperm. If the product is heat treated, it may be designated as heat treated or toasted.

hemicellulose extract. A by-product of the manufacture of pressed wood. It is the concentrated soluble material obtained from the treatment of wood at elevated temperature and pressure without use of acids, alkalis, or salts. It contains pentose and hextose sugars and has a total carbohydrate content of not less than 55%.

hominy feed. A mixture of corn bran, corn germ, and part of the starchy portion of either white or yellow corn kernels or mixture thereof, as produced in the manufacture of pearl hominy, hominy grits, or table meal, and must contain not less than 4% crude fat. If prefixed by the words white or yellow, the product must correspond thereto.

linseed meal. Linseed meal, mechanical extracted, is the product obtained by grinding the cake or chips which remain after removal of the most of the oil from flaxseed by a mechanical extraction process. It must contain no more than 10% fiber.

Linseed meal, solvent extracted, is the product obtained by grinding the flakes that remain after removal of most of the oil from flaxseed by a solvent extraction process. It must contain no more than 10% fiber.

malt sprouts. Obtained from malted barley by the removal of the sprouts that may include some of the malt hulls, other parts of malt, and foreign material unavoidably present. It must contain not less than 24% crude protein. The term malt sprouts when applied to a corresponding portion of other malted cereals must be used in qualified form, i.e., rye malt sprouts, wheat malt sprouts, etc.

meat meal. The dry rendered product from mammal tissues, exclusive of hair, hoof, horn and hide trimmings, manure, and stomach contents, except in such amounts as may occur unavoidably in good factory practice. It shall not contain added blood meal. It shall contain not more than 14% pepsin indigestible residue. Not more than 11% of the crude protein in the product shall be pepsin indigestible. It must be designated according to its protein content. If the product bears a name descriptive of its kind, composition, or origin, it must correspond thereto.

meat-and-bone meal. The dry-rendered product from mammal tissues, exclusive of hair, hoof, horn, hide trimmings, manure and stomach contents, except in such amounts as may occur unavoidably in good factory practice. It shall not contain added blood meal. It shall contain more than 4.4% phosphorus and not more than 14% pepsin indigestible residue. Not more than 11% of the crude protein shall be pepsin indigestible. It must be designated according to its protein content. If the product bears a name description of its kind, composition or origin it must correspond thereto.

meat meal tankage. The dry product from mammal tissues, exclusive of hair, hoof, horn, hide trimmings, manure, and stomach contents, except in such amounts as may occur unavoidably in good factory practice, especially prepared for feeding purposes by tankage under live steam or dry rendering. It shall contain not more than 14% pepsin indigestible residue. Not more than 11% of the crude protein in the product shall be pepsin indigestible. It must be designated according to its protein content. If the product bears a name descriptive of its kind, composition, or origin it must correspond thereto.

molasses, beet. A by-product of the manufacture of sucrose from sugar beets. It must contain not less than 48% total sugars, expressed as invert and its density determined by double dilution must not be less than 79.5° Brix.

molasses, starch. A by-product of the manufacture of dextrose from starch derived from corn or grain sorghums in which the starch is hydrolized by use of enzymes and/or acid. It must contain not less than 43% reducing sugars expressed as dextrose and not less than 50% total sugars expressed as dextrose. It shall contain not less than 73% total solids.

peanut meal. A ground product of the shelled peanuts, composed principally of the kernels, with such portion of the hulls, or fiber, and oil, as may be left in the ordinary course of manufacture by a mechanical or solvent extraction process. It must contain not more than 7% crude fiber.

poultry by-product meal. Consists of the ground, dry-rendered or wet-rendered clean parts of the carcass of slaughtered poultry, such as heads, feet, undeveloped eggs and intestines, exclusive of feathers, except in such trace amounts as might occur unavoidably in good factory practice. It must contain not more than 16% ash and not more than 4% acid-insoluble ash.

poultry feathers, hydrolized. The product resulting from the treatment under pressure of clean, undecomposed feathers from slaughtered poultry, free of additives and/or accelerators. Not less than 75% of its crude protein content must be digestible by the pepsin digestibility method.

rice bran. The pericarp or bran layer and germ of the rice, with only such quantity of hull fragments, chipped, broken or brewer's rice, and calcium carbonate as is unavoidable in the regular milling of edible rice. It must contain not more than 13% crude fiber. When the calcium carbonate exceeds 3%, the percentage must be declared in the brand name.

rice mill by-product. The total offal obtained in the milling of rice. It consists of rice hulls, rice bran, rice polishings, and broken rice grains. Its crude fiber content must not exceed 32%.

safflower meal. The ground residue obtained after extracting the oil from whole safflower seed, by a mechanical or solvent extraction process.

screenings. Obtained in the cleaning of grains that are included in the U.S. Grain Standard Act and other agricultural seeds. It may include light and broken grains and agricultural seeds, weed seeds, hulls, chaff, joints, straw, elevator or mill dust, sand, and dirt. It must be designated as grain screenings, mixed screenings, and chaff and/or dust.

No grade of screenings must contain any seeds or other material in amount that is either injurious to animals or will impart any objectionable odor or flavor to their milk or flesh. The screenings must contain not more than four whole prohibited noxious weed seeds per pound and must contain not more than 100 whole restricted noxious weed seeds per pound. The prohibited and restricted noxious weed seeds must be those named as such by the seed control law of the state in which the screenings is sold or used.

All grades of screenings must bear minimum guarantees of crude protein and crude fat and maximum guarantees of crude fiber and ash.

soybean meal. The product obtained by grinding the cake, chips, or flakes that remain after removal of most of the oil from soybeans by a mechanical or solvent extraction process. It must contain not more than 7% crude fiber. It may contain a nonnutritive inert, nontoxic conditioning agent to reduce caking and improve flowability in an amount not to exceed that necessary to accomplish its intended effect and in no case exceed 0.5%.

Soybean meal, dehulled, solvent extracted, is obtained by grinding the flakes remaining after removal of most of the oil from dehulled soybeans by a solvent extraction process. It must contain not more than 3.3% crude fiber.

soybean mill feed. Composed of soybean hulls and the offal from the tail of the mill that results from the manufacture of soy grits or flour. It must contain not less than 13% crude protein and not more than 32% crude fiber.

soybeans, heat processed. The product resulting from heating whole soybeans without removing any of the component parts. It may be ground, pelleted, flaked or powdered. It must be sold according to its crude protein content. Similar term: dry roasted soybeans.

sunflower meal. Obtained by grinding the residue remaining after extraction of most of the oil from whole sunflower seed, by mechanical or solvent extraction process. May also be obtained by grinding the residue after extraction of oil from dehulled sunflower seed by either process.

urea. Predominantly urea but may contain other nontoxic nitrogenous compounds that are present as by-products from the commercial synthesis and processing of urea. It must contain not less than 45% nitrogen (equivalent to 281.25% crude protein). If it contains less than 45% N but 41% or more N, it must be designated as urea and conditioner(s). If the name of the conditioner(s) does(do) appear in the product name, the ingredient listing must contain the specific name of the conditioner(s).

If the urea and conditioner(s) contribute more than 0.5% conditioner(s) to the mixed feed, the conditioner(s) must be named in the mixed ingredient list.

wheat bran. The coarse outer covering of the wheat kernel as separated from cleaned and scoured wheat in the usual process of commercial milling.

wheat mill run. Coarse wheat bran, fine particles of wheat bran, wheat shorts, wheat flour, and the offal from the "tail of the mill." This product must be obtained in the usual process of commercial milling and must contain not more than 9.5% crude fiber.

wheat middlings. Fine particles of wheat bran, wheat shorts, wheat germ, wheat flour, and some of the offal from the "tail of the mill." This product must be obtained in the usual process of commercial milling and must contain not more than 9.5% crude fiber.

wheat shorts. The fine particles of wheat bran, wheat germ, wheat flour, and the offal from the "tail of the mill." This product must be obtained in the usual process of commercial milling and must contain not more than 7% crude fiber.

Appendix 4

Composition of Feeds Commonly Used in Swine Diets

The following paragraphs are from p. 27 of National Research Council Publication *No. 2,* Nutrient Requirements of Swine (1979), and precede the tables on feed composition reproduced here from the same publication. Additional feed composition data are available in the National Research Council (1982), Joint U.S.–Canadian Tables of Feed Composition, National Academy of Sciences, Washington D.C.

In formulating diets to meet the recommended nutrient requirements of swine, it is necessary to know the nutrient composition and the bioavailability of nutrients in each ingredient used. Values in the tables have been compiled from several National Academy of Sciences–National Research Council reports.[1] Data were also obtained from the International Feedstuffs Institute, Utah State University,[2] the Subcommittee on Feed Composition of the N.R.C. Committee on Animal Nutrition,[3] and from individuals at universities, agricultural experiment stations, commercial feed companies, and the "Feedstuffs Yearbook," Vol. 48 (1976).

Individual ingredients may vary widely in composition because of the variation in species or variety, storage conditions, climate, soil moisture, and nutrient status. Variations in analytical procedures also affect values obtained. Therefore, the values given are an average and are subject to interpretation.

In this edition, short ingredient names are listed and only those commonly used in swine feeding are included. A six-digit International Feed Number is given for each ingredient. The first digit is the class designation.[4] In computer formulation this reference number may be used as the "numerical name" of an ingredient. This number is also listed after each Official Feed Definition in the Association of American Feed Control Officials *Handbook.*[5] Nutrient content is expressed on an "as-fed" or "as-is" basis.

[1]Atlas of Nutritional Data on United States and Canadian Feeds, *1971.* ISBN 0-309-01919-2; United States–Canadian Tables of Feed Composition, *1969.* ISBN 0-309-01684-3; Nutrient Requirements of Poultry, *1977.* ISBN 0-309-02725-X; Nutrient Requirements of Swine, *1973.* ISBN 0-309-02140-5.

[2]Lorin E. Harris, Director.

[3]Joseph H. Conrad, Chairman, J. R. Aitken, and Charles W. Deyoe, Lorin E. Harris, Paul W. Moe, Rodney L. Preston, and Peter J. Van Soest.

[4]Class designations: (1) dry forages and roughages; (2) pasture, range plants, and forages fed green; (3) silages; (4) energy feeds; (5) protein supplements; (6) minerals; (7) vitamins; and (8) additives.

[5]Epps, Ernest A., Jr., Division of Agricultural Chemistry, P.O. Box 16390-A, Baton Rouge, Louisiana 70803.

In addition to feedstuffs composition, the sources and composition of minerals frequently fed to swine are listed. The percentage of the mineral in each mineral source is given for the pure compound. The percentage of purity for technical and feed grade sources should, therefore, be multiplied by the listed percentage in this table to arrive at the percentage of the element in the source being used.

TABLE A4.1. Average Composition of Some Feed Ingredients Commonly Used in Swine Diets (Excluding Amino Acids

Line No.		International Feed Number[b]	Dry Matter (%)	Energy (kcal/kg)		Protein (%)	Ether Extract (%)	Crude Fiber (%)	Minerals			
				DE	ME				Calcium (%)	Phosphorus (%)	Potassium (%)	Chlorine (%)
01	Alfalfa meal, dehy., 17% protein	1-00-023	92	2580	2270	17.5	2.5	24.1	1.44	0.22	2.40	0.46
02	Barley	4-00-549	89	3086	2870	11.6	1.8	5.1	0.05	0.36	0.48	0.15
03	Barley, Pacific Coast	4-07-939	89	3130	2940	9.0	2.0	6.4	0.05	0.32	0.53	0.15
04	Beans, field (*Vicia faba*)	5-09-262	89	3263	3080	26.0	1.4	8.2	0.14	0.54	1.20	—
05	Beet pulp, dried	4-00-669	91	2866	2345	8.0	0.5	21.0	0.60	0.10	0.21	—
06	Blood meal, spray or ring dried	5-00-381	86	2690	1927	85.0	1.0	0.6	0.30	0.25	0.90	0.27
07	Brewers dried grains	5-02-141	92	1940	1710	25.3	6.2	15.3	0.29	0.52	0.09	0.12
08	Corn, dent yellow, grain	4-02-935	89	3525	3325	8.8	3.8	2.2	0.02	0.28	0.30	0.04
09	Corn and cob meal	4-02-849	85	3086	2500	7.8	3.0	10.0	0.04	0.21	0.45	0.04
10	Corn, gluten feed	5-02-903	90	3307	2400	22.0	2.5	10.0	0.40	0.80	0.57	0.22
11	gluten meal, 41%	5-02-411	91	3230	3069	41.0	2.5	4.0	0.23	0.55	0.31	0.11
12	Corn, distillers grain w/solubles, dehy.	5-02-843	93	3568	3390	27.2	9.0	9.1	0.35	0.95	1.00	0.17
13	Corn, distillers solubles, dehy.	5-02-844	92	3307	2900	28.5	9.0	4.0	0.35	1.33	1.75	0.26
14	Corn, hominy feed	4-02-887	90	3615	3365	10.0	6.9	6.0	0.04	0.50	0.67	0.05
15	Cottonseed meal, mechanical extracted	5-01-609	93	2954	2453	40.9	3.9	12.6	0.17	1.05	1.19	0.04
16	Cottonseed meal, solvent extracted	5-01-619	92	2689	2555	41.4	1.5	11.3	0.15	0.97	1.22	0.03
17	Feather meal	5-03-795	93	2778	2270	86.4	3.3	1.0	0.20	0.80	0.31	—
18	Fish meal, anchovy	5-01-985	92	3086	2450	64.2	10.0	1.0	3.73	2.43	0.90	0.29
19	herring	5-02-000	93	3086	2500	72.3	10.0	0.7	2.29	1.70	1.50	0.90
20	Menhaden	5-02-009	92	2734	2230	60.5	9.4	0.7	5.11	2.88	0.77	0.60
21	Fish solubles, condensed	5-01-969	51	3307	3190	31.5	4.0	0.2	0.30	0.50	1.74	2.65
22	Meat-and-bone meal, 50%	5-09-322	93	2866	2434	50.4	8.6	2.8	10.10	4.96	1.40	0.74
23	Meat meal, 55%	5-09-323	92	2998	2540	54.4	7.1	2.5	8.27	4.10	1.40	0.91
24	Molasses, beet	4-00-668	79	2460	2320	6.1	0.0	0.0	0.13	0.06	4.83	1.30
25	Molasses, cane	4-04-696	74	2469	2343	2.9	0.0	0.0	0.82	0.08	2.38	—
26	Oats	4-03-309	89	2866	2668	11.4	4.2	10.8	0.06	0.27	0.37	0.11
27	Oat groats (dehulled oats)	4-03-331	91	3690	3400	16.0	5.5	3.0	0.07	0.43	0.34	—
28	Peas	5-03-600	90	3527	3200	23.8	1.3	5.5	0.11	0.42	1.02	0.06
29	Peanut meal, expeller	5-03-649	90	3600	3200	45.0	7.3	12.0	0.16	0.55	1.12	0.03
30	Peanut meal, solvent	5-03-650	90	2845	2920	47.0	1.2	13.1	0.20	0.65	1.15	—
31	Rapeseed meal, solvent	5-03-871	94	2998	2670	35.0	1.8	12.4	0.66	1.09	0.80	—
32	Rice bran, solvent	4-03-930	91	3080	2200	12.9	0.6	11.4	0.07	1.50	1.35	0.07
33	Rice, broken	4-03-932	89	2513	2360	8.7	1.7	9.8	0.08	—	—	0.08
34	Rice, polishings	4-03-943	90	3792	3000	12.2	11.0	4.1	0.05	1.31	1.06	0.11
35	Rye, grain	4-04-047	89	3307	2712	12.6	1.8	2.8	0.08	0.30	0.46	—
36	Safflower meal, solvent	5-04-110	91	2960	2435	28.5	0.5	30.6	0.40	1.10	0.80	—
37	Sesame meal, expeller	5-04-220	93	3130	2560	42.0	7.0	6.5	1.99	1.37	1.20	0.06
38	Skim milk, dried	5-01-175	92	3792	3360	33.5	0.9	0.0	1.28	1.02	1.59	0.50
39	Sorghum, grain (Milo)	4-04-444	89	3439	3229	8.9	2.8	2.3	0.03	0.28	0.32	0.09
40	Soybeans, full-fat cooked	5-04-597	90	4056	3540	37.0	18.0	5.5	0.25	0.58	1.61	0.03
41	Soybean meal, dehulled, solvent	5-04-612	90	3860	3485	48.5	1.0	3.9	0.27	0.62	2.02	0.05
42	Soybean meal, expeller	5-04-600	90	3483	2990	42.6	4.0	6.2	0.27	0.61	1.83	0.07
43	Soybean meal, solvent	5-04-604	89	3350	3090	44.0	0.8	7.3	0.29	0.65	2.00	0.05
44	Sunflower meal, dehulled, solvent	5-04-739	93	2998	2605	42.0	2.9	12.2	0.37	1.00	1.00	0.10
45	Wheat bran	4-05-190	90	2513	2320	15.7	4.0	11.0	0.14	1.15	1.19	0.06
46	Wheat shorts	4-05-201	89	3175	2910	16.8	4.2	8.2	0.11	0.76	0.88	0.07
47	Wheat middlings	4-05-205	88	3050	2940	16.0	3.0	7.0	0.12	0.90	0.60	0.03
48	Wheat, hard, red winter	4-05-268	87	3483	3220	14.1	1.9	2.4	0.05	0.37	0.45	0.05
49	Wheat, soft, red winter	4-05-294	86	3659	3416	10.2	1.8	2.4	0.05	0.31	0.40	0.08
50	Whey, dried	4-01-182	93	3439	3190	13.6	0.8	1.3	0.97	0.76	1.05	1.50
51	Whey, low lactose	4-01-186	91	3307	2750	15.5	1.0	0.3	1.95	0.98	3.00	2.10
52	Yeast, brewers dried	7-05-527	93	3135	2707	44.4	1.0	2.7	0.12	1.40	1.70	0.12

[a] As fed basis.

[b] The first digit is the feed class, coded as follows: (1) dry forages and roughages; (2) pasture, range plants and forages fed green; (3) silages; (4) energy feeds; and (5) protein supplements.

The *abbreviations for terms used in the tables are as follows:* DE, digestible energy; dehy, dehydrated; kcal, kilocalories; kg, kilogram(s); ME, metabolizable energy; and w, with.

									Vitamins									
Line No.	Magnesium (%)	Sodium (%)	Sulfur (%)	Copper (mg/kg)	Iron (mg/kg)	Manganese (mg/kg)	Selenium (mg/kg)	Zinc (mg/kg)	Biotin (mg/kg)	Choline (mg/kg)	Folacin (mg/kg)	Niacin (mg/kg)	Pantothenic Acid (mg/kg)	Pyridoxine (mg/kg)	Riboflavin (mg/kg)	Thiamin (mg/kg)	Vitamin B_{12} (mg/kg)	Vitamin E (IU/kg)
01	0.26	0.08	0.21	8.2	310	28.0	0.60	17	0.30	1097	6.3	38	28.4	6.5	15.7	3.4	0.004	125
02	0.14	0.04	0.15	7.5	50	8.0	0.10	17	0.08	990	0.5	63	9.2	3.0	1.2	4.0	—	36
03	0.12	0.02	—	7.7	60	16.3	0.10	15	0.15	1034	0.5	48	7.0	2.9	1.6	5.5	—	36
04	0.13	0.80	—	4.1	70	8.4	—	42	0.09	1670	—	22	3.0	—	1.6	5.5	—	1
05	0.27	0.32	0.20	12.5	300	35.0	—	0.7	—	800	—	20	0.8	—	1.1	0.2	—	—
06	0.22	0.33	0.32	8.1	3000	6.4	—	306	0.30	749	0.3	22	1.1	4.4	1.3	0.5	0.440	—
07	0.16	0.15	0.31	21.1	250	37.8	0.70	98	0.96	1723	7.1	29	8.0	0.7	1.4	0.5	—	25
08	0.12	0.02	0.08	3.4	35	5.0	0.04	10	0.11	530	0.2	34	7.5	7.0	1.0	3.5	—	22
09	0.13	0.01	0.18	6.7	70	7.7	0.07	9	0.05	393	0.3	17	4.0	5.0	0.9	—	—	19
10	0.29	0.95	0.22	47.9	460	23.8	0.10	48	0.33	1518	0.3	66	17.0	15.0	2.4	2.0	—	15
11	0.05	0.07	0.40	28.3	400	8.9	1.00	20	0.18	330	0.2	50	10.0	7.9	1.7	0.2	—	20
12	0.35	0.90	0.30	44.7	280	30.0	0.39	80	0.30	3400	0.9	80	11.0	2.2	8.6	3.5	—	40
13	0.64	0.26	0.37	82.7	560	73.7	0.33	85	1.40	4842	1.1	116	21.0	10.0	11.6	6.9	—	55
14	0.24	0.10	0.03	13.3	70	14.5	—	3	0.13	1500	0.3	46	8.0	11.0	2.2	7.9	—	—
15	0.42	0.04	0.40	18.6	160	22.9	0.90	57	0.60	2753	2.7	38	7.7	5.3	4.2	9.7	—	15
16	0.40	0.04	—	17.8	110	20.2	—	—	0.55	2933	2.7	40	9.9	3.0	4.0	7.7	—	15
17	0.20	0.71	—	—	—	21.0	—	—	0.04	891	0.2	27	10.0	—	2.1	0.1	0.600	—
18	0.24	1.10	0.54	9.3	220	9.5	1.36	103	0.23	5100	0.2	135	20.0	4.0	7.1	0.1	0.352	6
19	0.15	0.61	0.69	4.5	80	4.7	1.93	132	0.20	5306	0.5	142	22.0	4.0	9.9	0.1	0.588	17
20	0.16	0.41	0.45	10.8	440	33.0	2.10	147	0.15	3056	1.0	55	9.0	4.0	4.9	0.2	0.150	7
21	0.02	3.10	0.12	44.9	30	14.4	2.00	38	0.18	4028	—	169	35.0	12.2	14.6	5.5	0.347	—
22	1.12	0.72	0.26	1.5	490	14.2	0.25	93	0.14	1996	0.6	46	4.1	12.8	4.4	0.2	0.070	0.8
23	1.13	0.73	0.26	1.5	440	12.3	0.25	103	0.14	2077	0.6	57	5.0	3.0	5.5	0.2	—	0.8
24	0.23	—	0.48	17.7	70	4.7	—	14	0.70	880	—	48	4.0	—	2.1	—	—	4.4
25	0.35	0.90	0.35	59.6	200	42.2	—	—	0.70	660	—	45	39.0	—	2.3	0.9	—	4.4
26	0.16	0.06	0.21	5.9	70	43.2	0.30	1	0.30	1100	0.4	15	29.2	1.0	1.1	6.0	—	20.0
27	0.09	—	0.20	6.4	90	28.6	—	—	0.20	1232	0.3	18	11.0	—	1.3	6.8	—	15.0
28	—	0.04	—	—	50	—	—	30	0.18	642	0.4	17	4.6	1.0	0.8	1.8	—	—
29	0.32	—	0.28	—	—	24.8	—	—	0.39	1640	—	165	46.8	—	5.1	7.1	—	2.9
30	0.40	0.10	—	—	—	29.9	—	—	0.39	1980	—	165	50.6	—	11.0	6.6	—	3.0
31	0.51	0.50	—	7.0	180	43.0	0.98	66	—	6464	—	153	9.0	7.0	3.7	1.7	—	19.1
32	0.95	0.07	0.18	13.0	190	138.0	—	30	4.20	1135	—	293	23.0	14.0	2.5	22.5	—	59.8
33	0.11	0.07	0.06	—	—	18.0	—	17	0.08	800	0.2	46	8.0	—	0.7	—	—	14.5
34	0.65	0.10	0.17	—	160	—	—	—	0.61	1237	—	520	47.0	—	1.8	19.8	—	90.0
35	0.12	0.02	0.15	7.8	100	66.9	—	31	0.60	—	0.6	16	9.2	—	1.5	4.4	—	15.0
36	0.37	0.06	—	10.8	560	19.8	—	44	1.56	2247	0.5	60	43.8	—	11.3	2.8	—	0.9
37	0.86	0.04	0.43	—	—	47.9	—	100	0.34	1690	—	30	6.0	12.5	3.6	2.8	—	—
38	0.11	0.44	0.31	11.5	50	2.0	0.12	40	0.33	1250	0.6	12	33.0	3.9	22.0	3.5	0.010	9.1
39	0.20	0.01	0.09	14.1	40	12.9	—	14	0.09	678	0.2	41	12.0	3.2	1.1	4.0	—	12.0
40	0.21	0.28	0.22	15.8	80	29.8	0.11	16	0.27	2420	3.5	22	15.6	10.8	2.6	6.6	—	0.9
41	0.27	0.34	0.43	36.3	120	27.5	0.10	45	0.32	2850	0.7	22	15.0	5.0	2.9	1.7	—	3.3
42	0.26	0.27	0.33	18.0	140	30.7	0.10	60	0.33	2703	0.5	37	14.0	—	3.7	1.7	—	6.1
43	0.27	0.34	0.43	36.3	120	29.3	0.10	27	0.32	2794	0.5	60	13.3	8.0	2.9	1.7	—	2.1
44	0.75	2.00	—	3.5	30	22.9	—	—	1.45	2894	—	220	10.0	16.0	3.1	—	—	11.0
45	0.52	0.05	0.22	10.2	170	100.0	0.50	95	0.10	980	1.8	321	31.0	7.0	3.1	8.0	—	10.8
46	0.26	0.07	0.23	12.1	100	115.0	0.50	106	0.10	930	1.4	100	17.6	11.0	2.0	19.9	—	29.9
47	0.29	0.60	0.16	4.4	40	43.0	0.80	64	0.10	1100	0.6	53	13.0	9.0	2.2	18.9	—	—
48	0.17	0.04	0.12	10.6	50	62.2	0.06	14	0.04	1090	0.4	56	13.5	3.4	1.4	4.5	—	12.6
49	0.10	0.04	0.12	9.7	40	51.3	0.06	14	0.04	788	0.4	48	11.0	4.0	1.2	4.3	—	13.2
50	0.13	2.00	1.04	40.0	130	6.1	0.06	—	0.34	1980	0.8	10	44.0	4.0	27.1	4.1	0.015	0.2
51	0.25	1.50	—	—	—	14.0	0.06	—	0.64	4392	1.4	19	69.0	4.0	29.9	5.7	0.015	—
52	0.23	0.07	0.38	32.8	120	5.2	1.00	39	1.05	3984	9.9	448	109.0	42.8	37.0	91.8	—	—

TABLE A4.2. Average Amino Acid Composition of Some Commonly Used Feedstuffs

	International Feed Number[b]	Protein (%)	Arginine (%)	Histidine (%)	Isoleucine (%)	Leucine (%)	Lysine (%)	Methionine (%)	Cystine (%)	Phenylalanine (%)	Tyrosine (%)	Threonine (%)	Tryptophan (%)	Valine (%)
Alfalfa meal, dehy., 17% protein	1-00-023	17.5	0.8	0.3	0.8	1.3	0.73	0.2	0.2	0.8	0.6	0.70	0.28	0.8
Barley	4-00-549	11.6	0.6	0.3	0.5	0.8	0.40	0.2	0.3	0.6	0.3	0.42	0.14	0.6
Barley, Pacific Coast	4-07-939	9.0	0.5	0.2	0.4	0.6	0.29	0.1	0.2	0.5	0.3	0.30	0.12	0.5
Beans, field (*Vicia faba*)	5-09-262	27.4	2.5	0.7	1.1	1.9	1.72	0.2	0.2	1.2	0.7	3.96	0.24	1.2
Beet pulp, dried	4-00-669	8.0	0.3	0.2	0.3	0.6	0.60	0.01	0.01	0.3	0.4	0.40	0.10	0.4
Blood meal, spray or ring dried	5-00-381	85.0	4.1	5.5	1.0	12.7	8.10	1.5	1.5	7.3	3.0	4.90	1.10	9.1
Brewers dried grains	5-02-141	25.3	0.8	0.6	1.4	2.5	0.90	0.6	0.4	1.5	1.2	0.98	0.34	1.7
Corn, dent yellow, grain	4-02-935	8.8	0.5	0.2	0.4	1.1	0.24	0.2	0.2	0.5	0.5	0.39	0.05	0.4
Corn and cob meal	4-02-849	7.8	0.4	0.2	0.4	1.0	0.18	0.1	0.1	0.4	—	0.35	0.07	0.4
Corn, gluten feed	5-02-903	22.0	1.0	0.7	0.7	1.9	0.63	0.5	0.5	0.8	0.6	0.89	0.10	1.0
gluten meal, 41%	5-02-411	40.6	1.4	1.0	2.2	7.2	0.78	1.0	0.7	2.9	1.0	1.40	0.21	2.2
Corn, distillers grain w/solubles, dehy.	5-02-843	27.2	1.0	0.7	1.0	2.6	0.60	0.6	0.3	1.2	0.7	0.92	0.19	1.3
Corn, distillers solubles, dehy.	5-02-844	28.5	1.1	0.7	1.3	2.1	0.90	0.5	0.4	1.3	1.0	1.00	0.30	1.4
Corn, hominy feed	4-02-887	10.0	0.5	0.2	0.4	0.8	0.40	0.1	0.1	0.4	0.5	0.40	0.10	0.5
Cottonseed meal, mechanical extracted	5-01-609	40.9	4.3	1.1	1.6	2.5	1.51	0.6	0.6	2.2	1.1	1.38	0.55	2.0
Cottonseed meal, solvent extracted	5-01-619	41.4	4.6	1.1	1.3	2.4	1.71	0.5	0.6	2.2	1.0	1.32	0.47	1.9
Feather meal	5-03-795	86.4	3.9	0.3	2.7	6.7	1.10	0.4	3.0	2.7	6.3	2.80	0.50	4.6
Fish meal, anchovy	5-01-985	64.2	3.7	1.5	3.0	5.0	5.10	1.9	0.6	2.7	2.2	2.68	0.74	3.4
herring	5-02-000	72.3	4.8	1.7	3.2	5.3	5.70	2.1	0.7	2.8	2.3	3.00	0.81	4.4
Menhaden	5-02-009	60.5	3.8	1.5	2.9	5.0	4.83	1.8	0.6	2.5	2.0	2.50	0.68	3.2
Fish solubles, 50% solids	5-01-969	31.5	1.6	1.6	0.7	1.9	1.73	0.5	0.3	0.9	0.4	0.86	0.31	1.2
Meat-and-bone meal, 50%	5-09-322	50.4	3.6	1.2	1.4	3.2	2.60	0.7	0.3	1.5	0.8	1.50	0.28	2.3
Meat meal, 55%	5-09-323	54.4	3.7	1.3	1.6	3.3	3.00	0.8	0.7	1.7	1.8	1.74	0.36	2.6
Oats	4-03-309	11.4	0.8	0.2	0.5	0.9	0.40	0.2	0.2	0.6	0.5	0.43	0.16	0.7
Oat groats (dehulled oats)	4-03-331	16.0	0.7	0.3	0.5	1.0	0.60	0.2	0.3	0.7	0.9	0.50	0.18	0.7
Peas	5-03-600	23.8	1.4	0.7	1.1	1.8	1.60	0.3	0.2	1.3	—	0.94	0.24	1.3
Peanut meal, expeller	5-03-649	45.0	4.7	1.1	1.8	3.6	1.55	0.4	0.7	2.6	—	1.40	0.46	2.6
Peanut meal, solvent	5-03-650	47.0	4.9	1.2	2.1	3.7	1.76	0.4	0.8	2.8	2.0	1.45	0.48	2.8
Rapeseed meal, solvent	5-03-871	35.0	1.9	1.0	1.3	2.3	2.10	0.7	0.4	1.4	0.8	1.53	0.45	1.8
Rice bran, solvent	4-03-930	12.9	0.9	0.3	0.4	0.9	0.59	0.2	0.1	0.6	0.7	0.48	0.15	0.6
Rice, broken	4-03-932	8.7	0.6	0.2	0.3	0.5	0.24	0.1	0.1	0.3	—	0.27	0.10	0.5
Rice, polishings	4-03-943	12.2	0.8	0.2	0.4	0.8	0.57	0.2	0.1	0.5	0.6	0.40	0.13	0.8
Rye, grain	4-04-047	12.6	0.5	0.3	0.5	0.7	0.49	0.2	0.2	0.6	0.3	0.86	0.12	0.6
Safflower meal, solvent	5-04-110	28.5	3.7	1.0	1.7	2.5	1.30	0.7	0.7	1.9	—	1.35	0.60	2.3
Sesame meal, expeller	5-04-220	42.0	4.2	1.1	2.1	3.3	1.30	1.2	0.6	2.2	2.0	1.65	0.80	2.4
Skim milk, dried	5-01-175	33.5	1.1	0.8	2.2	3.2	2.40	0.9	0.4	1.6	1.1	1.60	0.44	2.3
Sorghum, grain (Milo)	4-04-383	8.9	0.4	0.3	0.5	1.4	0.22	0.1	0.2	0.4	0.4	0.27	0.10	0.5
Soybeans, full-fat cooked	5-04-597	37.0	2.8	0.9	2.0	2.8	2.40	0.5	0.6	1.8	1.2	1.50	0.55	1.8
Soybean meal, dehulled, solvent	5-04-612	48.5	3.7	1.3	2.6	3.8	3.18	0.7	0.7	2.1	2.0	1.91	0.67	2.7
Soybean meal, solvent	5-04-604	44.0	3.3	1.2	2.4	3.5	2.93	0.7	0.7	2.3	1.3	1.81	0.62	2.3
Sunflower meal, dehulled, solvent	5-04-739	42.0	3.3	1.4	2.8	3.9	1.70	0.7	0.7	2.9	1.2	2.13	0.71	3.2
Wheat bran	4-05-190	15.7	1.0	0.3	0.6	0.9	0.59	0.2	0.3	0.5	0.4	0.42	0.30	0.7
Wheat, hard, red winter	4-05-268	14.1	0.6	0.2	0.6	0.9	0.40	0.2	0.3	0.7	0.6	0.37	0.18	0.6
Wheat middlings	4-05-205	16.0	1.8	0.4	0.6	1.1	0.69	0.2	0.3	0.6	0.5	0.49	0.20	0.7
Wheat shorts	4-05-201	16.8	1.2	0.5	0.6	1.1	0.81	0.2	0.3	0.7	0.5	0.61	0.19	0.8
Wheat, soft, red winter	4-05-294	10.2	0.4	0.2	0.4	0.6	0.31	0.2	0.2	0.5	0.4	0.32	0.12	0.4
Whey, dried	4-01-182	12.0	0.3	0.2	0.8	1.2	0.97	0.2	0.3	0.3	0.3	0.89	0.19	0.7
Whey, low lactose	4-01-186	15.5	0.7	0.1	0.3	0.2	1.47	0.6	0.6	0.1	0.2	0.50	0.18	0.3
Yeast, brewers dried	4-05-527	44.4	2.2	1.1	2.1	3.2	3.23	0.7	0.5	1.8	1.5	2.06	0.49	2.3

[a]As-fed basis.

[b]The first digit is the feed class, coded as follows: (1) dry forages and roughages; (2) pasture, range plants and forages fed green; (3) silages; (4) energy feeds and (5) protein supplements.

TABLE A4.3. Common Mineral Sources for Swine

Mineral	*Source*	*Chemical formula*	*Mineral content*[a]			
Calcium	Calcium carbonate		40% Ca	0.02% Na		
	Limestone		38% Ca	0.05% Na	0.01% F	
Calcium and phosphorus	Bone meal		24% Ca	12.6% P	0.37% Na	0.05% F
	Phosphate, Curaçao		36% Ca	14% P	0.3% Na	0.54% F
	Defluorinated		30–34% Ca	18% P	5.7% Na	0.16% F
	Dicalcium		18–24% Ca	18.5% P	0.6% Na	0.14% F
	Mono and dicalcium		16–19% Ca	21% P	0.6% Na	0.20% F
	Soft rock		17% Ca	9% P	0.1% Na	1.2% F
	Sodium tripoly		0	25% P	31.2% Na	0.03% F
Sodium and chlorine	Sodium chloride		39.3% Na	60.7% Cl		
Iron	Ferrous sulfate	$FeSO_4\ H_2O$	32.9% Fe			
	Ferrous sulfate	$FeSO_4\ 7H_2O$	20.1% Fe			
	Ferric ammonium citrate		16.5–18.5% Fe			
	Ferrous fumarate	$FeC_4H_2O_4$	32.9% Fe			
	Ferric chloride	$FeCl_3\ 6H_2O$	20.7% Fe			
	Ferrous carbonate	$FeCO_3$	48.2% Fe			
	Ferric oxide	Fe_2O_3	69.9% Fe			
	Ferrous oxide	FeO	77.8% Fe			
Copper	Cupric carbonate	$CuCO_3\ Cu(OH)_2$	57.5% Cu			
	Cupric chloride	$CuCl_2\ 2H_2O$	37.3% Cu			
	Cupric hydroxide	$Cu(OH)_2$	65.1% Cu			
	Cupric oxide	CuO	79.9% Cu			
	Cupric sulfate	$CuSO_4\ 5H_2O$	25.4% Cu			
Manganese	Manganese carbonate	$MnCO_3$	47.8% Mn			
	Manganous chloride	$MnCl_2\ 4H_2O$	27.8% Mn			
	Manganous oxide	MnO	77.4% Mn			
	Manganese sulfate	$MnSO_4\ 5H_2O$	22.7% Mn			
	Manganous sulfate	$MnSO_4\ H_2O$	32.5% Mn			
Zinc	Zinc carbonate	$5ZnO\ 2CO_34H_2O$	56.0% Zn			
	Zinc chloride	$ZnCl_2$	48.0% Zn			
	Zinc oxide	ZnO	80.3% Zn			
	Zinc sulfate	$ZnSO_4\ 7H_2O$	22.7% Zn			
	Zinc sulfate	$ZnSO_4\ H_2O$	36.4% Zn			
Iodine	Calcium iodate	$Ca(IO_3)_2$	65.1% I			
	Potassium iodide	KI	76.4% I			
	Cuprous iodide	CuI	66.6% I			
	Penta calcium orthoperiodate	$Ca_5(IO_6)_2$	39.3% I			
Selenium	Sodium selenite	Na_2SeO_3	45.6% Se	26.6% Na		
	Sodium selenate	Na_2SeO_4	41.8% Se	24.3% Na		

[a]Actual mineral levels in technical grade sources may vary.

Appendix 5

Sample Diets for Swine

TABLE A5.1. Examples of Suitable Starter (Creep) Diets for Use in Temperature or Tropical Environments[a]

Feedstuff	*A, %*	*B, %*	*C, %*	*D, %*	*E, %*
Ground barley (or grain sorghum)	—	—	—	—	12.0
Ground yellow maize	32.7	57.5	41.0	61.6	—
Rolled oats	31.0	—	16.0	—	40.0
Cane sugar (or beet)	8.0	—	10.0	—	12.0
Dried skim milk	10.0	12.0	—	—	10.0
Dried whey	—	3.0	—	—	—
Soybean meal	8.0	16.0	22.0	35.5	—
Meat meal	3.5	7.5	—	—	—
Fish meal	3.5	—	5.0	—	20.0
Dried brewers' yeast	—	—	1.0	—	6.0
Lard (with antioxidant)	—	—	2.5	—	—
Ground limestone	0.4	1.0	—	1.1	—
Dicalcium phosphate	2.0	2.0	1.0	1.3	—
Steamed bone meal	0.4	0.5	1.0	—	—
Salt (with trace minerals)	0.5	0.5	0.5	0.5	—
Vitamin supplement	To be added at levels recommended by supplier				
Antiobiotic supplement	To be added at levels recommended by supplier				

[a]Proportions of feedstuffs may be changed or other feedstuffs substituted to accommodate particular cost relationships. The number of combinations of ingredients compatible with adequate nutrition is almost unlimited.

TABLE A5.2. Examples of Suitable Growing Pig Diets (to 55-kg Body Weight) for Use in Temperate or Tropical Environments[a]

Feedstuff	*A*, %	*B*, %	*C*, %	*D*, %	*E*, %	*F*, %
Ground barley (or grain sorghum)	77.5	—	—	—	—	—
Ground yellow maize	—	77.5	72.7	58.5	—	—
Ground opaque-2 maize	—	—	—	—	—	91.0
Ground oats	—	—	—	19.0	—	—
Cassava	—	—	—	—	66.0	—
Soybean meal	18.0	18.0	19.0	10.0	21.5	5.6
Fish meal	—	—	—	—	3.0	—
Meat meal	—	—	—	5.0	—	—
Cottonseed meal	—	—	—	—	7.0	—
Alfalfa meal	—	—	5.0	5.0	—	—
Steamed bone meal	—	—	2.0	2.0	2.0	2.6
Dicalcium phosphate	2.8	2.8	—	—	—	—
Ground limestone	1.2	1.2	0.8	—	—	0.3
Salt (with trace minerals)	0.5	0.5	0.5	0.5	0.5	0.5
Vitamin supplement	To be added at levels recommended by supplier					
Antibiotic supplement	To be added at levels recommended by supplier					

[a]Proportions of feedstuffs may be changed or other feedstuffs substituted to accommodate particular cost relationships. The number of combinations of ingredients compatible with adequate nutrition is almost unlimited.

TABLE A5.3. Examples of Suitable Finishing Pig Diets (56–100-kg Body Weight) for Use in Temperate or Tropical Environments[a]

Feedstuff	*A*, %	*B*, %	*C*, %	*D*, %	*E*, %	*F*, %
Ground barley (or grain sorghum)	87.5	—	—	—	—	—
Ground yellow maize	—	85.4	87.5	65.0	—	—
Ground opaque-2 maize	—	—	—	—	96.0	—
Ground oats	—	—	—	19.0	—	—
Cassava	—	—	—	—	—	73.5
Soybean meal	6.0	8.0	8.0	5.0	—	15.0
Meat meal	5.0	—	—	5.0	—	4.0
Cottonseed meal	—	—	—	—	—	5.0
Alfalfa meal	—	5.0	—	5.0	—	—
Steamed bone meal	1.0	0.7	3.0	0.5	—	2.0
Dicalcium phosphate	—	—	0.5	—	2.0	—
Ground limestone	0.5	0.4	0.5	—	1.5	—
Salt (with trace minerals)	—	0.5	0.5	0.5	0.5	0.5
Vitamin supplement	To be added at levels recommended by supplier					

[a]Proportions of feedstuffs may be changed or other feedstuffs substituted to accommodate particular cost relationships. The number of combinations of ingredients compatible with adequate nutrition is almost unlimited.

TABLE A5.4. Sample Diets for Sows[a,b]

Feedstuff	*A, %*	*B, %*	*C, %*	*D, %*	*E, %*
Barley (or sorghum)	79.5				
Corn (maize)		54.5	45.0		
Opaque-2 maize				91.0	
Oats		25.0	15.0		
Cassava					62.5
Soybean meal	10.0	7.0	8.0	5.6	20.0
Fish meal	2.0				3.0
Meat meal		7.0	6.0		
Cottonseed meal					7.0
Alfalfa meal	5.0	5.0	10.0		5.0
Wheat middlings			15.0		
Bone meal		1.0	0.5	2.6	2.0
Calcium phosphate	3.0				
Limestone				0.3	
Salt (with trace mineral)	0.5	0.5	0.5	0.5	0.5
Vitamin supplement	+	+	+	+	+

[a]Proportions of feedstuffs may be changed or other feedstuffs substituted to effect lower costs. The number of combinations compatible with adequate nutrition is almost unlimited.
[b]Hand fed (1.8–2.2 kg/day) in pregnancy and self-fed in lactation.

TABLE A5.5. Sample Diets for Sows[a,b]

Feedstuff	*A, %*	*B, %*	*C, %*
Corn (maize)	36.0	23.0	—
Oats	—	25.0	—
Soybean meal	10.0	7.5	—
Meat meal	—	3.0	—
Alfalfa meal (or hay)	50.0	—	97.1
Maize cobs (ground)	—	15.0	—
Calcium phosphate	2.3	1.0	—
Limestone	0.2	—	—
Salt (with trace mineral)	0.5	0.5	0.5
Vitamin supplement	+	+	+
Sodium tripolyphosphate	—	—	2.4

[a]Proportions of feedstuffs may be changed or other feedstuffs substituted to effect lower costs. The number of combinations compatible with adequate nutrition is almost unlimited.
[b]Self-fed in pregnancy.

Index

See pages 680–704 for glossaries of terms used in swine production and feed processing.

H

P

T

U

V

W

Related AVI Books

BASIC ENGINEERING PRINCIPLES, 2nd Edition *Merkel*
BREEDING FIELD CROPS, 2nd Edition *Poehlman*
COMMERCIAL CHICKEN PRODUCTION MANUAL, 3rd Edition *North*
EGG SCIENCE AND TECHNOLOGY, 2nd Edition
Stadelman and Cotterill
FOOD ENGINEERING SYSTEMS, Vol. 1 *Farrall*
FOOD PRODUCTS FORMULARY
Vol. 1. *Long, Komarik and Tressler*
Vol. 2. *Tressler and Sultan*
Vol. 3. *Tressler and Woodroof*
Vol. 4. *Inglett and Inglett*
FOOD SCIENCE, Revised 3rd Edition *Potter*
FUNDAMENTALS OF ENTOMOLOGY AND PLANT PATHOLOGY
Pyenson
LABORATORY MANUAL FOR ENTOMOLOGY AND PLANT PATHOLOGY *Pyenson and Barke*
LEAF PROTEIN CONCENTRATES *Telek and Graham*
LIVESTOCK AND MEAT MARKETING, 2nd Edition *McCoy*
MEAT HANDBOOK, 4th Edition *Levie*
POULTRY: FEEDS AND NUTRITION, 2nd Edition *Patrick and Schaible*
POULTRY PRODUCTS TECHNOLOGY, 2nd Edition *Mountney*
PRACTICAL MEAT CUTTING AND MERCHANDISING, Vols. 1 and 2 *Fabbricante and Sultan*
PRINCIPLES OF ANIMAL ENVIRONMENT, Textbook Edition *Esmay*
PRINCIPLES OF FARM MACHINERY, 3rd Edition
Kepner, Bainer and Barger
PROCESSED MEATS, 2nd Edition *Kramlich, Pearson and Tauber*
PROCESSING AGRICULTURAL AND MUNICIPAL WASTES *Inglett*
WORLD FISH FARMING, 2nd Edition *Brown*